environmental
change and challenge

environmental
change and challenge

A CANADIAN PERSPECTIVE

SECOND EDITION

PHILIP DEARDEN AND BRUCE MITCHELL

OXFORD

UNIVERSITY PRESS

To Kate and Theresa Dearden, and in memory of George and Jean Mitchell

OXFORD
UNIVERSITY PRESS

70 Wynford Drive, Don Mills, Ontario M3C 1J9
www.oup.com/ca

Oxford University Press is a department of the University of Oxford.
It furthers the University's objective of excellence in research, scholarship, and education by publishing worldwide in

Oxford New York
Auckland Cape Town Dar es Salaam Hong Kong
Karachi Kuala Lumpur Madrid Melbourne
Mexico City Nairobi New Delhi Shanghai Taipei
Toronto

With offices in
Argentina Austria Brazil Chile Czech Republic
France Greece Guatemala Hungary Italy Japan
Poland Portugal Singapore South Korea Switzerland
Thailand Turkey Ukraine Vietnam

Oxford is a trade mark of Oxford University Press
in the UK and in certain other countries

Published in Canada
by Oxford University Press

Copyright © Oxford University Press Canada 2005

The moral rights of the author have been asserted

Database right Oxford University Press (maker)

First published 2005
Photos on pages 36, 144, 188, 463, 498 from PhotoDisc: Environmental Concerns/Getty Images; 41 (top), 69, 104, 148, 165, 277 (top), 427 PhotoDisc: Panoramic Landscapes/Getty Images.

Cover image: Eric Meola/Getty Images.
Cover and text design: Brett Miller

Library and Archives Canada Cataloguing in Publication

Dearden, Philip
        Environmental change and challenge : a Canadian perspective / Philip

Dearden and Bruce Mitchell. — 2nd ed.

Includes bibliographical references and index.
ISBN-13: 978-0-19-541927-6.—
ISBN-10: 0-19-541927-8

        1. Environmental management--Canada—Textbooks. 2. Human ecology—Canada—Textbooks.
3. Nature—Effect of human beings on—Textbooks.
4. Global environmental change—Textbooks.
Mitchell, Bruce, 1944- II. Title.

GF511.D42 2005   333.7'0971
C2005-900910-1

1 2 3 4 - 08 07 06 05
This book is printed on permanent (acid-free) paper ∞.
Printed in Canada

CONTENTS OVERVIEW

Preface ix
Acknowledgements xi

PART A   INTRODUCTION 1
Chapter 1   Environment, Resources, and Society 3
             GUEST STATEMENT: PERSPECTIVE ON THE TAR PONDS, TIM BABCOCK

PART B   THE ECOSPHERE 37
Chapter 2   Energy Flows and Ecosystems 41
             GUEST STATEMENT: LANDSCAPE ECOLOGY, DENNIS JELINSKI
Chapter 3   Ecosystem Change 70
             GUEST STATEMENT: CONTROLS ON HIGH-ELEVATION TREELINE IN WESTERNMOST CANADA,
                 ZE'EV GEDALOF & DAN J. SMITH
Chapter 4   Ecosystems and Matter Cycling 104

PART C   PLANNING AND MANAGEMENT: PHILOSOPHY, PROCESS, AND PRODUCT 145
Chapter 5   Planning and Management: Philosophy 148
             GUEST STATEMENT: RESOURCES AND MANAGEMENT: THE NEED FOR AN INTEGRATED
                 PERSPECTIVE AND BALANCED PHILOSOPHY, DAN SHRUBSOLE
Chapter 6   Planning and Management: Process, Method, and Product 164

PART D   RESOURCE AND ENVIRONMENTAL MANAGEMENT IN CANADA 183
Chapter 7   Climate Change 186
             GUEST STATEMENT: TRADITIONAL ECOLOGICAL KNOWLEDGE, SCIENCE, AND CLIMATE
                 CHANGE, ELISE HO
Chapter 8   Oceans and Fisheries 226
             GUEST STATEMENT: PUBLIC AWARENESS AND OCEAN MANAGEMENT, SABINE JESSEN
Chapter 9   Forests 275
             GUEST STATEMENT: REFORMING FOREST TENURE, KEVIN HANNA
Chapter 10   Agriculture 325
             GUEST STATEMENT: CANADA FEEDING THE WORLD? PETER SCHROEDER
Chapter 11   Endangered Species and Protected Areas 371
             GUEST STATEMENT: CANADA'S GREAT BEAR RAINFOREST, IAN MCALLISTER
Chapter 12   Water 423
             GUEST STATEMENT: HOW BECOMING A HERITAGE RIVER CAN INFLUENCE WATER
                 MANAGEMENT, BARBARA VEALE
Chapter 13   Minerals and Energy 459
             GUEST STATEMENT: MINERALS, MINING, AND IMPACT AND BENEFIT AGREEMENTS,
                 MICHAEL VITCH

PART E   ENVIRONMENTAL CHANGE AND CHALLENGE REVISITED 494
Chapter 14   Making It Happen 496
             GUEST STATEMENT: HOW STUDENTS BROUGHT SUSTAINABILITY INTO CAMPUS PLANNING AT
                 THE UNIVERSITY OF VICTORIA, GRAHAM WATT-GREMM

             Appendix: Conservation Organizations 519
             Glossary 530
             Index 543

DETAILED CONTENTS

Preface ix
Acknowledgements xi

PART A INTRODUCTION 1

Chapter 1 Environment, Resources, and
            Society 3
Introduction: Change and Challenge 3
Defining Environment and Resources 5
Alternative Approaches to Understanding
    Complex Natural and Socio-economic
    Systems 6
Science-based Management of Resources and
    Environment 7
The Sydney Tar Ponds 9
The Global Picture 15
The Canadian Picture 22
Jurisdictional Arrangements for Environmental
    Management in Canada 23
Measuring Progress 24
Implications 29
Summary 33
Key Terms 34
Review Questions 34
Related Websites 34
References and Suggested Reading 35

PART B THE ECOSPHERE 37

Chapter 2 Energy Flows and Ecosystems 41
Introduction 41
Energy 42
Energy Flows in Ecological Systems 45
Ecosystem Structure 56
Implications 65
Summary 67
Key Terms 67
Review Questions 68
Related Websites 68
References and Suggested Reading 68

Chapter 3 Ecosystem Change 70
Introduction 70
Ecological Succession 70
Ecosystem Homeostasis 79
Population Growth 87
Evolution, Speciation, and Extinction 90
Biodiversity 94
Implications 98

Summary 100
Key Terms 101
Review Questions 101
Related Websites 101
References and Suggested Reading 102

Chapter 4 Ecosystems and Matter Cycling 104
Introduction 104
Matter 104
Biogeochemical Cycles 105
The Hydrological Cycle 116
Biogeochemical Cycles and Human Activity 125
Implications 140
Summary 140
Key Terms 141
Review Questions 141
Related Websites 142
References and Suggested Reading 142

PART C PLANNING AND MANAGEMENT: PHILOSOPHY, PROCESS, AND PRODUCT 145

Chapter 5 Planning and Management:
            Philosophy 148
Introduction 148
Planning and Management Components 149
Implications 160
Summary 161
Key Terms 162
Review Questions 162
Related Websites 162
References and Suggested Reading 163

Chapter 6 Planning and Management:
            Process, Method, and Product 164
Introduction 164
Collaboration and Co-ordination 164
Stakeholders and Participatory Approaches 165
Communication 168
Adaptive Management 169
Impact and Risk Assessment 170
Dispute Resolution 174
Regional and Land-Use Planning 177
Implications 178
Summary 179
Key Terms 180
Review Questions 180
Related Websites 180
References and Suggested Reading 180

PART D RESOURCE AND ENVIRONMENTAL MANAGEMENT IN CANADA 183

Chapter 7 Climate Change 186
Introduction 186
Nature of Climate Change 187
Scientific Evidence Related to Climate
 Change 188
Modelling Climate Change 193
Scientific Explanations 195
Case Studies: Rise of Sea Level in PEI and
 Winter Tourism in Ontario 197
Communicating Global Change 207
Kyoto Protocol 209
Moving Forward 217
Summary 220
Key Terms 222
Review Questions 222
Related Websites 222
References and Suggested Reading 223

Chapter 8 Oceans and Fisheries 226
Introduction 226
Oceanic Ecosystems 227
Ocean Management Challenges 233
Global Responses 241
Canada's Oceans 242
Fisheries 248
Aboriginal Use of Marine Resources 256
Hydrocarbon Development 258
Pollution 260
Some Canadian Responses 261
Aquaculture 265
Implications 269
Summary 270
Key Terms 271
Review Questions 271
Related Websites 272
References and Suggested Reading 272

Chapter 9 Forests 275
The Boreal Rendezvous 275
An Overview of Canada's Forests 276
Forest Management Practices 287
Environmental and Social Impacts of Forest
 Management Practices 299
New Forestry 311
Canada's National Forest Strategies 314
Implications 318
Summary 320
Key Terms 321
Review Questions 322

Related Websites 322
References and Suggested Reading 322

Chapter 10 Agriculture 325
Introduction 325
Agriculture as an Ecological Process 330
Modern Farming Systems in the Industrialized
 World 331
Trends in Canadian Agriculture 338
Environmental Challenges for Canadian
 Agriculture 339
Sustainable Food Production Systems 359
Sustainable Agriculture in Action: Organic
 Farming 363
Implications 366
Summary 367
Key Terms 368
Review Questions 368
Related Websites 368
References and Suggested Reading 369

Chapter 11 Endangered Species and Protected Areas 371
Introduction 371
Valuing Biodiversity 372
Main Pressures Causing Extinction 376
Vulnerability to Extinction 389
Responses to the Loss of Biodiversity 392
Protected Areas 397
Implications 416
Summary 419
Key Terms 420
Review Questions 420
Related Websites 420
References and Suggested Reading 420

Chapter 12 Water 423
Introduction 423
Human Interventions in the Hydrological
 Cycle: Water Diversions 424
Water Quality 430
Water Security: Protecting Quantity and
 Quality 437
Water as Hazard 441
Heritage Rivers 448
Implications 454
Summary 455
Key Terms 455
Review Questions 455
Related Websites 456
References and Suggested Reading 457

Chapter 13 Minerals and Energy 459
Introduction 459
Framing Issues and Questions 460
Non-renewable Resources in Canada: Basic
  Information 463
Remediating Mined Landscapes: Sudbury,
  Ontario 465
Developing a Diamond Mine: Ekati, NWT 472
Energy Resources 480
Implications 488
Summary 489
Key Terms 490
Review Questions 490
Related Websites 491
References and Suggested Reading 491

**PART E ENVIRONMENTAL CHANGE AND
          CHALLENGE REVISITED 494**

Chapter 14 Making It Happen 496
Introduction 496
Global Perspectives 496
National Perspectives 501
Personal Perspectives 504
The Law of Everybody 511
Implications 514
Summary 515
Key Terms 516
Review Questions 516
Related Websites 517
References and Suggested Reading 517

Appendix: Conservation Organizations 519
Glossary 530
Index 543

PREFACE

On 3 November 2004, as we were putting the finishing touches to this book, a press release was issued by the Canadian Council of Ministers of Environment reporting on their latest meeting. The press release proudly told us of the main achievement. The ministers had agreed to meet again in another six months to agree on a draft framework that will serve as a basis for consultations towards a framework agreement on sustainability, which in turn is to be reviewed by the ministers before they take the framework for discussion to the provincial and territorial levels.

Only the week before, two scientific reports had been released. One, the product of four years of work by 250 scientists in eight countries, examined the rate of change in Arctic ecosystems as a result of global climatic change. The results surprised even the scientists, who found out that changes were taking place approximately twice as quickly as had been predicted. Increased melting of Arctic sea ice is expected to raise sea levels one metre above current levels during this century. This would inundate the homes of over one billion people and some of the most productive agricultural land on Earth. Polar bears and other Arctic species adapted to current conditions are unlikely to survive.

A second report provided further confirmation that climate change models and their predicted responses were going in the right direction. A US-based study documented the increased northern migration of a butterfly, Edith's checkerspot, as the southern part of its range became too hot and dry and how it was moving into higher elevations and latitudes to seek cooler climes. The butterfly was able to move in its search for appropriate conditions for its survival. Many other species will not be so lucky.

Global climatic change is not an isolated academic concept. It is happening here and now. Yet our societal leaders, far from being in the vanguard of action, are often busy formulating drafts for frameworks for consultations that may lead to plans to actually do something. They fiddle as Rome burns. And Canada falls further and further behind in terms of the actions that need to be taken to address critical environmental problems. If Britain's Prime Minister can proudly promote Britain's goal of cutting greenhouse gas emissions by 60 per cent in the next 50 years, why can't Canada formulate a plan to cut 6 per cent in the next six years?

It is not just global change that needs urgent attention. Canada, as described in both Chapters 1 and 14 of this new edition of *Environmental Change and Challenge*, is getting increasingly poor ratings for environmental management in international comparisons. We were rated 28 out of 29 OECD countries, for example, in a 2002 comparison of environmental indicators.

But the situation is not hopeless. We have a huge and beautiful country, one of the most magnificent places on Earth. We are a rich country. In general, our citizens have a high quality of life and value the environment. But changes need to take place. And those changes need to take place far more quickly than is currently the case. Our so-called leaders are only willing to make those changes if they perceive support for them. That support hinges on having a well-informed and active populace.

We believe it is critical that university students leave our universities when they graduate with a greater understanding of the planetary ecosystems that support life and of their impacts on ecosystems, as well as an awareness of what society and individuals can do to help improve the situation. If all university graduates came out thus informed and acted on this knowledge to create change in their own lifestyles and society, the prognosis for the future would be a little more optimistic.

This book was written to help contribute towards these changes. It was written for students taking a first course in environment to impart an understanding of the biosphere's function, and to link basic environmental management principles to environmental and resource problems in a Canadian context. The book provides both a basic background for those who will go on to specialize in fields other than environment and also a broad platform upon which later, more detailed courses on environment can build.

Part A provides an overall introduction to environment, resources, and society and the role of science, both social and natural, in helping understand the relationship among them. This relationship is illustrated in more detail by a well-known

Canadian case study, the Sydney Tar Ponds. We also provide a global and national context for environmental management and describe some approaches for assessing current progress in dealing with environmental challenges. If we don't know how we are doing, we can hardly judge with any degree of accuracy the severity of the problem or map out suitable strategies to address the problem

Part B (Chapters 2–4) provides a basic primer on the environmental processes that constitute the Earth's life support system. Primary emphasis is on energy flows, biogeochemical cycles, and biotic responses, with reference to Canadian examples wherever possible. A strong emphasis is placed both here and in subsequent sections on making explicit links between these principles and examples illustrating the principles in action.

Part C (Chapters 5 and 6) reviews what are referred to as different philosophies, processes, and products that should characterize high-quality resource and environmental planning and management. Some refer to such attributes as elements of 'best practice'. Our hope is that by the end of these two chapters you will be able to develop a mental checklist of the attributes you would expect to see used in planning and management, and that you would advocate either as a team member addressing resource and environmental issues or as a member of civil society.

Part D (Chapters 7–13) takes the basic science of Part B and the management approaches of Part C and puts them together by focusing on environmental and resource management themes: climate change, oceans and fisheries, forests, agriculture, endangered species and protected areas, water, and minerals and energy. In each chapter we provide an overview of the current situation in Canada and the main management challenges. Text boxes highlight particular case studies of interest.

The final section (Chapter 14) concludes the book with views from three perspectives—global, national, and personal. Here we emphasize the actions that individuals can take in moving towards a more sustainable society and introduce the 'Law of Everybody', suggesting that if everyone took a few conservation actions these would add up to a massive contribution to the overall changes required.

ACKNOWLEDGEMENTS

We express our appreciation to the authors of the guest statements, as well as to Tim Babcock and Elise Ho, who provided photographs taken by themselves, to Wendy Bullerwell of the Beaton Institute of Cape Breton Studies at the University College of Cape Breton, who helped arrange our use of archival material from the Beaton Institute, and to Jean Andrey, who provided a cartoon related to climate change.

Dennis Jelinski provided valuable comments on Chapter 2. Elise Ho and Patricia Fitzpatrick did excellent background research at the University of Waterloo, and Karen Topelko, Melissa Hausser, and Heather MacDonald provided valuable assistance at the University of Victoria. Frances Hannigan at the University of Waterloo worked diligently to assemble chapters, tables, figures, boxes, and photographs into a manuscript.

We would also like to express our appreciation to the people at Oxford University Press Canada, in particular Phyllis Wilson, for their work on this project, as well as our copy editor, Richard Tallman.

Philip Dearden

I grew up in Britain. Even though home was in one of the wilder parts of Britain, I was always struck with the biological impoverishment of my homeland and dreamed of living in a country where wild nature still existed. My dream was realized when I first came to Newfoundland as a graduate student in the early 1970s. Since that time I have travelled all over Canada, and most of the rest of the world, and have a strong appreciation of the beauty and grandeur of the Canadian landscape.

My main interest is in conservation and I have taught courses and undertaken research on this topic, based at the University of Victoria, for 25 years. Throughout this period I have taught large introductory classes in society and environment and loved every minute of it. I have a strong belief that the power of individual actions can help to make a better environmental future and that we need to support NGOs working in this area. I have held many positions in the Canadian Parks and Wilderness Society, including Chair of the British Columbia chapter, and am currently a Trustee Emeritus.

My main field of research is protected areas, and I maintain active research programs in Canada and in Asia on this topic. I am a member of the IUCN's World Commission on Protected Areas and have advised many international bodies on protected area management. Author of over 160 articles and nine books and monographs, including (with Rick Rollins) *Parks and Protected Areas in Canada: Planning and Management* (Oxford, 2002), I have also been recognized for excellence in teaching with an Alumni Outstanding Teacher Award and Maclean's Popular Professor recognition at the University of

Victoria. I am Leader of the Marine Protected Area Research Group at the University of Victoria and like nothing better than to be out on the water with my wife Jittiya and children Kate and Theresa.

Bruce Mitchell

I was raised in Prince Rupert, a small community on the northwestern coast of British Columbia whose economy was strongly based on natural resources, especially forests and fish. Thus, from an early age, I recognized the importance of 'resources' and the 'environment'. As a graduate student, I focused on water resources, having become more and more aware of how critical this resource is for natural systems and humankind. Now, in the early twenty-first century, it is apparent that water is a key resource at a global scale.

As a high-school student and an undergraduate, I worked summers as a shoreworker in a fish-processing plant and then as a deckhand on a troller, fishing for salmon. That provided first-hand experience with a resource-harvesting industry. But, more importantly, it made me aware of how knowledgeable people could be who had not finished their formal education in the school system. This showed me that what would later be called 'traditional ecological knowledge' is a remarkable source of understanding and insight. This became a lifelong lesson: experiential learning deserves respect, and those pursuing science should continuously look to such knowledge to enhance what they think they know based on science.

Studying geography provided me with a solid background to understand natural systems, as well as the way in which humans interact with or use them. As time passed, I became more and more convinced that many 'resource problems' were often 'people problems'. As a result, I oriented my work towards planning and management, while always remaining mindful of the need to draw on scientific research. I have published about 130 articles, and 25 books, have served as President of the Canadian Water Resources Association, been a visiting professor at nine universities in various countries, and received the Award for Scholarly

Distinction from the Canadian Association of Geographers as well as a Distinguished Teacher Award from the University of Waterloo.

As a faculty member, I have had the opportunity to conduct research and consult both in Canada and in developing countries. The work in countries such as Indonesia has made me aware of how fortunate most Canadians are, and how our system of governance and civil society allows us all to become engaged in societal and environmental issues. Barriers and challenges exist, but compared to the people of many nations, Canadians have considerable latitude to participate in and shape our shared future.

Introduction

The relationship among environment, resources, and society is one of the most important challenges, if not the most important challenge, currently faced by humans on Earth. For many of Earth's human inhabitants, this relationship is an ongoing reality as they try to meet their everyday needs of food, water, and shelter. They do not need to be reminded of how important it is. To ignore this relationship is to perish.

For others, usually urban dwellers, this reality seems distant. Food comes from the supermarket, water is piped into homes, even work and home environments have controlled temperature through heating and air conditioning. It is not until disruptions in these delivery systems—by floods, tsunamis, droughts, ice storms, earthquakes, hurricanes, insect infestations, and similar forces of nature—that many people realize that they, too, are dependent on the environment for survival, as has been true since before the dawn of human civilization.

This first section introduces some basic concepts regarding the relationship of environment, resources, and society and the ways in which we try to understand complex natural and socio-economic systems. There are many ways of knowing about environment. Here we concentrate mainly on the contribution of the natural and social sciences.

Science, especially environmental science, is also becoming an increasingly collaborative undertaking. This involves not only workers from one discipline, but those from many disciplines coming together to contribute their understanding of a particular phenomenon. Understanding of acid precipitation, for example, necessarily requires the input of chemists, biochemists, climatologists, geologists, hydrologists, geographers, biologists, health specialists, economists, and political and legal experts, to name a few. Each discipline has its own expertise and methods of approach and these can be combined together in different ways to yield more effective answers to environmental problems.

The need for a collaborative approach to environment is clearly illustrated by a Canadian case study, that of the Sydney Tar Ponds in Nova Scotia. Many of the problems resulting from the steelmaking and subsequent pollution of the Muggah Creek estuary came about through failure to use effectively scientific information that was known at the

◄ *(Philip Dearden)*

time. Even with the use of science, there can be high levels of uncertainty. How we can try to deal with uncertainty and change is one of the main themes of this book.

The Sydney Tar Ponds also help to illustrate some of the complexity involved in trying to address one environmental problem in one small place. Needless to say, the global situation is infinitely more complex. The next two sections of Chapter 1 provide an overview of the global and Canadian situations with regard to environment and society. What are some of the main trends that indicate future directions? Although there are disagreements about the rate and severity of change, few claim that overall environmental conditions are improving. One indicator, the Living Planet Index, suggests that the overall ecological health of the Earth has been reduced by 35 per cent since 1970.

There is also some good news. Global population growth predictions, for example, have declined to below 9 billion people by the year 2050. One important dimension that has shown only growth, however, is resource use, fuelled mainly by the demands of consumers in developed countries—we are reminded of the old comic strip, 'Pogo', in which the title character, a possum living in a swamp, famously proclaimed that 'We have found the enemy, and it is us.' If there is one fundamental message that we would like to convey, it is the power of individuals to make decisions on a daily basis that can reduce these pressures. Canadians have much to contribute in this regard, as we are among the most profligate consumers of energy and water in the world and are also among the most prolific producers of solid waste. Our society has developed into one of the most wasteful on the planet. Only we can turn that around.

The jurisdictional and institutional arrangements for environmental management in Canada are one critical aspect influencing our relationship with the environment and resources. Such arrangements are often little taken into account by scientists and environmentalists, but they can be the most important factors when considering how

and when a particular problem is going to be addressed. Canada is a large country and the various levels of government are complex, and, for political reasons rather than for the good of all Canadians, they often work poorly together, or even work against each other.

Whether the context is global, national, or regional, we are interested in measuring our progress in addressing environmental change. As pointed out above, however, the situation is very complex, with far more variables and changes than we can possibly measure. The next section of this first chapter provides some background on how we try to address this situation through the use of indicators, and outlines the various kinds of indicators and their strengths and weaknesses. Environment Canada, for example, has devised a system of national environmental indicators with the goal of keeping Canadians informed on the overall state of the environment.

The chapter ends with the presentation of a simple framework that summarizes the process of environmental management. Throughout the book, we return to this framework to illustrate deficiencies in understanding or lack of connection between different elements of the framework.

This first part of the book provides an overall introduction to environmental change and challenge with reference to the global, national, and regional levels. Most of the remainder of the book is concerned with Canada, although we return to a global perspective in the final chapter. The next part, Part B, provides an overview of the main environmental processes with which we need to be familiar to understand many environmental problems. Part C discusses some of the philosophical dimensions and best practices for various aspects of resource management. This is followed by Part D, in which we discuss various thematic aspects of resource management, such as fisheries, water, and climate change. The final section, Part E, draws together some of these themes, returns to global and national summaries of current trends, and points out some of the things that individuals can do to effect change for the better in the environment of tomorrow.

Chapter 1

Environment, Resources, and Society

Learning Objectives

- To appreciate different perspectives related to environment and resources.
- To understand different approaches to analyzing complex environmental and socio-economic systems.
- To understand the implications for change, complexity, uncertainty, and conflict relative to environmental issues and problems.
- To appreciate various aspects that must be addressed to bring 'science' to bear on environmental and resource problems.
- To understand institutional arrangements related to environmental management in Canada.
- To understand that Canada's natural environment and society are part of a global system.
- To describe different ways of looking at progress among nations on environmental matters over time.

INTRODUCTION: CHANGE AND CHALLENGE

'It has never rained this hard before in the Lower Mainland [Vancouver] since we've been keeping records', said Environment Canada meteorologist David Jones on 21 October 2003. Rain records in the wettest area of Canada were easily surpassed with 600 mm (that's two feet) in four days in some areas, with record intensities of 40 mm/hour. The rains produced extensive flooding with roads and railway lines washed out, communities cut off from the outside world, and loss of life.

At the other end of the country, just weeks before, Halifax and the east coast of Nova Scotia had faced a hurricane with wind speeds that reached over 158 km/hr. Hurricane Juan left over 300,000 homes and businesses in Nova Scotia without power for weeks afterwards. Homes were

Hurricane Juan damage on McNabs Island, located in Halifax Harbour (*R. Guscott, Nova Scotia Department of Natural Resources*).

destroyed, boats sank, and lives were lost. The Canadian Hurricane Centre announced that the last hurricane comparable in force was 110 years ago.

Five months after Hurricane Juan hit, on 19 February 2004, Nova Scotia and Prince Edward Island were struck by a fierce 'nor'easter' that dumped up to 95 cm of snow and provided gusts up to 100 km/hr on Halifax and other areas. Combined with powerful winds, the snow knocked out power for 12,000 residents of PEI and 2,000 Nova Scotians by the afternoon of the storm. Both provinces called a state of emergency, a first time this had been done in Nova Scotia, and for two nights people in Halifax were ordered to stay in their homes and off the roads so that snow clearing could be done. The snow and windstorm was dubbed 'White Juan' after Hurricane Juan.

Natural systems change. They have always changed and always will change. There is strong evidence, discussed later in this book, that human activities have now become a main driving force behind environmental change. Whatever the reason, it seems that changes are happening more abruptly and with greater magnitude than previously. They threaten societal well-being, and society must respond and respond quickly.

Changes also occur in societies due to shifts in human values, expectations, perceptions, and attitudes, which may have implications for future interactions between those societies and natural systems. The value of the world economy has increased more than six times in the last 25 years. This is not merely the result of population growth; the chief cause has been increased consumption. Expectations have changed. Things that were seen as luxuries 30 years ago, such as TVs and automobiles, are now found in some of the most remote societies on Earth. They have become necessities to many.

Changes in natural and human systems generate challenges. If we wish to protect the integrity of biophysical systems, yet also ensure that human needs are satisfied, questions arise about how to determine ecosystem integrity and how to define basic human needs. Such questions force us to think about conditions both *today* and in the *future*. Such questions also remind us that an understanding of environmental and resource systems requires both natural and social sciences. Neither alone provides sufficient understanding and insight to guide decisions. Finally, such questions pose fundamental challenges as to whether we can realistically expect to manage or control natural systems, or whether we should focus on trying to manage human interactions with natural systems.

In this chapter, we begin by explaining what we mean by 'environment', 'resources', and 'society', and then consider alternative ways to understand systems, issues, and problems. This is followed by a case study of the Sydney Tar Ponds in Nova Scotia to illustrate opportunities and challenges regarding resource and environmental systems, as well as the importance of using both science and social science to inform public decisions and policy-making. The Sydney Tar Ponds case study illustrates vividly that decisions often are made in the context of changing conditions, incomplete knowledge and understanding, conflicting interests and values, trade-offs, and uncertainty. These same conditions apply not only to the Sydney Tar Ponds or, indeed, to Canada generally, but also to the global picture. In this context, we provide an overview of some

Even in the most remote locations, television now seems to be a necessity. Here the TV antenna is on a hut in a remote community in Kerala, India (*Philip Dearden*).

major environmental trends and the main issues that arise. Canada is placed within this global context with special consideration of how environmental change and response can be measured through ecological footprints and indicators, and of the institutional aspects of environmental management in Canada. We conclude the chapter by identifying some key considerations regarding how scientific understanding and insight can be used to inform resource and environmental management and decisions.

DEFINING ENVIRONMENT AND RESOURCES

The **environment** is the combination of the atmosphere, hydrosphere, cryosphere, lithosphere, and biosphere in which humans, other living species, and non-animate phenomena exist. As an analogy, the environment is the habitat or home on which humans and others depend to survive. In contrast, **resources** are more specific, and normally are thought of as such things as forests, wildlife, oceans, rivers and lakes, minerals, and petroleum.

Some consider resources to be only those components of the environment with utility for humans. From this perspective, coal and copper were part of the environment, but were not resources, until humans had the understanding to recognize their existence, the insight to appreciate how they could be used, and the skills or technology to access and apply them. In other words, in this perspective, elements of the environment do not become resources until they have value for humans. This is considered to be an **anthropocentric view**, in the sense that value is defined relative to human interests, wants, and needs.

In contrast, another perspective is that resources exist independently of human wants and needs. On that basis, components of the environment, such as temperate rain forests and grizzly bears, have value regardless of their immediate value for people. This is labelled as an **ecocentric** or **biocentric view** because it values aspects of the environment simply because they exist and accepts that they have the right to exist.

In this book, we are interested in resources both as they have the potential to meet human needs, and also with regard to their own intrinsic value. Whichever category is emphasized, we often encounter change, complexity, uncertainty, and conflict. For example, different attitudes at different times may lead to an area being logged or used for mining, or designated as a protected area. People and other living beings drink water to live, and at the same time urban areas may compete with farmers to have access to water. An area considered to be of significant value for its biodiversity or ecological integrity may be designated as a national or provincial park—but then might change into an ecosystem of less intrinsic value to humans due to ecological processes. For instance, if a fire sweeps through an old-growth forest, the question arises as to whether the fire should be allowed to burn because it is a natural part of ecosystem processes, or whether humans should intervene to put it out.

The person who sees a manta ray underwater will never forget such a sight or dispute the right of such species to survive. However, a purely anthropocentric resource view will tend to look more at their value in terms of making money as material for bags (*Philip Dearden*).

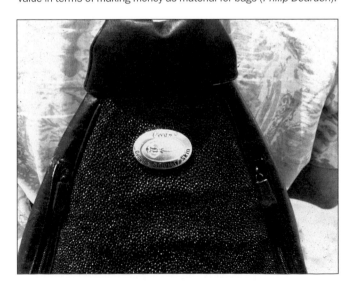

Thus, recognizing anthropocentric and biocentric perspectives does not automatically resolve all the problems that scientists and managers face when dealing with the environment and resources. However, being aware of such viewpoints should help to make us more aware about the positions that individuals or groups take with regard to what is appropriate action.

ALTERNATIVE APPROACHES TO UNDERSTANDING COMPLEX NATURAL AND SOCIO-ECONOMIC SYSTEMS

Systems have environmental, economic, and social components. Even if we focus on the environmental component, this can be subdivided into aspects requiring expertise in disciplines such as biology, zoology, chemistry, geology, and geography. However, while humans have organized knowledge into disciplines for convenience and manageability, the 'real world' is not organized in that way, nor does it recognize disciplinary boundaries. Thus, it is important to understand alternative ways of creating and applying knowledge. These include at least the following:

1. *Disciplinary*. Disciplinary understanding is organized around the concepts, theories, assumptions, and methods associated with an academic discipline. Disciplines have emerged and evolved in the belief that specialization will result in more in-depth understanding, and this is correct. However, since systems of interest to environmental scientists and managers have many components, the danger of a disciplinary approach is that important connections with parts of the system not considered by any one disciplinary specialist will not be considered.

2. *Multidisciplinary*. In order to get the in-depth insight of the disciplinary specialist but also gain the benefits of a broader view by drawing on specialists from various disciplines, the multidisciplinary approach emerged. In this approach, different specialists examine an issue, such as biodiversity, from their disciplinary perspectives, such as ecology, economics, and law. The specialists work in isolation, or only with others from the same discipline or profession, and provide separate reports, which are submitted to one person or group who then draws upon them to synthesize the findings and insights. In this manner, both depth and breadth are achieved, through synthesis of the findings of different specialists *after* they have completed their analyses.

3. *Cross-disciplinary*. While specialists in a multidisciplinary team work in isolation from one another, in cross-disciplinary research a disciplinary specialist 'crosses' the boundaries of other disciplines and borrows concepts, theories, and methods to enhance his or her disciplinary perspective. However, while in this approach the specialist deliberately and systematically crosses disciplinary boundaries to borrow from other disciplines, he or she does not actively engage with specialists from the other disciplines but simply draws on their ideas and approaches. This approach allows the investigator to make connections throughout an investigation that would not occur in a disciplinary or multidisciplinary approach, and this can be very positive. At the same time, it also can result in misunderstanding of the borrowed material, using theories, concepts, and methods out of context, and overlooking contradictory evidence, tests, or explanations in the discipline from which borrowing is being done.

4. *Interdisciplinary*. To overcome the limitations of the previous three approaches, interdisciplinary investigations involve disciplinary specialists crossing other disciplinary boundaries and engaging with other specialists from the very beginning of a research project. The objective is to achieve the benefits of both depth and breadth from the outset, as well as synthesis or integration, rather than at the end of the process as occurs in the multidisciplinary approach.

   This approach requires more time than the second and third approaches, as a team of disciplinary specialists must meet regularly throughout a study. In addition, this approach requires a high degree of respect, trust, and mutual understanding among the disciplinary specialists, since it is common for one disciplinary specialist to question basic beliefs or assumptions that another specialist takes for granted. This approach also requires patience, as disciplinary specialists have to be prepared to make the effort to learn the 'jargon' of other specialists so that clear communication can occur. Finally, an inter-disciplinary approach requires team members to have considerable self-confidence and a willing-ness to acknowledge strengths and weaknesses of their disciplines, as inevitably their disciplinary views will be challenged by others.

It is not uncommon for research projects initiated as interdisciplinary ventures to fail, and consequently to evolve into multidisciplinary or cross-disciplinary projects, because participants are unwilling to allocate the necessary extra time or do not have the patience to learn about and receive critiques from other specialists. However, when a group of disciplinary specialists over-comes these obstacles and mutual respect and trust are achieved, then the opportunity for enhanced understanding through benefiting from depth, breadth, and synthesis is achievable.

5. *Transdisciplinary.* Transdisciplinary research is similar to interdisciplinary research, except that this term is used when the issue or problem under investigation is generally recognized not to be in the domain or be the 'property' of any one discipline. A good example is the concept of sustainable development (discussed in more detail in Chapter 5), which did not emerge from any one discipline yet reflects ideas from envi-ronmental, economic, and social fields of inquiry.

In this book we examine many complex systems, and their understanding often is based on the knowledge derived from more than one discipline.

SUSTAINABLE DEVELOPMENT

Sustainable development is development that meets the needs of the present without compromising the ability of future generations to meet their own needs. It contains within it two key concepts:
- *the concept of 'needs', in particular the essential needs of the world's poor, to which overriding prior-ity should be given; and*
- *the idea of limitations imposed by the state of tech-nology and social organization on the environment's ability to meet present and future needs.*
  — *World Commission on Environment and Development (1987: 43)*

As an individual, you should be aware of what insight you can contribute from a disciplinary or interdisciplinary/transdisciplinary base and what knowledge you should have to be a valued member of an interdisciplinary or transdisciplinary team. As an individual, you can approach research from a disciplinary or cross-disciplinary perspec-tive. To participate in a multidisciplinary, interdis-ciplinary, or transdisciplinary project, you will need to be part of a group of people from different disciplines or professions.

SCIENCE-BASED MANAGEMENT OF RESOURCES AND ENVIRONMENT

In this book, we consider how understanding and insight from science can be used to inform management and decision-making. The nature of science is discussed in more detail in the introduc-tion to Part B of the text. In that regard, it is helpful to consider the view of Mills, Quigley, and Everest (2001: 11) regarding the nature of science-based natural resource decisions. In their words, a science-based decision 'must be made with full consideration and correct interpretation of all rele-vant science information, and the scientific under-standing must be revealed to all interested parties. The decision, however, must be left to the appro-priate decision-maker(s) with authority to make the decision.' Based on this interpretation, they provide five guidelines for contributions by scien-tists to effective management of resources and the environment.

Guidelines for Science-based Management of Resources and the Environment

1. *Focus the science on key issues, and communicate it in a policy-relevant form.* If science is to have value for management decisions, it must address pertinent management issues and research must be conceived in a manner rele-vant to such issues. This does not preclude scientific research from addressing basic or fundamental questions. However, to be perceived as relevant to the needs of managers, scientific work must be focused on and be timely and relevant to the needs of managers. In that regard, while scientists can provide important input into establishing management

goals, this task is properly in the domain of the value-laden process of decision-making and is not part of scientific research per se.

In their research, scientists should be describing key relationships within and between systems being managed. Improved understanding of environmental, economic, social, political, and legal systems affected by management is essential to appreciate basic scientific information that might be presented.

2. *Use scientific information to clarify issues, identify potential management options, and estimate consequences of decisions.* A basic challenge for management is to determine if a problem has been defined in an appropriate manner. Sometimes, due to complexity and uncertainty, managers may be unaware of questions that should be asked. In that regard, science, by helping to clarify relations and trends in systems, can contribute to clarifying known issues and to identifying issues previously overlooked or unknown. Science also can help to calculate the implications of different options related to an issue or problem.

3. *Clearly and simply communicate key scientific findings to all participants.* While it is important for scientists to publish their results in peer-reviewed journals, and that is to be encouraged, if their work is to be relevant for management then scientists must go beyond that to share findings in forums and formats accessible and understandable to a wide variety of readers.

Mills et al. suggest that scientific findings can be broadly shared in one or more of the following ways: (1) publishing a summary of key scientific findings in lay terms within an engaging format, (2) condensing the most policy-relevant scientific findings into a few clear bullets for electronic or print media, and (3) organizing conferences intended for mixed audiences, and ensuring that presentations are geared to a wide diversity of participants.

4. *Evaluate whether or not the final decision is consistent with scientific information.* Relevant scientific information being available and accessible is necessary, but not enough. It must be considered and incorporated into decision-making. One way to ascertain if that is happening, and to put pressure on decision-makers to do so, is

to conduct systematic and formal evaluations of decisions to determine to what extent they have relied on science.

Many benefits can emerge as a result of evaluations. First, an evaluation is a formal mechanism to demonstrate if decision-makers understood and used scientific information appropriately. Second, evaluation encourages scientists to continue to provide their results in a way understandable to non-scientists. Third, as a review is being conducted, an opportunity is presented for decision-makers to be sure they have understood the scientific information, and, if not, to improve their understanding before taking decisions that are practically irreversible.

Mills et al. recommend that a first step in an evaluation should be to identify the principal elements of a decision for which scientific information was available, and then, for each element, determine (1) if all of the relevant scientific information had been addressed and acknowledged in the decision, (2) if the scientific information had been correctly interpreted, based on current understanding, and then accurately presented, and (3) if projected outputs, outcomes and impacts, and risks associated with the decision had been considered and shared. In their view, if any element does not conform to these three considerations, then the decision cannot be said to have been consistent with science.

5. *Avoid advocacy of any particular solution.* There is much debate regarding this final guideline, since some scientists believe that they should

PROFESSIONAL JUDGEMENT

Since there seldom is time to conduct new research in the middle of a major policy debate, there always will be holes, sometimes big ones, in the science information. The scientists will be asked to at least hypothesize relations that might fill those holes and that will require significant personal judgement. Often, tight time frames will not permit the sort of multiple rounds of peer review that are desirable and typical in the science arena. In these circumstances, faith in the objectivity and independence of the scientists is particularly important.

— *Mills et al. (2001: 14)*

be advocates for solutions when their knowledge leads them to a preferred conclusion. Others maintain that if scientists, and their evidence and interpretations, are to be credible, they should not be seen to be favouring any particular position or solution. For example, if a scientist is a known supporter of the use of herbicides and pesticides to enhance agricultural production, would that person's evidence supporting the use of herbicides and pesticides be viewed as credible? Even if scientists can separate their basic values from their scientific understanding, there is danger that they will be perceived to favour a particular viewpoint, leaving doubt in some people's minds as to whether data, interpretations, conclusions, and recommendations from such a scientist have been 'contaminated' by those values.

The dilemma, of course, is that nobody is value-free or value-neutral, so to suggest that scientists can or should be value-free or 'objective' is difficult to sustain. However, there is a difference between being perceived to be open-minded in defining a problem or identifying alternative solutions, as opposed to being known for supporting a particular view or position and consistently producing findings that only support that one particular view or position. A further complication occurs when

there is insufficient evidence to support a conclusion, and so scientists are asked to provide a professional opinion. In this type of situation, a strong argument can be made that the scientist will be viewed as more credible if there is not a record of advocacy of a particular perspective, as highlighted in the quote in the accompanying box.

The following case study of the Sydney Tar Ponds illustrates the different kinds of knowledge required to understand resource and environmental problems. It also shows that the ideal of having 'science' inform policy decisions is often fraught with challenges.

THE SYDNEY TAR PONDS

Background

Sydney, the third largest city in Nova Scotia, is located on the northern part of Cape Breton Island (Figure 1.1). Extensive deposits of coal and iron ore in the Sydney area resulted in a long history of coal mining and steel production. These resources, in addition to a coastal fishery, forestry, and a striking natural landscape, have been the economic base for what one commentator has characterized as 'a culturally rich but economically poor and government-dependent community' (Rainham, 2002: 26).

In the late nineteenth century, the Industrial

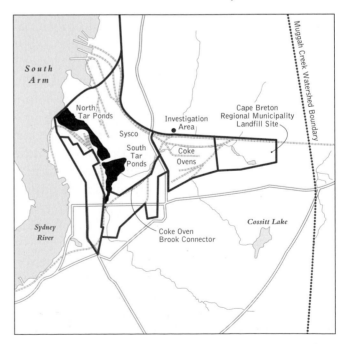

Figure 1.1 Location of Sydney, Nova Scotia, and the Tar Ponds. SOURCE: Rainham (2002: 27), from Joint Action Group, *Sydney, Cape Breton* (1999).

STEEL PRODUCTION IN SYDNEY

By 1912 Cape Breton was the source of nearly half of all the steel produced in Canada, and Sydney's Dominion Iron and Steel Company had the largest share of the pie. For half a century business boomed.

— *Lahey (1998: 38)*

Revolution was being powered by coal. Geologists were aware that Cape Breton Island had substantial coal deposits in the areas of Inverness, Glace Bay, North Sydney, Sydney Mines, and Dominion. Shafts were sunk and men and boys began the hard work of digging out the coal. Production steadily grew, and shortly after 1891, when a rail line connected Sydney to Halifax, annual coal production had grown to 1.5 million tonnes. By 1893, the numerous small coal mines joined together as the Dominion Coal Company (DOMCO), and output continued to grow—to more than 6 million tonnes annually by 1913.

The steady expansion of railroad lines in North America led to a high demand for steel, leading the American owner of DOMCO to form a partnership to create a new corporation, the Dominion Iron and Steel Company (DISCO). This new company received a free federal grant of 500 acres (200 hectares) of land along the Sydney waterfront, as well as a 30-year tax holiday from the provincial and municipal governments. On 1 July 1899, construction started on the new steel plant located along Muggah Creek on the 200-hectare site. The aim was to build 400 coke ovens, four 250-tonne capacity blast furnaces, and 10 open-hearth furnaces. A supply of coal was readily available, but more iron ore was needed, and this was procured from the iron ore mines at Wabana, Newfoundland.

Making high-quality steel requires high-quality ingredients. If coal deposits contain too much sulphur, then inferior coke is the result. Low-grade iron ore leads to the need for larger quantities of limestone to remove the impurities. To assess the quality of the basic inputs to the steelmaking process, science is essential. However, such science was not drawn upon when arrangements were being made for the basic raw materials. This was the beginning of what would become a pattern. As Barlow and May (2000: 11–12) observed:

neither Arthur Moxham, now vice-president and general manager of DISCO, nor Henry Whitney [the owner of DISCO] stopped to test the quality of their ingredients. In a rush to begin full operation, they failed to run the most basic tests on their coals and ores. DISCO tapped its first steel on New Year's Eve 1901 and right from the start there were problems.

The coal from Cape Breton seams was very high in sulphur, so far more coal had to be baked to produce usable coke. The iron ore from Wabana was full of impurities, such as silica and rock, so far more limestone was required to pull out the impurities as slag. The unusually large amounts of limestone required in the blast furnaces caused the furnace linings to deteriorate rapidly....

The poor quality of the basic ingredients led to higher costs, less marketable and inferior products, and far more waste. In what had been Muggah Creek, the slag would eventually create a mountain range of waste, stretching hundreds of feet high and reducing the mouth of the estuary by nearly a mile. With more coal being baked to make less coke, the mills were producing larger quantities of tarry sludge, benzenes (volatile liquid used as solvent), ammonium sulphates, naphthalene (a white crystalline substance used to make mothballs) and PAHs [polycyclic aromatic hydrocarbons,

Sydney, Nova Scotia, Coke Ovens, 1987. Aerial view of the coke ovens, showing the quenching plant with steam at left, coal pocket and batteries between exhaust stacks, centre, and conveyor leading from pocket to blending plant, right. By-product building is large brick structure, foreground. Smaller buildings are pump house, carpenter shop, and oil house (*G. Langille, Coke Ovens, SYSCO, 1987, 90–221–19653, Beaton Institute, University College of Cape Breton, Sydney, NS*).

BOX 1.1 HEALTH CHALLENGES RELATED TO THE TAR PONDS

Cancer rates are far above the national average, and higher than the rates in nearby Glace Bay or New Waterford. Cervical cancer in women in Sydney is 134 per cent higher than the provincial rate, which itself exceeds the national rate. All cancers in men and women are higher in Sydney, including brain cancer, breast cancer, stomach cancer, lung cancer, and salivary gland cancer, Alzheimer's, multiple sclerosis, heart disease, and birth defects are all much higher in Sydney.
SOURCE: May and Barlow (2001: 7).

the largest group of cancer-causing chemicals in the world]. These by-products and wastes were simply dumped into Muggah Creek....

The coal and steel operations continued for almost a century under various owners. The coke ovens, in which coal was baked at high temperatures to produce a higher-quality fuel source for the steel mill's open-hearth blast furnaces, were closed in 1988 (see photo). The steel plant stopped operating in 2001. The closure of these operations was a serious blow to the economy of Sydney. However, even though the coke ovens closed in 1988, the legacy of the uncontrolled dumping of millions of tonnes of toxic sludge remained a threat to adjacent residents and the environment, and significant industrial pollution in the Muggah Creek watershed remained.

For decades, air pollution originating from the coke ovens and open-hearth furnaces had been clearly evident. Nevertheless, the implications of the air pollution for land, waterways, and the Muggah Creek watershed were in some cases unknown and in other cases only poorly understood.

The area labelled as the 'tar ponds' was created from the deposit of chemical by-products from the coking process, runoff of water used for cooling in the coke ovens and the steel mill, leaching from contaminated soil at the coke and steel-making plants, an adjacent garbage dump site in the upper part of the watershed, and discharge of raw sewage from residential and commercial areas of Sydney. The outcome was a chemical- and bacteria-laden

river system, including the **estuary** full of contaminated sediments, the latter referred to as the tar ponds, 'a two-kilometre-long stretch of contaminated water and sludge which federal officials refer to as the largest chemical waste site in Canada' (Lahey, 1998: 37). Various government surveys of this ecosystem identified polycyclic aromatic hydrocarbons (PAHs), polychlorinated biphenyls (PCBs), and other chemicals and metals among the pollutants. Muggah Creek and the estuary empty into Sydney harbour, and discovery of PAHs in those waters led to the closure of the lobster fishery in the harbour and its approaches. Beyond the contamination of the aquatic system, the studies determined that the soil on the coke ovens site was highly contaminated. The site also contained storage tanks holding chemical waste.

The tar ponds are in the lower part of the Muggah Creek watershed, and since Dominion Iron and Steel built its plant in 1899, what was once an active, navigable waterway and a habitat for fish and birds became a highly contaminated narrow tidal outlet. A causeway and bridge divide what used to be an estuary into a north pond and a south pond. Surveys by governments and consultants have revealed that the two ponds contain about 700,000 tonnes of sediment contaminated with PAHs, including some 45,000 tonnes also containing PCBs at concentrations above 50 ppm.

Challenges for Epidemiological Studies

Various studies started to highlight that citizens of Sydney were encountering serious health problems. Unreleased government studies revealed significantly higher cancer rates in Sydney compared to the rest of Canada. For example, between 1977 and

HIGH LEVELS OF UNCERTAINTY

The interactions among claims and interpretations of evidence provided by empirical study and scientific assessment indicate there indeed exists an inherent complexity and uncertainty surrounding the understanding of health effects from chemical exposures from environmental contamination. Epidemiology, historically viewed as a parsimonious social science, seems to lose credibility and power when questions of causality arise, and its ability to generate meaningful change in Sydney is questionable.

— Rainham (2002: 31)

1980, residents in Sydney had 347 cancer deaths per 100,000 people, while the national average was 192. Health and Welfare Canada expressed concerns about the possible links to the toxic wastes from the coke ovens. More recent studies provide evidence that health issues in Sydney deserve attention. One study in the late 1990s found that the life expectancy for male and female residents in Sydney was as much as five years less than for the Canadian population as a whole. The primary contributions to reduced life expectancy were significantly higher levels of cancer and cardiovascular disease. Another study in the late 1990s showed increased incidences of cancer in the Sydney population over a 45-year period. Yet another study indicated that rates of major birth anomalies are significantly higher among Sydney residents relative to the rest of Nova Scotia.

In the mid-1980s, Health and Welfare Canada alerted the Environment Canada Atlantic regional office about health concerns related to the coking process, and the regional official alerted his provincial counterpart. In contrast, the epidemiologist for Nova Scotia took the position that the hazard depended on long-term exposure, and that, balanced against social and economic benefits of the coke ovens, it was reasonable to allow the coke operations to continue. The province then conducted its own investigation and concluded that unhealthy lifestyles (smoking, alcohol, poor diet, such as fatty foods and high salt intake) were more likely to be causes of the higher incidences of poor health.

As Rainham (2002: 30) noted, 'The Sydney issue is characterized by two core groups: stakeholders who believe that the community contamination poses serious risks to human health and stakeholders who believe that the community contamination is both contained (a necessary result of supposed economic prosperity) and negligible (in terms of risk to human well-being).' An important challenge is that evidence based on environmental epidemiology research is often not conclusive because many variables (smoking and other lifestyle factors) can affect human health, and there is uncertainty related to estimation of risks. Even though a large body of literature indicates a link between human morbidity and steelmaking, mining, and related industries, others always can argue that other variables are as likely to be the cause of morbidity.

Other factors can confound our understanding. For example, Goodarzi and Mukhopadhyay (2000:

369–70) have reviewed the possible source and level of concentrations of PAHs and metals from a sample of bore holes and lakes providing water to the Sydney Regional Municipality. They concluded that 'the origin of the PAHs and the priority metals in the drinking water of the Sydney basin is not known.' They identified at least three possible origins for the PAHs and metals: (1) from contamination through leaching of the combustion residuals from the Sydney coke oven site (the tar ponds) or from the disposal site for the two power plants of the Nova Scotia Power Authority; (2) from contamination through leakage or flow of paleowater from the sedimentary strata that contain hydrocarbon reservoirs and source rocks; and (3) from leaching of underground coal-bearing seams, which are ubiquitous in the Sydney Basin. These three possible explanations illustrate that it is not uncommon during scientific investigations for alternative explanations to emerge. It is not always possible to state definitively which is correct.

Problems When Science Is Not Used To Inform Decisions

As noted above, basic science was not used when arrangements were made for the basic raw materials (coal, iron ore) for the steel mill. This pattern continued when efforts began to address the industrial pollution in the tar ponds. Once it was acknowledged that the estuary was severely polluted and posed a health threat, alternative actions were identified, including incinerating, burying, and/or removing the waste.

Serious thinking about remediation followed a 1980 federal survey of lobsters in Sydney harbour. The survey revealed the lobsters were contaminated with the cancer-causing PAH chemicals, as well as with mercury, cadmium, and lead. This led to closure of the lobster fishery in 1982 in the south arm of the harbour. Testing by Environment Canada indicated the obvious source—the steel-making operations of what was then SYSCO, or the Sydney Steel Corporation, an agency of the Nova Scotia provincial government that had taken control of the failing private-sector operations in late 1967.

In April 1984 the consulting firm Acres International received a contract to determine the scope of the pollution and recommend options. Initial testing indicated that the tar ponds contained the equivalent of 540,000 tonnes (dry weight) of toxic waste, including from 4.4 to 8.8 million pounds of PAHs. The sludge on the bottom of the estuary was judged to be between one and four metres deep. The report by Acres focused on the challenges represented by the PAHs.

Although PCBs had been identified in the earlier study of lobsters, these were not considered to be a problem because they were not a by-product of making coke or steel. The source of the PCBs was waste transformer fluid used in the steel mill. Another source of PCBs was the railway yard operated by Canadian National Railways on the opposite side of Muggah Creek.

The consulting firm conducted random sampling by drilling bore holes in the estuary, and, on that basis, discovered only small quantities of PCBs. From those results and its analysis, Acres identified three options: (1) leave the polluted sludge in place and cover it; (2) remove the sludge and store it somewhere else; and (3) remove the sludge and incinerate it. Acres estimated that incin-eration would destroy 99.99 per cent of the PAH. However, PCBs are virtually indestructible at extreme temperatures and would only have been changed into airborne dioxins and other poisons if incinerated. Given the three options, along with the estimate of the high proportion of PAHs that would be destroyed by incineration and the almost 1,500 person years of work to be generated by incineration, the provincial government selected the incineration option.

In November 1987, the federal Minister of the Environment, along with the Nova Scotia Ministers of Environment and Development, met in Sydney to announce a $34.3 million package that would result in the excavation of the toxic waste in the tar ponds and then its incineration. The workers in the coke ovens would have the first opportunities for employ-ment in the cleanup project. At the press confer-ence, the ministers stated that the tar ponds were the worst toxic waste site in Canada and the second worst in North America. Nevertheless, Environment Canada decided after a preliminary screening that a full environmental impact assess-ment would not be required for the excavation and incineration.

The incinerator was supposed to be opera-tional by 1990. However, as 1992 began, the project was behind schedule and over budget, testing of the systems was still ongoing, and repairs were being made. It also had been decided to conduct further testing of the tar ponds to better understand the contaminants. And, in October 1992, a surprise was encountered. The additional testing by Acres identified a 'hot spot' of PCBs in the south tar pond. It was estimated that this hot spot had 4,000 tonnes of sludge contaminated with PCBs. Canadian law requires PCBs over 50 ppm to be incinerated at a minimum temperature of 1,200° Celsius, and the sample from the south tar pond revealed concentrations up to 633 ppm. The incinerator, designed to destroy only PAHs, had a capacity up to 900° Celsius. At the same time, serious problems were being encountered with the pumping and dredging systems for the sludge and with operating the incinerator.

By the fall of 1994, the problem created by the PCBs was not resolved, although the incinerator and dredging equipment were working (see photo). The originally announced budget of $34 million had climbed to more than $55 million. In Decem-ber 1994, the province decided that the incinerator option was not going to do the job. It called for tenders for new approaches to deal with the con-taminated sludge, and all the bids were above $100 million. The province rejected all the proposals as too expensive. It subsequently invited one Nova Scotia consulting firm, Jacques Whitford, to deter-mine what could be done for $20 million or less. In January 1996, the Nova Scotia Minister of Supply and Services announced he had accepted the Jacques Whitford plan to use the slag piled up next to the tar ponds to fill in the ponds. Once that work was done, grass and trees would be planted to create a park. This proposal was greeted with surprise and anger by the people of Sydney, including the mayor, none of whom had participated in the develop-ment of this 'solution'.

Notwithstanding the opposition, Jacques Whitford started the first phase of its work, at a cost of $5 million, which was a thorough sampling of the tar ponds to determine the extent of the PCBs. By this time 10 years had passed since the federal and provincial governments had announced the cleanup, and under federal law PCBs cannot be buried. The intent was to identify the PCB-contaminated sludge and remove it to a disposal site in Quebec. Throughout the spring the

Incinerator plant, November 2003 (*Timothy Babcock*).

sampling continued, and by midsummer the estimate was that 45,000 tonnes of PCB-contaminated sludge existed, leaving Jacques Whitford to express reservations about its proposal. The outcome was that the 'encapsulation' option was rejected. To this point, $60 million had been spent, and a viable solution had not emerged.

The two levels of government announced that they would pursue a more open and participatory approach, and would establish a community-government committee to develop a cleanup plan. The committee was named the Joint Action Group (JAG). In September 1998 an agreement was reached to clarify the relationship of JAG with the three levels of government and $62 million was committed to complete studies, designs, and other preparations for the cleanup, and in 2000 another consulting firm, Conestoga Rovers and Associates, was hired to manage the agreement intended to lead to the cleanup of the tar ponds as well as the coke ovens site. In the next few years, the following actions were taken: (1) building of a sewer system to divert tonnes of raw sewage flowing daily into the tar ponds; (2) demolition and removal of the derelict structures on the coke oven site; and (3) closure and capping of the old Sydney landfill.

Then, in May 2004, a $400 million cleanup of the tar ponds was announced by the federal and provincial governments. The first stage of the work involved construction of a dam during the summer of 2004 to seal off the flow of the contaminants from the site into the Sydney harbour. The next stage will be the completion of an environmental impact assessment, expected to take at least until 2005. Most of the cleanup work will then be completed between 2006 and 2010, with the federal government contributing $280 million and the Nova Scotia provincial government adding $120 million. The intent is to destroy the worst contaminants using existing proven technologies, treat the remaining material, and then encapsulate both sites using an engineered containment system.

What has been learned from this experience so far? At least three lessons stand out. First, when basic science is not used from the outset to inform policy decisions related to environmental issues, the probability is high that funds will not be allocated to effective solutions. Second, even when science is used, understanding can be incomplete, and decisions will have to be taken in the face of considerable uncertainty. Third, when local stakeholders are not included in the process, challenges can be expected to proposed solutions. In this book, our goal is to help you to enhance your understanding about scientific aspects of ecosystems and how that can be combined with 'best practices' (discussed more in Chapters 5 and 6) related to management of resources and the environment.

The Sydney Tar Ponds situation is a vivid example of challenges that can be faced from industrial legacies of pollution, but it is not unique. Other communities with similar challenges exist: the Lubicon First Nation in Alberta, surrounded by sour gas wells; the White Dog Reserve downstream from Dryden, Ontario, experiencing mercury contamination from a paper mill in Dryden; Deline, NWT, and its abandoned uranium mine; Yellowknife, NWT, with an abandoned gold mine; Port Hope, Ontario, with waste from a refinery for radium and uranium; Newcastle, NB, and Transcona, Manitoba, with abandoned wood preservative plants; and Squamish, BC, with heavy

LESSONS

It would appear that the residents of Sydney ... have suffered and fought in isolation and perhaps in vain. We appear not to have learned one thing from their ordeal. We continue to talk about the 'trade-off' between jobs and the environment; jobs and health. The tar ponds saga should have taught us that such trade-offs are wrong economically, environmentally and morally.

— *Barlow and May (2000: 200)*

metal pollution from an old chloralkali plant, and the nearby abandoned Britannia mine site on Howe Sound. All of these communities face formidable challenges, and the development and implementation of solutions will be major, and often expensive, undertakings. These community-based problems also illustrate that resource and environmental problems can be located within or adjacent to settlements and, indeed, in highly urbanized landscapes. Later, in Chapters 7, 12, and 13, other examples of environmental challenges in more urbanized situations will be presented.

THE GLOBAL PICTURE

Our home, planet Earth (photo), is different from all the other planets we know. Hurtling through space at 107,200 kms per hour, an apparently infinite supply of energy from the sun fuels a life support system that should provide perpetual sustenance for Earth's passengers. Unfortunately, this seems not to be the case. Organisms are becoming extinct at rates unsurpassed for 65 million years. These extinctions cover all life forms and probably represent the largest orgy of extinction experienced in the 4.5-billion-year history of the planet. Our seas are no longer the infinite sources of fish we thought they were. Our forests are dwindling at unprecedented rates. Even the atmosphere is changing in composition and making the spectre of significant climatic change a reality. Every raindrop that falls on this planet bears the indelible stamp of the one organism bringing about these changes—you and us (Box 1.2).

Recent estimates are showing the extent of human transformation of natural ecosystems. Vitousek et al. (1998) calculated that humans now appropriate more than 40 per cent of the gross primary productivity (see Chapter 2) of the planet for their own use. Tropical forests are being eroded at a rate of 0.8 per cent/year, marine fisheries by 1.5 per cent, freshwater ecosystems by 2.4 per cent, and mangroves by 2.5 per cent. Large but difficult to quantify losses are also occurring on coral reefs, croplands, and rangelands (Balmford et al., 2002). The World Wildlife Fund has developed an index, called the Living Planet Index (www.pandaorg/news_facts/publications/general/livingplanet/index.cfm), that quantifies the overall state of planetary ecosystems. The index shows a 35 per cent reduction overall in the planet's ecological health since 1970. Meanwhile, economic development has more than doubled (Figure 1.2).

Why are these changes taking place? One simple answer is that humanity is literally and figuratively lost. We don't know where we are in terms of our relationship with the planet and our life support systems. We don't know where we are going or where we will end up if we continue as we have been doing. We haven't asked if we want to go there or if we are on the right road. We haven't examined alternative destinations and how we might be able to head towards any of these instead. We are lost.

To answer these questions, we must know where we are in the first place. As Environment Canada (2003) stated recently:

Our home, planet Earth (© 1996. *The Living Earth Inc*).

Guest Statement

Perspective on the Tar Ponds

Tim Babcock

I moved to the outskirts of Sydney (on the *other* side of town from the tar ponds) in the fall of 2003, after two decades of working in Indonesia on issues related to resource management. What contrasts between the two places, and yet what similarities!

One of the first things that visitors to Sydney will likely notice, if they arrive by car along the Trans-Canada Highway from the mainland, is a sign on the outskirts (still) proclaiming Sydney as the 'Steel Center of Eastern Canada' (see photo). But this is only a small example of the types of disjunctions one soon comes across in the area. There is no sign, perhaps on the other side of the highway, proclaiming the existence of the most polluted industrial site in Canada.

Sydney: Steel Center of Eastern Canada, March 2004 (*Timothy Babcock*).

I very soon learned, however, that the tar ponds cleanup (for which a whole government agency has been set up) continues to be a major topic of discussion—and a source of frustration and anger—among the residents of the area. It is embedded, of course, in a much larger issue— the dismantling of the steel plant itself, the appropriate disposition and use of the land on

which it stands, and, far more seriously, the devastating effects of the plant's closure on the economy, and on the very fabric of society, of Cape Breton Island. Cape Breton, the recently amalgamated Regional Municipality has declared, is in a 'state of crisis'. Much uncertainty, at least in the public mind, still exists concerning the choice of cleanup measures, the amount of money available, and even when the program will get into high gear. Argument, not all of high quality, still rages concerning the appropriate means to clean up the site among the Sierra Club, the still extant though no longer funded Joint Action Group, members of the public, and the cleanup agency. We are no further ahead than we were in 1996, opine some informed individuals; others, with intimate knowledge, claim that progress is now 'rapid'.

At this point, many tens of millions of dollars have been spent to obtain what one presumes are the best scientific analyses and solutions currently available, and many more dollars have been spent on various forms of 'citizen participation' to discuss and propose a range of solutions. Yet the studies continue to be debated, and confusion, and much unhappiness, reigns. The question

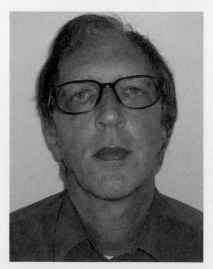

Tim Babcock

remains: were the processes flawed, or are they the 'state of the art', the best that available Canadian money can buy?

Over the past several decades Canada has spent hundreds of millions of other dollars providing what is hopefully good advice to developing countries on how to better manage their environments and natural resources. But are we following 'best practice' here in Canada? Would we wish such a lengthy and complex process as the tar ponds case demonstrates on a poor country with equally serious pollution problems, probably affecting far more people, but with far fewer scientific, educational, and financial resources to deal with them? And we have so far left politics out of our analysis: this adds a further, almost unlimited layer of complexity and uncertainty to the situation, where individuals and groups will inevitably, and selectively, use the results of 'science' to further their own sincerely, or cynically, held positions. There are clearly no simple answers.

Tim Babcock has a Ph.D. in anthropology, and for over 25 years worked on various development projects in Indonesia. He teaches at Cape Breton University College and does consulting work.

BOX 1.2 GLOBE FACTS

- Half the world's forest cover has gone and another 30 per cent is degraded.
- About 170,000 km^2 of tropical forest are destroyed every year. Almost half the original area has already been cleared.
- Amazonian deforestation was 41 per cent higher in 2002 than in 2001.
- More than half the world's wetlands have been converted to other uses.
- Topsoil is disappearing at a rate faster than accumulation on one-third of the world's croplands.
- Production of crops has been reduced by 13 per cent over the last 50 years as a result of environmental degradation.
- World economic activity has grown an average of 3 per cent per year since 1950. At this rate, by the year 2050, global output will be five times larger than it is today.
- Land degradation costs the world about $42 billion a year in lost crop and livestock output.
- 60,000 km^2 of new desert are formed every year, mainly as a result of overgrazing.
- Since pre-industrial times the concentration of carbon dioxide in the atmosphere has increased by 31 per cent and methane by 144 per cent.

- A heat wave in France in 2003 resulted in almost 15,000 deaths.
- Sea level has risen 10–20 cm in the last century.
- Over 3 per cent of the stratospheric ozone layer has been depleted, leading to increases in ultraviolet light striking the Earth's surface of over 6 per cent.
- The ozone hole over the Antarctic reached a record 26 million km^2 in 2003, and scientists predict it could increase further.
- More than 3,000 children die every day in Africa from malaria.
- Industrial fishing has killed off more than 90 per cent of the world's biggest and economically most important fish species.
- The number of large dams has increased from 5,000 to 45,000 since 1950.
- A billion people do not have access to safe water supplies.
- Two out of five people do not have access to adequate sanitation.
- About three-quarters of the world's population have access to one or more televisions.

SOURCES: Worldwatch Institute (2004a, 2004b).

We do not yet have a powerful suite of measures that show the extent to which economic activity is impacting the environment. Over the coming years, as better indicators of the relationship between the economy and environment are developed, we will be able to track how rapidly our economy is embracing environmental values and whether or not economic growth is depleting our natural capital.

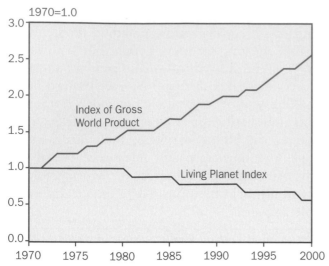

Figure 1.2 Changes in economic activity and ecosystem health, 1970–2000. SOURCE: Worldwatch (2004b).

One main variable that affects our impact on the planetary life support system is the number of passengers being supported. Although countless billions of passengers—from insects and fish to the great blue whale—are on board planet Earth, we are mainly concerned with those who seem to be having most impact on the system, humans or *Homo sapiens sapiens*. This species, along with a few others such as rats and cockroaches, has experienced a staggering increase in population numbers

What the future holds in terms of population growth depends on the reproductive decisions taken by today's children as they grow up. Children in the central highlands of Vietnam (*Philip Dearden*).

over the last century (see Box 1.3). The United Nations forecasts an increase to 8.9 billion people by the year 2050. This is lower than earlier estimates of 9.3 billion but still represents some 76 million additional people per year or 240,000 people per day to feed.

Much of this increase will occur in lesser developed countries, where populations are predicted to grow by 55 per cent, compared with only 4 per cent in developed countries. By 2050 the United Nations has forecast that the populations of the world's 48 least-developed countries will triple. China's massive population would continue to grow until 2025, when economic growth would trigger reductions in fertility. India is predicted to overtake China as the most populous country on Earth by 2050, with a population of 1.6 billion. Nigeria would also experience rapid growth, with the population tripling to 307 million, while Bangladesh would double to 280 million.

The different trajectories of the developed and developing countries are epitomized by Nigeria and Japan. On average, Japanese women give birth to just over one child in their lifetimes. Fourteen per cent of the Japanese population are younger than 15, and 19 per cent are older than 65. In contrast, 44 per cent of Nigeria's population are younger than 15 with only 3 per cent over 65. A typical Nigerian woman gives birth to six children over her lifetime.

In all these predictions there are great uncertainties. Much depends on trends in infant mortality rates and attempts to curb the growth of AIDS in the developing world. The largest generation in history, 1.2 billion people are now between the ages of 10 and 19, and the child-bearing decisions made by this generation will also be critical. If the world's women have just half a child more than predicted, then the population would swell to 10.6 billion by mid-century. And although the average annual growth rate has fallen from over 2 per cent, as it was from the 1950s to the 1990s, to less than 1.3 per cent today, that rate is being applied to a much larger and still increasing population.

The steep curve of population increase, shown in the figure in Box 1.3, coincides with the time that humans learned how to exploit the vast energy supplies of past *photosynthetic* activity laid down as coal and oil in the Earth's crust. Until then, energy supplies had been limited by daily

BOX 1.3 POPULATION AND EXPONENTIAL GROWTH

In early 1798, a British clergyman, Thomas Malthus, pointed out the enduring truth that population growth was geometric in nature (i.e., 2, 4, 8, 16, 32, 64, etc.), whereas the growth in food supply was arithmetic (i.e., 1, 2, 3, 4, 5, etc.). This will inevitably, said Malthus, lead to famine, disease, and war. This was not a popular viewpoint in his day when population growth was considered very beneficial. For many years this Malthusian view was ignored. The opening up of new lands for cultivation in North America and the southern hemisphere and later the development of Green Revolution techniques (Chapter 9) allowed food supplies to increase rapidly.

Increasing numbers of experts, watching the declines in food supplies per capita over the last few years (see Chapter 9) and the increases in population, particularly in less-developed countries, now feel that the Malthusian spectre is here once more. The figure below illustrates how global population has grown over the centuries and millennia. On the other hand, other pundits, particularly economists, feel that more population simply furnishes more resources—human resources—upon which to build increases in wealth for the future. Indeed, there are concerns that some developed countries will start losing population in the future and this will have negative impacts on their economies. For example, Japan is predicted to lose 20 per cent of its popu lation by 2050, with declines also taking place in Germany, Russia, and Italy. Both the US and Canada are predicted to have trends in the opposite direction, largely through immigration. By 2050 it is estimated that Canada's population will have increased by 14 per cent to 37 million through immigration, despite a fertility rate of 1.5 children per couple.

Political leaders from some of the lesser developed countries that are experiencing the most rapid population growth rates have also argued that population growth per se is not a problem, and that the main problem is overconsumption in the more developed countries. This distributive concern is echoed by women's groups—which also are wary of coercive birth control programs—that think that most progress could be achieved by improving the status of women. Women with more education have smaller, healthier families and their children have a better chance of making it out of poverty. Yet two-thirds of the world's 876 million people who can neither read nor write are women, and a majority of the 115 million children not attending school are girls. Women who have the choice to delay marriage and child-bearing past their teens also have fewer children than teen brides. Yet over 100 million girls will be married before their eighteenth birthdays in the next decade.

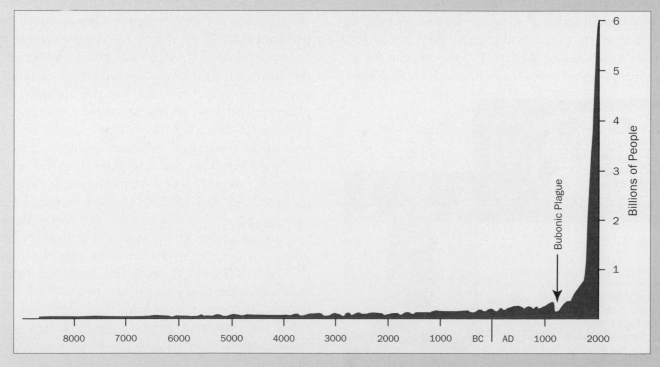

The growth of human population over time

inputs from the sun. The discovery of this new treasure house of energy allowed humans to increase food supplies dramatically and improve and greatly speed up the processing and transportation of materials (see also Chapter 13). Global energy consumption increased more than twelve-fold between 1850 and 1970, while population tripled. Energy use has increased another 73 per cent since 1970. There are now over six billion humans drawing upon the planetary life support system for sustenance; before the Industrial Revolution there were fewer than a billion. Another result of increased energy consumption is the pollution that now chokes this life support system and is causing unprecedented human-induced changes in global climate.

As we noted above, the number of passengers on planet Earth is not the only concern, for each passenger does not have the same impact on the life support system. Some passengers, those in first class, get special meals, three times a day, wine included; those in the economy section are lucky if they get one meal and must buy their own water, if it is available. Indeed, the top 20 per cent of the global population consume about 80 per cent of all resources used on the planet. The 12 per cent of the global population in North America and Europe account for almost 60 per cent of private consumption. This compares with the third of the global population in sub-Saharan Africa and South Asia that account for 3 per cent. In terms of metal use, for example, the 15 per cent of the population living in the US, Canada, Japan, Australia, and Western Europe account for 61 per cent of aluminum use, 60 per cent lead, 59 per cent copper, and 49 per cent steel. The average North American uses 22 kilograms of aluminum a year, the average African less than 1 kg (Worldwatch Institute, 2004b).

Energy consumption is also very unequally distributed, with the people in the wealthiest countries using 25 times more per capita than the world's poorest people. Over a third of the global population does not have access to electricity, but demands are growing. During the 1980s, the demand for electricity in China, for example, increased by more than 400 per cent (Sawin, 2004). There are also large differences in energy consumption among developed countries, with Canadians and Americans consuming 2.4 times as much energy as the average person in Western Europe.

Canadians are among the top per capita consumers of energy in the world (Figure 1.3), as discussed in more detail in Chapter 13. Each Canadian consumes as much energy as 60 Cambodians. Government policies encourage us to be wasteful by subsidizing cheap energy costs, and we as individuals do not resist. Energy is a good index of our planetary impact, reflecting our abilities to process materials and disrupt the environment through pollution such as *acid precipitation* (Chapter 4) and the production of *greenhouse gases* (Chapter 7).

Obviously, very different kinds of passengers share our planet, and the differences among them have increased rather than been reduced by increased wealth over the last 20 years. **Gross national product** (GNP) is an index that economists use to compare the market value of all goods and services produced for final consumption in an economy during one year. Over the last two decades, the planetary GNP has risen by $20 trillion, but only 15 per cent of this has trickled down to the 80 per cent of the passengers in the economy section of the spaceship. The rest has made the rich even richer. In contrast, over 2.4 billion people exist on under $2 per day and half of these people are in extreme poverty, as defined by the World Bank, and live on under $1 per day (www.development goals.com/poverty.html). Yet each year the poor countries of the world pay to the developed industrial/post-industrial world more than four times in interest payments the amount they receive in so-called aid programs. This has been likened to a hospital where the sick give blood transfusions to the healthy.

Although much is being done throughout the world to curb population growth, as seen in this sign-board in Vietnam encouraging couples to have only one child, consumption knows no bounds and we are constantly exhorted to buy more (*Philip Dearden*).

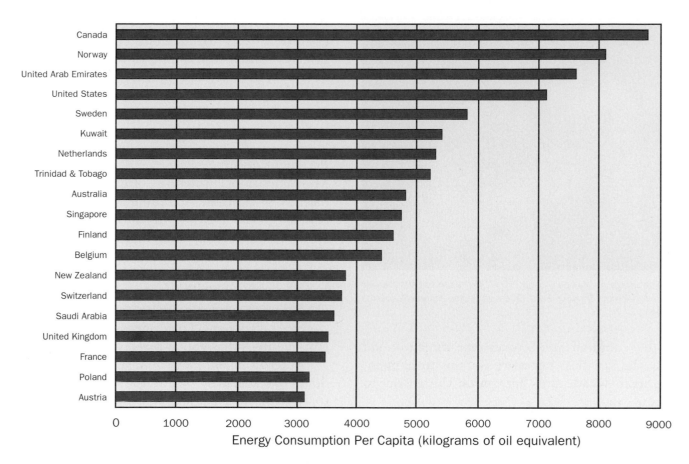

Figure 1.3 Energy consumption per capita in selected countries. The graph includes the countries with the highest energy consumption per capita. SOURCE: Resources Futures Inernational, *Global Change and Canadians* (Ottawa: Canadian Global Change Program, 1993): 12.

The stresses being put on the planetary life support system are a result of these factors of over-consumption and the resulting pollution, as well as overpopulation and the resulting poverty. Together,

Families in Lesotho are still large, but the planetary impact of this entire family will be a fraction of that of one Canadian child (*Philip Dearden*).

they result in pressures on the **carrying capacity** of the planet at all different scales. Although in the past there have been many examples of cultures violating the carrying capacities of their local environments with dire results, never before have we approached these limits at the global scale.

If we return to our original metaphor, although we may not be able to see every twist and turn of the path upon which we tread, we can at least get a fairly clear look at the direction in which we are heading. The prospects are not good. Some of these trends were brought to the public stage at the Earth Summit held in Rio de Janeiro in 1992 and at the follow-up meeting, the World Summit on Sustainable Development (WSSD) held in Johannesburg 10 years later. Details of some of the outcomes of these meetings are discussed in individual chapters, with an overall summary provided in Chapter 14. The main question remains, having glimpsed the direction of the road, whether the international community can come together

Satellite image of Canada (*Toutin, 1997. Processed at the Canada Centre for Remote Sensing, Earth Sciences Sector, Natural Resources Canada.*)

for the common good, to change direction and make the sacrifices necessary for this to happen. The next decade will determine the answer to this question.

THE CANADIAN PICTURE

Where does Canada, one of the most privileged nations in the world and one of the highest per capita consumers of energy, fit within this global picture? Our land is vast, about 13 million km^2, and our population small, about 32 million people, as shown in Figure 1.4 (Statistics Canada, 2003).

Population density is 0.04 people per hectare, compared with Bangladesh at 8 people per hectare. Canada would have to have a population of over 8 billion to equal this density. However, some people feel that Canada is already overpopulated. Between 1971 and 1991 the population increased from 22 million to 28 million, an increase of 27.7 per cent, and from 1951 to 2001 the population increased by 122 per cent (ibid.). Most population growth in Canada has occurred recently as a result of immigration rather than natural increase. Does this reflect concurrence with the economists' view that the more people the better? Are more people good for Canada when we consider there is more to Canada than just the economy? Does Canada have a moral obligation to accept migrants from overcrowded countries elsewhere? These are some of the important questions that policy-makers must consider.

In terms of numbers alone, there can be little doubt that Canada is not overpopulated compared to virtually any other country on Earth. Canada is the second largest country in the world in terms of area and includes 20 per cent of the world's wilderness, 24 per cent of its wetlands, 10 per cent of its forests, 9 per cent of its fresh water, and has the longest coastline in the world. However, as discussed above, it is not simply numbers of people, but rather the impacts of those people that are critical. Canadians are among the top producers per capita in the world of industrial and house-

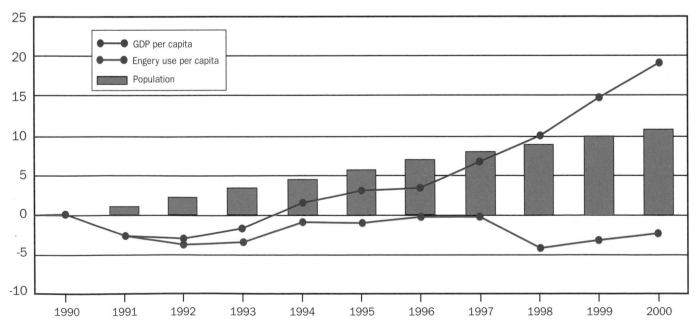

Figure 1.4 Change in population, GDP per capita, and energy use per capita (% change since 1990). SOURCE: Environment Canada (2003: vi)

hold garbage, hazardous wastes, and greenhouse gases (Box 1.4). Some point to the size of the country, the cold in winter, and heat in summer as reasons behind our remarkable energy consumption, but it is clear that Canadians can contribute substantially in reducing impacts on the planetary life support system. Suggestions on some of the ways that you can help contribute to the changes that need to be made are given throughout the text, and this is a theme that will be returned to in Chapter 14.

BOX 1.4 CANADA FACTS

- We generate about 360 kilograms of solid waste per capita per year, ranking seventh in the world.
- We generate almost 6 tonnes of hazardous waste for each US $1 million of goods and services produced; Japan generates less than a quarter of a tonne.
- We have one of the highest per capita uses of water in the world, about 15 cubic metres/day, roughly three times that of Sweden and Japan.
- We have one of the highest per capita energy uses in the world.
- We use our cars nearly 10 per cent more than residents of other industrialized countries.
- We emit 2 per cent of the world's greenhouse gases, with 0.5 per cent of the world's population, and rank eighth in global production of greenhouse gases per capita. The five top countries are all small oil-producing countries where gas leakage is high; among comparable countries, only the US and Australia surpass Canada.

JURISDICTIONAL ARRANGEMENTS FOR ENVIRONMENTAL MANAGEMENT IN CANADA

As noted in Mitchell (2004), no one government has total authority or responsibility for natural resources and the environment in Canada. Instead, under the Canadian Constitution, authority is divided between the federal and provincial governments, with territorial and municipal governments increasingly having a role. First Nations also are increasing their role, commensurate with their being recognized as another level of government. In addition, Canada is involved in bilateral arrangements with the United States to address environmental problems such as air pollution and to deal with shared water bodies such as the Great Lakes, and in multilateral arrangements with other nations or international organizations related to resources such as fisheries, migratory birds and animals, and minerals on or under the ocean floor.

Federal, Provincial, and Municipal Roles

Canada is a federated state, with power and authority shared between federal and provincial governments, and with municipal governments receiving their power and authority from provincial legislatures. The Constitution Act, 1982 differentiates between proprietary rights and legislative authority relative to natural resources and the environment. Ownership of all Crown lands and natural resources not specifically in private ownership is given to the provinces, except for the Canadian North (north of 60 degrees latitude), where the federal government has proprietary rights to land and resources until territories receive such power, and for resources found on or under seabeds off the coasts of Canada (but some provinces are challenging this right).

Legislative authority is mixed between the federal and provincial governments, often becoming a significant source of conflict. The federal government has jurisdiction over trade and commerce, giving it substantial authority over both interprovincial and export trading of resources (oil and natural gas, water). Alberta, Saskatchewan, and British Columbia, which have oil and natural gas, often object to the federal government becoming involved in setting prices and determining buyers, arguing that such matters are within provincial authority due to their responsibility for property and civil rights, and also because these resources are a matter of local interest. The federal government has used its legislative authority for navigation and shipping and for fisheries to create water pollution regulations—even though water within provinces is under the authority of the provinces. Thus, there are ambiguities and inconsistencies regarding jurisdiction over resources and the environment. One consequence is that it has been difficult to establish *national approaches* (combined federal and provincial) to deal with resource and environmental issues.

In the early to mid-1990s, many provincial governments began to download selected responsibilities, which they had traditionally held, to municipalities. The provinces argued that downloading was consistent with the principle of **subsidiarity**, which stipulates that decisions should be taken at the level closest to where consequences are most noticeable. While such an argument is rational, others have suggested that the primary motive for downloading was the desire of provincial governments to shift the cost of many responsibilities to lower levels of government as part of a strategy to reduce provincial debts and deficits. Whatever the motivation, the outcome has been that municipalities have become much more significant players in natural resource and environmental management, since, in many instances, provinces have withdrawn from related management activities.

There are instances of effective partnerships between provincial and municipal governments. One of the best and most enduring examples is the Ontario Conservation Authorities, watershed-based organizations established by statute in 1946 to manage many renewable resources within river basins. Individual authorities were established when two or more municipalities in a watershed petitioned the provincial government to establish one. When a majority of the municipalities in a watershed agreed that they would work collaboratively, then the province established a Conservation Authority. Today, 36 authorities exist, primarily in the more settled parts of the province.

While the provincial government would not impose a Conservation Authority, it initially provided a strong incentive for local governments to form one by offering funds not available to municipalities on their own but available after a Conservation Authority was established. This cost-sharing arrangement was a powerful stimulus for municipalities to agree to establish authorities, and for many years the cost-sharing was 50/50 between the province and the municipalities. This arrangement meant that, as Lord (1974: x) reported, 'an authority can flourish only when the local people have enough enthusiasm and conviction to support it financially. It has also meant that the authority does not exceed the financial

resources of its jurisdiction.' However, in the mid-1990s, the provincial government significantly reduced its proportion of the funding to the Conservation Authorities, as part of a drive to reduce government activities.

MEASURING PROGRESS

Some idea of global comparisons among countries is provided by comparisons of **ecological footprints** (Box 1.5). Some countries show ecological demands greatly in excess of their capabilities (Figure 1.5), and they import ecological capital from elsewhere to make up for this deficit. Although trade between nations is to be expected, this excess of ecological footprint over capacity allows nations to live beyond their ecological means. Canada has one of the largest available

When you have your morning cup of coffee or tea with sugar your ecological footprint is reaching out to the tropics where both drinks and sugar are grown. Tea plantation, Sri Lanka (top). Sugar cane fields, South Africa (bottom) (*Philip Dearden*).

BOX 1.5 YOUR ECOLOGICAL FOOTPRINT

Your 'ecological footprint' is the land base required to provide your needs, including all energy and material requirements, and also to dispose of your wastes. In fact, most of a person's footprint is caused by the space needed to absorb waste from energy consumption, especially carbon dioxide. Can nature provide enough of these services on an ongoing basis to meet the needs of an individual, community, or nation?

A 2004 study undertaken for the Federation of Canadian Municipalities (see www.fcm.ca/ english/communications/eco.pdf) examined the ecological footprints of 20 municipalities, that is, the area of land and sea throughout the world that is required to produce the amount of food, energy, and other materials the citizens use. The Canadian average was 7.25 hectares, third in the world rankings. The lowest and highest footprints were both in Ontario. Sudbury was the lowest at 6.87 hectares with York the highest at 10.33. One reason behind Sudbury's success is through the efforts of the public works engineer who is in charge of heating and sewage. He has introduced many innovative programs especially

related to using local energy sources such as the power of the wind, sun, and the Earth's heat that have reduced both the cost of heating and resource consumption. It is a striking example of how the initiative and energy of one person can make a significant improvement in planetary resource use.

York-Durham, on the other hand, uses 43 per cent more resources than the average Canadian. This is mainly a result of the fuel consumption generated by the sprawling suburban developments near Toronto that were based on car transportation with few other options.

On a global scale, only 1.6 hectares are available per person, and this amount is shrinking every year, largely as a result of population growth. Yet the collective global footprint is 2.2 hectares per person, with a North American average double that for Europeans and seven times greater than it is in Asia or Africa. To provide for everyone at Canadian standards would require three Earths, not one. By the year 2030, estimates suggest that only 0.9 ha will be available per person.

ecological capacities; we also have one of the largest ecological footprints per capita.

Another way to help assess environmental trends is through **indicators**. Earlier we talked of looking down the road to see which direction we are heading regarding planetary conditions. There

is a great consensus among scientists that the prognosis is not good. There are, however, some differences in opinion about the speed we are travelling along this road to environmental degradation. Indicators provide a speedometer to help assess the situation.

Indicators are not new. Doctors for many years have been using body temperature, measured easily by thermometer, as an indicator of the health of the human body. Gross domestic product has been used as an indicator of economic performance, as has the Dow Jones industrial average. These indicators tell us something of the current state of a particular system, but they do not help us understand why the system is in that state. One framework that helps develop causal linkages between indicators is described in Box 1.6. Over the last 15 years, there has been growing awareness of the need to develop indicators that would also gauge the health of other aspects of societal well-being, including the environment. Indicators are often used to supply information on environmental problems to enable policy-makers to value their seriousness; to support policy development and

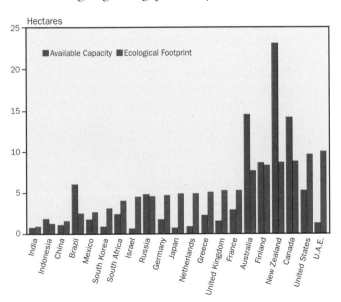

Figure 1.5 Ecological footprint per person in selected nations, 1999. SOURCE: Worldwatch (2004).

BOX 1.6 THE DPSIR INDICATOR FRAMEWORK

The most widespread framework for classifying environmental indicators is the Drivers-Pressures-State-Impact-Response (DPSIR) framework developed by the Organization for Economic Co-operation and Development (OECD) and adopted by all EU countries, the US, Canada, Australia, Japan, and many developing countries (e.g., Malaysia) and international organizations (such as the Commission for Sustainable Development of the UN, the United Nations Environment Program [UNEP], and the World Bank). The framework, as shown in the figure below, is popular due to its organization around key causal mechanisms of environmental problems.

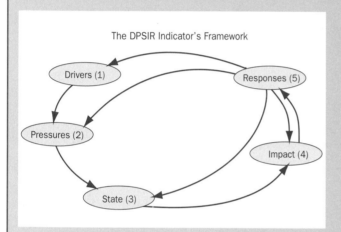

The DPSIR Indicator's Framework

Driving forces (1). Drivers are the underlying forces causing environmental change. They describe social, demographic, and economic developments in societies and corresponding changes in lifestyles, overall levels of consumption, and production patterns. Examples of drivers include population pressures and the demand for various consumer goods and services (e.g., cars, demand for red meat, increasing travel, etc).

Pressure indicators (2). These are the resulting pressures on environment from the drivers. Examples would include emission of pollutants, use of resources, use of land for roads, water withdrawals, deforestation, fisheries catch, etc. Initial interest in indicators focused further down the causal chain in the state indicators, to be described next. However, there is now widespread realization

that these are mere reflections of changes further up the chain, promoting much greater interest in drivers and pressures. One example of the link between drivers and pressures would be for a driver such as the number of cars. Not only can we address the driver (by seeking to limit the number of cars through better public transportation, raising the price of gas, or imposing special taxes, licensing fees, or tolls on those people who insist on driving their cars into downtown urban areas or cities during the daytime), but we can also try to reduce the pressure by making vehicles more fuel efficient and less polluting.

State indicators (3). State indicators describe the quantity and quality of physical phenomena (e.g., temperature), biological phenomena (e.g., fish stocks, extinctions), and chemical phenomena (e.g., CO_2 concentrations, phosphorus loading) and tell us the current state of a particular environmental system. They are often tracked over time to produce a *trend*.

Impact indicators (4). The changes in the state of the environment described by the state indicators result in societal impacts. For example, rises in global temperatures (a state indicator) have impacts on crop productivity, fisheries values, water availability, flooding, and so on.

Response indicators (5). Response indicators measure the effectiveness of attempts to prevent, compensate, ameliorate, or adapt to environmental changes and may be collective or individual efforts, both governmental and non-governmental. Responses may include regulatory action, environmental or research expenditures, public opinion and consumer preferences, changes in management strategies, and provision of environmental information. Examples include the number of cars with pollution control or houses with water-efficient utilities, the percentage of waste that communities and households recycle, use of public transport, passage of legislation, etc. Response indicators are critical to assess the effectiveness of policy interventions but are often the most difficult to develop and interpret.

the setting of priorities by identifying key factors that cause pressure on the environment; to monitor the effects of policy responses; and to raise public awareness and generate support for government actions.

A useful example of the DPSIR approach described in Box 1.6 is provided by the joint Environment Canada–US Environmental Protection Agency (2003) series of indicators on the state of the Great Lakes, which is broken down into pressure, state, and response indicators. It is interesting to note that two of the richest nations in the world found that insufficient data exist on many of the 80 desired variables. On the basis of the 43 indicators used, the conclusion is that the overall trend is 'mixed', i.e., some indicators show improving conditions, others deterioration.

Environment Canada (2003) has created a set of national environmental indicators. These are also a good source of information for monitoring Canada's progress, and are used in this book where appropriate. A description of the meter approach taken by Environment Canada to communicate the essential information shown by each indicator is provided in Box 1.7. Indicators can also be misleading, however. For example, in Environment Canada's 2003 report, key indicators are used to highlight the country's progress on various issues. For 'Biodiversity and Protected Areas', the highlight indicator is the amount of protected area over time in Canada. This shows an impressive 70 per cent rise since 1992 and leads to an overall meter rate of 70 per cent improvement. Yet, the number of species on Canada's endangered list has constantly risen over the last decade (Chapter 11) and now includes over 400 species. When these species have been reassessed to determine whether their status has improved as a result of conservation programs, only 16 per cent have improved over time, as shown through another indicator (Environment Canada, 2003). Had the highlight indicator been based on actual threats to endangered species, there would have been a far different outcome than the 70 per cent improvement shown under the selected highlight indicator. Clearly, the indicators chosen and highlighted can have a major influence on how progress is perceived.

Another difficult issue to resolve is the degree of aggregation of information included in an indicator. An almost infinite amount of information

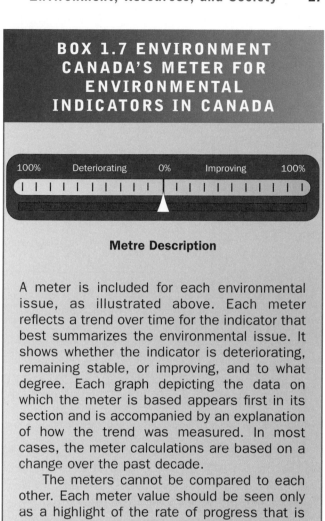

BOX 1.7 ENVIRONMENT CANADA'S METER FOR ENVIRONMENTAL INDICATORS IN CANADA

Metre Description

A meter is included for each environmental issue, as illustrated above. Each meter reflects a trend over time for the indicator that best summarizes the environmental issue. It shows whether the indicator is deteriorating, remaining stable, or improving, and to what degree. Each graph depicting the data on which the meter is based appears first in its section and is accompanied by an explanation of how the trend was measured. In most cases, the meter calculations are based on a change over the past decade.

The meters cannot be compared to each other. Each meter value should be seen only as a highlight of the rate of progress that is occurring in the issue. They do not allow comparisons of the relative importance of issues, and they do not show change with respect to specific science-based thresholds. Furthermore, the meters provide a national roll-up and therefore do not represent regional variation.

SOURCE: Statistics Canada (2003).

could be collected on environmental systems (Figure 1.6). Much of this information might be useful in understanding the basic nature of the system, but it is not necessary for decision-making. Research scientists and line agencies may be involved in the routine collection of such data, and without such data meaningful indicators cannot be constructed. At a higher level of sophistication, these raw data may form an *integrated database*, such as the integration of social and biophysical data as a basis for integrated watershed management planning. However, synthesis of these data into indica-

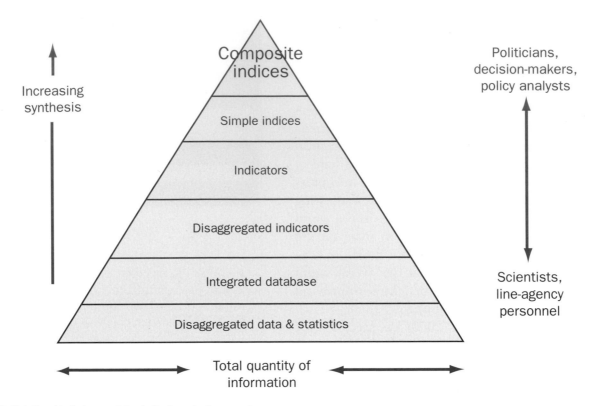

Figure 1.6 Relationship between data, indicators, indices, and users.

tors is often most useful to decision–makers, and indicators themselves may show greater or lesser degrees of aggregation, especially in a spatial sense. Higher levels of thematic aggregation produce *indices.* Simple indices are composed mainly of indicators that are all similar. The well-known Dow Jones industrial average, for example, combines changes in market processes for 30 bluechip stocks listed on the New York Stock Exchange.

Figure 1.7 shows an example of the various components of the Land Quality Index (LQI) that has been under design by a joint program sponsored by the World Bank, the Food and Agricultural Organization (FAO) of the United Nations, the UN Environment Program (UNEP), the UN Development Program (UNDP), and CGIAR (Consultative Group on International Agricultural Research). Each of the indicators forming the LQI may, in turn, be broken down into more specific indicators, as is shown for land degradation (see

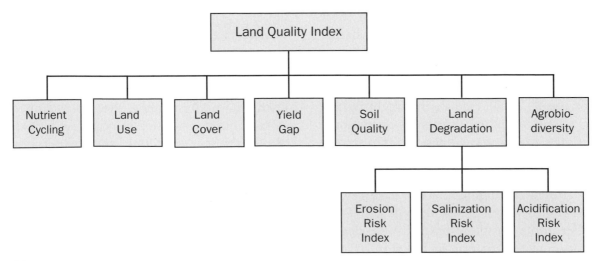

Figure 1.7 The Land Quality Index (LQI).

Dumanski, 2000; World Bank, 1997). The Living Planet Index, mentioned earlier, is another example.

Composite indices are often the most useful for decision-makers and represent the highest level of aggregation. They are few in number and incorporate many, often very different, sub-variables. The Human Development Index (HDI) created by the UNDP, the Environmental Sustainability Index (ESI) of the United Nations, and GNP are examples of aggregate indices. Many countries are developing their own composite indices to represent their progress towards environmental sustainability. The United Kingdom has a system of 15 headline indicators, for example, while the European Union has called upon 2,300 experts to boil down 60 indicators into a Sustainable Development Index of 10 themes. Many jurisdictions also use composite indicators to represent overall air and water quality.

In Canada, the National Round Table on the Environment and Economy (NRTEE) was charged by the Minister of Finance in 2002 with developing indicators to supplement the main economic indicators with information related to natural and human capital (NRTEE, 2003). They developed indicators related to water quality, forest cover, air quality, greenhouse gas emissions, wetlands cover-

Even if we could determine the value of insect pollinators to Canada's agriculture adequately, money could never buy the services they provide (*Philip Dearden*).

age, and educational attainment, but stopped short of the intended step of synthesizing these into a composite index to represent the total capital of Canada (economic, social, and environmental) that could be tracked on an ongoing basis to assess sustainability. Part of the rationale for not proceeding was the unease with aggregating together such different dimensions of capital and assuming that they were substitutable. For example, economists have derived ways of assessing the monetary value of ecosystem services, such as that provided by forests absorbing rain to reduce the damage caused by flooding, or the value of wild insects for pollinating agricultural crops. However, because we think we can measure the value of these services in money does not mean that money can buy these services. They are not substitutable.

Composite indices are highly attractive because they convey a lot of information and are useful for making macro-level policy decisions. However, these highly aggregated indicators also carry risks. They often tell us what is happening at the macro level but add little to explaining why. They may mask the complex detail that decision-makers require to make informed decisions. Composite indices must be highly transparent and able to be disaggregated to facilitate understanding of why change is occurring.

Ecological footprints and indicators provide some basis for assessing change and comparison among countries, but they raise the question about the role of science in environmental decision-making. This is discussed in more detail in the next section.

IMPLICATIONS

We appear to be violating global thresholds related to the carrying capacity of the life support system of the planet. We have ceased to live off the interest and are consuming the capital at such a rate that it threatens the future viability of the system. Many species reach such carrying capacity limits with their environment, overshoot them, and have their numbers drastically reduced by environmental factors. So far, we have been able to avoid this process because of human technological abilities, which have increased carrying capacities. But can we continue to increase our numbers and our habits of consumption indefinitely? Or must even humans accept some limits to activities and numbers?

If the answer to the latter question is yes, then identification of the changes that need to be made in general terms is not that difficult. We need to balance birth and death rates, restore climatic stability, protect our atmosphere and waters from excessive pollution, curb deforestation and replant trees, protect the remaining natural habitats, and stabilize soils. The challenge, however, is charting the course necessary to fulfill these objectives. The Soviet Union before its demise had possibly the most stringent and comprehensive environmental protection regulations in existence, and yet still ended up as one of the most polluted environments on Earth. The regulations were simply not enforced. The secret is having a course charted that not only addresses the goals mentioned above, but is actually able to achieve these goals through strategies that can be implemented.

In this book, we aim to provide some background as to how this can be achieved, with particular reference to the Canadian situation. In this chapter we started by discussing the characteristics of the 'environment' and 'resources'. We also examined different approaches to understanding complex systems and considered issues related to the use of 'science' in decision- and policy-making. The case study of the Sydney Tar Ponds dramatically illustrates the complexity of the environmental challenges being faced, even at the local level. They are characterized by uncertainty, rapid change and conflict, and the need to appreciate both the scientific and technical aspects of a problem and the social dimensions. This book attempts to provide an introduction to both of these aspects. Part B outlines some of the main processes of the *ecosphere*, the basic functionings of the planetary life support system, and the ways in which we are interrupting them. Part C details some of the main

planning and management approaches that have evolved within the Canadian context to address environmental challenges. Part D provides a thematic assessment of the challenges associated with particular activities such as fisheries, forestry, agriculture, wildlife use, water, energy production, and mineral extraction.

The relationship among these different aspects is illustrated in Figure 1.8. Natural systems form the basis for all human activity. A system is a recurring process of cause-and-effect pathways (Box 1.8). They range in scale from the giant atmospheric and oceanic circulation systems through to the processes underway in a single living cell. The pollination of a flower by a bee, the melting of a glacier, and the biological fixation of nitrogen from the atmosphere are all parts of such systems. These systems are infinitely complex, and there is a great deal of uncertainty as to how they function. One of the goals of natural science is to try to understand this complexity. We do this by constructing simplified models of how we think they work (Box 1, Figure 1.8). The models presented in Chapters 2 and 4 on how energy flows through the biosphere and on the nature of biogeochemical cycles are examples of these kinds of simplified representations of natural systems.

We do not know all the facts relating to these systems. We do not know all the components, let alone their functional relationships (Box 2, Figure 1.8). Many species, especially insects, even in well-explored temperate countries such as Canada, yet

BOX 1.8 SYSTEMS

You are in the educational system. You use the transportation system to go to your college or university that is warmed by a heating system. What is a system?

Systems are composed of sets of things—e.g., educational institutions, buses, heating components—that are all related and linked together in some sort of functional way. Between these different components there is the flow of material, such as students, passengers, or heat, subject to some driving force—a thirst for knowledge, the need to get somewhere, or a need to get or stay warm. Systems are generalized ways of looking at these processes.

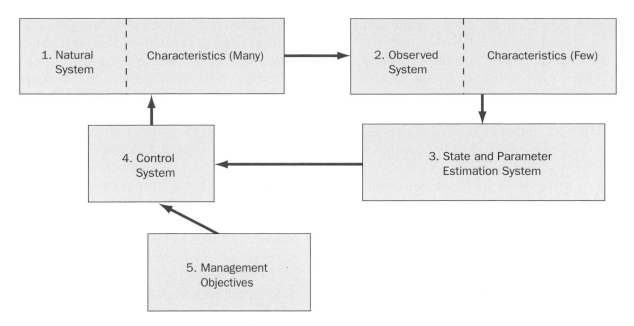

Figure 1.8 Simplified model of interaction of biophysical and social systems in resource management.

await discovery. Of those we do know about, we have to select which ones we think are important and worth representing in our simplified models. Only recently, for example, have we become aware of the critical role played by various lichens in the circulation and retention of nitrogen in temperate rain forests (Chapter 4). Furthermore, not all characteristics are measurable, even if we are aware of their existence. Thus, the simplified models that we use to understand natural systems are fraught with uncertainty.

On the basis of these models, we try to estimate the status of a given system (Box 3, Figure 1.8). How many fish spawn in a certain river? What proportion of the landscape supports commercial tree growth? What soil characteristics are suitable to support a given crop? If we understand the current status of the system, we can also start to ask what the result will be if certain parameters are changed. What would happen to the system, for example, if we took a certain number of fish from the river before they spawned, if we removed tree growth from a portion of the landscape, or if we grew a given crop in the same soil for a particular time period? In other words, we try to assess the impacts of various changes to the system. Quite formal processes of impact assessment have arisen in many jurisdictions that contain not only impacts on natural systems, but also on social systems, as described in Chapter 6.

On the basis of this understanding we try to

replace natural systems with control systems in which the main decision regulators are humans rather than nature (Box 4, Figure 1.8). Instead of natural forces determining the number of fish that reach the spawning grounds, or the age of trees before they are replaced by other trees, or what species will grow in a particular location, people make these decisions as we modify the environment to our own advantage. These control systems are considered under topics such as forestry, water, energy, and agriculture in Part D of the book.

Control systems are implemented on the basis of the social, economic, technological, and management constraints of a society (Box 5, Figure 1.8). These factors influence the demands for various outputs from the system and the speed of extraction. The environmental management strategies

The moral I labor toward is that a landscape as splendid as that of the Colorado Plateau can best be understood and given human significance by poets who have their feet planted in concrete—concrete data—and by scientists whose heads and hearts have not lost the capacity for wonder. Any good poet, in our age at least, must begin with the scientific view of the world; and any scientist worth listening to must be something of a poet, must possess the ability to communicate to the rest of us his sense of love and wonder at what his work discovers.

— Abbey (1977: 87)

discussed in Part C of the book outline some of the main approaches to mediating between the social and economic demands of the society and the productive capacity of the system. As with the natural system, these strategies are characterized not only by complexity and uncertainty, but also by *conflict* among different societal groups regarding the rate of outputs and distribution of benefits. Simply deciding which groups in society have a legitimate interest in a particular environmental issue is quite complex, as described in Chapter 11. Various dispute resolution mechanisms (Chapter 6) have emerged to address the conflicts arising from resource allocation decisions.

As we can see already, the challenges faced in environmental management are complex indeed, and consequently it is necessary to employ an integrated approach, such as the ecosystem approach described in Chapter 5, to understand this complexity. Both natural and social systems are fraught with uncertainty, making an adaptive approach (Chapter 6) a necessity, with strong adherence to the *precautionary principle*, as discussed in Chapter 5. Furthermore, our present predicament is largely the result of modification of natural systems before we had invested the time and effort—or had sufficient data or scientific expertise—to understand the consequences of our actions. The fisheries on both the Pacific and Atlantic coasts of Canada are in trouble (see Chapter 8) because our simplified system models were inadequate as a basis for decision-making. However, in some instances, even where the long-term implications of an activity on the future viability of a resource are understood, they still continue for political and economic reasons. The overharvesting of timber (Chapter 10) is a good example.

Perhaps the most important message underlying the environmental challenges we face is the need for a fundamental change in how we view our relationship with nature, as discussed in Chapter 14. These changes must take place at all levels, from international agencies such as the World Bank, through to national and regional governments, down to household and individual initiatives. Part of the goal of this book is to motivate the reader to become more involved in making these changes happen, both locally and globally.

That people want to become involved in helping to make this planet we call 'home' a better place was demonstrated dramatically by the overwhelming response worldwide to the massive earthquake off the coast of Sumatra and subsequent tsunamis that devastated coastal areas throughout South Asia and reached as far as East Africa, on 26 December 2004, killing over 220,000 people. Yet, when the chief scientist for the World Conservation Union (IUCN) suggested in an interview with Agence France Press that tourist and aquaculture development in South Asia had contributed to the extent of the loss of life and destruction, some media pundits for whom 'economic progress' is most important accused him of being, among other things, an 'environmental fundamentalist' who was part of the '"hate people" propaganda movement' (<www.brandmalaysia.com> and <www.dinocrat.com>).

In fact, caring about the environment means caring about people, as an editorial from the *Bangkok Post* (see box) suggests. The environmental damage from the tidal waves was extensive, with sea water reaching kilometres inland in some places. Some farmers in Sri Lanka, for example, lost everything and estimate some areas cannot be planted for up to two years, and as much as 80 per cent of the Sri Lankan fishing fleet was destroyed. Nesting areas of endangered sea turtles were demolished. In Sri Lanka's largest wildlife reserve, Yala National Park, human-built structures were destroyed and lives were lost, but the animals, apparently sensing the oncoming wave, fled to higher ground and survived. The United Nations initially allocated $1 million for assessment of the environmental damage in the region. In view of this global tragedy, we might do well to assess the environmental damage we cause in our daily lives.

A village on the coast of Sumatra following the tsunami of 26 December 2004 (*US Navy Photo by Photographer's Mate 2nd class Philip A. McDaniel* [Released]).

AN ENVIRONMENTAL VIEW OF THE ASIAN TSUNAMI

Jeff McNeely, chief scientist with the World Conservation Union (IUCN) based in Switzerland, hit the nail on the head when he attributed the enormous loss of life from Sunday's killer tidal waves ... to the human intrusion on natural shorelines. The IUCN environmentalist has lived for several years in Indonesia and Thailand. He no doubt saw for himself a lot of unhealthy development along these countries' coastlines and so had no difficulty in explaining why the tidal waves created by the earthquake in northwest Sumatra could have caused such widespread disaster among communities along the Indian Ocean rim.

Mr McNeely was simply stating the facts when he described Sunday's catastrophe as 'nothing new for nature'. He also was absolutely correct when he said that what made this natural phenomenon a disaster was that people now were occupying part of the geography that they should not.

Fifty years ago, the coastline rimming the Indian Ocean was occupied only by fishermen, not huge tourist hotels and associated attractions. As the tourist facilities mushroomed, and shrimp farms and other such 'developments' also competed for space, the coral reefs and mangrove forests which provided a natural barrier against heavy seas were cleared away. This is very much the case along Thailand's Andaman coast, especially in Phuket, Phangnga and Krabi—the provinces hit hardest by the killer tidal waves ... Much of the coastline where many of the tourist facilities stood before that great leveller, Mother Nature, struck was once public land and forest reserves.

The tourist establishments were allowed to put down roots using dubious legal documents on land which should be preserved in perpetuity as protected coastline or as offshore natural resources. The authorities were happy to turn a blind eye to these ill-conceived and often illegally developed facilities while they continued to attract freespending tourists from around the world to Thailand's beach resorts....

It will take years to rebuild—if this is considered wise—what was lost or damaged. There is clearly much to be learned from what took place on Sunday. It would make no sense at all to embark once again on development patterns that contributed to such heavy losses.... [B]efore any redevelopment goes ahead, the public and private sectors must realize that exploiting our coastal resources for tourism can carry a heavy price tag. Unchecked encroachment on our beaches and other areas of coastline to build tourist resorts perpetuates a disastrous trend of destroying our natural resources and confirms... that we are undermining our own natural protection from the likes of Sunday's disaster.

This disaster is a national crisis but, as many are so quick to point out, from crisis comes opportunity—in this case, the opportunity to correct what has gone wrong with our tourism development.

— Bangkok Post, 29 December 2004, accessed at: <www.forests.org/articles/print. asp?linkid=37689>

SUMMARY

from the outset to inform policy decisions related to environmental issues, the probability is high that effective solutions will not be found; (2) even when science is used, understanding can be incomplete, and decisions will have to be taken in the face of considerable uncertainty; and (3) when local stakeholders are not included in the process, challenges can be expected to proposed solutions. The third point reminds us that while science is important, it often is not sufficient by itself.

7. At the global scale, there is undeniable evidence of unprecedented environmental degradation as a result of human activities. Growing global population is a main challenge, as are the consumer demands of people in the wealthier countries.

8. Conditions continue to deteriorate in many poorer countries.

9. Canada is one of the most privileged countries, covering some 13 million km^2 and having a popula-

tion of some 32 million people. However, our environmental impacts are considerable. Our per capita consumption of water and energy is among the highest in the world. We also have some of the highest production per capita of waste products, including greenhouse gases.

10. Responsibility for the environment and natural resources is divided between the federal and provincial governments, with the territories and municipalities taking on increasingly important roles. First Nations also are much more involved. The shared responsibility often requires collaboration and partnerships, which can create tensions because of different interests and perspectives.

11. It is possible that many of the environmental problems could become acute over the next couple of decades. This will challenge our abilities to understand the Earth as a life support system and our management of the human activities causing changes.

KEY TERMS

anthropocentric view	ecocentric (biocentric)	environment	indicators
carrying capacity	view	estuary	resources
	ecological footprint	gross national product	subsidiarity

REVIEW QUESTIONS

1. What information is available in your municipality or province regarding environmental hazards? Could a Sydney Tar Ponds occur in your community?

2. Who should be responsible for dealing with environmental hazards resulting from earlier resource use and environmental standards no longer acceptable today?

3. If you were hired to provide recommendations related to the Sydney Tar Ponds, what information would you need to make a decision about the potential risk to health for people living in the community? How would you place a monetary value on any potential risk?

4. Outline the main arguments for considering population growth as a threat to global carrying capacity or as a building block for future economic growth.

5. What moral obligations, if any, do Canadians have to assist people in the developing world whose standards of living do not meet basic human needs?

6. Is population growth or environmental degradation the major problem in the less-developed countries? Which is cause and which is effect?

7. What are some of the main initiatives of Canadian governments to address environmental problems?

8. What is a system? Outline the components of a system that you use on a regular basis.

RELATED WEBSITES

SELECTION OF INDICATORS FOR GREAT LAKES:
   http://www.binational.net
FRASER RIVER BASIN INDICATORS:
   http://www.fraserbasin.bc.ca/indicators.html
COMMUNITY-BASED INDICATORS:
   http://www.sustainable.org

LIVING PLANET INDEX:
   http://www.pandaorg/news_facts/publications/general/ livingplanet/index.cfm
NATIONAL ROUNDTABLE ON ENVIRONMENT AND ECONOMY (NRTEE): http://www.nrtee-trnee.ca/
WORLDWATCH INSTITUTE:
   http://www.worldwatch.org

REFERENCES AND SUGGESTED READING

Abbey, E. 1977. *The Journey Home: Some Words in Defense of the American West.* New York: Dutton.

Balmford, A., et al. 2002. 'Economic reasons for conserving wild nature', *Science* 297 (Aug.): 950–3.

Barlow, M., and E. May. 2000. *Frederick Street: Life and Death on Canada's Love Canal.* Toronto: HarperCollins.

Dumanski, J. 2000. 'Land quality indicators: research plan', *Agriculture, Ecosystems and Environment* 81: 93–102

Environment Canada. 2003. *Environmental Signals: Canada's National Environmental Indicator Series 2003.* Ottawa: Environment Canada.

Furimsky, E. 2002. 'Sydney Tar Ponds: some problems in quantifying toxic waste', *Environmental Management* 30, 6: 872–9.

Goodarzi, F., and M. Mukhopadhyay. 2000. 'Metals and polyaromatic hydrocarbons in the drinking water of the Sydney Basin, Nova Scotia, Canada: a preliminary assessment of their source', *International Journal of Coal Geology* 43: 357–72.

Guernsy, J.R., R. Dewar, S. Weerasigne, S. Kirkland, and P.J. Veugelers. 2000. 'Incidence of cancer in Sydney and Cape Breton county, Nova Scotia 1979–1997', *Canadian Journal of Public Health* 91, 4: 285–92.

International Monetary Fund. 2002. *Economic Outlook Database.* Washington: IMF.

Joffres, M.R., T. Williams, B. Sabo, and R.A. Fox. 2001. 'Environmental sensitivities: prevalence of major symptoms in a referral center: the Nova Scotia Environmental Sensivities Research Center Study', *Environmental Health Perspectives* 109, 2: 161–5.

Lahey, A. 1998. 'Black lagoons: They're ugly. They stink. But are the tar ponds really killing the people of Sydney, Nova Scotia?', *Saturday Night* 113, 8 (Oct.): 37–40.

Lord, G.R. 1974. 'Introduction', in A.H. Richardson, *Conservation by the People: A History of the Conservation Authority Movement to 1970.* Toronto: University of Toronto Press, ix–xi.

Lubechenco, J. 1998. 'Entering the century of the environment: a new social contract for science', *Science* 279: 491–7.

May, E., and M. Barlow. 2001. 'The Tar Ponds', *Alternatives Journal* 27, 1: 7–11.

Mills, T.J., T.M. Quigley, and F.J. Everest. 2001. 'Science-based natural resource management decisions: what are they?', *Renewable Resources Journal* 19, 2: 10–15.

Mitchell, B., ed. 2004. *Resource and Environmental Management in Canada: Addressing Conflict and Uncertainty*, 3rd edn. Toronto: Oxford University Press.

National Round Table on Environment and Economy 2003. *2003 Environment and Sustainable Development Indicators for Canada.* Ottawa: NRTEE.

O'Leary, J., and K. Covell. 2002. 'The tar pond kids: toxic environments and adolescent well-being', *Canadian Journal of Behavioural Science* 34, 1: 34–43.

Parson, E.A. 2000. 'Environmental trends and environmental governance in Canada', *Canadian Public Policy* 26 (Supplement): S123–S143.

Rainham, D. 2002. 'Risk communication and public response to industrial chemical contamination in Sydney, Nova Scotia: a case study,' *Journal of Environmental Health* 65, 5: 26–32.

Sawin, J.L. 2004. 'Making better energy choices', in Worldwatch Institute (2004b: 24–43).

Statistics Canada. 2003. *Human Activity and the Environment: A Statistical Compendium.* Ottawa: Minister of Supply and Services Canada.

Toutin, T. 1997. 'Quantitative Aspects of Chromo-Stereoscopy for Depth Perception', *Photogrammetric Engineering and Remote Sensing* 63 (2): 193-203. (http://ess.nrca.gc.ca/esic/ccrspub-cctpub/pdf/1530.pdf) http://www.ccrs.nrcan.gc.ca/ccrs/data/showcase/showcase_e.html.

United Nations 2003. *2003 World Population Prospects: The 2002 Revision.* New York: UN.

Vitousek, P., et al. 1997. 'Human alteration of the global nitrogen cycle: causes and consequences', *Issues in Ecology* 1: 2–16.

Wackernagel, M. 1994. 'How big is our ecological footprint?', *Videas Journal* (Dec.): 2–3.

———— et al. 1993. *How Big Is Our Ecological Footprint? A Handbook for Estimating a Community's Appropriate Carrying Capacity.* Vancouver: University of British Columbia, Department of Family Practice.

World Bank. 1997. *Expanding the Measure of Wealth: Indicators of Environmentally Sustainable Development.* Washington: Environmentally Sustainable Development Studies and Monographs Series 17.

————. 1999. *Environmental Performance Indicators.* Washington: Environmental Economics Series No. 17.

World Commission on Environment and Development. 1987. *Our Common Future.* Oxford: Oxford University Press.

Worldwatch Institute. 2004a. *Vital Signs: The Trends That Are Shaping Our Future.* New York: W.W. Norton.

————. 2004b. *State of the World 2004.* New York: Worldwatch Institute/W.W. Norton.

The Ecosphere

On Spaceship Earth,
there are no passengers;
we are all members of the crew.
— Marshall McLuhan

The first section of the book provided an overview of the global and Canadian situations regarding the relationship between humans and the Earth. As the Canadian communications theorist Marshall McLuhan noted, we are not passive bystanders in terms of this interaction. We are members of the crew, helping to direct what happens to our planet. As crew members, we need to have some idea of the workings of our 'Spaceship Earth', and particularly of the nature of its life support system. Imparting such an overview is the prime purpose of this next section. Here we describe the natural systems through simplified models, following the conceptual framework outlined in Chapter 1 (Figure 1.5). This will provide the background needed to understand many of the environmental challenges faced by society today.

Most of what follows is derived from just one form of environmental knowledge, natural science (see Box B.1). Natural science tries to find order in and to understand nature so that accurate predictions can be made regarding the outcome of given changes that may occur. An important underlying assumption is that patterns in nature can be discerned if we approach things in the right manner with the right tools. The *scientific method* lays the foundation for how scientists approach this task. Another important assumption of the scientific method is that the same results will be obtained by different scientists if they repeat the experiment or observations in the same manner as the original observations. This is one reason why scientists must give detailed descriptions of their methodology and represents an attempt to be as objective as possible. What use would a thermometer be, for example, if its readings of temperature varied according to who took the measurement?

Although the scientific method strives to minimize the effects of different scientists, it is a myth to believe that science is totally 'objective'. Scientists work in a social environment and are greatly influenced by their peers and by society. The research topics selected and the ways in which research questions are

BOX B.1 TRADITIONAL ECOLOGICAL KNOWLEDGE

Most of the concepts presented in this book are the result of the 'scientific approach' to understanding different phenomena. There are other approaches, however, and one that is gaining increasing attention is **traditional ecological knowledge (TEK)**. Scientists around the world have found that indigenous peoples often have a detailed knowledge of their local environments, not to be entirely unexpected for peoples gaining their sustenance directly from that environment. They have also found that indigenous peoples tend to undertake the same kinds of tasks as Western scientists, such as classification and naming of different organisms and studies of population dynamics, geographical distributions, and optimal management strategies. Unlike Western science, however, this knowledge is rarely recorded in written form but is handed down orally from generation to generation.

Only now are we becoming more aware of this body of knowledge. In Canada, this has come about particularly as a result of increasing industrial interest in northern regions and the potential impacts of resource extraction on Native communities. Inevitably, this has given rise to discussions about which form of ecological knowledge, Western or traditional, is the 'best'. Both have their advantages and disadvantages. Modern science is informed by developments around the world, but is limited in terms of its knowledge of changes over time in a particular place, an aspect where traditional knowledge is particularly rich. Scientists also tend to concentrate on information that can be tested by replication and ignore idiosyncratic and individual behaviour, which is given substantial weight by indigenous hunters.

Management systems also differ. The traditional system is self-regulating, based on communal property arrangements. Conservation practices, such as rotation of hunting areas, were commonly practised, as described by Feit (1988) for moose hunting and beaver trapping by the James Bay Cree. However, the system is not infallible, especially under the onslaught of outside influences and commercialization. Similarly, the modern system of private property rights and state allocation of harvesting rights does not always work. This was discovered all too clearly with the North Atlantic cod fishery, where, if scientists and policy-makers had given more credence to the local ecological knowledge of inshore fishers in Newfoundland outport communities, the destruction of the fishery might have been averted. Scientists and indigenous peoples are now realizing the benefits provided by both systems of knowledge and management approaches, and are trying to use both though co-management arrangements.

framed are key components of the scientific method and yet are heavily influenced by the individual scientist. However, the fact that such social biases and values influence science should not be an excuse for not trying to maintain as value-free an approach as possible and always making sure that biases are explicit and documented.

Scientists collect data, or *facts* (observations that are widely accepted as truthful) about the environment, and then try to make some order out of those facts. This order is called *theory*. The theory of natural selection, first outlined by Charles Darwin and Alfred Russel Wallace in 1858, is a good example. When there is universal acceptance of the theory a *scientific law* may be established. A scientific law represents the most stringent form of understanding and lays down a universal truth that describes in all cases what happens in certain circumstances. In this next section, you will be introduced to some of these laws. They are useful because they provide firm building blocks upon which we can build our scientific understanding.

The scientific method links, with minimum error, the testing of *hypotheses* with the existing body of knowledge. It prescribes a series of steps that, over time, scientists have found are most likely to provide understanding about phenomena under study, irrespective of the observer undertaking the observations or experiment. The process starts with a question often derived either from observations of the environment or from existing literature. A hypothesis is generated to explain the phenomenon of interest and experiments or observations are designed to disprove the hypothesis. If the hypothesis is not rejected, then further experiments may be designed in further efforts to disprove it. If all efforts fail, the hypothesis becomes incorporated into theory and is accepted as a valid answer to the question. However, all natural scientists do not have to follow one scientific method. Many important scientific insights have come about through quite irregular approaches. Nonetheless, it is wise to learn from previous experience and to know the kinds of procedures usually followed in any given area of inquiry.

This said, we should not feel that any of the hypotheses, theories, or even laws advanced by science are unquestionable. The whole purpose of science is to ask questions, and science advances by continually changing and modifying previous knowledge. 'Be kind to scientists (and teachers!) but ruthless in your questions' is not a bad dictum to adhere to. Debate and disagreement in science are normal. In the following chapters, for example, the ideas on some topics are changing rapidly, and even some concepts (e.g., keystone species, climax vegetation) are being questioned by some scientists. This active debate is often misunderstood by those outside the academic community to imply that scientists do not know anything. It can also be used for political purposes to support inaction on measures that might be unpopular, such as limits on industrial emissions to reduce acidic precipitation. Debate and dispute, however, are signs of the vitality and strength of scientific thinking—and of the urgent need to get closer to truth—rather than the converse.

When we think of science we often tend to think of it within the context of natural and physical sciences, and most of the next three chapters focuses on this understanding of 'science'. However, social science is also critical to understanding and addressing environmental management. Just as natural scientists hope to provide greater understanding of the natural world, social scientists address the same need for social dimensions. How can we understand how individuals think, how to change their behaviour, how governments and the economy operate, and how legislation is formed and enforced? What can we learn from societies elsewhere and how they have interacted with their environmental surroundings in the past? These are some of the kinds of questions that we look to disciplines such as geography, anthropology, psychology, sociology, political science, and economics to address.

The methods of social science can also have much in common with the natural sciences, with the testing of hypotheses as a means to build theory and establish laws that will help provide generic understanding of seemingly disparate events. Yet, the great variability in social systems, the difficulties with isolating and controlling many of the variables under study in a social context (as opposed to natural phenomena or laboratory work), and the challenges of repeatability (it is often impossible to duplicate the same circumstances for repeated experiments) make adherence to strict scientific procedures impossible. These difficulties mean that social scientists must apply a very broad range of approaches to gaining social understanding. Thus, although theory formation is a critical part of the social sciences, the formation of universal laws is very rare.

Furthermore, although both natural and social scientists strive for *quantitative* (numerical) data due to the great precision that can be obtained, *qualitative* (non-numerical) data are sometimes all that can be obtained or even all that can or should be sought, depending on the particular research question. A case in point: federal fisheries scientists over many years tended to dismiss the qualitative data from inshore Newfoundland fishers as 'anecdotal' and chose to rely largely on their own estimates of stock biomass, which were based on offshore data from the trawler fishery and from their own survey boats. As it turned out, of course, the anecdotal observations and warnings based on generations of knowledge and experience were far more accurate than the quantitative data the scientists chose to believe. Both kinds of data are important. For example, if someone tells you that the weather was 'cool', that does not convey as much information as telling you that the temperature was $-25°$ C. Data on the beauty of a certain landscape is less amenable to quantitative assessment, although some scientists claim to have developed numerical ways to approach such problems!

The three chapters in Part B provide a basic overview of the main processes that maintain the planetary life support system. A simple model of the planet would look like the layers of an onion (Figure B.1). We are most concerned with the outer layer, the ecosphere, which consists of three main layers:

- The **lithosphere**, which is the outer layer of the Earth's mantle and the crust, and contains the rocks, minerals, and soils that provide the nutrients necessary for life.
- The **hydrosphere**, which contains all the water on Earth.

• The **atmosphere**, which contains the gases surrounding the lithosphere and hydrosphere. It can be further divided into three main sublayers, the innermost layer or **troposphere**, which contains 99 per cent of the water vapour and up to 90 per cent of the air, and is responsible for our weather. Two gases, nitrogen (78 per cent) and oxygen (21 per cent), account for 99 per cent of the gaseous volume. This layer extends on average to about 17 km before it gives way to the second layer, the **stratosphere**, wherein lies the main body of ozone that blocks out most of the ultraviolet radiation from the sun. At about 50 km from the Earth's surface is the **mesosphere** and above that the **thermosphere**. As distance from Earth increases, the pressure and density of the atmosphere decreases as it melds into space.

These three layers combine to produce the conditions necessary for life in the **ecosphere,** stretching from the depths of the ocean trenches up to the highest peaks, a layer some 20 kms in width, no larger than the peel of an apple, that contains some 30 million different organisms. The following chapters impart some idea of the main environmental processes of the ecosphere and describe how human activities interrupt these processes.

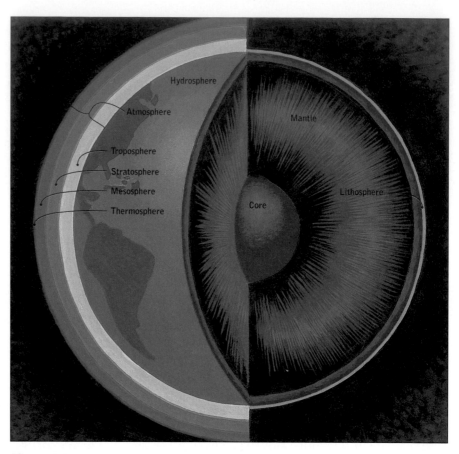

Figure B.1 A simplified model of the Earth showing the ecosphere.

KEY TERMS

atmosphere	lithosphere	thermosphere	troposphere
ecosphere	mesosphere	traditional ecological	
hydrosphere	stratosphere	knowledge (TEK)	

REFERENCE

Feit, H.A. 1988. 'Self-management and state-management: forms of knowing and managing northern wildlife', in M.M.R. Freeman and L.N. Carbyn, eds, *Traditional* *Knowledge and Renewable Resource Management in Northern Regions.* Edmonton: Boreal Institute, 72–91.

Chapter 2

Energy Flows and Ecosystems

Learning Objectives

- To know the nature of energy and the laws that govern its transformation.
- To understand the way energy flows through the ecosphere and links ecosystem components.
- To outline the main influences on the structure and nature of ecosystems.
- To learn the importance of reducing energy use in society.

INTRODUCTION

The annual arrival of the capelin to the beaches of Newfoundland to spawn on the first full moon of June had long been a bounty not only for many animal species, but also for the settlers who collected the fish for consumption and application to their gardens as fertilizer. The capelin is a small fish of the North Atlantic and an important food supply for many other species, including cod, salmon, halibut, mackerel, seals, various whale species, and many species of seabirds such as puffins and murres. In the late 1970s and early 1980s the numbers of fish coming to spawn declined markedly.

Normally 80 to 90 per cent of the Atlantic puffin's diet consisted of capelin. In the early 1980s, scientists noticed a large decline in this proportion, down to 13 per cent, resulting in severe malnutrition of puffin chicks and subsequent declines in the population owing in part to starvation. Their numbers fell as a direct result of the removal of their food base, the capelin. The energy flow between the species had been interrupted through the opening of an offshore capelin fishery that had removed the capelin from the food chains that nourish so many other marine species. The puffins were one noticeable victim of this appropriation, but other species feeding on the capelin suffered the same consequence. These species in turn would affect the abundance of other species at all levels in the food web as the numbers

The colourful bill is the most striking feature of the Atlantic puffin, which breeds among the rocks of sea islands (*Newfoundland and Labrador Parks and Natural Areas/Ned Pratt*).

Spawning capelin on a beach in Newfoundland (*Philip Dearden*).

Traditional societies in many lesser developed countries, such as the Lisu people in the highlands of northern Thailand, still rely on wood as the main source of energy. Although this can lead to deforestation if demands are greater than growth, it also limits the amount of energy that can be transformed by these societies (*Philip Dearden*).

of some species are controlled mainly by their predators. The example illustrates the importance of understanding how energy links species and flows through ecosystems. Changing the energy available at one part of the food chain will have repercussions throughout the ecosystem.

Reading this book, taking notes in class, even snoozing at home all require energy. That energy is coming ultimately from the radiant energy of the sun, transformed into chemical energy in the form of food supplies before being converted to mechanical energy in terms of physical exertion and activity. In this chapter you will gain an appreciation of energy in relationship to such transformations, how energy flows through ecosystems, and the ecosystem consequences that result.

ENERGY

Energy is the capacity to do work and is measured in calories. A **calorie** is the amount of heat necessary to raise one gram or one millilitre of water one degree Celsius (C), starting at 15 degrees. Energy comes in many forms: radiant energy (from the sun), chemical energy (stored in chemical bonds of molecules), as well as heat, mechanical, and electrical energy. Energy differs from matter in that it has no mass and does not occupy space. It affects matter by making it *do* things—work. Energy derived from an object's motion and mass is known as **kinetic energy**, whereas **potential energy** is stored energy that is available for later use. The water stored behind a dam is potential energy that

becomes kinetic energy as it pours over the dam. The gas in a car is potential energy before it is poured into the engine to create mechanical energy for propulsion.

Most of the energy available for use is termed **low-quality energy**, which is diffuse, dispersed, at low temperatures, and difficult to gather. The total energy of all moving atoms is referred to as **heat**, whereas temperature is a measure at a particular time of the average speed of motion of the atoms or molecules in a substance. The oceans, for example, contain an enormous amount of heat, but it is very costly to harness this energy for use. They have high heat content but low temperature. On the other hand, **high-quality energy**, such as a hot fire or coal or gasoline, is easy to use, but the energy disperses quickly. Much of our economy and technology are now built around the transformation of low-quality energy into high-quality energy for human use. It is important that we match the quality of the energy supply to the task at hand. In other words, the aim is not to use high-quality energy for tasks that can be undertaken by low-quality supplies. Heating space, such as your house, for example, requires only low-temperature heat, yet many homes are heated through the conversion of high-quality energy sources that entail significant energy losses in generation, transport, and application. Nuclear energy involving high-quality heat at several thousand degrees, which is then converted to high-quality electricity transmitted to homes and used in resistance heating, is very inefficient. The most efficient way to provide space heating is through having super-insulated houses and passive solar heating.

All organisms, including plants, require energy for growth, tissue replacement, movement, and reproduction. To gain a comprehensive perspective on life we must understand energy and how it is transformed and used. Box 2.1 provides an introductory definition of life.

Laws of Thermodynamics

Two laws of physics (or physical laws) describe the way in which trillions of energy transformations per second all over the globe take place. These are known as the *laws of thermodynamics*. The first one, the **law of conservation of energy**, tells us that energy can neither be created nor destroyed; it is merely changed from one form into another (nuclear is a form of potential energy—the energy

BOX 2.1   WHAT IS LIFE?

We have said that energy is essential for all life, but what is life? Living organisms have a number of common characteristics, including:

- They use energy to maintain internal order.
- They increase in size and complexity over time.
- They can reproduce.
- They react to their environment.
- They regulate and maintain a constant internal environment.
- They fit the biotic and abiotic requirements of a specific habitat.

We think we have a fairly good idea of what constitutes life, but there is still a lot of debate as to how life developed on Earth. Over 80 years ago two scientists proposed a theory, called the Big Bang theory, explaining the origin of the universe as the result of a massive explosion occurring some 15 billion years ago. The solar system came from the resulting matter. As the chunks of matter grew in size they heated up. As the Earth cooled, warm seas formed and precipitation occurred that helped to create a nutrient-rich environment. Over time, the constant bombardment of this nutrient-rich soup by high energy levels from the sun created chemical reactions producing simple organic compounds, such as amino acids. Scientists have managed to recreate several organic compounds necessary for life from inorganic molecules by bombarding them with energy.

Over billions of years larger organic molecules came to be synthesized until the first living cells, probably bacteria, developed between 3.6 and 3.8 billion years ago. These cells passed through several stages over billions of years, with increasingly complex development. This activity took place in the ocean environment, protected from ultraviolet (UV) radiation. Between 2.3 and 2.5 billion years ago, a major change occurred when photosynthetic bacteria developed that emitted oxygen into the atmosphere as they manufactured carbohydrates from the carbon dioxide in the atmosphere. Over time, the oxygen reacted with the abundant and poisonous methane in the atmosphere, reducing levels of this gas and leading to the atmosphere we know today. Some oxygen was also converted to ozone in the lower stratosphere that protected evolving life from UV radiation and allowed the emergence of life from deeper to shallower waters, and eventually onto land itself.

---

is simply held in the nucleus of an atom). Organisms do not create energy; rather, they obtain it from the surrounding environment. When an organism dies, the energy of that organism is not 'lost'. It flows back into the environment and is transformed into different types of energy, the total sum of which adds up to the original amount. Similarly, we all know that cars obtain their energy from gasoline. As the fuel gauge goes from full to empty this does not indicate that energy has been consumed; it has merely been transformed from chemical energy into other forms of energy, including the mechanical energy to move the car.

The second law of thermodynamics tells us that when energy is transformed from one form into another, there is always a decrease in the quality of usable energy. In any transformation, some energy is lost as lower-quality, dispersed energy that is dissipated to the surrounding environment, often as heat. The amount of energy lost varies depending on the nature of the transformation. In a coal-fired generating station, for example, at the most 35 per cent of the coal's energy is converted into electricity. The rest is given off as waste heat to the environment.

In a car only about 10 per cent of the chemical energy of the gasoline is actually converted into mechanical energy to turn the wheels. The remainder is dispersed into the environment. Put your hand onto the hood of a car that has just stopped running. The heat you feel is a result of this second law of energy or the *law of entropy*. **Entropy** is a measure of the disorder or randomness of a system. High-quality, useful energy has low entropy. As energy becomes dispersed through transformation, the entropy increases.

For organisms, the second law is particularly important because they must continuously expend energy to maintain themselves. Whenever energy is used, some of that energy is lost to the organism, creating a need for an ongoing supply that must exceed these losses if the organism is to survive. If losses exceed gains for an extended period of time, then the organism dies.

There are also many other important ramifications of this law. It tells us, for example, that energy cannot be recycled. It flows through systems in a constantly degrading manner. We think of 'advanced' societies as being energy consumers. Large dams and nuclear power stations, for example, are visible signs of a modern economy. As we become more economically developed we find new ways to transform energy. Cars, telephones, electric can openers, blenders, microwaves, hot tubs, computers, and compact disc players are all energy transformers. Yet, as more energy is transformed, more is dispersed into the atmosphere because entropy increases. This dispersion can be likened to a bar of soap in a bowl of water. As the soap is used over time it dissolves into the water, making it less and less useful. Energy may be thought of in a similar fashion as use gradually disperses useful and concentrated forms into the atmosphere.

Some of the principal transformations that have to take place to achieve a sustainable society are: to view high energy consumption as undesirable, to reduce energy waste, and to switch from the non-renewable sources of energy that now dominate (coal and oil particularly) to renewable sources, such as those discussed in Chapter 13. Until the Industrial Revolution, the speed of processing raw materials was limited by the energy available, supplied largely by human and animal labour combined with wood, wind, and water power. These sources were in turn limited by the input of solar energy over a relatively short time period. The use of coal and later oil to fuel steam engines removed these limitations and made accessible a vast storehouse of potential energy created by the sun over millions of years through the remains of compressed plants. Acid rain, climatic change, and many of our current environmental problems result directly from this transformation of the energy base of society from a renewable to a non-renewable one. In geological terms, we have released the energy input of millions of years in the blink of an eye—the last 250 years. Many problems are a result of this increase in entropy.

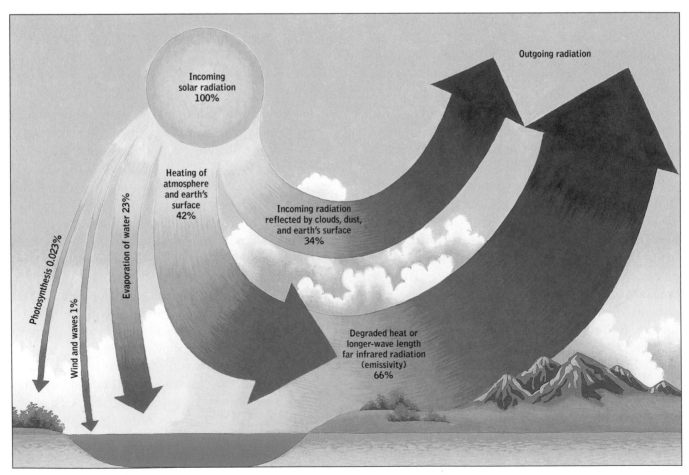

Figure 2.1 The Earth's energy input and output, a good example of the first law of thermodynamics.

ENERGY FLOWS IN ECOLOGICAL SYSTEMS

Energy is the basis for all life. The source of virtually all this energy is the sun. Over 150 million km away, the sun, a giant fireball of hydrogen and helium, constantly bombards the Earth with *radiant energy*. This energy, although it is only about 1/50 millionth of the sun's output, fuels our life support system, creates our climate, and powers the cycles of matter to be discussed in Chapter 4. About a third of the energy received is reflected by the atmosphere back into space (Figure 2.1). Of the remainder, about 42 per cent provides heat to the Earth's surface, 23 per cent causes evaporation of water, and less than 1 per cent forms the basis for our ecological systems. When we think of solar energy it is important to remember not just the direct heat from the sun but also these indirect forms of energy created by heat input.

Figure 2.1 illustrates the law of conservation of energy. The total amount of energy received by the Earth is equal to the total amount that is lost. One of the changes being caused by human activity is to delay the amount of heat lost back to space by trapping it in the atmosphere through increased levels of heat-trapping gases such as carbon dioxide and methane. The implications of this are discussed more fully in Chapter 7.

Producers and Consumers

The sun's energy is transformed into matter by plants through the process of **photosynthesis** (photo = light, synthesis = to put together). Through this process, plants combine carbon dioxide and water, using energy from the sun, into high-energy carbohydrates such as starches, cellulose, and sugars. Green pigments in the plants, called **chlorophylls**, absorb light energy from the sun. Photosynthesis also produces oxygen, some of which is used by plants in various metabolic processes. The rest goes into the atmosphere. Hundreds of millions of years of evolution have served to produce the oxygen in the atmosphere that we depend on for life.

Organisms with the ability to capture energy and manufacture matter are known as **autotrophs** (auto = self, trophos = feeding) or **producers**. All other organisms obtain their energy supply through eating other organisms, and are known as

Plants that grow on the forest floor have differing strategies to obtain enough light to survive. Most, such as many ferns, can survive on relatively low light levels. Some, such as the devil's club shown here, grow very large leaves (over 40 cm wide for the devil's club) in order to expose as much photosynthetic surface as possible to the low light levels. The devil's club is a member of the ginseng family and well known among Native peoples in western North America for its medicinal properties (*Philip Dearden*).

heterotrophs (hetero = different) or **consumers**. There are two kinds of autotrophs, **phototrophs** and **chemoautotrophs**. Phototrophs obtain their energy from light; chemoautotrophs gain their energy from chemicals available in the environment. Although most of us are aware of the critical role played by phototrophs (plants) in our life support system, the chemoautotrophs play an equally critical yet not so visible role (see Box 2.2). Most of them are bacteria and play a fundamental role in the **biogeochemical cycles**, to be discussed in more detail in Chapter 4.

Phototrophs convert the light energy of the sun into chemical energy, using carbon dioxide and water to produce carbohydrates. The second law of thermodynamics instructs us that some energy will be lost in this transformation; indeed, the efficiency rate is only between 1 and 3 per cent. In other words, 97–99 per cent of the energy will be lost. Nonetheless, this conversion is sufficient to produce billions of tons of living matter, or **biomass,** throughout the globe.

Besides photosynthesis, *cellular respiration* is another essential energy pathway in organisms. In both plants and animals, this involves a kind of reversal of the photosynthesis process in which energy is released rather than captured. High-energy organic carbohydrates are broken down

BOX 2.2 CHEMOAUTOTROPHS IN THE DEEP

We think of the deep sea floor as a biological desert. In the 1970s, however, scientists discovered that rich biological communities were supported at hydrothermal vents on the sea floor, mainly bacteria that derive their energy from sulphide emissions. It has now been discovered that similar kinds of chemoautotrophic-based communities are on the remains of whale skeletons found at depth, nourished by sulphides produced as the carcasses decay. Discoveries of fossils suggest that dead whales may have provided dispersal stepping stones for these communities for over 30 million years. The question then becomes—what was the impact on these communities when whales were virtually eliminated from the oceans by whalers? Scientists do not yet have the answer to this question.

through a series of steps to release the stored chemical bond energy. In other words, the potential energy is now realized as kinetic energy in the way described above. This produces the inorganic molecules, carbon dioxide, water, heat (because of the law of entropy), and energy that can be used by the organism for various purposes such as growth, feeding, seeking shelter, communicating with one another, producing seeds, and maintaining basic physiological functions such as constant body temperature and breathing. Since we are unable to obtain energy from photosynthesis or

through chemotaxis, this is how we, and all other organisms unable to fix their own energy, get our energy supplies.

For cellular respiration to occur, most organisms must have access to oxygen or they will die. Such organisms are known as **aerobic** organisms. Some species, anaerobic organisms, such as some bacteria, can survive even without oxygen. This makes them useful in the breakdown of organic wastes, such as sewage.

Food Chains

Some of the energy captured by autotrophs is subsequently passed on to other organisms, the consumers, by means of a **food chain** (Figure 2.2). **Herbivores** eat the producers and are, in turn, the source of energy for higher-level consumers or **carnivores**. Decomposers will feed upon all these organisms after they die. Each level of the food chain is known as a **trophic level**. A giant Douglas fir tree on the Pacific coast and a minute Arctic flower on Baffin Island are on the same trophic level—autotrophs. Herbivores, on the second level, range in size from elephants to locusts. The role in energy transformation, rather than the size of the organism, is the important factor in determining trophic level.

Some organisms, such as humans, raccoons, sea anemones, and cockroaches, are **omnivores** and can obtain their energy from different trophic levels. When we eat vegetables we are acting as **primary consumers**; when we eat beef we are at the second trophic level acting as **secondary consumers**; and when we eat fish that have

Animals as different as the grasshopper and the elephant are on the same trophic level (*Philip Dearden*).

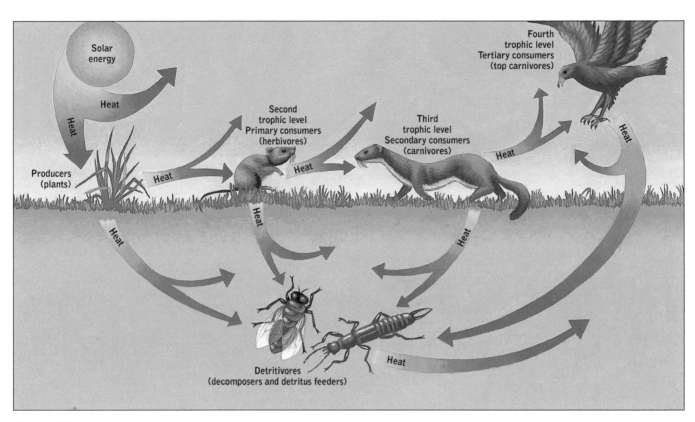

Figure 2.2 A food chain.

derived their energy from eating smaller organisms we may be **tertiary consumers** at the top of the food chain. The level at which food energy is being obtained has some important implications, to be discussed later.

We tend to concentrate on these grazing food chains, but equally important are the decomposer food chains (Figure 2.3). These are based on dead organic material or **detritus**, which is high in potential energy but difficult to digest for the consumer organisms described above. However, various species of microorganisms, bacteria, and fungi are able to digest this material as their source of energy. Indeed, many large grazing animals such as cows and moose have such bacteria in their stomachs to help break down the cellulose in plant material. These decomposers (or saprotrophs) derive their energy from dead matter (sapro = putrid). They are joined by consumers such as earthworms and marsh crabs, known as detritivores, which may consume both plant and animal remains.

These **decomposer food chains** play an integral role in breaking down plant and animal material into products such as carbon dioxide, water, and inorganic forms of phosphorus and nitrogen and other elements. For example, fungi that consume simple carbohydrates, such as glucose, first break down dead wood. Following this other fungi, bacteria, and organisms such as termites break down the cellulose that is the main constituent of the wood. Were it not for these organisms, wood and other dead organisms would accumulate indefinitely on the forest floor.

The relative importance of grazing and detrital food chains varies. The latter often dominate in forest ecosystems, where less than 10 per cent of the tree leaves may be eaten by herbivores. The remainder dies and becomes the basis for the detritus food chain. In the coastal forests of British Columbia, for example, there are some 140 different species of birds, mammals, and reptiles through which energy can flow. By way of contrast, there are over 8,000 known, as well as many unknown, species involved in breaking down the soil litter. The same is often true in freshwater aquatic systems where there may be relatively little plant growth but abundant detritus from overhanging leaves and dead insects. However, the converse is true in marine ecosystems (see Box 2.4), where 90 per cent of the photosynthetic **phytoplankton** (phyto = plant, plankton = float-

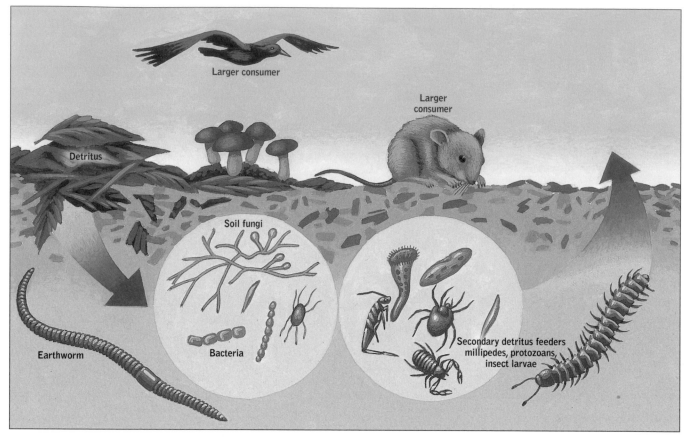

Figure 2.3 Detritus-based food chain.

BOX 2.3   CARNIVOROUS PLANTS

All plants are not autotrophs. Carnivorous plants, such as the pitcher plant, the floral emblem of Newfoundland, gain their energy from ingesting the bodies of insects that become trapped in their funnel-shaped leaves (see photo). The plant, which grows in boggy areas across Canada, has no photosynthetic surfaces. Instead, the leaves act as 'pitchers' to hold a soapy liquid from which a hapless insect cannot escape. It is now thought that the plant may be aided in the decomposition of dead insects by other insects that have developed immunities to the decomposing enzymes produced by the plant. In fact, the plant plays host to several insects that seem to thrive on the environment provided by the pitcher plant. These are examples of mutualism, where both species benefit from a relationship.

The carnivorous pitcher plant, the provincial flower of Newfoundland, grows in abundance (*John Maunder*).

ing) may be grazed by the primary consumers, the **zooplankton**.

In general, ecological theory suggests that the more species in the ecosystem, the more alternative pathways are available for energy flow and the more able the ecosystem is to withstand stress. In the Arctic, for example, a simple food chain might be phytoplankton to zooplankton to cod to ringed seal to polar bear. All these species are heavily dependent on the species at the preceding trophic level. Were one of these species to be drastically reduced in number or made extinct, the chances of the role of that species being compensated for by other species is quite low and the whole food chain may well collapse. This situation can be compared to that at the other extreme, such as a tropical forest, where there are many times more species and the chance of other species combining to fulfill the ecological role of a depleted one is much higher. In practice, however, many other factors are involved, such as the relative degree of specialization of the various organisms, and examples of relatively unstable complex systems abound, so care must be exercised before generalizing the theory to all ecological systems.

Rarely are food chains organized in the simple manner shown in Figure 2.2. Usually there are many competing organisms and energy paths. Thus, scientists tend to think of food chains more as complex **food webs** in which numerous alternative routes exist for energy to flow through the ecosystem (Figure 2.4). Ecologist Bristol Foster has described an unusual example of this breadth when on the BC coast he witnessed a garter snake foraging in a tidal pool and swallowing a small fish, while at the same time being held by a green sea anemone.

The number of species increases—from the poles to the tropics—as conditions become more amenable for life (Figure 2.5). In the Arctic, for

BOX 2.4 OCEANIC ECOSYSTEMS

From space, the Earth appears to be a blue, not a green, planet, reflecting the fact that 71 per cent of the Earth's surface is covered by oceans. Life originated in this blueness, perhaps 3.5 billion years ago, and only came onto land some 450 million years ago. Hence much of our biological ancestry lies within these waters. Although we know about more different species on land than in the oceans, the number of *phyla*, distinguished by differences in fundamental body characteristics, is higher in the oceans. Of the 33 different animal phyla, for example, 15 exist exclusively in the ocean and only one is exclusively land-based. We share the same phyla as the fishes, the *chordata*, characterized by a flexible spinal cord and complex nervous system.

Through their photosynthetic activity, the early bacteria that started in the oceans helped create the conditions under which the rest of life evolved. Current photosynthetic activity is no less important to our survival. Scientists estimate that the phytoplankton in the **euphotic zone** of the oceans produce between one-third and one-half of the global oxygen supply. In doing so they also extract carbon dioxide from the atmosphere. Some 90 per cent of this is recycled through marine food webs, but some also falls into the deep ocean as the detritus of decaying organisms, and is stored as dissolved carbon dioxide in deep ocean currents that may take over 1,000 years to reappear at the surface. The oceans contain at least 50 times as much gas as the atmosphere and are playing a critical role in helping delay the so-called greenhouse effect, discussed in more detail in Chapters 7 and 8.

These phytoplankton, so important to atmospheric regulation, are also the main autotrophic base for the marine food web. From tiny zooplankton through to the great whales, almost every marine animal has phytoplankton to thank for its existence. Phytoplankton flourish best in areas where ocean currents return nutrients from the deep ocean back to the euphotic zone. This occurs in shallow areas, such as the Grand Banks, where deep ocean currents meet the coast or where two deep ocean currents meet head on. Such areas are the most productive in what is generally an unproductive ocean, and they are the best sites for fisheries. Ninety per cent of the marine fish catch comes from these fertile nearshore waters. Unfortunately, these are also the sites of greatest pollution. The blueness of most of the rest of the ocean is a visible symbol of the low density of phytoplankton. That is why the sea is blue, not green.

Continued

BOX 2.4 OCEANIC ECOSYSTEMS CONTINUED

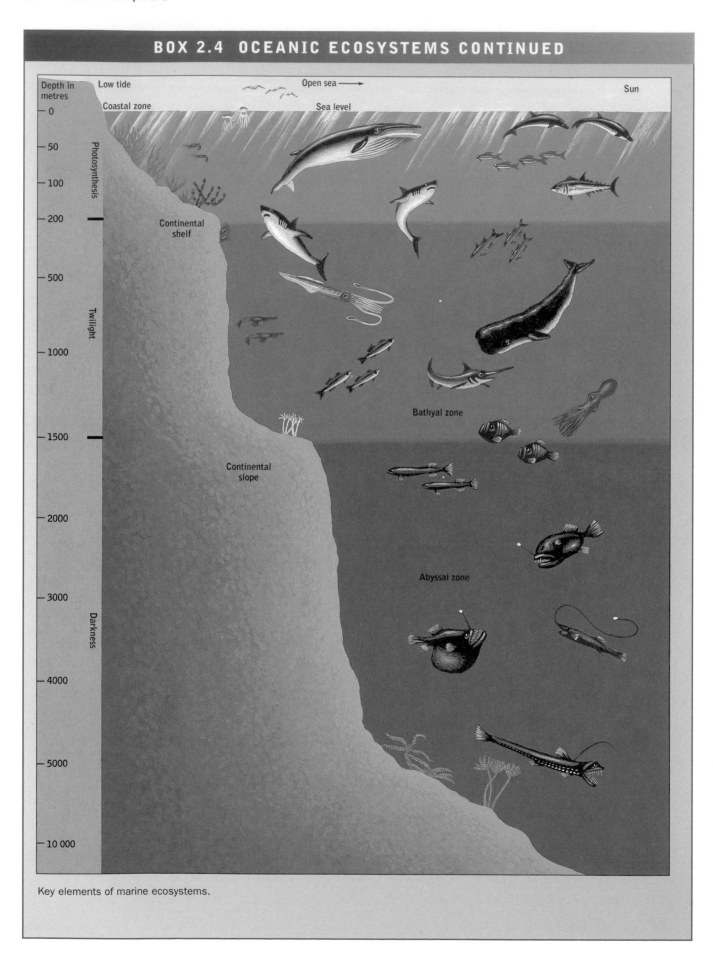

Key elements of marine ecosystems.

example, there are relatively few species, and therefore relatively few alternative pathways for energy flow. If a prey species, such as the Arctic hare, decreases in number, then so will the organism dependent on it higher in the food chain, such as the lynx, because there are few other species upon which these organisms can feed. This gives rise to the familiar population cycles in the North as predator numbers closely reflect the availability of dominant prey species (Figure 2.6).

This relationship might not be quite so simple though, as discussed in Box 2.5, and ecologists have long debated over whether ecosystems are mostly controlled by predators (*top-down control*) or by prey populations (*bottom-up control*). In the former, predators restrict the size of the prey population. This seems to occur, for example, where wolves control deer, elk, or moose populations or when sea otters control urchin populations. Conversely, in some systems the quality of available forage limits the number of herbivores, which in turn limits the number of predators. The predator numbers are essentially limited by the energy flow through the previous trophic levels. If herbivore populations fall as a result of disease or lack of forage, this will result in a drop in predator populations because of a lack of food. Ecosystems where controls are dominantly bottom-up tend to have marked limits on plant productivity through abiotic factors such as low nutrient supply, lack of water and similar factors, or very

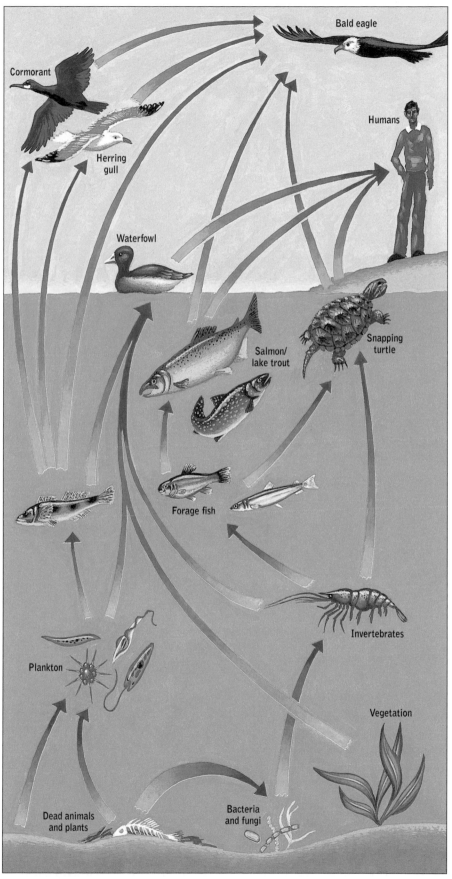

Figure 2.4 A simplified Great Lakes food web. SOURCE: Environment Canada, *Toxic Chemicals in the Great Lakes and Associated Effects* (Toronto: Department of Fisheries and Oceans; Ottawa: Health and Welfare Canada, 1991).

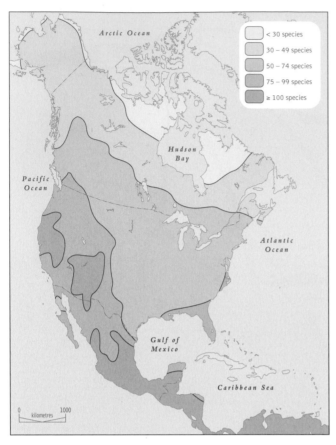

Figure 2.5 The number of mammal species per latitude. SOURCE: After Simpson (1964).

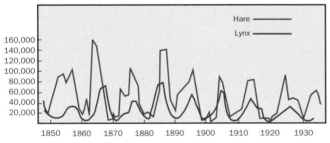

Figure 2.6 Lynx and hare cycles. Data based on pelts received by Hudson's Bay Company. SOURCE: D.A. MacLulich, University of Toronto Studies, Biological Series No. 43, 1937.

close relationships between a specific plant, a herbivore, and a carnivore. Ecosystems reflecting top-down control typically lack these features. However, as with most ecological phenomena, these are general guidelines—most ecosystems contain elements of both top-down and bottom-up control.

Biotic Pyramids

The second law of thermodynamics describes how energy flows from trophic level to trophic level, with a loss of usable energy at each succeeding transformation. In natural food chains, the *energy*

efficiency or amount of total energy input of a system that is transformed into work or some other usable form of energy may be as low as 1 per cent. In general, we expect about 90 per cent of the energy to be lost at each level (Figure 2.7). Similar losses may be experienced in biomass and numbers of organisms at each trophic level. This explains why there are fewer secondary than primary consumers and fewer tertiary than secondary. Carnivores must always have the lowest numbers in an ecosystem in order to be supported by the energy base below. The case of the Atlantic puffins, as described in the introduction to this chapter, provides a good example. The biomass of carnivores (puffins) could no longer be supported by the energy from the preceding trophic level, the capelin.

Some ecosystems, however, may display an inverted biomass pyramid. In natural grasslands such as those found in southern Saskatchewan, the dominant species, such as grasshoppers, are small-bodied and do not have a large biomass. In contrast, in this system many herbivores, such as antelope and mule deer, are large-bodied and long-lived with a large total biomass. The same situation exists in the oceans, discussed in more detail in Chapter 8. However, in both cases the productivity of the plant base is much greater than that of the herbivores.

There are several reasons for the low energy efficiencies of natural food chains. First, not all the biomass at each trophic level is converted into food for the next trophic level. Many organisms have developed characteristics to avoid getting eaten by something else. For example, many plant species have thorns or produce secondary chemicals to deter herbivores. Others have low nutritive levels. Generally, only between 10 and 20 per cent of the biomass of one trophic level is harvested by the next level. Furthermore, of that which is consumed, not all of it is digested. Humans, for example, are not well equipped to break down and consume the bones or fur of animals, nor are they equipped, compared to moose and other members of the deer family, to break down woody tissue. The proportion of ingested energy actually absorbed by an organism is the *assimilated food energy*. Finally, as cellular respiration occurs to liberate energy for the growth, maintenance, and reproduction of the organism, energy is further released as heat.

The longer the food chain, the more inefficient it is in terms of energy transformation, reflecting

BOX 2.5 ARCTIC POPULATION FLUCTUATIONS

Populations of many animals in the North show distinctive fluctuations in numbers over regular time periods. A three- to four-year cycle is recognized for smaller animals, such as lemmings and meadow voles, and a longer one of nine to ten years exists for larger animals, such as snowshoe hares and several species of grouse and ptarmigans. The predators of these herbivores show similar fluctuations in numbers as their energy source becomes critically depleted.

One of the themes that this book illustrates is the uncertainty surrounding many aspects of environmental management. In this case, there is a fair amount of certainty about the dates of fluctuations in these populations, but there is considerable uncertainty as to why these fluctuations occur. Four main ideas have been advanced:

1. The seasonality of the Arctic environment means that plants grow for only a few months every year, yet herbivores must eat throughout the year. Thus, when population levels rise, overexploitation of the food supply can occur rapidly and for a relatively long period, but the plants will take a long time to recover. The relationship between the herbivores and their food supply is the determining factor, and the predator numbers simply reflect those of the prey.
2. A second hypothesis builds on this idea, but suggests that as populations of small animals such as lemmings increase, then more nutrients vital to plant productivity become tied up in this higher trophic level. This lack of nutrients causes reductions in plant productivity and quality, leading to starvation for the herbivores.
3. Other ideas postulate more of an interaction between the predators and prey. Keith et al. (1984) studied the ecology of snowshoe hares in northern Alberta to test a food supply-predation hypothesis. This suggests that food supply shortages halt populations of herbivores that are subsequently caught by predators, until numbers fall enough to permit plant recovery. Keith et al. reported that malnutrition of hares was evident, but that predators caused 80 to 90 per cent of deaths, thereby supporting the hypothesis.
4. The final idea suggests that food supplies play a negligible role in population cycles and that these cycles would not occur in the absence of predators (Trostel et al., 1987). Scientists have found that sudden drops in hare populations occur despite supplementary feeding programs, a finding that would appear to support this hypothesis (Smith et al., 1988).

In short, ecological systems are not easy to understand. In many cases, several factors may contribute to changes such as population cycles.

the second law of thermodynamics. An Arctic marine food chain that starts from the producers (phytoplankton) to primary consumers (zooplankton) that are subsequently grazed by the largest animals ever to exist on earth, whales, is very efficient because it is so short, with only three energy transformations wherein energy is lost. Longer food chains involve a proportionately larger loss of energy due to the greater number of energy transformations. Entropy dictates that long food chains with five or six trophic levels, such as that supporting a killer whale, for example, are very scarce.

The *energy pyramid* also has important implications for humans. For example, it takes between 8 and 16 kg of grain to produce one kg of beef. This means that more land must be cultivated to provide people with a diet high in meat as opposed to a diet based on grains. As humans are one of the species that can access food energy at several different trophic levels, in terms of energy efficiency it would be better to operate as low on the food chain as possible, that is, as primary consumers or vegetarians. This topic is discussed in more detail in Chapter 10.

Productivity

Productivity in ecosystems is measured by the rate at which energy is transformed into biomass, or living matter, and is usually expressed in terms of kilocalories per square metre per year. **Gross primary productivity** (GPP) is the overall rate of biomass production, but there is an energy cost to capturing this energy. This cost, *cellular respiration* (R), must be subtracted from the GPP to reveal the **net primary productivity** (NPP). This is the amount of energy available to heterotrophs.

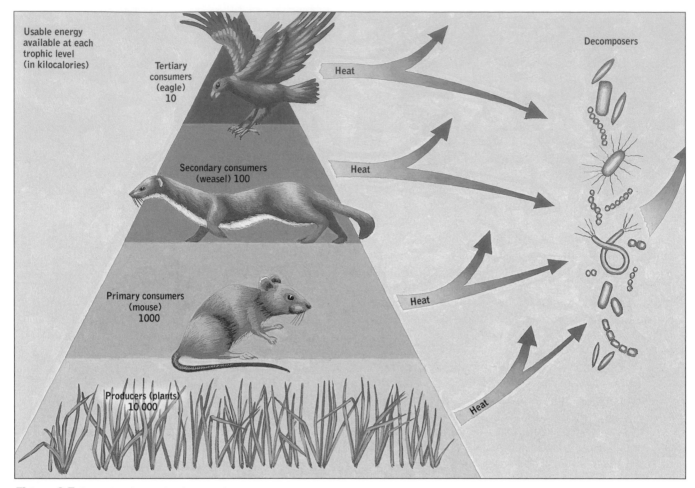

Figure 2.7 Generalized pyramid of energy flow.

All ecosystems are not the same in terms of their ability to fix biomass. Light levels, nutrient availability, temperature, and moisture, among other factors, regulate the rates of photosynthesis. The most productive ecosystems per unit area are estuaries, swamps and marshes, and tropical rain forests (Figure 2.8). Recent data indicate that the temperate rain forests, such as those that grow in the Pacific maritime ecozone, are just as productive as the tropical forests. Other ecosystems are more limited due to deficiencies in one or more of the characteristics noted above. A desert, for example, lacks water, the Arctic lacks heat, and the ocean lacks nutrients.

Unfortunately, in Canada some of our most highly industrialized and polluted lands are adjacent to estuaries. There is not one sizable estuary on the east coast of Vancouver Island, for example, that is not used by the logging industry either as a mill site or for log storage. The estuary of the Fraser River has been extensively dyked, industrialized, and polluted. Paradoxically, relative little

data exist on the affects of these intrusions on our most productive ecosystems.

Humans already take about 40 per cent of terrestrial NPP for their own use. The remainder supports all the other organisms on Earth, which in turn maintain the environmental conditions that keep us alive. The human population is projected to increase by about 50 per cent over the next 50 years. It is highly doubtful that the Earth's systems could withstand a concomitant increase in the amount of NPP being appropriated for human use. Once again, this is indicative of the carrying capacity challenge that we face.

In addition to primary productivity, measurements also can be made of *net community productivity* (NCP), including heterotrophic and autotrophic respiration. Measurements indicate that as communities mature, although GPP and NPP rise, an increasing proportion of the energy of the community is devoted to heterotrophic respiration (Table 2.1). In mature communities, the amount of respiration may be sufficient to account for all the energy

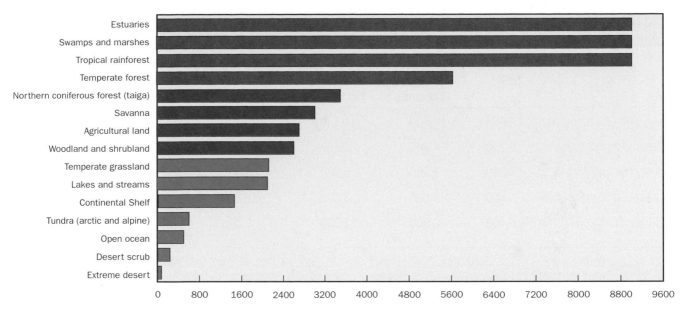

Figure 2.8 Estimated annual average net productivity of producers per unit of area in principal types of life zones and ecosystems. Values are given in kilocalories of energy produced per square metre per year.

being fixed by photosynthesis. There is thus no net gain, leading to the characterization of such communities by some foresters as 'decadent' because they are mainly interested in the productivity of the autotrophs.

Over time, natural systems mature towards maximization of NCP. On the other hand, humans are often concerned with maximizing NPP. This is an example of the 'decision regulators' discussed in Chapter 1. Natural forest system decision regulators may allow trees to achieve ages of several hundred to over a thousand years old before they die. The control system exerted by forest management determines that the life of the trees will be that which maximizes NPP, before considerable amounts of energy become devoted to heterotrophic respira-

tion (Figure 2.9). The age of the trees in systems managed for forestry will hence be much younger than is the case in natural systems.

Auxiliary energy flows allow some ecosystems and sites to be especially productive. For example, tidal energy in an estuary is a form of auxiliary energy flow that helps to bring in nutrients and dissipate wastes, so that organisms do not have to expend energy on these tasks and can devote more energy to growth. Agriculture, as discussed in Chapter 10, relies extensively on the inputs of auxiliary energy in the form of pesticides, fertilizer, tractor fuel, and the like to supplement the natural energy from the sun to augment crop growth. In many cases this subsidy, mostly derived

Estuaries are among the most productive ecosystems. Unfortunately, they are also very convenient sites for industrial activity, such as the log boom storage shown here, which inhibits productivity (*Philip Dearden*).

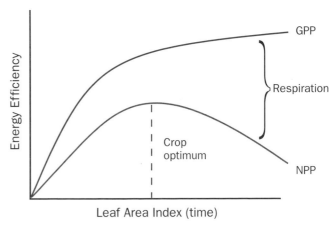

Figure 2.9 Diagram to illustrate general relationship between productivity and time as a forest matures. Foresters might consider the optimal stage of the forest to be at maximum NPP, even though GPP continues to increase over time.

Table 2.1	Production and Respiration as (kcal/m²/yr) in Growing and Climax Ecosystems

	Alfalfa Field (US)	Young Pine Plantation (England)	Medium-Aged Oak–Pine Forest (NY)	Large Flowering Spring (Silver Springs, Fla)	Mature Rain Forest (Puerto Rico)	Coastal Sound (Long Island, NY)
Gross primary production	24,400	12,200	11,500	20,800	45,000	5,700
Autotrophic respiration	9,200	4,700	6,400	12,000	32,000	3,200
Net primary production	15,200	7,500	5,000	8,800	2,500	2,500
Heterotrophic respiration	800	4,600	3,000	6,800	13,000	2,500
Net community production	14,400	2,900	2,000	2,000	Little or None	Little or None

SOURCE: Adapted from E.P. Odum, *Fundamentals of Ecology*, 3rd edn (Toronto: Holt, Rinehart and Winston, CBS College Publishing, 1971, 46). Copyright © 1971 by Saunders College Publishing, reproduced by permission of the publisher.

from fossil fuels, is in excess of the amount of energy input from the sun. Without this subsidy, productivity would be much reduced. There is a cost to this subsidy, however, in terms of high energy costs and the environmental externalities created as the subsidy disperses into the environment in the form of pollution.

ECOSYSTEM STRUCTURE

The energy flows described above are all part of the ecosphere. The ecosphere can be broken down in size to smaller units. At the smallest level is the individual *organism*. A group of individuals of the same species is a **population**. When taken together with other organisms in a particular environment, this is known as a *community*. The **ecosystem** is a collection of communities interacting with the physical environment. However, ecosystems represent a somewhat abstract conceptualization of the environment that can range greatly in scale. Due to the highly interactive nature of the relationship between organisms and their environment, it is often difficult to define precisely the boundary of an ecosystem. Furthermore, ecosystems are *open systems* in that they exchange material and organisms with other ecosystems. Ecosystems and communities thus provide useful abstractions for the study of the environment, but should not be taken to be very precise categories that will be agreed upon by all scientists.

Many ecosystems taken together and classified according to their dominant vegetation and animal communities form a **biome**. Canada, due to its size, has as many biomes as any country in the world (Figure 2.10). The main factors that control biome distribution are water availability and temperature. How these influence biomes at the global scale is summarized in Figure 2.11.

Abiotic Components

The food chains described above constitute the living or **biotic components** of ecosystems. The **abiotic components** have an important role in determining how these biotic components are distributed. Important abiotic factors include light, temperature, wind, water, and soil characteristics such as pH, soil type, and nutrient status. All these factors influence different organisms in various ways. The interaction among these characteristics and the organisms and between the organisms themselves determines where each organism can grow and how well it may grow.

Soils are critical in determining the vegetation growth of an area (Box 2.6). Soil is a mixture of inorganic materials such as sand, clay, and pebbles, decaying organic matter, such as leaves, and water and air. This mixture is home to billions of microorganisms that are constantly modifying and developing the soil. In the absence of these organisms, earth would be a sterile rock pile rather than the rich life-supporting environment it now is. Most of these organisms are in the surface layer of the soil, and one teaspoon may contain hundreds of millions of bacteria, algae, and fungi. In addition, many larger species—roundworms, mites, millipedes, and insects—play vital roles in this complex ecology.

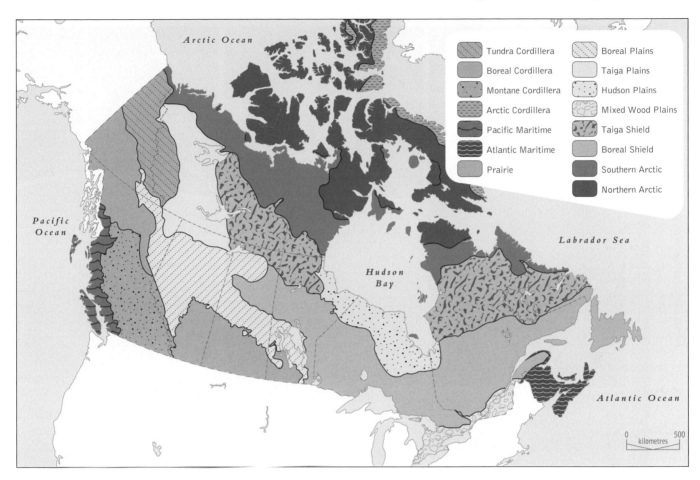

Figure 2.10 Terrestrial ecozones of Canada. SOURCE: E. Wiken, *Terrestrial Ecozones of Canada* (Ottawa: Environment Canada, 1986).

Most soils form from the *parent material* where they are found. This may originate from the weathered remains of bedrock or where sediments have been deposited from elsewhere by water, ice, landslides, or the wind. Over centuries, ongoing physical and chemical weathering and organic activities modify this mixture. As the parent material breaks down, inorganic elements such as calcium, iron, manganese, phosphorus, etc. (Chapter 4) are released. The amount of nutrients in the material and the speed of breakdown are a major influence on the fertility of the resulting soil. Hard rocks, such as granite, break down slowly and yield few nutrients. Different soils will result depending on the location. These different processes result in different layers forming in the soil, called **soil horizons**. A view across these horizons is called a *soil profile*. Figure 2.12 shows a generalized profile. However, not all soils have all these different horizons.

Time is also a critical factor in soil development. Soils that have been exposed to millions of years of chemical and physical weathering, such as

many tropical soils, have often lost their entire nutrient content. Conversely, where glaciation has scraped all the soil away, as in much of Canada as recently as 10,000 years ago, then the hard rocks, such as the granite of the Canadian Shield, have had little opportunity for weathering. They are also infertile. However, soils can be very fertile where ice sheets as they retreated deposited large quantities of clay rich in nutrients, as on the Prairies.

Soils also differ in their *texture*, or sizes of different materials. Clay is the finest, followed by silt, sand, and then gravel, the coarsest. Soils that contain a mixture of all these with decomposed organic material, or *humus*, are called *loams* and often make the best soils for vegetation growth. Texture is a main determinant of *soil permeability*, or the rate at which water can move through the soil. Water moves very slowly through soils composed mainly of the smallest particles, clay, and the soil easily becomes waterlogged. On the other hand, the large spaces between particles of sand or gravel lead to rapid drainage, and the soils may be too dry to support good vegetation growth. Plants obtain their nutrient supply necessary

Figure 2.11 Influence of temperature and rainfall on biome.

for growth from ions dissolved in the soil water, and so permeability is critical.

Soil has many different chemical characteristics. One of the most important is the pH value (see Chapter 4) measuring the acidity/alkalinity of the soil, which helps determine which minerals are available and in what form. Different plants have different mineral requirements. Farmers often try to change the acidity of their soils, for example, by adding lime if the soil is too acidic or sulphur if the soil is too alkaline.

Just as the laws of thermodynamics explain energy flows, some useful principles help us understand how organisms react to different abiotic influences. The first of these is known as the **limiting factor**. This tells us that all factors necessary for growth must be available in certain quantities if an organism is to survive. Thus, a surplus of water will not compensate for an absence

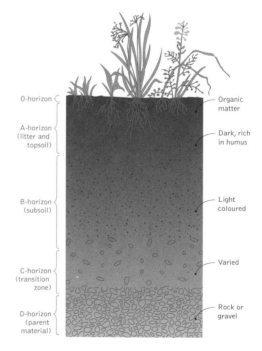

Figure 2.12 Generalized soil profile.

BOX 2.6 SOILS IN CANADA

Just as we can define ecozones, soil scientists can define soil zones that group together soils that are relatively similar in terms of their measurable characteristics. A glance at the soil map of Canada (below) will reveal a close resemblance to the ecozone map, as at this scale both tend to reflect the gross climatic and geological conditions of the region. The Canadian System of Soil Classification includes nine orders, the largest category of classification:

Brunisols cover 8.6 per cent, and are brown soils found mainly under forests.

Chernozems cover 5.1 per cent, occur under grasslands, and are some of the most productive soils.

Cryosols are the dominant soils in Canada, covering some 40 per cent of the country's land mass, and are found in association with permafrost.

Gleysols only cover 1.3 per cent and are found in areas that are often waterlogged.

Luvisols cover 8.8 per cent and occur in a wide variety of wooded ecosystems. They have higher clay content than brunisols.

Organics, which cover 4.1 per cent of Canada, form in wetland ecosystems where decomposition rates are slow.

Podzols are found beneath heathlands and coniferous forests, are relatively nutrient poor, and cover 15.6 per cent of Canada.

Regosols cover under 1 per cent of Canada and vary little from their parent material.

Solonets are saline soils, covering 0.7 per cent of Canada, and are found mostly in grassland ecosystems.

Soil zones of Canada.

of an essential nutrient or adequate warmth. In other words, a chain is only as strong as its weakest link. The weakest link is known as the *dominant limiting factor.* A major goal of agriculture is to remove the effect of the various limiting factors. Thus, auxiliary energy flows are employed to ensure that a crop has no competition from other plants (weeding), or that water supply is adequate (irrigation), or that the plant has optimal nutrient supply (fertilizer) (Chapter 10).

The corollary of the above is that all organisms have a range of conditions that they can tolerate and still survive. This is known as the **range of tolerance** for a particular species. This range is bounded on each side by a zone of intolerance for which limiting factors are too severe to permit growth (Figure 2.13). There may, for example, be too much or too little water. As conditions ameliorate for the particular factor, certain individuals within the population can tolerate the conditions, but because the conditions still are not optimal, relatively few individuals can exist. This is known as the **zone of physiological stress**. Still further amelioration creates a range where conditions are ideal for that species, the *optimum range.* Here, in theory, barring other factors, there will be the highest population of the particular organism.

The concepts of limiting factors and range of tolerance can be illustrated by an example from Saskatchewan, where smallmouth bass are introduced every year into lakes for sport fishing. These hatchery-raised fish survived from year to year (if they were not caught), proving that they were within their range of tolerance for survival as individuals. They did not, however, breed successfully, because the hatchlings needed a slightly warmer temperature to develop than did the mature fish to survive. Thus, different levels of tolerance exist for species at different life stages. The dominant limiting factor for the smallmouth in this environment was the low temperatures during hatching.

Water availability is often the critical factor that determines differences between communities. Where precipitation is in excess of about 1,000 mm per year, for example, trees will usually dominate the landscape if other factors are suitable. Below 750 mm, the range of tolerance for trees is exceeded and grasses will dominate because they have a much lower range of tolerance for water stress, in the order of 100 mm per annum. Below that level even grasses run into their zone of intolerance and then cacti, sagebrush, and other drought-resistant species dominate.

We must keep in mind that organisms are not reacting just to one abiotic factor, such as water availability, but to all the factors necessary for growth. Sometimes the optimal range for one factor will not overlap with the optimal range for other factors, which would place the organism in the zone of physiological stress for that factor, and it would become the dominant limiting factor. Organisms may also be out-competed for a particular factor in their optimum range by another organism with a greater tolerance to that environmental factor and again be forced into a zone of physiological stress. In other words, the simple single-factor model represented in Figure 2.13 is more complicated because of the numerous abiotic and biotic influences that must also be taken into account. That model does, however, provide a useful conceptual tool to help understand the spatial distribution of organisms.

Biotic Components

Other species also have an important role in influencing species distributions and abundance. Species

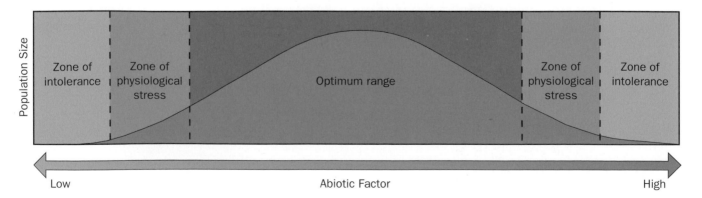

Figure 2.13 Range of tolerance.

In northern climates vegetation structure is very simple as both low temperatures and low rainfall result in growth conditions under which few species can survive. In the short Arctic summer, however, richer areas are ablaze with brightly coloured flowers, as on Herschel Island off the north coast of Yukon (*Philip Dearden*).

interact in several ways, including competition for scarce environmental resources. Each species needs a specific combination of the physical, chemical, and biological conditions for its growth. This is known as the **niche** of that species. Where the species lives is known as the **habitat**.

The **competitive exclusion principle** tells us that no two species can occupy the same niche in the same area. Most species have a *fundamental niche*, representing the potential range of conditions that it can occupy, and also a narrower *realized niche*, representing the range actually occupied. The physical conditions for growth exist throughout the fundamental niche, but the species may be out-competed in parts of this area through the overlapping requirements of other species. **Specialist** species have relatively narrow niches and are

generally more susceptible to population fluctuations as a result of environmental change. The panda is a classic example of such a specialist, with a total concentration on one plant, bamboo, as a source of food. Whenever the bamboo supply falls, as it does after it flowers, then this specialist species has few suitable alternative sources of food. Historically, when bamboo was abundant this did not particularly matter, as the pandas simply moved to a new area. However, as the animals have become increasingly restricted to smaller and more isolated reserves, this has become a major problem.

In Canada, specialist species include many of those discussed in later chapters, such as the burrowing owl and the whooping crane. **Generalist** species, on the other hand, like the black bear and coyote, may have a very broad niche, where few things organic are not considered a potential food item. Such generalist species have adapted most successfully to the new environments created by humans.

Competition

Intraspecific competition occurs among members of the same species, whereas **interspecific competition** occurs between different species. Both forms of competition are a result of demands for scarce resources. Intraspecific competition occurs particularly where individual species densities are very high. Interspecific competition occurs where species niches are similar. Competition may be reduced through **resource partitioning**, where the resources are used at different times or in different ways by species with overlap of fundamental niches. Hawks and owls, for example, both

The panda is a classic example of a specialist species (*Philip Dearden*).

Some species, such as the black bear, are very adaptable and have a relatively broad range of tolerance. On the Pacific coast, black bears are frequent scavengers of the intertidal zone where many items are considered potential food. The year-round availability of an abundant and varied food source results in very large individuals (*Philip Dearden*).

hunt for similar types of prey, but at different times, as owls are mainly nocturnal.

Intraspecific competition may lead to the domination of specific areas by certain individuals; the area is known as a *territory* and may be aggressively defended against intruders. Grizzly bears establish such territories, which may be as large as 1,000 km^2 for dominant males, although the possibility of defending such a large territory at all times from intruders is remote. During the breeding season male robins also establish and defend nesting territories, the boundaries of which are advertised in song. This kind of behaviour tries to establish sufficient resources for breeding pairs to be successful. Ultimately, intraspecific competition contributes to regulation of population size in areas where favourable habitat is limited, as those individuals unable to defend territories are outcast to less favourable areas where their likelihood of success is limited.

Biotic Relationships

There are other kinds of relationships between species besides competition. In predation, for example, a **predator** species benefits at the expense of a **prey** species. The lynx eating the hare and the osprey eating the fish are familiar examples of this kind of relationship, although in a broader sense we should also consider the herbivore eating the plant. Predation is a major factor in population control and usually results in the immediate death of the prey species. A predator must be able to kill and overwhelm prey on a regular basis without getting hurt. Usually predators are bigger than their prey, and often target weaker members of the prey population to avoid getting injured. They may also hunt as a group to improve the likelihood of a kill and minimize the possibility of a debilitating injury.

One theory that addresses the relationship between the benefit of making a kill and feeding against the cost of the energy expended to make the kill is **optimal foraging theory**. The theory recognizes that there is a point of compensation between the benefit of obtaining the prey and the costs of doing so, and that the predator's behaviour adjusts to optimize the benefits. It may be more worthwhile, for example, to hunt a smaller prey more often, even though it will result in less food intake, if the smaller prey can be dispatched with

This rather stunned looking red fox has just escaped from the clutches of a large female golden eagle that managed to lift it some 3 m from the ground before one of the authors happened on the scene and unwittingly rescued one carnivore from becoming the prey of another one at the next trophic level (*Philip Dearden*).

little fear of injury and eaten quickly so another predator cannot steal it. Optimal foraging theory also suggests that most predators, as one type of prey becomes scarce, will switch prey if they can. Several examples of this kind of behaviour are discussed within the marine context in Chapter 8.

Prey species have evolved many strategies to avoid being transferred along the food chain. Some plants develop physical defences such as thorns, while others may evolve chemical defences such as poisons, to deter their predators. The chemicals manufactured by plants provide the raw material for many of our modern medicines, such as aspirin, which comes from willows. Animal species employ a wide variety of predator avoidance strategies ranging from camouflage, alarm calls, and grouping to flight (Box 2.7).

A special kind of predator-prey relationship is **parasitism**, where the predator lives on or in its prey (or host). In this case the predator is often smaller than the prey, and gains its nourishment from the prey over a more extended time period that may lead to the eventual death of the host. This may cause the death of the parasite, too, although some parasites, such as dog fleas and mosquitoes, can readily switch hosts. Tapeworms, ticks, lamprey, and mistletoe are all examples of parasites.

Not all relationships between species are necessarily detrimental to one of the species. **Mutualism** is the term used to describe situations where the relationship benefits both species. These benefits may relate to enhanced food supplies, protection, or transport to other locations. The

Guest Statement

Landscape Ecology
Dennis Jelinski

The problem of relating phenomena across scales is the central problem in biology and in all of science
— S. Levin, 1992 Presidential Address to the Ecological Society of America

'Space' has been declared the 'final frontier' in ecology. Geographers have known this for years and over the last two decades ecology and geography have combined to spawn the new subdiscipline of landscape ecology, which deals explicitly with the effect of space on ecological processes and patterning. A *landscape* is simply a heterogeneous land area, such as a river valley in Waterton Lakes National Park, Alberta, a patchwork of logged areas in western Ontario, or the intertidal zone in Clayoquot Sound, BC. Landscape ecology seeks to understand the relationship between patterns and processes, explaining the spatial variation in landscapes at multiple scales as affected by natural causes and human society. In other words, landscape ecologists are interested in understanding how *heterogeneity* in the structure of the landscape might affect processes such as the movement of pollinators, the location of food supplies, refugia from predators, denning and calving areas, spawning beds, and shelter from inclement weather.

Landscape ecology, both as a science and especially as a way of examining the natural world, was instrumental in creating a major paradigm shift in ecology from the *equilibrium* view of ecosystem function to the *non-equilibrium* view. The former view embraced the notion of a 'balance of nature', a theory that had been front and centre in ecology for most of the twentieth century. The paradigm had at its core a belief that ecosystems were closed, integrated and self-regulated, and able to respond to positive feedbacks in accordance with the mechanistic principles of cause and effect, all directed towards achieving equilibrium or 'balance'. By zooming out in scale and examining systems at the coarser landscape scale, students of ecology discovered that in fact ecosystems were subject to a great degree of openness wherein energy, materials, and organisms were frequently exchanged across system boundaries. They were also subject to frequent disturbance and a great deal of chance (more formally termed *stochastic* events). This flux in nature is most apparent at landscape scales.

Dennis Jelinski

Over the last 12 years or so my research and that of my students has focused on landscape ecology in a number of different environments. For example, the effect of *habitat fragmentation* on birds has been a major focus of research in avian conservation. Much research has focused on identifying several area-sensitive species. On wet meadow grasslands flanking the Platte River, Nebraska, Chris Helzer and I studied a combination of habitat area and shape and found that a perimeter-area ratio of habitat patches was a more effective determinant of the presence and richness of grassland bird species than patch area alone (Helzer and Jelinski, 1999). In other words, some grassland bird species avoid habitat edges and habitat patches with high perimeter-to-area ratios. This means that ecosystem restoration efforts must try to ensure that habitat patches are as compact as possible so that brood parasites and generalist predators cannot gain ready access to all portions of the habitat, as is possible when patches are elongated.

In a very different way, Lindsay Mulkins, myself, and others show how landscape boundaries are very permeable. We (Mulkins et al., 2001) studied the flow of dissolved carbon from coastal old-growth forests in Clayoquot Sound, BC, and found that in nearby marine waters small zooplankton called mysids, which are an important component of the diet in grey whales, were seasonally affected by this importation, or *spatial subsidy*, of carbon. Thus, grey whales that in turn feed on these mysids are fuelled in part by carbon originating in terrestrial forests. This is a profound illustration of how open ecosystems really are!

In a more recent study in Barkley Sound, BC, we examined the effect of wave-tossed marine algae (such as kelps) on the nitrogen budget of beach plant communities. Nitrogen is in short supply in most ecosystems, but especially on these soil-poor beaches. The algae plus marine carrion, diatoms, and drift-wood that is also washed ashore are collectively termed *wrack*. This spatial subsidy of nutrients on otherwise low-nutrient cobble beaches elevates the level of nitrogen in a species of beach grass. The wrack also is important in the diets of terrestrial isopods (you can find them under most decaying logs), which migrate from nearby forest and forage on the dried algae and beach grass. Thus, both plant and animal members of beach and nearby forest communities benefit from materials that originate in the ocean.

Combined, these studies show that plants, animals, their ecological communities, and indeed entire ecosystems are strongly affected by the landscape in which they are embedded. Processes in the landscape are affected by spatial patterning at a great many scales, from those in a small patch of a square metre of beach, to patches of grasslands, to the juxtaposition of marine and terrestrial ecosystems. Landscape ecology as a management tool holds tremendous potential in addressing a great many environmental problems that are played out at landscape and regional scales, including threats to biodiversity, pollution, habitat fragmentation, parks management, and sustainable resource development.

Dennis Jelinski is Director of Bamfield Marine Sciences Centre on Vancouver Island and an Associate Professor in the Department of Geography, University of Victoria.

relationship between the nitrogen-fixing bacteria and their host plants, described in Chapter 4, is an example of such a relationship that results in enhanced nutrition for both species. Other examples include the relationships between flowering plants and their pollinators, which results in the transport of pollen to other plants, and the protection offered by ants to aphids in return for the food extracted from plants by the aphids. Interactions that appear to benefit only one partner but not harm the other are examples of **commensalism**. The growth of *epiphytes*, plants that use others for support but not nourishment, is one example.

Keystone Species

Species with a strong influence on the entire community are known as **keystone species**. They are named after the final wedge-shaped stone laid in an arch. Without the keystone, all the other stones in the arch will collapse. In Canada, our national symbol, the beaver, is a good example of such a species (Box 2.8). Beavers can have a profound impact on their environments through the dams they build that raise and lower water levels. This, in turn, affects the limits of tolerance of other species in the community that may suddenly find themselves submerged under a beaver dam or facing lower water levels downstream. Different species will have different reactions to this change, depending on, for one thing, their range of toler-ance relating to water. However, when a keystone species is removed there is generally a cascading

Rubber is a natural product made by rubber trees to help deter herbi-vores. Rubber plantation in southern Thailand (*Philip Dearden*).

BOX 2.7 CANADA'S OLYMPIC CHAMPION: THE PRONGHORN ANTELOPE

Like the bison (see Chapter 11), the pronghorn are also migratory, moving north in summer and south in winter, at which times they may gather together in large numbers. Their main predators (other than humans)—the plains wolf and grizzly—have become extirpated from their range, leaving the coyote and, further south, the bobcat as predators. Unlike all other members of the bovid family (sheep, oxen, and goats), pronghorns shed their horns after the rut. Populations of pronghorn are now protected in Grasslands National Park in Saskatchewan.

Pronghorn are reckoned to be the fastest middle-distance runners on Earth, having been clocked at speeds up to 98 km per hour for over 24 kilometres. They are superbly adapted for speed with small stomachs and big lungs, long strides, small feet, flexible spines like the cheetah, and additional ligaments and joints, all designed to make this animal the fastest on the prairie. Speed is the logical defence mechanism for a vulnerable animal easily seen by predators on the open prairies. Confidence in their escape mechanism made them rather curious, and early settlers found them easy to attract and shoot. Initial herds estimated at between 20 and 40 million animals in North America were soon reduced to fewer than 30,000 animals as a result of hunting pressure and land-use change. There are now some half a million pronghorn remaining. They are the sole survivors of a family of North American antelopes, unrelated to the African antelopes, that at one time contained over two dozen different species.

effect throughout the ecosystem as other species are affected. Interestingly, the same species may be a keystone is some communities and not in others, depending on the community composition in that particular locale.

It is especially significant when a keystone species is removed from an area, or **extirpated**, by human activity. Such changes may take some time before they become noticeably manifest. Changes to soil characteristics created by the extermination of major herbivores, such as bison from the prairie, may take centuries before they become noticeable,

and are generally not reversible. The same is true for the other large grazers, such as the great whales (discussed in Chapter 8), which have been greatly decimated over the last couple of centuries.

IMPLICATIONS

There are important implications for society and species distributions from the above discussion.

- All of the Earth's inhabitants are interlocked in environmental systems that depend on one another for survival. Perturbations in part of the system will have impacts on other parts of the system.
- The basic scientific laws that govern the transformation of matter and energy dictate that sooner or later society must transform itself from a throw-away society built on processing ever-increasing matter and energy flows to one where energy efficiencies are improved and matter flows are reduced.
- A species may have a wide range of tolerance to some factors but a very narrow range for others.
- Species with the largest ranges of tolerance for all factors tend to be the most widely distributed. Cockroaches and rats, for example, enjoy virtually global distribution.
- Many weed and pest species are successful because of their large range of tolerance. Eurasian water milfoil, a significant nuisance in many

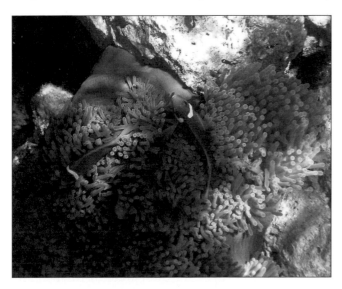

These Maldivian anemone fish live in a mutualistic relationship with the anemone, which provides protection from potential predators with its stinging tentacles while benefiting from the food scraps brought by the fish (*Philip Dearden*).

BOX 2.8 CANADA'S NATIONAL SYMBOL—THE BEAVER

The beaver is found all the way from Mexico to the Arctic and from Vancouver Island to Newfoundland. However, the beaver is mostly associated with the northern woods and their waterways, where it is well known for its capabilities for water engineering. Many different species of beaver could once be found throughout the northern hemisphere. A Eurasian counterpart remains in small populations, but it is the North American beaver that has flourished and become one of the continent's most successful mammals. It also played a critical role, as did the sea otter on the west coast, in attracting the colonial gaze of Europeans to the resources of North America.

The beaver is a rodent. In fact, it is the second largest rodent in the world. All rodents are distinguished by their sharp incisor teeth, designed to gnaw though bark, crack nuts, or attack any other edible vegetable matter in a similarly efficient manner. The success of this strategy is attested to by the proliferation of rodents, which comprise nearly 40 per cent of all mammal species. The specialty of the beaver, of course, is its ability to fell trees (some as large as a metre in diameter), which can then be used as a food supply and as a means to build its familiar dams and lodges. Trees, particularly hardwoods such as poplars, are felled close to the water's edge so that they can be dragged into the water, which is the beaver's preferred medium. With their broad flat tails, sleek coats, powerful and webbed hind feet, beavers are well equipped to undertake their aquatic construction activities. Their dams impede the flow of water, giving them greater access to trees and making them less vulnerable to terrestrial predators. They can use the pond so created as a low-energy way of transporting food to their lodge, which, surrounded by water, is virtually impregnable to predators.

Probably no other animal, except humans, has the ability to cause such a radical and deliberate change to the environment. Beaver dams benefit not only beavers but also other water-oriented organisms such as waterfowl, otters, muskrats, and frogs and other amphibians. With an estimated population of over 60 million animals prior to European settlement in North America, their ecological impact on the landscape would have been substantial. Although they were trapped out of large areas, they are now starting to recolonize as a result of conservation activities. Industrious, clever, effective, resilient, but not spectacular—is this the Canadian character?

The beaver, the national animal of Canada, can have a very significant impact on other species through its dam-building activities (*Philip Dearden*).

waterways in Canada, is an alien that can grow in conditions from Canada to Bangladesh.

- Response to growth factors is not independent. Grass, for example, is much more susceptible to drought when nitrogen intake is low.
- Tolerance for different factors may vary through the life cycle. Critical phases often occur when organisms are juveniles and during the time of reproduction.
- Some species can adapt to gradually changing conditions for some factors, up to a point. However, after this **threshold** of change is reached, the population will collapse.

From this discussion it should be apparent that ecosystems are very complicated. A complex set of interrelationships exists among organisms and between organisms and their environment. A change in part of this matrix will often result in corresponding changes throughout. Humans are now such a dominant influence on global environmental conditions at all scales that significant changes are underway as a result of human activities. There is considerable uncertainty as to how ecosystems and the entire life support system of this planet will react to these changes. Yet, even under natural conditions, ecosystems are not static. The next chapter will focus on how ecosystems change over time.

SUMMARY

1. Energy is the capacity to do work. Energy comes in many forms, including radiant energy (from the sun), chemical energy (stored in chemical bonds of molecules), and heat, mechanical, and electrical energy. Energy differs from matter in that it has no mass and does not occupy space.

2. Understanding energy flows is critical to the understanding of the ecosphere and environmental problems. The laws of thermodynamics explain how energy moves though systems. The first law states that energy can neither be created nor destroyed, but merely changed from one form to another. The second law informs us that at each energy transformation some energy is converted to a lower-quality, less useful form.

3. Energy is the basis for all life. Through the process of photosynthesis, certain organisms transform carbon dioxide and nutrients in the presence of radiant energy from the sun into organic matter. This matter forms the basis of the food chains by which energy is passed from trophic level to trophic level. At each transference, the second law of thermodynamics dictates that some energy is lost, typically as much as 90 per cent.

4. Productivity is a measure of the abilities of different communities to transform energy into biomass. The most productive communities are found in estuaries, wetlands, and rain forests.

5. The ecosphere is the thin, life-supporting layer of the Earth characterized by interactions between the biotic and abiotic components. It can be further subdivided into communities, ecosystems, and biomes.

6. The concepts of limiting factors and range of tolerance help us understand the interaction between the biotic and abiotic components of the ecosphere.

7. Each species needs a specific combination of physical, chemical, and biological conditions for its growth. This is the niche of that species.

8. The principle of competitive exclusion tells us that no two species can occupy the same niche in the same area at the same time.

9. Species compete for scarce resources in any given habitat. However, there are many other forms of relationship between species, such as predation, parasitism, mutualism, and commensalism.

10. Species with a strong influence on the entire community are known as keystone species.

KEY TERMS

abiotic components
aerobic
autotrophs
biogeochemical cycles
biomass
biome
biotic components
calorie
carnivores
chemoautotrophs
chlorophylls
commensalism
competitive exclusion
  principle
consumers
decomposer food chains
detritus

ecosystem
energy
entropy
euphotic zone
extirpated
food chain
food web
generalist
gross primary productivity
  (GPP)
habitat
heat
herbivores
heterotrophs
high-quality energy
interspecific competition
intraspecific competition

keystone species
kinetic energy
law of conservation of
  energy
limiting factor
low-quality energy
mutualism
net primary productivity
  (NPP)
niche
omnivores
optimal foraging theory
parasitism
photosynthesis
phototrophs
phytoplankton
population

potential energy
predator
prey
primary consumers
producers
range of tolerance
resource partitioning
secondary consumers
soil horizons
specialist
tertiary consumers
threshold
trophic level
zone of physiological
  stress
zooplankton

REVIEW QUESTIONS

1. What are the main biotic and abiotic components of ecosystems?
2. How do the laws of thermodynamics apply to living organisms?
3. How do the laws of thermodynamics apply to environmental management?
4. What are chemoautotrophs, and what role do they play in ecosystem dynamics?
5. On what trophic level is a pitcher plant? Why? Are there plants on the same trophic level in your area? What are they and where do they grow?
6. In what kinds of ecosystems do detritus food chains dominate?
7. What roles do phytoplankton play in maintaining ecospheric processes?
8. What are the management implications of recognizing concepts such as specialist, generalist, and keystone species? Can you think of any examples in your area?
9. What is optimal foraging theory?
10. What do you think the dominant limiting factors are for plant communities in your area?
11. Draw a cross section across (E–W) and down (N–S) your province or territory, and show the main environmental gradients and the vegetational response.
12. What are some of the main transformations that have to take place in society to reflect the implications of the laws of thermodynamics and law of conservation of matter?

RELATED WEBSITES

ENVIRONMENT CANADA, ECOTOXICOLOGY:
http://www.pyr.ec.gc.ca/EN/Wildlfie/migratory/ecotoxic.shtml

ENVIRONMENT CANADA, ECOSYSTEM INFORMATION:
http://ecoinfo.ec.gc.ca/index_e.cfm

REFERENCES AND SUGGESTED READING

Allan, J.D., M.S. Wipfli, J.P. Caouette, A. Prussian, and J. Rodgers. 2003. 'Influence of streamside vegetation on inputs of terrestrial invertebrates to salmonid food webs', *Canadian Journal of Fisheries and Aquatic Sciences* 60, 3: 309–20.

Cebrian, J. 2004. 'Role of first-order consumers in ecosystem carbon flow', *Ecology Letters* 7, 3: 232–40.

Colinvaux, P. 1980. *Why Big Fierce Animals Are Rare.* London: Penguin.

Elmhagen, B., M. Tannerfeldt, and A. Angerbjörn. 2002. 'Food-niche overlap between arctic and red foxes', *Canadian Journal of Zoology* 80, 7: 1274–85.

Gende, S.M., and T.P. Quinn. 2004. 'The relative importance of prey density and social dominance in determining energy intake by bears feeding on Pacific salmon', *Canadian Journal of Zoology* 82, 1: 75–85.

Hodges, K.E., and A.R.E. Sinclair. 2003. 'Does predation risk cause snowshoe hares to modify their diets?', *Canadian Journal of Zoology* 81, 12: 1973–85.

Keith, L.B., J.R. Cary, O.J. Rongstad, and M.C. Brittingham. 1984. 'Demography and ecology of a declining snowshoe hare population', *Wildlife Monographs* 90: 1–43.

Krebs, C.J. 1994. *Ecology*, 4th edn. New York: Harper-Collins.

———, R. Boonstra, S. Boutin, and A.R.E. Sinclair. 2001. 'What drives the 10 year cycle of snowshoe hares?', *Bioscience* 51: 25–35.

——— et al. 2003. 'Terrestrial trophic dynamics in the Canadian Arctic', *Canadian Journal of Zoology* 81, 5: 827–43.

McKinstry, M.C., P. Caffrey, and S.H. Anderson. 2001. 'The importance of beaver to wetland habitats and waterfowl in Wyoming', *Journal of the American Water Resources Association* 37, 6: 1571–7.

Methratta, E.T. 2003. 'Top-down and bottom-up factors in tidepool communities', *Journal of Experimental Marine Biology and Ecology* 299, 1: 77–96.

Mills, E.L., et al. 2003. 'Lake Ontario: food web dynamics in a changing ecosystem (1970–2000)', *Canadian Journal of Fisheries and Aquatic Sciences* 60, 4: 471–90.

Moore, P.G. 2002. 'Mammals in intertidal and maritime ecosystems: Interactions, impacts and implications', *Oceanography and Marine Biology* 40: 491–608.

Mulkins, L.M., D.E. Jelinski, J.D. Karagatzides, and A. Carr. 2001. 'Carbon isotope composition of mysids at a terrestrial-marine ecotone, Clayoquot Sound, British Columbia, Canada', *Estuarine, Coastal and Shelf Science* 54: 669–75.

Odum, E.P. 1989. *Ecology and Our Endangered Life-Support Systems*. Sunderland, Mass.: Sinauer Associates.

Palmer, T.M., M.L. Stanton, and T.P. Young. 2003. 'Competition and coexistence: Exploring mechanisms that restrict and maintain diversity within mutualist guilds', *American Naturalist* 162, 4: S63–S79.

Schulze, E.D., and H.A. Mooney, eds. 1993. *Biodiversity and Ecosystem Function*. New York: Springer-Verlag.

Scott, G.A.J. 1995. *Canada's Vegetation: A World Perspective*. Montreal and Kingston: McGill-Queen's University Press.

Simpson, C.G. 1964. 'Species density of North American recent mammals', *Systematic Zoology* 13: 15–73.

Smith, C.R. 1992. 'Whale falls: Chemosynthesis on the deep sea floor', *Oceanus* 35: 74–8.

Smith, J.N.M., C.J. Krebs, A.R.E. Sinclair, and R. Boonstra. 1988. 'Population biology of snowshoe hares II: Interactions with winter food plants', *Journal of Animal Ecology* 57: 269–86.

Stephenson, R. 1982. 'Nunamiut Eskimos, wildlife biologists and wolves', in F.H. Harrington and P.C. Paquet, eds, *Wolves of the World: Perspective of Behavior Ecology and Conservation*. Park Ridge, NJ: Noyes Publishers, 434–9.

Thomson, J. 2003. 'When is it mutualism?', *American Naturalist* 162, 4: S1–S9.

Trostel, K., A.R.E. Sinclair, C.J. Walters, and C.J. Krebs. 1987. 'Can predation cause the 10 year hare cycle?', *Oecologia* 74: 185–92.

Zimmer, K.D., M.A. Hanson, and M.G. Butler. 2003. 'Relationships among nutrients, phytoplankton, macrophytes, and fish in prairie wetlands', *Canadian Journal of Fisheries and Aquatic Sciences* 60, 6: 721–30.

Chapter 3

Ecosystem Change

Learning Objectives

- To understand the nature of ecosystem change and its implications for environmental management.
- To describe the process of primary and secondary succession and the ways in which humans change these processes.
- To outline what is meant by ecosystem homeostasis.
- To discuss examples of invasive species and their impacts and management.
- To understand the ecological implications of species removal.
- To describe the main factors affecting population growth in a species.
- To discuss the nature of evolution, speciation, and extinction.
- To understand the importance of biodiversity.

INTRODUCTION

The interaction between the ecosystem components, the abiotic and biotic factors discussed in the preceding chapter, is a dynamic one. Communities and ecosystems change over time. The speed of change depends on the factors creating the change and the response of the organisms through their range of tolerance to the new environment. Some changes are very rapid, such as a landslide that totally destroys the existing ecosystem. Others, such as climate change, occur over long time periods and allow communities to adjust slowly to the new conditions. As vegetation communities change, so, too, do the heterotrophic components dependent on plants for food. Similarly, if the components of the food web change, this may well cause a change in vegetation.

In this chapter, several aspects of change in ecosystems will be examined, starting with the process of ecological succession. This will be followed by a discussion of the concept of ecosystem homeostasis and its contributing factors, as well as different aspects of population growth. Finally,

longer-term changes such as evolution, speciation, extinction, and biodiversity will be considered.

ECOLOGICAL SUCCESSION

Ecological succession is an example of slow adaptive processes. It involves the gradual replacement of one assemblage of species by another, as conditions change over time. **Primary succession** is the colonization and subsequent occupancy of a previously unvegetated surface, such as when a glacier retreats or a landslide destroys the previous ecosystem (Figure 3.1). Little soil exists, and the first species to occupy the area, known as *primary colonizers*, must be able to withstand high variability in temperatures and water availability and a difficult-to-access nutrient supply from rocks. Few species can tolerate such conditions. Lichens are often the first.

Lichens can exist on bare rock surfaces (Box 3.1). Over time, lichens, in combination with other physical and chemical processes, break down rocks. They trap water and nutrients. Biomass increases. Over centuries, these changes make it possible for

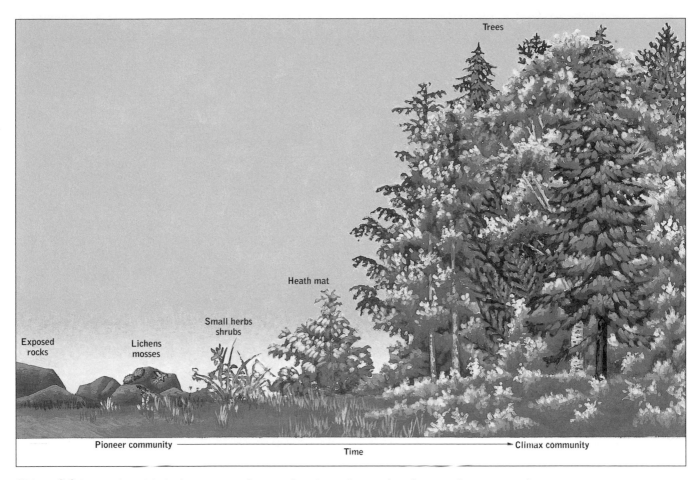

Trees

Heath mat

Small herbs
shrubs

Exposed
rocks

Lichens
mosses

Pioneer community ──────────────────── Time ────────────→ Climax community

Figure 3.1 A general model of primary succession over time, from a bare rock surface to a forest community.

other species to colonize, moss species in particular. Mosses grow faster than lichens, resulting in greater accumulation of biomass and soil over time. The lichens are eventually out-competed by the faster-growing mosses.

The next stage is invasion by herbaceous plants, such as grasses and species that we often

Only 10,000 years ago, most of Canada was covered in a thick layer of ice (*Philip Dearden*).

think of as weeds. Such species are able to colonize a wide range of habitats and have efficient reproductive strategies to allow them to do so. Fireweed is a good example, and the numerous wind-born seeds of this species are a common sight in many parts of Canada in the autumn.

Over time, the growth and decay of these early herbaceous species allow certain hardy shrubs to become established, which in turn further ameliorate conditions until more demanding tree species become established. In areas where precipitation and temperature are adequate, tree growth will often represent the final stage of this successional process. Each stage along the way is known as a seral stage, with the final stage known as the **climax community**. Although it is not possible to predict which species will occupy different areas undergoing succession, it is often possible to predict the type of plant that will dominate.

The foregoing is not an inevitable linear progression. It is a guideline to help understand the changes that may take place in communities. In some instances, for example in recently glaciated

BOX 3.1 LICHENS

Some environments have such harsh growing conditions that virtually nothing can survive. Lichens are one of the few types of organism that can be found in such places. Lichens are partnerships between fungi and photosynthetic algae that serve to benefit both partners in a mutualistic relationship. The fungi are able to cling to rocks or trees with their filaments and soak up water. In turn, the algae produce food for both partners through photosynthesis. This may include the fixing of nitrogen from the atmosphere by *cyanobacteria*. This combination is able to survive intense cold and drought and has been evolving for over one billion years, making lichens one of the most primitive of living organisms. Individual lichens may be over 4,000 years old. Over centuries, sufficient growth of lichens may occur so that the thinnest of soils is produced, allowing other species able to tolerate harsh conditions to colonize. Lichens are therefore very important *primary colonizers*.

Over 18,000 species of lichens have been described throughout the world, and undoubtedly many others have yet to be discovered. They have different life forms—dust, crust, scale, leaf, club, shrub, and hair. The best known in Canada include the leafy variety found growing on trees, an encrusting variety that grows on rocks and sometimes trees, and the so-called (and misnamed) reindeer mosses found throughout northern Canada. Besides providing an essential food supply for caribou and other animals, lichens have also been used by humans as flour (when dried and ground up) and as a dye for wool and other fabrics.

Lichens, due to their adaptive ability to absorb mineral requirements directly from the air, are very efficient accumulators of pollution. They can concentrate pollutants to exceed their own tolerance levels, and hence are excellent *indicator species* for air pollution, as lichens will be absent from heavily polluted areas. Wong and Brodo (1992) analyzed current against past collections of lichens in Ontario and found that of the 465 species collected prior to 1930, 42 had not been collected since that date and another 10 not since 1960.

Lichen (*Philip Dearden*).

Fireweed, seen here growing alongside the Tatshenshini River in northern British Columbia, is a common herb in early successional sites throughout Canada (*Philip Dearden*).

The term 'tree line' is used to describe areas where vegetation communities dominated by trees give way to those dominated by other types of vegetation, such as herbs and grasses. Rarely, however, is there a sharp line but rather an ecotone where patches of both tree- and grass-dominated communities exist together (*Philip Dearden*).

terrain, very hardy species of trees, such as willows and alders, may become established in favoured sites with little previous colonization having occurred. *Cyclic succession* may also occur where a community progresses through several seral stages but is then returned to earlier stages by natural phenomena such as fire (Box 3.2 and Figure 3.2) or intense insect attack. The different seral stages are also not discrete but may blend from one into another. These blending zones tend to be the areas with the highest species diversity since they contain species from more than one community. These zones are known as **ecotones** and occur as relatively richer zones between communities.

Sand dune succession is also a common form of primary succession, where the primary colonizers are not lichens but grasses that have the ability to withstand not only the high variability in temperature and water, but also the constantly shifting sand. The grasses help to stabilize the sand until mat-forming shrubs invade. Later, conditions may become suitable for hardy trees, such as pines, that may in turn be replaced by other tree species such as oaks.

Climax should be thought of as a relative rather than absolute stage. Communities do not change up to the climax and then cease to change. However, the nature of the species assemblage is more constant over time, once a climax stage is reached. Some ecologists prefer the term of *mature community* to describe this stage rather than climax, as the latter has stronger implications of mature

Sand dunes are a good place to observe the successional changes over time, as shown here at Shallow Bay on the west coast of Newfoundland. With increasing distance from the sea, the communities change to those in later seral stages representing the buildup and colonization of the sand (*Philip Dearden*).

communities being static over time. Even in mature communities, future changes in pathogens, predation, and climate will generate ongoing changes.

The climax vegetation for most areas is strongly influenced by the prevailing climate, and is therefore known as a *climatic climax*. In some areas other factors such as soil conditions may be more important than climate in determining community composition and structure. These are known as *edaphic climaxes* (Box 3.3).

In addition to the primary succession described above, successional processes also occur on previously vegetated surfaces such as abandoned fields or avalanche tracks or following a fire, where soil is already present. This process is known

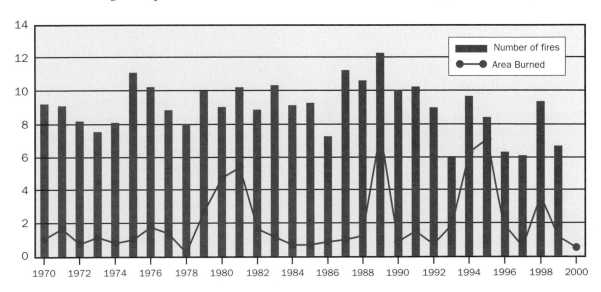

Figure 3.2 Number of forest fires in Canada (*thousands*) and area burned (*millions of hectares*). SOURCE: Environment Canada (2003: 46); www.nrcan-rncan.gc.ca/cts-sct/science/prodserv/firereport.

BOX 3.2 FIRE, MANAGEMENT, AND ECOSYSTEM CHANGE

In many areas, fire is a natural occurrence that has a profound impact on plant and animal communities. In some communities it may be the dominant influence, and if suppressed by human interference those communities will change in composition. Fire has been used as a tool by humans since earliest times to manipulate ecosystems to produce desired effects, such as removing forests to facilitate agriculture, burning grasslands to generate new grass growth, and scaring animals into running in certain directions so they can be more readily hunted. Weber and Taylor (1992) outline five uses of fire in forest management: hazard reduction, silviculture, insect and disease control, wildlife habitat enhancement, and range burning.

Fire has several important ecological and social implications:

- It favours the growth of certain species over others. Some species are quite fire-resistant (such as the Douglas fir), while the heat from fire may aid in the germination of other species (e.g., many pine species). Fire may result in the death of other species. Thus, where fire is common it may be the dominant influence on the composition and structure of some communities.
- It tends to increase the diversity of species in a community over the short term. Fire releases nutrients from the biomass into the soil and atmosphere; some may be lost from the site, while the remainder helps stimulate growth of some species.
- It stimulates the growth of various grasses and herbs that provide fodder for herbivores, which may in turn increase carnivore populations.
- Soil temperatures are increased, not only during the fire, but also afterwards—the site has a lower albedo and is more open to the sun. This also influences chemical and biological properties of the soil, stimulating microbial activities and enhancing decomposition.
- Fires that are too intense or frequent may cause sufficient nutrient impoverishment of a site to preclude further growth of trees, and the vegetation may become dominated by grasses and low shrubs. This is known as a *plagioclimax*. Many of the heathlands of Northern Europe were created in this manner, and clear-cutting and fire in nutrient-poor black spruce forests in Canada can have the same effect.

Fire is also a highly emotive topic for management. Early concepts of forestry and conservation encouraged policies of total fire suppression, with little accord taken of the role of fire in various ecosystems. This led to unanticipated changes in some ecosystems and a buildup of organic debris. Managers of protected areas, such as parks, now realize that if fire is a natural part of an ecosystem, fire suppression policies are altering the ecosystem in unnatural ways. This has led to prescribed burning programs in many parks (Chapter 11). If the buildup of fuel becomes too great over time, then fire can jump from the forest floor up to the canopy and kill many trees. For this reason, fire managers often prefer to burn frequently to control the accumulation of fuel. Whether fire should be suppressed or not should be a reflection of knowledge of ecosystem characteristics. Some fires may be ecologically appropriate. Others may be a reflection of human carelessness or lack of ecological understanding. Furthermore, it is not possible to ignore the potentially destructive effects of fires on human livelihoods.

Since the mid-1980s, an average of 2.38 million hectares of forest have gone up in flames every year in Canada. Before this time, the average was one million hectares. Many scientists believe that the increase may be due to global warming caused by the emission of greenhouse gases (see Chapter 7). The burning of millions of tons of carbon that is biologically fixed in the biomass of the trees releases carbon dioxide, further exacerbating the buildup of greenhouse gases in the atmosphere. This is an example of a positive feedback loop. The hotter it gets, the drier it gets, the more fires we have, the more carbon dioxide is released, and the hotter it gets. Scientists predict that temperature rise associated with global warming could be between 4–6° C in another 40 years in the boreal forest zone. They also predict lower rainfalls. This will lead to greater drying of the land surface and increased frequency and area of fire. Overall, forest ecosystems will show a high degree of disturbance. Species dependent on old-growth characteristics, such as woodland caribou, will be put under increasing pressure.

Guest Statement

Controls on High-Elevation Treeline in Westernmost Canada

Ze'ev Gedalof and Dan J. Smith

One important problem facing scientists and resource managers is predicting how forests will respond to global environmental change. Trees are typically long-lived, and they reproduce relatively slowly. Unlike most animals, they cannot change their location in response to changing environmental pressures. As the Earth's climatic and chemical properties change, forests are being exposed to different selective pressures than existed at the time of their establishment. As trees become increasingly stressed, they will become more vulnerable to wildfire, insect attack, fungal pathogens, drought, and disease. Understanding how these vulnerabilities will change over time will allow scientists to be proactive in their efforts to preserve forest biodiversity.

One method of predicting how the distribution of trees might change in response to the changing climate is to evaluate how individual trees have responded to climatic variability in the past. In temperate regions, such as Canada, trees produce growth rings each year. The width of these rings often varies, depending on the availability of heat, moisture, nutrients, light, or other limiting factors. At particularly severe locations, such as high elevation, there will often be a single limiting factor (for example, summer temperature). Variability in the width of tree rings at these locations can therefore be used to understand how the species might respond to future changes in this limiting factor, and also to reconstruct how the factor varied over the lifetime of the tree—which can be considerably longer than the interval over which people have kept records. In Canada, the oldest known living tree is a yellow cedar from Vancouver Island, which is over 1,200 years old (Laroque and Smith, 1999).

In westernmost

Dan J. Smith

Ze'ev Gedalof

Canada, mountain hemlock (*Tsuga mertensiana*) is the most common treeline species. By collecting core samples from mountain hemlock trees growing in Strathcona Provincial Park, on Vancouver Island, we have been able to develop a tree-ring chronology that extends back to AD 1412 (see Figure 1). Annual radial growth at this site is limited by low summer temperatures and enhanced by low winter precipitation. These results are consistent with the hypothesis that the total length of the growing season is the most critical factor influencing growth at treeline environments. Deep, persistent snowpacks keep soil temperatures low and tend to delay the initiation and rates of metabolic processes, reducing growth. In contrast, warm temperatures during the growing season melt snow more quickly, increase root and shoot temperatures, enhance metabolic rates, and increase photosynthesis, thereby increasing radial growth rates.

It is possible to quantify the relationship between temperature, precipitation, and radial growth and then forecast the expected changes to growth under various global warming scenarios. However, the relationship is complicated. Using a response surface model to visualize the relationship between these parameters in three dimensions reveals that the relationship is complex and

Figure 1 An example of a tree-ring sample as seen by a high-resolution scanning/analysis system.

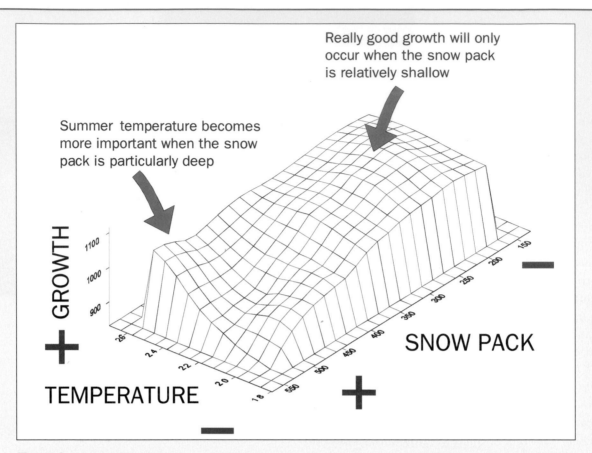

Figure 2 A response surface graph, showing the relationship between summer temperature, the depth of the April 1st snowpack, and annual radial growth of mountain hemlock.

highly non-linear (Figure 2). From this figure it is apparent that the species' response to summer temperature is highly dependent on the amount of snow that has accumulated. In years with a deep spring snowpack, radial growth is highly sensitive to variability in summer temperatures. In contrast, when the snowpack is shallow good growth occurs regardless of summer temperature.

The future health of mountain hemlock forests therefore depends not only on changes to temperature, but also on changes to precipitation—changes that are much less certain in the current generation of climate models! If the future turns out to be warmer and drier, mountain hemlock may be well adapted. However, mountain hemlock's downhill neighbours, western hemlock (*Tsuga heterophylla*) and Pacific silver fir (*Abies amabilis*) are probably better adapted to a warmer, less snowy climate and may ultimately out-compete mountain hemlock. Different tree species exhibit distinct responses to climatic variability, and only by studying these individual responses can scientists hope to understand the potential impacts of global environmental change on forest ecosystems.

Ze'ev Gedalof is an Assistant Professor in the Department of Geography at the University of Guelph. Dan Smith is Head of the Tree Ring Laboratory in the Department of Geography at the University of Victoria.

as **secondary succession**. The earlier soil-forming stages of primary succession do not have to be repeated, so the process is much shorter, with the dispersal characteristics of invading species being a main factor in community composition. Annual weeds again play a main role until perennial weeds, such as goldenrod, start to become established. Eventually the community will be invaded by shrub and ultimately tree species. A major challenge for agriculture and forest

BOX 3.3 EDAPHIC CLIMAX: TABLE MOUNTAIN, NEWFOUNDLAND

The west coast of Newfoundland (like most of the rest of the island) is dominated by the boreal forest (Chapter 10). In Gros Morne National Park (Figure 3.3), however, and at other locations on the west coast, this greenery (white spruce, paper birch, balsam fir) is punctuated by practically treeless orange-coloured outcrops that bear little if any similarity to the surrounding vegetation. These outcrops result from the distinctive chemical composition of the bedrock, known as 'serpentine'. Along with three other serpentine outcrops in western Newfoundland, the Table Mountain massif in Gros Morne was formed on the floor of the Atlantic Ocean millions of years ago and rafted up to its present position through the process of continental drift.

This geological history has given the serpentines a distinctive chemical composition characterized by high levels of nickel, chromium, and magnesium and low levels of calcium. Most of the climax species of the surrounding forests cannot tolerate these conditions; if they grow at all, they are stunted. Instead, the serpentines are host to relict communities of tough arctic-alpine species that have survived since the retreat of the glaciers and have not been displaced through the process of succession like the arctic-alpines on the surrounding bedrock. These serpentine communities are edaphic climaxes, where the underlying geology has been more important in determining plant cover than climate.

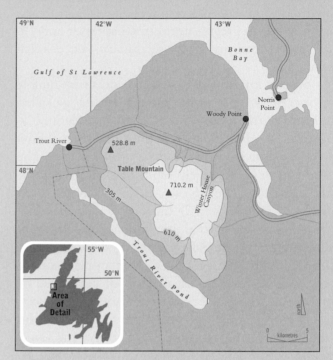

Figure 3.3 Location of Table Mountain.

The inhospitable soil chemistry has allowed rare species, such as this *Lychnis alpina*, to continue to grow in the area as relicts from the ice age (*Philip Dearden*).

The difference between the dominant vegetation of the edaphic climax of Table Mountain and the surrounding boreal forest can be clearly seen along the geological boundary (*Philip Dearden*).

managers is to prevent this natural recolonization taking place by species that may not yield the required products. As a result, chemical herbicides, as discussed in greater detail in Chapters 9 and 10, are often used to arrest secondary successional processes.

Similar kinds of processes also occur in aquatic environments. Here, the natural aging is called **eutrophication** (eu = well, trophos = feeding) as nutrient supplies increase over time with inflow and the growth and decay of communities. The process can be relatively rapid in shallow lakes as the nutrients (one of the auxiliary energy flows discussed in Chapter 2) promote increased plant growth that leads to more biomass and nutrient accumulation. The lake becomes shallower over time, with less surface area of water and the aquatic communities may eventually be out-competed by marsh and ultimately terrestrial plants. This process is a good example of a **positive feedback loop** (the shallower the lake gets, the stronger the forces become to make it shallower), which will be discussed in more detail in the next section. Eutrophication may also constitute a significant management problem as the species being replaced often have higher values to humans than the species replacing them. This problem is discussed in more detail in Chapter 4.

Indicators of Immature and Mature Ecosystems

As successional changes take place in communities, several trends emerge. For example, productivity declines as the slower growing species move in, and diversity increases as more specialized species come to dominate the community and more finely subdivide the resources of the particular habitat. However, the increase in diversity will not continue indefinitely, according to the intermediate disturbance hypothesis (Figure 3.4). This hypothesis suggests that ecosystems need a certain amount of disturbance to maintain diversity, or all the pioneer species would be gradually eliminated. Competitive exclusion reduces diversity. Disturbance occurs at different scales and might take the form of forest fire in an old-growth forest, floods, volcanic eruptions, or climate change.

Despite the flexibility of the successional concept, certain differences between mature and immature systems are generic (Table 3.1). In general, mature ecosystems tend to have a high level of community organization among a large number of larger plants and have a well-developed trophic structure. Decomposers dominate food chains, with a high efficiency of nutrient cycling and energy use. Net productivity is low. Immature ecosystems tend to have the opposite of these characteristics.

Effects of Human Activities

Humans have a profound influence on ecological succession. Many activities are directed towards keeping certain communities in early seral stages. In other words, humans seek to maintain the characteristics of the immature ecosystems shown in Table 3.1, as opposed to those of the mature ecosystems that would result if natural processes were allowed to proceed. Agriculture, for example, usually involves large inputs of auxiliary energy flows to ensure that succession does not take place as weeds try to colonize the same areas being used to grow crops. The same can be said for commercial forestry. Maintaining ecosystems in early successional stages has several implications:

- The productivity of early successional phases is often higher than later phases.
- Nutrient cycling, to be discussed in more detail in the next chapter, is often more rapid in early stages. Trees, for example, not only hold nutrients in their mass for a longer time than herbaceous plants, but they also protect the soil from high temperatures. High temperatures result in more rapid breakdown of organic material and release of nutrients to the environment. Water uptake and storage by plants are also much reduced. Consequently,

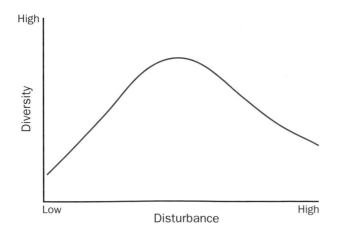

Figure 3.4 The intermediate disturbance hypothesis.

Table 3.1	Characteristics of Immature and Mature Ecosystems	
Characteristic	**Immature Ecosystem**	**Mature Ecosystem**
Food chains	Linear, predominantly grazer	Weblike, predominantly detritus
Net productivity	High	Low
Species diversity	Low	High
Niche specialization	Broad	Narrow
Nutrient cycles	Open	Closed
Nutrient conservation	Poor	Good
Stability	Low	Higher

SOURCE: Modified from E. Odum, 'The Strategy of Ecosystem Development', *Science* 164 (1969): 262–70. Copyright © 1969 by the American Association for the Advancement of Science.

disturbance may result in a significant loss of nutrient capital from a site through losses in soil water to streams.

- Biodiversity overall tends to be reduced.
- The species most adversely affected are often highly specialized ones at higher trophic levels.
- The species that benefit most are usually pioneer species (weeds and pests) that have broad ranges of tolerance and efficient reproductive strategies for wide dispersal.

ECOSYSTEM HOMEOSTASIS

In the early 1970s, residents of the Okanagan Valley in British Columbia began to complain to the government about excessive weed growth in some of the lakes in the valley. Several popular beaches were becoming virtually unusable due to the weeds, and the extent of the invasion appeared to be growing rapidly. This was of considerable concern to the residents not only because of the impacts on their recreational activities but also because of the impact on the economy of this tourist area for which water-based recreation was the main attraction.

Eurasian water milfoil had arrived in the area and over the next couple of decades was to spread not only to all the lakes in the Okanagan but also to many other lakes in southern BC and other provinces. The government spent significant amounts of money trying to control the spread of the species, but to no avail. Originating in Eurasia, the milfoil had reached the eastern shores of this

Signs warning of the spread of Eurasian water milfoil were placed at boat loading ramps throughout British Columbia, but did little to stem the colonization (*Philip Dearden*).

In the tourist economy of the Okanagan Valley where resorts rely on water-based activities to attract clientele, there was considerable conflict among different stakeholders regarding the most appropriate means of controlling the spread of milfoil (*Philip Dearden*).

BOX 3.4 THE GAIA HYPOTHESIS

As humans became capable of probing deeper and deeper into space, and thus able to view Earth from this unique perspective, it became increasingly clear that our planet was significantly different from all the millions of others. It seems like a happy coincidence that of the vast range of temperatures that could be experienced, those on Earth are just right for life, between the freezing and boiling points for water, even though the energy output of the sun has increased by over 30 per cent during the last 3.6 billion years. The gaseous composition of the atmosphere is also just what we need to breathe.

One hypothesis, the *Gaia hypothesis*, named after the ancient Greek goddess of the Earth by the originator of the idea, James Lovelock, postulates that the Earth acts as one giant self-regulating superorganism to help maintain these conditions necessary for life. Organisms act like cells in a body, with integrated functions to promote the health of that body or, in this case, optimum conditions for life. Active, automatic feedback processes among the atmosphere, lithosphere, hydrosphere, and ecosphere maintain this homeostasis.

There is no doubt that since the beginning of life on Earth, organisms have adapted to existing conditions and also modified these conditions in ways that are beneficial to life. However, few scientists believe that Earth is a superorganism with all living things interacting to maintain an equilibrial environment. The historical record clearly shows that organisms do not regulate natural cycles. Furthermore, to extend this thinking to assume that Earth will adapt and compensate in ways that are beneficial to humans for all of the changes that are now being initiated by human activity is mere wishful thinking.

continent probably a century ago and since that time had spread across the continent, replacing native aquatic plants in many water bodies.

This biophysical event, the spread of a Eurasian plant into North America, also illustrates the dynamic relationship among the biophysical, socio-economic, and management systems that is the main focus of this book. In BC, for example, the dependence of local economies such as that of the Okanagan Valley on water-based tourism triggered a strong response to milfoil that involved the use of the chemical 2,4-D. This created considerable conflict between different stakeholders regarding the relative impacts of the plant versus the control mechanism. Critics claimed that management had failed to consider the broader perspectives that would have been included with an ecosystem-based approach to the problem, and that management failed to adapt to the changing parameters of the situation. Chapter 6 discusses various approaches to these kinds of resource management issues in greater detail.

Situations such as this are not uncommon. We tend to think of ecosystems as having relatively constant characteristics, of being in a balance where internal processes adjust for changes in external conditions. This is known as **ecosystem homeostasis** and implies not a static state but one of dynamic equilibrium. James Lovelock (1988) postulated the *Gaia hypothesis*, which claims that the ecosphere itself is a self-regulating homeostatic system in which the biotic and abiotic components interact to produce a balanced, constant state (Box 3.4). This is an example of a highly integrated system, in which there is a strong interaction between the different parts of the system. Other systems may not be so highly dependent on one another. Cells in a colony of single-celled organisms, for example, may be removed and have little effect on the remainder due to the low integration of the system.

Not all ecosystems are equal in their abilities to withstand perturbations. **Inertia** is the ability of an ecosystem to withstand change, whereas **resilience** refers to the ability to recover to the original state following disturbance. Ecosystems can have low inertia and high resilience or any other possible combination. In terms of human usage it is best to work with systems that have both high inertia and high resilience. This means that they are relatively difficult to disturb, and even when disturbed will recover quickly. Such systems are relatively stable. The best growth sites for forestry—alluvial sites in nutrient-rich areas at low elevations—would fit into this category. Many tropical and Arctic sites would fit into the opposite combination where sites are readily disturbed and recover only very slowly, if at all.

Ecosystems have various attributes, such as feedback, species interactions, and population

BOX 3.5 PURPLE LOOSESTRIFE: ALIEN INVADER

Purple loosestrife was introduced to North America from Europe in ballast water from ships over a century ago. An aggressive invader of aquatic systems, this 'weed' has spread through thousands of hectares of wetlands in Quebec, Ontario, and Manitoba. It is estimated that 190,000 hectares of land in North America are invaded by purple loosestrife each year. After its woody root systems have become established, native plants and the animals that depend on them for food are forced out.

At the University of Guelph, experiments on test sites with the *Galerucella pusilla* beetle have showed promising results in controlling this invader plant. The beetle has a voracious appetite for purple loosestrife. The beetles eat the metre-high plant at such a rate that the plant's capacity to produce seed (about 2.5 million per plant per year) has been cut by up to 99 per cent.

Use of the beetles to control purple loosestrife looks promising, as previous control efforts that have relied on pulling out the plant, or burning, mowing, dusting, and spraying, provided negligible results. Furthermore, the beetles do not seem to have an appetite for native plant species.

A roadside sign describes the knapweed problem (*Philip Dearden*).

dynamics, that tend to maintain this equilibrium. Some scientists feel that there is no such thing as equilibrium and that ecosystems are in a permanent state of non-equilibrium. In some cases, as with the milfoil described above, this is obviously true. In the case of the milfoil, this involved the replacement of a variety of native aquatic species with a monospecific stand of the alien species. Similar effects are common with other non-native invaders, such as the purple loosestrife (Box 3.5), sea lamprey, and zebra mussels, which escape the normal control factors in their habitats and violate ecosystem homeostasis. Such is the scale of these changes today that this might prove to be the rule rather than the exception.

Alien or Invader Species

Organisms found in an area outside their normal range, such as Eurasian water milfoil and purple loosestrife, are considered **alien** or **invader**

species. Many species transported to a new environment do not survive. Others multiply rapidly, crowd out native species, and change native habitats. They may occupy a wide range of environments and disperse easily, and are typically difficult to eradicate. Their impacts are often irreversible. Scientists suggest that invasive species are second only to habitat destruction as a leading cause of biodiversity loss.

Canada has felt the impact of thousands of non-native (alien) species introductions (Figure 3.5). It is estimated that over 500 species of alien plants in Canada have developed into agricultural weeds that cost farmers millions of dollars every year to control (Table 3.2). One example is the various species of knapweed introduced to Canada and the US from the Balkan states, probably in

Cumulative number of species

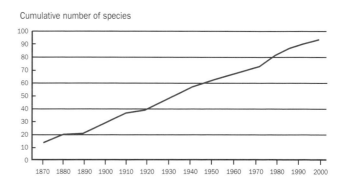

Figure 3.5 Alien agricultural and forest pests introduced in Canada (1870–2000). SOURCE: Auditor General (2002: 4:5). Reproduced with the permission of the Minister of Public Woks and Government Services, 2004.

Table 3.2	Some Introduced Agricultural and Forestry Pests and Diseases in Canada			
Species	**Place and Date of Discovery**	**Current Area**	**Impacts/Notes**	
Asian long-horned beetle *Anaplophora glabripennis*	Brooklyn and Amityville, NY: 1966; Chicago, Ill.: 1998; no known established population in Canada	Intercepted in warehouses in both Ontario and British Columbia	Poses serious threat to Canadian hardwood forests (maple being the preferred host); expensive eradication effort (over $5 million to date in New York alone) undertaken in Brooklyn, Amityville, and Chicago after detection of established populations	
Brown spruce longhorn beetle *Tetropium fuscum*	Likely arrived in Port of Halifax, Nova Scotia; found in Point Pleasant Park around 1990	Halifax Regional Municipality, Nova Scotia	No chemical pesticide currently registered to control the beetle; 5,309 trees are confirmed positive for removal in the Halifax Regional Munici-pality, 2,645 of these are in Point Pleasant Park; native to Europe and Asia	
Colorado potato beetle *Lepinotarsa decemlineata*	Extended its range during 1850s	Across Canada	Significant pest of potato crops; native to the western United States but fed on another species prior to the arrival of the potato in the 1850s; after discovering new host, quickly followed the potato path back across North America; good example of a species spread by habitat change	
Dutch elm disease *Ophiostoma ulumi* and *O. novo-ulmi*	Quebec: 1944 (first discovery; introduction was probably before 1940)	Present in all provinces except Newfoundland and Labrador, Alberta, and British Columbia	A fungus spread by the native elm bark beetle (*Hylurogopinus rufipes*) and the European elm bark beetle (*Scolytus multistriatus*); controlled through public education campaigns, surveillance , insecticides, and sanitation (removal of diseased elms)	

SOURCE: Statistics Canada (2003: 75).

shipments of alfalfa. The diffuse knapweed causes the most problems; it has a wide range of tolerance and a very effective seed dispersal system that it has used to colonize vast areas of rangeland in western Canada. It is also thought that this species may be *allelopathic*, that is, it can directly inhibit the growth of surrounding species through production of chemicals in the soil. The species displaces native species and considerably reduces the carrying capacity of the rangelands. Cattle will only eat it as a last resort, and the nutritive content is under 10 per cent of that of the displaced native species. Initial control efforts relied on chemical sprays. A more integrated approach is now being taken using biological control and attempting to limit its spread through stricter controls on vehicular access to rangelands, one of the main means of seed distribution.

Besides plants, many other species have also proved troublesome. Two fungi, chestnut blight and Dutch elm disease, for example, have had significant impacts on the landscape of central and eastern Canada. Both attack native trees that at one time were conspicuous parts of the deciduous forests. The American chestnut was attacked by an Asian pathogenic fungus that was introduced on stocks of Japanese chestnuts last century, and the elm by a European fungus transmitted between trees by beetles. Over 600,000 elm trees were killed in Quebec alone, and 80 per cent of Toronto's elms died in one year in the 1970s. Researchers at the Universities of Manitoba and Toronto are now exploring the potential of a natural toxin called *mansonone* that occurs in some elms and might be used to breed seedlings that are more resistant to Dutch elm disease.

Another fungus, the white pine blister rust, illustrates the complexity of the impacts of invasive species. The fungus, originating from Eurasia, attacks five-needled pines and causes extensive mortality. Whitebark pine is a key component of the subalpine ecosystems of the Canadian Rockies. It has a mutualistic relationship (Chapter 2) with Clark's nutcracker, a type of crow that eats the seed. Unlike many pines, the cones are not opened by fire but by animal activity. The seeds cannot be carried by wind and rely on the nutcracker for dispersal. The bird caches the seeds in forest openings for easy retrieval, creating perfect conditions for germination of the seed. Stuart-Smith et al.

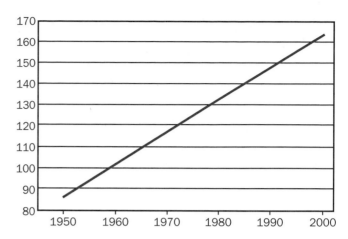

Figure 3.6 Cumulative number of aquatic invasive species in the Great Lakes, 1950–2000. SOURCE: Auditor General (2002: 4:14). Reproduced with the permission of the Minister of Public Works and Government Services, 2004.

(2002) measured mortality rates of the pine in excess of 20 per cent in some areas of the national parks as a result of fungus attack. There is concern that if mortality rates increase this will also lead to population declines in the Clark's nutcracker.

Many of the most serious invasions are in aquatic ecosystems. One notable recent invasion is the spread of the zebra mussel into the waters of the Great Lakes. The mussel, named for its striped shell, joins a long line of aliens in the Great Lakes, including the sea lamprey, alewife, and rainbow smelt (Figure 3.6). It is thought that the mussel, a native of the Black and Caspian Seas, was introduced from the ballast of freighters in the mid-1980s; it was first found in 1988 in a sample of aquatic worms collected from the bottom of Lake St Clair at Windsor–Detroit, which connects Lakes Erie and Huron. Evidence from Europe indicated that the species was an aggressive colonizer able to displace most native species. It displaced 13 species from Lake St Clair and caused the near-extinction of 10 species in western Lake Erie.

The mussels start their three- to six-year life as free swimming larvae before attaching to a hard surface, usually in the top three to four metres of the water, although they can live as deep as 30 m. By the end of 1988 the mussels had colonized half of Lake St Clair and two-thirds of Lake Erie at densities as high as 30,000 per square metre. The mussel has now spread extensively through the Great Lakes, where it appears capable of colonizing any hard surface. It has encrusted water intakes and discharges, severely reducing their efficiency

Zebra Mussel infestation. (*John Mitchell/Valan Photos*).

and necessitating significant expense to remove them. Water intakes may be reduced by as much as 50 per cent. Many different approaches are being tried to screen out the mussels, but they appear to be able to pass through most physical barriers. At the moment, chlorination is the most common measure, but this also raises problems due to the potential for formation of toxic organochlorines. Ontario Power Generation has spent over $20 million installing and maintaining chlorine applicators at its Great Lakes facilities and inland facilities and another $13 million on research to reduce chlorine use. The mussels also colonize spawning sites for other fish, with as yet undetermined impacts on their populations.

Other impacts on population levels of other species are also likely to be caused more indirectly through food-chain effects. The mussels are filter feeders that remove phytoplankton from the water, thereby affecting all the species higher in the food chain, such as walleye, bass, trout, and perch. In some European locations, invasion by the mussels has led to clearer water as a result of removing the phytoplankton. These changes may also benefit some species, even fish species. Bottom feeders, such as carp and whitefish, and invertebrates such as crayfish may benefit as more nutrients are returned to the lake bottoms, either in the form of mussels themselves or mussel feces.

The mussels do not remove all species of phytoplankton equally, though. This is leading to problems with blooms of blue-green algae, such as the toxic *Microcystis aeruginosa*, that are not ingested by the mussels.

It remains to be seen whether species higher in the food chain, such as scaup and other waterfowl, can help control the spread of the mussel. Already, numbers of some of these species, which stop over to feed during their migration, appear to have risen considerably. Realistically, it appears that the ducks may have some impact, as they have in Europe, but that the infestation will be too large and the number of ducks too small for the problem to be controlled in this manner.

If controlling the zebra mussels seems impossible, what about trying to ensure that the same kind of problem doesn't arise again? Many invasive aquatic species arrive in their new habitat courtesy of ocean freighters, which take on water for ballast in one part of the world and release it in another. Over 3,000 species are being transported around the world every day through this process. Scientists have known for a long time that ballast water is a major threat to aquatic communities. In 1980, sampling of 50 freighters entering the Great Lakes found 56 different species of exotic aquatic invertebrates as well as over 100 phytoplanktons. Human pathogens are also transported. In 1998 a study in the Great Lakes found a wide range of pathogens in ballast water, including giardia, salmonella, fecal coliforms, and *Vibrio cholerae*. Given the magnitude of these introductions, it is inevitable that from time to time one or two of these species will not only find a tolerable home in their new location, but also explode into great numbers. The US requires freighters to discharge ballast water offshore. Canada now asks freighters to swap their ballast water for seawater before entering the St Lawrence. Unfortunately, as is so often the case in Canada, the program is voluntary, there is no monitoring to encourage compliance, and it is geographically restricted. Scientists are now considering ways in which unwelcome organisms can be killed before they are released, such as by chemicals or heating. As yet, a satisfactory solution has not been found.

In 2004 a new international convention to prevent the potentially devastating effects of the spread of harmful aquatic organisms carried by ships' ballast water was adopted by the International Maritime Organization (IMO), the United Nations agency responsible for the safety and security of

> *... the federal government has no means to determine the greatest threats to Canada's ecosystems from invasive species; to set national priorities for prevention, control or eradication; and to allocate its scarce resources to areas of greatest risk.... There is a bias toward continuing dialogue and consensus-building and a lack of practical action to prevent invasive species from harming Canada's ecosystems, habitats or native species.*
>
> — *Auditor General (2002)*

shipping and the prevention of marine pollution from ships. This convention will require all new ships to implement a Ballast Water and Sediments Management Plan. They will have to carry a Ballast Water Record Book and will be required to carry out ballast water management procedures to a given standard. Existing ships will be required to do the same, but after a phase-in period.

Although the impact of the zebra mussels on the Great Lakes has been enormous, fisheries biologists feel that a much greater threat is on the horizon, again created by human activities. This is the Asian big head carp. The carp has a voracious appetite, eating half its body weight every day and reaching almost a metre in length. The carp escaped from fish farms in the southern US in the 1990s and invaded the Illinois and Mississippi River systems. Only a canal in Chicago protected by an electric fence that connects to Lake Michigan is preventing the fish from entering the Great Lakes. Biologists feel that it is only a matter of time until the carp enter the lakes, which would lead to the demise of the entire fishery. One possible means of entrance is through the live fish trade. Carp are brought live from the fish farms in the US to markets and restaurants in Toronto. The water, subsequently, is discarded into the drainage system along with any fingerlings.

Canada already has legislation and programs that deal with the problem of invasive species, especially those that may damage agricultural and forest crops or pose a danger to human health. Under the terms of the United Nations Convention on Biological Diversity, discussed in more detail in Chapter 11, Canada is also committed to containing invasives that threaten biodiversity. However, an audit of progress in this regard undertaken by the Auditor General (2002) discerned no real progress in addressing the problem.

Species Removal

Just as the introduction of species to new habitats can disturb ecosystem homeostasis, so, too, can the removal of species from food webs. The reduction of some species, the so-called keystone species discussed in the last chapter, may be particularly disruptive. One well-known example relates to the extirpation of the sea otter from the Pacific coast.

When James Cook anchored at Nootka Sound on the west coast of Vancouver Island in 1778 he reported that the fur of the sea otter 'is softer and finer than that of any others we know of; and, therefore, the discovery of this part of the continent of North America, where so valuable an article of commerce may be met with, cannot be a matter of indifference.' Indeed, it was not. The British, seeking trading goods to barter with the Chinese in exchange for tea, discovered that sea otter pelts were in great demand in China, and thus made every effort to ensure that the west coast became British (rather than Spanish or Russian!) Columbia.

The sea otter is a large sea-going weasel of the outer coasts, flourishing in the giant kelp beds. They lack a protective layer of blubber but have a very fine fur that traps air and insulates them from the cold Pacific waters. They also need a lot of food (up to nine kilograms per day) to fuel the fast metabolism that counteracts energy loss to the environment. Favourite prey are sea urchins, crabs, shellfish, and slow-moving fish.

The otters were easy to catch and Russian, American, and Spanish hunters, aided by local Native populations, finished off what the British had begun. Within 40 years, populations were reduced from over half a million to 1,000–2,000. On the coast of British Columbia, it is likely that they were completely extirpated. However, relict populations remained both to the south, around Monterey in northern California, and to the north in the Aleutian Islands. Individuals from this latter population have now been reintroduced to the coast of British Columbia, where small but vibrant populations thrive again.

Scientists discovered the otters' key role in maintaining ecosystem homeostasis after studying two groups of islands off Alaska. They noticed that although the two groups were very similar in terms of location and physical conditions, one group had much more life—bald eagles, seals, kelp beds, and otters—than the other. Otters play a

A sea otter off Vancouver Island (*World Wildlife Fund Canada/ © Mark Hobson*).

critical role in controlling sea urchin populations (Estes et al., 1989). Sea urchins are voracious eaters of kelp (large, brown seaweed), which may be the world's fastest-growing plant, with increments of up to 60 centimetres per day. Given the support of the ocean, kelp does not need to invest much energy in heavy support structures, leaving more energy for growth (another example of the auxiliary energy flow discussed in Chapter 2).

Kelp plays a major role in coastal communities. It provides food and habitat for many other species. Diatoms, algae, and microbes grow on the fronds of the kelp, along with colonies of filter-feeding bryozoans and hydroids. Predators abound. Fish come to feed off the colonists or seek protection from open-water predators such as seals, sea lions, and killer whales. When overgrazed by sea urchins, this productive habitat disappears. The urchins eat through the holdfasts that anchor the kelp to the ocean floor and the kelp is soon washed away into the open ocean or onto land. As the kelp disappears, so do the species dependent on it. On the two islands in Alaska, one island had managed to escape the fur rampage that eliminated otters elsewhere, and this one displayed the rich coastal community that should extend all along the outer coast of the North Pacific. Otter populations, through their control of the urchin population, are therefore critical in maintaining the productivity of the entire community, right up to bald eagle populations. The fact that the fashion tastes of Chinese mandarins 200 years ago, met by traders from the other side of the world who wanted to enjoy afternoon tea, is still reflected in bald eagle

populations 7,000 km away on the BC coast indicates the complex interactions between biophysical and human systems.

Feedback

Feedback is an important aspect of maintaining stability in ecosystems whereby information is fed back into a system as a result of change. Feedback initiates responses that may exacerbate (**positive feedback**) or moderate (**negative feedback**) the change. There is, for example, considerable debate regarding the role of feedback loops in global climate change, as discussed in more detail in Chapter 7. One positive feedback loop that may have a strong influence in Canada is the effect of increased temperatures in the North. This would increase the area of snow-free land in summer and is known as the **albedo effect**. Snow has a high albedo; in other words, it reflects rather than absorbs much of the incoming radiation. As temperatures rise, the area covered in snow will be replaced by areas free of snow, uncovering rocks and vegetation with lower albedo values. This will cause more heat to be absorbed, which in turn will contribute to global warming. A similar contribution of forest fires was noted in Box 3.2.

On the other hand, negative feedback loops may also be in operation, and serve to counteract such positive feedback loops. One possibility regards the role of phytoplankton in global warming. Phytoplankton produce a gas called dimethyl sulphide. When sea water interacts with the gas, sulphur particles formed in the atmosphere serve as condensation nuclei for cloud droplets. As the planet heats up, the productivity of the phytoplankton should increase, leading to an increase in the amount of gas and cloud droplets produced. This will have the effect of increasing cloud cover and reflecting away solar radiation that could lead to cooling of the Earth, maintaining a dynamic equilibrium.

Almost all the examples in this chapter can be used to illustrate some aspect of feedback mechanisms. The allelopathic qualities of the diffuse knapweed, for example, show a positive feedback loop that helps the spread of the species. The more the species spreads, the more conditions are created into which only *it* can spread. The sea otters provide a negative feedback loop on the sea urchin-kelp relationship. If the urchins become too numerous and overgraze the kelp beds, increases in otter populations will help reverse this imbalance. When this negative feedback loop was removed from the system, there was nothing to maintain the dynamic homeostasis of the system.

Similar examples of feedback loops occur at all scales, even down to the regulation of temperatures in individual organisms. Sometimes these feedback messages can be rapid, as in the case of organism thermoregulation. In other cases, considerable time delays can occur between the stimulus for change and the resulting feedback response. Unfortunately, as the example of the positive feedback loop and snowmelt described above indicates, sometimes the delay between the stimulus and the response may be so long that we are not aware of it. By the time we are aware, it may be too late to try to moderate the stimulus and a powerful positive feedback loop may already, albeit slowly, have been set in motion. This is one reason why many scientists support immediate actions to cut down on the emission of greenhouse gases (see Chapter 7), even though we do not yet have a clear understanding of all the relationships involved.

We are also becoming more aware of the chaotic nature of many systems whereby a slight perturbation becomes greatly enhanced by positive feedback. The so-called *butterfly effect*, for example, traced how the turbulence of a butterfly flapping its wings in South America might, through cascading effects on airflows, influence the weather in North America (see www.imho.com/grae/chaos/chaos.html). Further research has revealed the existence of similar phenomena in many different systems where very small changes in systems can have great influence on outcomes. Chaos theory tries to discern pattern and regularity in such systems and allow for greater predictability.

Synergism

Synergism is also an important characteristic that may influence change in ecosystems. A synergistic relationship occurs when the effect of two or more separate entities together is greater than the sum of the individual entities. Thus, for example, in Chapter 4 some attention is devoted to the problem of acid deposition. The effects of acid deposition are often exacerbated by the presence of other pollutants, such as ground-level ozone. Individually, both these forms of pollution may cause a certain amount of damage to an ecosystem. In combination, however, their effects are magnified.

At an individual scale we can see the same

effects for smokers. The potential for getting lung cancer is increased both by smoking and by the inhalation of radon gas. However, if someone who smokes also inhales radon gas the chances of getting lung cancer are greater than the sum of the individual exposures to each hazard. This is because the particles of tobacco ash that accumulate in the lungs as a result of smoking also act as anchors for radon particles. Together, the two exposures are far more dangerous than the sum of the individual exposures.

POPULATION GROWTH

The number of individuals of a species is known as the **population**. When calculated on the basis of a certain area, such as the number of sea otters per hectare, it becomes *population density*. The number of organisms in a population is important because low numbers will make a species more vulnerable to extinction. Changes in population characteristics are known as *population dynamics*.

Populations change as a result of the balance among the factors promoting population growth and those promoting reduction. The most common response is through adjustments in the birth and/or death rates to the factors shown in Figure 3.7, although emigration and immigration can be important factors in some species. Population change is calculated by the formula $I = (b - d) N$, where I is the rate of change in the number of individuals in the population, b is the average birth rate, d is the average death rate, and N is the number of individuals in the population at the present time. As long as births are greater than deaths, then a population will increase exponentially over time (Figure 3.8) until the environmental resistance of the factors shown in Figure 3.7 begin to have an inhibiting effect that will serve to flatten out the curve.

The **carrying capacity** of an environment is the number of individuals of a given species that can be sustained in a given area indefinitely given a constancy of resource supply and demand. Most species will grow rapidly in numbers up to this point and then fluctuate around the carrying capacity in a dynamic equilibrium (Figure 3.9). The carrying capacity is not one fixed figure, however, but will vary along with changes in the other abiotic and biotic parts of the ecosystem. In the puffin-capelin example in Chapter 2, the carrying capacity of the North Atlantic waters to support the puffin population was severely reduced due to a reduction in their food supply caused by competition from another organism, humans. Management inputs, such as provision of supplementary feeding or other habitat requirements, are often used to change the capacity of an area to meet human demands.

Organisms that demonstrate the kind of S-shaped growth curve of Figure 3.9 are *density-dependent*, and as the population density increases, the rate of growth decreases. In other words, the larger the population, the smaller the growth rate. This view is in accord with the equilibrium view of ecosystems discussed earlier. Some organisms, however, are *density-independent*, and the population operates with a positive feedback loop—the more individuals there are in the population the more that are born, and the population grows at an increasing rate to demonstrate a J-shaped curve. At some point this population meets environmental resistance, causing the population to crash back to,

or below, the carrying capacity. The algae blooms that you will notice on ponds in the late spring or early summer are a result of this kind of growth. In reaction to the increased nutrient availability after winter, spectacular growth can occur until this food supply is exhausted and the population crashes.

In some locations in Europe, this is what has happened with the zebra mussels discussed earlier. Mussel populations have soared up to a peak and stayed there for a few years before exhausting the food supply and crashing to between 10 and 40 per cent of the original numbers. However, there are other locations, such as Sweden, where the expected crash has yet to occur. Given the enormous food supplies of the Great Lakes and the low

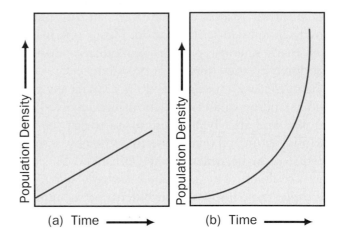

Figure 3.8 Geometric (a) and arithmetic (b) growth patterns.

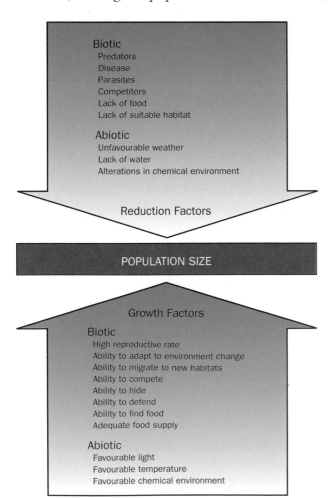

Figure 3.7 Factors affecting population growth.

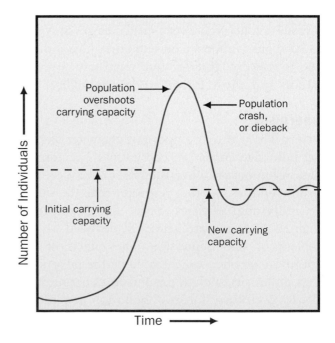

Figure 3.9 Carrying capacity and population growth rates.

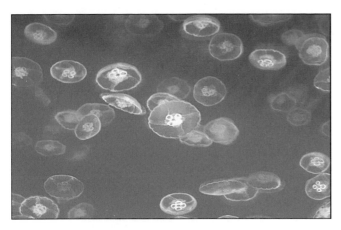

These jellyfish and Pacific white-sided dolphins are good examples of marine r and K species respectively (*Philip Dearden*).

numbers of predators, such as waterfowl, it may be a very long time before any natural population crash may happen there.

The capacity of species to increase in number is known as their **biotic potential**, the maximum rate at which a species may increase if there is no environmental resistance. Different species, however, have different reproductive strategies. Some species, such as zebra mussels, are known as *r-strategists*, which produce large numbers of young early in life and over a short time period, but invest little parental energy in their upbringing. Most of their energy is spent on reproduction and they have few resources left to devote to maintaining a longer lifespan. Such species are usually small and short-lived and can respond to favourable conditions through rapid reproduction. They are opportunists and their reproductive strategy is essentially based on quantity. Such species tend to dominate the

early seral stages of the successional process.

K-strategists, on the other hand, produce few offspring but devote considerable effort into ensuring that these offspring reach maturity. Their strategy is based on quality. Individuals live longer and are usually larger. Populations of K-strategists often reach the carrying capacity of an environment and are relatively stable compared to r-strategists, which may experience large variation in population size. The characteristics of these different strategists are summarized in Table 3.3.

Typical examples of r-strategists include insects, rodents, algae, annual plants, and fish. A mature female codfish, for example, may produce over 9 million eggs in one season. However, fewer than 5 per cent of these may mature and last the first year. Most K-strategists are larger organisms, such as the larger mammals (including humans). Their lower biotic potential and ability to disperse

Table 3.3	Characteristics of K-Strategists and r-Strategists
K-Strategists	**r-Strategists**
Few, larger young	Early reproductive age
More care of young	Many small young
Slower development	Little care of young
Later reproductive age	Rapid development
Greater competitive ability	Limited competitive ability
Longer life	Short life
Larger adults	Small adults
Live in generally stable environments	Live in variable or unpredictable environments
Emphasis on efficiency	Emphasis on productivity
Stable populations usually close to carrying capacity	Large population fluctuations usually far below carrying capacity

BOX 3.6 MUSKOX

Perhaps no animal epitomizes the Arctic more than the prehistoric-looking muskox. With its great shaggy head and flowing mane facing into an Arctic storm, it seems to symbolize the determination of life to survive in even the most inhospitable environments. Muskox are, in fact, superbly adapted to the rigours of the North. They have a long, coarse outer coat covering a fine under layer of soft wool; all of their extremities (e.g., tail and ears) are covered by this coat. Their keen sense of smell allows them to find food in the winter darkness, they have low metabolic rates, and they minimize energy loss by remaining relatively inactive.

Muskox are not widely distributed throughout the Arctic—they concentrate in areas where vegetative conditions are best, where they can graze and browse on most Arctic herbs, grasses, and shrubs. In these areas they stay in bands of females with a single bull that takes the leadership role in finding routes and repelling predators, mainly Arctic wolves. It is against such predators that their well-known defensive strategy evolved of forming a circle around the calves with their massively horned heads pointing outward. Although effective against wolves, this strategy proved much less effective against humans with guns, who could stand at a distance and shoot this rather large target at will. Over 16,000 hides were shipped out of Canada

Muskox (*Eastcott. Momatina/Valan Photos*).

between 1864 and 1916. In addition, Arctic explorers used them as a main source of meat. Consequently, with their low birth rate (one calf every second year) and a few hard winters, numbers fell drastically. They are now protected, but may be hunted by licence and by indigenous hunters. About 8,000 muskox are now left in the Canadian Arctic, ranging from the Arctic islands south to the Thelon game sanctuary and the coast of Hudson Bay.

---

often means that they are more restricted to the later seral stages of succession. Many endangered species (see Chapter 11) are K-strategists. The great whales (Chapter 8), with perhaps only one offspring every three years, are a good example. When the conditions to which they have become accustomed, and under which they evolved their reproductive strategy, change dramatically, such as with the introduction of new predators (humans), they have little capacity to respond in terms of increasing their reproductive rate. Muskox are another example (Box 3.6).

In addition to the factors outlined above, chance also plays an important role in determining population size. Severe winters, disease outbreaks, fires, droughts, and similar factors often have a major impact on populations that continually find themselves recovering from one natural catastrophe or another. Peary caribou, for example, exist north

of the seventy-fourth parallel by digging under the snow to feed on vegetation. In 1974–5, heavy snows and freezing rains led to high mortality as the herd starved to death, unable to reach their food source. Unfortunately, these are the very same weather conditions predicted to become more common as a result of global climate change. In 1993 there were over 3,000 caribou on Bathurst Island in the High Arctic. By 1997, as a result of repeated bad winters, the number was down to 75.

EVOLUTION, SPECIATION, AND EXTINCTION

Over the long term, populations adapt to changing conditions through **evolution**, a change in the genetic makeup of the population with time. Within any population there will be some variation in the genetic composition that may predispose a certain segment of the population to adapt to certain conditions. If conditions change to favour those certain conditions, then the success of the part of the population that is genetically better adapted to the new conditions will be improved. In this way, over time, *natural selection* can lead to changes in the characteristics of a population. *Phyletic evolution* is the process in which a population has undergone so much change that it is no longer able to interbreed with the original population, and a new species is formed. The evolution of the polar bear from the grizzly bear over time is an example. Genetic diversity helps protect species from extinction. The resilience of a species depends partly on the magnitude of the environmental change, how rapidly it takes place, and the capacity of the gene pool of the species to respond to these changes. In general, the broader the gene pool is, the greater the capacity to adapt to change.

Changes in the abiotic environment are not alone in promoting evolutionary change. Species may also change through **coevolution**, whereby changes in one species cause changes to occur in another. Each species may become an evolutionary force affecting the other. A typical case is where a prey species evolves to be more effective in avoiding a predator. In turn, the predator may evolve more efficient hunting techniques to detect the prey. Many such relationships have evolved in the tropical forests, especially between specific plants and animals, due to the long period of evolutionary change that has taken place in such environments. Canada also has many examples, particularly relating to pollination, where various insects, birds, and bats have evolved to pollinate flowering plants, and, in turn, a great diversity of plant shapes, sizes, and colours have developed as a direct response to the activities of the pollinators.

Evolution results in the formation of new species as a result of divergent natural selection responding to environmental changes. This is the process of **speciation**. It occurs most often when members of the same species become geographically isolated so that they can no longer interbreed. If conditions differ in the respective environments of the different breeding groups, then over time natural selection will favour those individuals best fit to those conditions. Thus, it is thought that the polar bear evolved as a separate species from the grizzly bear some 10,000 years ago. Bears that had characteristics that aided hunting seals on ice flows,

Many tropical orchids are products of coevolution. The flower has evolved to imitate the female wasp of the species that pollinates the flower. The male is deceived into thinking that it is a female, flies into the flower and in so doing picks up pollen that is subsequently taken to the next imitator, and pollination occurs (*Philip Dearden*).

such as lighter-coloured fur and greater strength, would be relatively more successful in the Far North as opposed to in the rest of the range where a brown pelt and greater mobility might be an advantage. In this way, a single bear species became two bear species through adaptation to different environments and the process of natural selection. This process of local adaptation and speciation is known as *adaptive radiation*.

Extinction is the opposite of this process and represents the elimination of a species that can no longer survive under new conditions. The fossil record leads scientists to estimate that perhaps almost 99 per cent of the species that have lived on Earth are extinct. The fact that we perhaps still have up to 50 million species, more than have ever existed before, indicates that speciation has exceeded the extinction level. However, speciation takes time. Even with r-strategists it may take hundreds and thousands of years; with K-strategists this may be extended to tens of thousands of years. Evidence suggests that the activities of humanity have tipped the scale recently in favour of extinction over speciation (Box 3.7), as will be discussed in more detail in Chapter 11. Table 3.4 gives some examples of species that at one time existed in Canada but are now extinct.

Extinction, like speciation, should not be considered just a smooth, constant process, but one punctuated by relatively sudden and catastrophic changes. It appears that multicellular life, for

BOX 3.7   HUMANS AND EXTINCTION

Extinction, as pointed out in the text, is a natural process. Scientists can try to estimate the average rate of species extinction by examining the fossil record, which suggests that extinctions among mammals might be expected to occur at the rate of about one every 400 years, and among birds, one every 200 years. Current extinction rates are difficult to estimate because we do not have a full inventory of species, and so we do not know what we are losing. Based on current rates of habitat destruction for tropical forests, figures of over 100,000 extinct species per year are often quoted. Many of these extinct species are likely to be undescribed arthropods, as these comprise the majority of species in tropical forests. More conservative estimates suggest that somewhere between 2 and 8 per cent of the planet's species will become extinct over the next 25 years, many more than what would have occurred in pre-human times.

Although there may be disagreements over the rate of extinction, there is widespread consensus that humans have vastly increased this rate. In 1996, over 7,000 scientists working with the World Conservation Union (IUCN) completed the most comprehensive evaluation of species status ever attempted. The report concluded that one-quarter of the world's mammal species are threatened with extinction, and that half of those may be gone within the next decade. Over a third of global primate populations are threatened. In North America, 30 per cent of bird species are declining in number.

Evidence suggests that humans have had a major impact on biodiversity for quite some time. Paul Martin (1967), for example, was one of the first to suggest that humans may have been a major factor in causing the extirpation of several species of large mammals from North America at the end of the last ice age, some 10,000 years ago. At this time at least 27 genera comprising 56 species of large mammals, two genera and 21 species of smaller mammals, and several large birds became extinct. The extinctions included 10 species of horse, four species of camel, two species of bison, a native cow, four elephant species, the sabre-toothed tiger, and the American lion. Although this was also a time

It seems difficult to believe that what we now consider primitive weapons, such as this fishing spear from the Warao people of the Orinoco delta, may have enabled humans to hunt many other species to extinction (*Philip Dearden*).

of global climatic change, no such extirpations were associated with the same time period in Eurasia. This period also saw a substantial immigration of humans from the Asian continent that began to prey on animals unfamiliar with, and therefore not adapted to, human hunting. This hunting, combined with the environmental stresses experienced through habitat alteration and repercussions through the food chain, were sufficient to extirpate the species.

Charles Kay (1994) has also conducted research into the subsequent impact of Native Americans on ungulate populations before the arrival of European influences. He concludes that even then humans were the main limiting factor on ungulate populations in the intermountain West, and that elk, in particular, were overexploited. The people had no effective conservation strategies and hunted to maximize their individual needs, irrespective of environmental impacts. Thus the image of North America as a vast wilderness unaffected by human activities before the coming of the Europeans appears to be a myth. Even that mightiest symbol of the wild, the grizzly bear, was apparently under pressure from Aboriginal hunters in Alaska (Birkedal, 1993). Human impacts on biodiversity will be discussed in more detail in Chapter 11.

Table 3.4	Some Canadian Vertebrate Species That Are Now Extinct		
Species	Distribution	Last Recorded	Probably Causes
Great auk (*Alca impennis*)	Canada, Iceland, UK, Greenland, Russia	1844	Hunting
Labrador duck (*Camptorhynchus labradorius*)	Canada, US	1878	Hunting, habitat alteration
Passenger pigeon (*Ectopistes migratorius*)	Canada, US	1914	Hunting, habitat alteration
Deepwater cisco (*Coregonusjohannae*)	Canada, US	1955	Commercial fishing, introduced predators
Longjaw cisco (*Coregonus alpenae*)	Canada, US (Great Lakes)	1978	Commercial fishing, introduced predators

example, has experienced five major and many minor mass extinctions. Scientists think that the age of the dinosaurs, a remarkably successful dynasty that effectively relegated mammals to minor ecological roles for over 140 million years, was brought to an end 65 million years ago by the impact of a large extraterrestrial object. And then the mammals took over. Perhaps the dinosaurs, through the traditional Darwinian process of evolution and speciation, managed to outcompete the mammals for a long period of time, and, were it not for the chance impact of the asteroid, might still be the dominant animal life. However, this chance occurrence not only led to the demise of the cold-blooded dinosaurs, but also favoured the survival of the rodent-like mammals with their smaller body size, less specialization, and greater numbers. Small body size was likely a sign of the mammals' inability to challenge the dinosaurs during the normal evolutionary process; however, small body size became a positive feature favouring survival under the new conditions.

There are other examples of this non-random impact of mass extinction on life. The features that make some species successful during ordinary times may be completely unrelated to the new conditions, making life's pathway somewhat chaotic and unpredictable, rather than the smooth path that evolutionary theory might suggest.

Finally, as emphasized in Box 3.8, we should not think that evolution is fundamentally a story that demonstrates the benefits of complexity over simplicity. Humans—the most complex organ-

isms—are not the pinnacle of life's achievement nor the most successful. For that accolade we have to travel to the other end of the complexity spectrum. From the beginning of the fossil record until present times, bacteria have provided the most stable presence. Says Stephen Jay Gould (1994: 87):

> Bacteria represent the great success story of life's pathway. They occupy a wider domain of environments and span a greater range of biochemistries than any other group. They are adaptable, indestructible and astoundingly diverse. We cannot even imagine how anthropogenic intervention might threaten their extinction, although we worry about our impact on nearly every other form of life. The number of *Escherichia coli* cells in the gut of each human being exceeds the number of humans that has ever lived on this planet.

Dinosaur Provincial Park in Alberta is a World Heritage Site where more species of extinct dinosaurs have been found and identified than anywhere else in the world (*Philip Dearden*).

BOX 3.8 THE BURGESS SHALES

The site of the Burgess Shales World Heritage Site on the slopes of Mount Wapta in Yoho National Park (*Philip Dearden*).

Stephen Jay Gould, the well-known Harvard paleontologist, called the Burgess Shales in Yoho National Park in British Columbia the single most important scientific site in the world. The reason for this superlative is the extensive bed of fossils that can be found high on the flanks of Mount Wapta. They are fossils from the Cambrian era, some 530 million years ago, when there was a great flourishing of diverse life forms. The special feature of the site is that the fossils from this era are preserved in great detail, even down to the soft body parts, such as stomach contents.

The story revealed by this detail is one of great diversity at this time when all but one phylum of animal life made a first appearance in the geological record. The site also contains many body patterns for which there are no current counterparts. Thus it seems as if life could be characterized as three billion years of unicellularity, followed by this enormously diverse Cambrian flowering in a brief five-million-year period, and a further 500 million years of variations on the basic anatomical patterns set in the Cambrian period. Why, or how, this flowering took place is uncertain. It would seem to require a combination of explanations. First, there was literally an open field available for colonization—an environment ripe to support life, but with little life in it. Therefore, species did not have to be particularly good competitors to survive. Virtually anything could survive. Since this time, even after mass extinctions, sufficient species have remained to make it pretty tough competition for any newcomers. Second, it seems as if the early multicellular animals must have maintained flexibility for genetic change and adaptability that declined as greater specialization arose and organisms concentrated on refining the successful designs that had already evolved. Furthermore, we have little idea why most of these early experiments in life died out and yet others remained. There seem to be no common traits shared by the survivors to indicate they were the victors of Darwinian strife. Perhaps just the lucky ones survived.

Gould, in his fascinating book, *Wonderful Life*, suggests that this challenges our established view of evolution as an inevitable progression over time from the primitive and few to the sophisticated and many. It also radically challenges our view of ourselves as being the logical end point of evolutionary change, the rightful inheritors of the world. In Gould's words: 'If humanity arose just yesterday as a small twig on one branch of a flourishing tree [of evolution], then life may not, in any genuine sense, exist for us or because of us. Perhaps we are only an afterthought, a kind of cosmic accident, just one bauble on the Christmas tree of evolution' (Gould, 1989: 44). In other words, we should be humble!

BIODIVERSITY

Over billions of years, interaction between the abiotic and biotic factors through the process of evolution has produced many different life forms. The main classification is the **species**, life forms that resemble one another and can interbreed successfully. Two species are created from one as a response of natural selection to changes in environmental conditions, as explained in the previous section. **Biodiversity** is

BOX 3.9 CAROLINIAN CANADA

Carolinian Canada is the wedge of land stretching from Toronto west to Windsor that contains 25 per cent of the country's population. It also contains the highest number of tree species in the country as the mixing zone between the eastern deciduous forests to the south and mixed coniferous-deciduous forests to the north. Being the southernmost part of the country, with the mediating effects on climate of the southern Great Lakes, allows semi-tropical tree species such as the cucumber and sassafras to spread up into this land of ice and snow. After Vancouver and Victoria, Windsor ranks as the third warmest city in Canada, and it is the most humid city in the country. It is little wonder that in summer the humidity and southern vegetation can give the appearance of being much further south.

Besides the distinctive vegetation, the Carolinian zone also supports a noteworthy bird population. Point Pelee is noted as one of the top birding spots in North America. Not only do birds (exhausted from the crossing of Lake Erie on their migrations north in winter) like to rest here, but it is also part of the Carolinian forest and the nesting habitat for many species that are unusual for Canada, particularly warblers. Of the 360 bird species seen, about 90 stay to nest, and in spring there may be 25 to 30 different warblers spotted on a good day. In the fall, the birds are joined by thousands of monarch butterflies as they, too, pause here before heading south on their 3,600 km journey to the Gulf of Mexico for the winter.

Point Pelee is a national park. Most of the rest of the Carolinian forest is not so well protected and is heavily fragmented by agriculture and urban development. As a result it is estimated that close to 40 per cent of Canada's rare, threatened, and endangered species are primarily Carolinian. The Carolinian Canada Program, started in 1984, has co-ordinated the efforts of government agencies and private landowners to try to protect the remaining forest. About half of the 38 targeted sites now have some degree of protection, but biologists still worry that these fragments are too small and isolated to be able to protect this most diverse area of Canada.

---

the sum of all these interactions, and is usually recognized at three different levels:

- *genetic diversity*, the variability in genetic makeup among individuals of the same species. Individuals generally vary in the genetic information encoded in their DNA. However, some individuals do not. Genetically uniform *clones* may exist in separate individuals. One example is the trembling aspen. Through vegetative propagation genetically identical individuals can cover an area of up to 50 hectares. In general, genetic diversity in a population increases the ability to avoid inbreeding and withstand stress.

- *species diversity*, the total number of species in an area. The number of different species in an area is often known as *species richness*.

- *ecosystem diversity*, the variety of ecosystems in an area. Some ecosystems are more vulnerable to human interference than others (Box 3.9). Estuaries and wetlands, for example, are highly productive but are often used for industrial and agricultural activities. As these ecosystems are replaced by human-controlled ecosystems, natural diversity at the landscape level is reduced.

Scientific knowledge of biodiversity is very primitive (Box 3.10). There may be up to 50 million species, although most scientific estimates suggest between 3 million and 30 million, of which we have identified some 1.8 million (Figure 3.10). Some 56 per cent of these are insects, 14 per cent are plants, and just 3 per cent are vertebrates such as mammals, birds, and fish. Even new mammals are still being discovered, such as the giant muntjac and saola discovered recently on the borders of Vietnam and Laos. However, most species awaiting discovery are probably invertebrates, bacteria, and fungi from the tropics. We also know relatively little about the ocean (Chapter 8). Only about 15 per cent of described species are from the oceans. Most biologists agree that there are fewer species to be found there than on land. On the other hand, there are 32 phyla in the oceans, compared with only 12 on land.

Species identification is only the first building block in biodiversity. We need also to understand the differences in genetic diversity within species and how species interact in ecosystems to really understand how the life support system of the planet works (Box 3.11).

BOX 3.10 COUNTING CRITTERS

Since actually discovering and describing new species is a very slow process, scientists have devised several means to help estimate just how many species we might have on Earth.

Species-area curves are one of the most popular approaches. As an area increases in size, the number of species that may be found there also increases, but the number of new species gradually levels off as the same species are encountered repeatedly. This relationship has enabled scientists to construct species-area curves that show the number of species likely to be found in areas of different size and hence predict how many species might be found in larger unsampled areas.

Rain forest insect samples. Most of the species we have yet to describe are probably rain forest insects. Studies indicate that there are up to 1,200 beetle species in the canopy of one single tree in Peru. Since we know that 40 per cent of insects are beetles, this suggests over 3,000 insect species in the canopy alone. About half this number will be found lower down and on the trunk, leading to estimates of 4,500 species on one tree alone. Given that there are over 50,000 species of tropical tree, the numbers that may be found in total will be vast. Some researchers predict that there are over 30 million insect species.

Ecological ratios can be used to predict the populations of little-known groups from their relationships with better-known groups. For example, the ratio of fungus to plant species in Europe is 6:1. If this holds worldwide then there should be over 600,000 species of fungus. At the moment fewer than 70,000 have been described.

None of these approaches (these are just three of the more popular ones) can deliver an absolute estimate of the numbers of species on the planet. Each one, however, is consistent in indicating that we have a long way to go before we can claim that we truly know the nature of life on this planet.

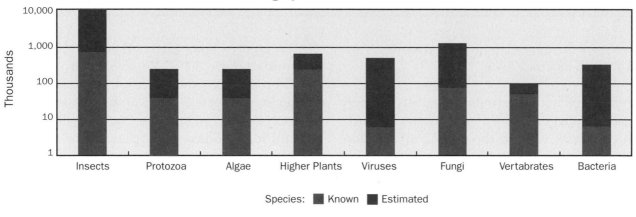

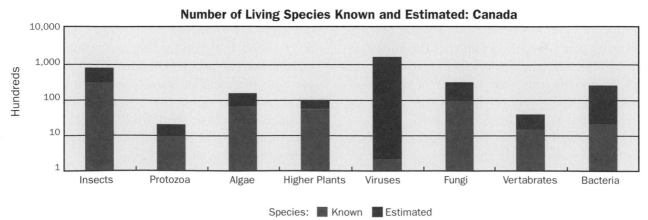

Figure 3.10 Numbers of known and estimated living species in the world and in Canada. SOURCE: B. Groombridge, *Global Biodiversity: Status of the World's Living Resources* (London: Chapman and Hall, 1992), 17.

Biodiversity in Canada

Biodiversity is not evenly distributed around the world. Some biomes, mainly tropical forests, are extremely diverse (Box 3.12), but in temperate latitudes there is much less diversity. Overall, as discussed in Chapter 2, species numbers decline in a gradient from the tropics to the poles. Latin America, for example, is home to over 85,000 plant species. North America has 17,000, of which only 4,000 occur in Canada. A similar gradient for

Most of the world's species are insects, and many more await discovery (*Philip Dearden*).

birds is shown in Table 3.5. Several reasons have been advanced to account for the latitudinal gradient in species richness. Rohde (1992), following a review of these different reasons, concludes that many factors contribute at different scales, but the primary cause appears to be the effect of solar radiation (i.e., temperature) that increases evolutionary speed at lower latitudes.

Including flora, fauna, and microorganisms but not viruses, it is estimated that Canada has over 71,000 different species. Specialists estimate that there are slightly fewer than this number of organisms (68,000) yet awaiting discovery. The taxonomic groups containing the most numbers of species are shown in Figure 3.11. Mosquin (1994) points out that although the groups represented in this graph are not as well known as other groups, such as birds and mammals, they undertake key functions in ecosystems, often functions that we are only just becoming aware of and that support the more familiar and larger organisms. Beneficial insects, for example, fertilize flowers and control pests, crustaceans provide food for fish, bacteria recycle nutrients, and bread, beer, and penicillin all come from fungi.

Another important element of biodiversity is the concept of endemism. **Endemic species** are

BOX 3.11 BIOLOGICAL UNCERTAINTY

Attention has been drawn to the high degree of biological uncertainty regarding the numbers of species in Canada and elsewhere, not to mention the ecological functions of each species. Few things in the natural world are absolutes. Even gender differences blur. We have known this for some time for more primitive species, such as slugs and earthworms, which are *hermaphrodites*; that is, one individual has both male and female sexual functions. Still, it was somewhat of a surprise to Charles Francis, a biologist with the Canadian Wildlife Service undertaking field research in Malaysia, to find the first free-ranging wild male mammals that lactate. He was collecting Dayak fruit bats when he found that several males also had breasts and milk. He speculates that this may have evolved among males that are monogamous. Another theory suggests that something in the bats' diet contains steroids that mimic female hormones.

There is also uncertainty regarding many of

the basic ecological principles described in these last two chapters. Concepts such as succession, ecosystems, and communities have been found unsatisfactory for addressing many real-life ecological problems. Ecosystems often exhibit many different states that can be reached by various paths and that may produce no identifiable local climax (Pickett et al., 1992). Furthermore, species distributions along environmental gradients may overlap broadly (Levin, 1989) and be subject to many natural and human-induced disturbances. Heterogeneity is the norm rather than the exception. These observations have changed how ecologists look at ecological systems to recognize the openness of systems, the importance of episodic events, and the numerous possibilities for intervention in ecological processes. Nonetheless, the classical concepts provide a useful background against which to organize this new, more flexible, 'non-equilibrium' approach.

Table 3.5	Changes in the Numbers of Breeding Birds in Areas of Comparable Size with Latitude	
Location	Approx. Median Latitude	No. of Species of Breeding Birds
Greenland	70°N	56
Labrador	55°N	81
Newfoundland	49°N	118
New York State	43°N	195
Guatemala	15°N	469
Colombia	5°N	1,525

SOURCE: E.O. Wilson, *The Diversity of Life* (Cambridge, Mass.: Harvard University Press, 1992), 196. Copyright © 1992 by E.O. Wilson. Reprinted by permission of Harvard University Press.

ones that are found nowhere else on Earth. In Canada, we have relatively few endemic species compared, for example, to southern Africa, where some 80 per cent of the plants are endemic, or southwest Australia, where 68 per cent are endemic. In Canada it is estimated that between 1 and 5 per cent of our species may be endemic. Examples include the Vancouver Island marmot (Canada's only endangered endemic mammal species), the Acadian whitefish, and 28 species of plants from the Yukon. Reasons for our low endemism include the recent glaciation over most of the country that effectively wiped out localized species and the wide-ranging nature of many of our existing species. In terms of protecting biodiversity, it is especially important that endemic species are given consideration.

IMPLICATIONS

This chapter has emphasized that ecosystems are dynamic entities that change over time. Without such change we would not have evolved and the dinosaurs would not have become extinct. The main implication of this is to accept and try to understand the nature of these changes and be able to distinguish between those that are essentially the results of natural processes and those that are the result of human activities. We cannot impose static management regimes on dynamic ecosystem processes without causing ecological disruptions. A visible reminder of this was the fire-suppression policy characteristic of many national park services, which often ignored the natural role of fire in these ecosystems. When fires did start in such ecosystems, the buildup of fuel was often so great as to cause a major and very damaging fire, as happened in Yellowstone National Park in the US in 1988. Most park services have now abandoned such practices for a more dynamic approach that tries to mimic the role of natural fires through prescribed burning programs.

Unfortunately, the temporal and spatial scales of ecosystem change are often so great that they are very difficult to observe in the human lifespan. Scientists are only now beginning to unravel the mysteries of some of these dynamic interactions between the different components of the ecosphere.

The Vancouver Island marmot is Canada's only endangered endemic mammal species (*Philip Dearden*).

BOX 3.12 THE TROPICAL FORESTS

Charles Darwin, who described the mechanisms for evolution in *On the Origin of Species* (1859), originated most of his ideas in the tropics. It was in the tropics—where life is speeded up through high energy inputs and abundant moisture, where adaptation is at its most complex and intricate, and where the struggle for survival is most dramatic—that evolution could most readily be appreciated.

The diversity of the tropical forests is astounding—estimates suggest that at least half of the world's species are within the 7 per cent of the globe's surface that the tropics cover. For example: in 100 square metres in Costa Rica, researchers found 233 tree species; one tree in Venezuela was home to at least 47 different species of orchids; there are 978 different species of beetles that live on sloths; and over 1,750 different species of fish live in the Amazon basin. In general, the rain forests of South America are the richest in species, followed by Southeast Asia and then Africa. Several factors account for this abundance.

1. The tropical rain forests have been around for over 200 million years, since the time of the dinosaurs and before the evolution of the flowering plants. At this time it is thought that there was just one gigantic landmass, before continental drift started to form the continents as we now know them. The vegetation of many areas was subsequently wiped out by succeeding glacial periods, which had minimum impact on the rain forests. Hence, evolutionary forces and speciation have had a long time to operate in the tropics.

2. Over the long period of evolution there is a kind of positive feedback loop. As more species have developed and adapted, this has caused further adaptations as more species seek to protect themselves against being eaten and also to improve their harvesting of available food supplies. It is thought, in particular, that plant diversity has been partly the result of the need to adapt defences against the myriad of insects that graze on them. As the plants develop their defences, so do insects constantly adapt to the new challenge. The very high biodiversity of these groups is due to the speed of these evolutionary processes. In a system where most plants are immune to most insects, but highly susceptible to a few, it pays to be a long way from a member of your own species. Successful trees are hence widely distributed, which allows more opportunity for speciation to occur.

3. The tropics receive a higher input of energy from the sun than other areas of the globe. Not only are they closer to the sun, they also have little or no winter. The flux in solar input at the equator between the seasons is 13 per cent, but at latitude 50 degrees the variation is 400 per cent.

4. Tropical rain forests receive a minimum of 2,000 mm of precipitation that is evenly distributed throughout the year. Moisture is therefore not a limiting factor, allowing for constant growth. There is a strong correlation between diversity and rainfall.

5. The tropical rain forests are the most diverse ecosystems that have evolved on Earth. They are also characterized by examples of coevolution and mutualism, where two species are absolutely codependent on one another. Over 900 species of wasp, for example, have evolved to pollinate the same number of fig trees. Each wasp has adapted to just one species of fig. Should anything destroy the food supply in such finely tuned systems, then the other species will also meet its demise.

While the evolutionary process has benefited from most of these characteristics, the soils have suffered. They have been exposed to weathering processes for a very long time with no renewal and remixing from glaciation. The warm temperatures and abundant moisture are perfect for chemical weathering to great depths and most tropical soils have long since had their nutrients washed out. A fundamental difference between tropical and temperate ecosystems is that in the tropics, unlike more temperate climes, most of the nutrients are stored in the biomass and not in the soils. When tropical vegetation is removed, by logging, for example, then this removes most of the nutrients.

Species that evolved among the complexity of tropical forests have developed many adaptations to protect themselves. The camouflage of the leaf insect in this photo gives it some protection from predators (*Philip Dearden*).

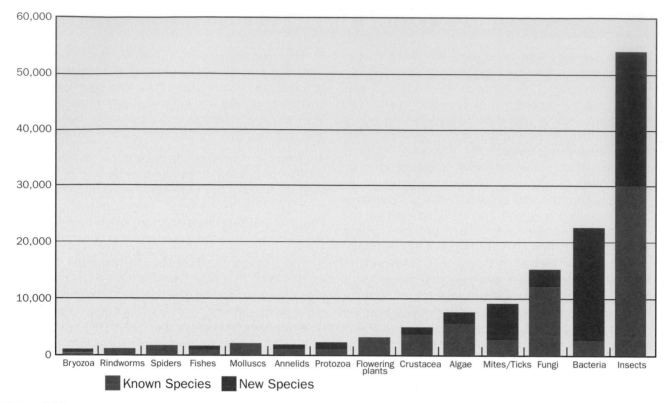

Figure 3.11 Groups with the most species in Canada (excluding viruses). SOURCE: T. Mosquin, P.G. Whiting, and D.E. McAllister, *Canada's Biodiversity: The Variety of Life, Its Status, Economic Benefits, Conservation Costs and Unmet Needs* (Ottawa: Canadian Museum of Nature, 1995), 58.

There are complicated feedback loops and synergistic relations. In some cases, positive feedback loops are strengthened to accelerate undesirable changes that underlie some of the most serious environmental challenges facing humanity, such as global warming, a topic that will be discussed in more detail in Chapter 7. When faced with such dynamic ecosystem changes, we must use equally dynamic thinking to face the challenges of the future.

SUMMARY

1. Ecosystems change over time. The speed of change varies from very slow, over evolutionary time scales, through to rapid, caused by events such as landslides and volcanic eruptions.

2. Ecological succession occurs as a slow adaptive process involving the gradual replacement of one assemblage of species by another as conditions change over time. Primary succession occurs on surfaces that have not been previously vegetated, such as surfaces exposed by glacial retreat; secondary succession occurs on previously vegetated surfaces, such as abandoned fields. Fire is an important element in ecosystem change. Some ecosystems, such as much of the boreal forest, have evolved in conjunction with periodic fires. Fire suppression in such ecosystems can be detrimental to these natural processes.

3. Ecosystem homeostasis is a state of dynamic equilibrium in which the internal processes of an ecosystem adjust for changes in external conditions. Not all ecosystems are equal in their abilities to withstand perturbations. Inertia is the ability of an ecosystem to withstand change; resilience is the ability to recover to the original state following disturbance. Both contribute to the stability of the system.

4. Important causes of loss of ecosystem homeostasis include the introduction of alien species and the removal of native keystone species.

5. Feedback mechanisms exist in ecosystems that may either exacerbate (positive feedback loops) or mitigate (negative feedback loops) change.

6. Population change occurs as a result of the balance between factors promoting growth (e.g., increase in birth rates or reduction in death rates) and those promoting reduction (e.g., declines in birth or survival rates or increase in death rates).

7. Different species have different reproductive strategies. K-strategists produce few offspring but devote considerable effort to ensure that these offspring reach matu-

rity. In comparison, r-strategists produce large numbers of young starting early in life and over a short time period, and devote little or no energy to parental care.

8. Populations adapt to changing conditions over the long term through evolution. Evolution results in the formation of new species as a result of divergent natural selection responding to environmental change.

This is speciation. Extinction results in the elimination of species that can no longer survive under new conditions.

9. Biodiversity involves the variety of life at three different scales: genetic, species, and landscape. Estimates suggest that Canada has a total of 71,000 known non-viral species, with 68,000 yet awaiting discovery.

KEY TERMS

albedo effect
alien or invader species
biodiversity
biotic potential
carrying capacity
climax community

coevolution
ecological succession
ecosystem homeostasis
ecotones
endemic species
eutrophication

evolution
extinction
inertia
negative feedback
population
positive feedback loop

primary succession
resilience
secondary succession
speciation
species
synergism

REVIEW QUESTIONS

1. What are the different kinds of succession? Can you identify different seral stages in your area?

2. What is an edaphic climax? Can you find some local examples and identify the dominant limiting factor?

3. How does the concept of succession relate to environmental management?

4. How important was fire in the development of vegetation patterns in your region? Is there a fire management plan in your region? If so, what are its management goals?

5. Identify the main plant and animal species in your region that are not native. What effect are they having on the local ecosystems? What are the implications for management?

6. Can you think of any other examples of negative and positive feedback loops in the ecosphere besides those mentioned in the text?

7. Are K-strategists or r-strategists most vulnerable to environmental change?

8. How does genetic diversity help protect a species from extinction?

9. What is coevolution? Can you think of any examples of coevolution among species in Canada?

10. Where in Canada has been called the most important scientific site in the world and why?

11. What is endemism and why does Canada have relatively few endemic species?

RELATED WEBSITES

INVASIVE SPECIES:
  www.invadingspecies.com
  www.oag-bvg.gc.ca/domino/reports.nsf.html
  http://www.imo.org/ome.asp?topic_id=548
  http://globallast.imo.org/
  http://cws-scf.ec.gc.ca/habitat/ramsar/
  http://www.iucn.org/biodiversityday/100booklet.pdf
  http://www.invasivespecies.gov/geog/canada.shtml

BIODIVERSITY CONVENTION OFFICE:
  http://www.bco.ec.gc.ca/en/default.cfm

BIODIVERSITY IN CANADA:
  http://www.cbin.ec.gc.ca/default_e.cfm

CANADIAN BIODIVERSITY STRATEGY:
  http://www.bco.ec.gc.ca/documents/CBS_E.pdf

IUCN RED LIST OF THREATENED SPECIES:
  http://www.redlist.org

WILDLIFE AND NATURE, ATLANTIC REGION:
  http://www.atl.ec.gc.ca/wildlife/index.html

REFERENCES AND SUGGESTED READING

Auditor General. 2002. 'Chapter 4: Invasive Species', in *The 2002 Report of the Commissioner of the Environment and Sustainable Development*. Ottawa: Minister of Supply and Services.

Balmford, A., R.E. Green, and M. Jenkins. 2003. 'Measuring the changing state of nature', *Trends in Ecology and Evolution* 18, 7: 326–30.

Biodiversity Science Assessment Team. 1994. *Biodiversity in Canada: A Science Assessment for Environment Canada*. Ottawa: Environment Canada.

Birkedal, T. 1993. 'Ancient hunters in the Alaskan wilderness: human predators and their role and effect on wildlife populations and the implications for resource management', in W.E. Brown and S.D. Veirs Jr, eds, *Partners in Stewardship: Proceedings of the 7th Conference on Research and Resource Management in Parks and on Public Lands*. Hancock, Mich.: The George Wright Society, 228–34.

Carlton, J.T., and J.B. Geller. 1993. 'Ecological roulette: the global transport of non-indigenous marine organisms', *Science* 198: 394–6.

Dalpe, Y. 2003. 'Mycorrhizal fungi biodiversity in Canadian soils', *Canadian Journal of Soil Science* 83, 3: 321–30.

Darwin, C.R. 1859. *On the Origin of Species*. London. John Murray.

Dearden, P. 1979. 'Some factors influencing the composition and location of plant communities on a serpentine bedrock in western Newfoundland', *Journal of Biogeography* 6: 93–104.

———. 1983. 'An Anatomy of a biological hazard: *Myriophyllum spicatum* L. in the Okanagan Valley, British Columbia', *Journal of Environmental Management* 17: 47–61.

Donlan, C.J., and P.S. Martin. 2004. 'Role of ecological history in invasive species management and conservation', *Conservation Biology* 18, 1: 267–9.

Dulloo, M.E., S.P. Kell, and C.G. Jones. 2002. 'Impact and control of invasive alien species on small islands', *International Forestry Review* 4, 4: 277–85.

Environment Canada. 2003. *Environmental Signals: Canada's National Environmental Indicator Series 2003*. Ottawa: Environment Canada.

Estes, J.A., D.O. Duggins, and G.B. Rathbun. 1989. 'The ecology of extinctions in kelp forest communities', *Conservation Biology* 3: 252–64.

Gedalof, Z., and D.J. Smith. 2001. 'Dendroclimatic response of mountain hemlock (*Tsuga mertensiana*) in Pacific North America', *Canadian Journal of Forest Research* 31: 322–32.

Gillis, P.L., and G.L. Mackie. 1994. 'Impact of the zebra mussel, *Dreissena polymorpha*, on populations of Unionidae (*Bivalvia*) in Lake St. Clair', *Canadian Journal of Zoology* 72: 1260–71.

Gould, S.J. 1989. *Wonderful Life: The Burgess Shale and the Nature of History*. New York: W.W. Norton.

———. 1994. 'The evolution of life on the earth', *Scientific American* 271: 84–91.

———. 2002. *I Have Landed: The End of the Beginning in Natural History*. New York: Harmony Books.

Goulet, H. 2003. 'Biodiversity of ground beetles (*Coleoptera: Carabidae*) in Canadian agricultural soils', *Canadian Journal of Soil Science* 83, 3: 259–64.

Grigorovich, I.A., R.I. Colautti, E.L. Mills, K. Holeck, A.G. Ballert, and H.J. MacIsaac. 2003. 'Ballast-mediated animal introductions in the Laurentian Great Lakes: retrospective and prospective analyses', *Canadian Journal of Fisheries and Aquatic Sciences* 60, 6: 740–56.

Hoddle, M.S. 2004. 'Restoring balance: Using exotic species to control invasive exotic species', *Conservation Biology* 18, 1: 38–49.

Jablonski, D. 2004. 'Extinction: past and present', *Nature* 427, 6975: 589.

Jones, G.A., and G.H.R. Henry. 2003. 'Primary plant succession on recently deglaciated terrain in the Canadian High Arctic', *Journal of Biogeography* 30, 2: 277–96.

Kay, C.E. 1994. 'Aboriginal overkill: the role of Native Americans in structuring western ecosystems', *Human Nature* 5: 359–98.

Kerr, R.A. 2003. 'Mass extinction: Extinction by a whoosh, not a bang?', *Science* 302, 5649: 1315.

Laroque, C.P., and D.J. Smith. 1999. 'Tree-ring analysis of yellow cedar (*Chamaecyparis nootkatensis*) on Vancouver Island, British Columbia', *Canadian Journal of Forest Research* 21: 115–23.

Lesieur, D., S. Gauthier, and Y. Bergeron. 2002. 'Fire frequency and vegetation dynamics for the south-central boreal forest of Quebec, Canada', *Canadian Journal of Forest Research* 32: 1996–2002.

Levin, S.A. 1989. 'Challenges in the development of a theory of community and ecosystem structure and function', in J. Roughgarden, R.M. May, and S.A. Lewis, eds, *Perspectives in Ecological Theory*. Princeton, NJ: Princeton University Press, 242–55.

Lovelock, J.E. 1988. *The Ages of Gaia*. New York: W.W. Norton.

McNeely, J.A. 2001. *The Great Reshuffling: Human Dimensions of Invasive Alien Species*. Gland, Switz.: IUCN.

———, H.A. Mooney, L.E. Neville, P. Schei, and J.K. Waage. 2001. *A Global Strategy on Global Invasive Species*. Gland, Switz.: IUCN.

Martin, P.S. 1967. 'Prehistoric overkill', in P.S. Martin and H.E. Wright Jr, eds, *Pleistocene Extinctions: The Search for a Cause*. New Haven: Yale University Press, 75–120.

Mosquin, T. 1994. 'A conceptual framework for the ecological functions of biodiversity', *Global Biodiversity* 4: 2–16.

——— and P.G. Whiting. 1992. *Canada Country Study of Biodiversity*. Ottawa: Canadian Museum of Nature.

Neave, D., E. Neave, T. Rotherham, and B. McAfee. 2002. 'Forest biodiversity in Canada', *Forestry Chronicle* 78, 6: 779–83.

Noonburg, E.G., B.J. Shuter, and P.A. Abrams. 2003. 'Indirect effects of zebra mussels (*Dreissena polymorpha*) on the planktonic food web', *Canadian Journal of Fisheries and Aquatic Sciences* 60, 11: 1353–68.

Perrings, C. 2002. 'Biological invasions in aquatic systems: The economic problem', *Bulletin of Marine Science* 70, 2: 541–52.

Picket, S.T.A., V.T. Parker, and P.L. Fiedler. 1992. 'The new paradigm in ecology: implications for conservation biology above the species level', in P.L. Fiedler and S.K. Jain, eds, *Conservation Biology*. New York: Chapman and Hall, 66–88.

Quammen, D. 1988. *The Flight of the Iguana: A Sidelong View of Science and Nature*. New York: Touchstone Books.

Reice, S.R. 1994. 'Nonequilibrium determinants of biological community structure', *American Scientist* 82: 424–35.

Rohde, K. 1992. 'Latitudinal gradients in species diversity: the search for the primary cause', *Oikos* 65: 514–27.

Statistics Canada. 2003. *Human Activity and the Environment: Annual Statistics 2003*. Ottawa. Ministry of Industry.

Stuart-Smith, J., B. Wilson, R. Walker, and E. MacDonald. 2002. 'Conserving whitebark pine in the Canadian Rockies', *Research Links* 10: 11–14.

Weber, M.G., and S.W. Taylor. 1992. 'The use of prescribed fire in the management of Canada's forested lands', *Forestry Chronicle* 68: 324–33.

Wong, P.Y., and I. Brodo. 1992. 'The lichens of southern Ontario', *Syllogeus* 69: 1–79.

Chapter 4

Ecosystems and Matter Cycling

Learning Objectives

- To understand the nature of matter.
- To describe why human intervention in biogeochemical cycles is a fundamental factor behind many environmental issues.
- To outline the main components and pathways of the phosphorus, nitrogen, sulphur, and carbon cycles.
- To identify the main components of the hydrological cycle and the nature of human intervention in the cycle.
- To discuss the causes, effects, and management approaches to eutrophication and acid deposition.

INTRODUCTION

The collapse in the Atlantic puffin population, described in Chapter 2, was a result of human interference with energy flow through the ecosystem. There are implications, however, for other aspects of ecosystem functioning. Puffins and most other seabirds play an important role in recycling nutrients, particularly phosphorus, from marine to terrestrial ecosystems. If these systems are disturbed, then the efficiency of the recycling mechanisms can be greatly reduced. As the phosphorus cycle has very limited recycling capabilities from aquatic to terrestrial systems, the impact of interfering with it in this way could be substantial. This chapter explains how matter, such as phosphorus, cycles in the ecosphere and some of the implications of disturbing these cycles.

MATTER

Everything is either matter or energy. However, in contrast to the supply of energy, which is virtually infinite, the supply of matter on Earth is limited to that which we now have. **Matter**, unlike energy,

has mass and takes up space. Matter is what things are made of and is composed of the 92 natural and 17 synthesized chemical elements such as carbon, oxygen, hydrogen, and calcium. *Atoms* are the smallest particles that still exhibit the characteristics of the element. Subatomic particles include *protons*, *neutrons*, and *electrons* that have different

Water is the only substance that occurs in all three phases of matter at the ambient temperatures of the Earth's surface (*Philip Dearden*).

electrical charges. At a larger scale, the same kinds of atoms can join together to form molecules. When two different atoms come together they are known as a **compound**. Water (H_2O), for example, is a compound made up of two hydrogen atoms (H) and one oxygen atom (O). Four major kinds of organic compounds—carbohydrates (sugars and starches), fats (lipids, hormones, etc.), proteins (enzymes, etc.), and nucleic acids (DNA, RNA, etc.)—make up living organisms.

Matter also exists in three different states, solid, liquid, and gas, and can be transformed from one to the other by changes in heat and pressure. At the existing temperatures at the surface of the Earth we have only one representative of the liquid state of matter, water. We can also readily see water in its other two states as ice (solid) or vapour (clouds).

Just as the laws of thermodynamics explain energy flow, the **law of conservation of matter** helps us understand how matter is transformed. This law tells us that matter can neither be created nor destroyed, but merely transformed from one

form into another. Thus, matter cannot be consumed so that it no longer exists; it will always exist, but in a changed form. When we throw something away it is still with us, on this planet, as matter somewhere. There *is* no 'away'. All pollution stems from this law. The huge super-stacks on large smelters such as at Inco in Sudbury do not dispose of waste; they just disperse those wastes over a much larger area. The matter dispersed is the same, and ultimately falls as acid deposition somewhere else. The same is true for all the wastes that we wash down our sinks. They do not disappear, but collect in larger water bodies and create pollution problems.

BIOGEOCHEMICAL CYCLES

For millions of years, matter has been moving among different components of the ecosphere. These cycles are as essential to life as the energy flow described in Chapter 2. About 30 of the naturally occurring elements are a necessary part of living things. These are known as **nutrients**, and may be further classified into **macronutrients**, which are needed in relatively large amounts by all organisms, and **micronutrients**, required in lesser amounts by most species (Table 4.1). About 97 per cent of organic mass is composed of six nutrients: carbon, oxygen, hydrogen, nitrogen, phosphorus, and sulphur. These nutrients are cycled continuously among different components of the ecosphere in characteristic paths known as **biogeochemical cycles**.

Figure 4.1 shows a generalized model of such a cycle. Like all the subsequent diagrams of cycles in this chapter, it exemplifies the types of simplifying models that scientists construct to try to represent the vast complexity of Earth processes, as described in Chapter 1. Nutrients can be stored in the different compartments shown in Figure 4.1 for varying amounts of time. In general, there is a large, relatively slow-moving abiotic pool that may be in the

According to the law of matter, emissions from stacks such as these do not simply disappear but end up somewhere else, often with undesirable consequences, such as acid deposition or global warming (*Philip Dearden*).

We know from studies of chemistry that our bodies are re-organized star-dust, recycled again and again, so that, truly, our bones are of corals made.
— *Stan Rowe (1993)*

Table 4.1	Relative Amounts of Chemical Elements That Make Up Living Things				
Major Macronutrients (>1% dry organic weight)		Relatively Minor Macronutrients (0.2–1% dry organic weight)		Micronutrients (<0.2% dry organic weight)	
Name of Element	Symbol	Name of Element	Symbol	Name of Element	Symbol
Carbon	C	Calcium	Ca	Aluminum	Al
Hydrogen	H	Chlorine	Cl	Boron	B
Nitrogen	N	Copper	Cu	Bromine	Br
Oxygen	O	Iron	Fe	Chromium	Cr
Phosphorus	P	Magnesium	Mg	Cobalt	Co
		Potassium	K	Fluorine	F
		Sodium	Na	Gallium	Ga
		Sulphur	S	Iodine	I
				Manganese	Mn
				Molybdenum	Mo
				Selenium	Se
				Silicon	Si
				Strontium	Sr
				Tin	Sn
				Titanium	Ti
				Vanadium	V
				Zinc	Zn

SOURCE: C.E. Kupchella and M.C. Hyland, *Environmental Science: Living within the System of Nature* (Toronto: Allyn and Bacon, 1989), 45.

atmosphere or lithosphere, and is chemically unusable by the biotic part of the ecosystem or is physically remote. There is a more rapidly interacting exchange pool between the biotic and abiotic components. Nutrients move at various speeds from the biotic to abiotic pools. For example, very rapid exchange takes place through respiration as carbon and oxygen move rapidly between the biotic and atmospheric components. The elements that now make up your body have undergone millions of years of recycling through these various compartments. You are a product of recycling!

Ecosystems also vary substantially in terms of the speeds of cycling and the relative proportions of nutrients in each compartment. Some systems have nutrient-poor soils, for example, and have developed different mechanisms to store nutrients in other compartments. Tropical forest ecosystems

Slash-and-burn agriculture is a common way to transfer nutrients from the biomass to the soil to increase agricultural productivity. This photograph shows several swiddens (fields cut in the forest) by the hill tribe people of northern Thailand. The soils in the swiddens rapidly lose fertility caused by the burning of the biomass, and they are then abandoned for secondary succession to occur (*Philip Dearden*).

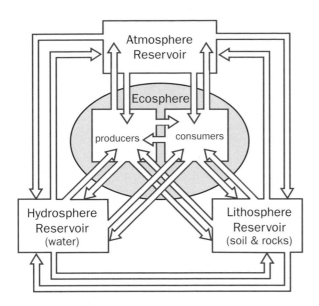

Figure 4.1 Generalized model of the biogeochemical cycle.

are classic examples. Most of the nutrients are stored in the biomass as opposed to the soil system (Table 4.2). When leaves fall to the ground they are rapidly mined for nutrients by plant roots before those nutrients have a chance to be leached out of the system. In contrast, many temperate forests have soils of high fertility. Removal of the nutrients in the biomass, through logging for example, does not remove as high a proportion of the site nutrient capital as removal in tropical ecosystems. This is discussed in more detail in Chapter 9.

Speed of cycling may also change within a cycle, depending on season. For the carbon cycle, for example, there is greater uptake of CO_2 in spring and summer as deciduous trees grow leaves. In fall there is a correspondingly greater release as the leaves fall off and decompose (see Box 4.1). On average, a carbon dioxide molecule stays in the

atmospheric component of the cycle from five to seven years. This is known as the *residence time*. It takes, on average, 300 years for a carbon molecule to pass through the lithosphere, atmosphere, hydrosphere, and biotic components of the carbon cycle. By way of contrast, it may take a water molecule two million years to make a complete cycle. The speed of cycling is influenced by such factors as the chemical reactivity of the substance. Carbon, for example, participates in many chemical reactions. It also occurs as a gas. In general, a gaseous phase allows for a speeding up of a cycle because gas molecules move more quickly than molecules in the other states of matter.

Cycles can be classified according to the main source of their matter. **Gaseous cycles**, as the name would suggest, have most of their matter in the atmosphere. The nitrogen cycle is a good example. **Sedimentary cycles**, such as the phosphorus and sulphur cycles, hold most of their matter in the lithosphere. In general, elements in sedimentary cycles tend to cycle more slowly than those in gaseous cycles and the elements may be locked into geological formations for millions of years.

Under natural conditions, recycling rates between components achieve a balance over time, where inputs and outputs are equal. Human activity serves to change the speed of transference between the different components of the cycles. Many of our pollution problems result from a human-induced buildup in one or more components of a cycle that cannot be effectively dissipated by natural processes.

In addition to the biogeochemical cycles, some attention will also be given in this chapter to the hydrological cycle. This cycle is critical in all other

Table 4.2	**Approximate Distributions of Carbon and Nitrogen in Temperate and Tropical Rain Forests**

	Tropical Rain Forest	Temperate
Carbon in vegetation	75%	50%
Carbon in litter and soil	25%	50%
Nitrogen in biomass	50%	6%
Nitrogen in biomass above ground	44%	3%

BOX 4.1 THE DECOMPOSERS

In Chapter 2, attention was drawn to the importance of decomposer organisms and detritus food chains. These are the main means by which nutrients in the biotic component of the ecosphere are returned to the abiotic, so that plants can once again use them. Photosynthesis has been described as the process of making a complicated product out of simple components; decomposition is the reverse process of making simple components out of that complicated product.

Decomposer organisms such as fungi may attack leaves while still on the plant; the fungi release products such as sugars, which are then washed to the ground by rainfall. Once leaves fall to the ground they are broken down progressively by various groups of organisms. Larger organisms such as earthworms, slugs, snails, beetles, ants, and termites help break up the leaf material initially. Many gardeners are fully aware of the abilities of slugs, for example, to devour green leaves in great quantities.

Fungi and heterotrophic bacteria further break down the organic matter, releasing more resistant carbohydrates, followed by cellulose and lignin. The humus—the organic layer in the soil—is composed mainly of products that can resist rapid breakdown. A chemical process, oxidation, is mainly responsible for the decay of this material.

As everyone knows who has witnessed leaf decay in autumn, the process can occur quite rapidly. The speed varies depending on the environment. Warm environments tend to promote more rapid microbial activity. Leaf decay in the

Slugs play an important role in breaking down vegetable matter. In the wet west coast forests the biomass of slugs is greater than that of any other animal in the ecosystem (*Philip Dearden*).

tropics takes place in a matter of weeks. In the boreal forest, however, where conditions are cold and the leaves, such as spruce and pine needles, quite resistant, recycling of the nutrients held in the leaves may take decades. Overall, the average recycling time for organic material in the wet tropics is five months; in the boreal forest it is 350 years. The high amounts of lignin found in leafs of needle-leafed trees help protect the trees against freezing conditions but offer little food value to decomposers, so decay is slow. In comparison, deciduous trees, such as maple, have a high reward for decomposers: high nitrogen levels and little protective lignin, so they decay very quickly.

cycles since water plays a major role in the mobilization and transportation of materials. The energy for this, as with all other aspects of the cycles, ultimately comes from the sun. Photosynthesis powers the biotic aspects of the cycles, and atmospheric circulation, fuelled by the sun's energy, controls the water power that is so important for weathering and erosional processes.

Sedimentary Cycles

Sedimentary cycles mobilize materials from the lithosphere to the hydrosphere and back to the lithosphere. Some, such as sulphur, may involve a gaseous phase, while others, such as phosphorus, do not. These cycles rely essentially on geological

uplift over long time periods to complete the cycle. Lack of a gaseous phase means that the cycle is missing one potential route for more rapid recycling, which can lead to problems when mobilization rates are increased through human activity. Phosphorus and sulphur will be discussed here, but other elements, such as calcium, magnesium, and potassium, follow similar pathways.

Phosphorus (P)

Phosphorus, a macronutrient incorporated into many organic molecules, is essential for metabolic energy use. It is relatively rare on the Earth's surface compared with biological demand; it is therefore essential that phosphorus cycles efficiently between components. Many organisms have devised means to store this element preferentially in their tissues and phosphorus moves very readily within plants from older tissues to more active growth sites. Deciduous trees may recirculate up to 30 per cent of their phosphorus back to their more permanent components before the leaves fall in an effort to preserve this nutrient.

Under natural circumstances, phosphorus is a prime example of a nutrient held in a tight circulation pattern between the biotic and abiotic components. Replenishment rates through weathering and soil availability are limited, thus the amount retained by the biomass is quite critical. The residence time of phosphorus in terrestrial systems can be up to 100 years before it is leached into the hydrosphere. Phosphorus is often the dominant limiting factor (Chapter 2) in freshwater aquatic systems and for plant growth in terrestrial soils. Agricultural productivity relies heavily on

augmenting this supply (auxiliary energy flow) through fertilizer application. Box 4.2 outlines the impact humans have on the phosphorus cycle.

The availability of phosphorus in the soil is influenced by soil acidity. Acidity is measured on the pH scale, which is discussed in more detail later in this chapter. Below pH 5.5, for example, phosphorus reacts with aluminum and iron to form insoluble compounds. Above pH 7, the same thing happens in combination with calcium. Obviously, things that change soil pH, such as acid precipitation, can have a critical effect on phosphorus availability. This is an example of the kind of synergistic reaction discussed in Chapter 3 in which the combination of either high or low pH values with low phosphorus availability, due to chemical reactions, can have a stronger effect than the sum of the two individually.

Rocks in the Earth's crust are the main reservoir of phosphorus (Figure 4.2). Geological uplift and subsequent weathering (Box 4.3) make phosphorus available in the soil where it is taken up by plant roots. Phosphate ions are the main source of phosphorus for plants and are released from slowly dissolving minerals such as iron, calcium, and magnesium phosphates. Many higher plants have a mutualistic relationship with soil fungi or mycorrhizae, which help them gain improved access to phosphorus in the soil. Once incorporated into plant material, the phosphorus may be passed on to other organisms at higher trophic levels.

Animal wastes are a significant source of phosphorus return to the soil. All organisms eventually

---

BOX 4.2 HUMAN IMPACTS ON THE PHOSPHORUS CYCLE

Humans intervene in the phosphorus cycle in several ways that serve to accelerate the mobilization rate:

- mining of phosphate-rich rocks for fertilizer and detergent production, creating excessive runoff into aquatic environments;
- biomass removal, leading to accelerated erosion of sediment and solutes into streams;
- concentration of large numbers of organisms such as humans, cattle, and pigs, creating heavy burdens of phosphate-rich waste materials; and
- removal of phosphorus from oceanic ecosystems through fishing, with the phosphorus returned to fresh water again and ultimately the marine system, through the dissolution of wastes.

The major implication of all these interventions is for excessive phosphorus accumulation in freshwater systems, resulting in eutrophication. Human activity is now estimated to account for about two-thirds of the phosphorus reaching the oceans. The environmental impacts of this nutrient enrichment will be discussed in more detail later.

The Head-Smashed-In World Heritage Site in southern Alberta is rich in phosphorus as Native peoples used the cliff to kill stampeding bison (*Philip Dearden*).

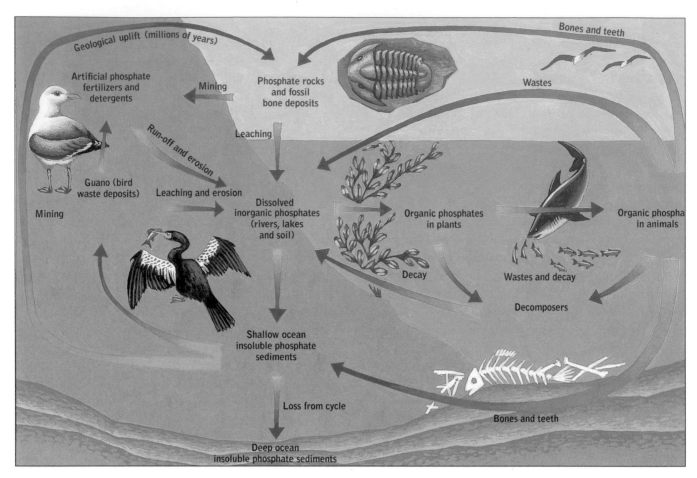

Figure 4.2 The phosphorus cycle.

die and the organic material is broken down by the decomposer food chains. This may take some time as a considerable amount of the phosphorus is within animal bones. In the past, farmers have used concentrated sources of animal bones, such as the bison jumps used by Native peoples on the prairies, as a source of phosphate fertilizer.

Following breakdown in the soil, the phosphorus is then either taken up again by plants or removed by water transport. Bacteria mineralize the returned organic phosphorus into inorganic forms so that plants can take it up once more. Most of the water transport occurs in particulate form by streams, and is one reason to be concerned about excessive sedimentation occurring through land-use activities such as agriculture and logging, described in Chapters 9 and 10.

Stream transport ultimately ends up in the ocean. Estuaries have such high productivities, as discussed in Chapter 2, in part because of this nutrient input from upstream. The circulation patterns within estuaries tend to trap nutrients, but some

phosphorus finds its way into the shallow ocean areas of the coastal zone. It may be fixed in biomass by phytoplankton or other aquatic plants in the euphotic (eu = well, photos = light) zone and once again incorporated into the food chain. The coastal zones, with this plentiful supply of nutrients and photosynthetic energy from the sun, cover less than 10 per cent of the ocean's surface but account for over 90 per cent of all ocean species.

Beyond the coastal zone and the continental shelves, water depth increases into the open ocean. Phosphorus and other nutrients that have not been incorporated into food chains, plus elements from the death of oceanic organisms, filter through to the bathyal and ultimately the abyssal zones (see figure with Box 2.4). Here, uptake by organisms is extremely limited, and the nutrients must either be moved back to the euphotic zone by upwelling currents or wait to be geologically uplifted over millions of years to move into another component of the cycle. Where such upwellings occur, such as off the west coasts of Africa and South America,

BOX 4.3 WEATHERING, THE ROCK CYCLE, AND PLANT UPTAKE

The weathering of the rocks of the Earth's crust plays an important role in supplying long-term inputs to biogeochemical cycles. Weathering is part of the rock cycle whereby rocks that have been uplifted are eroded into different constituents. The rock cycle involves the transformation of rocks from one type to another, such as when volcanic rocks are eroded and washed into the ocean. Over millions of years the resulting sediments are turned into sedimentary rocks. In turn, these sedimentary rocks may be compressed within the Earth's crust and altered by heat and pressure before once more being uplifted through the process of continental drift.

Weathering involves numerous different processes. In Canada, mechanical weathering involves the physical breakup of rocks as a result of changing temperatures. The action of water is important. Chemical processes, such as hydration and carbonation, further the process by removing elements in solution. Secondary clay minerals are produced from primary rock minerals by hydrolysis and oxidation. These clays are very important in terms of holding the nutrients in the soil. The soil can be thought of as a giant filter bed in which each particle is chemically active. As water percolates through, containing many different nutrients in solution, some of these are held by the clays and become available for plant uptake.

Plants constantly lose moisture from their leaves. This creates a moisture gradient within the plant that serves to draw water up to replace that which has been lost. Water then moves from the roots in replacement, and more nutrient-laden water is taken in by the roots. It is the job of the roots to keep the plant supplied with water. As nutrients are removed from the soil water around the plants, new nutrients move within the soil water to replace them.

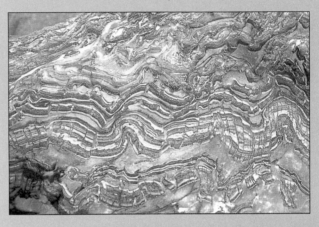

These sedimentary rocks along the Alsek River in the Yukon have been compressed and folded as part of the rock cycle (*Philip Dearden*).

plentiful fisheries are found due to the combination of high nutrient and energy levels. Some fish species, such as salmon, are **anadromous,** spending part of their lives in salt water and part in fresh water, where they die after spawning. When they die the nutrients they have collected during their ocean phase are returned to the freshwater system, resulting in a significant input of nutrients, including phosphates (Chapter 8).

Two other recycling mechanisms also occur. One is the biotic one described earlier as marine birds, such as puffins, cormorants, and other fish-eating birds, return phosphorus to land in the form of their droppings, representing the phosphorus that has concentrated through the marine food chain. This phosphorus, known as guano, constitutes the largest source of phosphorus for human use and is heavily mined for fertilizer production. A small amount of phosphorus is also returned to land through the atmosphere as sea spray.

Sulphur (S)

Like phosphorus, sulphur is a sedimentary cycle, but it differs from phosphorus in two important ways. First, it has an atmospheric component and,

The nutrients that have sustained this salmon are now being recycled (*Philip Dearden*).

therefore, better recycling potential. Sulphur is not often a limiting factor for growth in aquatic or terrestrial ecosystems. Second, like most of the other cycles but unlike phosphorus, it has strong dependencies on microbial activity. Sulphur is a necessary component for all life and is a building component of proteins.

Sulphur is unavailable in the lithosphere and must be transformed into sulphates to be able to be absorbed by plants. Bacteria are critical here, changing sulphur into various forms in the soil (Figure 4.3). The exact form depends on factors such as the presence (aerobic) or absence (anaerobic) of oxygen, which is usually a reflection of the relationship of the particular site of transformation to the water table and the presence of other elements such as iron. From these microbial transformations by chemoautotrophs (discussed in Chapter 2), gases such as hydrogen sulphide (H_2S) may be released directly to the atmosphere (giving the familiar 'rotten eggs' smell we associate with marshlands) or sulphate salts (SO_4) are produced. Through their roots, plants can then absorb the sulphates, sulphur enters the food chain, and the same processes occur as in the biotic components of the other cycles.

The complexity of these cycles is illustrated further by some of the interactions that occur between cycles. For example, the phosphorus cycle benefits when iron sulphides are formed in sediments and phosphorus is converted from insoluble to soluble forms, where it becomes available for uptake.

The upward movement of the gaseous phase of the sulphur cycle is also important, as significant quantities of sulphur are returned to the atmosphere, thereby shortening the long sediment uplift time that characterizes the phosphorus cycle. This is fortunate since average ocean residence times are quite long, and sulphur is continually lost to the ocean floor. From the upper reaches of the oceans, sulphur can be returned to the atmosphere by phytoplankton or photochemical reactions. However, unlike phosphorus, a relatively small proportion of sulphur is fixed in organic matter and availability is not usually a problem. As with phosphorus, human intervention in the sulphur cycle (Box 4.4) is significant.

Gaseous Cycles

Nitrogen

Nitrogen is a colourless, tasteless, odourless gas required by all organisms for life. It is an essential

---

BOX 4.4 HUMAN IMPACTS ON THE SULPHUR CYCLE

Humans intervene in the sulphur cycle mainly through:

- the burning of sulphur-containing coal, largely to produce electricity; and
- the smelting of metal ores that also contain sulphates.

Almost 99 per cent of the sulphur dioxide and about one-third of the sulphur compounds reaching the atmosphere come from these activities. These sulphur compounds react with oxygen and water vapour to produce sulphuric acid (H_2SO_4), a main component of acid deposition, as discussed later in this chapter.

---

component of chlorophyll, proteins, and amino acids. The atmosphere is over 78 per cent nitrogen gas (N_2), and also contains other forms of gaseous nitrogen such as ammonia (NH_3), nitrogen dioxide (NO_2), nitrous oxide (N_2O), and nitric oxide (NO). Excess quantities of these other forms are involved in many of our most challenging environmental problems such as acid deposition, ozone depletion, and global climate change (see Box 4.5).

Most organisms cannot gain access to the necessary nitrogen from the atmosphere. The nitrogen is instead obtained from the soil as nitrates. The main way in which the atmospheric reservoir is linked to the biotic components of the food chain is through **nitrogen fixation** and **denitrification**, both mediated through microbial activity (Figure 4.4).

Biological nitrogen fixation occurs as bacteria transform atmospheric nitrogen (N_2) into ammonia (NH_3). The most important fixers are bacteria of the *Rhizobium* family that grow on the root nodules of certain plants, such as members of the pea or legume family, like peas, beans, clover, and alfalfa. The bacteria and roots of the plant communicate through chemical stimuli that result in the bacteria infecting root cells. Once infected, the cells swell into the nodules that you can see on the roots of the peas or beans in your garden. In a remarkable example of coevolution, the plant supplies the products of photosynthesis to the relationship, and the bacteria transform the atmos-

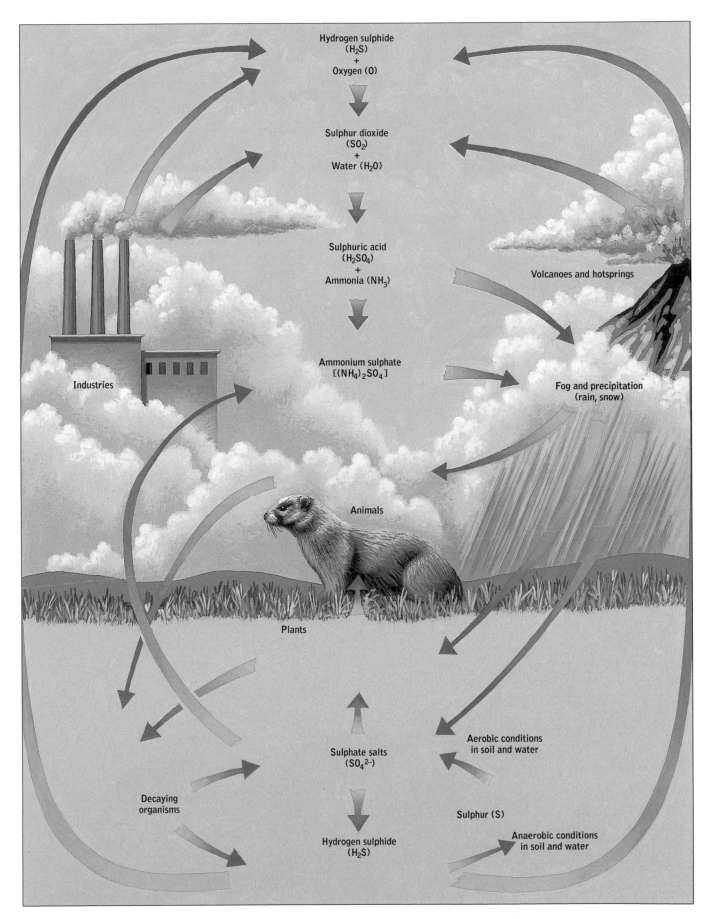

Figure 4.3 The sulphur cycle.

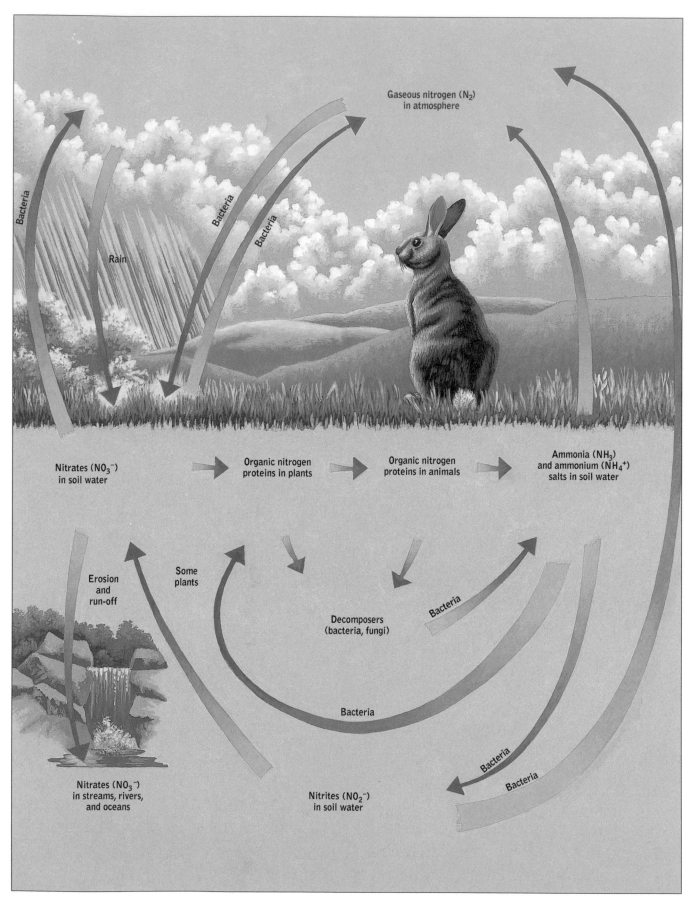

Figure 4.4 The nitrogen cycle.

BOX 4.5 HUMAN IMPACTS ON THE NITROGEN CYCLE

Humans disrupt the nitrogen cycle in many ways:

- Chemical fixation to supply nitrates and ammonia as fertilizer. The amount fixed is estimated to be about 120 million tonnes per year compared with the 170 million tonnes produced by natural processes. The main impacts are through runoff of excess fertilizer (contributing to eutrophication) and denitrification (contributing to climatic change). Agricultural fertilizers are implicated in both. Eutrophication will be discussed in more detail in the next section. Denitrification transforms nitrogen fertilizers into nitrous oxide, a green house gas (Chapter 7), and is also involved in the catalytic destruction of the ozone layer. There are also health concerns related to excessive nitrate levels from fertilizers running into water supplies (Box 4.8).
- Removal of nitrate and ammonium ions from agricultural soils through the harvesting of nitrogen-rich crops.
- High-temperature combustion produces nitric oxides (NO), which combine with oxygen to produce nitrogen dioxide (NO_2), which reacts with water vapour to form nitric acid (HNO_3), a main component of acid deposition.

pheric nitrogen into nitrates. It is one of the few examples known where two organisms co-operate to make one molecule.

This resulting enrichment is why farmers grow crops such as alfalfa and clover as part of a crop rotation to help build up nitrates in the soil. About one-half of the nitrogen circulating in agricultural ecosystems comes from this source. The increasing costs of fertilizer worldwide have focused more attention on biological nitrogen fixation (BNF) as a part of meeting the global food challenges of the future (see Chapter 10). Through genetic engineering, for example, it may be possible to inject other crops, such as cereals, with similar symbiotic unions between plants and nitrogen-fixing bacteria. It may not be that simple, however. For example, research indicates that species involved in di-nitrogen fixation may also be particularly susceptible to phosphorus deficiencies, given their high P and energy requirements.

Some wild species such as alder, lupines, and vetch also have similar bacteria and hence play a valuable role when they act as primary colonizers in the successional process or when they help recolonize sites that have been logged (Chapter 9) or otherwise disturbed. These relationships are mutualistic in that both organisms gain. The plant receives enhanced nutrient supply and the organisms find a home in which the plant supplies them with various sugars. One of the most celebrated examples of such a partnership is that between the fungi and algae in lichens, discussed in more detail in Chapter 2.

Other bacteria and algae that fix nitrogen are not attached to specific plants. In the Arctic, where nitrogen is a limiting factor on plant growth, an important source of nitrogen fixation is by cyanobacteria, or blue-green algae, such as *Nostoc commune*. Sometimes combinations of various cyanobacteria and other organisms form crusts over the soil surface, known as cryptogamic crusts, which are important in facilitating successional processes. Other important free-living nitrogen-fixing microorganisms are also found in the ocean. Estimates suggest that in terrestrial ecosystems, about double the amount of nitrogen is fixed by mutualistic relationships as by these free-floating relationships. The major supply, however, comes from the breakdown of existing biomass by decomposer food chains. Nitrogen is tightly circulated in most ecosystems between the dead and living biomass.

In addition to these biological mechanisms, some nitrogen is also fixed through atmospheric fixation that occurs largely during thunderstorms. Lightning causes extremely high temperatures that unite oxygen and nitrogen to form nitric acid (HNO_3), which is subsequently carried to earth as precipitation and converted into the nitrates (NO_3). These nitrates can then be taken up by plant roots. Estimates on the importance of atmospheric fixation vary, but 10 per cent of total fixation would be a maximum figure and most estimates are about half this.

Once fixed in the soil, ammonia and nitrates are incorporated into plant matter and then through the food chain. When the plants die, the biomass is converted back to ammonia (NH_3) and ammonium salts (NH_4) by bacterial action and returned to the soil. This is known as **mineralization**. Some ammonia and ammonium is changed to nitrites and then to nitrates by chemotrophic bacteria such as *Nitrosomonas* and *Nitrobacter* in a process known as nitrification. There are also other bacteria—anaerobic bacteria—which through the denitrification process convert nitrates into nitrogen gas and back into the atmosphere (Figure 4.4). Denitrification occurs in anaerobic conditions, especially where there are large amounts of nitrates available, such as on flooded agricultural fields.

Nitrates are highly soluble in water and if not held tightly may be lost to the ecosystem by surface runoff. Ammonia is also susceptible to loss by soil erosion, as it tends to adhere to soil particles. Like phosphorus, nitrogen is often a limiting factor for growth. When excessive concentrations occur in water, it is a major contributor to the process of eutrophication. Unlike phosphorus, however, nitrogen is not immobilized in deep-ocean sediments but has an effective feedback mechanism to the atmosphere from the ocean through microbial denitrification.

Carbon

Although carbon dioxide gas (CO_2) constitutes only 0.03 per cent of the atmosphere, it is the main reservoir for the carbon that is the building block for all necessary fats, proteins, and carbohydrates that constitute life. Plants take up carbon dioxide directly from the atmosphere through the process of photosynthesis, and at the same time emit oxygen. The carbon becomes incorporated into the biomass and is passed along the food chain. Residence times can vary greatly, but older forests constitute a significant repository for carbon for centuries. Respiration by organisms transforms some of this carbon back into carbon dioxide (Figure 4.5), and the cellular respiration of decomposers helps to return the carbon from dead organisms into the atmosphere. Most of this is in the form of CO_2, but also methane (CH_4) in anaerobic conditions. Thus, the cycling of carbon and the flow of energy through food chains are intimately related.

Besides this relatively rapid exchange, some carbon can also be stored in the lithosphere for extended periods as organisms become buried before they decompose. This is particularly true under relatively inefficient anaerobic decay conditions such as in peat bogs. Through geological time, millions of years of photosynthetic energy have been transformed into 'fossil fuels' by this process as a result of heat and compression. The highly productive forests and marine environments of the distant past have become the coal, oil, and natural gas that fuel the world's economy today (see Box 4.6).

Some of the carbon dioxide is dissolved into the shallower ocean before re-entering the atmosphere. Residence time is in the order of six years in these shallower waters, but much longer (up to 350 years) when mixed with deeper waters. These residence times are now of considerable scientific interest due to the rising levels of carbon dioxide in the atmosphere and the potential for the oceans to absorb these increases (see Chapter 8).

Large amounts of carbon are also stored for much longer periods in the ocean. When marine organisms die, their shells of calcium carbonate ($CaCO_3$) become cemented together to form rocks such as limestone. Over millions of years, the limestone may be uplifted to become land and then is slowly weathered to release the carbon back into the carbon cycle.

THE HYDROLOGICAL CYCLE

Water, like the nutrients discussed above, is necessary for all life. You are 70 per cent water. Although other planets, such as Venus and Mars, have water, only on Earth does it occur as a liquid. It also occurs in a fixed supply that cycles between various reservoirs driven by energy from the sun. By far the largest reservoir is the ocean, containing over 97 per cent of the water on Earth. Most of the rest is tied up in the polar ice caps, with only a small amount readily available as the fresh water

Large amounts of carbon are stored in the lithosphere, such as in these coal-beds, the product of millions of years of photosynthetic activity (*Philip Dearden*).

Figure 4.5 The carbon cycle.

BOX 4.6 HUMAN IMPACTS ON THE CARBON CYCLE

As human populations have increased, two major changes to the carbon cycle have occurred:

- Natural vegetation, usually dominated by tree growth, has been replaced by land uses, such as urban and agricultural systems that have reduced capacity to uptake and store carbon.
- For the last 200 years or so, human activity, particularly industrial activity, has mobilized large amounts of fossil fuels from the lithospheric component of the cycle to the atmospheric component. Estimates suggest that this mobilization represents the release of one

million years of photosynthetic activity every year. In 2002, over 6.44 billion tons of carbon were released into the atmosphere by human activities, an increase of 1 per cent over the previous year. The current atmospheric concentration is 372.9 parts per million (ppm), an increase of 18 per cent since 1960 and 31 per cent since the onset of the Industrial Revolution in 1750. These concentration levels have been experienced before, but not for at least 420,000 years and probably not for 20 million years (Chapter 7).

Coral reefs, such as here in the Maldives, store large amounts of carbon from the remains of thousands of years of coral growth (*Philip Dearden*).

that sustains terrestrial life (Table 4.3). Water travels ceaselessly between these various reservoirs through the main processes of evaporation and precipitation known as the **hydrological cycle** (Figure 4.6).

Average residence times in each reservoir vary greatly (Table 4.3). In the deep ocean, it may take 37,000 years before water is recycled through evaporation into the atmosphere, whereas once in the atmosphere, average residence time is in the order of 9–12 days. These figures have special relevance in regard to the effects of pollution. Although many major rivers have suffered from critical pollution incidents, the flushing action of rivers, combined with the short residence time of the water, means that a relatively rapid recovery is often possible. This is not the case, however, with

groundwater pollution, especially deep groundwater pollution.

The hydrological cycle involves the transport of water from the oceans to the atmosphere, through terrestrial and subterranean systems, and back to the ocean, all fuelled by energy from the sun. Eighty-four per cent of the water in the atmosphere is evaporated directly from the ocean surface. The remainder comes from evaporation from smaller water bodies, from the leaves of plants (**transpiration**), or from the soil and plants (**evapotranspiration**). As it evaporates, water leaves behind accumulated impurities. The most common dissolved substance in the ocean is sodium chloride, or table salt, which also contains many other elements in trace amounts. Evaporation acts as a giant purification plant until further pollutants are encountered in the atmosphere.

Once in the atmosphere the water vapour cools, condenses around tiny particles called condensation nuclei, forms clouds, and is precipitated to earth as either rain, snow, or hail. The warmer the air is, the more water it can hold. Moisture content can be expressed in terms of relative humidity, the amount of moisture held compared with how much could be held if fully saturated at a particular temperature. At a relative humidity of 100 per cent, the air is saturated and cloud, fog, and mist form. Clouds are moved around by winds and continue to grow until precipitation (Box 4.7) occurs and the water is returned to earth.

About 77 per cent of precipitation falls into the ocean. The remainder joins the terrestrial part of the cycle in ice caps, lakes, rivers, groundwater,

Table 4.3	Global Water Storage	
Reservoir	**Average Renewal Rate**	**Per cent of Global Total**
World oceans	3,100 years	97.2
Ice sheets and glaciers	16,000 years	2.15
Groundwater	300–4,600 years	0.62
Lakes (fresh water)	10–100 years	0.009
Inland seas, saline lakes	10–100 years	0.008
Soil moisture	280 days	0.005
Atmosphere	9–12 days	0.001
Rivers and streams	12–20 days	0.0001

BOX 4.7 PRECIPITATION

Precipitation occurs in several forms: rain, snow, hail, dew, fog, and rime ice (frost). It occurs when the accumulated particles of condensed water or ice in clouds become large enough that they overcome the uplifting air currents and fall to earth as a result of gravity. Some of this precipitation may never reach the ground. Lower air layers may be warmer and drier and re-evaporation may occur as the precipitates pass through these layers, giving an excellent example of the speed of some of these mini-cycles that occur as parts of the larger Earth cycles. Distribution of precipitation is one of the main factors influencing the nature and location of global biomes and the ecozones of Canada. In Canada, precipitation varies from almost nothing in the Arctic to over 3,000 mm on the west coast (Figure 4.7).

Differences in precipitation occur for various reasons. At the global scale, heating of equatorial regions causes air to rise. As it rises it cools and condenses, clouds form, and precipitation occurs. As a result, equatorial regions tend to have consistent high rainfall. Where this air falls as it cools, over subequatorial regions, it tends to be dry, such as in the Sahara desert.

In Canada, much of the precipitation comes from low pressure systems, large cells of rising air that form along the boundary between warm and cold air masses. The main factor influencing relative precipitation levels in Canada is where moisture-laden winds cross oceans and are forced to rise as a result of mountain barriers. The most extreme example is where westerlies coming across the Pacific meet the western Cordillera and create the highest precipitation levels in the country. As the air warms up in its descent from the mountains it can hold more moisture, and precipitation levels fall considerably to produce a **rainshadow effect** that accounts for the small amounts of precipitation across the Prairies, with as little annual precipitation as 300–400 mm. Precipitation levels rise again in central Canada due to disturbances bringing moisture from the Gulf of Mexico and Atlantic Ocean. In southern Ontario, precipitation increases to 800 mm and over 1,000 mm in the lee of the Great Lakes, a major source of moisture for downwind localities. In Atlantic Canada, exposure to maritime influences increases once more, with up to 1,500 mm of precipitation falling annually on the south coast of Newfoundland.

The Arctic is very dry because the prevailing winds from the north are very cold and therefore have little capacity for carrying moisture, they pass over terrain that has relatively few sources of water evaporation (sources often remain frozen for a good proportion of the year), and there is an absence of low-pressure systems in winter. Topography is also an important determinant of whether the precipitation falls as snow or rain.

More localized precipitation may be produced by convection as warmer and lighter air rises and cooler, heavier air sinks. This mechanism is important for localized storms in summer when the air is heated by the warm ground and may send columns of moist, warm air to great elevations, resulting in thunderstorm activity.

Desert-like conditions occur in some areas of Canada largely due to rainshadow effects. Although we associate the Fraser River with the high rainfalls of the west coast, the interior of British Columbia is quite dry (*Philip Dearden*).

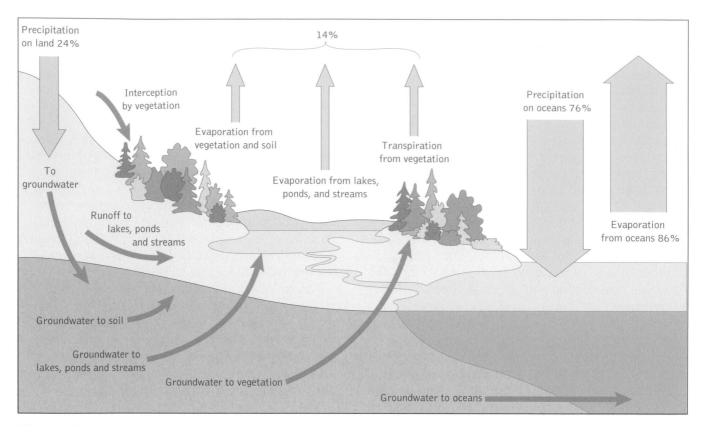

Precipitation
on land 24%

14%

Interception
by vegetation

Evaporation from
vegetation and soil

Transpiration
from vegetation

Precipitation
on oceans 76%

To
groundwater

Evaporation from lakes,
ponds, and streams

Evaporation
from oceans 86%

Runoff to
lakes, ponds
and streams

Groundwater to soil

Groundwater to
lakes, ponds and streams

Groundwater to vegetation

Groundwater to oceans

Figure 4.6 The hydrological cycle.

and transport between these compartments. Gravity moves water down through the soil until it reaches the water table, where all the spaces between the soil particles are full of water. This is the **groundwater** (Box 4.8). Lakes, streams, and other evidence of surface water occur where the land surface is below the water table. Surface water is a major factor in sculpting the shape of the surface of the Earth. At greater depth, the groundwater may penetrate to occupy various geological formations, known as aquifers.

As mentioned in Box 4.9, water is unique in that at the ambient temperatures and pressure of the Earth's surface it is the only substance that exists in all three phases of matter (solid, liquid, vapour). Water is stored in all three forms within the hydrological cycle and moves among these forms by the processes shown in Figure 4.8. *Sublimation* is the process for direct transfer between the solid and vapour phases of matter, regardless of direction. This explains why on bright sunny winter days when the air is dry, snowbanks may

Water in the solid phase of ice may be in these large glaciers of the St Elias range in western Yukon for thousands of years before melting and flowing to the ocean (*Philip Dearden*).

Relative humidity is high most of the year in Atlantic Canada and produces some beautiful atmospheric effects (*Philip Dearden*).

BOX 4.8 GROUNDWATER

Groundwater is found within spaces between soil and rock particles and in crevices and cracks in the rocks below the surface of the Earth. Above the water table is the *unsaturated zone* where the spaces contain both water and air. In this zone water is called *soil moisture*. Groundwater moves the same as surface waters, downhill, but rarely as quickly, and not at all through impermeable materials such as clay. Permeable materials allow the passage of water, usually through cracks and spaces between particles. An **aquifer** is a formation of permeable rocks or loose materials that contains usable sources of groundwater. They vary greatly in size and composition. Porous media aquifers consist of materials such as sand and gravel through which the water moves through the spaces between particles. Fractured aquifers occur where the water moves through joints and cracks in solid rock. If an aquifer lies between layers of impermeable material, it is called a confined aquifer, which

may be punctured by an *artesian* well, releasing the pressurized water to the surface. If the pressure is sufficient to bring water to the surface the well is known as a flowing artesian well.

Areas where water enters aquifers are known as *recharge* areas; *discharge* areas are where the water once more appears above ground. Such discharge areas can contribute significantly to surface water flow, especially in periods of low precipitation. Groundwater, of course, is a very significant part of the Canadian water supply. Dependencies range from 100 per cent in Prince Edward Island down to 17 per cent in Quebec, with Ontario having the largest total consumption.

An example of the interaction between groundwater and biogeochemical cycle disruption is provided by contamination of the groundwater of the Abbotsford Aquifer in the Lower Mainland of BC. Agriculture is the main land use. Changes in agricultural products over the last 25 years have led to high concentrations of nitrogen.

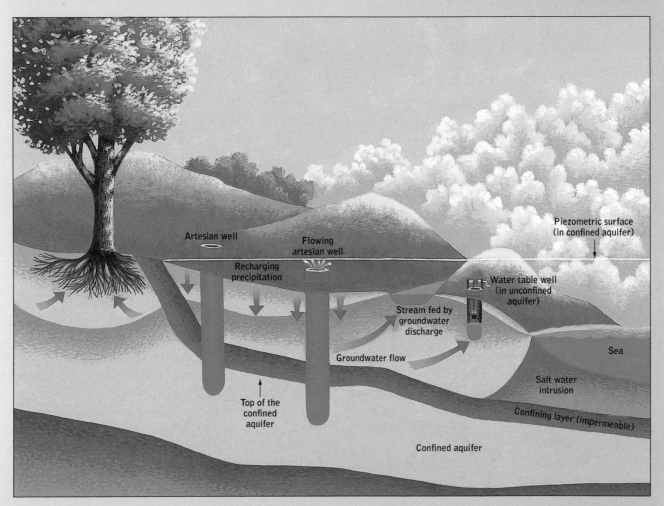

Groundwater flow.

These changes include a shift in animal production from locally fed dairy/beef cattle to poultry production based on outside feed. Crops have changed from grass and hay to raspberry production, which requires less nitrogen. Poultry manure is used to fertilize the raspberries. Nitrogen use is less than one-half of that produced. The remainder infiltrates to the groundwater. Of 2,297 groundwater samples taken, 71 per cent have exceeded the Guideline for Canadian Drinking Water Quality (10mg/L), with some as high as 91.9 mg/L. This is cause for concern. These nitrates may be transformed to nitrites in the digestive systems of babies, which in turn may lead to an oxygen deficiency in the blood, known as methemoglobinemia, or blue-baby syndrome. High nitrate levels have also been linked to cancer. More information on this can be found at Environment Canada's 'ecoinfo' Web site, listed at the end of this chapter.

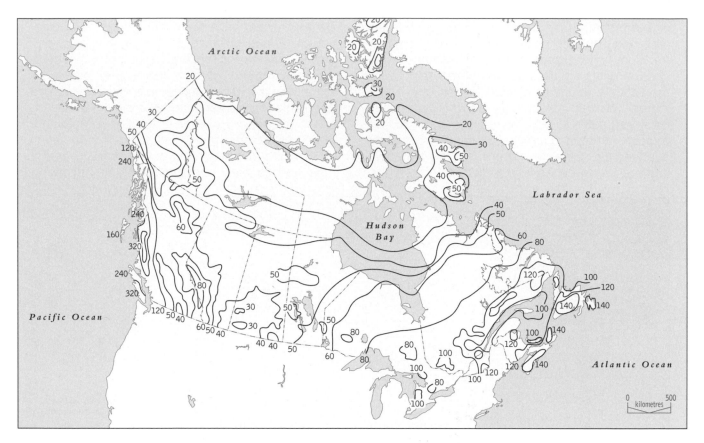

Figure 4.7 Average annual rain and snow for Canada (cm). SOURCE: After D. Phillips, *The Climate of Canada* (Ottawa: Minister of Supply and Services Canada, 1990), 210.

lose size without any visible melting. About 75 per cent of the world's fresh water is stored in the solid phase, and it may stay in this phase for a long time. Measurements in the Antarctic, for example, indicate that some of the ice is over 100,000 years old.

Although over the short term there are relatively constant amounts of water in the different storage compartments, over the long term these can change markedly. For example, large amounts of water are evaporated from the oceans and precipitated on land as snow during glacial periods. Over time, the snow accumulates and builds ice fields that may be over a kilometre thick. This effectively removes water from the oceanic component, causing the sea level to fall. During glacial times, then, the area of land will increase relative to the ocean, and the area covered by ice may increase by up to 300 per cent.

In Canada the solid phase of water is particularly important. Canada is estimated to have perhaps up to one-third of the world's fresh water, but most of this is held in a solid state. Compared to the surface supply in Canada, the country's 100,000 glaciers contain over 1.5 times the volume of fresh water. Furthermore, over 95 per cent of the country is snow-covered for part of the

BOX 4.9 SOME IMPORTANT PROPERTIES OF WATER

Water has several properties that make it unique:

- Water is a molecule (H_2O) that can exist in a liquid, gaseous, or solid state.
- These molecules have a strong mutual attraction, promoting high surface tension and high capacity to adhere to other surfaces; these properties allow water to move upward through plants.
- Water has a high heat capacity, meaning that it can store a great deal of heat without an equivalent rise in temperature; this is the reason why the oceans have such a moderating influence on climate.
- It takes a lot of heat to change water from liquid to gaseous form; this is why evaporation results in a cooling effect.
- Few solids do not undergo some dissolution in water; this allows water to carry dissolved

nutrients to plants, but it also means that water is easily polluted.

- Unlike other substances, when water passes from a liquid to a solid state it becomes less rather than more dense; this is why ice floats on top of water and permits aquatic life to exist in cold climates.

Most of these properties spring from the fact that although the water molecule is electrically neutral, the charges are distributed in a bipolar manner. In other words, a positive charge is at one end of the molecule and a negative charge is at the other. This means that water molecules have a strong attraction for each other, and also explains why water is such a good solvent as the charges increase the chemical reactivity of other substances.

winter. Spring melt is hence a critical part of the hydrological cycle in Canada as water moves from the solid to liquid phase. This creates a runoff regime for many Canadian rivers, characterized by low late winter flows and a high spring melt flow that slowly diminishes over the summer into the winter lows, as water becomes stored in the solid phase once more. This marked seasonality is one of the reasons why Canada has developed considerable expertise in the construction of water storage facilities. At the same time, human impacts on the hydrological cycle are significant (Box 4.10).

Canada also has abundant storage of fresh water in lake systems, covering almost 8 per cent of the area of the country. These lakes are replen-

ished by river flow that contains approximately 9 per cent of the total river discharge in the world (Table 4.4). About 60 per cent of the discharge in Canada drains north to the Arctic Ocean (Figure 4.9), whereas 90 per cent of the Canadian population lives within 300 km of the US border, creating the potential for water deficits in this water-rich country (see Chapter 12).

As water demands grow, we have increasingly become dependent on groundwater sources. Once

BOX 4.10 HUMAN IMPACTS ON THE HYDROLOGICAL CYCLE

Human activities have also impacted the hydrological cycle. Changes include:

- the storage and redistribution of runoff to augment water supplies for domestic, agricultural, and industrial uses;
- the building of storage structures to control floods;
- the drainage of wetlands;
- the pumping of groundwater;
- cloud seeding;
- land-use changes such as deforestation, urbanization, and agriculture that affect runoff and evapotranspiration patterns;
- climatic change caused by interference with biogeochemical cycles.

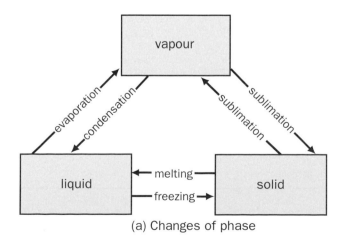

(a) Changes of phase

Figure 4.8 Changes of phase in the hydrological cycle,

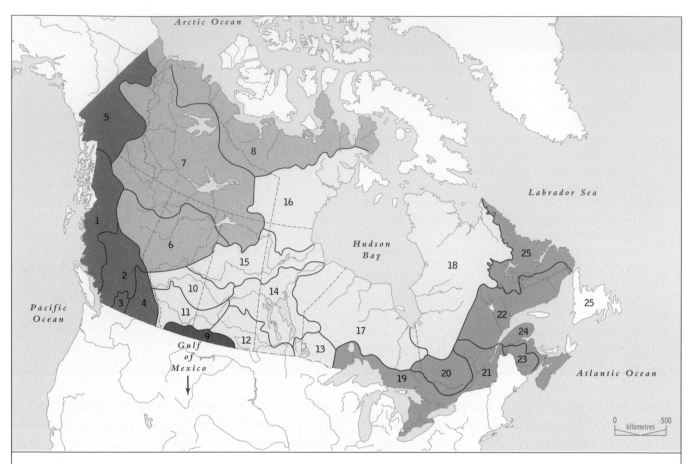

Ocean Basin Region		River Basin Region	Area in 000s km²
Pacific	1	Pacific Coastal	352
	2	Fraser-Lower Mainland	234
	3	Okanagan-Similkameen[a]	14
	4	Columbia[a]	90
	5	Yukon	328
Arctic	6	Peace-Athabasca	487
	7	Lower Mackenzie	1300
	8	Arctic Coast-Islands	2025
Gulf of Mexico	9	Missouri[a]	26
Hudson Bay	10	North Saskatchewan	146
	11	South Saskatchewan[a]	170
	12	Assiniboine-Red[a]	190
	13	Winnipeg[a]	107
	14	Lower Saskatchewan-Nelson	363
	15	Churchill	298
	16	Keewatin	689
	17	Northern Ontario	694
	18	Northern Quebec	950
Atlantic	19	Great Lakes[a]	319
	20	Ottawa	146
	21	St Lawrence[a]	116
	22	North Shore-Gaspé	403
	23	St John-St Croix[a]	37
	24	Maritime Coastal	114
	25	Newfoundland-Labrador	376
CANADA			9974

Figure 4.9 Drainage regions of Canada. [a]Canadian portion only; area on US side of international basin regions excluded from total. SOURCE: Environment Canada, *Currents of Change: Final Report, Inquiry on Federal Water Policy* (Ottawa: Environment Canada, 1985), 35. Reproduced with the permission of the Minister of Public Works and Government Services Canada, 1997.

Table 4.4	Mean Annual Stream Discharge to the Oceans for Selected Canadian Rivers	
River	**Area (km^2)**	**Discharge (m^{3}s^{-1})**
Saguenay	90,100	1,820
St Lawrence	1,026,000	9,860
Churchill	281,300	1,200
Nelson	722,600	2,370
Albany	133,900	1,400
Koksoak	133,400	2,550
Yukon (at Alaska border)	297,300	2,320
Fraser	219,600	3,540
Columbia (at Washington border)	154,600	2,800
Mackenzie	984,195	10,800

SOURCE: D. Briggs et al., *Fundamentals of Physical Geography*, 2nd Canadian edn (Toronto: Copp Clark Pitman, 1993), 206.

they become polluted with agricultural biocides or industrial wastes, however, they may be unsuitable for human use for centuries. The importance of this is underscored by the fact that Canada is estimated to have 37 times the amount of water in underground sources as in surface sources, and one-quarter of the Canadian population relies on groundwater for domestic use, while some communities, such as Fredericton, are almost totally dependent on groundwater.

BIOGEOCHEMICAL CYCLES AND HUMAN ACTIVITY

Despite the apparent sophistication of human society, the humble fact remains that society could not exist without biogeochemical cycles and those unpretentious bacteria that make them work. Yet all the cycles are susceptible to perturbations by human activity. Such is the scale of human actions that the major transfers taking place between some of the reservoirs in the cycles are human-induced. Some of the most notable and difficult environmental challenges now faced by society spring from these transfers. The purpose of this section is to discuss two of these in more detail: eutrophication and acid deposition. A third example, global climatic change—largely resulting from disruptions in the carbon and nitrogen cycles—is of such significance that Chapter 7 is devoted to it.

Eutrophication

Eutrophication is a natural process of nutrient enrichment of water bodies that leads to greater productivity. Phosphorus and nitrogen are often the two main limiting factors for plant growth in aquatic ecosystems. Systems with relatively low nutrient levels, **oligotrophic** ecosystems, have quite different characteristics from those with high levels (**eutrophic**), as summarized in Table 4.5. Natural terrestrial ecosystems are relatively efficient in terms of holding nutrient capital. The progression from an oligotrophic to eutrophic condition, through the process of succession discussed in Chapter 3, may take place over thousands of years. This rate is influenced by the geological makeup of the catchment area and the depth of the receiving waters. Catchments with fertile soils will progress more quickly than those with soils lacking in nutrients. Depth is important because shallower lakes tend to recycle nutrients more efficiently.

How Is Eutrophication Caused?

Cultural eutrophication (eutrophication caused by human activity) may speed up the natural eutrophication process by several decades, mainly through the addition of phosphates and nitrates to the water body. As lakes become shallower as a result of this input, nutrients are used more efficiently, productivity increases, and eutrophication progresses.

Table 4.5 — Characteristics of Oligotrophic and Eutrophic Water Bodies

Characteristic	Oligotrophic	Eutrophic
Nutrient cycling	low	High
Productivity (total biomass)	low	High
Species diversity	high*	Low
Relative numbers of undesirable species	low	High
Water quality	high	Low

*Lakes that are extremely non-productive (e.g., high mountain lakes) will have low species diversity.

This is a classic example of a positive feedback loop as change in the system promotes even more change in the same direction. Additional phosphates and nitrates come from many different sources (Table 4.6), and in accord with the law of conservation of matter discussed earlier, they do not simply disappear but accumulate in aquatic ecosystems.

What Are the Effects?

This enrichment promotes increased growth of aquatic plants, particularly favouring the growth of floating phytoplankton over **benthic** plants rooted in the substrate. As the benthic plants become out-competed for light by the phytoplankton, they produce less oxygen at depth. Oxygen is critical for the maintenance of more diverse, oxygen-demanding fish species such as trout and other members of the salmonid family, which also start to decline in

Animal feedlots are a major source of nutrients, such as phosphates and nitrates, which speed up eutrophication (*Philip Dearden*).

number. The oxygen produced by photosynthesis by the phytoplankton tends to stay in the shallower water, escaping back to the atmosphere rather than replenishing supplies at greater depths.

Oxygen depletion is further exacerbated by the decay of the large mass of phytoplankton produced. Dead matter filters to the bottom of the lake where it is consumed by oxygen-demanding decomposers. Once broken down, nutrients may be returned to the surface through convection currents and provide more food for more phytoplankton and algae. Blue-green algae replace green algae in eutrophic lakes, which further exacerbates the problem, as most blue-green algae are not consumed by the next trophic level, the zooplankton.

These effects of oxygen depletion in a water body also occur whenever excess organic matter is added. Under natural conditions, water is able to absorb and break down small amounts of organic matter, with the amount depending on the size, flow, and temperature of the receiving water body. The greater the size and flow are and the lower the temperature, the greater the ability to absorb organic materials and retain oxygen levels.

When organic wastes are added to a body of water, the oxygen levels fall as the number of bacteria rises to help break down the waste. This is known as the **oxygen sag curve** (Figure 4.10), and is measured by the **biological oxygen demand (BOD)**, the amount of dissolved oxygen needed by aerobic decomposers to break down the organic material in a given volume of water at a certain temperature over a given period. At the discharge source, the oxygen sag curve starts to fall and there is a corresponding rise in the BOD. As distance from the input source increases and the bacteria

Table 4.6	Main Nutrient Sources Contributing to Cultural Eutrophication
Run-off from	fertilizers (N and P)
	feedlots (N and P)
	land-use change, such as cultivation, construction, mining natural sources
Discharge of	detergents (P)
	untreated sewage (N and P)
	primary and secondary treated sewage (N and P)
Dissolved nitrogen oxides	(from internal combustion engines)

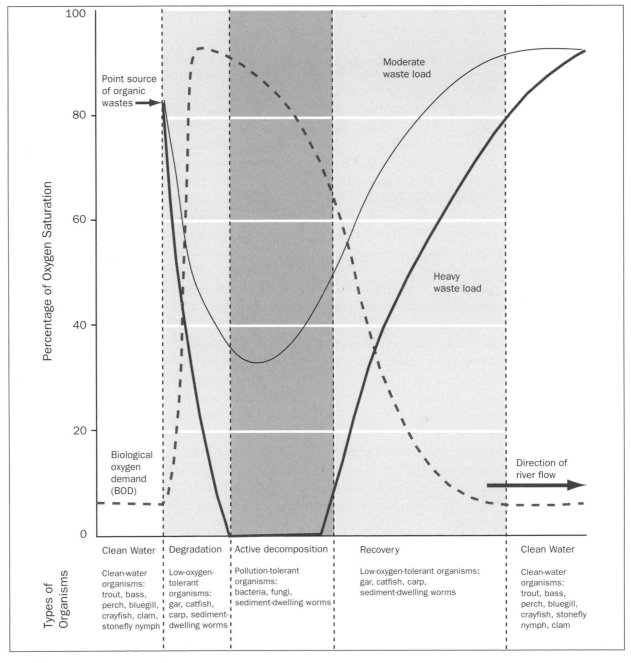

Figure 4.10 Oxygen sag curve and biological oxygen demand (BOD).

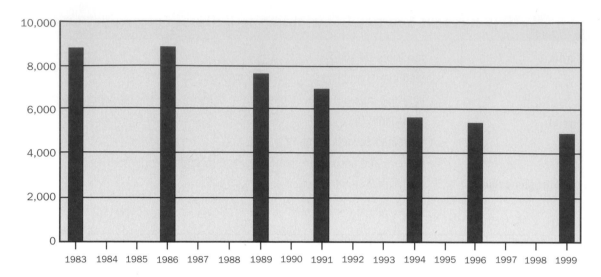

Figure 4.11 Phosphorus loadings in municipal wastewater effluents, 1983–1999. SOURCE: Environment Canada, 2003b. Reproduced with the permission of the Minister of Public Works and Government Services, 2005.

digest the wastes, then the oxygen content returns to normal and the BOD falls.

Major sources of nitrates and phosphates, such as runoff from feedlots and sewage discharge, also contain large amounts of oxygen-demanding wastes. Heat is another source of oxygen stress. The overall result is a progression to a less useful and less healthy water body. The composition of the fish species changes to those less dependent on high oxygen levels, species that are generally less desirable for human purposes. Populations of waterfowl may fall as aquatic plants die off. The water becomes infested with algae, aquatic weeds, and phytoplankton, making swimming and boating unpleasant and giving off unpleasant odours. Water treatment for domestic or industrial purposes becomes more expensive.

What Can We Do about It?

The main way to control eutrophication is through limiting the input of nutrients into the water body (Table 4.6). Domestic and animal wastes must be treated to remove phosphates (Figure 4.11). Advanced treatment can remove up to 90 per cent of these wastes. More difficult problems occur with diffuse, **non-point sources** such as runoff from urban areas and agricultural land. Such flows really have to be controlled at the source since they enter the water body, by definition, in so many different locations. In the past, measures to control water pollution have been largely directed towards **point sources** of pollution, or single discharge points, such as effluent discharges from sewage plants or indus-

trial processes. By and large, due to the high visibility of such sources, they are easy to identify and monitor, and pollution from such sources has fallen as a result. Increasing attention is now being directed towards the non-point sources (see Chapter 12).

Eutrophication used to be considered a problem of smaller water bodies, but now entire areas of the world's largest water bodies, the oceans, are becoming so eutrophic they are being described as 'dead zones'. Over 150 such zones have been recognized by the United Nations (http://www.unep.org/geo/yearbook/), some as large as 70,000 km² in size. One of the largest and best-known areas occurs in the Gulf of Mexico, which receives all the excess fertilizers brought down from the Mississippi watershed. However, similar areas are now found on every continent and researchers expect that their numbers and size will increase as global climatic change generates greater rainfall and greater runoff in many areas. The next section explores one of Canada's most notable eutrophication challenges, Lake Erie, and points to some of the ways to address the problem.

Lake Erie: An Example of Eutrophication Control

In all parts of the country, there are many examples of eutrophic lakes. Lake Erie is a particularly well-known case due to its size and importance. Lake Erie is the second smallest and also the shallowest of the Great Lakes, which contain almost 20 per cent of the world's fresh water. Erie has experienced considerable changes in fish species composition since the early explorers described a highly diverse community. Gone are the lake sturgeon, cisco, blue pike, and lake whitefish as the human population in the basin has increased and water quality has declined. There are some 11 million people in the Erie drainage basin, and 39 per cent of the Canadian and 44 per cent of the American shore is taken up by urban development. There is also considerable industrial use around the shore and intensive agricultural use throughout the basin.

In the past, up to 90 per cent of the bottom layer of the central zone of the lake became oxygen deficient in the summer. Huge algae mats over 20 metres in length and a metre deep became common. Beaches were closed. The natural eutrophication that might have taken thousands of years was superseded by cultural eutrophication in the space of 50 years.

In 1972 the Great Lakes Water Quality Agreement was signed by Canada and the US to try to

come to terms with this problem. The signing of the 1985 Great Lakes Charter, in which the two countries agreed to take a co-operative and ecosystem-based approach to the lakes, further strengthened international efforts. Since the 1970s, phosphorus controls, implemented under the Canada Water Act, have led to significant reductions in the phosphorus concentration of the water (Figure 4.12). Phosphate-based detergents were banned and municipal waste treatment plants upgraded.

These measures have led to improvement of water quality, but significant problems still remain. The controls are largely on point-source pollution, discharges that have a readily identifiable source, such as waste treatment plants and industrial complexes. However, much of the remaining nutrient load comes from non-point sources, such as runoff from agricultural fields, lawn fertilizer, and construction sites that are much more difficult to regulate. Figure 4.12 shows that phosphorus levels are still in excess of the guidelines established in the Great Lakes Water Quality Agreement and levels are likely to increase with future growth in population. Furthermore, the levels of nitrite and nitrate concentrations have been rising, leading to increased eutrophication due to other nutrients.

Acid Deposition

In 1966, fisheries researcher Harold Harvey was puzzled to find that the 4,000 pink salmon he had introduced to Lumsden Lake in the La Cloche Mountains southwest of Sudbury, Ontario, the previous year had all disappeared. Their passage upstream and downstream of the lake had been blocked. To unravel the mystery he began to take more measurements of the lake and look into its past history. The results were startling—not only had the salmon disappeared; many other species of fish indigenous to the lake had gone missing as well (Table 4.7). The reason soon became apparent. Between 1961 and 1971, Lumsden Lake had experienced a hundredfold increase in the acidity of its waters, as had many other lakes in the same region (Table 4.8). The changes had shifted the

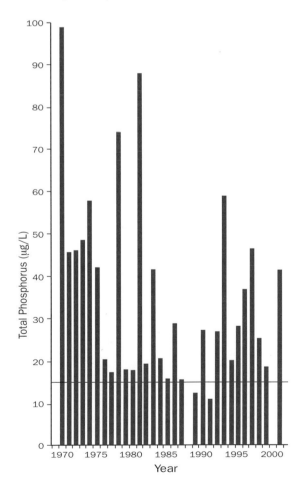

Figure 4.12 Phosphorus concentrations in western basin of Lake Erie. *Note:* Blank indicates no sampling. Horizontal line on graph represents the phosphorus guideline as listed in the Great Lakes Water Quality Agreement. SOURCE: Environment Canada and US Protection Agency (2003: 66).

Although significant reductions in nutrient loadings have been achieved, phosphorus concentrations in Lake Erie appear to be increasing again.

— Environment Canada and US Environmental Protection Agency (2003: 17)

Table 4.7	Disappearance of Fish from Lumsden Lake
1950s	Eight species present
1960	Last report of yellow perch
1960	Last report of burbot
1960–5	Sport fishery fails
1967	Last capture of lake trout
1967	Last capture of slimy sculpin
1968	White sucker suddenly rare
1969	Last capture of trout-perch
1969	Last capture of lake herring
1969	Last capture of white sucker
1970	One fish species present
1971	Lake chub very rare

SOURCE: H. Harvey, unpublished speech, based on R. Beamish and H. Harvey, 'Acidification of the La Cloche Mountain Lakes, Ontario, and Resulting Fish Mortalities', *Journal of the Fisheries Research Board of Canada* 29, 8 (1972): 1135. Reprinted with the author's permission.

lakes outside the limits of tolerance of the species, as discussed in Chapter 2. The indigenous fish species, and many of the species upon which they depended for food, simply could not tolerate the new conditions and perished. They were victims of the effects of **acid deposition**.

What Is Acid Deposition?

Acids are chemicals that release hydrogen ions (H+) when dissolved in water, whereas a *base* is a chemical that releases hydroxyl ions (OH–). When in contact, acids and bases neutralize each other as they come together to form water (H_2O). Acidity is

Table 4.8	Lake Acidification in the La Cloche Mountains, 1961–71	
Lake	pH 1961	pH 1971
Broker	6.8	4.7
David	5.2	4.3
George	6.5	4.7
Johnnie	6.8	4.8
Lumsden	6.8	4.4
Mahzenazing	6.8	5.3
O.A.S.	5.5	4.3
Spoon	6.8	5.6
Sunfish	6.8	5.6
Grey (1959)	5.6	4.1
Tyson (1955)	7.4	4.9

SOURCE: H. Harvey, unpublished speech, based on R. Beamish and H. Harvey, 'Acidification of the La Cloche Mountain Lakes, Ontario, and Resulting Fish Mortalities', *Journal of the Fisheries Research Board of Canada* 29, 8 (1972): 1135. Reprinted with the author's permission.

a measure of the concentration of hydrogen ions in a solution, and is measured on a scale, known as the pH scale, that goes from 0 to 14 (Figure 4.13). The midpoint of the scale, pH 7, represents a neutral balance between the presence of acidic hydrogen ions and basic hydroxyl ions. The pH scale is logarithmic. A decrease in value from pH 6 to pH 5 means that the solution has become 10 times more acidic. If the number drops to pH 4 from pH 6, then the solution is 100 times more acidic.

Precipitation, either as snow or rain, tends to be slightly acidic, even without human interference, due to the chemical reaction as carbon dioxide in the atmosphere combines with water to form carbonic acid. Generally, a pH value of 5.6 is accorded to 'clean' rain. Acid rain is defined as deposition that is more acidic than this. In Canada, rainfall has been recorded with pH levels much lower than this. Acidic deposition is a more generic term that includes not only rainfall, but also snow, fog, and dry deposition from dust.

How Is Acid Deposition Caused?

The increases in acidity reflected in the pH levels of the lakes in Table 4.8 are due to human interference in the sulphur and nitrogen cycles. The largest sources are through the smelting of sulphur-rich metal ores and the burning of fossil fuels for energy. These processes change the distribution of the elements between the various sources shown in Figures 4.3 and 4.4, and consequently natural processes are inadequate to deal with the buildup of matter. Increased amounts of sulphur and various forms of nitrogen are ejected into the atmosphere, where they may travel thousands of kilometres before being returned to the lithosphere as a result of depositional processes. Human activities account for more than 90 per cent of the sulphur dioxide and nitrogen oxide emissions in North America.

Excessive sulphur is produced when ore bodies, such as copper and nickel, are roasted (smelted) at high temperatures to release the metal. Unfortunately, such ores often contain more sulphur than metal, and the sulphur is released into the atmosphere as a waste product of the process. A similar effect is created when sulphur-containing coal is burned as the energy source in power plants. In Canada, the smelting of metal ores accounts for half of the SO_2 emissions east of Saskatchewan, with power generation and other sources accounting for 20 and 30 per cent respectively. In the US, electrical utilities are the largest source, accounting for 67 per cent of emissions. The effects of these emissions became obvious fairly early on around smelting plants such as those at Inco in Sudbury and in Trail, BC. Trees were destroyed over large areas. Now, there are very encouraging signs of rehabilitation, as discussed in Chapter 13.

These obvious signs of ecological damage were ignored for many years. However, as the ecological implications became better known, governments responded by encouraging industry to build higher stacks to eject the waste further into the atmosphere. Inco, for example, built a 381-metre superstack in 1972. This improved matters somewhat at the local scale but merely served to create problems elsewhere, especially in Quebec, as entire air masses became acidified and dropped their acid burdens over a larger area. Weather patterns are not

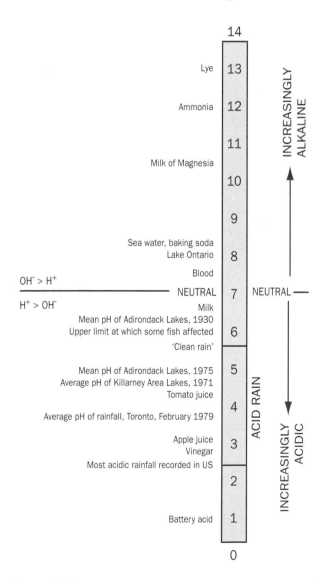

Figure 4.13 The acid (pH) scale.

random, and so acidified air masses tend to travel in the same kinds of patterns. In central and eastern Canada, as air masses travel from southwest to northeast, they bring a heavy pollution burden from the heavily industrialized Ohio Valley in the US, which falls mostly in Canada. It is estimated that approximately half of the sulphate falling in Canada originates in the US.

These point sources of pollution are, however, easier to monitor and control than the other main source of acids—nitrogen emissions—where some 35 per cent comes from various means of transport as a result of high-temperature combustion. The remainder is split between emissions from thermo-electric generating stations and other industrial, commercial, and residential combustion processes.

What Are the Effects of Acid Deposition?

Aquatic Effects

The effects of acid deposition on the fish of Lumsden Lake and other aquatic ecosystems are one visible sign of some of the impacts of acid deposition. Other species are also affected as the pH of the water body declines. Indeed, as can be seen in Figure 4.14, fish are often not the most sensitive species, and are really more of an indicator of the damage that has already occurred. As insects such as mayflies become eliminated, species higher in the food chain that feed on them become affected through food depletion. The same is true of fish-eating birds, such as loons, whose young have been shown to have a lower chance of survival on acidified lakes due to starvation.

Unfortunately, some of these impacts are permanent. Examination of historical angling records in Nova Scotia, for example, indicate that of 60 main salmon rivers, 13 runs of salmon are extinct and a further 18 are virtually extinct. It is estimated that over half of the total salmon stock has been lost as a result of declining pH levels. It has also been suggested that the endangered Acadian whitefish is threatened by declining pH levels.

Even where fish themselves manage to survive, they may be grossly disfigured with twisted backbones and flattened heads because their bones have been deprived of the necessary nutrients for strength. Reproductive capacities may be sufficiently impaired to lead to eventual population declines. Generally, the time of reproduction is the most sensitive part of the life cycle. The critical factor is often the lower pH level of the water as the snow

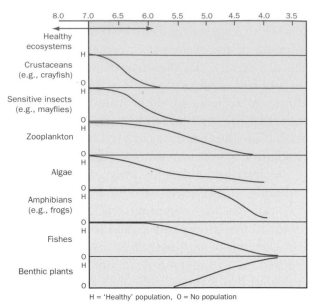

Figure 4.14 Sensitivity of various aquatic organisms to pH level. SOURCE: Environment Canada, *The State of Canada's Environment* (Ottawa: Minister of Supply and Services Canada, 1991).

melts. At this time the buildup of acids over the winter can result in even higher acidity than experienced through the rest of the year. This pulse of acidity is called the **acid shock** and it may also be one of the causes of stress on amphibious creatures, such as frogs, that often use small temporary pools of water for breeding in the spring following runoff. At Lumsden Lake, for example, spring runoff produced a pH as low as 3.3, over 100 times more acidic than the 1961 levels.

More acidic water and the subsequent food chain effects are not the only concern. Other chemical changes also occur. The increased acidity, for example, releases large amounts of aluminum from the terrestrial ecosystem into the rivers and lakes. Here it forms a toxic scum, lethal to many forms of aquatic life.

Terrestrial Effects

Terrestrial effects of acid deposition first became visible around emission sources, such as at Trail, BC, and Sudbury, Ontario, as trees began to die. Before joining the soil, the acids eat away at the sensitive photosynthetic surfaces of the leaves. Broadleaved trees, such as sugar maples, the source of Quebec's $40-million annual maple syrup industry, are thought to be particularly susceptible due to the large surface area of their leaves.

Once in the soil, the acids leach away the nutrients required for plant growth, leading to nutrient deficiencies. The high levels of aluminum released by the acids also help to inhibit the uptake of nutrients. The bacteria so critical to the workings of many biogeochemical cycles, as described earlier, are also adversely affected and cause changes in natural soil processes. Decomposition and humus formation are retarded. The actual physical contact of the acids with the plant roots can also inhibit growth and lower resistance to disease.

These impacts are now visible over much wider areas than those surrounding sources of high emissions. Extensive areas of damage have been recorded in Europe and in the eastern United States. At these larger scales, it is often difficult, however, to single out one cause, such as acid deposition. It is likely that other factors—climate change and other pollutants, such as high levels of ozone brought about by excessive nitrogen oxides—are also important in placing stress on these communities in synergistic reactions.

In central and eastern Canada, 89 per cent of the high-capability forest land receives in excess of 20 kg/ha of acid deposition annually, the figure originally defined as being an 'acceptable' level of deposition. Severe impacts have been noticed. A survey of white birch around the Bay of Fundy, for example, found 10 per cent of the trees already dead and virtually all others showing signs of damage (RMCC, 1990).

The impacts on plant life are not restricted to natural ecosystems. Significant changes can also occur on agricultural lands as direct damage to crops or more indirect changes through changes to soil chemistry. The growth of crops such as beets, radishes, tomatoes, beans, and lettuce is inhibited, and biological nitrogen fixation is diminished at a pH of 4. In central and eastern Canada, 84 per cent of the most productive agricultural lands receive more than the 20 kg/ha. In some areas the application of lime to the soil has become a routine agricultural procedure in an attempt to neutralize the acids by adding more basic ions.

Heterotrophs can also be affected. As the forest cover diminishes, so does the habitat for many species. Toxic metals such as cadmium, zinc, and mercury, released from the soil by the acids, may be concentrated by certain species of plants and lichens and accumulated in the livers of species eating them, such as moose and caribou.

Sensitivity

Not all ecosystems are equally sensitive to the effects of acid deposition. Some areas have a high capacity to neutralize the excess acids due to the high base capacity of the bedrock and soils. For example, the Prairies are not as sensitive to the effects of acid deposition, as underlying carbonate-rich rocks, deep soils, and other factors combine to provide high *buffering* capacity. On the other hand, areas with difficult-to-weather rocks with low nutrient content (e.g., granite) and with thin soils following glaciation often have very low buffering capacities. Much of central and eastern Canada and coastal British Columbia fit under this latter case (Figure 4.15). The provinces most vulnerable to acid deposition in terms of amount of area ranked as highly sensitive are Quebec (82 per cent), Newfoundland (56 per cent), and Nova Scotia (45 per cent). The spatial coincidence of low buffering capacities and high deposition rates explains why most attention in Canada has centred on the central and eastern parts of the country.

Socio-economic Effects

The environmental effects described above have socio-economic implications, and there are also direct effects of acid deposition on human health. In aquatic ecosystems, for example, declines in fish populations obviously have implications for those involved in fishing, whether commercially or for sport. The impacts of acid deposition on tree growth and the forest industry are also of considerable concern. Some studies indicate that tree

The smelter at Trail caused extensive damage to the surrounding vegetation (*Philip Dearden*).

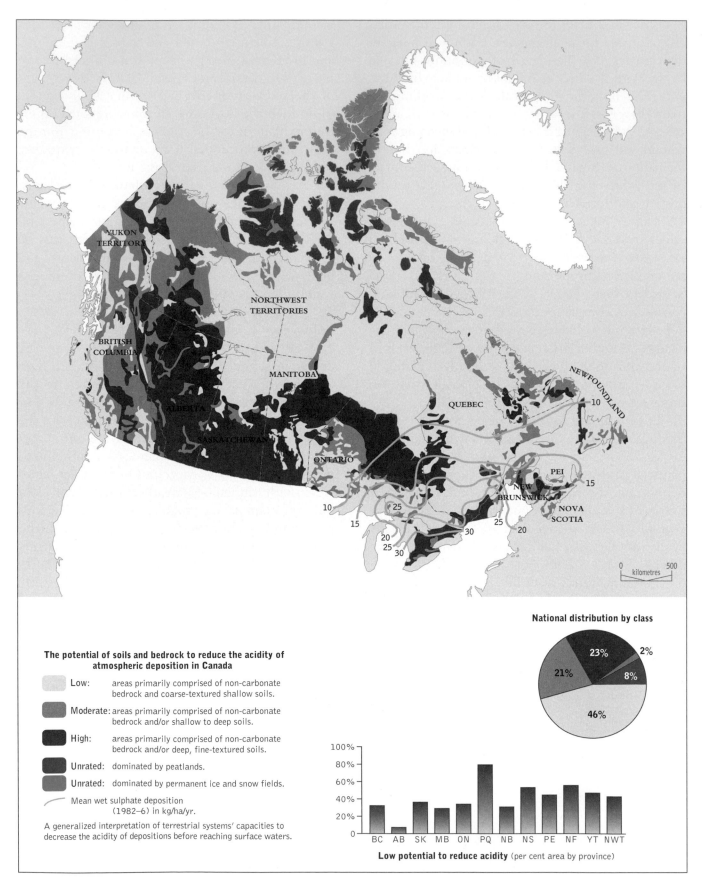

The potential of soils and bedrock to reduce the acidity of atmospheric deposition in Canada

Low: areas primarily comprised of non-carbonate bedrock and coarse-textured shallow soils.

Moderate: areas primarily comprised of non-carbonate bedrock and/or shallow to deep soils.

High: areas primarily comprised of non-carbonate bedrock and/or deep, fine-textured soils.

Unrated: dominated by peatlands.

Unrated: dominated by permanent ice and snow fields.

Mean wet sulphate deposition (1982–6) in kg/ha/yr.

A generalized interpretation of terrestrial systems' capacities to decrease the acidity of depositions before reaching surface waters.

National distribution by class

23%   2%
21%   8%
46%

Low potential to reduce acidity (per cent area by province)

Figure 4.15 The potential of soil and bedrock to reduce the acidity of atmospheric deposition in Canada. SOURCE: Environment Canada, 'Acid Rain National Sensitivity Assessment', Environment Fact Sheet 88–1 (Ottawa: Environment Canada, 1988).

growth reductions of up to 20 per cent might be experienced. The full effects, however, are likely to be much greater than this. European research indicates that a time lag of some 20 to 30 years is likely before the effects of acid rain are reflected in reduced tree growth. As outlined in Box 4.11, relatively little is known about many of the complex relationships in the soil community that may be affected by increased acidity.

Many values, however, are difficult to express in monetary terms. Thousands of Canadians, for example, maintain lakeside cottages. It is challenging, and some would say impossible, to ascribe a value to the changes that might occur as the lakes become devoid of life. Common loons, for example, are for many the quintessential symbol of the Canadian wilderness, yet populations have been found to be quite sensitive to lake acidification. Below pH levels of 4.5, loons do not seem able to find enough food to feed their young.

It is not only the natural environment that is damaged as a result of acid deposition, but also the human-built environment. The acids eat away at

Loons, with their beautiful colouration and haunting cries, are a wilderness symbol for many Canadians, but their breeding success has been reduced by the effects of acid rain (*Wayne Lankinen/Valan Photos*).

certain building materials. The effects have been most damaging in Europe where there are many old monuments, but the effects are also seen on the Houses of Parliament and nearby statues in Ottawa. It has been estimated that acid rain causes $1 billion worth of damage in Canada every year (Environment Canada, 2003a).

The most direct impacts on human health are thought to arise from inhalation of airborne acidified particles, which can impair respiratory processes and lead to lung damage. There appear to be significant relationships between air pollution levels in southwestern Ontario, for example, and hospital admissions for respiratory illnesses. Comparative studies between heavily polluted areas in Ontario and less polluted areas in Manitoba found diminished lung capacity in about 2 per cent of the children in the more polluted areas. Further studies have confirmed this relationship, although whether high sulphate or ozone levels are responsible has yet to be conclusively determined. Calculations by the US government suggest that southern Ontario will reap health benefits of over $1 billion per year by the year 2010 as a result of current efforts to reduce emissions in the US.

Humans can also be affected by ingestion of some of the products of acid rain. In drinking water, for example, in some areas where older delivery systems are in place, the increased acidity of water can corrode pipes and fittings and result

BOX 4.11 SOIL ORGANISMS AND ACID DEPOSITION

Soil contains many organisms, including mites, insects, worms, bacteria, fungi, and many others involved in the critical ecological functions of biogeochemical cycling and energy flow. For example, there may be over 100 million bacteria and several kilometres of fungal hyphae in a single gram of healthy soil. What happens to these organisms when they are subject to increased acidity is one of the key questions about the impacts of acid deposition.

Mycorrhizal fungi, for example, are very important to the growth of many plants because they help transport nutrients from the soil water into the roots. Browning and Hutchinson (1991) undertook research on Jack pine to try to understand the potential impacts of increased acidity on these mutualistic relationships. They found that changes in calcium-to-aluminum ratios caused by increased acidity influenced the succession of mycorrhizal fungi on tree root systems. The potential effects of these kinds of changes on tree growth, the ecological health of the community, and forest yields are still highly uncertain.

in elevated levels of lead in the water supply. This is one reason why in certain areas, such as Victoria, it is recommended that schools flush their water fountains early in the morning to spill the water that has been held overnight and that might show elevated lead levels.

Acidified water may also hold other substances in solution that are deleterious to human health. Excess levels of aluminum have already been noted. When metals such as mercury, chromium, and nickel are leached from the substrate into water, they may be taken up and concentrated along the food chain, and eventually cause a human health problem.

What Can We Do about It?

Had emission levels continued at 1980 levels, it was predicted that 600,000 lakes in Canada would eventually be virtually lifeless due to excess acidity. Thankfully, this has not happened (Figure 4.16) because of various measures to combat the problem (see Chapter 13). Critical early measures were to undertake more scientific research to ascertain the scale and dimensions of the problem, and then to raise public and political awareness to implement the necessary changes.

One of the main challenges associated with acid deposition is that the areas generating the emissions and causing the problem are not the sole recipients of the deposition. In Canada, estimates suggest that over half of the acid deposition originates from the US. Furthermore, the acidic fog that forms in the Canadian Arctic in spring is largely a result of emissions from Europe and Russia. To address such problems, international efforts are required, as is a greater concern on the part of individuals (see Box 4.12).

Canada has addressed the problem through national, bilateral, and multilateral efforts. In 1983, the Canadian Council of Resource and Environment Ministers agreed on an annual target deposition of 20 kg/ha as an acceptable goal, taking political and economic costs into account. Two years later an agreement was reached among all the provinces east of Saskatchewan (the area defined as 'eastern' Canada within the context of acid rain), which set specific emission reductions to reach this target. Progress was rapid. The area of eastern Canada receiving 20 kg/ha or more of wet sulphate per year declined by nearly 59 per cent from 0.71 million km^2 in 1980 to about 0.29 million km^2 in 1993 (Figure 4.17). By 1994, SO_2 emissions were down to 2.7 million tonnes, a 41 per cent reduction from the 1980 level of 4.6 million tonnes.

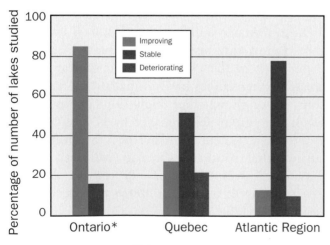

*73 percent of lakes studied in Ontario are in the Sudbury region.

Figure 4.16 Trends in lake acidity 1981–1997. SOURCE: Environment Canada (2003b: 15).

BOX 4.12   WHAT YOU CAN DO

Many of the challenges and problems discussed in this book are international in scope and require the co-ordinated efforts of different levels of government, industry, and individuals. The following are some of the ways in which individuals can try to have a positive influence.

1. Acid deposition is profoundly influenced by the personal decisions we all make regarding our use of fossil fuels in transport and electricity consumption. Think about your decisions and how you can minimize use.
2. Many consumer items, such as TVs, cellphones, computers, and other electronics, contain materials such as lead and nickel that contain sulphur. As you buy and dispose of these items you are helping to increase acid deposition.
3. Use chemical fertilizers sparingly on your gardens.
4. Eat less meat.
5. Let your political representatives know that you are in favour of mandatory measures to curb sulphur emissions and treat livestock wastes, even if this costs you more money.

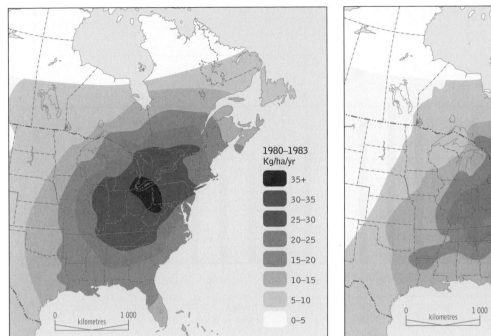

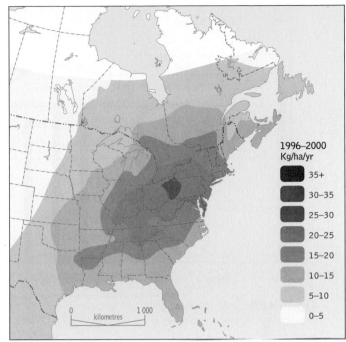

Figure 4.17 Wet sulphate deposition since 1980s. Source: Environment Canada (2003b), reproduced with the permission of the Minister of Public Works and Government Services, 2005.

Nonetheless, over 800,000 km² in eastern Canada continued to receive wet sulphate deposition in excess of the critical load (Figure 4.16). Critical loads were defined as the amount of sulphate that could be deposited on an area and still maintain 95 per cent of lakes in that region at pH 6 or over. In 1998, the provinces, territories, and federal government signed the Canada-Wide Acid Rain Strategy for Post-2000, committing them to further actions to deal with acid rain. The new targets for eastern Canada emissions are shown in Table 4.9. Under this scenario, and assuming a

Table 4.9	Sulphur Dioxide Reduction Targets for Ontario, Quebec, New Brunswick, and Nova Scotia		
	Former Eastern Canada Acid Rain Program Cap	**New Target under The Canada-Wide Acid Rain Strategy**	**Timeline for New Target**
Ontario	885 kt	• 442.5 kt (50% reduction)	2015[a]
Quebec	500 kt	• 300 kt (40% reduction)	2002
		• 250 kt (50% reduction)	2010
New Brunswick	175 kt	• 122.5 kt (30% reduction)	2005
		• 87.5 kt (50% reduction)	2010
Nova Scotia	189 kt	• 142 kt (25% reduction)	2005
		• 94.5 kt (50% cumulative reduction goal)[b]	2010

a Ontario has proposed and is consulting on the proposal to advance this timeline to 2010.
b Ninety-four and a half kilotonnes is a reduction target and not a cap. Nova Scotia's commitment is to reduce SP₂ emissions by 25 per cent from the existing cap by 2005 and to further reduce emissions to achieve a cumulative reduction goal of 50 per cent by 2010 from existing sources.
Source: Environment Canada, at: <www.ccme.ca/assets/pdf/acid_rain_e.pdf>.

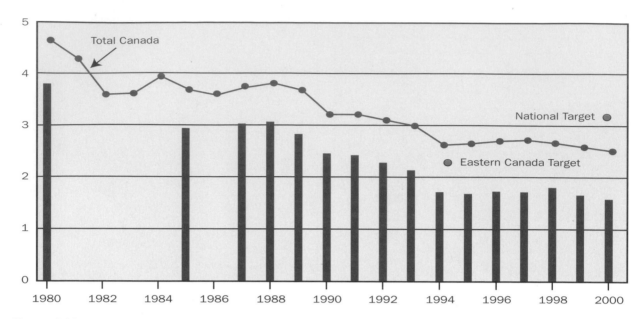

Figure 4.18 Sulphur dioxide emissions for eastern Canada, 1980–2000 (million tonnes). SOURCE: Environment Canada, 2003b. Reproduced with the permission of the Minister of Public Works and Government Services, 2005.

further 50 per cent reduction in US emissions over and above those already committed to, the area receiving more than the critical load would be reduced to 247,000 sq. km. However, scientists estimate that a further 75 per cent reduction in SO_2 emissions beyond current commitments will be needed to address this situation.

Bilaterally, early efforts to convince decision-makers in the US to control emissions met with little success. However, in 1990 intense lobbying efforts came to fruition. The US Clean Air Act was revised to cut in half the 1980 sulphur emission levels by the turn of the century. Whether the US will actually meet its target of a 9.1-million-tonne reduction by the year 2010 is open to question. However, overall, the amount of acid rain falling on eastern Canada originating in both countries is down an estimated 33 per cent since controls were implemented and total North American emissions are now 40 per cent of 1980 levels (Figure 4.18). Over half the eastern Canadian emission reductions have occurred at the smelters at Sudbury, Ontario, and Rouyn-Noranda, Quebec.

Efforts to control the emissions of nitrogen oxides have not been quite so successful. Between 1980 and 1990 both Canadian and US emissions remained fairly constant and show an increase to 2000 (Figure 4.19). Canada was committed under the terms of the Canada–US Air Quality Agreement to a 10 per cent reduction from stationary sources by the year 2000. Annual emissions in the

US should also have been reduced by 1.8 million tonnes below the 1980 level by the year 2000. Mobile emission reductions are being sought by introducing more stringent performance standards on exhaust emissions from new vehicles. Programs announced in 1996 aimed to cut hydrocarbon emissions by about 30 per cent and nitrogen oxide by 60 per cent on 1998 model cars.

Multilaterally, Canada also has been active, signing the 1979 Convention on Long-Range Trans-Boundary Air Pollution and then taking international leadership in the signing of the 1985 Helsinki Protocol that committed member countries to a 30 per cent reduction in emissions by 1994, a target that was met. Canada is also a signa-

Automobile emissions reduce the air quality in many urban centres (*P. Ivy/Ivy Images*).

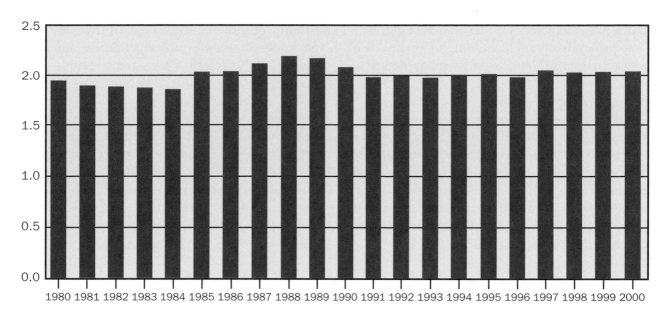

Figure 4.19 Nitrogen oxide emissions in Canada (million tonnes). SOURCE: Environment Canada (2003b: 15).

tory to the UN Economic Commission for Europe sulphur and NO_X protocols, which set national and regional targets that Canada has met.

Canada has met all its goals and commitments, some with considerable time to spare, in terms of acid deposition reductions. Quite rightly, Environment Canada's indicator meter (Figure 4.20) shows signs of improvement. Yet, despite reductions in emissions, acid deposition has not declined as expected. Rain in Ontario's Muskoka–Haliburton area is still pH 3.9–4.4, 40 times more acidic than normal. This seems to be because neutralizing salts, such as calcium and magnesium, also seem to have declined in the rain. Scientists are not sure why this is.

A study in the Ontario, Quebec, and the Atlantic region shows that of 202 lakes monitored since 1983, 33 per cent have improved, 56 per cent have not changed, and 11 per cent experienced increased acidity. However, only in the Sudbury region have the majority of monitored lakes showed some improvement, attributable to the reduced emissions from the Inco smelter. Elsewhere there was little change. While these studies

confirm the relationship between declining emissions and lowering acidity levels, they do not indicate an end to the acid deposition problem for several reasons.

- It is still uncertain whether clean air regulations will be met in the US. Utilities that surpass their emission targets are allowed to trade their excess polluting capacities to heavier polluting facilities. How effective this will be in overall controls remains to be seen.
- Unlike the US, Canada does not have a Clean Air Act that can be enforced by law, creating the possibility of further non-compliance with targets.
- Emission controls have focused largely on point-source control of sulphate emissions. It is more difficult to address the more diffuse nitrogen derivatives coming mainly from the transportation sector. The pattern of wet nitrate deposition has changed little over the last decade.
- Concern has focused on eastern Canada, but pockets of acidity exist all across the country.
- Although some lake pH levels are showing signs of recovery, it is still too early to assess the biotic implications in terms of community health.
- Emission targets have been consistently set below levels required to ensure biological recovery. Scientists (e.g., Jeffrie et al., 2003) have suggested that a further 75 per cent emission

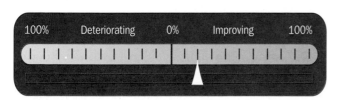

Figure 4.20 Environment Canada's metre for acid deposition. SOURCE: Environment Canada (2003b: 15).

cut will be necessary for some areas to recover.
- The pH levels are still low enough in many areas to have damaging effects on sensitive species.

IMPLICATIONS

Understanding the nature of matter and the way that nutrients cycle in the ecosphere is fundamental to appreciating many of the more challenging environmental issues that society currently faces. Acid deposition, eutrophication, and global climate change all have their root in disruption of biogeochemical cycles. Science is only just starting to unravel some of the secrets of these cycles, but we know enough to understand their significance. Research also clearly shows their complexity and their interconnectedness. We add nitrogen to soils to try to boost productivity and produce more food. However, this also results in eutrophication, depletion of the ozone layer (Chapter 7), and other problems. Clearly, we have to understand the basic science of this interconnectivity. However, this also has to be linked to our abilities to manage the situation. The following chapters in Part C discuss some of the main approaches in environmental planning and management.

SUMMARY

1. Matter has mass and takes up space. It is composed of 92 natural and 17 synthesized chemical elements. The law of conservation of matter states that matter can neither be created nor destroyed, but merely transformed from one form into another. Matter cannot be consumed.

2. Elements necessary for life are known as nutrients. They cycle between the different components of the ecosphere in characteristic paths known as biogeochemical cycles.

3. Humans disturb these cycles through various activities resulting in environmental problems such as acid rain and global warming.

4. Cycles can be classified into gaseous or sedimentary, depending on the location of their major reserves.

5. Phosphorus is an example of a sedimentary cycle. The main reservoir of phosphorus is the Earth's crust. Phosphates are made available in the soil water through erosional processes and are taken up by plant roots and passed along the food chain. There is no atmospheric component to the cycle, making it especially vulnerable to disruption. The main human use for phosphorus is as fertilizer. It is a main cause of eutrophication.

6. Sulphur is also a sedimentary cycle, but differs from phosphorus in that it has an atmospheric component. Like phosphorus, it is an essential component for all life. Bacteria enable plants to gain access to elemental sulphur by transforming it to sulphates in the soil. Sulphur is a main component of acid deposition.

7. Nitrogen is a gaseous cycle. Almost 80 per cent of the atmosphere is composed of nitrogen gas, yet most organisms cannot use it as a source of nitrates. Instead, various bacteria help transform nitrogen into a form that can be used by plants. As with the other cycles, these nitrates are then passed along the food chain. Nitrates are used as fertilizers and contribute to eutrophication. Various nitrous oxides also contribute to acid deposition and the catalytic destruction of ozone.

8. Carbon dioxide constitutes only 0.03 per cent of the atmosphere, but it is the main source of carbon—the basis for life—through the process of photosynthesis. Carbon becomes incorporated into the biomass and is passed along the food chain. Respiration by organisms transforms some of this carbon back into carbon dioxide and the cellular respiration of decomposers helps to return the carbon from dead organisms into the atmosphere. Carbon dioxide emissions from burning fossil fuels are a main contributor to global climatic change.

9. Water travels between the different components of the ecosphere by means of the hydrological cycle fuelled by energy from the sun. Ninety seven per cent of water is in the oceans. Less than 1 per cent is readily available for human use.

10. Canada has up to one-third of the world's fresh water. Most of this is held in the solid phase as ice. Canada also has high storage capacities for liquid water, with lakes covering an estimated 8 per cent of the country. These are replenished by a river flow containing approximately 9 per cent of the total river discharge in the world.

11. Major pollution problems, such as eutrophication, acid deposition, and global warming, are the result of human disruption of biogeochemical cycles.

12. Eutrophication is the nutrient enrichment of water bodies over time. Although a natural process, human disruption of the phosphorus and nitrogen cycles has caused a marked acceleration in the rate of eutrophication. This promotes excessive plant growth that leads to oxygen depletion when the plants die and start to decay. Over time this leads to changes in the composition of fish species and makes water treatment more expensive. The rate of eutrophication can be slowed down by limiting the inputs of nutrients into the water body.

13. Lake Erie is a classic example of eutrophication. Although phosphorus loadings have been reduced considerably, they are still above mandated guidelines and appear to be increasing.

14. Precipitation tends to be naturally acidic. However, as a result of disturbances in the sulphur and nitrogen cycles, the acidity has increased dramatically over much of Canada during the last few decades. The largest impacts are caused by the burning of sulphur-rich fossil fuels or through the smelting of sulphur-rich metal ores. The resulting sulphur dioxide mixes with water in the atmosphere to produce sulphuric acid. Emissions of various nitrogen oxides as by-products of high-temperature combustion account for most of the remainder. The increase in acidity has a damaging effect on aquatic and terrestrial ecosystems as well as upon human health.

15. Emission controls have been agreed on to try to limit these impacts, and targets have been exceeded. However, concomitant improvements in reducing acid rain and the recovery of aquatic systems have not been seen. Many scientists believe that more stringent measures are called for.

KEY TERMS

acid deposition	compound	law of conservation of	nutrients
acid shock	denitrification	matter	oligotrophic
anadromous	eutrophic	macronutrients	oxygen sag curve
aquifer	eutrophication	matter	point sources
benthic	evapotranspiration	micronutrients	rainshadow effect
biogeochemical cycles	gaseous cycles	mineralization	sedimentary cycles
biological oxygen demand	groundwater	nitrogen fixation	transpiration
(BOD)	hydrological cycle	non-point sources	

REVIEW QUESTIONS

1. Summarize some of the key differences and similarities between energy and matter.
2. Why is life dependent on biogeochemical cycles?
3. Explain why decomposer organisms are important in biogeochemical cycling.
4. What are some of the important implications of biogeochemical cycling for forestry and agricultural activities?
5. Outline the main characteristics of the hydrological cycle in Canada.
6. Which biogeochemical cycles are most responsible for eutrophication, and in what ways are they changed by human activities?
7. Are there any eutrophic lakes in your region? If so, what are the main inputs causing eutrophication and where do they come from?
8. How can eutrophication be controlled? Discuss one example.
9. What is the pH scale, and what is it used for?
10. Which biogeochemical cycle is most responsible for acid deposition, and how do disruptions occur as a result of human activities?
11. What are the effects of eutrophication and acid deposition on aquatic ecosystems? What do you think might be the impacts of them both together?
12. Are all areas equally sensitive to the impacts of acid deposition? If not, what influences the relative vulnerability of different areas?
13. What are the main socio-economic impacts of acid deposition likely to be?

RELATED WEBSITES

ENVIRONMENT CANADA, ACID RAIN:
http://www.ec.gc.ca/acidrain

ENVIRONMENT CANADA, AIR POLLUTANT EMISSIONS:
http://www.ec.gc.ca/pdb/ape/cape_home_e.cfm

ENVIRONMENT CANADA, CLEAN AIR:
http://www.ec.gc.ca/air/introduction_e.cfm

ENVIRONMENT CANADA, SUSTAINING THE
ENVIRONMENT AND RESOURCES FOR CANADIANS:
http://www.environmentandresources.gc.ca/default.
asp?lang_En

ENVIRONMENT CANADA, WATER:
http://www.ec.gc/water_e.html

NITRATE LEVELS:
http://www.ecoinfo.ec.gc.ca/env_ind/region/nitrate/
nitrate_e.cfn

REFERENCES AND SUGGESTED READING

Blodau, C. 2002. 'Carbon cycling in peatlands—A review of processes and controls', *Environmental Reviews* 10, 2: 111–34.

Briggs, D., P. Smithson, T. Ball, P. Johnson, P. Kershaw, and A. Lewkowicz. 1993. *Fundamentals of Physical Geography*, 2nd Canadian edn. Toronto: Copp Clark Pitman.

Browning, M.H.R., and T.C. Hutchinson. 1991. 'The effects of aluminum and calcium on the growth and nutrition of selected ectomycorrizal fungi of jack pine', *Canadian Journal of Botany* 69: 1691–9.

Dalton, H. 2003. 'Nitrogen: The essential public enemy', *Journal of Applied Ecology* 40, 5: 771–81.

Environment Canada. 2003a. Atmospheric Science: Acid Rain. At: <http://www.ns.ec.gc.ca/msc/as/acidfaq.html>.

———. 2003b. *Environmental Signals: Canada's National Environmental Indicator Series, 2003*. Ottawa, Canada. Cat. En40-775/2002E. <http://www.ec.gc.ca/soer-ree/English/Indicator_series/default.cfm>.

——— and US Environmental Protection Agency. 2003. *State of the Great Lakes, 2003*. Ottawa and Washington.

Galloway, J.N., and E.B. Cowling. 2002. 'Reactive nitrogen and the world: 200 years of change', *Ambio* 31, 2: 64–71.

Government of Canada. 2002. *2001 Annual Progress Report on the Canada-Wide Acid Rain Strategy for Post 2000*. Ottawa: Environment Canada. Available at: <http://www.ccme.ca/assets/pdf/acid_rain_e.pdf>.

Hendrickson, O. 2003. 'Influences of global change on carbon sequestration by agricultural and forest soils', *Environmental Reviews* 11, 3: 161–92.

Janzen, H.H., et al. 2003. 'The fate of nitrogen in agroecosystems: An illustration using Canadian estimates', *Nutrient Cycling in Agroecosystems* 67, 1: 85–102.

Jeffries, D.S., et al. 2003. 'Assessing the recovery of lakes in southeastern Canada from the effects of acidic deposition', *Ambio* 32: 176–83.

———, T.G. Brydges, P.J. Dillon, and W. Keller. 2003. 'Monitoring the results of Canada/USA acid rain control programs: Some lake responses', *Environmental Monitoring and Assessment* 88, 1–3: 3–19.

Keller, W., J.H. Heneberry, and S.S. Dixit. 2003. 'Decreased acid deposit and the chemical recovery of Killarney, Ontario, Lakes', *Ambio* 32: 183–7.

Lennihan, R., D.M. Chapin, and L.G. Dickson. 1994. 'Nitrogen fixation and photoynthesis in high arctic forms of *Nostoc commune*', *Canadian Journal of Botany* 72: 940–5.

Magee, J.A., M. Obedzinski, S.D. McCormick, and J.F. Kocik. 2003. 'Effects of episodic acidification on Atlantic salmon (*Salmo salar*) smolts', *Canadian Journal of Fisheries and Aquatic Sciences* 60, 2: 214–21.

Malley, D.F., and K.H. Mills. 1992. 'Whole-lake experimentation as a tool to assess ecosystem health, response to stress and recovery: The Experimental Lakes Area experience', *Journal of Aquatic Ecosystem Health* 1: 159–74.

Matson, P., K.A. Lohse, and S.J. Hall. 2002. 'The globalization of nitrogen deposition: Consequences for terrestrial ecosystems', *Ambio* 31, 2: 113–19.

Ohmura, A., and W. Wild. 2002. 'Climate change: Is the hydrological cycle accelerating?', *Science* 298, 5597: 1345–6.

Pilgram, W., T.A. Clair, J. Choate, and R. Hughes. 2003. 'Changes in acid precipitation–related water chemistry of lakes from Southwestern New Brunswick, Canada, 1986–2001', *Environmental Monitoring and Assessment* 88, 1–3: 39–52.

Rabalais, N.N. 2002. 'Nitrogen in aquatic ecosystems', *Ambio* 31, 2: 102–12.

RMCC (Federal/Provincial Research and Monitoring Co-ordinating Committee). 1990. *The 1990 Canadian Long-Range Transport of Air Pollutants and Acid Deposition Study*. Downsview, Ont.: Environment Canada, Atmospheric Environment Service.

Rowe, S. 1993. 'In search of the holy grass: How to bond with the wilderness in nature and ourselves', *Environment Views* (Winter): 7–11.

Tao, F., M. Yokozawa, Y. Hayashi, and E. Lin. 2003. 'Terrestrial water cycle and the impact of climate change', *Ambio* 32, 4: 295–301.

Vile, M.A., S.D. Bridgham, R.K. Wieder, and M. Novak. 2003. 'Atmospheric sulfur deposition alters pathways of gaseous carbon production in peatlands', *Global Biogeochemical Cycles* 17, 2: article no. 1058.

Vitousek, P.M., et al. 1997. 'Human alteration of the global nitrogen cycle: causes and consequences', *Issues in Ecology* 1: 2–16.

———, S. Hattenschwiler, L. Olander, and S. Allison. 2003. 'Nitrogen and nature', *Ambio* 31, 2: 97–101.

Winter, J.G., P.J. Dillon, M.N. Futter, R.H. Nicholls, W.A. Scheider, and L.D. Scott. 2002. 'Total phosphorous budgets and nitrogen loads: Lake Simcoe, Ontario (1990 to 1998)', *Journal of Great Lakes* 28, 3: 301–14.

Planning and Management: Philosophy, Process, and Product

The chapters in Part B focused mainly on the natural science pertinent to resources and the environment in Canada. In the following two chapters, attention is given to planning and management related to philosophies, processes, and products that can be applied in resource and environmental management. A key point, however, needs to be made. Often it is inappropriate to think that humans 'manage' the environment or natural resources. Instead, we often attempt to manage the *interaction* between humans and the environment. This is why resource and environmental management involves more than application of 'science' or 'technical expertise'. It also requires sensitivity to various, and often different, values, interests, needs, and wants.

Chapter 5 focuses on the 'philosophy' of planning and management related to resources and environments, while Chapter 6 considers 'processes and products'. Together, they provide an overview of concepts, approaches, and methods that can be drawn upon.

Chapter 5 begins by considering what is meant by the concept of *best practice*. It then examines the importance of *context* for a problem-solving situation and the need to be able to design solutions to fit specific situations. In this regard, we face a basic challenge. The unique characteristics of a given place and time suggest that it is best to develop approaches and solutions specific to a situation. However, when this is done, a potential problem is that some people may perceive other people or regions receiving preferential treatment, since different arrangements are being applied to them. As a result, there often is pressure, in the spirit of being equitable, to use the same or a similar approach in all places, regardless of differences among them. As a result, there is debate over the merits of using a standard approach versus allowing discretion to develop different approaches, depending on circumstances prevailing in a particular place.

While being able to appreciate the importance of context is necessary, best practice suggests that we all need to understand the distinction between most probable, desirable, and feasible futures, and why a *vision* or *direction* needs to be established to help in making choices about what is the right thing to do. Too often, it seems, those concerned with or responsible for environmental problems give most of their attention to what is the most likely or probable future and how we should prepare to deal with it. Certainly, understanding what is likely to occur in the future is very important, but it is also important to recognize that choices can often be made and that the most probable future is not necessarily the most desirable future. Thus, best practice indicates that we should have a clear sense of what kind of desirable future we aspire to, so that we then can judge, knowing what is likely to occur, whether by intervening it is possible to move closer towards what is desired. A fundamental challenge, of course, is that societies are not homogeneous, and therefore at any given time there may be competing visions about what kind of future is desirable. Thus, a major task is to identify, and achieve, a shared vision.

Various possibilities for what could constitute a desired future have been identified, and attention in Chapter 5 is given to several of them: *sustainable development, sustainable livelihoods*, and *environmental justice*. Sustainable development, popularized in 1987 with publication of *Our Common Future*, the report of the World Commission on Environment and Development, has been the cause of much debate and disagreement, as different groups interpret it in ways that favour their values and interests. Despite conflicting views about what sustainable development means, it frequently appears in policies related to the environment and natural resources. Thus, it is important to have a critical appreciation of its strengths and limitations. More recently, attention is being given to two other concepts—sustainable livelihoods and environmental justice—as alternative possible directions for the future. Our intent here is to ensure that you understand what these concepts mean, what they offer, and what their weaknesses are. From such understanding, we hope that you will be in a better position to determine the attributes of a desired future.

In determining what might be a desirable future, it is important to appreciate that basic values shape perspectives. In that spirit, we examine the difference between *biocentric* and *anthropocentric* views. As part of this discussion, the guest statement by Dan Shrubsole draws attention to different interpretations about what constitutes a 'resource', and how that can influence thinking about what are appropriate decisions to take.

A basic concept associated with best practice is a *systems perspective*, and how that can be applied as an ecosystem approach. In Chapter 5, discussion focuses on how the ecosystem approach can be translated from concept to practice, which is not always easy, especially since most often administrative or political boundaries do not reflect or respect ecosystem boundaries. Another best practice element is the need for thinking simultaneously in the short, medium, and long terms. Too often, our society expects instant results or gratification, which places emphasis on the short term. In addition, elections at municipal, provincial, and federal levels usually occur within five years, and sometimes as frequently as every two or three years, which further drives attention towards the immediate and short term. The point is not that we should think only in the long term, but that we should be thinking simultaneously at several time scales and also have the patience and understanding to recognize that results may take some time to emerge.

Aspects of planning and management continue to be the focus in Chapter 6, but there the emphasis shifts to processes and products. For example, if a systems perspective, noted in Chapter 5, is to be used, this requires considerable *collaboration* and *co-ordination* among governments and First Nations, public agencies, the private sector, and non-government organizations. These are needed to overcome what is often termed the *silo effect*, or the propensity of organizations or individuals to focus only on their own interests and responsibilities, and not to consider those of others. One means to achieve collaboration and co-ordination is through *stakeholder* and *participatory approaches*, and by continuously seeking to enhance *communication* among participants. In Chapter 6, alternative approaches to facilitate participation and communication are examined. None is perfect, and each has different strengths and weaknesses. The challenge is to determine which mix of approaches to use, so that the strengths of one can offset the limitations of another.

Best practice increasingly includes *adaptive management*, *impact and risk assessment*, and *dispute resolution*. These processes all explicitly recognize that there are high levels of change, complexity, and uncertainty in resource and environmental management, and that conflicts often occur. Therefore, the challenge is not to eliminate or avoid change, complexity, uncertainty, and conflict, but instead to manage within the reality of their presence. *Adaptive management* provides one process through which to monitor experience, in order to make systematic adjustments as a result of experience. *Impact and risk assessment* encourages us to be proactive, that is, to look ahead and anticipate positive and negative results from management actions, so that we can enhance the positive and mitigate the negative. *Dispute resolution* methods and processes have been created to allow us to deal with the presence of fundamentally different values, needs, interests, and behaviour, which need to be reconciled or at least managed.

A challenge in resource and environmental management is that strategies, programs, and plans have to gain credibility or legitimacy. This is normally realized through having a statutory or legislative base, political commitment, or administrative endorsement. The more of these that are in place, the more visibility and authority there will be for strategies, programs, or plans. One way to enhance their credibility is to link them explicitly to other management tools that have a statutory base, such as *regional and land-use plans*. When connections are made with such tools or methods, the probability of initiatives being sustained and implemented usually goes up markedly.

The discussion of philosophies, processes, and products in Part C provides an overview of elements of best practice related to resource and environmental management. The intent of presenting them is twofold. First, being aware of them will allow you to examine ongoing initiatives to determine if they reflect best practice. Where that is not the case, then we must begin to consider what would have to be added or changed for best practice to be achieved. Second, understanding the elements of best practice will allow you to incorporate them into solutions you want to develop relative to emerging or new issues.

Planning and Management: Philosophy

Learning Objectives

- To appreciate the significance of different planning and management approaches.
- To understand the importance of context for a problem-solving situation and the need to be able to design solutions to fit particular contexts.
- To distinguish between most probable, desirable, and feasible futures, and why it is important to identify a vision or direction to help in making choices about what is the right thing to do.
- To understand the strengths and limitations of sustainable development, sustainable livelihoods, and environmental justice as the basis for a vision or desirable future.
- To know the difference between biocentric and anthropocentric perspectives.
- To understand the significance of a systems perspective and how that can be applied as an ecosystem approach.
- To realize the importance of thinking simultaneously in the short, medium, and long terms.

INTRODUCTION

Improved resource and environmental management is likely to be achieved if two aspects are considered. First, and as already discussed in Chapter 1, using science to inform decision-making is desirable. Second, management and decisions should reflect what are accepted as the best planning and management approaches in terms of concepts, processes, and methods. In this chapter, we focus on what might be termed *philosophy* and consider basic concepts widely accepted around the world when addressing resource and environmental problems. In Chapter 6, attention will turn to *processes, methods,* and *products*, or those processes (such as public participation and community-based approaches), methods (impact and risk assessment), and outcomes (effective, efficient, equitable) generally deemed to be desirable.

Being aware of different approaches allows us to assess what is being done in any given situation

MANAGEMENT CHALLENGES

Central challenges in successful governance of the environment over the next few decades will involve developing more effective ways to integrate high quality, objective scientific and technical assessment with key decision needs; learning more effective processes for managing under uncertainty and responding adaptively to advances in knowledge; and effectively coordinating inevitably shared authority and capacity across multiple levels of government and between diverse public and private actors.

— Parson (2000: S123)

and to judge whether the approach is likely to be appropriate. In that way, the ideas reviewed in Chapters 5 and 6 serve as a framework or template to see if the way a problem is being addressed reflects what many professionals around the world consider to be the most suitable approaches.

BOX 5.1 CUSTOM-DESIGNING TO DEAL WITH AGRICULTURAL DIFFUSE POLLUTION

Stonehouse (1994) reviewed Canadian experience regarding adoption and use of soil conservation practices, with implications for non-point-source pollution control. He concluded that both cultural and biophysical factors influence conservation decisions by farmers, and that each set of factors operates at a localized, individual farm or land-unit level and at a broader regional or national scale. As a result, he recommended that a *targeted approach* is the best way to apply public policies to help solve degradation problems, in contrast to the dominant *universal application approach* normally used in Canada. The latter approach assumes that farmers and landowners are homogeneous, and that they contribute to degradation in more or less the same way. In contrast, he argued that farmers are extremely heterogeneous with respect to their personal characteristics, economic circumstances, and managerial abilities. Furthermore, their farms vary in size, natural resource characteristics, and enterprise specializations. Such differences result in wide variations in conservation effort, leading to the need for differential treatment of farmers regarding remedial or preventative public policy measures.

Stonehouse suggested farmers can be identified as innovators, early adopters, mainstream adopters, laggards, or apathetic types, in descending order of conservation effort. In a targeted approach, the type of farmer is matched with different mixes of policy instruments (ranging from voluntary to enforced compliance). For example, for *innovators* (or those most likely to adopt new ideas), some form of public recognition of their stewardship and public leadership is appropriate, along with, where necessary, information and extension assistance for newly emerging conservation practices. *Early adopters*, in contrast, would most likely be better assisted by information and extension assistance. *Mainstream adopters* would need some positive financial incentives to induce adoption and use, particularly for practices whose profitability was unclear or whose risk-loading was significant. For *laggards*, a group with much lower inclination to adopt conservation efforts, much stronger incentives are required due to their lack of problem awareness, low management skills, economic problems, and other reasons. Thus, for laggards, the most effective policy approach will be positive financial incentives and necessary information and extension support. Members in the final group, the *apathetic*, are often aware of degradation problems but are unwilling to take remedial action. For them, strong measures must be used, and the possibilities are regulation and control of farming activities associated with resource depletion, supported where needed by litigation, taxes, and penalties. Key messages from Stonehouse's evaluation of Canadian experience are the need to custom-design solutions, to apply a mix of policy instruments, and to use a targeted rather than a universal approach.

PLANNING AND MANAGEMENT COMPONENTS

Context

The **context**—i.e., specific characteristics of a time and place—needs to be systematically considered when developing a strategy, plan, or approach for a resource or environmental management problem. Biophysical, economic, social, and political conditions frequently differ from place to place and from time to time, highlighting that it usually is inappropriate to proceed as if one model or approach will be sufficient in every situation. Instead, it often is necessary to custom-design solutions relative to the conditions in different places and at different times. Thus, the basic message or lesson is that 'one size does not fit all situations' (Box 5.1).

Accepting the importance of context reaffirms the desirability of including local people when developing a strategy or implementing initiatives, as they often have extra insight into the conditions of a region or place (discussed further in Chapter 6). Understanding the significance of context also reminds us that this can change as conditions change, requiring capacity and willingness to modify strategies and initiatives to ensure that they remain relevant. This is discussed further in Chapter 6 with reference to the concept of adaptive management.

While context is important, in practice it is not unusual for public agencies to prefer a standardized approach to problem-solving. The rationale is that a standardized approach is easiest to defend or justify. If every area or region is treated the same, then no region can be perceived to be receiving special or preferential treatment. In contrast, if a specific approach is designed for one region or place, there is the possibility that people in other regions will conclude that such a region is receiving preferential treatment.

Our suggestion is that, notwithstanding the possibility of criticism about **custom-designed solutions**, managers should recognize the specific conditions of a place and time, and design accordingly. It is better to develop an approach especially designed for the needs of a place and accept criticism about perceived favouritism than to use a standardized, cookie-cutter approach that avoids such criticism but does not really fit the conditions or needs in any one place or time.

Context in the Big Picture

Experience since the Rio Summit in June 1992 and the Summit in Johannesburg during 2002 has confirmed that management of natural resources and the environment involves many organizations and jurisdictions that frequently have overlapping and/or conflicting legal mandates and responsibilities, numerous and often conflicting interests regarding access and rights to environmental systems, and growing skepticism about the formal mechanisms of governments—local, state, national, regional, international—to deliver services effectively, efficiently, and equitably.

Behind such challenges is the reality that governance of resources and the environment occurs in situations defined by high levels of *complexity* and *uncertainty*. Furthermore, managers function in situations characterized by rapid *change*, and at the same time often need to become agents of positive change. Finally, managers increasingly deal with *conflict* due to many different interests related to resources and the environment. As Homer-Dixon (2000: 1) has observed, 'the complexity, unpredictability, and pace of events in our world, and the severity of global environmental stress, are soaring.' In his view, such complexity, unpredictability, and rapid change create an ingenuity gap, or growing disparities between those who adapt well and

those who do not. Homer-Dixon (2000: 3) defines ingenuity as 'ideas for better institutions and social arrangements, like efficient governments and competent governments'.

Experience in the last decade suggests that four other contextual aspects also are important for understanding progress related to managing resources and the environment wisely. One is the preoccupation that many national governments have shown about debt and deficit reduction. Since Rio in 1992, many national governments have reduced significantly their expenditures on environmental infrastructure and services, which usually has had a serious negative impact on agencies responsible for resources and the environment. For example, in Canada, during the mid-1990s the federal agency responsible for water management had its budget reduced by 35 per cent. In the province of Ontario, the Ministry of Environment had its operating budget cut by 44 per cent and the Ministry of Natural Resources lost 42 per cent of its budget between 1995–6 and 1997–8—an approach that Kreutzwiser (1998: 138) characterized as a 'scorched earth retreat' by the provincial government, relative to its normal responsibilities for resources and the environment. Similar decisions have been taken in Alberta and British Columbia.

Second, and emerging from the concern about debt and deficit reduction, many national and state governments have been: (1) downloading responsibilities for environmental services to lower levels of government, which usually have not had the human or financial resources to maintain levels of service; (2) commercializing such services; or (3) privatizing these services. In reference to the Ontario government's decision to close provincial water testing laboratories in 1996, Kreutzwiser (1998: 139) concluded that the provincial government made such cutbacks 'without any consideration of the capacity in the private sector, or among municipalities, to test municipal water supplies'.

Such initiatives usually have been justified with reference to the principle of *subsidiarity* (allocating responsibilities to levels of government closest to where the services are used or received) or *efficiency* (providing services at least cost). However, some believe that these decisions are at least as much ideological as they are rational, reflecting a belief in market-based economies as the most effective way to allocate scarce societal resources, and that less government involvement is desirable. Thus, while the rationale for the subsidiarity principle is sensible—allowing people closest to decisions to participate directly in decisions that affect them—it also requires willingness to maintain, create, or enhance human and financial capacity at local levels to deal with resource and environmental management issues. When senior levels of government have downloaded their traditional management responsibilities to local levels without regard for the existing capacity to handle those responsibilities or without consideration as to how the necessary capacity can be created, subsidiarity has been presented as the basis for decisions that in fact may have been driven more by financial cost-cutting considerations. Different ideological perspectives present quite different views of the environment and resources, and these different views can affect planning and management dramatically. One view is presented by Krajnc in the accompanying box.

Third, while many governments have embraced a position favouring less government intervention, reliance on the private sector and market forces to deliver products and services efficiently, acceptance of the value of globalization, and modified management processes to reflect a 'business model' in which efficiency, results-based management, and tangible products have been emphasized, they have shown less interest in using systematic and thorough consultation processes regarding development and implementation of policy. Indeed, in some jurisdictions, consultation that is transparent and accessible to the general public has almost disappeared. When this has occurred, while the senior levels have identified subsidiarity as a desired principle to guide decisions, in practice their actions have not reflected consideration of all the preconditions required for subsidiarity to function effectively.

Fourth, since the late 1990s, many governments have been steadily backing away from concern for, or commitment to, environmental issues, and instead have been emphasizing strategies for economic growth. An often cited example in that regard is the decision by President George W. Bush to have the United States withdraw from the 1997 Kyoto Protocol for reducing global warming (see also Chapter 7) and to refocus

NEO-CONSERVATISM

A series of sharp cuts to the Ontario Ministry of the Environment's (MOE) budget in the 1990s ... left it with fewer resources at the turn of the century than it controlled in the mid-1970s—shortly after the ministry was created and when it was performing a fraction of the functions it later accumulated in its ever-expanding mandate.

...The most recent wave of cuts and policy changes represents not a minor but a major retrenchment, and can best be attributed to a particular brand of conservative ideology—namely neo-conservatism.

— Krajnc (2000: 112)

energy policies in the US on supply solutions with little or no attention to demand management strategies. There is some irony in this situation, given that many polls indicate that the general public over the past five years has showed increasing concern about environmental issues.

Thus, much of what is (or is not) happening in resource and environmental management around the world may be attributable to the shifting political power of different ideologies. To be an informed citizen, planner, or manager, individuals should aim to be informed about the basic values

PLACEBO POLICIES

*The (Ontario) government's approach to agenda management was the development of 'placebo policies' for high profile issues. **Placebo policies** are designed to play down the salience of environmental concerns and to side-step issues by addressing the symptoms of a problem rather than its causes. Such 'smoke and mirror' policies give the appearance of action through symbolic gestures rather than the necessary substantive policy changes.*

A case in point was the government's attempt at manipulating public opinion on the issue of urban smog.... A highly publicized smog patrol program was introduced, as well as summer gasoline volatility limits to reduce volatile organic compound (VOC) emissions by 2 per cent. Overall VOC emissions were nonetheless expected to rise by 10 per cent per decade as a result of other government policies that promote urban sprawl, and decrease support for public transit systems, thus increasing reliance on automobiles.

— Krajnc (2000: 124)

BOX 5.2　THINKING BEYOND THE NEAR AND NOW

When did we become so focused on the present moment and our immediate situation? It seems to me there was a time not so long ago when we were focused as a society on building a better future for our children and our grandchildren. We had a broader sense of the connections between the landscape, the communities of Ontario and the economy.... And we valued that. We worked for a better tomorrow, but at the same time we didn't forget about the past, about those who had made sacrifices and worked hard to create the opportunities we enjoyed.

Somehow the awareness of the past and concern about what the future might become seem to be missing from current public discourse and decision-making. In the same way, there is little concern about how other areas of the province are impacted by decisions made in the urban centres. We accept as inevitable that the population of the GTA [Greater Toronto Area] will continue to swell by 100,000 new residents every year. Yet, have we really come to grips with what that will mean for our remaining forests, watercourses, air quality, and our most productive agricultural land? Do we care about our countryside?

Alas, life moves too quickly to linger on such considerations. We must grow at an expanding rate. We can't worry about those things now. Approve those subdivisions—they're needed

now. Build more highways—otherwise we can't get to work. Just get the garbage to Michigan—and make sure it's nowhere near us. Do something now about that new crisis in this morning's newspaper.

It's rather like speeding down a dark northern Ontario highway on a moonless June night with only your low beams on. You have that confident, comfortable feeling because there is no one on the road and you're making good time. But if you just click on your high beams you'll see the moose standing there only a few seconds in front of you....

The analogy of a speeding car works quite well in a discussion about the 'near and now', because as most drivers know, we have a tendency to focus on objects and surroundings just in front of the car. It takes training and discipline to bring your eyes up to the distant horizon where you become aware of events and objects far ahead—in both time and space. But only then can you acquire the capacity to anticipate, plan and react to future hazards. And so it is with public decision-making that affects the environment. We all must keep our eyes on the horizon.

SOURCE: Environmental Commissioner of Ontario (2003: 4).

and assumptions of various ideologies so that they can determine whether arguments realistically present the rationale for policies and actions. In Box 5.2, the Environmental Commissioner of Ontario identifies some basic challenges and poses fundamental questions that deserve our attention. While he refers to Ontario, his comments have general applicability to the entire country.

Vision

Ends need to be distinguished from means. In other words, before deciding how to deal with resource and environmental management problems or opportunities, managers should determine what ends or desirable future conditions are sought, as highlighted in Box 5.2. This consideration is often referred to as a need to have a clear vision or sense of direction. The vision for Environment Canada, for example, is given in Chapter 14.

According to Nanus (1992: 8–17), a **vision** represents a realistic, credible, and attractive future for a region, community, or group. It is ideal to have a shared vision, one to which many people are committed. Achieving a shared vision is challenging, however, since many interests exist in a society, some of which are mutually exclusive. If a shared vision about a desirable future is to be achieved, it is important to involve many stakeholders in the

A VISION FOR YOUR AREA OR COMMUNITY

What would be the key features of a vision for resource and environmental management for your college or university, or community, region, or province? Does one exist now?

management process, a matter discussed further in Chapter 6.

Visions can be developed in various ways. One is to ask three questions: (1) What is likely to happen? (2) What ought to happen? (3) What can happen? In much work in resource and environmental management, the focus is on the first question. An answer to this question helps to establish the likelihood of some future state, assuming continuation of current conditions, or estimating changes in some current conditions. Answers to this question are useful because they help to establish the most probable future. However, this question does not help determine if the most probable or likely future is also the most desirable. To deal with the issue of desirability, we also need to ask about what ought to happen and thereby consider what would be desirable future conditions for a society or place. These two questions (what is most likely, what is most desirable) reflect the difference between forecasting and backcasting, as explained by Tinker in the adjacent box. Finally, the third question imposes discipline by requiring us to consider what is feasible or practical. We need to address all three of these questions instead of relying only on the first, which is the most usual practice.

Sustainable Development

Sustainable development emerged in the late 1980s as one possible vision. Sustainable development offers a major challenge to Canada and societies everywhere since it requires a re-examination of current values, policies, processes, and practices. **Sustainable development** has three strategic aspects. At one level, it represents a *philosophy*, in that it presents a vision or direction regarding the nature of future societies. In sustainable societies, attention is given to meeting basic human needs, realizing equity and justice, achieving self-empowerment, protecting integrity of biophysical systems, integrating environmental and economic considerations, and keeping future options open.

As a *process*, sustainable development emphasizes a system of governance and management characterized by openness, decentralization, and accessibility. It accepts the legitimacy of local or indigenous knowledge and seeks to incorporate such understanding with science-based knowledge when developing strategies and plans. It also recognizes that conditions change and that it is

FORECASTING VERSUS BACKCASTING

Forecasting takes the trends of yesterday and today and projects mechanistically forward as if humankind were not an intelligent species with the capacity for individual and societal choice. Backcasting sets itself against such predestination and insists on free will, dreaming what tomorrow might be and determining how to get there from today. Forecasting is driving down the freeway and, from one's speed and direction, working out where one will be by nightfall. Backcasting is deciding first where one wants to sleep that night and then planning a day's drive that will get one there.

— Tinker (1996: xi)

necessary to be flexible and adaptable, thereby allowing for policies and practices to be modified as experience accumulates. As a *product* related to specific places or resource sectors, sustainable development seeks to ensure that economic, environmental, and social aspects are considered together and that trade-offs are made in a way that is visible and transparent to those affected.

The concept of sustainable development has generated both enthusiasm and frustration. The enthusiasm has come from those who believe that it provides a vision for the twenty-first century, one in which more attention will be given to longer-term implications of development and to balancing economic, social, and environmental considerations. The phrase 'think globally and act locally' reminds us that while ultimately the planet is a single system in which actions in one part often have implications for other parts, resolution of problems also requires significant action at the local level, thereby stimulating self-empowerment, partnerships, and co-operative approaches to management and development (see Chapter 6).

The frustration has come from those who believe that 'sustainable development' is so vague that it can be defined in ways to suit different and often conflicting interests. It has been argued that the general acceptance of sustainable development occurs only because different interests can use it to justify their often incompatible positions. Thus, developers like the concept because they can argue that growth must continue if basic human needs are to be met and if standards of living are to continue to rise. In contrast, environmentalists support the concept because they can use it to argue that environmental integrity must be given priority if there is to be long-term and equitable development. The response of federal government departments to establish a sustainable development strategy for Canada is discussed further in Chapter 14.

Livelihood Approach

Based on experience up to the mid- to late 1990s, sustainable development often is greeted today with skepticism. Many observers conclude that the concept is used to justify whatever someone wants to achieve, whether economic development or environmental protection. Thus, for many, the concept is seen as incapable of providing practical guidance. As a result, in some countries and regions, interest has shifted to the concept of **sustainable livelihoods**, which is seen to be more realistic and focused. In turn, interest in sustainable livelihoods has attracted criticism from those who view it as too anthropocentric. Such a concern is legitimate and needs ongoing debate.

Nonetheless, the concept of sustainable livelihoods does highlight the concerns of the World Commission on Environment and Development (1987: 43) that 'Poverty is a major cause and effect of global environmental problems', that 'many present development trends leave increasing numbers of people poor and vulnerable', and that sustainable development contains within it two key concepts, one of which is 'the concept of "needs", in particular the essential needs of the world's poor, to which overriding priority should be given'. The significance of poverty as one cause of environmental degradation was highlighted in Chapter 1.

In the sustainable livelihood literature, attention is directed towards ways for local people to meet basic needs (food, housing), as well as other needs related to security and dignity, through meaningful work, at the same time minimizing environmental degradation, rehabilitating damaged environments, and addressing concerns about social justice. Strategies for sustainable livelihoods usually aim to create diverse opportunities, efficiency, and sufficiency relative to basic needs, while also achieving social equity and sensitivity regarding environmental integrity.

Environmental Justice

The environmental justice movement combines values and goals from the civil rights, poor peoples', occupational safety and health, and grass-roots environmental movements, and 'the vision that emerges is a qualitative challenge to existing business and government agendas, as well as to the reform agendas of mainstream environmentalism' (Higgins, 1993: 292). These challenges emerged in the late 1970s and early 1980s, and focused on institutional responses related to toxic waste disposal, public facility siting (from airports to electric power transmission lines), species and wilderness preservation, nuclear power and weapons, and the political positions and growing professional orientation of mainstream environmentalism.

Those working within the framework of environmental justice advocate policies and practices for ecologically and economically appropriate development, and argue that pollution *prevention* rather than pollution *control* should be sought. This emphasis on prevention places the environmental justice movement in direct opposition to strategies based on control of pollution, usually reflected in market-based approaches (permits to pollute, user-pay regulations) that are favoured by the neo-conservative ideology mentioned earlier.

Another defining characteristic of environmental justice is a belief in *empowerment* as an end in itself, as well as a means to achieve substantive improvements. For environmental justice activists,

---

DEFINING ENVIRONMENTAL JUSTICE

Environmental justice involves the right to a safe, healthy, productive, and sustainable environment for all living things, in which 'environment' is viewed in its totality, and includes ecological (biological), physical (natural and built), social, political, aesthetic, and economic components. Environmental justice refers to the conditions in which such a right can be freely exercised, through which individual and group identities, needs, and dignities are preserved, fulfilled, and respected in a way that provides for self-actualization and personal and community empowerment.

Adapted from: http://www-personal.umich.edu/~jrajzer/nre/definitions.html

BOX 5.3 ABORIGINAL PERSPECTIVES ON ENVIRONMENT AND RESOURCES

First Nations peoples use the expression Turtle Island to refer to North America, which is thought of as a shell of a turtle surrounded by oceans. The images of the protective shell jutting out and the living creature within make a powerful metaphor for the connection with and respect for the land that all First Nations cultures share.

— Mary Ellen Turpel

We have always been here on this land we call Turtle Island, on our homelands given to us by the Creator, and we have a responsibility to care for and live in harmony with all of her creations. . . . There is harmony in the universe, among our relatives, among the animals and among all creatures. . . . we do not believe in competition, in the survival of the fittest. We believe all should be cared for in our Nations, that caring and sharing, not self-interest, must be our overriding aims. . . .

There are many resources in this country, and Canadians respond to this with massive resource development. We believe that people have forgotten the importance of protecting the environment for future generations. The peoples that I represent, those with an indigenous philosophy, have a world view that is different from that of the corporate mainstream. We have a view of the environment that does not stop all forms of development, but allows it to proceed in a way that respects the environment and ensures that it is protected for future generations. . . . But to respect Mother Earth as a living entity is not easy, particularly when the pre-occupation of economic development may well be to exploit natural resources rather than to preserve or sustain them. That endangers our common survival and the survival of future generations who are relying on us to preserve the planet for them.

— Ovide Mercredi

SOURCE: Mercredi and Turpel (1993: 14, 16, 44, 45, 155–6)

government agencies are viewed as too frequently uncommunicative, unresponsive, and openly hostile (ibid.). As a result, environmental justice advocates promote open, transparent, and accessible decision-making processes. In that context, they believe an important task is to ensure that technical and/or scientific documents should be transparent, i.e., understandable, to citizens who wish to become informed about environmental issues and to participate in environmental planning and decision-making.

Environmental justice is a very broad concept, and incorporates numerous dimensions, as shown in the adjacent box. The concept emphasizes that a safe, healthy, and productive environment should be a right, not a privilege, and that individuals and groups should be heard when decisions are being taken that may alter the quality of the environment and, thereby, human welfare. Box 14.2 in the final chapter also discusses the potential for conflict between the environmental and human justice lobbies.

Ethics and Values

To ensure that a shared vision is endorsed by a group or society, it must be consistent with, and

reflect, basic ethics and values. Alternatively, a vision may outline a desirable future significantly different from the present situation, and for it to be achieved there will have to be a shift in fundamental values. The vision can help to clarify what different values will have to prevail for the different desirable future to be realized.

O'Riordan (1976) identified two sets of values. At one end of a continuum, he identifies **ecocentric values**, which include a belief that there is a natural order governing relationships between living things, and that there is a harmony and balance reflecting a natural order that humankind tends to disrupt through ignorance and presumption. Other key values include reverence for, humility towards, responsibility towards, and stewardship of nature. Those with ecocentric values are not against technology per se but favour application of low-impact technology, oppose bigness and impersonality in all forms, and advocate behaviour consistent with ecological principles of diversity and change. The comments by Mary Ellen Turpel and Ovide Mercredi in Box 5.3 highlight some values of Aboriginal peoples, which often align closely with an ecocentric outlook.

In contrast, a **technocentric perspective** is based on the assumption that humankind is able to understand, control, and manipulate nature to suit its purposes, and that nature and other living and non-living things exist to meet human needs and wants. While ecocentrics are concerned about choosing appropriate ends and using consistent means, technocentrics focus more on means because confidence in human ingenuity and rights causes less concern about assessing the moral aspects of activities or consequences. Technocentrics admire the capacity and power of technology, and believe technology and human inventiveness will be able to overcome possible resource shortages, as well as remediate or rehabilitate environmental degradation. Most, but not all, technocentrics live in major urban centres.

While O'Riordan believes recognizing a spectrum of values and associated ethics is important if resource and environmental managers are to understand why ideas are supported or opposed, he cautions that we should not assume that the values held by a group or society can be neatly allocated into these two categories. The boundaries are often blurred and indistinct, and many people will support certain aspects in both categories, depending on the conditions. In that regard he shares a story about a man who asked a socially conscious friend what he would do if he had two houses. The friend replied that he would keep one and give one to the state. The man then asked what his friend would do if he had two cows. Again, the friend answered that he would keep one and give one to the state. The man next asked what he would do if he had two chickens. The friend responded that he would keep them both. When asked why, the friend stated: 'Because I have two chickens.' This story highlights that circumstances (referred to earlier as 'context') can be very important in shaping outlooks about what is needed, important, and desirable. Thus, it is important for resource and environmental managers to be aware of dominant and secondary values in a society, so that they can determine how these can be used, or may have to be altered, if a vision is to be defined and then achieved. In the guest statement (opposite), Dan Shrubsole provides further insight into the shades of meaning that can be associated with perspectives reflecting different basic values.

Systems and Ecosystem Perspective

The Saskatchewan Round Table on Environment and Economy, in its *Conservation Strategy* published in 1992, stated that 'ecosystems consist of communities of plants, animals and micro-organisms, interacting with each other and the non-living elements of their environment.' In the same year, the final report of the Royal Commission on the Future of the Toronto Waterfront argued that an ecosystem approach emphasizes that human activities are interrelated, and that decisions made in one area affect all others. The Royal Commission concluded that 'dealing effectively with ... environmental problems ... requires a holistic or ecosystem approach to managing human activities.'

Such interpretations of ecosystems and an ecosystem approach reflect the ideas presented in

Grasslands National Park, Saskatchewan (*Parks Canada, Grasslands National Park*).

Toronto waterfront (*Andrew Leyerly*).

Guest Statement

Resources and Management: The Need for an Integrated Perspective and Balanced Philosophy

Dan Shrubsole

Resources can be understood as stocks of material that exist in the natural environment and are valued by humans. However, it is more appropriate in the context of resource and environmental management to highlight their functions and interrelationships with humans, as well as their non-human values. This integrative perspective, at least its human dimensions, was advocated as early as 1933 by Zimmermann, who provided the following definition of the word 'resource'.

> [It] *does not refer to a thing nor a substance but a function which a thing or substance may perform or to an operation in which it may take part*, namely, the function or operation of attaining a given end, such as satisfying a human want. In other words, the word 'resource' is an abstraction reflecting human appraisal and relating to a function or operation. As such, it is akin to words such as food, property, or capital but much wider in its sweep than any of these. (Zimmermann, 1964: 8; emphasis in original)

Resources, resource problems, and solutions need to be understood in the context of the nature of the interrelationships among the biophysical environments, cultures, societies, technologies, laws, economies, and politics of a region. The more recent concept of sustainable development supports this view, although it advocates a focus on people's needs (current and future) rather than 'wants', which were noted by Zimmermann. It also advocates greater consideration of non-human values.

In considering how to solve resource problems, the functions and interrelationships between and within human and natural systems must be considered carefully. In some instances, a philosophy fails to provide adequate treatment to these aspects and may contribute to undesirable and/or unintended outcomes and impacts. For instance, in response to the view that economic activities are a main cause of environmental degradation, some ecocentrics might advocate the adoption of zero growth policies. While these might maintain the current levels of resource consumption, they could contribute to increased conflicts over resources (even wars) because in

Dan Shrubsole

some instances the real problems reflect inequitable distribution of food and wealth. From another philosophical perspective, technocentrics often use data, such as life expectancy, that indicate the environmental crisis has been overstated. Caution must be applied to reaching this conclusion because of the usual selectivity and short duration of the supporting data. This conclusion can also misdiagnose fundamental resource problems that reflect inequitable distributions of wealth and resource availability. Thus, overgeneralization and oversimplification of problems must be avoided. It is important to consider how an environmental philosophy allows us to effectively embrace the integrative aspects of resource and environmental management.

This is easier said than done—many of us inherit rather than consciously select our philosophy. Inheriting a philosophy can happen quite subtly. For instance, we have traditionally labelled resource managers according to their knowledge about resource stocks, such as 'water managers', 'foresters', 'agronomists', 'wildlife managers', or 'land managers'. These labels imply an orientation that overemphasizes functional and biophysical processes relative to the interrelationships between natural and human systems and within human systems. Ensuring that there is an appropriate balance of knowledge (i.e., integrative knowledge) between the human and biophysical systems for resource and environmental managers is, in part, a question of educational philosophy. It is also a question of personal philosophy, needs, wants, and talents. Consciously reflecting on the need for an integrative resource perspective, for both its human and non-human values, may assist you in better defining your own environmental philosophy and environmental problems, and implementing effective solutions.

Dan Shrubsole is an Associate Professor, Department of Geography, University of Western Ontario. His research focuses on sustainable water management, flood hazards, and risk management.

Chapters 2, 3, and 4, in which emphasis was placed upon systems, interrelationships or linkages, energy flows, and ongoing change. In this section, attention turns to the characteristics, opportunities, and challenges provided by the systems or ecosystem approach for resource and environmental management.

Characteristics of the Ecosystem Approach

The Royal Commission on the Future of the Toronto Waterfront identified the following characteristics of an ecosystem approach, noting that it:

- includes the whole system, not just parts of it;
- focuses on the interrelationships among the elements;
- recognizes the dynamic nature of the ecosystem, presenting a moving picture rather than a still photograph;
- incorporates the concepts of carrying capacity, resilience, and sustainability—suggesting that there are limits to human activity;
- uses a broad definition of environments—natural, physical, economic, social, and cultural;
- encompasses both urban and rural activities;
- is based on natural geographic units such as watersheds, rather than on political boundaries;
- embraces all levels of activity—local, regional, national, and international;
- understands that humans are part of nature, not separate from it;
- emphasizes the importance of species other than humans, and of generations other than the present;
- is based on an ethic in which progress is measured by the quality, well-being, integrity, and dignity it accords natural, social, and economic systems.

Opportunities through the Ecosystem Approach

If the above attributes are accepted as defining an ecosystem approach, they present some basic challenges to contemporary environmental and resource management. First, in a Western, industrialized society such as Canada, most humans believe they have a dominant role relative to nature, that the environment and natural resources exist to satisfy human needs and wants. This is an anthropocentric perspective, in contrast to the ecocentric world view described earlier. By emphasizing that humans are part of nature rather than separate from it, and by recognizing the inherent value of non-human species and things, the ecosystem approach questions such a belief.

Second, by taking a holistic perspective focusing on interrelationships, the ecosystem approach reminds us of the need to consider management problems and solutions in the context of linked 'systems'. It forces us to appreciate that decisions made about one system, such as land, can have consequences for other systems, such as water or wildlife, and vice versa. Thus, many flooding problems often are associated with land-use practices such as removing vegetation, which accelerates runoff, or with allowing development in areas subject to regular inundation. Concentrating only on the aquatic system to reduce flood damages is unlikely to be effective, since many land-based activities exert some influence on this system.

In contrast, the conventional approach to environmental or resource management often has focused on systems in isolation from one another, as reflected by one government agency being responsible for forestry, another for wildlife, another for water, another for agriculture, and another for urban development. All of these agencies have a shared interest in resources such as wetlands, but if they each focus only on their own goals or interests they can undermine the activities or values of other agencies.

Third, the ecosystem approach demands that the links between natural and economic or social systems be considered. Such a focus also is one of the basic thrusts behind sustainable development. When such linkages are recognized, it becomes apparent that certain thresholds normally exist in natural systems, and exceeding these thresholds leads to deterioration and degradation. For example, agricultural production can be increased by adding chemical inputs (fertilizers, pesticides, herbicides). However, the cumulative effects of agrochemicals may eventually cause safety problems for human consumption of the product being grown (e.g., fruit or vegetables). Other concerns also may be created through introducing chemicals into adjacent environments, resulting in eutrophication and pollution, as discussed in Chapters 4 and 10. This reminds us that while sustainable development accepts the need for development to meet basic human needs, some kinds of growth are not sustainable if they lead to degradation of natural, economic, or social systems.

Fourth, the holistic perspective reminds us that decisions made at one place or scale can have implications for other places or scales. If a community deposits untreated sewage into an adjacent river, people and communities downstream will bear some costs related to that action. Or, if a community, state, or nation is unwilling to reduce emissions from factories, some of the costs will be borne by people, states, or nations downwind, since air pollutants often are carried well beyond the borders of the community or state in which the pollutants are generated, as is the case with acid deposition, discussed in Chapter 4.

Fifth, given the impacts of decisions in one place for people and activities in other places, the ecosystem approach raises questions regarding what is the most appropriate areal or spatial unit for planning and management. The conventional management unit usually has been based on political or administrative boundaries, such as municipal, regional, provincial, or national boundaries. In contrast, the ecosystem approach suggests that areas identified on the basis of other units, such as watersheds or airsheds, may have more functional value. For example, in managing for migratory birds that travel from the Gulf of Mexico to the Canadian Arctic, national boundaries have little relevance. The management area in this situation transcends at least three nations (Canada, United States, Mexico). If managers in each nation develop plans and strategies without regard to what is being done in other jurisdictions, the effectiveness of their separate initiatives is likely to be modest, as discussed in Chapter 11.

Sixth, an ecosystem approach highlights that systems are dynamic or continuously changing. An ecosystem, whether a local wetland, prairie grassland, boreal forest, or urbanizing area, is not static. In addition to daily, seasonal, and annual variations, ongoing longer-term changes occur, as illustrated by the transition of natural grasslands to cultivated cropland, or of farmland to urban land use. This aspect of change also emphasizes why managers and management strategies must be capable of adapting or adjusting to evolving situations. This latter aspect has led to interest in *adaptive management*, discussed further in Chapter 6.

In summary, the ecosystem approach incorpo-

SPATIAL UNITS FOR ECOSYSTEM-BASED MANAGEMENT

At the most basic level, problems derive from the definition of the units for which planning and management are undertaken. Management units often bear no relationship to the realities of ecological problems (even the home range of the species for which protection is sought), their connections to economic and social processes, or local people's cultural and political identity. Instead, they are arbitrary units defined by lines drawn on the map—lines often drawn by someone who has never been to the region and who, for example, decides a river would make a good boundary.

— Slocombe (1993: 616)

rates the key ideas that humans are part of nature rather than separate from it, that a perspective emphasizing interrelationships is needed, and that critical thresholds exist. When these aspects are combined, it can be appreciated why the Royal Commission on the Toronto Waterfront (1992: 31–2) concluded that:

the ecosystem approach is both a way of doing things and a way of thinking, a renewal of values and philosophy. It is not really a new concept: since time immemorial, aboriginal peoples around the world have understood their connectedness to the rest of the ecosystem—to the land, water, air, and other life forms. But, under many influences, and over many centuries, our society has lost its awareness of our place in ecosystems and, with it, our understanding of how they function.

The need and rationale for adopting an ecosystem approach was spelled out some time ago by the Conservation Authorities of Ontario (1993: 2):

The fundamental problem that exists in resource management today is not financial constraint. It is that the current body of legislation, agency structures and mandates do not recognize the concept of ecosystem-based management. The overlapping of mandates between the Ministries of Natural Resources, Environment and Energy, Agriculture and Food and Municipal Affairs and Conservation Authorities and Municipalities is evident to everyone.

The situation has evolved over time as public agencies reacted to specific problems with specific solutions. This issue-by-issue approach results in a situation that, when viewed from an ecosystem perspective, borders on the ludicrous. Thus, many reasons can be identified as to why an ecosystem approach should be used more frequently in Canada for environmental management.

Long-term View

In resource and environmental management it is important to have a long-term view (more than 15 years), while also being able to identify actions that should be taken in the short term (less than 5 years) and middle term (5–15 years). The rationale is that systems often change slowly and that a significant period of time may be required to shift values, attitudes, and behaviour.

Furthermore, many of our environmental problems have emerged after many decades, or even centuries, so it is unrealistic to assume that they can be reversed or 'fixed' in a few years. For example, changes in the aquatic ecology of the Great Lakes (see Chapters 3 and 4) reflect decades of different kinds of land use (agriculture, industry, settlement) on lands adjacent to the lakes.

Unfortunately, our society usually wants 'instant results', and does not often show the patience required to deal with problems that have been created over a long time period. There are many reasons for the short-term perspective. One is the relatively short time between elections (up to five years for federal and provincial elections, and two to three years for municipal elections). Politicians normally want to have tangible results so they can show what has been accomplished during their term of office and why they should be re-elected. In contrast, those running for office for the first time emphasize how little has been accomplished by those previously elected and why they should be given opportunity to demonstrate what they could accomplish. This mindset drives decisions focusing on short-term, tangible results and usually results in low priority being given to long-term strategies involving intangible outcomes.

In many instances, changes are required in basic values, attitudes, and behaviour, such as having Canadians modify basic patterns of activity so that the air quality targets identified for Canada in the Kyoto Protocol can be achieved (see Chapter 7).

There was much outcry over the years 2002–4 against targets committed to by the federal government under the Kyoto agreement, for fear that it would cause economic hardship or disadvantage. The Premier of Alberta was one of the most aggressive in criticizing the targets, since they could eventually lead to less reliance on a 'petroeconomy', the backbone of the Alberta economy. Thus, his criticism, while focused on the possible negative short-term consequences for the Alberta economy, reflected the reality that, when changes occur, some gain and some lose. Those who might lose are understandably not usually very enthusiastic.

The focus on tangible results is also why, when budget cutbacks are experienced in the public education system, one of the first 'casualties' usually is outdoor education and field trips, which are viewed as a luxury. And yet, as supporters of outdoor education and field trips argue, if young students are to appreciate the role of natural systems and change their attitude towards them, what better way to accomplish that than by exposing the students directly to these systems through field trips or outdoor education programs. Unfortunately, the 'product' or result of such programs—changed values and attitudes—is difficult to document in a tangible way, and it often takes many years before behaviours are influenced and changed in order to have significant positive consequences for the environment.

Thus, while a long-term perspective is required, accompanied by patience, perseverance, determination, and commitment, often most of our society is more preoccupied with shorter-term and visible results. This creates many difficulties for those believing that solutions will require a commitment of funds and human resources over an extended and sustained period of time.

IMPLICATIONS

The concepts, or 'philosophy', that reflect best practice for management of natural resources and the environment are becoming more clearly identified. Important concepts include (1) recognizing the contextual factors that characterize a problem situation and being willing to design solutions that address specific attributes of the problem; (2) establishing a vision that identifies a desirable future so that appropriate means can be identified to achieve the desired end; (3) appreciating the

BOX 5.4 WHAT YOU CAN DO

1. Ask elected officials and designated experts to explain how solutions to problems have been designed with regard to the particular conditions in the area of concern.
2. Watch for, and challenge, placebo policies and practices.
3. Identify a desirable future consistent with ideas presented in this chapter and book, and then ensure that your behaviour is consistent with this vision.
4. Urge your college or university to include ideas and practices consistent with ideas in this chapter in its curriculum and practices.
5. Strive to incorporate more ecocentric values into your personal outlook on living.
6. Expect public agencies to use a 'systems approach' when dealing with resource and environmental problems, and challenge them when that does not seem to be happening.
7. Advocate both short- and long-term thinking and action related to environmental problems.

strengths and limitations of potential desirable futures, such as sustainable development, sustainable livelihoods, and environmental justice; (4) clarifying underlying values that influence attitudes and behaviour, and developing ethical principles or guidelines consistent with the desired future; (5) adapting a systems perspective to ensure the interactions of various environmental and human subsystems are considered; and (6) looking beyond the present and immediate future to consider the longer term. In the spirit of transforming these ideals into action, Box 5.4 outlines initiatives that you can undertake as an individual or as part of a group.

SUMMARY

1. Context refers to the specific conditions related to a time and place. Since context can vary significantly from place to place and time to time, it is desirable to design solutions to fit specific environmental problems. A challenge in doing so is that some people may conclude that a region or group is receiving special consideration.

2. Attention to context reinforces the importance of incorporating experiential knowledge of local people with scientific knowledge when seeking to understand problems and develop solutions. Based on living in an area for many years, local people often have insight and understanding that scientists do not have.

3. Important aspects of context include rapid change, high complexity and uncertainty, and significant conflict. We should not be surprised to encounter such aspects when dealing with environmental issues.

4. Neo-conservatism has become an increasingly important ideology shaping the context for environmental problem-solving. Characteristics include reducing scope of government, relying on market mechanisms to allocate scarce resources, and accepting globalization. This ideology has driven many governments to be preoccupied with deficit and debt reduction and to commercialize or privatize management functions previously taken on by governments.

5. The subsidiarity principle, which stipulates that people closest to problems should participate directly in decisions affecting them, encourages involving local people in decisions. It also is used to justify approaches based in a neo-conservative ideology because decisions to download or privatize responsibilities are justified using the subsidiarity principle.

6. A vision represents a realistic, credible, and attractive future for a region, community, or group. It is ideal to have a shared vision, one to which many people are committed. Achieving a shared vision is challenging because many interests exist in a society, some of which are mutually exclusive. Without a vision or a well-defined end point, managers, their political masters, and society itself have difficulty in determining the most appropriate means to apply to achieve the desired ends.

7. Sustainable development, sustainable livelihoods, and environmental justice each are possible visions to guide environmental management.
8. Ecocentric and technocentric world views reflect different basic values and interests. Their existence is one explanation for why conflicts arise when decisions must be made related to the environment or resources.
9. Ecosystems consist of communities of plants, humans, animals, and micro-organisms interacting with each other and the non-living elements of their environment, and their management requires a systems or holistic perspective.
10. A long-term view (15 years of more) needs to be maintained, but, at the same time, there must be the capability to deal with immediate or short-term issues. Too often, only a short-term perspective is taken.

KEY TERMS

context	ecocentric values	sustainable development	technocentric
custom-designed solutions	environmental justice placebo policies	sustainable livelihoods	perspective vision

REVIEW QUESTIONS

1. Can science be 'objective' and provide unbiased input to inform management decisions related to natural resources and the environment? What are the reasons for your position in that regard?
2. Why do many believe it is desirable to custom-design solutions to the specific conditions of a problem, rather than design approaches to apply uniformly throughout a region?
3. What is the 'ingenuity gap'?
4. Explain why the principle of subsidiarity generates debate.
5. What are the implications of neo-conservatism for resource and environmental management?
6. Explain what a placebo policy is. Are any of these used by your local or provincial governments, or by the federal government, related to the environment?
7. Why do many consider establishing a vision to be important for resource and environmental management?
8. In what ways are sustainable development and sustainable livelihoods different? Which do you believe is more useful?
9. What would a strategy that incorporated principles of environmental justice look like if it were to be developed for your community?
10. Do you believe you are primarily ecocentric or technocentric in your values and behaviour? What are the implications of this view for how you interact with the natural environment?
11. What are the major obstacles to effective implementation of an ecosystem-based approach? What would have to be the first things to change if implementation were to improve?
12. If there were to be a model of an ideal ecosystem approach, what characteristics would it have?
13. Why is there a predisposition to favour short-term rather than long-term strategies related to resources and the environment? What is the best example of a long-term strategy in your community or province?

RELATED WEBSITES

ENVIRONMENTAL COMMISSIONER OF ONTARIO:
  http://www.eco.on.ca
ENVIRONMENTAL JUSTICE:
  http://www.epa.gov/compliance/environmentaljustice
NATIONAL ROUND TABLE ON ENVIRONMENT AND
  ECONOMY: http://www.nrtee-trnee.ca

SUSTAINABLE LIVELIHOODS:
  http://www.undp.org/sl
  http://www.undp.og/sl/Documents/documents
WORLD SUMMIT ON SUSTAINABLE DEVELOPMENT,
  JOHANNESBURG, 2002:
  http://www.un.org/esa/sustdev/index

REFERENCES AND SUGGESTED READING

Amalric, R. 1998. 'Sustainable livelihoods: entrepreneurship, political strategies and governance', *Development* 41: 31–41.

Bernstein, S., and B. Cashore. 2000. 'Globalization, fourth paths of internationalization and domestic policy change: the case of ecoforestry in British Columbia, Canada', *Canadian Journal of Political Science* 33: 67–99.

Carney, D., M. Drinkwater, T. Rusinow, K. Neefjes, S. Wanmall, and N. Singh. 2000. 'Livelihood Approaches Compared', draft prepared Nov. 1999, revised Feb. 2000. Available at: <www.undp.org/sl/documents>.

Conservation Authorities of Ontario. 1993. *Restructuring Resource Management in Ontario: A Blueprint for Success*. Mississauga, Ont.: Credit Valley Conservation Authority.

Coward, H., R. Ommer, and T. Pitcher. 2000. *Just Fish: Ethics and Canadian Marine Fisheries*. Social and Economic Papers No. 23. St John's: ISER Books.

Dwivedi, O.P., P. Kyba, P.J. Stoett, and R. Tiessen, eds. 2001. *Sustainable Development and Canada: National and International Perspectives*. Peterborough, Ont.: Broadview Press.

Environmental Commissioner of Ontario (Gord Miller). 2003. *Environmental Commissioner of Ontario 2002–2003 Annual Report*. Toronto: Environmental Commissioner of Ontario.

Higgins, R.R. 1993. 'Race and environmental equity: an overview of the environmental justice issue in the policy process', *Polity* 26: 281–300.

Homer-Dixon, T. 2000. *The Ingenuity Gap*. New York: Alfred A. Knopf.

Krajnc, A. 2000. 'Wither Ontario's environment? Neoconservatism and the decline of the Environment Ministry', *Canadian Public Policy* 26: 111–27.

Kreutswizer, R. 1998. 'Water resources management: the changing landscape in Ontario', in R.D. Needham, ed., *Coping with the World around Us: Changing Approaches to Land Use, Resources and Environment*, Department of Geography Publication No. 50. Waterloo, Ont.: University of Waterloo, 135–48.

McKenzie, J.I. 2002. *Environmental Politics in Canada: Managing the Commons into the Twenty-first Century*. Toronto: Oxford University Press.

Mercredi, O., and M.E. Turpel. 1993. *In the Rapids: Navigating the Future of First Nations*. Toronto: Viking.

Nanus, B. 1992. *Visionary Leadership*. San Francisco: Jossey-Bass

Ommer, R. 2000. *Just Fish: Ethics and Canadian Marine Fisheries*. St John's: Institute of Social and Economic Research, Memorial University of Newfoundland.

O'Riordan, T. 1976. *Environmentalism*. London: Pion.

Parson, E.A. 2000. 'Environmental trends and environmental governance in Canada', *Canadian Public Policy* 26 (Supplement): S123–43.

Reed, M.G. 2003. *Taking Stands: Gender and the Sustainability of Rural Communities*. Vancouver: University of British Columbia Press.

Royal Commission on the Future of the Toronto Waterfront. 1992. *Regeneration*. Ottawa and Toronto: Minister of Supply and Services Canada and Queen's Printer of Ontario.

Saskatchewan Round Table on Environment and Economy. 1992. *Conservation Strategy for Sustainable Development in Saskatchewan*. Regina: Saskatchewan Round Table on Environment and Economy.

Scoones, I. 1998. *Sustainable Rural Livelihoods: A Framework for Analysis*. Brighton, UK: University of Sussex, Institute of Development Studies.

Singh, N., and S. Wanmali, 1998. 'Sustainable Livelihoods Concept Paper'. United Nations Development Program, Sustainable Livelihoods Documents, at: <http://www.undp.org/sl.Documents.html>.

Slocombe, S.D. 1993. 'Implementing ecosystem-based management: development of theory, practice and research for planning and managing a region', *BioScience* 43: 612–22.

Stewart, J.M. 1993. 'Future state visioning—a powerful leadership process', *Long Range Planning* 26: 89–98.

Stonehouse, D.P. 1994. 'Canadian experiences with the adoption and use of soil conservation practices', in T.L. Napier, S.M. Camboni, and S.A. El-Swaify, eds, *Adopting Conservation on the Farm*. Ankeny, Iowa: Soil and Water Conservation Society, 369–95.

Tinker, J. 1996. 'Introduction', in J.B. Robinson et al., eds, *Life in 2030: Exploring a Sustainable Future for Canada*. Vancouver: University of British Columbia Press, ix–xv.

World Commission on Environment and Development. 1987. *Our Common Future*. Oxford: Oxford University Press.

Zimmermann, E. 1964 [1933]. *Introduction to World Resources*. New York: Harper & Row.

Chapter 6

Planning and Management:
Process, Method, and Product

Learning Objectives

- To understand the strengths and limitations of key methods and processes related to resource and environmental management.
- To know the distinction between collaboration and co-ordination.
- To appreciate the benefits and limitations of participatory approaches.
- To understand the obstacles inhibiting communication of scientific results and conclusions to policy-makers and the lay public.
- To examine why adaptive management has been designed to expect uncertainty and surprise.
- To gain an understanding of the connection between risk and impact assessment.
- To identify alternative approaches to resolving conflicts.
- To realize the relationship between regional and land-use planning, and resource and environmental management.

INTRODUCTION

In Chapter 5, we argued that planning and management can reflect basic philosophies, processes, methods, and products. In that chapter, the emphasis was on concepts reflecting philosophy. In this chapter, attention turns to various processes, methods, and products with regard to resource and environmental management. This chapter provides a lot of information. To make it less 'abstract', we suggest that after reading each section you pause and consider the implications of

what you have just read to the Sydney Tar Ponds case study in Chapter 1, as well as what this reading might mean relative to an issue or problem in your own community.

COLLABORATION AND CO-ORDINATION

In Chapter 5, it was noted the systems or ecosystem approach emphasizes that different components of resource and environmental systems are interconnected, and that therefore a holistic perspective is required. However, for practical reasons, public agencies often focus on a subset of resources or the environment. Hence, we have departments of Agriculture, Environment, Forestry, Water, and so on. If a holistic approach is to be taken, there is a need for collaboration among the various agencies and with other stakeholders.

For Himmelman (1996: 29), **collaboration** involves exchanging information, modifying activ-

COLLABORATION AND CO-ORDINATION

Collaboration: working together
Co-ordination: harmonious adjustment

RATIONALE FOR COLLABORATION

Many of our problems appear insoluble because our ... conceptions about how to manage in an increasingly interconnected world are limited. Our current problem-solving models frequently position participants as adversaries, pit them against one another, and leave them to operate with an incomplete appreciation of the problem and a restricted vision of what is possible.

The impasses are typically characterized as technical or economic. I argue ... that the roadblocks are as much conceptual and organizational as they are technical and economic. Many of our problems are unsolvable because our systems of organizing are not geared for a highly interdependent environment.

— Gray (1989: xvii–xviii)

ities in light of others' needs, sharing resources, and enhancing the capacity of others to achieve mutual benefit and to realize common goals or purposes. Selin and Chavez (1995) added other dimensions, arguing that collaboration involves joint decision-making to resolve problems, with power being shared and stakeholders accepting collective responsibility for their actions and the outcomes.

Collaboration is needed within, between, and among organizations. Once collaboration is agreed to, then **co-ordination**—the effective or harmonious working together of different departments, groups, and individuals—can be sought through various arrangements. For example, interdepartmental committees or task forces often provide a way to co-ordinate the activities of different agencies and stakeholders. Such interdepartmental mechanisms provide the means through which effective collaboration can be achieved. Different public participation processes, addressed in the next section, also can be used to facilitate collaboration.

Collaboration is increasingly accepted as desirable because the complexity and uncertainty associated with resource and environmental issues create a challenge for any individual or organization to have sufficient knowledge or authority to deal with them. Indeed, this is why multi-, inter-, and transdisciplinary research teams are often used, as explained in Chapter 1. Furthermore, different values and interests contribute to conflict, providing another reason for various stakeholders to come together to determine how they can meet

their respective needs. Done well, collaboration involves sharing of information and insight to achieve multiple goals. When this is accomplished, collaboration can contribute to more open, participatory processes, and to solutions to which different stakeholders feel committed. In that manner, the likelihood of finding acceptable solutions that lead to effective implementation is increased.

However, collaboration should not be viewed as being always accepted or endorsed by everyone. Some individuals or organizations may decide not to reach out to other groups because they are determined to satisfy their own interests and goals, regardless of what others want. And when such stakeholders are powerful, they often choose to ignore other interests and concerns, as they single-mindedly pursue their own interests. Another potential disadvantage of collaboration is that all parties may compromise their principles or interests to reach a 'common denominator' that may not always represent a good, long-term decision. As a result, we should always remember that collaboration and co-ordination are means to an end, and not an end in themselves. We should seek to apply collaborative approaches with the purpose of achieving an accepted vision for a desirable future. A notable success in that regard is outlined in Box 6.1

STAKEHOLDERS AND PARTICIPATORY APPROACHES

The Manitoba Round Table on Environment and Economy (1992) argued some time ago that one of 10 basic principles for sustainable development should be *shared responsibility* and that one of six guidelines to achieve sustainable development should be *public participation*. The Manitoba Round Table concluded that it was essential to 'encourage and provide opportunity for consultation and meaningful participation in decision-making processes by all Manitobans', as well as to ensure 'due process, prior notification and appropriate and timely redress for those affected by policies, programs, decisions and developments'. This principle and guideline continue to be relevant today for all jurisdictions across the country.

Here we focus on the nature of public participation as a means to reallocate *power* among participants, on alternative ways to facilitate *empowerment* of people relative to the environmental manage-

BOX 6.1 BOUNDARY WATERS TREATY, 1909

Canada and the United States share many lakes, rivers, aquifers, and wetlands along their 8,840-km border. Indeed, more than half of the border passes through water bodies. Examples include the Saint John and Saint Croix rivers in New Brunswick, the Great Lakes–St Lawrence River system in Ontario and Quebec, the Souris River in Saskatchewan and Manitoba, the Red River in Manitoba, the St Mary and Milk rivers in Alberta, and the Columbia River in BC.

The two countries signed the Boundary Waters Treaty in 1909 to provide a framework for a collaborative approach and resolution of disputes over water resources shared by both countries. For example, there were concerns during the 1970s in Canada about the Garrison Diversion project, proposed to transfer water from the upper reaches of the Missouri River into North Dakota to irrigate over 100,000 hectares of farmland, increase municipal water supplies, and support fish and wildlife management. Other benefits to the state would be flood control, land drainage, and stream flow regulation. However, the 'return flow' water, which would leave North

Dakota through the Souris and Red rivers to flow northward into Manitoba, was at issue. Of particular concern was the possibility of foreign biota, especially some species of trash fish, entering Lake Winnipeg and destroying the commercial fishery there. Projects such as the Garrison Diversion are examined through the International Joint Commission, set up under the Boundary Waters Treaty.

The International Joint Commission, or IJC, has six commissioners, three from Canada and three from the United States. The IJC has power to adjudicate regarding issues associated with water development proposals in one country that could affect water levels or flows in the other country and to serve as a commission of inquiry when matters are referred to it by the two federal governments. The commissioners examined the proposed Garrison Diversion, which was not completed.

The six commissioners are expected to consider what would be best for both countries, rather than being advocates for their own country, and the record indicates that they have effectively done exactly that.

ment process. The idea of sharing power has introduced a significant new dimension to public participation.

Degrees of Sharing in Decision-making

In Chapter 5, basic characteristics of sustainable development and environmental justice were outlined. Several characteristics—reducing injustice, increasing self determination—provide the rationale for sharing power in environmental management. However, redistribution of power for environmental management from government agencies and private firms to First Nations people and other members of the public can challenge vested interests. As a result, it is not always easily or readily accepted. Furthermore, power-sharing raises challenging questions about accountability and responsibility for decisions.

Arnstein (1969) provided a helpful perspective regarding the issue of power redistribution by identifying what she called 'rungs' on the ladder of citizen participation (Table 6.1).

As long ago as the late 1960s and early 1970s in Canada, public involvement programs had

moved up to Arnstein's rungs of informing, consultation, and placation. In other words, environmental managers often incorporated public participation into environmental management

TENSIONS EXIST FROM USE OF PARTICIPATORY APPROACHES

Useful and legitimate synthesis of expert knowledge with democratically accountable deliberation and decision-making poses grave challenges, both conceptual and practical, to the design of policy processes and institutions. At a conceptual level, the domains of science and of democratic politics have different goals, standards of merit, norms of participation, and procedures for resolving differences. At a practical level, desired knowledge is often unavailable, and available relevant knowledge is often not adequately used. Knowledge is often inadequate to give high confidence in the consequences of decisions, and decisions sometimes cannot be delayed until high confidence is obtained. Uncertainty is thus unavoidable and pervasive in environmental governance....

— *Parson (2000: S128)*

Table 6.1	Rungs on the Ladder of Citizen Participation	
Rungs	**Nature of Involvement**	**Degrees of Power**
8. Citizen control		
7. Delegation	Citizens are given management responsibility for all or parts of programs	Degrees of citizen power
6. Partnership	Trade-offs are negotiated	Degrees of tokenism
5. Placation	Advice is received from citizens but not acted on	
4. Consultation	Citizens are heard but not necessarily heeded	
3. Information	Citizens' rights and options are identified	Non-participation
2. Therapy	Powerholders educate or 'cure' citizens	
1. Manipulation	Rubberstamp committees	

SOURCE: Adapted from Arnstein (1969).

initiatives. However, because of a belief that public agencies were accountable for resource allocation decisions and expenditure of public funds, the position usually taken by public agencies was that information and advice received through public participation was only one of several sources to be considered, and that the public agency would retain decision-making authority.

Resource and environmental managers believed that they had the legal mandate, responsibility, and power to decide which trade-offs best reflected societal needs and interests and to make final decisions. This viewpoint usually was reinforced from the conviction that no one public interest existed, but rather that many different interests were present and frequent conflicts among them were common. Giving responsibility for decision-making to citizens was often viewed by public agencies as dangerous, as this could too easily evolve into a form of anarchy in which no one was responsible or accountable for decisions.

During the 1980s, increasing dissatisfaction emerged with the process, methods, and products associated with many resource and environmental management decisions. Growing numbers of Canadians rejected the idea that 'technically correct' answers always could be found. Instead, the prevailing view became that such decisions ultimately depended on weighing conflicting goals, aspirations, and values. In these situations, technical or scientific expertise was a legitimate input, but only one of several. Furthermore, as decisions seemed to become more complex, with more considerations and trade-offs, it became less credible that any one expert or organization could have all of the necessary insight to reach a 'wise' decision.

From these considerations arose the idea that 'stakeholders' had a right to participate in decisions. **Stakeholders** are those who should be included because of their direct interest, including (1) any public agency with prescribed management responsibilities, (2) all interests significantly affected by a decision, and (3) all parties who might intervene in the decision-making process to facilitate, block, or delay it. Because more and more decisions of an increasingly complex nature had to be made, the traditional forum for public participation—the political process with elected representatives consulting and then reflecting constituents' views—no longer seemed adequate. As a result, various individuals and non-governmental organizations (NGOs) began pressing for public involvement to move higher up on Arnstein's ladder. While not many politicians and public servants believed that total 'citizen control' was feasible or desirable, expectations were that 'partnerships' and 'delegated power' that gave effective power to the public were desirable and achievable.

As an example, by the early 1990s the British Columbia Ministry of Environment, Lands and Parks (1992: 4) was committed to 'shared responsibility and partnerships in reaching the goals of environmental protection and sustainability; consultation and discussion; . . . strong representation and advocacy for environmental protection and enhancement'. For its part, the Saskatchewan Round Table (1992) commented that there was 'value [in] a full and open declaration of diverse interests, early in the consultative process', which could be used to develop trust and build consensus. However, the Saskatchewan Round Table also recognized that such an approach can be 'time-consuming, costly and sometimes frustrating'. Nevertheless, it concluded that unilateral decisions that superficially appear efficient very often end up in 'costly lengthy appeals or court action'.

While the above statements support the use of participatory and stakeholder approaches and endorse the sharing of power with members of the public, in practice this is not always achieved. For example, an environmental impact assessment of a hog-processing plant and associated waste treatment plant in Brandon, Manitoba, did not achieve many of these ideals. Maple Leaf Pork decided to locate a $120-million hog-processing plant in Brandon after a nine-month search that considered proposals from 42 communities in western Canada. The final decision to locate in Brandon involved active participation in the negotiations with Maple Leaf Pork by the Premier of Manitoba and the mayor of Brandon, and together the two levels of government offered almost $20 million in subsidies. For the provincial government, the motive was to promote the hog industry in Manitoba as part of a policy of agricultural diversification for the province. For the city, there was the prospect of 2,100 new jobs and an overall population increase of 7,000 by 2007.

The impact assessment was conducted under provincial legislation. As the process began, the business community, led by the Chamber of Commerce, supported the project. A citizens' group was formed by those who opposed the project for various reasons, the most important of which were opposition to public subsidies to private corporations, concern about negative envi-

> ### EMPOWERMENT IS NOT ALWAYS ACHIEVED: ENVIRONMENTAL ASSESSMENT OF A HOG-PROCESSING PLANT IN BRANDON, MANITOBA
>
> *This was a classic case of 'decide-announce-defend' that saw little or no input from members of the community beyond the political and business leadership. In many respects, the EA [environmental assessment] became a sidelined issue that further marginalized the already disempowered project opponents.*
>
> *. . . The EA process in this case was at best legitimating and certainly not participatory, empowering, or equitable.*
>
> — *Diduck and Mitchell (2003: 360)*

ronmental impacts, and unhappiness about lack of transparency in the process. However, the relevant ministry decided not to hold public hearings, despite demands for them. Instead, the public was given opportunity to participate by commenting through a public registry, comment periods when the public was invited to send in comments regarding the proposed project at selected periods of time and asked to respond by a specified time, and access to appeal procedures. Ten public information meetings also were organized.

Despite situations such as this, in which participation is very limited, the idea of **partnerships** among governments, private companies, and the general public has become increasingly popular. The partnership concept has been implemented through **co-management** initiatives and other approaches that reflect a genuine reallocation of power to citizens and away from elected officials or technical experts. Co-management arrangements have been developed particularly with First Nations peoples regarding management of forests, fish, and wildlife. In these situations, power is allocated to First Nations or other local people. Some of the challenges associated with power-sharing with First Nations are highlighted within the context of marine resources in Chapter 8.

COMMUNICATION

At an international conference focused on 'climate change communication', the organizers observed that communication has three main purposes: (1) raise awareness, (2) confer understanding, and (3) motivate action (Scott et al., 2000: iii). Furthermore, at the same meeting, Andrey and Mortsch (2000: WP1) argued that 'communication is thought to be effective only when these changes in awareness and understanding result in attitudinal adjustments and/or improve the basis upon which decisions are made.'

Carpenter (1995) identified important aspects to consider regarding how communication of scientific understanding should be conducted. These include:

• Much of the public does not understand science, or how scientific research is conducted.
• Other than for weather and gambling, much of the public does not understand 'probability', and the idea of risk as part of life is rejected by many persons. Association and causation are often confused or assumed to be the same.
• The media do not deal well with the 'ebb and flow' of scientific research, which progresses by new research disproving or challenging existing understanding. This creates confusion and doubt about the authority or credibility of scientists when different views exist.
• In the courts, expert witnesses appear for different sides in a case and present conflicting testimony. The same should be understood in relation to scientific 'findings'.

To overcome the communication challenge, we first must recognize that different target audiences exist, such as other scientists, planners and managers, elected decision-makers, and the general public. Consequently, we should be sure that messages are crafted with regard to who will be the recipient and what can reasonably be assumed to be their level of understanding. To guide the preparation of messages, Carpenter further recommended attention to four complementary questions, regardless of the audience:

1. What do we know, with what accuracy, and how confident are we about our data?
2. What don't we know, and why are we uncertain?
3. What could we know, with more time, money, and talent?
4. What should we know in order to act in the face of uncertainty?

With regard to the final question, Carpenter reminds us that risk assessment does not necessarily lead to risk reduction if the uncertainty is not understood and appropriate action is not taken. To illustrate, he tells an amusing tale about a prospective air traveller worried about riding on an airplane in case another passenger might be carrying a bomb. He thus called the airline office and asked what the probability was of a passenger being on board his flight with a bomb. The answer: a very low probability. He then asked about the probability of two passengers being on the same flight, each carrying a bomb. The answer: completely improbable. He then decided to carry a bomb himself on his flight!

The important message in this section is that while it is important to achieve understanding of natural and human systems and their interactions, it also is important to determine how this knowledge and insight can be shared with others who may not have the same scientific background but are still key stakeholders in terms of taking, facilitating, or thwarting action. To test yourself, consider the four questions above with regard to the Sydney Tar Ponds case study in Chapter 1 and determine how you would answer each question, and how you would craft your answer as the target audience changed from scientists, to elected decision-makers, to the general public.

ADAPTIVE MANAGEMENT

Surprise, Turbulence, and Change

Trist (1980) observed that there is no such thing as *the future*, but instead there are *alternative possible futures*. Which future actually occurs depends very much on choices made and on actions taken to implement those choices. He argued that 'the

---

COMMUNICATING UNCERTAINTY: CATCH-22

If environmental professionals are candid about uncertainties, the client/recipient will likely be disappointed and perhaps berate the professionals for not being helpful and just simply telling them what is going to happen and what to do. If the environmental professionals ignore or obscure the uncertainties and give unambiguous predictions and advice, events may very well show them to be substantially wrong. Then their credibility will be gone, and they will not be consulted in the future.

— Carpenter (1995: 129)

paradox is that under conditions of uncertainty one has to make choices, and then endeavour actively to make these choices happen rather than leave things alone in the hope that they will arrange themselves for the best.'

A challenge in making choices and taking initiatives is that what Trist referred to as *turbulent conditions* have become increasingly prevalent. For example, energy plans in Canada became obsolete in the early 1970s when OPEC countries rapidly quadrupled oil prices. Such an increase in prices had not been included in the forecasts and assumptions on which energy plans had been based. During the 1980s when decisions were being made about the east coast fishery, there was no expectation that in the early 1990s the cod fishery would be effectively closed and thousands of people in Atlantic Canada would lose their jobs (see Chapter 8). More recently, few anticipated the terrorist attack on the World Trade Center in New York City in mid-September 2001. These and other events came as surprises to most people. They created bewilderment and anxiety, and raised doubts regarding the capability of science, planning, and planners because decision-makers and their decision process apparently could not anticipate or adapt to rapid change.

Adaptive Environmental Management

Although awareness of the need for **adaptive environmental management** is not new, it was popularized by Holling (1978, 1986) and his colleagues. They concluded that policies and approaches should be able to cope with the uncertain, the unexpected, and the unknown. Holling and his co-workers observed that the customary way of handling the unknown has been through trial and error. Errors or mistakes provide new information to allow subsequent activity to be modified. As a result, 'failures' generate new information and insight, which in turn lead to new knowledge.

However, effective trial-and-error management has preconditions. The experiment should not destroy the experimenter. Or, at a minimum, someone must remain to learn and benefit from the experiment. The experiment also should not cause irreversible, negative changes in the environment. Furthermore, the experimenter should have the will and capability to begin over.

> ### COMMENTARY: THE NATURE OF ADAPTIVE MANAGEMENT
>
> Adaptive management *is an approach to natural resource policy that embodies a simple imperative: policies are experiments;* learn from them. *In order to live we use the resources of the world, but we do not understand nature well enough to know how to live harmoniously within environmental limits. Adaptive management takes that uncertainty seriously, treating human interventions in natural systems as experimental probes. Its practitioners take special care with information. First, they are explicit about what they expect, so that they can design methods and apparatus to make measurements. Second, they collect and analyze information so that expectations can be compared with actuality. Finally, they transform comparison into learning—they correct errors, improve their imperfect understanding, and change action and plans. Linking science and human purpose, adaptive management serves as a compass for us to use in searching for a sustainable future.*
>
> — Lee (1993: 9)

In Holling's view, a major challenge for the use of adaptive environmental management is that it can be difficult to satisfy the preconditions. For example, the concern about climate change, discussed in Chapter 7, reflects worry that we may not be able to reverse such change before serious problems have occurred. Moreover, even when errors are not in theory irreversible, the magnitude of the original capital investment, as well as personal and political investment in a particular decision or course of action, often makes reversibility unlikely. Many people in industrialized societies simply do not like to admit to or pay for past mistakes, but prefer to try to correct them. The outcome of trying to correct an inappropriate initiative often is increasing investment, increasing costs related to control and maintenance, and progressive loss of future options.

IMPACT AND RISK ASSESSMENT

Environmental impacts, intended and unintended, positive and negative, are common to all development initiatives. The use of environmental impact assessment (EIA) has been a response to the increasing size and complexity of projects, greater uncertainty in predicting impacts, and growing demands by the general public and special interest groups to become more involved in planning and decision-making processes (Lawrence, 2003).

Many definitions of **environmental impact assessment** have been presented. For example, Beanlands and Duinker (1983: 18) defined it as 'a process or set of activities designed to contribute pertinent environmental information to project or programme decision-making'. In a slightly different way, the Canadian Environmental Assessment Research Council (1988: 1) defined it as a 'process which attempts to identify and predict the impacts of legislative proposals, policies, programs, projects and operational procedures on the biophysical environment and on human health and well-being. It also interprets and communicates information about those impacts and investigates and proposes means for their management.'

Initially, environmental impact assessment primarily emphasized the physical and biological resources that might be affected by a project, with attention to reducing negative consequences. However, partially in response to public pressure, the focus gradually broadened to incorporate social concerns, leading to the concept of social impact assessment. In addition, more attention has been given to basic policy questions, such as establishing the appropriateness of the objectives a project is designed to satisfy, considering alternative projects that also could satisfy the same objectives, and examining how compensation could be provided for impacts or losses that cannot be mitigated.

Smith (1993) suggests that three types of **impact assessment** can be identified:

1. *Technology assessment*. This most general type of assessment concentrates on the broad consequences of changing technology on a society, people, or region.
2. *Environmental impact assessment*. Initially, such assessments focused on the effects of development on natural ecosystems. Over time, the biophysical emphasis has been broadened to include social consequences of development.
3. *Social impact assessment*. Here, the main focus is on the impacts of development on human communities and their welfare. Initially, such assessments were done separately from environmental impact assessments. Today they

normally are included as an integral part of environmental impact assessments.

Since each of these elements—technology, environment, society—is important, Smith suggests that it is more appropriate to use the term 'impact assessment' than 'environmental impact assessment'. His approach is sensible, since such assessments also incorporate economic implications of proposals.

Risk assessment underlies impact assessment, since it focuses on determining the probability or likelihood of an environmentally or socially negative event of some specified magnitude, such as an oil spill, and the costs of dealing with the consequences. Of course, since risks have to be estimated, our calculations may be incorrect. Thus, at the Earth Summit of 1992, the **precautionary principle** was endorsed. This principle states that, in order to protect the environment, when there are risks of serious or irreversible damage, lack of full scientific certainty regarding the extent or possibility of risk shall not be used as a reason for postponing cost-effective measures to prevent environmental degradation. In other words, decision-makers and managers must always err on the side of caution.

Challenges in Impact Assessment

People conducting impact assessment have to juggle technical matters and value judgements. There often are not right or wrong answers, but rather different answers depending on the starting point for the assessment and the assumptions made.

The Types of Initiatives To Be Assessed

In Canada and most other countries, impact assessments have primarily been conducted for development and waste management *projects*, especially major capital projects such as dams and reservoirs, nuclear or other types of power plants, oil or natural gas drilling or pipelines, waste disposal facilities, and runway expansions at major airports. The rationale has been that such development usually has the potential for significant environmental and social impacts, and there are readily identifiable stakeholders—proponents, people, communities—to be affected.

However, it has been argued for some time

that impact assessments also could and should be completed for both *policies* and *programs*. The argument in favour of this approach, called 'strategic assessment', is that projects often are simply the means of implementing policies and programs. Thus, waiting to conduct impact assessments on projects results in such work occurring too late. At the policy or program level, decisions may already have been taken to preclude or eliminate possible alternatives—the problem of decisions sometimes becoming practically irreversible, as discussed above. A recent example of this occurred when BC Hydro proposed a pipeline to cross the Strait of Georgia to bring natural gas to Vancouver Island to generate power. The EIA was to be conducted on the specific impacts of the pipeline on the environment. However, environmental groups argued for a much broader EIA that would examine the need for the power and the alternatives available, rather than merely assess mitigation strategies for the pipeline.

While it is easy to accept the logic of having impact assessments completed for policies and programs as well as for projects, such assessments can be difficult to do. As Bregha et al. (1990: 3) have explained, policies can be general or specific, stated or implicit, incremental or radical, independent or linked to other policies.

Other challenges arise regarding which activities should have impact assessments completed. One issue is particularly difficult. On the one hand, society expects government regulations to be reasonable and efficient. In that regard, it is normal to have some lower limit or threshold below which assessments are not required. For example, it is unlikely that if one homeowner were to build an outdoor barbecue in her backyard that air quality in the community would be adversely affected. However, if every homeowner decided to build an outdoor barbecue, air quality possibly could be affected. The issue, then, is how to balance reasonableness and efficiency against the dilemma that the impact of many small developments *in aggregate* may have significant implications. This problem has been described as one of cumulative effects.

When Impact Assessments Should Be Done
The final report by the federal Environmental Assessment Panel (1991: 2) that reviewed the

Rafferty-Alameda Dam and reservoir projects in southeastern Saskatchewan stated that 'environmental impact assessment should be applied early in project planning. That is the intent of both provincial and federal processes.' The Panel concluded that the Rafferty-Alameda projects were 'well advanced, however, when both the first and this Panel became involved. This put some limits on the usefulness of the review.' Unfortunately, too often in Canada it appears that developments continue to be well advanced before environmental impact assessments are conducted.

The creators of environmental impact assessment had intended that EIA would be used jointly with other analyses to determine the appropriateness of development proposals and to help to design mitigating measures. However, as with Rafferty-Alameda, environmental assessments often have been conducted after the basic decision has already been taken regarding whether or not the project will be carried out. As a result, the impact assessment has rarely been used to determine whether or not a project should be approved; rather, it has become a tool to establish which mitigating measures could be used to reduce or soften negative impacts.

Determining the Significance of Impacts and Effects
As already discussed, a difficult challenge is to determine the significance or implications of impacts and effects. This is not a scientific or technical issue. Significance is influenced by the prevailing values in a society at a specific place and time. What might be considered significant in one place or at one time will not necessarily be viewed in the same way in another place or time. Furthermore, people from different cultural backgrounds or of different value systems and ideologies in the same place and at the same time may have quite different perspectives regarding what is important to protect or preserve. Accepting that judgements about significance are not only technical or scientific matters strengthens the rationale for extending partnerships in environmental management and for ensuring that key stakeholders participate in planning and assessments.

One of the major challenges in determining significance is that often the issues in a dispute do not lend themselves to a monetary valuation;

rather, they are characterized by intangible features. What is the value of an undisturbed stretch of whitewater in a river that might be inundated by a dam built to generate hydroelectricity? What are the values of a wetland that is to be drained to allow a subdivision to be built or to make it possible for a farmer to increase food production? What is the cost to wildlife from disturbance to their habitat due to a mine or a pipeline to transport oil? What is the value of a stand of old white pine or of Douglas fir? Questions such as these pose major dilemmas for those involved in impact assessment. There is usually no obvious answer, and people of different views and different vested interests may come to quite different conclusions about their significance and what action is appropriate.

Inadequate Understanding of Ecosystems

The scientific understanding of ecosystems often is incomplete, as discussed in Chapter 2. Furthermore, information about a specific ecosystem often is inadequate to allow estimates about what might be the effects from human intervention.

Even the most basic ecological concepts are not without problems. For example, some time ago McIntosh (1980) commented that ideas such as community, stability, succession, and climax were causing (and still do cause) major disagreements among ecologists. In some instances, scientists disagree over terminology and definitions, even after more than three-quarters of a century of research. Some of these difficulties already have been considered in Chapters 2, 3, and 4. With uncertainty and disagreement over basic ecological concepts, it is understandable why predictions are difficult to make with confidence, or why different scientific interpretations may be made from the same data, especially when such data are incomplete. This situation partially explains why it is not unusual for proponents and opponents at environmental hearings to each have their scientific experts who have reached opposite conclusions about impacts or risks.

To address in part the problem of incomplete information and understanding, Nakashima (1990) argued that more use should be made of **indigenous knowledge** or traditional ecological knowledge (TEK). He illustrated his argument by discussing the ecological understanding of seabirds by the Inuit in the coastal communities of Inukjuak

and Kuujjarapik, Quebec, and Sanikiluaq, NWT (now Nunavut), along southeastern Hudson Bay. He noted that Inuit hunters are well positioned to make a contribution to the protection of the environment since they have accumulated excellent information about the range and behaviour of animals as a direct outcome of their hunting lifestyle. Such knowledge can be invaluable in estimating and assessing the potential impacts of an oil spill on wildlife.

Incomplete understanding and inadequate data will continue to be the reality for those involved in impact assessments in Canada. For this reason, use of adaptive environmental management approaches is attractive, given its emphasis on learning by trial and error and its acceptance of uncertainty. In addition, these problems highlight the valuable contribution that indigenous knowledge can make to environmental management.

The Nature of Public Involvement

As discussed earlier in this chapter, several reasons are used to justify the role of public involvement. Garipey (1991) indicated that public involvement has at least three functions in impact assessment:

1. It opens up the decision-making process to the public, and in so doing contributes to that process being fair. Furthermore, by making decision-making accessible to the public, it enhances the credibility of the process. However, there is always danger of the public or individuals being 'co-opted' through such processes.
2. Public involvement helps to broaden the range of issues and alternative solutions considered, and allows the public to share in devising mitigation measures. In this manner, citizens share in establishing the conditions under which a proposal will be approved.
3. Public participation can contribute to social change. That is, by participating in the decision-making process, the public in a particular place become more aware about conditions in their own environment. Such enhanced awareness can lead to new initiatives within the community to identify and begin to address other problems, and also can lead to a greater likelihood that the government will appreciate the concerns of a community.

BOX 6.2 THE EXPERIMENTAL LAKES RESEARCH AREA

Generating enough electricity to fuel society's demands while minimizing environmental impacts is one of the main challenges we face today. Hydroelectric power has generally been regarded as preferable to coal- and oil-generating stations, which are linked to acid deposition, and to nuclear power plants with their associated difficulties in disposing of the wastes. It was somewhat of a surprise, therefore, when researchers at the Experimental Lakes Research Area (ELA) in northwestern Ontario discovered that reservoirs created for hydroelectric generation were responsible for releasing large amounts of carbon dioxide and particularly methane to the atmosphere. The emissions occur as a result of bacterial decomposition of flooded peat and forest biomass.

This was not the ELA's first finding of global significance. Since it was established in 1968,

researchers have also made significant contributions to the understanding of eutrophication and acid deposition. Fifty-seven small lakes and their watersheds are used as experimental sites to help determine the impacts of various environmental perturbations. Research on eutrophication at the ELA was instrumental in developing the phosphorus control strategies in the Great Lakes Water Quality Agreement. In 1987, research on acidification was initiated and contributed to new estimates regarding damage to aquatic ecosystems. Again, the results were used as the basis for international accords to limit emissions. The lakes have also provided a valuable function as monitoring sites to track changes in environmental conditions over time, such as global warming. In view of the national and global importance of the research at the ELA, it is unfortunate that government funding has been sharply cut back.

The Development of Monitoring

Chapters 2, 3, and 4 highlighted that many ecological processes unfold over *decades* and, therefore, that effects of some ecosystem changes may emerge only slowly over a lengthy time span. It also has been established that, due to incomplete ecological science, it often is difficult to predict which changes may occur in ecosystem structures or processes as a result of development. If we are to improve our knowledge of the resiliency, sensitivity, and recuperative powers of ecosystems, monitoring is needed. In addition, monitoring can help to ensure that mitigating measures recommended in an impact assessment have actually been implemented. Monitoring also can track public concerns or fears regarding a project and thereby help to ensure that they are recognized and addressed.

Too often in environmental and resource management in Canada, and in other countries, such monitoring is not conducted. It usually is time-consuming and expensive, and the results may not become useful until many years of monitoring have occurred. When financial and human resources are scarce, it is tempting for managers to reduce or eliminate monitoring and redirect scarce resources to new development activity, as has happened with the Experimental Lakes Area in northwestern Ontario (Box 6.2). However, by following that

route we reduce the opportunity to learn from previous experience and thus make it difficult to apply the adaptive management approach. In some instances the local community has been used to conduct monitoring activities, and such an approach seems very promising.

DISPUTE RESOLUTION

Conflicts and disputes occur for many reasons. They may emerge due to clashing or incompatible values, interests, needs, or actions. In an environmental context, conflicts may arise as a result of one or both of substantive or procedural issues. At a substantive level, disputes may arise about the effects of resource use or project development, about multiple use of resources and areas, about policies, legislation, and regulations, or over jurisdiction and ownership of resources. At a procedural level, conflicts may occur related to who should be involved, and at what times and in what ways.

However, conflict is not necessarily bad or undesirable. Conflict can help to highlight aspects in a process or system that hinder effective performance. Conflict can also be positive if it leads to clarification of differences due to poor information or misunderstandings. Approached in a constructive manner, conflict can result in creative and prac-

BOX 6.3 CLAYOQUOT SOUND, VANCOUVER ISLAND, BC

On 13 January 1993, a full-page advertisement appeared in the *New York Times* with the question, 'Will Canada do nothing to save Clayoquot Sound, one of the last great temperate rainforests in the world?' It was paid for by eight major international conservation groups. Six months later, Greenpeace International in London produced a 17-page colour booklet, 'British Columbia's Catalogue of Shame', outlining the background to the Clayoquot decision and demanding 'an end to the all clearcut logging in Clayoquot Sound, full inventory of all plants and animals to be carried out and outstanding native land claim issues to be settled'. Robert Kennedy Jr flew in as a lawyer representing the Natural Resources Defence Council of the US and promised the support of his organization. A resource and environmental management issue that had made nightly headline news in Canada also resulted in the arrest of over 800 protestors and captured worldwide attention.

At stake was the future of one of the greatest remaining temperate rain forests of the world and the largest remaining tract of old-growth forest on Vancouver Island, which was slated to undergo forest harvesting. The area is also under First Nations land claim negotiations, is coveted by the mining industry, is an arena for conflict between the rapidly expanding aquaculture industry and fishers and recreationists, and includes part of a national park.

The attention drawn to Clayoquot Sound as a result of the protests has resulted in several innovations. The area planned for complete protection was expanded. Part of the Sound is now included in an international biosphere reserve recognized by UNESCO. The reserve is helping to plan towards a more sustainable future. Some logging operations are now also joint operations between logging companies and First Nations, and thus are helping to address some of the chronic poverty found in some First Nation communities. Tourism has continued to increase in the area, again helping to combat poverty and providing more support for further preservation.

tical solutions to problems. On the other hand, conflict can be negative if it breeds lack of trust or misunderstanding, or reinforces biases. It can be negative as well if it is ignored or set aside and leads to an escalation of a problem or to creating stronger obstacles to be overcome later.

Conflicts over resource management have often become high-profile news issues in Canada. Some of these are highlighted in subsequent chapters, such as the conflicts over the seal hunt, Native whaling, and the lobster fishery in New Brunswick, discussed in Chapter 8. One of the largest conflicts in Canada over resource management focused on the temperate rain forests of Clayoquot Sound on the west coast of Vancouver Island (Box 6.3).

Whether conflict is positive or negative, it usually always is present. The reason for the persistence of conflict is because people see things differently, want different things, have different beliefs, and live their lives in different ways. The basic differences among people and their values, interests, needs, and activities create conflict. Such differences can be exacerbated by other factors. These include lack of understanding of other people or groups; people using different kinds and sources of infor-

mation; differences in culture, experience, or education; and different values, traditions, principles, assumptions, experiences, perceptions, and biases. Thus, conflicts are a normal part of life, and we need to devise ways to deal with them. We should accept the legitimacy of conflict and recognize that environmental planning and management is often a process for resolving conflicts.

Approaches to Handling Disputes

Disputes usually centre on three main aspects: *rights*, *interests*, and *power*. The traditional means to deal with disputes are political, administrative, and judicial. The latter is the most familiar to most citizens and involves using the courts. In litigation, the main issues of concern are *fact*, *precedent*, and *procedure*. Attention focuses on establishing a winner or on punishing an offender.

The strength of the judicial or litigation approach is that it uses a process that has evolved over centuries. Standards for procedure and evidence are well established. Accountability is ensured through appeal mechanisms and the professional certification of lawyers. On the other hand, the judicial process often is viewed as unduly adversarial, time-consuming, and expensive. Emond (1987) has noted that an adversarial and adjudicative process also encourages participants to exaggerate their private interests, to conceal their 'bottom lines', to withhold information, and to try to discredit their opponents. In addition, the courts do not always provide a level playing field between, for example, a small group of private citizens or a First Nation opposing a resource project and a large multinational corporation as the proponent. The greater financial resources of the corporation can pay for expensive legal expertise, not to mention assist in gaining access to and influence within the corridors of political power.

One alternative to the judicial approach is often referred to as alternative dispute or conflict resolution. **Alternative dispute resolution** (ADR) emphasizes the interests and needs of the parties involved. And, while the judicial approach finishes with either a winner or a loser, or by identifying a party to be punished, in alternative dispute resolution the focus is on reparation for harm done and on improving future conduct. Another key distinction between the two is that the judicial approach emphasizes *argument*, while the alternative dispute resolution approach stresses *persuasion*.

Attributes of Alternative Dispute Resolution

At least six strengths or advantages of alternative dispute resolution can be identified (Shaftoe, 1993):

1. an emphasis on the issues and interests rather than the procedures;
2. an outcome that normally results in a greater commitment to the agreement;
3. the attainment of a long-lasting settlement;
4. constructive communication and improved understanding;
5. effective use of information and experts;
6. increased flexibility.

These strengths of alternative dispute resolution also highlight some of the limitations of the judicial or court-based approach. However, the conclusion should not be drawn that legally based approaches are never appropriate for dealing with environmental issues. No approach is perfect, and alternative dispute resolution is no exception. Thus, while in some circumstances alternative dispute resolution may be more effective than litigation, the key is to recognize the strengths and weaknesses of each and to determine when one or the other will be the most effective.

Types of Alternative Dispute Resolution

The features of specific types of alternative dispute resolution approaches include public consultation, negotiation, mediation, and arbitration. Various aspects of *public consultation* were reviewed earlier, in the context of the ideas of partnerships and stakeholders. Public participation or citizen involvement has been used explicitly since the late 1960s for resource and environmental management in Canada. Initially, such consultation focused on having the public help in identifying key issues and in reviewing possible solutions. However, this public input was simply one of many inputs considered by the managers who ultimately determined which trade-offs were appropriate and then made the final decisions. The members of the public had no real power or authority in the management process.

By the mid-1980s, this approach had been modified as public participation moved towards the concepts of partnership and delegated power. Some co-management initiatives illustrated the shift to give real power to the public. However, public consultation is not normally considered to be one of the emerging types of alternative dispute resolution, as in the latter all decisions concerning the dispute are the exclusive domain of stakeholders.

Negotiation is one of the two main types of alternative dispute resolution technique. Its distinctive characteristics include two or more parties involved in a dispute joining in a voluntary, joint exploration of issues with the goal of reaching a mutually acceptable agreement. Because participation is voluntary, participants can withdraw at any time. Through joint exploration, the parties strive to identify and define issues of mutual concern and to

---

CONSENSUS

Traditional [indigenous] society was based upon the principle of consensus for government .... Consensus must be the most perfect form of democracy known because it means that there is no imposition of the rule of the majority. Everyone has input and no one is excluded.

— Ovide Mercredi, in Mercredi and Turpel (1993: 115)

develop mutually acceptable solutions. The normal procedure is to reach agreements by consensus.

Mediation is the second main type of alternative dispute resolution technique. The distinguishing feature is that mediation includes a neutral third party (called a mediator) whose task is to help the disputants overcome their differences and reach a settlement. The third party has no power to impose any outcome. The responsibility to accept or reject any solution remains exclusively with the stakeholders in the dispute. In addition, the third party has to be acceptable to all of the parties in the conflict.

Mediators have a variety of roles. They assist the parties in the dispute to come together. In this role, mediators act as facilitators. They also can help the parties with regard to fact-finding. The mediators do not necessarily have the expertise to provide the needed information, but they can help to identify what information is needed and then assist in finding the necessary data. During mediation, the mediator often first meets separately with the parties in the dispute and subsequently holds joint meetings of all parties. Mediators try to assist each stakeholder to understand the interests and objectives of the other stakeholders, to find points in common, and to settle differences through negotiation and compromise. Key roles for the mediator are to maintain momentum in the negotiations, keep the parties communicating with each other, and ensure that proposals are realistic.

Arbitration differs significantly from negotiation and mediation because the stakeholders normally accept a third party who has the responsibility to make a decision regarding the issue(s) in conflict. In mediation, the third party has no power to impose a settlement. In arbitration, normally the arbitrator's decision is binding on the parties. However, there are instances characterized as 'non-binding arbitration' in which the arbitrator makes a decision regarding the conflict but the stakeholders may accept or reject it. In some ways, the judicial or court-based approach contains many of the elements of binding arbitration since the judge reaches a decision that is then imposed on those involved in the dispute. The main difference is that in arbitration the stakeholders usually have a voice in selecting who will be the arbitrator. In the judicial procedure, the participants do not have a role in deciding which judge will hear their case.

As already noted, public consultation or public participation has been used in Canada for environmental management for about four decades, so there is considerable experience with it. Judicial or court-based approaches also have been used ever since the country was settled by people from Europe. The newly emerging approaches are negotiation and mediation, although aspects of negotiation certainly have been included as part of public consultation and judicial approaches. In Part D, some of the case studies provide further details regarding how negotiation and mediation are being used for addressing conflicts over the environment. At this point, you might consider how ADR methods could be used to deal with the different interests represented in the Sydney Tar Ponds situation.

REGIONAL AND LAND-USE PLANNING

Regional and land-use planning represents a process and a product, and, ideally, the product (or plan) reflects a vision about how development should occur in a region. Resource and environmental managers should be aware of arrangements for regional and land-use planning for several reasons. First, work undertaken in creating or updating a plan often is directly relevant for resource and environmental management because it can provide valuable information and insight. Second, regional and land-use plans frequently have a statutory or legislative basis and thus become part of the official documentation governing activity in an area. In contrast, many resource and environmental management plans, such as a watershed plan, usually do not have a legal basis, and this often creates difficulty in achieving effective implementation. It is common, once a resource management plan has been created, for various agencies to be responsible for implementing specific recommendations that fall under each of their jurisdictions. However, such agencies also have other responsibilities, and they have to determine what priority to give to recommendations from resource management plans. As a result, if resource managers can link their plans to official regional land-use plans, the probability becomes higher that recommendations will be acted upon.

The same argument can be made with relation to environmental impact assessment. As with regional and land-use plans, environmental impact assessment has a legal basis in federal and provincial governments in Canada. Thus, the likelihood of action being taken on recommendations within a resource management plan will be further enhanced if they are related to associated impact assessment statements and to regional and land-use plans. Put another way, if such links are not created, then it becomes too easy for decision-makers to overlook or ignore resource management plans since they usually do not have any legal underpinning.

In Chapter 5, in the section focused on system and ecosystem perspectives, it was noted that ecosystems are dynamic, thereby continuously changing. This is a reminder that the context, also discussed in Chapter 5, within which a resource management plan must function can be expected to change. One implication is that resource management plans need to be updated and modified to ensure that they remain relevant. The notion of monitoring and modifying plans in light of changing circumstances, new knowledge, and lessons learned is also embodied in adaptive environmental management, already discussed. In most jurisdictions, provision is made for reviewing, updating, and modifying regional and land-use plans. Such capacity related to this planning provides another reason to link resource management plans to them, as both can therefore be reviewed and updated at the same time.

IMPLICATIONS

The approaches discussed in this chapter represent what many would view as ideals for resource and environmental management regarding processes, methods, and products. Some of them provide fundamental challenges to us as individuals and society. For example, basic values in Western, industrialized societies frequently are self-interest and competition. It often is assumed or believed that scarce resources—natural and/or human—are allocated most efficiently and equitably through competition as each agent—individual citizens, interest groups, corporations, resource managers—pursues its own best interests, while governments try to mediate this process in the greater public interest. However, the concepts of *collaboration, co-ordination, partnerships*, and *stakeholders* suggest a different perspective, one based on a willingness to recognize the legitimacy of many interests and needs, and on a desire to try to satisfy them. In a similar manner, *alternative dispute resolution* is predi-

cated on the idea of groups working together for mutual gain, rather than determining winners and losers. In a different way, *adaptive management* rejects a belief that humans can completely understand and control natural systems. Instead, this approach accepts that our understanding will always be incomplete and limited, resulting in ongoing surprises that will require us to adapt and modify our policies and practices. *Impact* and *risk assessments* further reflect acceptance of uncertainty relative to resource and environmental systems, and encourage us to strive to anticipate and monitor outcomes—intended and unintended, desirable and undesirable—so that we can determine where and when adjustments are needed.

In Chapter 5, attention was given to the importance of a systems approach, and the discussion in this chapter related to *regional and land-use planning* highlights that resource and environmental management does not occur in a vacuum. Other management processes occur in parallel, making it desirable always to be watching to determine how connections can be made with such other processes, particularly when they have a legal basis.

Finally, *communication* reminds us that it is critically important to ensure that stakeholders share information, insights, needs, and priorities with each other in problem-solving situations. It seems as if there can never be too much time allocated to improving communication among participants in a management process.

Given the above, what can you do as an individual? By being aware of these characteristics of processes, methods, and products, you can first critically examine any initiative related to resource or environmental problems to determine if these approaches and their inherent values are being included. If one or more is absent, you can be a voice drawing attention to the need to incorporate them. Second, you can reflect on what you have been acculturated to believe are appropriate values to guide behaviour in problem-solving, such as self-interest and competition. Alternative approaches exist, and we should not always approach problems without questioning why and how we instinctively try to deal with them. Third, you can pay attention to related approaches, such as regional and land-use planning, and risk and impact assessment, to see how these can support or advance resource and environmental management. And fourth, you can strive to ensure that information and understanding you have is shared with others in the spirit of improving our understanding not only of structures and processes related to natural and human systems, but also of enhancing appreciation of the almost inevitable range of underlying values, assumptions, and attitudes that will shape behaviour. Awareness of the above processes, methods, and products, and sensitivity to them, will help generate diverse ways to define, frame, and solve problems.

SUMMARY

1. If a systems approach is to be used, collaboration and co-ordination are required.

2. Collaboration involves the exchange of information, modification of activities in light of others' needs, sharing of resources, enhancement of the capacity of others in order to achieve mutual benefit and to realize common goals or purposes, and joint decision-making to resolve problems, during which power is shared and stakeholders accept collective responsibility for their actions and the outcomes.

3. Collaboration is increasingly accepted as desirable because the complexity and uncertainty associated with resource and environmental issues create a challenge for any individual or organization to have sufficient knowledge or authority to deal with them.

4. Participatory approaches aim to incorporate insight from individuals and groups affected by decisions or with responsibility for issues, and ideally lead to sharing of power and authority. Not all groups welcome power-sharing, viewing any relinquishment of power as undermining their role in environmental management.

5. Stakeholders are those who should be included in decision-making because of their direct interest, including (1) any public agency with prescribed management responsibilities; (2) all interests significantly affected by a decision; and (3) all parties who might intervene in the decision-making process or block or delay the process.

6. Co-management has been one innovative approach to sharing power, in which local people are allocated responsibility and authority for certain aspects of resource and environmental management. Co-management has been applied with regard to forestry and wildlife.

7. Effective communication of science and local knowledge is essential, and communications should be prepared with regard to who will be the recipient, and what can reasonably be assumed to be their level of understanding.

8. Effective communication should address four complementary questions: (1) What do we know, with what accuracy, and how confident are we about our data? (2) What don't we know, and why are we uncertain? (3) What could we know, with more time, money, and talent? (4) What should we know in order to act in the face of uncertainty?

9. The concept of adaptive environmental management accepts that (1) surprise, uncertainty, and the unexpected are normal; (2) it is not possible to eliminate them through management initiatives; and (3) management should provide allowance for them. As a result, management is viewed as an experiment, requiring systematic monitoring of results so that we can learn from experience.

10. Environmental impact assessment is used to identify and predict the impacts of legislative proposals, policies, programs, projects, and operational procedures on the biophysical environment and on human health and well-being. It also interprets and communicates information about those impacts and investigates and proposes means for their management.

11. Risk assessment focuses on determining the probability or likelihood of an event of some specified magnitude, as well as the likelihood of the associated consequences. It is recognized that since risks have to be estimated, our calculations may be incorrect. For this reason the precautionary principle is used. This principle states that, in order to protect the environment, when there are risks of serious or irreversible damage, lack of full scientific certainty shall not be used as a reason for postponing cost-effective measures to prevent environmental degradation.

12. Disputes usually centre on three main aspects: rights, interests, and power. The traditional means to deal with disputes are political, administrative, and judicial. Increasingly, attention is being given to alternative dispute resolution, in which information and understanding are shared and efforts are made to find solutions that address needs of all stakeholders.

13. Regional and land-use plans can be valuable because they often have a statutory or legislative basis and become part of the official documentation governing activity in an area. In contrast, many resource and environmental management plans do not have a legal basis, and this often creates difficulty for effective implementation. By connecting resource and environmental management plans to regional and land-use plans, statutory authority can be gained.

KEY TERMS

adaptive environmental management	collaboration	impact assessment	precautionary principle
	co-management	indigenous knowledge	risk assessment
alternative dispute resolution	co-ordination	mediation	stakeholders
arbitration	environmental impact assessment	negotiation	
		partnerships	

REVIEW QUESTIONS

1. Explain the difference between collaboration and co-ordination. Why are both needed in resource and environmental management?
2. What does the word 'stakeholder' imply for resource and environmental management? How would you go about identifying stakeholders in a specific problem-solving situation?
3. Why is it often difficult to communicate results from scientific research to the lay public? What can be done to improve such communication?
4. Do you believe that scientists are objective when they conduct research?
5. What were the motivations for people to develop the concept of adaptive management? What are the greatest strengths and weaknesses of the approach? How might it be applied to an environmental

problem in your community or province?
6. Why is the precautionary principle considered important? Has it been effective in practical terms?
7. What are the distinctive features of alternative dispute resolution? In what kinds of situations might it be a better way to deal with conflicts than the judicial approach?
8. What is the benefit of connecting or relating resource and environmental management to regional and land-use plans, and to environmental impact assessments? Is that being done in the community or province in which you live?
9. To what extent did collaboration and co-ordination occur during the process used to deal with the Sydney Tar Ponds problem discussed in Chapter 1? What initiatives should be taken to achieve a more collaborative and co-ordinated approach in the future?

RELATED WEBSITES

CANADIAN ENVIRONMENTAL ASSESSMENT AGENCY:
    http://www.ceaa.gc.ca

CLAYOQUOT SOUND:
    http://www.forests.org

CO-MANAGEMENT: www.pcffa.org/fn-aprov
    http://www.indiana.edu/~iascp/coman2003

ENVIRONMENTAL JUSTICE:
    http://www.epa.gov/compliance/environmentaljustice

EXPERIMENTAL LAKES RESEARCH AREA:
    http://www.dfo-mpo.gc.ca/regions/central/science/
        enviro/ela-rle

INTERNATIONAL JOINT COMMISSION:
    http://www.icj.org/en/home/main-accueil

MANITOBA ROUND TABLE ON ENVIRONMENT AND ECONOMY:
    http://www.iisd1.iisd.ca/worldsd/canada/prov.manrt

RAFFERTY-ALAMEDA DAMS:
    http://www.swa.sk.ca/watermanagement/damsandreservoirs

SUSTAINABLE LIVELIHOODS:
    http://www.undp.org.sl
    http://www.undp.org/sl/Documents/documents

REFERENCES AND SUGGESTED READING

Andrey, J., and L. Mortsch. 2000. 'Communicating about climate change: challenges and opportunities', in D. Scott, B. Jones, J. Andrey, R. Gibson, P. Kay, L. Mortsch, and K. Warriner, eds, *Climate Change Communication: Proceedings of an International Conference*. Waterloo, Ont.: University of Waterloo and Environment Canada, Adaptation and Impacts Research Group, WP1–WP11.

Arnstein, S. 1969. 'A ladder of citizen participation', *Journal of the American Institute of Planners* 35, 4: 216–24.

Beanlands, G.E., and P.N. Duinker. 1983. *An Ecological Framework for Environmental Impact Assessment in Canada*. Halifax: Institute for Resource and Environmental Studies, Dalhousie University.

Bregha, F., J. Benidickson, D. Gamble, T. Shillington, and E.

Weick. 1990. *The Integration of Environmental Considerations into Government Policy*. Prepared for the Canadian Environmental Assessment Research Council. Ottawa: Minister of Supply and Services Canada.

British Columbia Ministry of Environment, Lands and Parks. 1992. *New Approaches to Environmental Protection in British Columbia: A Legislation Discussion Paper*. Victoria: Ministry of Environment, Lands and Parks.

Canadian Environmental Assessment Research Council. 1988. *Evaluating Environmental Impact Assessment: An Action Perspective*. Ottawa: Minister of Supply and Services Canada.

Carpenter, R.A. 1995. 'Communicating environmental science uncertainties', *Environmental Professional* 17: 127–36.

Diduck, A., and B. Mitchell. 2003. 'Learning, public involvement and environmental assessment: a Canadian case study', *Journal of Environmental Assessment Policy and Management* 5: 339–64.

Emond, D.P. 1987. 'Accommodating negotiation/mediation within existing assessment and approval processes', in *The Place of Negotiation in Environmental Assessment*. Prepared for the Canadian Environmental Assessment Research Council. Ottawa: Minister of Supply and Services Canada, 45–52.

Environmental Assessment Panel. 1991. *Rafferty-Alameda Project: Report of the Environmental Assessment Panel*. Ottawa: Federal Environmental Assessment Review Office.

Garipey, M. 1991. 'Toward a dual-influence system: assessing the effects of public participation in environmental impact assessment for Hydro-Québec projects', *Environmental Impact Assessment Review* 11, 4: 353–74.

Gray, B. 1989. *Collaborating: Finding Common Ground for Multiparty Problems*. San Francisco: Jossey-Bass.

Himmelman, A.T. 1996. 'On the theory and practice of transformational collaboration: from social service to social justice', in Huxham (1996: 19–43).

Holling, C.S., ed. 1978. *Adaptive Environmental Assessment and Management*. Chichester: John Wiley.

Holling, C.S., 1986. 'The resilience of terrestrial ecosystems: local surprise and global change', in W.C. Clark and R.E. Munn, eds, *Sustainable Development in the Biosphere*. Cambridge: Cambridge University Press, 292–317.

Huxham, C., ed. 1996. *Creating Collaborative Advantage*. Thousand Oaks, Calif.: Sage.

Ingles, A.W., A. Musch, and S. Qwist-Hoffman. 1999. *The Participatory Process for Supporting Collaborative Management of Natural Resources: An Overview*. Rome: United Nations Food and Agriculture Organization.

Lawrence, D., ed. 2003. *Environmental Impact Assessment: Practical Solutions to Recurrent Problems*. Toronto: John Wiley.

Lee, K.N. 1993. *Compass and Gyroscope: Integrating Science and Politics for the Environment*. Washington: Island Press.

McIntosh, R.P. 1980. 'The relationship between succession and the recovery process in ecosystems', in J. Cairns, ed., *The Recovery Process in Damaged Ecosystems*. Ann Arbor, Mich.: Ann Arbor Science Publishers, 11–62.

McManus, P.A. 2000. 'Beyond Kyoto? Media representation of an environmental issue', *Australian Geographical Studies* 38: 306–19.

Manitoba Round Table on Environment and Economy. 1992. *Sustainable Development: Towards Institutional Change in the Manitoba Public Sector*. Winnipeg: Manitoba Round Table on Environment and Economy.

Mercredi, O., and M.E. Turpel. 1993. *In the Rapids: Navigating the Future of First Nations*. Toronto: Viking.

Nakashima, D.J. 1990. *Application of Native Knowledge in EIA: Inuit, Eiders and Hudson Bay Oil*. Prepared for the Canadian Environmental Assessment Research Council. Ottawa: Minister of Supply and Services Canada.

Noble, B.F. 2004. 'Applying adaptive environmental management', in B. Mitchell, ed., *Resource and Environmental Management in Canada: Addressing Conflict and Uncertainty*. Toronto: Oxford University Press, 442–66

Palerm, J.R. 2000. 'An empirical-theoretical analysis framework for public participation in environmental impact assessment', *Journal of Environmental Planning and Management* 43: 581–600.

Parson, E.A. 2000. 'Environmental trends and environmental governance in Canada', *Canadian Public Policy* 26 (supplement, Aug.): S123–43.

Saarikoski, H. 2000. 'Environmental impact assessment (EIA) as collaborative learning process', *Environmental Impact Assessment Review* 20: 681–700.

Saskatchewan Round Table on Environment and Economy. 1992. *Conservation Strategy for Sustainable Development in Saskatchewan*. Regina: Saskatchewan Round Table on Environment and Economy.

Scott, D., B. Jones, J. Andrey, R. Gibson, P. Kay, L. Mortsch, and K. Warriner. 2000. *Climate Change Communication: Proceedings of an International Conference*. Waterloo, Ont.: University of Waterloo and Environment Canada, Adaptation and Impacts Research Group.

Selin, S., and D. Chavez. 1995. 'Developing a collaborative model for environmental planning and management', *Environmental Management* 19: 189–95.

Shaftoe, D., ed. 1993. *Responding to Changing Times: Environmental Mediation in Canada*. Waterloo, Ont.: The Network: Interaction of Conflict Resolution, Conrad Grebel College.

Sinclair, A.J., and A.P. Diduck. 2001. 'Public involvement in EA in Canada: A transformative learning perspective', *Environmental Impact Assessment Review* 21: 113–36.

Smith, L.G. 1993. *Impact Assessment and Sustainable Resource Management*. Harlow, UK, and New York: Longman Scientific and Technical and John Wiley.

Trist, E. 1980. 'The environment and system-response capability', *Futures* 12, 2: 113–27.

RESOURCE AND ENVIRONMENTAL MANAGEMENT IN CANADA

Parts B and C focused on science and management related to resources and the environment. In Part D we discover how ideas and methods from science and management can be applied in practical problem-solving situations.

Several comments about the structure in this section should be helpful. It may appear contradictory to argue for a systems or ecosystem approach in Parts B and C, and then organize Part D on the basis of specific 'resources' or attributes of the environment, such as agriculture, forestry, or wildlife issues. Indeed, it would be inappropriate to isolate components of resource or environmental systems and examine them on a sector-by-sector basis, as the chapter headings in Part D might suggest. However, it is appropriate to use one resource or environmental aspect as the starting point for a discussion, as long as attention is given to other elements of the ecosystem and their linkages. This is the approach used in this section of the book.

For example, in Chapter 12 we examine challenges related to flood-ing and human use of flood plains. Flooding is clearly a problem of too much water at a particular place at a given time. Understanding the process of flooding, however, requires attention not only to aquatic systems but also to connected terrestrial systems, since land-use practices can be a major variable influencing vulnerability to flooding. Likewise, Chapter 7 focuses on climate change and on changes in the atmosphere that can lead to the alteration of climatic conditions. Yet, for us to understand what causes climate change we need to consider other systems as diverse as fossil fuel resources and transportation indus-tries, as well as examine basic values and attitudes that result in many people deciding that sport utility vehicles are desirable to meet their transportation needs.

Three different types of resources or environments are considered in the following seven chapters. The first, addressed in Chapter 7, is the atmospheric system, with emphasis on climate change. The second, the focus of Chapters 8 to 12, covers various renewable resources—oceans, agriculture, forestry, wildlife, and water systems. The third type, considered in Chapter 13, deals with non-renewable resources—minerals and fossil fuels that are created (or renewed) over the course of a geological time span rather than human lifespans.

In all of these chapters, interest centres on examining how science can be used to inform analysis and management, and how elements of best practice in management have been or could be applied. Thus, after examining the nature of weather and climate, Chapter 7 turns to assessing the scientific evidence related to climate change and to the role of climate models in aiding our understanding of changes. On this basis, it is possible to assess critically the scientific conclusions and explanations being offered relative to climate change. We also find that while solid scientific understanding of environmental systems and how they may be changing is of critical importance, it is equally important to communicate that understanding to non-technical specialists. Consequently, this chapter also addresses the challenges encountered by researchers when they seek to communicate the scientific insight about climate change to policy-makers and the general public. These challenges are formidable, and too often they are not systematically thought about by scientists, who frequently are more interested in the 'purity' of their scientific work and in communicating their findings to peers. As in other chapters, an effort is made in this chapter to relate scientific understanding to 'real-world' situations. In that context, two detailed case studies—how the potential rise of sea level caused by climate change would affect Prince Edward Island, and the possible consequences for winter tourism in south-central Ontario resulting from milder winters—are presented. In addition, a detailed discussion is provided regarding the Kyoto Protocol. Besides examining basic issues and approaches represented by this international agreement, attention is given to the federal, provincial, and territorial government perspectives.

Case studies in other chapters provide further opportunity to learn about how science and management can be connected. For example, Chapter 8, on oceans, looks at the challenges involved in the depletion of the Pacific salmon fishery, the debate over the seal hunt off the east coast, and the closure of the groundfish fishery in Atlantic Canada. In later chapters, case studies cover the environmental impacts of the James Bay hydroelectric project in northern Quebec and of the development of diamond mines in the Northwest Territories and Nunavut, the implications of the Walkerton, Ontario, experience for water security, the lessons about managing natural hazards from the 1997 Red River flood in Manitoba, and the opportunities from using wind and other renewable sources of energy production.

The chapters in Part D do not always provide detailed scientific or management background, as these were covered in Parts B and C. For example, the case studies about water in Chapter 12 do not examine aspects of the hydrological cycle or contaminant pathways since these were addressed in Chapter 4. Furthermore, other case studies appear in Parts B and C to highlight the relevance of science and management concepts and methods.

The issues and examples considered here raise fundamental questions about humans' relationships with the environment and resources. We see a range of attitudes towards the environment and other living things, covering the spectrum from humans dominating nature to humans striving to live in harmony with nature. The extinction of species, such as those reported in Chapter 11, reminds us that we have been (and can still be) incredibly arrogant in believing that it was acceptable for us to eliminate some species forever. In contrast, as also shown in Chapter 11, conscious decisions are being taken to protect valued areas, in some instances because we believe it is important to protect examples of different biomes. Notwithstanding some significant and positive accomplishments, current lifestyles in Canada continue to be a major contributor to global climate change, as discussed in Chapter 7. The changes associated with climatic warming may confound attempts to identify and protect examples of biospheres if those change in the future because of different climatic conditions.

Finally, the case studies and examples illustrate how pervasive conflict can be in resource and environmental management. This reinforces our belief that resource and environmental management is not only a scientific or technical exercise.

To be effective, it must be able to recognize, identify, and incorporate different values and interests. For this reason, managers often spend significant time trying to resolve conflicts. Given such a reality, scientists must also develop a greater appreciation that their work will usually be used in situations in which values and emotions can be as important as, or of greater importance than, theories, models, and quantitative evidence.

The mix of case studies and examples in Part D and throughout this book will help you to appreciate the change, complexity, uncertainty, and conflict that are integral parts of resource and environmental management.

Chapter 7

Climate Change

Learning Objectives

- To understand the difference between weather and climate.
- To know the difference between climate change and global warming.
- To appreciate why the science of climate change is characterized by complexity and uncertainty.
- To understand the nature of scientific evidence regarding climate change.
- To understand the scientific explanation for climate change.
- To realize the implications of climate change for natural and human systems.
- To appreciate the challenges of sharing information and insight related to climate change.
- To comprehend the Kyoto Protocol, and different perspectives in Canada related to it.
- To discover what you can do as an individual to minimize the impacts of climate change.

INTRODUCTION

Climate is naturally variable. It is never exactly the same from one period to another. Sometimes it can shift dramatically within a few hundred or thousand years, as it does when ice ages begin and end. Usually it varies within much narrower limits. For most of the past 1,000 years, for example, the world's average temperature has remained within about half a degree of 14° C.

Over the past 100 or so years, however, the world's climate has changed noticeably. The world's average temperature was approximately 0.6° C warmer at the end of the twentieth century than it was at the beginning, and the 1990s were the hottest decade in 140 years of global climate records. Such changes may seem trifling, but the difference between global temperatures now and at the peak of the last ice age is a mere 5° C. Evidence of earlier climates suggests that global

UNCERTAINTY RELATED TO CLIMATE CHANGE CONSEQUENCES

Addressing climate change risk is a major environmental challenge, yet advocates and even experts disagree on the costs and lifestyle impacts of reducing greenhouse gas (GHG) emissions. One side argues that a concerted effort to reduce GHGs will launch a new era of technological innovation, productivity gains and job creation, while the other argues that economic output will decline and unemployment will rise. The public and politicians don't know whom to believe.

— *Jaccard et al. (2003: 29)*

temperatures have warmed more during the twentieth century than in any other century during the past 1,000 years. (Canadian Council of Ministers of the Environment, 2003: 5).

Figure 7.1 Kyoto and world governments. SOURCE: Malcom Mayes/artizan.com.

As Andrey and Mortsch (2000: WP1) observed, by early in the twenty-first century climate change had become part of the vocabulary of the North American public. Climate change cartoons and greeting cards had become common (Figure 7.1). In their opinion, perhaps the ultimate proof appeared during the popular TV game show *Who Wants To Be a Millionaire?* in March 2000 when the following question was posed: 'What gas is primarily responsible for the greenhouse effect? (a) hydrogen, (b) oxygen, (c) nitrogen, or (d) carbon dioxide.' The significance was not that such a question was posed, but the $1,000 dollar value for the answer. This relatively low amount highlighted that almost any contestant was expected to know the answer. On the other hand, as the quotation in the box on the previous page indicates, much complexity and uncertainty is associated with climate change.

NATURE OF CLIMATE CHANGE

The condition of the atmosphere at any time or place, i.e., the weather, is expressed by a combination of several elements, primarily (a) *temperature* and (b) *precipitation* and *humidity*, but to a lesser degree (c) *winds* and (d) *air pressure*. These are called the elements of weather and climate because they are the ingredients out of which various weather and climatic types are compounded. The **weather** of any place is the sum total of its atmospheric conditions (temperature, pressure, winds, moisture, and precipitation) for a *short* period of time. It is the momentary state of the atmosphere. Thus we speak of the weather, not the climate, for today or of last week.

Climate, on the other hand, is a composite or generalization of the variety of day-to-day weather conditions. It is not just 'average weather', for the variations from the mean, or average, are as important as the mean itself.

A distinction also should be made between climate change and global warming. Climate represents average day-to-day weather conditions as well as seasonal variations for a particular place or region. In that context, **climate change** is defined as 'a long-term shift or alteration in the climate of a specific location, a region, or the entire planet' (Hengeveld et al., 2002: 1). A shift is measured for variables associated with average weather conditions, such as temperature, precipitation, and wind patterns (velocity, direction). A change in variability of climate also is included as climate change. In contrast, **global warming**, often mentioned by the media, addresses changes only in average surface *temperatures*. It does not address whether conditions are becoming wetter or drier, for example. A frequent misunderstanding is that global warming means uniform warming throughout the world. Such an interpretation is not correct. An increase in average global temperatures drives alterations in atmospheric circulation patterns, which can contribute to some areas warming at higher rates, others at lower rates, and even others to become cooler.

This coral was found near the Arctic Ocean and illustrates how climates have changed in the past (*Philip Dearden*).

What are the 'causes' of climate? The Earth's surface and atmosphere are heated differentially by short-wave radiation from the sun. The differences in heat and pressure between the poles and the tropics fuel the global circulation system as heat and moisture are redistributed around the world. The temperature balance of the Earth is maintained through the return of the continually absorbed solar radiation back to space as infrared radiation, consistent with the first law of thermodynamics (Chapter 2). Long-term temperature changes are a result of shifts in the amount of energy received or absorbed. These may be caused over long cycles (100,000 years) by factors such as the shape of the Earth's orbit around the sun, wobbles of the Earth's axis, and the angle of tilt. Such a 100,000-year cycle of glaciation can be traced over 600,000 years, using evidence from sources such as glacier ice and the chemical characteristics of marine sediments. It is more difficult,

---

GREENHOUSE EFFECT

*The **greenhouse effect** describes the role of the atmosphere in insulating the planet from heat loss, much like a blanket on our bed insulates our bodies from heat loss. The small concentrations of greenhouse gases within the atmosphere that cause this effect allow most of the sunlight to pass through the atmosphere to heat the planet. However, these gases absorb much of the outgoing heat energy radiated by the Earth itself, and return much of this energy back towards the surface. This keeps the surface much warmer than if they were absent. This process is referred to as the 'greenhouse effect' because, in some respects, it resembles the role of glass in a greenhouse.*

— *Hengeveld et al. (2002: 2)*

---

however, to explain some of the shorter-term fluctuations that occur.

Natural events, such as the eruption of large volcanoes and changes in ocean currents such as El Niño, are known to have an influence. Volcanoes, illustrated by the eruption of Mount Pinatubo in 1991, eject large quantities of dust and sulphur particles into the **atmosphere**, which reduce the amount of solar radiation reaching the surface of the Earth. Changes in ocean currents can also be influential (Chapter 8). **El Niño** represents a marked warming of the waters in the eastern and central portions of the tropical Pacific, as westerly winds weaken or stop blowing, usually two to three times every decade. In normal years, the trade winds amass warm water in the western Pacific. As the winds slacken, this water spreads back eastward and towards the pole into the rest of the Pacific. This triggers weather changes in at least two-thirds of the globe, causing droughts and extreme rainfall in countries along the Pacific and Indian oceans, including Africa, eastern Asia, and North America.

However, it is increasingly apparent that climatic change may occur more rapidly than ever before due to human activities. These aspects are considered below.

SCIENTIFIC EVIDENCE RELATED TO CLIMATE CHANGE

In the context of the distinction among weather, climate, and global warming, the following statements are supported by solid scientific evidence:

1. The world has been warming, with the average global temperature at the Earth's surface having increased by about 0.6° C, with an error range of plus or minus 0.2° C, since the late nineteenth century (Figure 7.2). This increase is a global average, and in some areas, especially over continents, warming has been several times greater than the global average.
2. The increase in the average temperature for the northern hemisphere during the twentieth century was the largest of any century in the past 1,000 years (Figure 7.3).
3. **Greenhouse gas** concentrations, especially those of carbon dioxide, methane, nitrous oxide, and tropospheric ozone, have been rising for several decades (Figure 7.4). Carbon dioxide and methane are at higher concentra-

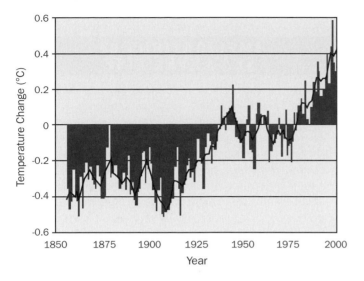

Figure 7.2 Variation in global average surface temperature between 1856 and 2000. SOURCE: Miller (2002: A-2), prepared by Professor Danny Harvey, Department of Geography, University of Toronto, using data in electronic form available from the UK Meteorological Office website <http://www.meo.gov.uk>.

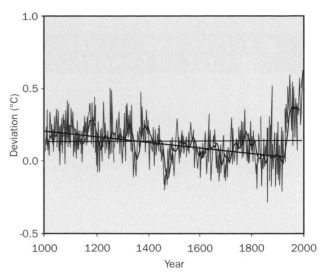

Figure 7.3 Variation in northern hemisphere average surface temperature. Based largely on ice core, tree ring, and coral reef data (thin light line). Also shown is the 20-year running mean of the annual paleoclimatic data and the 1000–1900 trend line (thick lines), and the directly observed temperature variation of Figure 7.2 (thin dark line). SOURCE: Miller (2002: A-2), prepared by Professor Danny Harvey, Department of Geography, University of Toronto, using paleoclimatic and historical data from UK website. (see Figures 7.2) and using data from the US National Oceanographic and Atmospheric paleoclimatology website <http://www.ngtc.noaa.gov/paleo>.

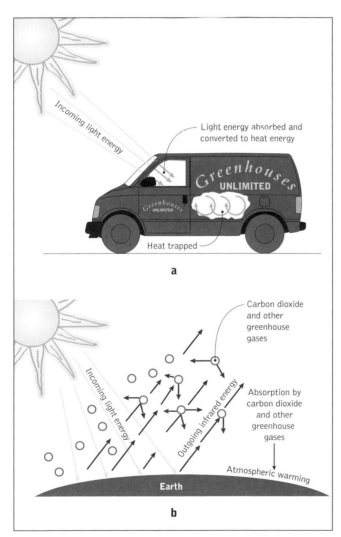

Figure 7.4 The greenhouse effect.

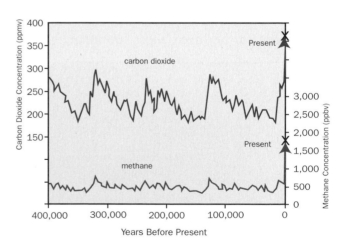

Figure 7.5 Variation in atmospheric concentrations of carbon dioxide and methane to 400,000 years before present. Measured from the Vostok ice core in Antarctica (thin lines) and during the past 200 years (heavy line). SOURCE: Miller (2002: A-1), prepared by Professor Danny Harvey, Department of Geography, University of Toronto, using data in electronic form obtained from the US National Oceanographic and Atmospheric Administration (NOAA) paleoclimatology website <http://www.ngdc.noaa.gov/paleo>.

GLACIERS AND CLIMATE CHANGE IN CANADA

Since 1950, the greatest warming in Canada has occurred in the West and Northwest. Most glaciers in these regions are also shrinking rapidly. The 1,300 or so glaciers on the eastern slopes of the Rockies, for example, are now about 25 per cent to 75 per cent smaller than they were in 1850. The area of warming also covers many of the High Arctic islands in Nunavut, where glaciers such as the Melville Island South Ice Cap have been shrinking gradually since at least the late 1950s. In eastern Nunavut, however, the situation is more complex: some glaciers are shrinking, while others are growing.

The melting of glaciers is a concern for Alberta, Saskatchewan and Manitoba. Farmers depend on glacier-fed rivers like the Saskatchewan and the Bow for irrigation water, and cities like Edmonton, Calgary and Saskatoon rely on them for municipal water supplies and recreation. At The Pas in Manitoba, reduced flows on the Saskatchewan could interfere with the native fishery and hydroelectric power generation.

— Canadian Council of Ministers of the Environment (2003: 20)

IMPLICATIONS OF SEA-LEVEL RISE

Rising sea levels threaten familiar shoreline environments. Coastal wetlands, which are important ecosystems, and barriers against shoreline erosion, gradually disappear. Bluffs and beaches are more exposed to erosion by waves, groundwater is more likely to become contaminated by salt water, and low-lying coastal areas may be permanently lost. In addition, wharves, buildings, roads and other valuable seaside property face a greater risk of damage as a result of flooding from storms.

— Canadian Council of Ministers of the Environment (2003: 13)

tions now than any time over the last 420,000 years (Figure 7.5). There is no scientific dispute about such increases.

4. In most parts of the world, glaciers since 1980 have lost, on average, more mass than they have gained. Furthermore, most mountain glaciers have been retreating during the past 100 years. Exceptions are found in Norway and New Zealand, where some glaciers have been advancing. The explanation for these exceptions is that they are associated with remarkable increases in precipitation rather than with decreases in temperatures.

5. In many areas of the world, reduced snow cover has been documented, as well as earlier spring melting of ice on rivers and lakes. For example, snow cover in the northern hemisphere has decreased by approximately 10 per cent since 1996. The cold temperatures and large amounts of snow in the winter of 2004 across most of Canada were an exception to this general pattern. The extent of Arctic sea ice has declined at a rate of 3 per cent per decade, and the summer minimum thickness has decreased by almost 40 per cent.

6. Measurements show that permafrost is warming in many regions.

7. The Intergovernmental Panel on Climate Change (2001a) reported that the average rate of sea level rise has increased from 0.1 to 0.2 mm per year during the last 3,000 years to 1 to 2 mm per year in the twentieth century, a tenfold increase. Caution is needed

The forest industry is a major contributor to the rising levels of carbon dioxide in the atmosphere not only through deforestation but also through emissions from processing plants (*Philip Dearden*).

Guest Statement

Traditional Ecological Knowledge (TEK), Science, and Climate Change
Elise Ho

Climate change is expected to affect many areas of the globe, particularly northern regions. Thus, the western James Bay region in Ontario is expected to experience drastic changes due to climate change. For the predominantly First Nations communities in this area, climate change may affect transportation networks, building infrastructure, recreational opportunities, traditional lifestyles, animal and plant populations, river and sea ice characteristics, water levels, and weather. Unfortunately, current General Circulation Models (GCMS) [discussed in the following section] are not able to predict future changes in climate at this regional scale. Furthermore, Hudson and James bays (which are mostly covered by ice during the winter) create complex feedback loops that confound many regional climate projections. Lastly, these relatively remote northern regions are not monitored to the same extent as southern regions in Canada, and weather stations are sparsely located.

As a graduate student, I investigated this problem for my Master's thesis. I examined how conventional scientific approaches to understanding climate change could be complemented by traditional ecological knowledge (TEK). TEK is 'a cumulative body of knowledge, practice and belief, evolving by adaptive processes and handed down through generations by cultural transmission, about the relationship of living beings (including humans) with one another and with their environment' (Berkes, 1999: 8). Regrettably, TEK is regarded by some to be anecdotal and less factual than what many people consider to be 'science', and it has not been given enough recognition as a valuable and insightful knowledge system in conventional environmental management and assessment practices.

For my research, I visited two First Nations communities in the western James Bay region, Fort Albany and Moose Factory. The Elders in these communities shared their knowledge of climate change with me, which I was able to compare with

the scientific evidence of climate change in the region. Since GCMS are not robust for this regional analysis, the only other scientific information with respect to climate change in the region was temperature and precipitation data from one weather station. I found that TEK was able to identify a

Elise Ho

larger range of climate change variables that contributed to a more complete understanding of how climate change has affected and might continue to affect the western James Bay area. TEK provided in-depth and specific insight regarding aspects such as changing weather characteristics, animal migration patterns and behaviour, and changing seasonal averages. Furthermore, while scientific analyses of aspects such as breakup dates for river ice are informative, TEK was able to complement this information through knowledge of the severity of the breakup events, changes in ice thickness, and other ice characteristics. I concluded that TEK and conventional scientific approaches are complementary in climate studies, and that scientific studies of climate change in this region were limited by number and scope.

My study is part of the growing research that seeks to understand how TEK and science can work as complementary knowledge systems, especially in northern climate studies. Hopefully, as we seek to better understand climate change and its effects on our environment, we will become increasingly open to different, yet also valuable, knowledge systems such as TEK.

Elise Ho completed her Master of Environmental Studies degree at the University of Waterloo, and now is studying for her Ph.D. at the University of Toronto.

Some countries, such as the Maldives, are so low-lying that they could be mostly flooded as early as 2050, if global sea levels change as predicted (*Philip Dearden*).

A view of the Moose River from Moose Factory Island. Traditional environmental knowledge is vital to understanding the complex changes that occur in regions such as Moose Factory and Moosonee. Climate change has not only affected average temperatures, but also animal and bird migration, weather patterns, and the freeze/thaw cycles of the Moose River. *(From the Cree Village Ecolodge site <www.creevillage.com> with permission of Randy Kapashesit).*

in interpreting such findings, however, as it is recognized that in some areas land is still rebounding from the weight of the last glaciation. Thus, data from tidal gauges must be interpreted in light of the combination of sea level rise and land rebound, which could mask the increase in sea level rise.

One attribute of 'good science' is the use of cross-checking data sources, to ensure that findings are not unduly influenced by measurement error

or limitations of any single data source. The findings above related to temperature, greenhouse gas concentrations, glaciers, snow cover, river and lake ice breakup, permafrost, sea-level rise, and traditional environmental knowledge all indicate that climate change is occurring. In the next section, attention turns to the reasons for this change.

BOX 7.1 MEASURING CLIMATE CHANGE

An essential step in attempting to assess climatic change is to see how present variations in climate compare with those of the past. Current data are largely instrument-based weather observations, i.e., the instrumental record. Even here there are difficulties. More modern and accurate data, for example, data on the upper atmosphere gathered by satellite, are available only for the last two decades or so. This period also coincides with the greatest human impacts on climate and does not provide any type of control for climate change in the absence of industrialization.

Former climates are reconstructed by scientists using proxy information from many different sources. For example, examination of historical records of climate-influenced factors such as the price of wheat in Europe over the past 800 years, the blooming dates of cherry trees in Kyoto, Japan, since AD 812, the height of the Nile River at Cairo since AD 622, the number of severe winters in China since the sixth century, examination of sailors' and explorers' logs, and other such

sources all contribute to building up a picture of past climates.

There are also climate-sensitive natural indicators such as tree rings and glacial ice. Cores obtained from ice in Greenland and Antarctica that go back tens of thousands of years have been analyzed using the ratio of two oxygen isotopes that indicate the air temperature when the original snow accumulated on the glacier surface. Tree rings are also very useful. Outside the tropics where there are noted differences in seasons, the width and density of tree rings reflect growth conditions, including climate. Some species, red cedar in coastal BC, for example, may live for well over 1,000 years and can provide valuable indicators as to past climates. The same kinds of rings also characterize the growth of many long-living corals in the tropics, which may be over 800 years old, and provide valuable evidence for previous El Niño events.

Once the pattern of and causes for climate change are understood, a foundation exists for considering the implications and for designing possible policies and actions.

MODELLING CLIMATE CHANGE

The uncertainty associated with global climate change is encouraging scientists to explore many different ways of assessing past and future climates (Box 7.1). One approach is **climate modelling**. While concern about climate change due to greenhouse gas emissions is relatively recent, climate modelling is not. The earliest global climate models date back to the 1950s, far ahead of when scientists were concerned about carbon dioxide emissions and their effects on the atmosphere. However, more recent concerns about global warming have propelled the science of climate modelling to the forefront.

Climate Models

All climate models consider some or all of five components in order to predict future climates.

- radiation—both incoming (solar) and out-going (absorbed, reflected);
- dynamics—the horizontal and vertical move-

CO₂ × 2, OR CO₂ × 4?

In June 2004, Mr G. Beauchemin, who is responsible for disaster preparedness for the government of Quebec, commented at the annual conference of the Canadian Water Resources Association in Montreal that the CO₂ × 2 scenario is the future we expect and are preparing for. However, he said, this is very likely to be the 'best-case' scenario. He argued that risk managers should be worrying about the 'worst-case' scenario, such as CO₂ × 4.

---

ment of energy around the globe;
- surface processes—the effects of the Earth's surface (snow cover, vegetation) on climate (i.e., albedo, emissivity);
- chemistry—the chemical composition of the atmosphere and its interactions with other Earth processes (i.e., carbon cycling);
- time step and resolution—the time step (minutes or decades) and the spatial scale (your backyard or the entire globe) of the model.

The nature of the Earth's climate and its complexity make comprehensive climate modelling difficult. The many components, interactions, and feedback loops in the global climate cannot be

---

BOX 7.2   FOUR TYPES OF CLIMATE MODELS

1. *Energy Balance Models (EBMs)*

   EBMs can be either non- or one-dimensional. In the first case, the Earth (or any point on the Earth) is treated as a single entity, and only incoming and outgoing radiation are modelled. In one-dimensional EBMs, temperature is modelled as a function of latitude and radiation balance.

2. *One-Dimensional Radiative-Convective (RC) Climate Models*

   In this model, the one-dimension is altitude. One-dimensional RC models take into account incoming and outgoing solar radiation, as well as convective processes that affect the vertical distribution of temperature. These models are useful for examining the vertical distribution of solar radiation and cloud cover, and are very useful for examining the effects of volcanic emissions on temperature.

3. *Two-Dimensional Statistical-Dynamic (SD) Climate Models*

   This type of model takes into account either the two horizontal dimensions or one horizontal dimension and the vertical dimension. The latter are most frequently modelled, thus combining the latitudinal EBMs with the vertical RC models. These models can examine wind speed, direction, and other horizontal energy transfers.

4. *General Circulation Models (GCMs)*

   The three types of climate models described above are still used for various purposes in climate research. However, since the 1980s, **general circulation models** (GCM) have largely taken over the field of climate modelling, and most model development is devoted to them. It is by far the most complex type of model, as the GCM takes into account the three-dimensional nature of the Earth's atmosphere, oceans, or both.

entirely represented by any mathematical model, and therefore all models simplify certain aspects of climate. There are four main types of climate models, each increasing in complexity. These are outlined in Box 7.2.

Unlike EBMs, GCMs attempt to examine all of the climatic elements and processes, making these models very complex. GCMs model the Earth's atmosphere and oceans under certain climate change scenarios, the most popular being $2 \times CO_2$. In this situation, the Earth's climate is modelled to indicate the changes that would occur if atmospheric concentrations of carbon dioxide were doubled from pre-Industrial Revolution levels, which many scientists believe will occur by 2050.

In a GCM, the Earth's surface is divided into a grid, in which case a larger grid results in a simpler model and a smaller grid requires more calculations. For each grid, a series of fundamental equations are solved at the surface of the grid (sea level) and for several layers of the atmosphere and subsurface layers (the vertical dimension). The fundamental equations deal with:

- conservation of momentum;
- conservation of mass;
- conservation of energy;
- ideal gas law.

Beginning with present-day or known values, the solutions for these equations are solved and repeated at each time step of the simulation, and then the results are interpolated between the grid points to cover the Earth's entire surface. The two key constraints, spatial resolution (grid size) and temporal resolution (length of time step), are compromised to meet current computational capacity. Most models operate at spatial resolutions of a few degrees latitude and longitude and at time steps of less than one hour. The vertical dimension is often divided into 10 layers, with two subsurface layers. Due to these simplifications, GCMs are best used for global or overall climate modelling, not for regional representations of climate change.

Other aspects of the global climate are also simplified, and thus limit the predictive capabilities of many GCMs. For example, known or present-day values are required to run many models, but in some areas of the world for some variables (i.e., temperature, sea ice cover, cloud cover) these values are unavailable or scarce. Therefore, assumptions are made to fill in these missing values, which may not be accurate. There are also many complicated feedbacks that cannot all be accounted for in GCMs, partly due to their complexity and the **uncertainty** of how they react under given circumstances. While the relationship between greenhouse gas emissions and temperature is a relatively straightforward positive feedback loop (where a positive change in one variable results in a positive change in the other), the relationship between increased temperature and cloud cover is relatively uncertain and relies on many other variables. Finally, many of the climatic interactions at the Earth's surface are difficult to model and are under-represented in many GCMs. For example, ocean layers and interactions are difficult to model, but their effects on regional and global climate can be quite significant.

In summary, while GCMs are becoming increasingly sophisticated, many complex aspects and interactions of the global climate need to be understood more fully, and computational facilities need to be better developed to support these new models.

Limitations of GCMs

While GCMs provide overall indications of future climates, their limitations for policy and planning need to be appreciated. Many scientists have recognized that the coarse spatial resolution, poor predictive capacity for precipitation, relatively weak simulation of oceans, lack of baseline data, and many other limitations cause GCM outputs to be highly variable. Some researchers have become discouraged by the dominance of GCMs in climate change sciences and by the scientific disregard of other climate models (i.e., EBM, RC, SD, or combinations). Others caution against the misinterpretation that GCMs are accurate and realistic models of global climate, and stress that much more improvement is needed so that GCMs may best represent the complex nature of the Earth's climate.

The Regional Climate Modelling Laboratory of the Université du Québec à Montréal (UQAM) is one of several international research institutions responding to the recurring critique of GCMs and their poor regional spatial resolution. The Canadian Regional Climate Model (CRCM) operates as a 'nested' model within a larger GCM, has a spatial resolution of 45 km between grid points, and runs

INTERGOVERNMENTAL PANEL ON CLIMATE CHANGE

The Intergovernmental Panel on Climate Change (IPCC) was established in 1988 by the World Meteorological Organization and the United Nations Environment Program.

The IPCC was created to assess scientific, technical, and socio-economic information related to understanding the risks from human-induced change to climate. The IPCC does not conduct original research, nor does it monitor climate data. Its assessments are based on peer reviewed and published scientific literature.

The IPCC has three working groups and a task force. Its First Assessment Report, published in 1991, had an important role related to the UN Framework Convention on Climate Change, which was adopted at the Earth Summit at Rio de Janeiro in 1992.

Its Second Assessment Report, published in 1995, became a significant input into negotiations that resulted in the Kyoto Protocol in 1997.

The Third Assessment Report, produced in 2001, was the product of the work of over 2,000 scientists from many disciplines from all around the world.

A Fourth Assessment Report is planned to be published in 2007.

at a time step of 15 minutes. This relatively high spatial resolution allows for the modelling of local clouds, thunderclouds, soil evaporation, and precipitation, which are too specific for most GCMs. While still developing, the UQAM and other RCs are promising tools for climate modellers.

SCIENTIFIC EXPLANATIONS

The Intergovernmental Panel on Climate Change in its most recent report in 2001 concluded that, as noted earlier in this chapter, worldwide trends in the twentieth century consistently and strongly reveal an increase in global surface temperature. There is now

virtually no scientific debate about this finding. Furthermore, as also noted above, greenhouse gas concentrations in the atmosphere have increased significantly. There is strong scientific consensus that the increase in greenhouse gases has been caused by human activities. The basis for this conclusion comes from data showing that: (1) the rate of increase of greenhouse gases in the atmosphere over the past century closely matches the rate of human-driven emissions into the atmosphere; (2) atmospheric oxygen has been falling at the same rate as fossil-fuel emissions of carbon dioxide have been increasing; and (3) significantly, the change in proportions of carbon isotopes in the atmosphere provides evidence that the atmosphere is being enriched with carbon from fossil-fuel sources rather than from natural sources. Regarding the third point, the Intergovernmental Panel on Climate Change (2001b) estimated that 70–90 per cent of the increase in carbon dioxide emissions is associated with the burning of **fossil fuels**, with the balance attributed to changes in land use, especially deforestation, and other greenhouse gases (Box 7.3). The National Research Council (2001) in the United States has reached the same conclusion.

Natural and human variables both contribute to climate change, but it is difficult to determine their relative contribution since they usually operate at the same time. Such variables alter the balance of incoming and outgoing energy in the

Although this rural scene may look bucolic, the rising numbers of cattle on the Earth contribute significantly to two of the problems discussed in this book: eutrophication and increased methane levels leading to global climatic change (*Philip Dearden*).

BOX 7.3 CONTRIBUTIONS OF DEFORESTATION, METHANE, AND NITROUS OXIDE TO CLIMATE CHANGE

Deforestation is a main cause of increasing CO_2 levels. Trees are a main repository for carbon. When they are cut down, removed, and processed, the carbon in the trees is also removed from that site. If the site is allowed to regrow trees, then over time the amount of carbon on the site will increase again. However, the rapid deforestation in this century, with many sites being used for other purposes, means that there is far less carbon now held by this biological component of the cycle than a century ago. Much of the carbon has been returned to the atmosphere as carbon dioxide. Forest fires have the same effect. Deforestation thus leads to increased amounts of CO_2 in the atmosphere and reduced uptake. Deforestation may account for up to one-third of the rise in CO_2 in the atmosphere.

Methane occurs in much smaller quantities relative to carbon dioxide (the single most important influence on rising temperature), but traps 25 to 30 times the heat per molecule of CO_2. Its major sources are organic, such as from organic decay in swamps and other anaerobic environments, and from the waste products of ruminants, such as sheep and cattle. Humans have encouraged increases in methane emissions, mainly through land-use changes, and by vastly increasing the number of domesticated ruminants and the amount of decay in waste disposal sites. Methane concentrations are increasing at a rate of between 0.75 and 1.0 per cent every year.

Nitrous oxide is produced naturally from soils and water. Major human influences are applications of nitrogen fertilizers, increases in agricultural land, and the burning of fossil fuels and biomass. When nitrogen fertilizers are added to agricultural land, some of the nitrogen is denitrified by bacteria and becomes nitrous oxide. Thus, global warming also involves disruption of the nitrogen cycle as well as the carbon cycle. This kind of interaction illustrates some of the complexities of the modern environmental challenge. Fertilizers are necessary to feed a burgeoning world population, but, although essential for food supplies in the short term, their use also contributes to global climatic change that may ultimately lead to much greater reduction in food supplies.

Earth's atmospheric system, and are usually referred to as 'radiative forcings'. Positive forcings produce warming; negative forcings lead to cooling. Greenhouse gases, aerosols, variations in solar output, and volcanic eruptions are viewed to be the most influential radiative forcings.

Figure 7.6, from the Intergovernmental Panel on Climate Change (2001b), shows the relative contribution of different radiative forcings since 1750. Greenhouse gases, and in particular carbon dioxide, have been the main forcing factor. In comparison, the effect of variation in solar output has been minor. Working in the opposite direction, sulphate aerosols, emissions from volcanic eruptions, biomass aerosols, and depletion of the stratospheric **ozone layer** have contributed to cooling at the Earth's surface. Figure 7.6 is derived from what is viewed to be a high level of scientific understanding about most greenhouse gases, based on excellent data regarding their concentrations in the atmosphere and good insight about their radiative forcing characteristics.

What about the role of humans and their activities related to climate change? The reports from the Intergovernmental Panel on Climate Change (2001a, 2001b, 2001c) are explicit that most warming since the mid-twentieth century is associated with human activities. This view is based on climate modelling that shows when only the influences of variation in solar output and emissions from volcanic eruptions are included, simulated temperature changes do not match the observed changes very well. In contrast, when the models include the forcing factors related to human activi-

INCREASING VULNERABILITY OF COASTAL AREAS

With a growing migration of human settlements to coastal areas for economic and aesthetic reasons, human society is becoming increasingly vulnerable to the possible coastal effects of climate change, such as rises in mean sea level, and an increase in the frequency of storm surges due to an increase in storminess.

— *Shaw (2001: 13)*

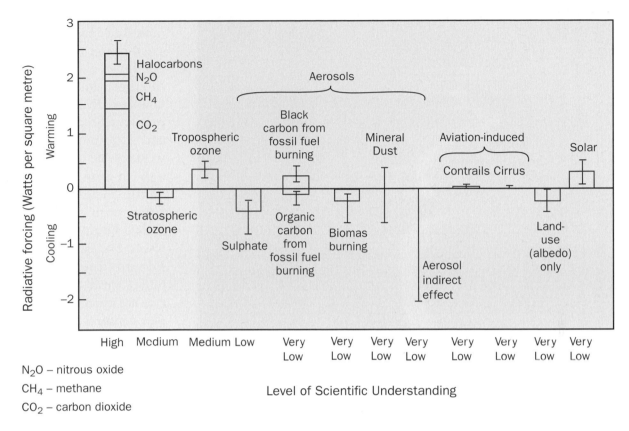

N$_2$O – nitrous oxide
CH$_4$ – methane
CO$_2$ – carbon dioxide

Figure 7.6 Relative contribution of different radiative forcings, 1750 to 2000. These radiative forcings arise from changes in the atmospheric composition, alteration of surface reflectance by land use, and variation in the output of the sun. Except for solar variation, some form of human activity is linked to each. The rectangular bars represent estimates of the contributions of these forcings—some of which yield warming, and some cooling. Forcing due to episodic volcanic events, which lead to a negative forcing lasting only for a few years, is not shown. The indirect effect of aerosols shown is their effect on the size and number of cloud droplets. A second indirect effect of aerosols on clouds, namely their effect on cloud lifetime, which would also lead to a negative forcing, is not shown. Effects of aviation on greenhouse gases are included in the individual bars. The vertical line about the rectangular bars indicates a range of estimates, guided by the spread in the published values of the forcings and physical understanding. Some of the forcings possess a much greater degree of certainty than others. A vertical line without a rectangular bar denotes a forcing for which no best estimate can be given owing to large uncertainties. The overall level of scientific understanding for each forcing varies considerably, as noted. Some of the radiative forcing agents are well mixed over the globe, such as CO$_2$, thereby perturbing the global heat balance. Others represent perturbations with stronger regional signatures because of their spatial distribution, such as aerosols. For this and other reasons, a simple sum of the positive and negative bars cannot be expected to yield the net effect on the climate system. The simulations of this assessment report indicate that the estimated net effect of these perturbations is to have warmed the global climate since 1750. SOURCE: Miller (2002: A-4), from IPCC (2001b); reprinted with permission of the Intergovernmental Panel on Climate Change. © Intergovernmental Panel on Climate Change.

ties (greenhouse gases, stratospheric ozone depletion, sulphate aerosols), better matches are found between the observed and simulated patterns of temperature change. Not surprisingly, the best match occurs when both natural and human forcing factors are combined.

There is no guarantee that greenhouse gases have caused the temperature increases at the Earth's surface, but the evidence is strong and credible. As the National Research Council in the US (2001) noted, the view that most of the observed increase in temperatures is likely due to an increase in greenhouse gas concentrations 'accurately reflects the current thinking of the scientific community on this issue.'

CASE STUDIES: RISE OF SEA LEVEL IN PEI AND WINTER TOURISM IN ONTARIO

So far in this chapter, we have concentrated on understanding the nature of climate change, the manner in which climate has been changing, and reasons for the change. Before we turn to the general implications and possible policies and actions, it is useful to examine the possible consequences of climate change 'on the ground'. In that regard, two case studies will help to illuminate what might happen in the future. The first consid-

Prince Edward Island coastline (*Al Harvey*/www.slidefarm.com).

However, with anticipated increased concentrations of greenhouse gases leading to continued climate change, the Intergovernmental Panel on Climate Change (2001b) has predicted that between 1990 and 2100 mean sea level at a world scale may increase by 0.09 to 0.88 metres, with an average increase of 0.5 metres. Changes in sea level at local scales could be higher or lower. For east-central Prince Edward Island, during the past 6,000 years, mean sea-level rise has been about 0.3 metres per century, but at a slower rate in the past 200 years. At Charlottetown, records indicate that since 1911 the rate of sea-level change has been 0.32 metres per century.

The coast of Prince Edward Island is one of the areas in Canada most vulnerable to sea-level rise (Figure 7.6). As Shaw and his colleagues (2001: 14) observed, this vulnerability is due to many factors, including presence of soft sandstone bedrock, a sandy and dynamic shore zone starved of sediment in places, an indented shoreline with many estuaries and marshes, low terrain inland from the shoreline leading to high potential for flooding, high rates of shoreline retreat, and submergence of the coast, which continues today. Against that background, Shaw and others analyzed the physical and socio-economic impacts of climate change and accelerated sea-level rise on the coast of PEI.

ers what would happen in Prince Edward Island related to rising sea levels if anticipated climate change continues. The second considers what would happen to winter recreation activities in south-central Ontario.

Case Study 1: Prince Edward Island and Rising Sea Level

At a global scale, records document that mean sea level has been rising at a rate of 0.1 to 0.2 metres per century during the last 100–200 years.

Sea-Level Rise

Figure 7.8 shows that since the last ice age (during the last 10,000 years) the relative sea level has gone up by 45 metres or more in central and eastern PEI. One result has been inundation of the land once connecting PEI to the Nova Scotia and

New Brunswick mainland. The relative rate of sea-level rise (an outcome of both land subsidence and rising sea level) during the last 6,000 years has averaged some 30 centimetres/century.

One reason for the change in relative sea level has been crustal subsidence, due to adjustments to different ice and water loads in the post-ice age period. It has been estimated that crustal subsidence has been about 20 centimetres each century in PEI, indicating the balance of 12 centimetres per century is due to global and regional sea-level rise. As sea levels increase, the impact of storm surges and ocean waves also increases because

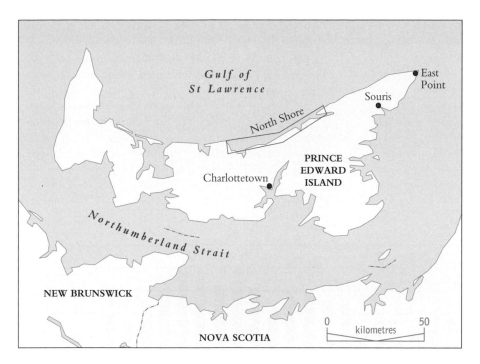

Figure 7.7 The coast of Prince Edward Island. SOURCE: Shaw et al. (2001: 15). Reprinted with permission of the Minister of Public Works and Government Services, Canada, 2005 and Courtesy of Natural Resources Canada, Geological Survey of Canada.

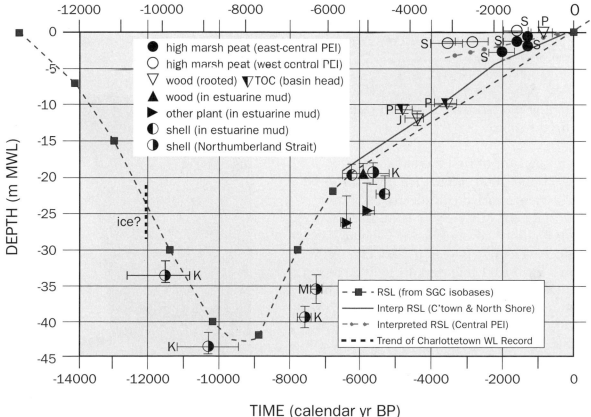

Figure 7.8 Relative sea level increase in coastal and eastern PEI since the last ice age. Note: This represents the combination of sea-level change and vertical land movement. The black broken line indicates the general trend and the solid line is the best estimate of trend for Charlottetown and the central North Shore over the past 6,000 years. SOURCE: Shaw et al. (2001: 35) Reprinted with permission of the Minister of Public Works and Government Services, Canada, 2005 and Courtesy of Natural Resources Canada, Geological Survey of Canada.

OSCILLATION

The Southern Oscillation is based on the monthly or seasonal fluctuation in the air pressure differences between Tahiti and Darwin, Australia, in the South Pacific Ocean. Sustained negative values often indicate El Niño episodes, associated with sustained warming of the central and eastern tropical Pacific Ocean. Positive values lead to La Niña episodes, usually leading to wetter than normal conditions in eastern and northern Australia. The North Atlantic Oscillation, also referred to by some as the Arctic Oscillation, relates to opposing atmospheric patterns in northern middle and high latitudes. Positive phases involve low pressure over the polar region and high pressure at mid-latitudes. Such conditions result in ocean storms being driven further north, leading to wetter weather in Alaska, Scotland, and Scandinavia and drier conditions in the western United States and the Mediterranean. In negative phases, the pattern is reversed.

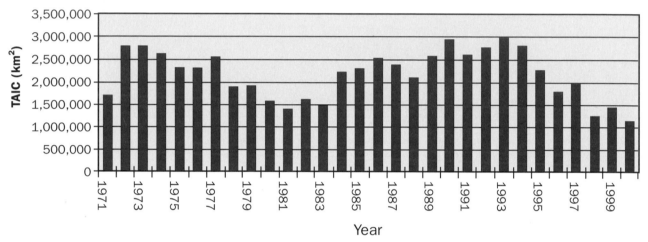

Figure 7.9 TAIC: Total accumulated ice coverage. Annual ice cover in the Gulf of St Lawrence, 1971–2000. SOURCE: Shaw et al. (2001: 46). Reprinted with permission of the Minister of Public Works and Government Services, Canada, 2005 and Courtesy of Natural Resources Canada, Geological Survey of Canada.

the water hitting the coastline is carried to higher elevations on the shore.

Ice Cover in the Gulf of St Lawrence

Between 1971 and 2000, annual ice cover in the Gulf of St Lawrence varied considerably, as shown in Figure 7.9. Extent of ice cover is measured by 'total accumulated ice coverage' (TAIC). Figure 7.9 indicates that over this 30-year period the TAIC varied from a low of 1.1 million km^2 in 2000 to a high of just under 3 million km^2 in 1993. The figure also suggests there may be a 15-year cycle in variability. Climate change is one potential cause of changes in ice cover, but scientists also recognize that other factors have an influence, including El Niño, Southern Oscillation, North Atlantic Oscillation, solar flux, and volcanic eruptions.

Calculations based on the Canadian Climate Change Model indicate that by 2050 the extent of sea ice in the northern hemisphere will be reduced

The 21 January 2000 ice ride-up moved this Charlotte-town lighthouse and damaged a golf course (*D.L. Forbes, Geological Survey of Canada Atlantic*).

(LeDrew, 2002). The Gulf of St Lawrence could become free of ice for the entire year. This latter estimate is important, because sea ice in the Gulf inhibits development of waves and therefore reduces erosion from winter storms on the coastline. If sea ice decreases or disappears in the Gulf, the size of waves can be expected to increase, with an associated increase in erosion. In addition, it is possible that storm surge impacts also could be increased, but this is less easy to determine.

Another consideration is that if wave action increases in frequency or magnitude, but there is still some ice in the Gulf adjacent to coastlines, there can be damage expected from ice ride-up or pile-up against the coastline. For example, on 21 January 2000, an ice ride-up during a storm pushed a lighthouse in Charlottetown off its foundation and also caused major damage to a golf course near Charlottetown. In various places along the coast on PEI, and other provinces in Atlantic Canada, the ice ride-up and pile-up damaged wharves and other coastal infrastructure, as well as some homes. On the other hand, on the north coast of PEI, damage to the shore from waves was minor, mainly due to the protection provided by sea ice along the shoreline, which bore the brunt of the energy from wind-driven waves.

Increased Storminess and Storm Surges

Earlier in this chapter, it was noted that climate change relates to more than changes in temperature and precipitation. It can also involve alterations in wind speeds and patterns, as well as new extremes of weather conditions.

Storm surges reflect the meteorological consequences on sea level in an area, and are the difference between an observed water level and what would be expected under predicted tidal conditions. A large storm surge combined with an exceptionally high tide creates conditions when flooding and damage are likely to occur. In contrast, a 'negative' storm surge can cause sea levels to be lower than predicted from tidal charts. Between 1960 and 1998, storm surges exceeding 60 centimetres occurred approximately eight times a year in Charlottetown, primarily associated with winter storms. Protective works for the Charlottetown harbour are more than adequate to protect it from surges causing water levels to increase up to 60 centimetres. However, storm surges in the

future accompanied by sea-level increases due to climate change have the potential to cause damage because the high water levels can be expected to exceed the design capacity of protective works. Furthermore, not only the severity of storm surges can be expected to increase—so can the frequency of higher storm surges, consistent with more variability due to climate change. Finally, if more storm surges to higher levels are accompanied by reduced ice cover in the Gulf of St Lawrence, conditions will have been created for significant damage potential.

Climate Change Future for Charlottetown and the North Coast

Three scenarios related to sea-level rise affecting Charlottetown were examined by Shaw and his colleagues: (1) a recreation of the storm of 21 January 2000 to test the accuracy of the model for flooding that occurred with a maximum water level of 4.23 metres above 'chart datum' or CD (the lowest predicted water level averaged over several years); (2) a lower storm surge than occurred in January 2000 but superimposed on anticipated future sea-level rise to give a flood level of 4.70 metres; and (3) a water level associated with the January 2000 storm combined with estimated sea-level rise of 0.7 metres by 2100 to give a flood level of 4.93 metres above CD.

Under Scenario 1, about 460 properties are either flooded or placed under risk of flooding, and the assessed property value at risk is calculated to be $172 million. In Scenario 2, the at-risk property value increases to $190 million, of which one-third is commercial property. In addition, some farmland becomes at risk. Scenario 3 results in the value of property at risk climbing to an estimated $202 million, with commercial property again representing one-third of property at risk.

A further complication for estimating damage potential is that Charlottetown has many designated heritage areas and structures, most in the central part of the city. Some 335 municipally designated properties are mostly located in the downtown core, as are about a dozen federally designated sites. Many designated sites and areas are within the probable flood plain identified in the modelling study, indicating that potential costs to heritage values are not trivial. Under Scenarios 2 and 3, the municipally designated heritage properties at risk of flooding are estimated to total $10.5 million and $11.3 million, respectively.

Damage to properties is not the only risk from flooding exacerbated by sea-level rise and storm surges. Municipal infrastructure, including sewer, water, and storm systems, also is vulnerable. Possible impacts on the storm/sewage system include: (1) a significant rise in water level would lead to a surcharge from outfalls and lift stations into the sewage lines, resulting in backups into residential and commercial areas; (2) water levels high enough to reach the lift station (on the waterfront, pumping sewage to the primary sewage treatment plant) could lead to sea water being pumped along with or in place of sewage to the main treatment plant, causing corrosion and possible lack of capacity to move sewage to the treatment plant; and (3) extended flooding of a lift station or the treatment plant could result in their becoming incapable of functioning due to temporary or permanent damage. The damage could range from just over $300,000 for a small lift station to over $25 million for the main sewage treatment plant. Other infrastructure at risk would include sidewalks and streets, estimated at a cost of at least $13 million.

Other risks to Charlottetown could occur through water-borne diseases, loss of employment, and disruption to education for students.

Along the north shore of PEI, it is estimated that heightened sea levels, reduced sea ice, and increased wave energy would create severe erosion damage and possibly reconfiguration of some coastlines. Since the mid-1930s, some coastal areas have already experienced serious erosion, while other areas have experienced shoreline recovery and growth of sand dunes. A long-term perspective indicates that the north coast has generally been retreating at a rate of at least 0.5 metres annually for several thousand years, which has led to property literally disappearing, wetlands experiencing encroachment, and coastal infrastructure being placed at risk.

Adaptation to Climate Change

Five types of adaptation are usually recognized: (1) prevent the loss by adopting measures that reduce vulnerability; (2) tolerate the loss by doing nothing and absorbing the cost of losses when they happen; (3) spread or share the loss by distributing the costs over a larger population, such as through

insurance; (4) change the affected activity by ceasing to do certain things or by changing to other activities; and (5) change the location of the activity by moving to a less vulnerable location. For PEI, Shaw and his colleagues identify three adaptation measures: protection, accommodation, and retreat.

Protection usually involves constructing structural measures to protect property, buildings, and infrastructure. Structural measures can involve individual initiatives or major public works projects, such as sea walls, revetments, and groynes designed to trap sediment. Shaw and his colleagues (2001: 55) noted that 'in general, protection is costly and may have limited long-term effectiveness in exposed locations, though it may be successful as flood protection where wave energy is limited.'

There also can be side effects from some structural measures. For example, groynes often are effective in trapping sediments and therefore in protecting the properties adjacent to the area in which sediment is trapped. However, locations downdrift of the groynes may experience reduced sediment deposition and thus become more vulnerable to erosion, slumping of shorelines, and eventually increased rate of shoreline loss.

Accommodation usually involves a mix of approaches—redesign of structures to reduce their vulnerability, zoning to guide appropriate land use involving low capital investment in vulnerable areas, and other measures such as rehabilitating coastal dune systems, renewing wetlands, nourishing beaches, and replacing causeways with bridges.

However, 'stabilizing' natural systems can undermine the natural functioning of ecosystems and can be counterproductive. For example, the endangered piping plover requires an active beach habitat and prefers environments including washover channels through dune systems. However, cottage and other property owners usually do not favour unstable shorelines, and as a result there can be conflict between various groups, such as bird watchers and cottagers, with different interests and priorities.

Retreat, the third general approach, seeks to avoid vulnerability. It usually involves recognizing the high risk or vulnerability of a place and taking a conscious decision to relocate buildings, other capital works, or infrastructure away from hazardous places. The initial cost of relocating is normally very high, but in the long term the costs are usually

DESIGNING A STRATEGY FOR PEI

An interesting project for you would be to design an appropriate strategy for Charlottetown and the north shore of PEI, using a mix of approaches. What would be your first priority for an overall approach? How might the strategy change if you addressed the challenge for another community vulnerable to sea-level rise? Do you live in a community that might become more vulnerable to flooding as a result of climate change? If so, what strategies might you design?

judged to be much lower than if money has to be spent after each damaging natural event to rebuild or repair damaged properties. This approach is the easiest when development has not already occurred and zoning can be used to keep development away from vulnerable places. However, many Canadians have chosen to live or work close to rivers, lakes, or coastlines, so relocation often would require a very substantial investment.

Case Study 2: Ontario and Winter Recreation

Scott and his colleagues (2002) have noted that the Canadian Tourism Commission reported total tourism expenditures in Canada during 2000 were more than $54 billion and supported 546,400 jobs, and that the contribution of tourism to the Cana-

dian economy exceeded that from agriculture, forestry, mining, and fisheries. They also observed that climate plays a significant role relative to tourism and recreation, as in some regions it is the key resource base for tourism, and furthermore the seasons influence the frequency and duration of outdoor recreation activities.

In Canada, there is a distinct seasonal pattern of tourism expenditures, with summer accounting for almost 40 per cent of annual domestic and over 60 per cent of annual international tourism expenditures. Length of operating season is therefore an important aspect for the success of tourism. Due to the dominance of the summer tourism season, any extension of the warm-weather recreation season could be beneficial in economic terms as it would allow a longer period for golfing, camping, and boating. However, such economic benefits need to be balanced against new costs, such as increased environmental deterioration due to more visitors for longer periods. There also could be intersectoral resource conflicts, such as between increased demand for water to irrigate golf courses and to irrigate agricultural crops.

While warmer weather could extend and enhance summer recreation activity, winter-based activities—such as downhill skiing, snowboarding, Nordic skiing, snowmobiling, ice fishing—that depend on snow and ice could be adversely affected. Research on winter-based recreation activities in the Great Lakes region indicates that such negative impacts are very likely. For example, a study in the early 1990s, using climate change scenarios available at that time, found that the ski season in the area north of the Great Lakes would likely be reduced by 30–40 per cent. The same study also found that the skiing season to the south of Georgian Bay would be reduced by 40–100 per cent, with the latter number leading to the elimination of a multi-million dollar industry. Two studies of the ski season in the lower Laurentians area of Quebec indicated a reduction of the ski season from between 34 and 49 per cent, or between 42 and 87

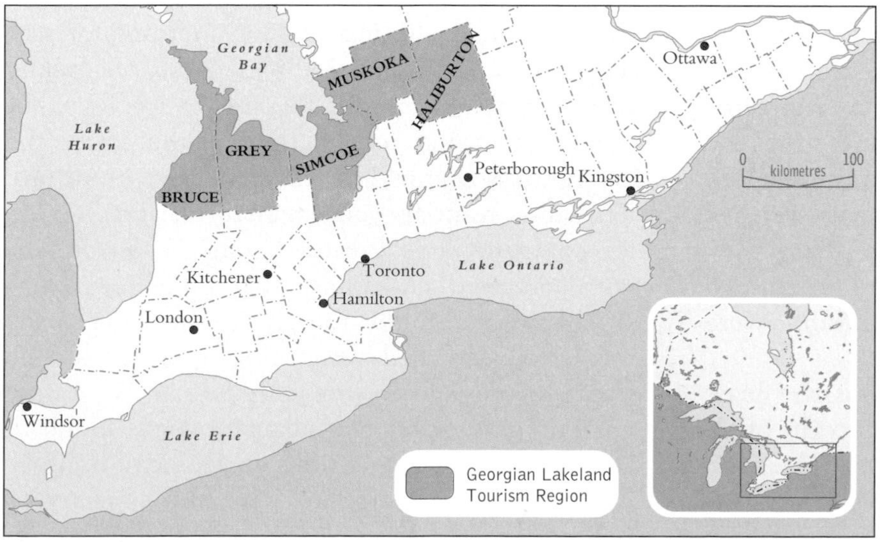

Figure 7.10 Georgian Lakeland Tourism Region. SOURCE: Scott et al. (2002: 8).

per cent. It was noted that whatever happens to the ski areas and seasons in the Great Lakes region is of critical importance, since the Great Lakes region is the primary feeder area for more challenging ski resorts in North America and even beyond.

To update the studies referred to above, Scott and his colleagues conducted the first 'integrated sectoral assessment' of four core winter recreation industries (alpine and Nordic skiing, snowmobiling, ice fishing) in the Georgian Lakeland Tourism Region in south-central Ontario (Figure 7.10) in order to improve understanding of different vulnerabilities of major winter recreation industries, the role of climate adaptation, the potential net impact of climate change in the study area, the relationship between climate change and regional development planning, and the capacity of communities to adapt to climate change.

The Georgian Lakeland Tourism Region, representative of Ontario's 'cottage country', was selected because the recreation-tourism sector is the second most important component in its economy, and climate has been recognized as a limiting factor for winter-based recreation. Inter-regional tourism is dominant, with tourists from within Ontario making up the largest portion of visitors (93 per cent) and associated expenditures. Furthermore, it is a major recreation destination for people living in the Greater Toronto Area. Many tourism-based businesses cannot survive on activities based on one season, so the income from the winter season activity is critical.

The analysis was conducted in several stages. Stage one involved construction of a historical climate data set for seven climate stations. Those data then were used to calibrate a climate-recreation simulation model. In the second stage, the recreation season simulation model was calibrated with observed data on recreation activity. Thresholds from the literature were used to define suitable climatic conditions, such as what constitutes a skiable day. Then, the recreation seasons were simulated for the 1961–90 period.

Alpine skiing, it was found, was the least vulnerable to current climate variability. In other words, from one year to another it experienced the least variability in length of season. The primary reason was the widespread use of snow-making technology, which, in some years, extended the

skiing season by almost 100 days. Although snow-making equipment required a multi-million dollar investment, those costs were repaid several times during mild or low snowfall winters during the 1990s and can be anticipated to prevent losses in the tens of millions of dollars annually under anticipated climate change scenarios.

By incorporating snow-making technology into the analysis, Scott and his co-workers concluded that the alpine skiing seasons would be reduced by between 21 and 34 per cent, in contrast to the studies in the early 1990s that had not considered the effect of snow-making and suggested that the season could be reduced by 40–100 per cent in the Georgian Lakeland Tourism Region under doubled CO_2 conditions. However, Scott et al. pointed out that considerable uncertainty exists about the extra costs of snow-making under warmer conditions, and therefore it is possible that the economic benefits of an extended skiing season might be offset by extra costs for snow-making.

Vulnerability was higher for the two trail-based activities. Nordic skiing and snowmobiling seasons were estimated to become more than 50 per cent shorter in some years. It was estimated that the season for snowmobiling could be shortened by 29–49 per cent and Nordic skiing by 39–55 per cent as soon as the 2020s. They noted that adjustments through snow-making technology had reduced the vulnerability of alpine skiing, but any comparable adjustments had not yet been considered for snowmobiling.

Ice fishing on Lake Simcoe was vulnerable. Estimated changes in average ice thickness over the lake could eliminate safe conditions for ice-fishing huts in some parts of the lake. Ice fishing could continue without the use of ice huts, but even then it was judged that by the 2020s the average season could be shortened by one-third. The study team also commented that the thinner ice could result in greater hazard for ice fishers, with increased probability that more ice fishers would break through the ice. In addition to the threat to individual well-being, there would be associated costs for rescue activities.

The most significant adaptation was found to be snow-making technology. Improved snow-making was estimated to have the capacity to reduce average seasonal losses in the alpine ski industry by 5–7 per cent in the 2020s, with estimated annual savings of $8.6 million to $15.5. million. These savings would become even higher under the projected climate change scenarios for the 2050s and 2080s.

Other adaptation strategies were identified. On the supply side, winter trail operators could modify snow-making practices to reduce their vulnerability. Alpine ski and trail areas also could modify design of slopes (develop north-facing slopes, develop at higher altitudes, smooth rough surfaces to reduce snow requirements for safe operations). Some recreation operators have diversified by becoming four-season facilities with multiple recreation activities in each season as a way to reduce vulnerability to climate variability.

Other Examples

The two case studies presented above provide insight into what might be anticipated in terms of impacts from climate change. However, there are many other possible consequences, some of which are noted below.

Terrestrial Systems

It is conceivable that within your lifetime many terrestrial systems, along with the associated fauna and flora, will change dramatically. For example, on the Canadian Prairies, boreal forests may shift anywhere from 100 to 700 km to the north, to be replaced by grasslands and more southern forest species. In the Arctic, the southern permafrost border could move 500 km northward and the tree line could move from 200 to 300 km to the north. These shifts are illustrated in Figure 7.11. In

CLIMATE CHANGE AND ENVIRONMENTAL REFUGEES

Future climate change is expected to have considerable impacts on natural resource systems, and it is well-established that changes in the natural environment can affect human sustenance and livelihoods. This in turn can lead to instability and conflict, often followed by displacements of people and changes in occupancy and migration patterns. Therefore, as hazards and disruptions associated with climate change grow in this century, so, too, may the likelihood of related population displacements.

— McLeman and Smit (2003: 6)

the Rockies, glaciers under 100 metres thick are vulnerable to melt and could disappear by 2030.

The consequences of change to terrestrial systems could be dramatic. For example, polar bears may no longer remain to breed in Wapusk National Park in Manitoba, and yet the park was created in 1996 to protect the habitat of polar bears. At the other extreme, the hoary marmot in the Rockies and other areas in the Western Cordillera is likely to thrive, as changed climate leads to more avalanches, which will expand its preferred habitat of open meadow. These examples illustrate that the rationale for national and provincial parks, which were created to protect representative ecosystems (Chapter 11), may disappear or dramatically change as the distinctive ecosystem currently protected by such parks evolves into something totally different.

Freshwater Systems

As a result of changes documented so far, every part of Canada except the southern Prairies has become wetter, with precipitation increasing from 5 per cent to 35 per cent since 1950. At the same time, generally higher temperatures cause higher rates of evapotranspiration. What might be the outcomes? On the west coast of British Columbia, increased cloud cover and more rain can be expected. As a result, water supplies should be relatively secure, but tourism may be adversely affected if potential tourists are not attracted to an area that already has a reputation of receiving an abundance of rainfall, sometimes referred to as 'liquid sunshine'. In other areas, agriculture operations may become more vulnerable, leading to pressure for expansion of irrigation systems, which may place high pressure on surface and groundwater systems. On the Great Lakes, the shipping season may be extended due to less ice on the lakes, but at the same time drier conditions may contribute to the drop in lake levels, requiring lakers to carry less freight in order to navigate locks and other passages with depth constraints.

Ocean and Coastal Systems

It appears as if both sea temperatures and sea levels will increase. The latter will have the effects outlined above for Charlottetown for various other coastal communities (e.g., Vancouver, St John's), the severity depending on the nature of the coastline and the amount of increase.

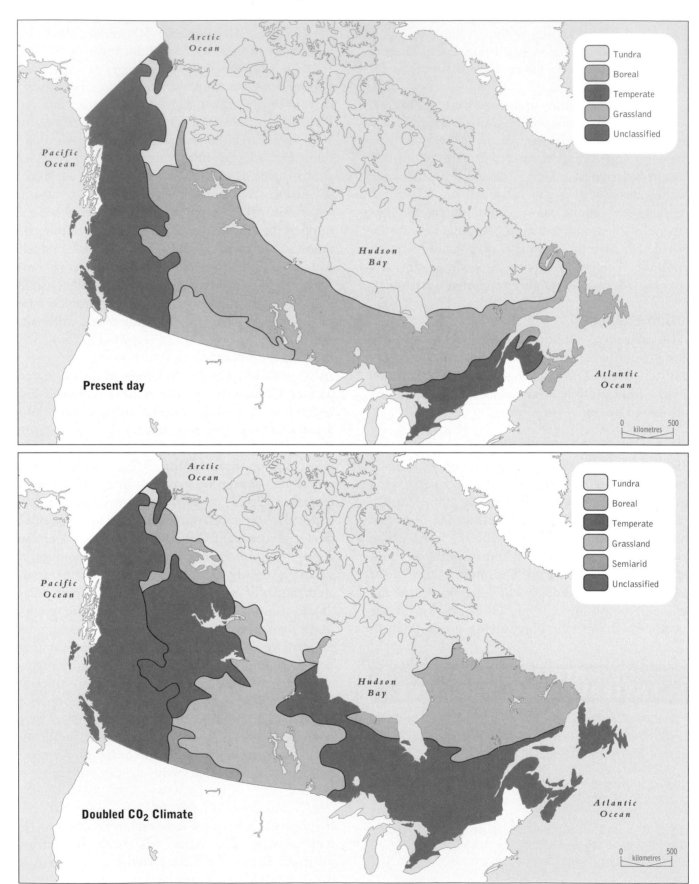

Figure 7.11 Changes in forest and grassland boundaries resulting from a typical doubled CO_2 climate. SOURCE: H. Hengeveld, *Understanding Atmospheric Change*, State of the Environment Report 91–2 (Ottawa: Environment Canada, 44. Reproduced with the permission of the Minister of Public Works and Government Services, 2005.

COMPLEXITY AND ALTERNATIVE EXPLANATIONS

The media have a hard time dealing with the ebb and flow of scientific research, which characteristically attempts to falsify one, or perhaps competing, hypotheses and moves, usually unevenly, toward a most strongly accepted description of reality. Ecosystem problems usually involve multiple causes, so more than one hypothesis may be correct, a source of even more confusion.

— Carpenter (1995: 128)

However, it is also clear that wave action may change, especially becoming more severe in areas previously covered by ice for part or most of the year. Wave action on shorelines will contribute to erosion and to changes in wetland complexes, both enhancing and damaging them.

COMMUNICATING GLOBAL CHANGE

In Chapter 6, communication was identified as one of seven attributes of best practice related to processes, methods, and products in resource and environmental management. In that regard, Andrey and Mortsch (2000) have identified several challenges for communicating information or understanding about global change.

1. *Global change is a complex issue.* The global climate system is enormously complex, mainly due to the many linkages and feedback mechanisms in the atmospheric system. Furthermore, the associated socio-economic system is complex and continuously changing.

A related complication is that while many people have heard about 'global warming' or 'global change', the level of understanding usually is poor. Polls consistently show that many Canadians have a poor understanding of the meaning, causes, or effects of global change. Few see the connections between energy use and deforestation, and climate change. Another complication is that the media often provide misleading or incorrect information, such as when the *Toronto Star* (19 Oct. 1999) reported that methane is one of the most dangerous of all greenhouse gases and is 32 times as dangerous to the ozone layer as carbon dioxide, or when the *Red Deer Advocate* (17 Sept. 1999) stated that scientists around the world believe carbon dioxide emissions are depleting the ozone layer. Although ozone-depleting substances (CFCs) are also greenhouse gases, ozone depletion by itself has a very minor effect on the global energy budget, and, indeed, is thought to have a net cooling effect (Box 7.4).

Another complication is that scientists and the general public do not often speak the same language related to global change. As Chalecki (2000: A2, 15) observed, 'Scientists often examine small pieces of larger environmental problems in great detail within the limits of their discipline, while most non-scientists have a somewhat fuzzy understanding of the larger issues, often fed by outdated knowledge and half-formed opinions.'

2. *Uncertainties exist regarding almost every aspect of the global change issue, and these increase when moving from natural to human systems.* Four main sources of uncertainty occur: (1) statistical randomness, or the variability in nature; (2) lack of scientific understanding about processes involved; (3) lack of or inadequate data; and (4) imprecision in risk assessment methods due to varying protocols for conducting research. All of these are relevant in global change research. They collectively contribute to significant uncertainty, which encourages a 'wait-and-see' attitude on the part of some policy-

Canada's North would experience major changes as a result of global warming as water levels change, as the amount and duration of snow cover change, and as permafrost melts. This could have drastic effects on areas such as the Mackenzie Delta (*Philip Dearden*).

BOX 7.4 OZONE DEPLETION

Ultraviolet radiation from the sun causes some oxygen molecules to split apart into free oxygen atoms. These may recombine with other oxygen molecules to form ozone (O_3) in the outer layer of the atmosphere, known as the stratosphere. This layer of ozone helps to filter out ultraviolet (UV) radiation from penetrating to the Earth's surface where it destroys protein and DNA molecules. Without this protective layer, it is doubtful whether life could have evolved at all on Earth.

Although there are natural causes of variation of ozone levels, recent observations have indicated that this layer is being broken down by the emission of various chemicals from the Earth. Since 1979, the amount of stratospheric ozone over the entire globe has fallen per decade by about 4 to 6 per cent in the mid-latitudes and by 10 to 12 per cent in higher latitudes. These decreases have led to average increases in exposure to ultraviolet-B (UV-B) of 6.8 per cent per decade at 55° N and 9.9 per cent in the same latitude in the southern hemisphere. In general, penetration of UV-B radiation increases by 2 per cent for every 1 per cent decrease in the ozone layer. UV-B radiation is responsible for various potentially negative health effects to humans and animals, mainly related to eyes, skin, and immune systems. Human vulnerability to UV-B depends on a person's location (latitude, altitude), duration and timing of outdoor activities, and precautionary behaviour (use of sunscreen, sunglasses, or protective clothing).

makers due to skepticism about the information and understanding provided by scientists (Fraser Institute, 1997a, 1997b, 1998, 1999). Such a view has been referred to as 'brownlash' by Ehrlich and Ehrlich (1997) because the intent is to 'minimize the seriousness of environmental problems' and 'help to fuel a backlash against "green" policies'.

3. *Impacts of global change will be disproportionately heavier on people in less-developed countries and on future generations.* Human-induced climate change impacts will fall mainly on future generations. Areas at greatest risk are those with limited fresh water, prone to drought, along coasts, and generally in less-developed nations. One consequence is that many people in developed nations, confronted with the various issues and problems in their lives, give less attention to global change challenges. Most give priority to issues with some immediacy or urgency, and global change does not fall into that category. As Andrey and Mortsch (2000) observe, the consequences of climate change are diffuse rather than concentrated, indirect rather than direct, unintended rather than intended, and affect statistical or anonymous people rather than identifiable individuals.

4. *The basic causes of global change are embedded in current values and lifestyles.* In the developed world, including Canada, relatively high standards of living and materialistic lifestyles rely on extensive use of energy based on fossil fuels. Much of this use results from residential heating and cooling and from personal transportation by car. It is the cumulative effect of millions of people going about their normal lives that contributes to global warming. It is easy for any one individual to conclude that by changing lifestyle he or she will make virtually no difference at a global scale. Thus, a dilemma is created for any individual, city, province, or country to take action, because the scale of the challenge requires unprecedented collaboration (see Chapter 6). Thus, individuals believe they are helpless on their own to make a difference, and for many people more immediate issues compete for attention and resources.

The above attribute flags an important cautionary note. We should not get so preoccupied about providing information and increasing understanding that we neglect to create a sense of ownership and empowerment relative to this issue. As is often argued, the purpose of communication is not only to provide knowledge to raise awareness and increase understanding, but also it should lead to changes in attitudes and behaviour.

In Chapter 6, four questions posed by Carpenter (1995) related to communication were considered. Here, each question is addressed relative to climate change.

1. *What do we know, with what accuracy, and how confident are we about the data?* Notwithstanding many uncertainties related to global change, some components are both well understood and documented, and therefore there is a high degree of confidence about data, interpretations, and conclusions. For example, our understanding is solid regard-

ing the increase in atmospheric carbon dioxide, alterations in the biogeochemical cycle of nitrogen, and broad-scale changes in land use and vegetation cover. However, the implications of the cause-and-effect sequence from such changes for impacts on ecosystems and human health are much less understood.

To improve our understanding, the scientific method is often applied, which involves making observations related to the patterns in nature and the outcomes of experiments, and leads to development of hypotheses to account for observed patterns and relationships. Findings are published so other investigators can replicate and confirm, or challenge, them. As more data are collected and analyzed, hypotheses may be modified or changed, and there is always some uncertainty due to awareness that more data could be assembled. As Carpenter (1995: 132) commented, such an approach 'is confusing to laypersons who, understandably, do not readily reconcile the claims of regularity and predictability with a certain amount of irreducible uncertainty.'

2. *What don't we know, and why are we uncertain?* As noted previously, we often are uncertain due to measurement problems because of insufficient observations and natural variability in ecosystems, challenges of extrapolating from one time and/or place to another time and place, and inadequacy of models and our basic understanding of structures and processes in ecosystems. Some environmental phenomena are both uncertain (because of difficulties of precise measurement) and variable (changeable). The latter factor often compounds uncertainty. We are aware that temperature and precipitation can vary on a daily, seasonal, or annual basis. Furthermore, impacts on humans and human activity are usually a function of exposure, or the amount of time spent in a particular place. The fact that variability is part of the normal behaviour of ecosystems needs to be documented and explained, and not become used as an explanation for not taking action due to 'uncertainty'.

3. *What could we know with more time, money, and talent?* Additional information can usually be acquired, but a decision to obtain more data must be taken with regard to the resources available for such an undertaking, and to what the 'value added' of more data will be. In some cases, it may well be that additional information would not provide significant new insight or lead to management decisions that would result in significant societal benefits.

There is also a dilemma faced in experimentation. It is difficult to know how an ecosystem will respond to stress except by placing it under stress. However, there could well be ethical questions related to creating such stresses if outcomes could include injury or mortality to other living species or significant costs imposed on some members of society. We also appreciate that ecosystem responses may be characterized by lags, thresholds, and rapid transformations or flips from one state to another— highlighted by the surprise associated with the discovery of the Antarctic ozone hole, despite years of monitoring CFCs and a sophisticated understanding of atmospheric chemistry.

4. *What should we know in order to act in the context of uncertainty?* Both action and non-action involve acceptance of risk. As a result, we often strive to conduct a risk assessment. In such a process, it is usual to identify choices or options, as well as the benefits, costs, and risks for each, plus how those will be distributed across individuals or groups in a society (Chapter 6). In conducting such an assessment, it is normal to consider the following: (1) whether an option will increase any risk beyond what could be expected if the status quo were continued; (2) different risks for alternative options to realize the same goal (an electric car would reduce pollutants relative to gasoline-based cars, but what would be the environmental impacts from having to generate much more electricity to charge batteries for electric cars?); (3) the types of risks created (immediate or delayed; low probability but high impact, or high probability but low impact; unequally distributed; accepted voluntarily or involuntarily); and (4) the benefits to be realized, set against the possibility of risks.

Risk assessment by itself cannot eliminate risk, but it can assist us in making more informed decisions. However, people can use information about risk in unpredictable ways, as illustrated by the anecdote from Carpenter (1995: 134) recounted in the previous chapter of the airline passenger who carried a bomb on board the plane as 'insurance'.

KYOTO PROTOCOL

During December 1997, representatives from more than 160 countries met in Kyoto, Japan. The outcome was an agreement, popularly referred to as the **Kyoto Protocol**, with targets for 38 developed nations as well as the European Community

GREENHOUSE EMISSIONS FROM CANADA

Canadians currently produce about 700 megatonnes of greenhouse gases per year, much of it from wasteful energy use. This is about 2 per cent of total global emissions, coming from a country with about half of 1 per cent of the world's population.

In other words, the average Canadian produces four times the global average level of emissions.

— *David Suzuki Foundation (2003: 2)*

to ensure that 'their aggregate anthropocentric carbon dioxide equivalent emissions of the greenhouse gases [e.g., carbon dioxide (CO_2); methane (CH_4); nitrous oxide (N_2O); hydrofluorocarbons (HFCs); perfluorocarbons (PFCs); sulfur hexafluoride (SF_6)]...do not exceed their assigned amounts ...with a view to reducing their overall emissions of such gases by at least 5 per cent below 1990 levels in the commitment period 2008 to 2012'

Table 7.1	Greenhouse Gas Emission Reduction Targets under the Kyoto Protocol for Selected Countries

Country	Reduction Commitment as Percentage of base year
Australia	108
Canada	94
European Community	92
France	92
Germany	92
Iceland	110
Japan	94
Netherlands	92
New Zealand	100
Norway	101
Russian Federation	100
Sweden	92
Ukraine	100
United Kingdom	92
United States	93

SOURCE: Annex B, Kyoto Protocol, 1997.

(Kyoto Protocol, 1997: Article 3). The Protocol will be legally binding when ratified by at least 55 countries accounting for at least 55 per cent of the developed world's 1990 emissions of carbon dioxide. The targets are shown for selected countries in Table 7.1. Developing countries, including China and India, were not included in the targets in the first time period because their per capita emissions were much lower than those of developed countries. Another reason was that their economies were judged to be much less able to absorb the costs of changing to cleaner fuels, since the main source of greenhouse gas emissions is carbon dioxide from use of fossil fuels.

Canada is to reduce greenhouse emissions to 6 per cent below 1990 levels by between 2008 and 2012. After much debate, considered later in this chapter, Canada ratified the Protocol in December 2002. However, at the end of March 2001, shortly after taking office US President George W. Bush stated that he opposed the Protocol, the US would not agree to it, and the US would develop its own approach. He argued that it was inappropriate for China and India, countries with the largest populations, not to be included in the Kyoto targets. Bush was quite correct that those two countries have the largest populations, but he ignored the fact that their per capita emissions of greenhouse gases are much lower than those of the United States, which has the worst record. With only 4 to 5 per cent of the world's population, the US accounts for between 20 and 25 per cent of the global emissions of greenhouse gases. Bush's position on the Kyoto Protocol was consistent with his stand on other international matters, ranging from a land mine agreement, an international court of justice, and terrorism—the United States will take unilateral action rather than participate in multilateral agreements.

Bush explained in February 2002 that the US would use a 'voluntary approach' related to greenhouse gas emissions, with the purpose to reduce 'greenhouse gas intensity' by 18 per cent over 10 years. Greenhouse gas intensity is the ratio of greenhouse gas emissions to economic output. For the United States, this translates into a goal to reduce greenhouse gas emissions from an estimated 183 mt per million dollars of gross domestic product in 2002 to 151 mt in 2012. Unlike the Kyoto Protocol, which requires an absolute reduction in greenhouse gas emissions, the American approach will result in emissions continuing to increase as its economy grows, but at a slower rate than without this arrangement.

Critics of Bush's position have noted that the United States introduced various voluntary measures following the **Rio Declaration** in 1992 in order to meet its pledge to reduce emissions to 1990 levels by 2000. While emission growth in the US was reduced by 2 per cent compared to what would have happened under 'business as usual', overall US emissions increased by 13.6 per cent between 1992 and 2002. In contrast, if the US accepted its target in the Kyoto Protocol, US emissions would have to be reduced by approximately 30 per cent below 1990 levels by the 2008–12 time period. Those supporting Bush's approach argue that accepting the target in the Kyoto Protocol would cause millions of Americans to lose their jobs and discourage long-term investments in clean energy technology.

The American position and approach are significant because they are based on determination to protect the US economy in the short term, to ensure that jobs are not lost due to costs associated with reducing emissions, and to maintain its international economic competitiveness. Because the US is such a dominant player in the global economy, its position is cited by people in other countries, including Canada, who believe it would be economically foolish to accept the Kyoto Protocol targets when the nation with the largest economy has decided not to do so.

With the United States government choosing not to ratify the Kyoto Protocol, attention turned to Russia since the other industrialized nations that had ratified the Protocol accounted for 44.2 per cent of 1990 emissions, and Russia accounted for 17.4 per cent. If Russia ratified, the total would be above the 55 per cent of 1990 emissions agreed to in 1997 in Japan. However, at a conference in Milan, Italy, in early December 2003, Russia announced it would not ratify Kyoto. The main reason was that the pollution cuts required of Russia under the agreement would slow down the economic growth that President Vladimir Putin had made a top priority. His goal is to have Russia double its gross domestic product by 2010. His top adviser commented that it would be unfair to Russia to have to curb emissions and jeopardize its growth while the United States and other nations, which account for the significant portion of global emissions, had refused to ratify. At the same confer-

ence, an American representative, Republican Senator James Inhofe, Chairman of the US Senate Committee on Environment and Public Works, reiterated the US opposition to Kyoto, stating that the Protocol and its policies were 'inconsistent with freedom, prosperity and environmental policy progress'. A significant shift occurred at the end of September 2004, however, when Russia's cabinet approved the Kyoto Protocol. The European Union had long lobbied Russia to ratify the pact, and in the spring of 2004 President Putin pledged to have Russia ratify in exchange for support from the EU for Russia to join the World Trade Organization. On 22 October 2004 the lower house of the Russian parliament, the Duma, approved ratification by a vote of 334–73; five days later Russia's upper house followed suit in a vote of 139–1, with one abstention.

Against this background, we now can turn to some of the specific aspects of the Protocol and then to the approaches of the federal and selected provincial governments in Canada.

Specific Features of the Kyoto Protocol

Legal Basis

Unlike the Framework Convention on Climate Change signed at the Earth Summit in Rio de Janeiro in 1992, which committed countries only to 'aim' to stabilize emissions at 1990 levels by 2000, the Kyoto Protocol commitments are legally binding on nations under international law.

Assigned Amounts

For the commitment period 2008–12, the Protocol states that overall average emissions are to be 94.8 per cent relative to 1990 levels. 'Assigned amounts' are identified for each developed nation. While the targets were set for different allowed emissions with reference to variables such as population, gross national product, and carbon intensity of economies, the final targets were determined politically. Canada is to reduce its emissions by 6 per cent. Australia is allowed to increase emissions by 8 per cent, and Iceland can increase by 10 per cent.

Greenhouse Gases

Six greenhouse gases are identified in the Protocol. Three are viewed as the main greenhouse gases produced by human activity: carbon dioxide, nitrous oxide, and methane. The other three—

hydrofluorocarbons, perfluorocarbons, sulphur hexafluoride—are released in small quantities but are long-lasting and significant contributors to climatic change.

Exclusion of Most Forest and Soil Sinks

The assigned emission amounts for most nations are a percentage of gross emissions in 1990. Gross emissions are the anthropocentric (human-caused) greenhouse gas emissions from energy, industrial processes, agriculture, and waste. However, they do not include carbon fluxes from forests, soil, and other carbon reservoirs.

When a nation calculates whether it is in compliance with its target emissions, it must count emissions and carbon flux changes due to afforestation, reforestation, and deforestation since 1990. In Canada, the view of the federal government is that it can interpret the Protocol to include loss of carbon from agricultural soil in calculating the balance between emissions and carbon flux removal. Indeed, the target for Australia of 108 per cent of 1990 emissions was partially based on arguments that it had positive net emissions related to land-use change and forestry in 1990. As a result, the 1990 allowable emissions were calculated to reflect the 1990 gross emissions plus the net emissions from land-use change and forestry.

Due to uncertainty and methodological challenges in measuring emissions from land-use change and forestry, some observers are concerned that countries will use forest and soil sinks to claim credits that will be difficult to verify. There are also problems in reaching agreement about the meaning of key terms such as reforestation, afforestation, and deforestation. In Canada, reforestation normally is interpreted to mean replanting and natural regeneration after logging, and afforestation to planting trees in areas traditionally forested. In contrast, the Intergovernmental Panel on Climate Change defines reforestation as planting of forests on lands historically forested but later converted to another use, and afforestation as planting of new forests on lands historically not forested. These different interpretations are important. If reforestation is interpreted to include planting trees after harvesting, as Canada does, then emissions from harvesting are not counted. As a result, only the credit from planting would be used in calculating the carbon reservoir credit.

At the Milan conference in December 2003, the signatories to the Kyoto Protocol reached agreement on how industrialized countries can earn credit towards their emission targets by preserving or establishing forests.

Clean Development Mechanism

Emission reduction commitments can be fulfilled through a clean development mechanism, allowing 'emission credits' in countries not given targets through the Protocol to be used by countries included in the Protocol targets. Initiatives are certified as satisfying the clean development mechanism when they involve voluntary participation by each party; real, measurable, and long-term benefits for mitigation of climate change; and emission reductions in addition to those that would have occurred without the initiative. Such initiatives also are to help the host nation. The Protocol allows countries with targets under the Protocol to meet their commitments for the 2008–12 period by claiming certified clean development emission reductions achieved between 2000 and 2007.

The major concern about the clean development mechanism is that emission credit may be given for projects that would have occurred without such a mechanism in place. The Protocol only requires reductions in emissions that would be in addition to what would have occurred without an initiative. This is referred to as 'emissions additionality'. However, the Protocol does not require an initiative to be one that would only have occurred if the clean development mechanism did not exist. This is referred to as 'project additionality'. A concern, therefore, is that credit will be given for initiatives that would have occurred regardless of the Protocol. This is a basic challenge in any system for identifying emission credits, since projects for which credit is given may have occurred for other reasons, such as reduction of air pollution to safeguard public health.

Emissions Trading

Under the Protocol, a country can meet its emission commitments by acquiring from other countries 'emission reduction units' related to initiatives that cause a reduction in emissions or enhancement of sinks incremental to what would otherwise have occurred without the initiative. When a nation buys some emission reduction units, these are added to its allowable emissions and subtracted from the allowable emissions of the selling country.

However, because developing countries are not given emission targets under the Protocol in order to allow them to develop their economies, they cannot sell emission credits to developed countries, even when such sales could benefit them economically. Developing countries also cannot agree to voluntary emission targets, which could benefit some if they could introduce low-cost emission reductions and then sell emission credits.

The theory of **emission trading** is based on the belief that it is more efficient for one country to purchase emission credits from another country that can generate credits in a less costly manner. In such a way, overall compliance goals can be reached at a reduced overall cost. However, as with all approaches, there are problems. For example, because the economies of Eastern European countries such as Russia and the Ukraine have collapsed, their emission allowances for 2008 to 2012, based on 1990 levels, will probably be higher than their actual emissions. As a result, they could sell their surplus emission rights (sometimes referred to as 'hot air') to other countries that then would be able to keep increasing their emissions. The outcome would be an overall increased emission of greenhouse gases, whereas if such 'hot air' credits were not transferable the overall levels would decrease.

Another concern is that the buyer of emission credits from another country does not have to be concerned about whether the selling nation will actually comply with its emission amounts after the sale. This is referred to as a 'seller beware' system, as opposed to the 'buyer beware' warning we are familiar with in economic transactions. The net effect, if the selling country cannot meet its emissions targets after selling some emission credits, will be an overall increase in emissions. Dealing with this issue is a challenge, as a sophisticated international monitoring and enforcement capacity is required.

Canada's Federal Government

On 10 December 2002 the House of Commons voted 195 to 77 in favour of a motion to ratify the Kyoto Protocol. The Liberal majority led those supporting the motion, while the Canadian Alliance and Progressive Conservative parties opposed it. Through this vote, Canada committed

---

CANADIAN FEDERAL GOVERNMENT PERSPECTIVE ON THE NEED TO ADDRESS CLIMATE CHANGE

In Canada the evidence of climate change is very real. Permafrost in the north is melting. Polar bears are having difficulty hunting because of shortened sea ice seasons. We are faced with lower water levels in the Great Lakes and rising sea levels and changing water temperatures on our coasts.

What does this mean to people? It means that the infrastructure of Canada's north is being undermined. It means that the lifestyles of Canada's aboriginal peoples are being threatened. It means that our water supply is decreasing. It means that sources of all Canadians' economic well-being are at risk as we bear the increasing costs associated with the impacts of climate change.

— Government of Canada (2001a)

---

itself to cut average greenhouse gas emission levels to 6 per cent below 1990 levels by 2008–12.

Prime Minister Chrétien signed Canada onto the Kyoto Protocol on 16 December 2002, and the next day the Minister of the Environment, David Anderson, travelled to New York to officially inform the United Nations. Canada's ratification brought the total to almost 100 countries that have committed to the Kyoto Protocol, accounting for about 40 per cent of 1990 emissions.

Canada's Approach to Implementing the Kyoto Protocol

The Canadian government ratified the Kyoto Protocol without a clear plan on how it would be implemented in Canada. The Prime Minister argued that details would be worked out. This position was supported by Anderson, who compared the challenge of global warming to World War II. He was often quoted as saying that 'If Winston Churchill had said in 1939 that we are not going to challenge the Nazis until we know exactly how much it will cost and how long the war will last, then we would have never won the war.' In contrast, Stephen Harper, leader of the then Canadian Alliance Party, argued that it was inappropriate to ratify the Protocol without providing a clear plan regarding how it would be achieved. His view was that 'I don't think we have any idea what the government is going to do' and that the

The role of the oceans in helping to mitigate the impacts of global warming through absorption of carbon dioxide is still uncertain, as is the oceanic response to warmer temperatures. Scientists are already detecting larger wave swells in many parts of the world that may be linked to these changes (*Philip Dearden*).

implementation would turn into a great disaster for Canada and Canadians.

The federal government allocated $2 billion in its 2003 budget for Kyoto Protocol initiatives over a five-year period. One specific initiative, the creation of Sustainable Development Technology Canada, was identified. This arm's-length foundation was targeted to receive $250 million to support technologies that are so new that they are not yet commercially viable. The remaining $1.75 billion were to be allocated to other initiatives once the details were worked out, with the budget indicating only that the money would be 'to support climate change science, environmental technology and cost-effective climate change measures and partnerships in areas such as renewable energy, energy efficiency, sustainable transportation and new alternative fuels'.

The components of a plan for implementing the Kyoto Protocol will include requiring major industrial emitters to reduce their greenhouse gas emissions, taxes on private vehicles such as sports utility vehicles, minimum requirements for fuel alcohol, subsidies to install energy-efficient windows, and an emissions-trading framework.

The federal government has compiled a list of large industrial emitters of greenhouse gases, such as oil and gas, mining, and pulp and paper companies, each of which generates large amounts of carbon dioxide per unit of product. Each industry will be required to reduce emissions extensively. However, the political nature of this approach was illustrated in January 2003 when the gigantic Ontario automotive industry was removed from the list of large industrial emitters as a result of lobbying from the auto assembly industry.

Alberta Government Perspective

On 3 September 2002, Alberta Premier Ralph Klein wrote to Prime Minister Chrétien expressing his concerns about the federal government's intent to ratify the Kyoto Protocol based on what, to him, was an ambiguous strategy. With regard to the federal strategy, Klein commented that 'It's like signing a mortgage for a property you have never seen and for a price that you have never discussed.' Klein argued at the annual Premiers' Conference in Halifax in August 2002 that if the federal government's approach was supported then his province's oil and gas industry would be devastated. Furthermore, the wealth of his province would be harmed significantly, meaning it would no longer be able to make the same level of contribution to the funding program that reallocates money from well-to-do provinces to poorer provinces through federal equalization payments.

The economy of Alberta has a strong basis in fossil fuels due to its petroleum and natural gas industries. In that context, the Alberta government has argued that it agrees Canada must be a responsible member of the global community, and Alberta is committed to reducing greenhouse gas emissions. However, it prefers a 'made in Alberta' approach to the Kyoto Protocol. On one of the Alberta government websites it is explained that 'Alberta's goal is to make the province more competitive by developing new economic and environmental opportunities while managing climate change. Immediate investment in new energy, emissions, and environmental technologies will result in environmental improvements with economic benefits.' Furthermore, the government argues that 'in order for Alberta and Canada to remain competitive, actions must be compatible with our largest trading partner, the US.'

A 'made in Alberta' approach is represented by Bill 32, the Climate Change and Emissions Management Act, introduced on 19 November 2002. Under this legislation, Alberta will identify (1) an overall emission target for Alberta, along with targets for specific sectors of the economy, to be established through negotiation and regulated agreements; (2) a framework for an emissions trading system; and (3) a fund to help different sectors meet performance targets and to encourage private-sector investment in research, technology, and energy conservation.

The legislation is one tool to implement the Alberta action plan on climate change, announced in mid-October 2002. Under that plan, Alberta will cut greenhouse gas emission intensity (the same concept followed in the United States, relating emissions to dollar of economic production) by 50 per cent below 1990 levels by the year 2020 through investment in energy research, innovation technology, sectoral agreements, energy efficiency, and conservation by consumers. Compared to the Kyoto Protocol, Alberta will aim to reach its target by 2020 rather than the 2008–12 period, and as long as its economy grows it will continue to produce more greenhouse gas emissions, but at a reduced rate. To counter criticism, the government states that 'Alberta recognizes that more significant emissions reductions will be required over the longer term. This plan is only the beginning of a 50-year initiative to dramatically reduce carbon emissions.'

Alberta has received support from governments in British Columbia, Saskatchewan, and Newfoundland, which also have fossil fuel resources in their provinces, and to some extent from Ontario, which has expressed concern about remaining economically competitive with firms in the United States and other industrial nations.

The Manitoba and Quebec Perspectives

The provincial governments of both Manitoba and Quebec have supported the Kyoto Protocol. Unlike Alberta, whose economy depends on fossil fuels, these two provinces rely primarily on hydroelectric power and so view Kyoto as a positive initiative. For example, Manitoba each year exports about $300 million of hydro-generated electricity to the United States, and different Quebec governments have based future economic growth on further development of hydro power in the northern areas of that province. Since hydro power is generally viewed as 'clean' in terms of greenhouse gas emissions, the governments of these two provinces view Kyoto as providing support for their development strategies.

Premier Gary Doer of Manitoba argued that if the Kyoto Protocol targets for Canada were not adopted the environment in the western provinces would be so damaged that parts of it would be unrecognizable. He commented that it was not only the boreal forests of Manitoba that were under threat, but also the grasslands of Alberta. Doer was quoted as saying that he did not want Manitoba 'to go from raising cattle to raising camels'.

The Nunavut Perspective

Prior to the ratification of the Kyoto Protocol, Premier Paul Okalik argued at the annual Premiers' Conference that the destruction of his territory's environment due to global change was not worth the money that would be received through Alberta's equalization payments. In his words, 'our custom is to pass on our traditional knowledge to our children—you can't put a price tag on that, so you can keep your money for one.' He further commented that the fishery resource in Nunavut was being damaged by climate warming.

Overview

The above discussion highlights the different perspectives on climate change across Canada. Different governments mount arguments that reflect the economic interests of their particular jurisdictions (Figure 7.12). Each presents what it claims to be a principled argument, but nearly always the arguments reflect economic self-interest and opportunities for development. Only occasionally is a long-term view or a 'big picture' perspective taken.

The Sierra Club of Canada (2004) is not a disinterested observer of the climate change debate, but nevertheless it is useful to look at the report card it issued in September 2004 for each government related to climate change (Table 7.2). This was the twelfth report card, which it began in 1993, the year after the Rio Summit meeting. The report card, first titled 'RIO' in reference to the Rio Summit and now an acronym for 'Report on International Obligations', is released annually to report progress on environment and development commitments. RIO reports on commitments to increase overseas development assistance, to reduce greenhouse gases, to enhance biodiversity, to review and reform pesticide and toxic policies, to improve environmental assessment, to make trade and environment mutually supportive, to conserve and sustain use of living marine resources, and to protect forests.

BOX 7.5 SIGNS OF RADICAL CHANGE IN ARCTIC ECOSYSTEM

J. Patrick Coolican
Seattle Times
29 Oct. 2003

Shrubs are appearing where before there were none; gray whales are venturing farther north; clams and their predators, diving sea ducks, are less plentiful. The ice is melting. These disparate phenomena are signs of a radical change in the Arctic ecosystem. Moreover, changes in the Arctic mean changes everywhere else. That consensus was part of a broad discussion among 400 Arctic scientists meeting in Seattle in late October 2003, as part of a new multimillion-dollar effort to study the far-reaching changes occurring in the far North.

The Study of Environmental Arctic Change (SEARCH), sponsored by the National Science Foundation and based at the University of Washington, has brought together climatologists, biologists, oceanographers, and social anthropologists to examine how warmer temperatures are altering life on the tundra and how the changes may affect the rest of the Earth.

Their findings add to a growing, and somewhat ominous scientific consensus.

The Arctic, which has seen temperatures increase 3 to 4 degrees Fahrenheit over the past century, is on track to become as warm as it has been in 130,000 years, according to Jonathan Overpeck, a University of Arizona paleoclimatologist, who studies the history of climate change. The research will help establish a key unknown, Overpeck said: What is the point of no return? The point at which a reduction in greenhouse emissions will no longer matter? When will life be inexorably changed? (Overpeck said that although there's a broad scientific consensus that human beings are responsible for the recent warming, there are dissenting scientists, who say the change is not caused by human activity.)

In an interview after the symposium, Overpeck laid out potential consequences of the warming: Warmer temperatures could lead to the melting of the vast field of ice that currently carpets about 80 per cent of the island of Greenland. (Arctic sea-ice thickness reached its lowest recorded level in September 2002). That could lead to a 6-meter rise in sea levels, flooding coastal areas. Scientists are still trying to find out how long this process could take, Overpeck

said. The melting also could disrupt the conveyor of warm ocean currents that make places like England habitable as opposed to barren, icy outposts.

Also, 30 per cent of the world's carbon is trapped under the icy tundra of the Arctic region, according to Matthew Sturm, a polar scientist with the US Army. A melting would allow the carbon to escape into the atmosphere as either carbon dioxide or methane, both greenhouse gases, which would lead to even warmer temperatures. A ratcheting up of heat would ensue. 'The Arctic will change the globe', Sturm said.

Researchers also are examining how the climate changes will affect people living in the Arctic. 'Ice is a portal of life, an extension of the land' for indigenous communities, said Caleb Pungowiyi, a social anthropologist who lives in the Arctic region. Although indigenous populations are wary of the effects of climate change, there could be good effects, like a larger caribou population and less populous insect swarms, Pungowiyi said.

The warmer Arctic is definitely changing life for plants and animals, according to Jacqueline Grebmeier, an oceanographer at the University of Tennessee. She noted plants growing on what was once a barren tundra, odd migratory patterns, disappearing species.

Studying changes in the Arctic will help scientists understand how future climate change will affect life in the more southern latitudes, said Mark Nuttall of the University of Alberta. 'The Arctic is the canary in the mine shaft. A barometer for the future', he said.

Implicit in a radical climate change, Overpeck asserted, is a mass species extinction. There have been five known mass species extinctions—meaning more than 50 per cent of the Earth's species were wiped out—and four were related to climate change, Overpeck said. Unlike in the past, however, when plants and animals could freely adapt to climate change, Overpeck said, this time human development stands in the way—big-box stores and factories and casinos and subdivisions that will prevent species from seeking new habitats. 'We're really playing with fire', he said.

SOURCE: <http://seattletimes.nwsource.com/html/localnews/2001777464_arctic29m.html>

Figure 7.12 The Kyoto Protocol. SOURCE: *Michael De Adder/artizans.com.*

MOVING FORWARD

While scientific evidence supports the occurrence of climate change and the often negative environmental and social consequences of this change, it is also apparent that there is opposition to taking action due to concerns about negative consequences for economic well-being and standards of living. Thus, if progress is to occur, it will be important to identify and implement initiatives that generate multiple benefits.

One example of such an initiative has been provided by Steve Hounsel of Ontario Power Generation (OPG). OPG, the successor to the former Ontario Hydro, has a mix of nuclear, hydroelectric, and fossil power. About 31 per cent of the energy mix in Ontario is provided from fossil fuels, notably coal, and greenhouse gas emissions from this source are significant. For example, in 2000, OPG fossil fuel generating stations emitted 38.5 million metric tonnes of CO_2. Being aware that such emissions are undesirable, OPG has set voluntary targets to stabilize net emissions of CO_2 at 1990 levels. This will include many initiatives, such as returning performance of nuclear generating plants to world-class standards, internal energy efficiency programs, switching to cleaner fuels, and a more aggressive move to green, renewable energy. All of these will reduce use of fossil energy. In parallel, OPG is seeking carbon offsets through emissions trading and has introduced a carbon sequestration and biodiversity management program. The latter program illustrates how two desirable goals can be achieved simultaneously, as outlined below.

Ontario Power Generation's Carbon Sequestration and Biodiversity Management Program

The program was announced in the spring of 2000, with an initial goal of planting at least 1.6 million native trees and shrubs in southern Ontario over a five-year period ending in 2005. It was estimated that this would sequester about 900,000 metric tonnes of CO_2 over the lifetime of the trees. The program has since been extended to

Table 7.2		Commitment to Reduce Greenhouse Gases, 2004
Jurisdiction	Grade	Comments
Canada	B	The B is for the last six months of the Chrétien government, and is down from an A in 2003. After ratification of the Kyoto Protocol, the government announced the laudable EnerGuide Home Retrofit Incentive, which led to a tripling of requests for home energy audits. It also announced $100 million to build a Canadian ethanol industry. Since February 2004, no new programs were announced and some elements of the Kyoto plan were placed on hold. Not enough time had passed to give a grade to the approach of the Paul Martin government to climate change.
Newfoundland and Labrador	C–	The grade slipped from C+ in 2003 because the province still did not have a plan to meet greenhouse gas reduction commitments. Opportunities related to wind energy have not been pursued, despite the strong potential. Energy retrofits have been completed on many government buildings and some solid outreach programs exist.
Nova Scotia	C–	Nova Scotia has worked effectively with the federal government to establish a GHG inventory, but it has done less well in making needed changes to meet Kyoto targets. The province continues to rely overly on fossil fuels.
Prince Edward Island	A–	PEI received the highest grade in Canada, up from a B+ in 2003. This high grade is for excellent progress in developing a provincial energy strategy, and because 5 per cent of the energy now is based on wind energy from the North Cape Wind Farm. The minus is due to lack of action on transportation.
New Brunswick	C	Down from a B in 2003, because of lost ground due to poor planning and energy choices that miss opportunities to meet Kyoto targets. The lower grade also reflects a decision to refurbish a 1,000-megawatt oil-fired power plant, responsible for one-fifth of the greenhouse gas emissions for the province.
Quebec	B+	Quebec's good grade reflects that it has consistently supported the Kyoto Protocol, continues to promote energy conservation, is expanding wind energy capacity, and is leading in development of an adaptation strategy. Points were lost from Hydro-Québec's decision to build a fossil fuel power plant over the border in New Brunswick.
Ontario	D+	The grade went up from an F in 2003 based on a commitment to phase out coal and to meet Kyoto obligations. The grade would be higher if the actions of the new government were consistent with election promises. The decision to invest in fixing old nuclear power plants has dropped the grade. The work towards an emissions trading system is flawed, and not consistent with a commitment to phase out emissions.
Manitoba	B–	The grade is down from a B+ in 2003. The Premier continues to be a strong supporter of Kyoto, but the provincial government has actively supported emission trading schemes rather than significant reductions in fossil fuel and other emissions in the province. There is no overall provincial energy plan.
Saskatchewan	C	The provincial government continues to support Kyoto but needs to show more action related to implementing measures consistent with commitments relative to Kyoto. The Office of Energy Conservation is a good idea, but is under-resourced, with only two staff.

Alberta	F	F is the same grade as for 2003. Alberta continues to make progress in reducing the intensity of GHG emissions, but this US-style approach means that the absolute amount of GHG emissions will continue to rise. In previous years, Alberta made progress in reducing flaring of gases, thus reducing toxics and GHG emissions. The 'Ride the Wind' renewable energy-powered bus system in Calgary is a model for the nation. The anti-Kyoto rhetoric has not toned down.
British Columbia	F	Like Alberta, an F grade as in 2003. The Campbell government has no policies to address climate change, but does have policies to encourage maximum development of provincial fossil fuel resources, including developing coal reserves and encouraging offshore oil and gas exploration. The Campbell government has supported the Alberta government in its opposition to Kyoto.
Northwest Territories	F	Down from a C in the previous year. The energy policy released in July 2003, if implemented, will result in staggering increases in GHG emissions due to fossil fuel industrialization in the Mackenzie Valley and Delta. The strategy does not commit the NWT to meet Kyoto targets.
Yukon	B–	Same grade as in 2003. The territorial government appears to be in a 'holding pattern' regarding climate change. No new initiatives have been started, and there seems to be strong support for more oil and gas development.
Nunavut	C–	No change in grade from last year. The government has several ongoing modest climate change research projects underway, especially related to collecting traditional knowledge on climate change.

SOURCE: Sierra Club of Canada (2004).

increase the goal to the planting of 2 million trees by the fall of 2006, which also will result in the addition of at least 900 hectares of new forest.

The program focuses on restoring forest habitat in southern Ontario, an area that has lost more than 80 per cent of its original forest cover. The Carolinian biotic zone of southwestern Ontario is considered to be the most ecologically threatened region in the country, and supports 71 nationally endangered or threatened species, 50 species and 37 vegetation types considered to be globally rare, more than 500 provincially rare plant and animal species, and over 65 provincially rare vegetation community types. Planting is targeted to expand key forested areas and to connect woodland patches to help promote recovery of declining wildlife at risk due to loss and fragmentation of habitat. Management plans at each site have been prepared, with explicit goals related to regional biodiversity conservation priorities. In the first two years of the program, OPG with its partners planted over one million native trees and

shrubs, matched to site conditions, on more than 450 hectares of land.

The program of **carbon sequestration** thus helps to enhance biodiversity, but it also contributes to the Greenhouse Gas Management Strategy of OPG. OPG has estimated that once the 2 million trees are planted, they will trap and offset about a million tonnes of carbon dioxide over the lifetime of the trees, and thereby will contribute to a reduction of climate warming. The program is viewed as unique in that it links a GHG emissions problem to a biodiversity solution, illustrating how working with nature can sequester some of the CO_2 emissions. In addition, many of the same plantings provide other ecosystem benefits, especially along riparian systems, by reducing erosion and enhancing stream water quality.

The OPG Carbon Sequestration and Biodiversity Management Program has been recognized by a number of external awards: (1) in 2000, Environment Canada and the US Environmental Protection Agency gave it the State of the Lakes Ecosystem

Conference Success Story Award; (2) in 2001, Carolinian Canada Conservation granted it the organization's Conservation Award. Carolinian Canada Conservation is a partnership of some 20 public and private agencies co-operating to conserve biodiversity in the Carolinian life zone of Canada; and (3) in 2002, the Wildlife Habitat Canada Forest Stewardship Recognition Program presented it with an award, providing national recognition.

The Carbon Sequestration and Biodiversity Management Program illustrates how environmental science from different fields (biodiversity, climate change) can be integrated to develop innovative solutions to environmental problems. While OPG recognizes that this program will not, by any stretch, solve greenhouse gas problems, it is a contribution to an overall solution and also provides many other benefits. We need to identify more such initiatives, which incrementally and cumulatively can move us in the direction of reducing the serious challenges presented by climate warming and global change.

What Else?

Clearly, there is considerable uncertainty regarding the precise effects of anticipated climate change. Furthermore, concerted and co-ordinated initiatives by provinces, states, and nations will be necessary to reduce the projected negative impacts. Some of the multilateral initiatives underway have been identified in this chapter, but it is obvious that some nations, especially the United States, are unwilling to participate in co-ordinated global activities if there might be economic disadvantage to them in the short term. But what can individuals do? Box 7.6 presents some options. A change in one person's behaviour will not have a significant impact. However, cumulatively, many individual actions can be significant. The challenge will be to decide whether we are prepared to make such modifications in our behaviour since, in many instances, we will not be the direct beneficiaries—those will more likely be received by one or two generations in the future, such as your grandchildren.

SUMMARY

1. The *weather* of any place is the sum total of its atmospheric conditions (temperature, pressure, winds, moisture, and precipitation) for a *short* period of time. It is the momentary state of the atmosphere. *Climate* is a composite or generalization of the variety of day-to-day weather conditions. It is not just 'average weather', for the variations from the mean, or average, are as important as the mean itself.

2. Scientific evidence confirms that the world has been warming, with the average global temperature at the Earth's surface having increased by about 0.6° C, with an error range of plus or minus 0.2° C, since the late nineteenth century.

3. The increase in the average temperature for the northern hemisphere during the twentieth century was the largest of any century in the past 1,000 years.

4. Evidence showing increases in greenhouse gases in the atmosphere, loss of mass in glaciers, reduction in permafrost and snow cover, and rises in sea level are all consistent with global temperature increases. Local knowledge from Aboriginal people provides further confirmation.

5. There are four basic climate models, the most commonly used being the General Circulation Model. GCMs are best used for global or overall climate modelling, but not for regional representations of climate change.

6. Coarse spatial resolution, poor predictive capacity for precipitation, relatively weak simulation of oceans, lack of baseline data, and many other limitations cause GCM outputs to be highly variable.

7. The reports from the Intergovernmental Panel on Climate Change are explicit that most warming since the mid-twentieth century is associated with human activities.

8. At a global scale, records document that mean sea level has been rising at a rate of 0.1 to 0.2 metres per century during the last 100 to 200 years. The coast of Prince Edward Island is one of the areas in Canada most vulnerable to sea-level rise

9. In the Georgian Lakeland Tourism Region in southern Ontario, it is estimated that the alpine skiing seasons would be reduced by between 21 and 34 per cent, in contrast to the studies in the early 1990s that had not considered the effect of snow-making and suggested that the season could be reduced by 40–100 per cent under doubled CO_2 conditions.

10. Various challenges for communicating information or understanding about global change exist, including the following: (1) Global change is a complex issue. (2) Uncertainties exist regarding almost every aspect of the global change issue, and these increase when moving from natural to human systems. (3) Impacts of global change will be disproportionately heavier on people in lesser-developed countries and on future generations. (4) The basic causes of global change are embedded in current values and lifestyles.

11. To improve communication about situations such as global change, which are characterized by high uncertainty, four aspects should be addressed: (1) What do we know, with what accuracy, and how confident are we about the data? (2) What don't we know, and why are we uncertain? (3) What could we know, with more time, money, and talent? (4) What should we know in order to act in the context of uncertainty?

12. During December 1997, representatives from more than 160 countries met in Kyoto, Japan, and reached an agreement, popularly referred to as the Kyoto Protocol, with targets for 38 developed nations as well as the European Community to reduce their overall emissions of greenhouse gases by at least 5 per cent below 1990 levels by the period 2008–12.

13. Canada, which ratified the Protocol in December 2002, is to reduce greenhouse emissions to 6 per cent below 1990 levels by between 2008 and 2012. In contrast, at the end of March 2001, US President George W. Bush stated that the US would not agree to it and would develop its own approach.

14. Provincial governments in Alberta, British Columbia, and Saskatchewan, producers of oil or natural gas, have criticized Canada's ratification of Kyoto, whereas the provincial governments of Manitoba and Quebec, producers of hydroelectricity, have supported this action.

15. Ontario Power Generation's Carbon Sequestration and Biodiversity Management Program is one example of how an initiative can contribute to reduced climate change and also enhance biodiversity.

KEY TERMS

atmosphere	El Niño	global warming	ozone layer
carbon sequestration	emission trading	greenhouse effect	Rio Declaration
climate	fossil fuels	greenhouse gas	stratosphere
climate change	general circulation models	Kyoto Protocol	uncertainty
climate modelling	(GCMs)	ozone	weather

REVIEW QUESTIONS

1. Explain the difference between weather and climate.
2. What are some of the key natural and human causes of climate change?
3. How credible is the evidence for climate change or global warming? For which aspects is there the most uncertainty?
4. Have you noticed any indication of climate change in the area in which you live? If so, what are these, and how confident are you that these are valid and reliable indicators of climate change?
5. What would be the best sources of traditional ecological knowledge about climate change in your area?
6. Why have general circulation models (GCMs) become dominant in the modelling research focused on climate change? What are the limitations of GCMs?
7. What is viewed as the main cause of increased carbon dioxide emissions to the atmosphere in the twentieth and twenty-first centuries?
8. The two case studies (sea-level rise and its impact on Prince Edward Island, and reduced snow cover and its impact on winter recreation) highlight possible impacts from climate change. What would be other appropriate Canadian case studies to understand the consequences for marine and coastal resources, freshwater resources, and terrestrial resources?
9. What are the basic communication challenges related to climatic change, and what, in your view, should be the first steps to overcome them?
10. What are the strengths and limitations of the Kyoto Protocol? If it did not exist, and you were to start with a 'blank sheet', what type of Protocol would you propose?
11. There have been different perspectives in Canada relating to the best way to reduce greenhouse gases. What do you think is the best strategy to resolve differences among federal, territorial, and provincial governments and among the provincial governments as a group?

RELATED WEBSITES

ENVIRONMENT CANADA, CLIMATE CHANGE SITE:
  http://www.ec.gc.ca/climate
ENVIRONMENT CANADA, CLIMATE TRENDS AND
  VARIATIONS BULLETIN:
    http://www.mscsmc.ec.gc.ca/ccrm/bulletin
ENVIRONMENT CANADA, STATE OF THE
  ENVIRONMENT INFOBASE:
    http://www.ec.gc.ca/soer-ree
GOVERNMENT OF CANADA:
  http://www.climatechange.gc.ca

INTERGOVERNMENTAL PANEL ON CLIMATE CHANGE:
  http://www.ipcc.ch
NATURAL RESOURCES CANADA:
  http://adaptation.nrcan.gc.ca/posters
UNITED NATIONS:
  http://www.unfcc.int
UNITED STATES ENVIRONMENTAL PROTECTION
  AGENCY:
    http://yosemite.epa.gov/OAR.globalwarming.nsf

REFERENCES AND SUGGESTED READING

Andrey, J., and L. Mortsch. 2000. 'Communicating about climate change: challenges and opportunities', in Scott et al. (2000: WP1–WP11).

Ashmore, P., and M. Church. 2001. *The Impact of Climate Change on Rivers and River Processes in Canada*. Geological Survey of Canada Bulletin 555. Ottawa: Natural Resources Canada.

Barnett, J. 2003. 'Security and climate change', *Global Environmental Change* 13, 1: 7–17.

Berkes, F. 1999. *Sacred Ecology: Traditional Ecological Knowledge and Resource Management*. Philadelphia: Taylor and Francis.

Bernstein, S., and C. Gore. 2001. 'Policy implications of the Kyoto Protocol for Canada', *Isuma* 2, 4: 26–36

Brown, R.D. 1993. 'Implications of global climate warming for Canadian East Coast sea-ice and iceberg regimes over the next 50–100 years', *Canadian Climate Digest*. Ottawa: Minister of Supply and Services.

Bruce, J.P. 2001. 'Intergovernmental Panel on Climate Change and the role of science in policy', *Isuma* 2, 4: 11–15.

Canadian Council of Ministers of the Environment. 2003. *Climate, Nature, People: Indicators of Canada's Changing Climate*. Winnipeg: Canadian Council of Ministers of the Environment.

Carpenter, R.A. 1995. 'Communicating environmental science uncertainties', *Environmental Professional* 17: 127–36.

Chalecki, E.L. 2000. 'Same planet, different worlds: the climate change information gap', in Scott et al. (2000: A2, 15–22).

Chiotti, Q. 1998. 'An assessment of the regional impacts and opportunities from climate change in Canada', *Canadian Geographer* 42: 380–93.

Cohen, S.J. 1997. 'What if and so what in Northwest Canada: could climate change make a difference to the future of the Mackenzie Basin?', *Arctic* 50: 293–307.

David Suzuki Foundation. 2003. *Climate Change: Kyoto Protocol*. At: <http:www.davidsuzuki.org/Climate_Change/Kyoto/Kyoto_Protocol.asp>.

Ehrlich, P., and A. Ehrlich. 1996. *Betrayal of Science and Reason: How Anti-Environmental Rhetoric Threatens Our Future*. Washington: Island Press.

Environment Canada. 2003. *Environmental Signals: Canada's National Environmental Indicator Series*. Ottawa: Environment Canada.

Filion, Y. 2000. 'Climatic change: implications for Canadian water resources and hydropower production', *Canadian Water Resources Journal* 25: 255–69.

Fleming, S.W., and G.K. Clarke. 2003. 'Glacial control of water resources and related environmental responses to climatic warming: empirical analysis using historical streamflow data from Northwestern Canada', *Canadian Water Resources Journal* 28: 69–86.

Fraser Institute. 1997a. *Global Warming: The Science and the Politics*. Vancouver: Fraser Institute.

Fraser Institute. 1997b, 1998, 1999. *Environmental Indicators for Canada and the United States*. Vancouver: Fraser Institute.

Gilmore, J.M. 2000. 'Ten illusions that must be dispelled before people will act on your global warming message', in Scott et al. (2000: B1, 1–5).

Good, J. 2000. 'The case study of climate change: the nature of risk and the risk of nature', in Scott et al. (2000: B2, 21–33).

Gough, W.A. 1998. 'Projections of sea-level change in Hudson and James Bay, Canada, due to global warming', *Arctic and Alpine Research* 30: 80–8.

——— and E. Wolfe. 2001. 'Climate warming scenarios for Hudson Bay, Canada from general circulation models', *Arctic* 54: 142–8.

Government of Alberta. 2003. *Kyoto Protocol: Our Position*. At: <http:www.gov.ab.ca/home/kyoto/Display.cfm?id'5>.

Government of Canada. 2001a. *Viewpoint: The Road Ahead*. At: <http://www.climatechange.gc.ca/english/whats_new/road_e.html>.

Government of Canada. 2001b. *Canada and the Kyoto Protocol*. At: <http:www.climatechange.gc.ca/english/whats_new.overview.html>.

Harvey, L.D.D. 2000. *Global Warming: The Hard Science*. Harlow, UK: Prentice-Hall.

———. 2003. 'Climatic change: addressing complexity, uncertainty, and conflict', in B. Mitchell, ed., *Resource and Environmental Management in Canada: Addressing Conflict and Uncertainty*, 3rd edn. Toronto: Oxford University Press, 132–65.

Hengeveld, H. 1991. *Understanding Atmospheric Change: A Survey of the Background Science and Implications of Climate Change and Ozone Depletion*. SOE Report No. 91–2, Ottawa: Minister of Supply and Services Canada.

———, E. Bush, and P. Edwards. 2002. *Frequently Asked Questions about Climate Change Science*. Ottawa: Minister of Supply and Services.

Intergovernmental Panel on Climate Change. 2001a. *Climate Change 2001: Synthesis Report. A Contribution of Working Groups I, II and III to the Third Assessment Report of the Intergovernmental Panel on Climate Change*, eds R.T. Watson and the Core Writing Team. Cambridge: Cambridge University Press.

————. 2001b. *Climate Change 2001: The Scientific Basis. Contributions of Working Group I to the Third Assessment Report of the Intergovernmental Panel on Climate Change*, eds J.T. Houghton, Y. Ding, D.J. Griggs, M. Noguer, P.J. van der Linden, X. Dai, K. Maskell, and C.A. Johnson. Cambridge: Cambridge University Press.

————. 2001c. *Climate Change 2001: Impacts, Adaptation, and Vulnerability. Contribution of Working Group II to the Third Assessment Report of the Intergovernmental Panel on Climate Change*, eds J.J. McCarthy, O.F. Canziani, N.A. Leary, D.J. Dokken, and K.S. White. Cambridge: Cambridge University Press.

Jaccard, M., J. Nyboer, and B. Sadownik. 2002. *The Cost of Climate Policy*. Vancouver: University of British Columbia Press.

————, ————, and ————. 2003. 'The cost of climate policy', *Horizons* 6, 1: 29–30.

Kerr, J., and L. Packer. 1997. 'The impact of climate change on mammal diversity in Canada', *Environmental Monitoring and Assessment* 49: 263–70.

Lane, P., and Associates. 1988. 'Preliminary study of the possible impacts of a one metre rise in sea level at Charlottetown, Prince Edward Island', *Canadian Climate Digest*. Ottawa: Minister of Supply and Services.

LeDrew, E. 2002. *A Uniquely Canadian Dimension to Climate Change: The Role of the Cryosphere in the Everyday Lives of Canadians*. Waterloo, Ont.: University of Waterloo, Faculty of Environmental Studies, Research Lecture 2002.

Lomborg, B. 2001. *The Skeptical Environmentalist: Measuring the Real State of the World*. Cambridge: Cambridge University Press.

Lonergan, S. 1998. *The Role of Environmental Degradation in Population Displacement*. Research Report No. 1. Victoria: University of Victoria, Global Environmental Change and Human Security Project.

McBain, G., A. Weaver, and N. Roulet. 2001. 'The science of climate change: what do we know?', *Isuma* 2, 4: 16–25.

McLeman, R., and B. Smit 2003. 'Climate Change, Migration and Security', *Commentary No. 86*. Ottawa: Canadian Security Intelligence Service.

McMichael, A.J., and T. Kjellstrom. 2002. 'Sustainable development, global environmental change and public health', *Isuma* 3, 2: 43–50.

Miller, G. 2002. *Climate Change: Is the Science Sound?* Special Report to the Legislative Assembly of Ontario. Toronto: Environmental Commissioner of Ontario.

Mote, P.W. 2003. 'Twentieth-century fluctuations and trends in temperature, precipitation, and mountain snowpack in the Georgia Basin–Puget Sound Region', *Canadian Water Resources Journal* 28, 4: 567–85.

National Research Council. 1993. *Issues in Risk Assessment*. Washington: National Academy Press.

————. 1994. *Science and Judgment in Risk Assessment*. Washington: National Academy Press.

————, Division of Earth and Life Studies, Committee on the Science of Climate Change. 2001. *Climate Change Science: An Analysis of Some Key Questions*. Washington: National Academy Press.

Reidlinger, D. 1999. 'Climate change and the Inuvialuit of Banks Island, NWT: using traditional environmental knowledge to complement Western science', *Arctic* 52: 430–2.

———— and F. Berkes. 2001. 'Contributions of traditional knowledge to understanding climate change in the Canadian Arctic', *Polar Record* 37: 315–28.

Rolfe, C. 1998. 'Kyoto Protocol to the United Nations Framework Convention on Climate Change: A Guide to the Protocol and Analysis of its Effectiveness'. At: <http:www.sierraclub.ca/national/climate/kyoto.html>.

Scott, D.N. 2001a. 'Carbon sinks and the preservation of old growth forests under the Kyoto Protocol', *Journal of Environmental Law and Practice* 10: 105–45.

————. 2001b. 'Looking for loopholes', *Alternatives Journal* 27, 4: 22–9.

————, B. Jones, J. Andrey, R. Gibson, P. Kay, L. Mortsch, and K. Warriner. 2000. *Climate Change Communication: Proceedings of an International Conference*. Hull, Que., and Waterloo, Ont.: Environment Canada and University of Waterloo, June.

————, ————, C. Lemieux, G. McBoyle, B. Mills, S. Svenson, and G. Wall. 2002. *The Vulnerability of Winter Recreation to Climate Change in Ontario's Lakeland Tourism Region*. Department of Geography Publication Series Occasional Paper No. 18. Waterloo, Ont.: University of Waterloo.

Shaw, R.W., and CCAF A041 Project Team. 2001. *Coastal Impacts of Climate Change and Sea-Level Rise on Prince Edward Island: Synthesis Report*. Climate Change Action Fund Project CCAF A041. Ottawa: Ministry of Supply and Services.

Sierra Club of Canada. 2003. *Rio + 11: The Eleventh Annual Rio (Report on International Obligations) Report Card, 2003*. Ottawa: Sierra Club of Canada, June.

————. 2004. *Rio + 12: The Twelfth Annual RIO (Report on International Obligations) Report Card, 2004*. Ottawa: Sierra Club of Canada, Sept.

Strong, M.A. 2000. 'A case study of the municipal climate change outreach strategies for the residential sector in the City of Ottawa', in Scott et al. (2000: B4, 8–16).

Tickell, C. 2002. 'Communicating climate change', *Science* 297: 737.

Torrie, R., R. Parfett, and P. Steenhof. 2002. *Kyoto and Beyond: The Low Emission Path to Innovation and Efficiency*. Vancouver and Ottawa: David Suzuki Foundation and Climate Action Network.

Trewartha, G.T. 1954. *An Introduction to Climate*. New York: McGraw-Hill.

Tynan, C.T., and D.P. DeMaster. 1997. 'Observations and predictions of Arctic climate change: potential effects on marine animals', *Arctic* 50: 308–22.

Van Kooten, G.C., and G. Hauer. 2001. 'Global climate change: Canadian policy and the role of terrestrial ecosystems', *Canadian Public Policy* 27: 267–78.

Vitousek, P.M. 1994. 'Beyond global warming: ecology and global change', *Ecology* 75: 1861–76.

Chapter 8

Oceans and Fisheries

Learning Objectives

- To understand the nature of oceanic ecosystems and their similarities and differences with terrestrial ecosystems.
- To know the main challenges facing the oceans.
- To learn about some of the global management responses to these challenges.
- To realize the main characteristics of Canada's ocean regions.
- To understand the reasons behind the collapse of Canada's east coast groundfish fishery.
- To appreciate the background to Aboriginal use of marine resources and examine some current conflicts.
- To gain an understanding of Canada's main management strategies for ocean management.
- To know some of the concerns regarding aquaculture.

INTRODUCTION

At the height of the Northwest Atlantic cod fishery in 1968 over 800,000 tonnes of northern cod were landed in Canada; by 2002, the amount landed had been reduced to 5,600 tonnes in one restricted opening. In response to this crisis, the federal government introduced a moratorium on cod fishing in 1992, but the moratorium has been totally ineffective in rebuilding stocks, which remain at historically low levels. Scientists have put forth a number of suggestions to explain the stock's recovery failure, including changes in oceanic temperatures and increased predation. But the truth is, scientists are very far from understanding the basics of the marine ecosystems that have been sustaining the people of the Maritime provinces for centuries.

Meanwhile, on Canada's west coast, the once mighty salmon runs have dwindled over the last decade to a trickle in many areas. Some stocks have become extinct, others have been placed on the endangered species list by the Committee on the Status of Endangered Wildlife in Canada (COSEWIC), and numbers of returning salmon have so decreased in many rivers (including the world's largest salmon river—the Fraser) that fishing has been banned or drastically reduced for many stocks.

When the fishing economies on both coasts collapsed, coastal economies collapsed with them. Politicians (who had consistently approved fishing quotas well in excess of those recommended by their own scientists) finally began to realize the gravity of the situation—there was no quick fix to set things right again. As a result, a way of life for countless Canadians would have to change.

Over the last decade, fisheries management in Canada has risen from relative obscurity to become a high-profile public issue. Society has finally realized that the planet on which we live is finite. For thousands of years the world's oceans—covering 72 per cent of the Earth's surface—have been considered vast, inexhaustible resources. But

'No Fishing' (*Victoria Times Colonist*).

this past decade has soundly dismissed this idea, as fisheries around the world collapsed. The United Nations Food and Agriculture Organization (FAO) estimates that 70 per cent of the world's fisheries have reached or surpassed their limits of exploitation. Some fisheries, such as for the North Atlantic cod, may never recover.

The collapse in these fisheries is not really the root of the problem; it is a symptom of the problem, just as a runny nose is a symptom of a cold. The real problem is that we do not understand our ocean ecosystems and we have yet to learn that they cannot be managed. Human use needs managing, not fisheries. The theme of this book, *change and challenge*, is perhaps best epitomized in our relationship with our oceans.

OCEANIC ECOSYSTEMS

The oceans and their well-being are integral to sustaining life on this planet. They are key components in global cycles and energy flows (Chapter 2). Marine ecosystems are home to a vast array of organisms displaying greater diversity of taxonomic groups than their terrestrial counterparts. Marine organisms help feed us, and they are also the source of many valuable medicinal products. The value of marine ecological goods and services is estimated to be $21 trillion annually, 70 per cent more than for terrestrial systems (Costanza et al., 1997). We use the seas to dump our waste products and to transport most of our goods around the world. The oceans enrich our cultures, and nations draw

The oceans sustain an amazing variety of life. Here are two fish with very different life strategies. The whale shark, the world's largest fish, cruises the world's oceans feeding on plankton. The stone fish rarely moves and sits on coral reefs, disguised as a stone, until it spots prey to ambush. It is highly poisonous to humans (*Philip Dearden*).

strength and inspiration from their links to the vital life-giving nature and awesome power of the seas.

One of the major difficulties mitigating against sustainable human use of the oceans is our lack of understanding of oceanic ecosystems. In 2003, a $1 billion, 10-year expedition announced that it had described 150 new species of fish and another 1,700 plants and animals in just the first three years of its travels, and anticipates that there may be as many as 5,000 species of fish waiting to be discovered. Not all these new discoveries are small and in distant locales. For example, one discovery was a new species of squid over nine metres in length in the Gulf of Mexico, another, a giant jellyfish in the heavily studied waters off Monterey, California. If we know so little about oceanic ecosystem components, it is even more difficult to understand the functional relationships among them.

Yet the general principles that govern life and energy flows and matter cycles discussed in Chapters 2 and 4 also hold true for oceanic ecosystems (see Box 2.5). But there are important differences. On land, water is the most common limiting factor for life. In the oceans, this is obviously not the case. Here it is nutrients. While the oceans cover over 70 per cent of the Earth's surface, they account for only 50 per cent of the global primary productivity. Much of the ocean surface is the marine equivalent to a desert, with productivity limited to the areas where nutrients are abundant. Generally, nutrient concentrations increase with depth due to the decomposition of organisms falling from the surface layer.

The most productive areas are in the coastal zones and in areas of the ocean where upwellings from the deep ocean return nutrients to the surface layers and where photosynthetic activity occurs. In terrestrial ecosystems, productivity generally increases from the poles to the tropics. In the ocean this is not true. Although there are productive areas in the tropics (Box 8.1), some of the most highly productive marine areas in the world occur off the coast of Canada, such as the Grand Banks and in the Arctic Ocean, where nutrient upwellings occur. These upwellings promote large phytoplankton populations, especially in the Arctic where there is virtually unlimited light in the summer. Many species of whales and birds migrate to these waters to take advantage of this abundance. The largest whale, the blue whale, and the bird with the largest wing span in the world, the albatross, are two good examples.

Besides nutrient availability, the other major ecological influences on marine life are temperature and light. Both temperature and light decrease with depth. This means that surface waters in the euphotic zone (Box 2.4) are both warmer and have higher light levels, resulting in higher productivity. There is usually a sharp transition in temperature between the warmer surface waters and the cooler waters underneath. This is known as the **thermocline** and generally occurs between 120 and 240 metres in depth, depending on latitude and ocean currents.

The deepest part of the ocean is over 9,000 metres deep, but over three-quarters is between

BOX 8.1 CORAL REEFS—THE RAIN FORESTS OF THE SEA

Coral reefs are in many ways oceanic analogues of the tropical rain forests. Found throughout tropical and subtropical seas, they are among the most diverse and productive ecosystems on Earth. Like the rain forests they also have an ancient evolutionary history, having first appeared over 225 million years ago, with some living reefs perhaps as old as 2.5 million years. With solar radiation providing the primary source of energy, these habitats are found predominantly within 30° north and south of the equator, at depths of less than 50–70 metres. Coral reefs are categorized into three main types: (1) fringing reefs that are found along the shorelines of continents and offshore islands; (2) barrier reefs that have been separated from land by lagoons; and (3) atolls that appear as their own coral islands, having developed around (now submerged) lagoons.

Atolls, the Maldives (*Philip Dearden*).

Coral reefs are made up of countless numbers of individual **coral polyps** and the calcium carbonate skeletons deposited by prior generations of corals and other reef-associated organisms (e.g., coralline red algae and molluscs). These limestone secretions serve as a substrate for live coral polyps to grow and flourish.

Many coral species are involved in symbiotic relationships with unicellular algae, or **zooxanthellae**, that live inside the coral's protective skeleton. These photosynthetic algae produce carbohydrates that serve as the primary food source for the corals in which they live. When water temperatures get too warm, the **zooxanthellae** are often expelled, leading to the eventual death of the corals. This is called coral bleaching and has been recorded over large areas of reef throughout the world over the last decade. Scientists fear that global climatic change may have a very destructive effect on coral reefs.

Acting as sources of food and refuge to an incredible diversity of sea life, coral reefs are vitally important habitats. These complex ecosystems provide a number of critical ecosystem services, including the regulation of environmental disturbances, the treatment of organic wastes, the production of food, and the creation of recreational opportunities, for example, for scuba divers. In monetary terms, coral reef habitats have been estimated to generate approximately US$375 billion annually (Costanza et al., 1997).

A coral reef that has been dynamited for fishing in Myanmar (*Philip Dearden*).

Despite their value, coral reefs are now highly threatened worldwide, having been plagued with the effects of destructive fishing practices, coastal erosion, marine pollution, and irresponsible tourism activities (McClanahan, 2002). Southeast Asia is the world's epicentre for coral reef diversity and over 80 per cent of the reefs here are considered threatened, with 50 per cent in the high-risk category. In fact, the cataclysmic tsunamis of 26 December 2004 would have

wreaked less devastation if many protective coral reefs had not been previously destroyed. Local, national, and international initiatives have now been launched to counter these effects, but consumers can help too. Coral has become a fashionable accessory; avoid purchasing jewellery with coral. You

Coral reefs are home to some of the most complex and colourful organisms on Earth. Here are the bangai cardinal fish from Indonesia, an endemic nudibranch from the Andaman Sea, and a clam and coral from the Maldives (*Philip Dearden*).

can also help by not purchasing coral as a souvenir or for your aquarium. If you keep tropical fish, ensure that they have been bred in captivity and in a sustainable way. Research shows that as fish species are removed from the reefs, the whole community becomes less stable and more likely to collapse (Carr et al., 2002).

4,000 and 6,000 metres. Most productivity is on the continental shelves at a depth of less than 200 metres and within the top 100 metres. Most fisheries are concentrated in these areas. In the 1970s, however, scientists discovered that rich biological communities, mainly made up of bacteria that derive their energy from sulphide emissions, were centred around hydrothermal vents on the sea floor. Rich communities of tube worms, clams, and mussels have now been documented at depths exceeding 2,000 metres at over 100 sites worldwide, including off the west coast of Canada. Scientists speculate that life on Earth originated in such hydrothermal vent systems, based on chemosynthesis. Similar kinds of chemoautotrophic-based communities are found on the remains of whale skeletons at depth, nourished by sulphides produced as the carcasses decay, as discussed in Box 2.2.

Another interesting difference between terrestrial and marine ecosystems is in the shape of the **biomass pyramids**, discussed in Chapter 2. In terrestrial ecosystems, you will recall, the pyramids stand upright. There is a broad base of primary producers and a reduced biomass at each subsequent tropic level. This arrangement is reversed in

oceanic pyramids because the biomass of the primary producers, the phytoplankton, alive at any one time is quite small compared to their predators. The food chains are still dependent on a broad energy base, but the turnover of biomass at the first trophic level is rapid. This constantly replenished, short-lived base supports long-living predators, such as the blue whale, that store energy over a much longer period.

The **carbon balance** in oceanic ecosystems is now receiving a lot of scientific research because of the crucial importance of the oceans in absorbing carbon dioxide, a greenhouse gas, and the role this may play in mitigating the impacts of global climate change (Chapter 7). The ocean surface takes up about 2 billion metric tonnes of carbon per year by gas exchange. This is driven by wind exchange and the imbalances between the amount of carbon in the atmosphere and the oceans. Marine primary producers get their carbon from the dissolved CO_2 in the water, as bicarbonate. There is a balance between the amount of CO_2 in the atmosphere and bicarbonate in the water. If there is too much in either compartment, then a gradient is created and the carbon migrates along the gradient between the ocean's surface and the atmosphere. However, carbon is also constantly being moved out of the surface layers of the ocean into deeper water, where it is stored in dead organisms, ocean sediments, and coral reefs. The more that is stored at depth, the greater will be the gradient pulling carbon out of the atmosphere and into the surface waters to compensate for these losses. However, the lag time in these movements is considerable, and too slow to hope that this process can compensate for all the extra CO_2 released into the atmosphere by human activities.

The carbon-saturated water does not stay where it is but moves around the globe, mainly as a result of differing water densities. This is known as **thermohaline circulation** and involves warm surface water that is cooled at high latitudes sinking into deeper basins with water close to 0° C. The key to this circulation is the high salt content of the sea water that allows the water

density to increase before it freezes and sinks at certain sites. The main sites for conversion are in the North Atlantic, the Arctic Ocean, and the Weddell Sea in the Antarctic. When the water sinks, it carries with it large quantities of carbon. This sinking is the main mechanism for the removal of atmospheric carbon by the oceans. The cold water is then carried along the ocean floor until it mixes with surface waters and is transported back by wind-driven currents to the conversion areas (Figure 8.1).

The system is in fact a series of interlinked and variable currents that serve to mediate the Earth's climate through the transport of heat and water around the globe. Its scale is vast, with the flow estimated at more than 100 Amazon Rivers and the heat delivered to the North Atlantic being about one-quarter of that received directly from the sun. One of the main concerns with global

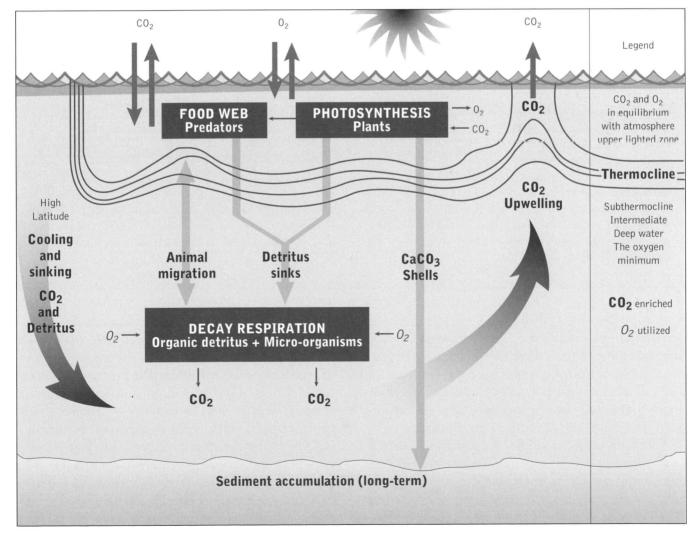

Figure 8.1   The ocean-atmosphere carbon cycle. SOURCE: *Oceans 2020* by John G. Field, et al. Copyright © 2002 Island Press. Reproduced by permission.

BOX 8.2 NEPTUNE PROJECT

NEPTUNE (North-East Pacific Time-series Undersea Networked Experiment) is a joint US–Canada venture that will feature 30 or more sea-floor 'laboratories' or nodes spaced about 100 km apart and connected over a 200,000 km² region of the Juan de Fuca plate. Land-based scientists in BC, Washington, and Oregon will control and monitor sampling instruments, video cameras, and remotely operated vehicles as they collect a wide range of real-time data from the sea floor to the waves at the surface.

NEPTUNE instruments are designed to measure and observe the complex physical, chemical, biological, and geological processes in the deep sea and immediately respond to events such as underwater volcanic eruptions, storms, fish migrations, earthquakes, tsunamis, and plankton blooms. This new knowledge will be applied to a range of global concerns, such as providing advanced earthquake and tsunami warnings, gaining more accurate means of fish conservation and stock estimates, creating regional climate change models, and researching safer methods of extracting energy resources.

The first test site for NEPTUNE is known as the VENUS (Victoria Experimental Network Under the Sea) project and is led by the University of Victoria. VENUS will provide real-time measurements, images, and sound data via fibre optic cables from three sea-floor locations off the southern BC coast: Saanich Inlet,

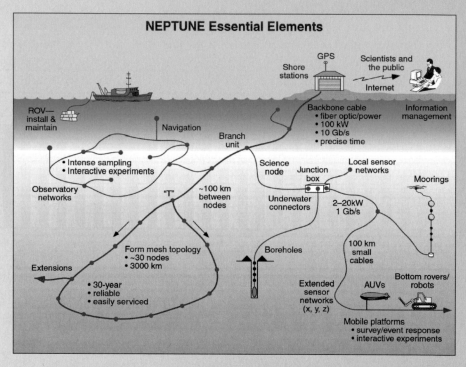

'Essential elements'. SOURCE: NEPTUNE Program, University of Washington.

the Strait of Georgia, and Juan de Fuca Strait. The first cable, which will be laid in Saanich Inlet, is expected to be complete in the spring of 2005.

warming (Chapter 7) is the impact it may have on thermohaline circulation. For example, with increased temperatures there will be increased fresh-water melting from the polar ice caps. Fresh water is less dense and will not sink to the same depth as super-cooled salty water, and therefore less carbon will be sequestered for shorter periods of time.

This is a classic example of a positive feedback loop (Chapter 3): the warmer the atmosphere gets due to increased CO_2 emissions, the more ice will melt resulting in less CO_2 being absorbed by the ocean, which will result in increased atmospheric warming ... and more melting. The cycle will

continue to perpetuate itself unless equally strong negative feedback loops come into play. One example of a potential negative feedback loop in this context is the plankton that produce dimethyl sulphide gas. Given enough nutrient supply, a warmer climate should produce a greater abundance of plankton. The gas is important in enhancing the formation of atmospheric sulphate aerosol particles and cloud condensation nuclei that would tend to screen out sunlight and hence produce a negative feedback loop for the effects of global warming. Unfortunately, scientists do not feel the effects of this negative feedback loop will be enough to counteract the effects of the numerous positive feedback loops likely to be triggered by global warming.

There is a great deal of scientific uncertainty about these global systems. Scientists know they exist and that they are of crucial importance in determining global climatic conditions. They also know that these are dynamic systems that change over time. However, due to a lack of good baseline data, it is often difficult to assess the dimensions of natural change and the mechanisms underlying it. Large spatial changes happen only rarely and slowly, requiring the collection of very long data sets of frequent measurements in order to understand them. Ocean scientists are now trying to establish these kinds of monitoring systems so that we can understand observed changes and reduce uncertainty (Box 8.2). Meanwhile, the physical evidence to support global climatic change continues to grow. In 2002, scientists reported that the Arctic's largest ice shelf had broken up.

OCEAN MANAGEMENT CHALLENGES

Oceanic ecosystems have been providing humans with sustenance since time immemorial. Coastal zones, including the continental shelf, occupy about 18 per cent of the Earth's surface, supply about 90 per cent of the global fish catch, and account for roughly 25 per cent of global primary productivity. Approximately 60 per cent of the world's population also lives in coastal zones, and this percentage is predicted to increase.

The oceans are crucial to the way in which planetary ecosystems work and to the functioning of human society. In 1883, Thomas Huxley, the great nineteenth-century biologist, voiced the common opinion 'that probably all the great sea fisheries are inexhaustible.' This has proven to be far from the truth. This section will review some of the main management challenges facing ocean ecosystems as a result of fisheries and other human activities (Figure 8.2).

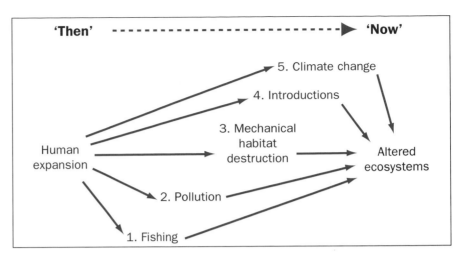

Figure 8.2   Historical sequence of human impacts on marine ecosystems. SOURCE: Roberts (2003: 170), from Jackson et al. (2001). Copyright © 2001 AAAS. <http://www.sciencemag.org>.

Fisheries

Fisheries provide some 20 per cent of the world's annual animal protein supply, but in some areas of Asia and Oceania fish provide virtually all the protein supply. Over one billion people rely on fish as their primary source of protein. This supply is tapped by fishers ranging from villagers using homemade canoes trying to feed their families through to multi-million dollar offshore factory ships owned by multinational corporations. Given this great scale and variation, it is only recently that fisheries scientists have started to understand more about what is happening in global fisheries.

There is clear evidence, obtained from bottom sediment cores going back hundreds of years, that fish populations fluctuated quite widely as a result

Small-scale fishers on the east coast of Sri Lanka sort the morning's catch (*Philip Dearden*).

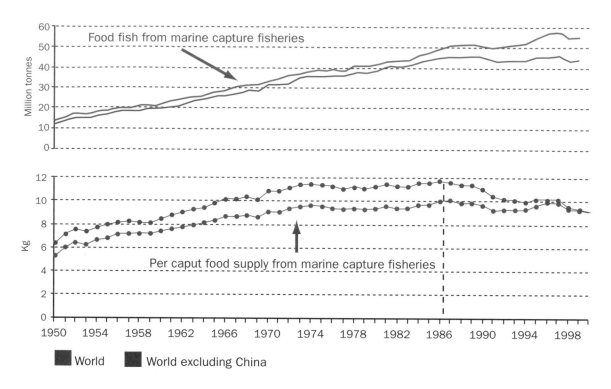

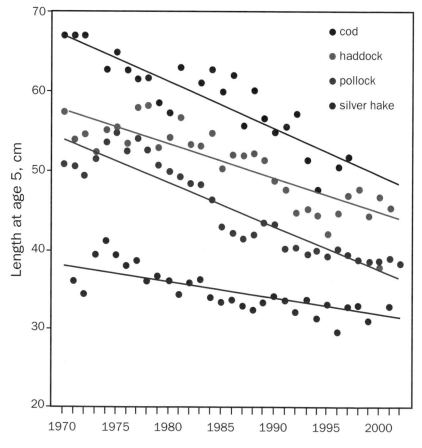

Figure 8.3 Absolute and per caput food fish production from marine capture for the world and the world excluding China, 1950–99. SOURCE: United Nations Food and Agriculture Organization, Fisheries Department, 2003, at: <http://www.fao.org/fi/statist/nature_CHINA/30JAN02.ASP>.

of changing environmental conditions even before the advent of modern fishing. However, the scale and speed of current changes are unprecedented. As stated earlier, over 70 per cent of global fisheries are now fully utilized or overexploited. The graph in Figure 8.3 shows a gradual levelling off in more recent years, but many scientists argue that this levelling off masks a major unprecedented collapse of ocean fisheries. Scientists at the Fisheries Research Centre at the University of British Columbia, for example, revealed that the catch data shown in the graph may be up to 25 per cent above actual catch data. This is suspected to be the result of the routine reporting of inflated catches by China (Watson and Pauly, 2001). The problem with this is that permitted catch levels are based largely on historical catch information. If the latter is inflated this leads to unsustainable catch levels in the future.

Figure 8.4 Fish size decline, 1970–2000. SOURCE: Fisheries and Oceans Canada, 2003. Reproduced with the permission of Her Majesty the Queen in Right of Canada, 2005.

Figure 8.5 Bycatch. Reprinted by permission of Adrian Raeside.

The fish being caught are also substantially smaller than they were historically (Figure 8.4). Recent research suggests that for top predators, current sizes are one-fifth to one-half what they used to be (Myers and Worm, 2003). These problems are not immediately obvious, as scientists tend to look only at the most recent data rather than comparing these with historical catches. This problem is known as a **shifting baseline**, where scientists have no other option than to take the current degraded state as the baseline rather than the historical ecological abundance. Says fisheries biologist Ransom Myers (ibid.), 'Since 1950, with the onset of industrialized fisheries, we have rapidly reduced the resource base to less than 10 per cent—not just in some areas, not just for some stocks, but for entire communities of these large fish species from the tropics to the poles.' Removal of virtually all the large predatory fish from oceanic ecosystems will have significant implications for the structure and functioning of marine ecosystems.

The total catch is also made up of many different species, which creates several problems. In theory, fishing should be self-regulating. As the catch of the target species declines, this should result in a reduced fishing effort as it becomes unprofitable. Unfortunately, this is not what happens. Instead, fishing fleets increase their effort towards the target species and then turn their efforts to the next most profitable species, until that too is depleted. Then they pursue the next most profitable species, etc. This is a familiar foraging behaviour in ecology, known as **prey switching**. In the fishery it leads to **serial depletion**, where one stock after another becomes progressively depleted, even if the total catch remains the same. This has been well documented in whales (Box 8.3). Unfortunately, in switching the target species, many fisheries also take some of the depleted species as bycatch (Figure 8.5), making it even more difficult for the stocks to recover.

Target species are not the only changes to the world's fishing activities. It has also been shown that we are progressively exploiting lower and lower trophic levels to derive our catch. Initially we targeted large, long-lived fish high in the food chain, but as these became depleted, smaller fish were sought until these, too, became depleted. Fisheries in many areas have now refocused on invertebrates, like the Atlantic crab fishery. The large-sized fish no longer exist. This is known as **fishing down the food chain** and can be clearly seen in Figure 8.6. The gains in fish catch shown in Figure 8.3 were mainly the result of reaching further down the food chain to previously under-exploited trophic levels.

This shark was caught by a 'ghost net' snagged on the bottom, but still doing the deadly function it was designed for (*Philip Dearden*).

BOX 8.3 LEVIATHAN

Canada has the longest coastline in the world as well as some of the richest waters. Cold oceanic waters from the north and deep upwelling currents mix with warmer waters from the tropics to create an abundance of life. Rich nutrient supplies accompanied by shallow seas and long daylight hours in the summer have created some of the richest waters off the east coast. Plankton flourish and provide the base for a diverse food web that supports three main groups of marine mammals, the Odontoceti or toothed whales, the Mysticeti or baleen whales, and the Pinnipedia or seals and walruses. The baleen whales all feed on plankton, small fish, and marine algae by means of plates of baleen in their mouths that serve to filter these organisms from the water. From these small food items, the baleen whales have evolved as the largest creatures on this planet. The blue whale is the largest of the baleen whales; the largest recorded was a female over 30 metres in length and weighing 140 tonnes. A calf is typically seven metres long at birth and may weigh almost three tonnes. Blue whales, as well as other baleen whales such as the fin and sei whales, were once found in abundance off the east coast, as was the largest toothed whale, the sperm whale. These whale populations were all decimated by hunting and are only now starting to recover.

In the late fifteenth century, Europe was finding itself increasingly short of the oil needed to light its lanterns. Marine mammals, with a thick layer of blubber, were the solution. Exposed to high heat, the blubber can be rendered down to oil. The Basques—a great seafaring people from northern Spain—discovered the rich whaling grounds off the east coast of Canada, and the slaughter began. The abundance of whales in these waters at this time is difficult to imagine. Whales could be harpooned from shore. Early mariners complained of whales as a navigational hazard because they were so numerous; one missionary in the Gulf of St Lawrence reported that the whales were so numerous and loud they kept him awake all night! The limiting factor was not the number of whales, but the ability to process them, and as increasing numbers of shore stations were established, another toehold of colonization began.

The rest of the story is fairly well known. As human numbers increased, whale numbers declined. First to go was the one hunted the

Humpback whales were once common off both the Pacific and Atlantic coasts of Canada (*Philip Dearden*).

most, the black right whale. The black right whale was targeted because it was slow, it floated when it was killed, and one whale could be rendered into over 16,000 litres of oil. Besides the oil, the baleen was used for other indispensable purposes such as clothing supports, brush bristles, sieves, and plumes for military helmets. A Basque shipowner could pay off his ship and all his expenses and still make a good profit in one year from such whales. Although the whales gained some respite when the Spanish Armada was destroyed by England, other nations finished off the job.

Other whales—sperms, humpbacks, blues, fins, seis, and minkes—soon joined the right whale as commercially, if not biologically, extinct. This is a classic example of serial depletion. By the early 1970s there were no commercially viable populations of large whales remaining. A global moratorium on whaling was announced in 1987 and still stands today, despite pressure applied by Japan and Norway, both of which continue to kill whales for 'research purposes'. Populations for most species have been slow to recover due to the slow reproductive rates of these K strategists (Chapter 3).

The removal of such a large biomass from the top of the food chain obviously has ecological repercussions for other organisms. Unfortunately, we have interfered so much with populations of other animals that the impacts are seldom clear. One possible implication is an increase in other krill eaters that would benefit from removal of these large and efficient

Whale harpoon in the Fisheries Museum of the Atlantic, Lunenburg, Nova Scotia (*Philip Dearden*).

previously used for nesting by birds such as the albatross. Did declining whale numbers also lead to a decline in albatross? We cannot say for sure, but the example does illustrate the complexities of changes in food webs.

Whaling has now been controlled throughout most of the world, and on all of Canada's coasts a resurgence is occurring, albeit very slowly for some species, such as the right whales. The largest whale, the blue, is also in trouble. Although numbers still exist, finding a mate is difficult for this wide-ranging species. Scientists have now found evidence that the blues are interbreeding with the more common fin whales, the second largest whale in the world, and fear that this hybridization will result in a loss of genetic identity for the blue, which could then disappear as a species. Unfortunately, the rapacious killing by our ancestors denied all future generations the spectacle of seeing our seas full of these mighty creatures. Are we denying future generations other such opportunities by our actions today?

competitors. For example, in the Antarctic it has been suggested that increases in the populations of krill-eating seals (e.g., fur and crab-eating seals) came about as a result of whaling. More breeding sites are required to support the higher populations of seals, and so seals colonize areas

Figure 8.6 Fishing down the food chain.

The rich waters off the BC coast attract many pelagic (open ocean) bird species, such as this black-footed albatross wheeling off the west coast of the Queen Charlotte Islands (*Philip Dearden*).

Besides the impact on target species, fishing activities have many other ecological repercussions. Of particular concern is the impact on non-target organisms, or so-called **bycatch**. Estimates suggest that 25 per cent of the world's catch is dumped because it is not the right species or size. Virtually all organisms dumped overboard die. The world's largest turtle, the leatherback turtle, is rated as critically endangered on the Red List of the International Union for the Conservation of Nature and Natural Resources (IUCN). In 1980, there were some 91,000 leatherbacks left in the Pacific; there are now fewer than 5,000. They are mainly the victims of the longline and gill-net fisheries. **Longline** fishing in all the world's deep

Model of a box trawl in the Fisheries Museum of the Atlantic, Lunenburg, Nova Scotia. The large weights are dragged along the sea floor, destroying everything in their path (*Philip Dearden*).

oceans kills some 40,000 sea turtles each year, along with 300,000 seabirds and millions of sharks. Bycatch of albatrosses, petrels, and shearwaters in longline fisheries is one of the greatest threats to these seabirds (Robertson and Gales, 1998; Tasker et al., 2000). All 21 of the world's species of albatross are now considered at risk of extinction, along with 57 species of sharks and rays.

One of the most destructive means of fishing is **bottom trawling**, where heavy nets are dragged along the sea floor scooping up everything in their path. This is a common method used for catching shrimp, and the ratio of shrimp to other organisms caught is generally around 10 per cent (i.e., one shrimp caught for every 10 organisms caught unintentionally). Elliott Norse of the Centre for Marine Conservation has characterized bottom trawling as 'hanging a huge net dragged from a blimp across a forest, knocking down the trees and scooping up the plants and animals, and then throwing away everything except the deer' (1998: http://www.sunsonline.org). The effects are worse than clear-cutting a forest, although no one will ever see the ocean's forests.

One noted victim of bottom trawling was the ancient sponge reefs found off the coast of BC. The sponge reefs, found at depths of between 165 and 240 metres, are over 9,000 years old. Previously they were known only from fossils in Europe dating from 146–245 million years ago, and they were thought to be extinct before being discovered in 1988. The federal government called for *voluntary* trawl restrictions, but in 2002, after documentation of extensive damage to the most pristine reefs by trawling, mandatory closures were finally implemented.

In addition to the direct destructive effects of fishing on non-target organisms, there are also indirect effects through food chain relationships. Steller's sea lion, for example, was once very abundant in the North Pacific with over 300,000 animals recorded in 1960. By 1990, this number had fallen to 66,000, and the US declared the sea

One of the deep sponges found off the BC coast.

This dovekie was killed by an oil spill. Oil penetrates through the feathers to the layer of down beneath and decreases the effectiveness of the feathers' insulating properties (*John W. Chardine*).

lion endangered. The main reason for the decline is thought to have been the decline in pollock, their chief food source. As the harvests for cod diminished elsewhere, the demand for pollock increased, leading to unprecedented catches and a subsequent decline in sea lion numbers. Scientists have recently suggested another factor that may have contributed to this. Earlier industrial whaling activities removed the bulk of the killer whales' main prey. In response, the killer whales started 'fishing down the food chain', eating sea lions in much greater numbers. Reaching even further down the food chain, killer whales are now implicated in the decline of sea otters in the North Pacific (Chapter 3). But sea lions and sea otters are not the only marine mammals to experience

declining numbers; two-thirds of all marine mammals are now classified as endangered on the IUCN Red List (Chapter 11).

Pollution

As the recipient of all the polluted water that flows off the land as well as airborne contaminants, the oceans are the ultimate sink for many pollutants. The scale of global pollution is now astounding—even the oceans are being rendered eutrophic (Chapter 4), in some cases from farming practices thousands of kilometres away. Scientists suspect that iron particles picked up into the atmosphere in Africa and blown across the Atlantic are a major factor in increasing the growth of smothering algae on the coral reefs of the Caribbean.

Marine pollutants take a number of forms, originate from many different sources, and have a wide range of effects. About 80 per cent of ocean pollution comes from activities on land. The remaining 20 per cent comes from activities at sea, such as waste disposal, oil spills, vessel traffic, oil and gas exploration, and mining. Although we are all familiar with major oil spills, such as that of the *Exxon Valdez*

Steller's sea lions, the world's largest sea lions, are now highly endangered in the North Pacific. These are at Cape St James, off the southern tip of the Queen Charlotte Islands (Haida Gwaii), also the windiest spot in Canada (*Philip Dearden*).

in Alaska, we think little of the many diffuse sources of oil, such as leaking car engines, that eventually seep into the oceans. In fact, the total amount from non-point sources of pollution is considerably greater than from point-source pollution. During the last 20 years, many governments have made considerable progress in monitoring and regulating point-source pollution, such as effluent discharge from factories. However, addressing non-point-source pollution is a much more challenging task, as it requires changes in behaviour from millions of people (see Chapter 12).

Chemical pollutants take two main forms: toxic materials and nutrients. We live in a chemical society, with over 100,000 chemicals used in manufacturing and released into the environment every year. Many of these chemicals are harmful to life; when they end up in the ocean, they may cause instant death if released in sufficient quantity, or they may have sub-lethal effects such as inhibition of reproduction. The chemicals are also subject to *bioconcentration*, discussed in more detail in Chapter 10. Synthetic organic chemicals and toxic metals tend to concentrate along two main interfaces in the ocean: the boundary between the seabed and water, and the boundary between the water and the atmosphere.

One rapidly emerging impact relating to pollution is that of **endocrine disruption**. The endocrine system consists of glands and hormones that control many bodily processes such as sex, metabolism, and growth. Many chemicals in everyday use have been found to mimic these processes and may stimulate, replace, or repress the natural processes. Over 50 such endocrine disrupters have been positively identified, and scientists suspect that there are hundreds more. Many of them are in commonly used products such as soaps and detergents. The major effect of these chemicals on marine life detected so far is the feminization of various aquatic species. Hermaphroditic fish are appearing all over Europe and also in the Great Lakes. All the fish in the River Aire in England, for example, show signs of feminization. Increasing numbers of

reports from the Arctic are also reporting hermaphroditic polar bears. Many endocrine disrupters, such as the pesticide DDT and other organochlorines, are vulnerable to both the grasshopper effect and biomagnification (as described in Chapter 10), and would logically find their way to the top Arctic predator. There is now extensive research underway in Europe on this issue, as it is also feared that these so-called 'gender-benders' may be at least partially behind the documented fall in human male sperm counts in industrialized societies over the past 50 years.

Oil and Gas Development

World demand for energy, particularly oil and gas, continues to rise (Chapter 13). Every time you jump into your car or go on an airline, you are sending a financial message to industry and the government that you support further development of fossil fuel sources. All of the world's cheap and accessible terrestrial sources have now been tapped. Many of the world's main oil fields, such as the North Sea in Europe and Hibernia off the coast of Newfoundland, occur in sedimentary basins under the oceans. More than 60 per cent of current global production comes from these sources. Offshore oil rigs are a source of chronic, low-level pollution caused by the disposal of drilling mud and drill cuttings, which smother the local environment and are often contaminated with oil or chemicals. Oil is pumped to the surface and loaded on tankers or piped ashore, where it is often stored in close proximity to the ocean before being refined and further distributed. There is

Offshore oil rig, South Africa (*Philip Dearden*).

potential for spillage at every stage. Many seabirds and marine organisms are highly vulnerable to oil pollution. Following the wreck of the *Exxon Valdez* in Alaska, for example, over 750 sea otters were killed. Peterson et al. (2003) argue, however, that the long-term effects of the spill, including mortalities of species such as sea otters, were much greater, because of ongoing impacts such as contaminated food chains. Oil is lethal to marine life by virtue of its physical effects and its chemical composition (Garshelis, 1997). Cleanup attempts following oil spills can also be damaging to many species.

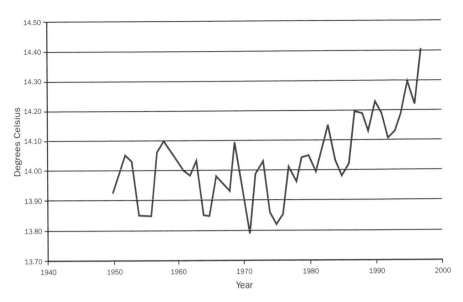

Figure 8.7 Global average sea surface temperature, 1950–97.

Most accessible ocean oil basins have already been developed, which has meant that exploration and development are pushing into increasingly challenging and fragile environments, such as the Arctic Ocean and North Pacific. The most effective way to reduce the effects of oil exploration and development in these regions is through reducing demand for oil products.

Coastal Development

Almost half of the world's population lives within 100 kilometres of the coast, and this is expected to increase to 75 per cent by mid-century. These regions also support the highest concentration of supporting infrastructure, industrial plants, energy use, food production, and tourism in the world. Half of the world's coastal wetlands have been filled in to support these developments. Meanwhile, as in Canada, environmental decision-making and management within the coastal zone are often highly fragmented among many different agencies. These extreme pressures on fragile ecosystems with ineffective management control have often led to highly degraded coastal environments in many countries. Half of the countries in the world do not have any coastal legislation in place to address this situation.

Climate Change

Global air temperature is expected to warm by 1.4 to 5.8° C this century, affecting sea-surface temperatures (Figure 8.7) and raising the global sea level by nine to 88 cm (IPCC, 2001). Higher sea

levels increase the impact of storm surges, accelerate habitat degradation, alter tidal ranges, exacerbate flooding, and change sediment and nutrient circulation patterns. A case study of coastal areas in Prince Edward Island in Chapter 7 highlights some of these challenges. Rising levels will displace approximately one billion people, many of them among the poorest in the world. Recent estimates suggest that an increase in mean sea-surface temperature of only 1° C could cause the global destruction of coral reef ecosystems (Hoegh-Guldberg, 1999). Increases in temperature may slow or shut down the thermohaline circulation (discussed earlier), causing widespread climatic change, changes in the geographic distributions of fisheries, and increased risk of hypoxia in the deep ocean.

GLOBAL RESPONSES

International Agreements

Strong international action can help address these problems, and there are many examples of international treaties concerning the oceans. The overall international legal framework is provided by the United Nations Convention on the Law of the Sea (UNCLOS) that entered into force in 1994. Activities under its jurisdiction include: protection and preservation, navigation, pollution, access to marine resources, exploitation of marine resources, conservation, monitoring, and research. One of the most important provisions was the agreement for coastal nations to establish **exclusive**

economic zones (EEZs). The Convention also established 45 per cent of the seabed as common property. Canada ratified the Convention in November 2003.

More specific agreements include the 1972 London Dumping Convention that has led to a gradual decline in the amounts of sewage sludge and industrial waste being dumped into the oceans. In terms of fisheries, the UN Moratorium on High Seas Driftnets was quite successful. The same kind of approach now needs to be taken to reduce the impacts of other highly damaging types of fishing gear, especially longlines and industrial trawlers with high rates of bycatch. The United States has already taken some steps to protect embattled marine species by closing the west coast to longlining altogether and by restricting the Hawaii longlining fleet from fishing for swordfish. There are also temporal and spatial closures for gill netting and moves to ban bottom trawling, which may be *the* most destructive fishing method of all.

Some strong ecologically based marine agreements have emerged over the last decade, including the UN Convention on Straddling and Highly Migratory Fish Stocks and the Global Program of Action for the Protection of the Marine Environment from Land-Based Activities. However, these new agreements in some instances are not yet fully implemented. For example, 10 of the world's top 15 fishing nations have not yet ratified the Straddling Stocks agreement, although Canada has done so.

In addition to these international agreements, important international programs are trying to create a more comprehensive, global approach to oceans management. Agenda 21, for example, arose from the UN Conference on Environment and Development in 1992 as a program of action to promote sustainable development. Chapter 17 of the Agenda outlines principles and objectives related to oceans. At the World Summit on Sustainable Development (WSSD) held in Johannesburg in 2002, governments committed to restore most of the major global fisheries to commercial viability by 2015.

Marine Protected Areas

Compared to the terrestrial environment, where ecological communities are often associated with areas defined by geographical features (such as mountains and rivers), precise boundaries of distinct communities or processes in oceans are rare. Geographic scales are large and biological processes are not self-contained within a given area. The water overlying the sea bed is a mobile, third dimension that provides nourishment for much of ocean life. These characteristics of the ocean environment mean surveys take time and are costly, and because of the high variability in sizes of marine populations, surveys are typically uncertain. Until recently, these difficulties prevented scientists from being able to assess the effectiveness of **marine protected areas (MPAs)**. However, in 2001, following two and a half years of rigorous scientific review, 161 of the world's top marine scientists signed a consensus statement (http://www.nceas.ucsb.edu/Consensus) that marine reserves:

- conserve both fisheries and biodiversity;
- are the best way to protect resident species;
- provide a critical benchmark for the evaluation of threats to ocean communities;
- are required in networks for long-term fishery and conservation benefits;
- are a central management tool supported by existing scientific information.

The establishment of MPAs has lagged substantially behind their terrestrial counterparts—currently under 0.5 per cent of the ocean has been designated as protected. However, with increasing political awareness of the importance of the oceans and their highly degraded state, MPAs are starting to receive some attention. The 2002 World Summit on Sustainable Development set 2012 as the target date for completion of an effectively managed, ecologically representative network of marine and coastal protected areas within and beyond areas of national jurisdiction. The latter occur in the 64 per cent of the oceans beyond the 200-nautical-mile limit of the EEZs of coastal states. These are known as the High Seas and, due to the lack of an established legal framework for their allocation and management other than UNCLOS, it will be necessary to take concerted international action to accomplish their protection.

CANADA'S OCEANS

Canada has the longest coastline of any country and the second largest continental shelf, equal to 30 per cent of Canada's land mass (Figure 8.8).

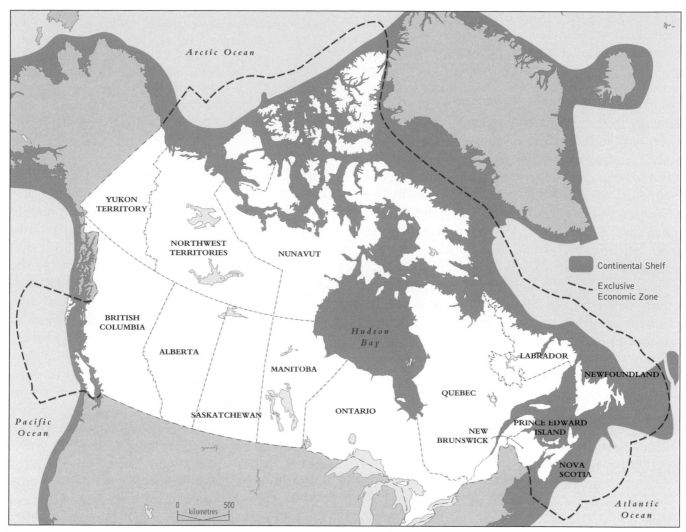

Figure 8.8 Canada's coastline and continental shelf.

The population of the southern pods of killer whales has fallen 20 per cent over the last 10 years and is now considered endangered (*Adrian Dorst*).

There are some 1,200 species of fish and many globally important populations of marine mammals, including grey, bowhead, beluga, right, minke, humpback, and killer whales. Unfortunately, several of these species are also on Canada's list of species at risk (see Chapter 11), including the beluga, bowhead, northern right, and Georgia Strait killer whales. Over seven million Canadians live in coastal communities.

The federal government largely holds jurisdiction for the marine environment below the high-water mark. Twenty-seven different federal agencies and departments are involved. The lead agency is the Department of Fisheries and Oceans (DFO), which has traditionally focused its efforts on commercial fishery management but is also responsible for all marine species (except seabirds, which come under Environment Canada). DFO is also the lead agency for Canada's Ocean Strategy, discussed later. Provincial governments have responsibility for shorelines, some areas of seabed, and some specific activities, such as aquaculture. Municipal governments also influence the coastal zone since they have responsibility for many land-based activities affecting the oceans. There are quite a few areas of overlap between these different levels of jurisdiction and, inevitably, conflicts have arisen. As a result, coastal and marine resource management is typically fragmented and often ineffective. In addition, the very size of Canada's coastal zone, ranging from the Mediterranean climate of BC's Gulf Islands through to the High Arctic, means that a wide range of conditions and activities exists. The next section provides an overview of the main marine ecozones in Canada.

Marine Ecozones

Pacific Marine

From the continental shelf, with depths of less than 200 metres, this zone descends to the Abyssal Plain offshore, with depths in excess of 2,500 metres. Sea temperatures and salinities are higher than in the Arctic Ocean, and this zone acts as a transition zone between cold polar waters and the temperate waters of mid-latitudes. The zone is of particular importance for the feeding, staging, and

The mammoth bones of the California Gray lie bleaching on the shores of those silvery waters, and are scattered along the broken coasts, from Siberia to the Gulf of California; and ere long it may be questioned whether this mammal will not be numbered among the extinct species of the Pacific.
— Captain Charles Scammon, 1874

wintering of waterfowl, seabirds, and shorebirds. Pollution is a constant concern. Ten pulp and paper mills discharge effluent into adjacent marine coastal waters, and large areas have been closed to shellfish harvesting as a result. One small oil spill off the Washington coast in 1988 resulted in the death of a least 56,000 birds on the BC coast.

The Pacific coastal waters also support large populations of sea lions and seals and a variety of whale species. However, many of these populations have yet to recover from hunting activities. The Pacific (or California) grey whale is the most abundant whale and can be readily seen on its yearly migrations from the calving grounds of Baja California to the feeding areas in the Arctic Ocean as it passes, and sometimes lingers and feeds, in BC coastal waters. The Pacific grey has recovered twice from excessive hunting pressure to become one of the most populous whales, with a current population of around 25,000 individuals (Box 8.4). Humpback whales are also sometimes seen, but the best-known sighting is of the killer whales that ply the waters around Vancouver Island, taking advantage of the salmon coming back to spawn. The sea otter was also once common on this coast, but was *extirpated* through hunting in the last century and is only now beginning to recover in numbers as a result of reintroductions (Chapter 3).

Arctic Basin

This zone covers the Arctic Ocean basin between Canada and Russia, Greenland, and Norway. Depths are mostly greater than 2,000 metres. It is largely covered in permanent pack ice. Biological productivity and diversity are low. Three species of whale, the bowhead, beluga, and narwhal, are found up to the edge of the pack ice, and the ringed seal, the smallest of all the pinnipeds or finfeet (seals, sea lions, walruses), is the most abundant marine mammal. The seals have claws on their front flippers to help scrape breathing holes through the ice, and in the winter their bodies are

A cormorant killed by an oil spill (*Philip Dearden*).

BOX 8.4 THE MAKAH AND THE WHALE HUNT

On 17 May 1999, members of the Makah Indian tribe in Neah Bay in the state of Washington returned to the whale-hunting practice of their ancestors with the help of powerboats and a high-calibre rifle. On that day a crew of tribal members set out in a traditional dugout canoe and proceeded to fire three harpoons into a migratory grey whale off the Olympic coast of Washington and across the international boundary from Victoria, the site of a thriving whale-watching industry. The wounded mammal then dragged the canoe further out to sea before a motorized support crew closed in and fired two high-calibre rifle shots to complete the kill. This day marked the first and only legal whale hunt by the Makah in over 70 years.

The hunting of grey whales in this region had been officially banned since the 1920s, when their numbers had plummeted and landed them on the endangered species list. After a strong recovery the grey whale was removed from the endangered species list in 1994, but only to be met with increasing pressure from the Makah to reinstate the whale hunt with passionate claims that the hunt is vital to the survival of the Makah culture. Eventually, the US government agreed to the Makah's requests and acknowledged their 'right' to hunt, which had been guaranteed in an 1855 treaty. In accordance with the International Whaling Commission, the US government authorized the Makah to resume whaling for 20 days each year and to harvest a maximum of four grey whales.

The decision to resume the hunt was met by fierce opposition from angered environmentalists and animal rights activists. Many believe that the Makah no longer have any real subsistence need and that allowing a renewed hunt will be the first step towards a disintegration of a world ban on whale hunting. Many activists opposed the hunt through extensive demonstrations at the reservation boundary and others by actively interfering on the water by scaring off the whales or irritating hunters.

In a more recent development, a US appeal court has temporarily reversed the decision to allow whale hunting until a more thorough impact assessment has been conducted. Although the Makah will not be able to continue whale hunting in the interim, this does not mark the end of this moral and cultural debate. However, it has allowed for more time to properly investigate the varied arguments and their implied consequences.

No reasonable person would deny that the Makah people, or the whale-hunting tribes of Vancouver Island, have endured severe cultural losses, or that the grey whale had a significant role in their way of life. The question is whether the act of killing whales is necessary for revitalizing Makah culture and whether this objective outweighs the numerous moral and political ramifications of this action. Will resumption of the hunt spur the efforts of some coastal indigenous people in British Columbia to resume hunting? Will the Makah hunt be used as a wedge to break down international protection against whaling?

One reason the conflict has had such a high profile is the multi-million dollar whale-watching industry that has developed on the west coast. The industry helps introduce thousands of people every year to the oceans, although there are some very real questions as to how much educational information people receive and attain, and whether or not the whale-watching experience (and other nature-based tourism experiences) can alter consumer behaviour (e.g., using fewer chemicals at home) to assist in ocean conservation (Malcolm, 2003). Nonetheless, many people have been whale-watching and do not take kindly to the graphic images of whales being shot, dragged up on a beach, and butchered, as shown on television.

In contrast, Canada has been licensing the Inuit to kill the much more endangered northern bowhead for many years, with relatively little public response. Every year the International Whaling Commission, of which Canada is no longer a member due to this practice, sends a note of protest to the Canadian government.

Whale-watching has become a thriving business on both the Pacific and Atlantic coasts (*Philip Dearden*).

> *Perhaps of greatest concern for the future state of the two polar marine systems must be the potential for cumulative impacts from multiple sources, combining global pressures in the form of climatic and ozone changes with local pressures in the form of pollution and unsustainable exploitation of living resources. While species may be able to withstand and adapt to one, or even several of these pressures, the cumulative impact of a number of impacts operating synergistically may be too much to bear. Moreover, should severe impacts occur at a critical point in the food chain, dependent and associated species are likely to follow suit. In this sense, the polar marine ecosystems are particularly vulnerable and exposed to risk.*
> — *Clarke and Harris (2003: 21)*

composed of up to 40 per cent blubber to insulate against the cold. Seals are mainly krill eaters, but also eat small fish. So well adapted are they to their ice environment that even mating occurs under the ice. In spring, the female finds a crevice through the pack ice and excavates a den for pupping in the snow above. Polar bears are the seals' main predator.

Arctic Archipelago

This maze of channels and islands extends from Baffin Island and Hudson Bay in the east through to the Beaufort Sea in the west. Ice cover is not permanent, and open water exists in most areas for two to three months every year. There are also areas of permanent open water, known as **polynyas**, that are biologically productive, allowing regionally significant concentrations of marine life. Resident birds such as black guillemots, thick-billed murres, ivory gulls, and dovekies congregate in large numbers at these sites in winter to obtain food in the open leads. The polynyas vary greatly in size from 60–90 metres in diameter to as large as the North Water polynya between Ellesmere Island and Greenland that may cover as much as 130,000 square kilometres. These areas remain ice-free as a result of various combinations of tides, currents, ocean-bottom upwellings, and winds, all of which promote constant water movement. Their attraction for marine mammals also made polynyas attractive sites for Inuit settlements and later for whalers who battled to get through to large polynyas (such as the North Water) as early as possible every year to kill the whales. To this day

there are still only a few hundred bowheads in the area, from pre-exploitation numbers of perhaps 11,000. It is doubtful whether the population will ever recover.

Perhaps contrary to expectation, Arctic marine mammals live in a less extreme environment than their terrestrial counterparts. Polar waters seldom fall below the freezing point. The temperature gradient between external and internal temperatures that animals living in this environment must cope with is therefore only about half that of land animals. Animals such as whales, seals, and walruses have virtually no hair, but instead rely on a thick layer of blubber to protect them from the cold. Their skin is exposed directly to the water, and they therefore need a very finely tuned thermal system to maintain a skin temperature warm enough to resist freezing but cool enough to avoid excessive waste of energy.

Inuit people engage in hunting and fishing activities throughout the zone. Potential also exists for increased exploitation of hydrocarbons, which could cause serious resource conflicts between traditional Inuit ways of life and the populations of wildlife upon which they depend, and the desire for economic development through exploitation of oil and gas reserves.

Northwest Atlantic

Extending from Hudson Strait and Ungava Bay down to the St Lawrence, this zone includes mostly continental shelf where depths are less than 200 metres, although depths in the Labrador Sea may reach 1,000 metres. In winter, ice formations move along the coast as far as the south coast of Newfoundland. Sea temperatures in this zone may be 20° C higher than in the Arctic due to warm ocean currents. Marine life is generally abundant. A deep trench channels cold oceanic water up the St Lawrence as far as the entrance to the Saguenay Fjord, creating a productive feeding zone for beluga and various baleen whales in the upwelling currents. Some 22 species of whales occur in the zone, as well as six species of seals.

This zone was the site of some of the most controversial international exposure for Canada because of the seal hunt that used to occur annually off these shores. Since the mid-eighteenth century, the harp seal had been the target of hunting, mainly for pelts. Between 1820 and 1860,

Icebergs, such as this one outside the harbour of St John's, are characteristic of the Northwest Atlantic (*Philip Dearden*).

for example, about half a million harp seals were killed every year. In 1831, over 300 ships and 10,000 sealers pursued the hunt; 687,000 pelts were taken. Publicity over the hunt in the early 1980s led to bans on the importation of sealskins into Europe. Subsequently, the hunt all but collapsed (Box 8.5).

Atlantic

This zone includes the highly productive waters of the Grand Banks, Scotian Shelf, St Lawrence Trough, and Bay of Fundy as well as the Northwest Atlantic Basin. Generally ice-free, this zone is a mixing ground for warmer waters moving north from the tropics and the cold water of the Labrador Current. Waters are generally shallow, with large areas less than 150 metres in depth. There is an abundant and diverse population of

BOX 8.5 THE SEAL HUNT

The annual hunt for seals off the east coast was probably Canada's biggest foray on the international stage regarding environmental matters. Famous celebrities, such as French actress Brigitte Bardot, were shown nightly on televisions across the world as they tried to protect seemingly helpless white-coated seal pups from being clubbed to death. Eventually, following bans on the import of sealskins by the US and the European Union, the Canadian government banned the hunt in 1987. However, it now appears that a new hunt will be promoted, ostensibly to help in the recovery of the endangered cod stocks.

Seals eat fish. On that everyone agrees. What is contentious, however, is the number of seals there are and how many fish of various species they are eating. Federal scientists estimate a harp seal population of about 5.2 million consuming some 140,000 tonnes of cod per year, enough to put a considerable dent in the recovery of the fish stock. Widespread disagreement exists concerning all of these figures, however. Some estimates of seal numbers are as low as two million. Federal researchers have indicated that virtually all the cod eaten by harp seals are the smaller, and not commercially fished, Arctic cod.

On 3 February 2003, Fisheries Minister Robert Thibault announced a new multi-year Atlantic Seal Hunt Management Plan that set the total allowable catch of harp seals at 975,000

over the next three years, which will provide income to about 12,000 sealers and their families. This quota is the highest permitted cull since the introduction of quota management in 1971. The management plan prohibits the hunting of harp seal pups (whitecoats).

International environmental groups have suggested that the real reason behind the desire to increase the seal hunt is not related to seal predation on cod, but more related to the need for a political scapegoat in economically depressed areas and the demand for seal penises on the Asian market.

A beater, like this harp seal pup, is one that has moulted its white fur (*International Fund for Animal Welfare* © IFAW, www.ifaw.org).

Thick-billed murres nest along sea cliffs (*Newfoundland and Labrador Parks and Natural Areas/Ned Pratt*).

Northern gannets range over the North Atlantic and are often seen well offshore (*Newfoundland and Labrador Parks and Natural Areas/Ned Pratt*).

marine mammals such as grey, hooded, and harp seals and Atlantic pilot, killer, and northern bottlenose whales. The fisheries of these coasts were at one time among the most abundant on Earth, and a large population of over 300 different seabirds was also supported by the marine bounty. These include the northern gannet, which almost followed the great auk into extinction (see Chapter 11) as a result of hunting before protective laws were invoked in the early part of the twentieth century. Other species include the Atlantic puffin, murres, great and double-crested cormorants, guillemots, great black-backed gulls, and many others. The region also contains Canada's only confirmed coral reef. Scientists on board a DFO research ship in 2003 discovered the reef between Cape Breton Island and Newfoundland. Unfortunately the reef, made up mainly of *Lophelia pertusa*, has already been badly damaged by fishing activities.

Scattered throughout the zone, major gas and oil discoveries have been made. The Hibernia field on the eastern edge of the Grand Banks, some 300 km east-southeast of St John's, was the first field put into production. In 2002 it was joined by the Terra Nova field, with a third, White Rose, coming into production in 2006.

FISHERIES

Canada has been blessed with some of the most productive marine environments in the world, and these environments were critical in sustaining populations of Aboriginal peoples and attracting European attention on both coasts. Unfortunately, the squandering of this rich biological heritage stands as one of the sharpest reminders of our inability to manage ourselves in a way that sustains resources over a long period of time. Since 1992— the year of the cod fishery closure—commercial landings have declined by 19 per cent overall, including cod by 82 per cent, haddock by 32 per cent, turbot by 35 per cent, redfish and rockfish by 69 per cent, and hake by 43 per cent. Bear in mind, too, that these percentages reflect a shifting baseline, as discussed earlier. The number of vessels in Canada's commercial fishing fleets has decreased by 31 per cent, but the average income earned within fleet sectors has doubled. In Atlantic Canada, for instance, the average vessel earned $18,400 in 2002, more than double the average income of $8,700 in 1992. Overall, the catch has doubled in value over the last decade. This is mainly due to the growth in shellfish catches, up by 77 per cent, which have been spurred by the closure of the Atlantic cod fishery.

Case Study: East Coast Fisheries

The marine fishery has been an essential component of the economy and culture of Atlantic Canada for centuries. After 1977, when Canada declared a 200-nautical-mile (370-km) fishing limit off its coasts, cod were the mainstay for more than 50,000 fishers and 60,000 fish-plant workers in Atlantic Canada. In Newfoundland and Labrador alone, about 700 communities depended entirely on the cod fishery, which had a 1991 value to fishers of over $226 million.

However, between July 1992 and December 1993, decisions to cut back on the rate of harvesting resulted in employment losses for 40,000 to 50,000 people in Newfoundland, the Maritime provinces, and Quebec. How could this dramatic collapse of a renewable resource occur in such a relatively short period of time? Why did fisheries scientists fail to anticipate the collapse? Why was action not taken earlier to avoid degradation of the fishery and economic disruption? Is it possible for the fishery to rebound and become a mainstay in the regional economy once again?

Small outport communities in Newfoundland have always relied heavily on harvesting marine products, from seals to fish (*Newfoundland and Labrador Tourism*).

The Nature of the Collapse

Table 8.1 presents a chronology of events and decisions that led to, and followed, the virtual closing of the Atlantic groundfish fishery by the end of 1993 (Figure 8.9). There are four main areas for this fishery in Atlantic Canada: the Scotian Shelf, the Gulf of St Lawrence, the Grand Banks, and the Labrador coast (Figure 8.10). The two areas most affected by the harvesting cutbacks have been the Scotian Shelf, which extends from the mouth of the Bay of Fundy to the northern tip of Cape Breton Island, and the Labrador coast.

These areas supported two different kinds of fishery. The fishery on the Scotian Shelf is readily accessible to the inshore fishers along the coast of Nova Scotia and New Brunswick, and includes a wide mix of species, including cod, haddock, flounder, pollock, hake, herring, redfish, crab, scallop, and lobster. In contrast, the Labrador coast fishery is dominated by the northern cod stock, and extends east of the Labrador coast and north

> *... northern cod off Newfoundland recently declined slightly, although TACs [total allowable catches] were reduced substantially when it was realized that previous exploitation rates had been higher than the target levels. The stock is now increasing as a result of improved recruitment.*
> — *Environment Canada (1991: 7–8)*

and east of Newfoundland. The northern cod traditionally yielded about half of Atlantic Canada's cod catch and one-quarter of all groundfish landings in the region. The northern cod has formed the backbone of the Atlantic fishery. This explains why stock depletion has been such a blow to regional economies, where fishing has provided a significant percentage of the jobs—and almost all jobs in small outport communities.

The northern cod were caught by larger inshore vessels, and especially by offshore draggers or factory trawlers, multi-million dollar boats that drag huge nets across the bottom of the ocean and stay on the fishing grounds for extended periods of time without landing. However, the northern cod migrate to the shores of Newfoundland in the summer, and in consequence also supported an inshore fishery that relied on much smaller boats using traps, hooks, and nets. The inshore fishery has been an important one. Until the late 1950s, the inshore catch was typically higher than 150,000 tonnes. By 1974, as a result of overfishing by boats from more than 20 nations, the inshore catch had fallen to 35,000 tonnes. After Canada declared its exclusive fishing zone in 1977 and banned fishing by foreign draggers in that area, the inshore catch increased. It peaked at 115,000 tonnes in 1982, but by 1986 the catch had fallen to 68,000 tonnes and inshore fishers were complaining that the fish caught

Table 8.1	The Path to the Collapse of the Atlantic Groundfish Fishery

Date	Event
1968–78	Following a peak high of 810,000 tonnes caught in 1968, the catch drops to 137,000 tonnes in 1978.
1977	Canada declares a 200-nautical-mile (370-km) exclusive fishing zone. The declaration requires foreign fleets to fish outside of the boundary. Stocks begin to rebuild.
1982	A task force to study the difficulties of the east coast fishery concludes that northern cod is a bright spot in an otherwise discouraging situation. The total catch of northern cod is expected to reach 380,000 tonnes by 1987.
1986	Catches of northern cod continue to increase, reaching 252,000 tonnes, almost twice what they were in 1978. Scientists are mainly optimistic about the stock, but inshore fishers in Newfoundland express concern about what they consider to be depletion of the stock.
1989	Based on new scientific advice, the federal Minister of Fisheries reduces the total allowable catch (TAC) for northern cod to 235,000 tonnes, down from 266,000 tonnes in 1988.
1990	
January	The TAC for northern cod is reduced further to 199,000 tonnes.
May	The federal government announces a $584 million, five-year program to help communities affected by the falling fish quotas.
October	A task force report recommends a further reduction in the TAC for northern cod to between 100,000 and 150,000 tonnes for 1991. The federal government decides the 1991 quota will be 185,000 tonnes.
1991	Northern cod catches drop sharply. Scientists agree that the stock is in serious trouble, but they do not know why.
1992	
February	The federal Fisheries Minister reduces the northern cod TAC to 120,000 tonnes, and curtails offshore fishing for six months.
July	The Minister of Fisheries announces a moratorium on northern cod fishing until May 1994, and explains that Ottawa will provide $500 million (later to become $772 million) to compensate the 20,000 fishers and plant workers expected to lose their jobs because of the moratorium.
	The federal government announces another $140 million to assist Atlantic fishers until 15 May 1994, at which time a new cod compensation program would take effect. Thus, the total cost in assistance from 1992 to May 1994 is $912 million.
1993	
July	The Fisheries Resource Conservation Council (FRCC), an advisory group for the federal minister, releases a report blaming the fisheries crisis on overfishing, changes in migration, harsh climatic conditions, poor feeding conditions, and seals and other predators. A subsequent report recommends cuts affecting another 9,000 to 12,000 workers and all but closing the cod fishery. The government bans cod fishing in five more areas and sharply reduces quotas for other valuable species. The result is a total loss of about 35,000–40,000 fisher and fish-plant jobs in Atlantic Canada since the closures began in July 1992.
November	The FRCC urges that the two-year fishing ban on northern cod off Newfoundland also include the Gulf of St Lawrence and areas off Nova Scotia.
December	The Task Force on Incomes and Adjustment in its Atlantic Fishery Report calls for halving the number of fishery workers by buying out fisher licences and fish-processing plants, offering early retirement packages, and retraining people. The federal Fisheries Minister closes all but one Atlantic cod fishery and curtails catches of other species. Another 5,000 people are estimated to have lost their jobs.
1994	
January	The government announces a ban in most of Newfoundland on catching cod for personal use because the fish stocks are so depleted that even this small hook-and-line fishery threatens them. The government suggests that this ban is necessary because too many Newfoundlanders were catching fish for private sale, often to restaurants.
February	The Northwest Atlantic Fisheries Organization (NAFO) votes to stop fishing for one year the declining stocks that congregate on the southeast tip or 'tail' of the Grand Banks, which lies beyond the 200-mile limit. Of the 11 NAFO members at the meeting, eight approve the ban and three (Denmark, Norway, European Union) abstain. The three members that abstain argue that there are enough young cod to allow fishing to continue.

Date	Event
February	The new Liberal federal government announces that Atlantic cod fishers will receive $1.7 billion over five years.
April	Canadian fisheries officials board and impound the *Kristina Logos*, a freezer trawler registered in Panama, carrying an all-Portuguese crew, but owned in Nova Scotia, for fishing in restricted waters beyond the 200-mile limit on the southern Grand Banks. Technically, Canada has no authority beyond the 200-mile limit, but the Fisheries Minister declares that Canada is prepared to seize ships operating just outside the limit in order to protect endangered species. The *Kristina Logos* had 107 tonnes of fish onboard, more than half of which was the protected cod and flounder.
	The federal government announces a new five-year, $1.9 billion aid package for Atlantic fishers and plant workers left unemployed due to closure of the groundfish fishery. The new aid package is designed to help some 30,000 people in Newfoundland, the Maritimes, and Quebec. The Fisheries Minister also announces that some people will have to leave the fisheries.
May	A bill is passed giving fisheries officers legal authority to arrest ships fishing beyond the 200-mile limit in areas in which agreements had been reached to curtail fishing. Ships fishing in contravention of the agreement can be boarded and towed to St John's, where the captain and owners can face prosecution and penalties, including fines up to $500,000 and seizure of the vessel.
July	Fisheries Minister Brian Tobin announces an easing of the fishery ban for personal use. Between 26 August and 30 September, Newfoundlanders are allowed to catch 10 groundfish a day on two days of every week.
July	HMCS *Fraser* and a fisheries patrol boat intercept two American boats suspected of fishing illegally about 15 km outside the 200-mile limit. The boats are escorted to St John's. The captains of the two vessels are charged with fishing without a Canadian licence, which could result in a maximum fine of $750,000.
August	Two other American fishing vessels are seized in separate actions for allegedly fishing in Canadian waters. The Fisheries Minister announces that Canada will continue to act unilaterally to protect depleted fish stocks until a strong international agreement is developed. Canada advocates a binding convention that will allow regional fisheries organizations to set total catch limits, create mechanisms to resolve disputes, and enforce rules. Japan and European Union countries resist a binding agreement, preferring a less stringent UN resolution.
October	The Fisheries Minister announces that $300 million will be allocated to allow the Department of Fisheries and Oceans (DFO) to purchase licences from fishers who caught the endangered cod, flatfish, and haddock stocks. DFO will also pay a pension for fishers retiring prior to age 65. The intent is to reduce the more than 20,000 Atlantic fishers who traditionally depend on the fish for their living. Even if the stocks recover, less than half of the present number of fishers can expect to make a living from the sea in the future.
1995	Some 3,500 former fishers and plant workers in Atlantic Canada, mostly in Newfoundland, are cut off from the federal aid package as eligibility rules are tightened.
1997	Some 5,000 fishers, mostly from Newfoundland but some from Quebec, become involved in a limited test fishery off southern Newfoundland and in the northern Gulf of St Lawrence. This is the first commercial fishery allowed since the closure of the cod fishery. The catch limit is set at 16,000 tonnes, and the purpose of the fishery is to allow officials to obtain more information regarding the state of cod stocks.
1998	The Atlantic Groundfish Strategy (TAGS)—a five-year income supplement program worth $1.9 billion— ends, leaving fisherman increasingly impatient to resume fishing practices.
2003	
February	Fisheries and Oceans Minister Robert Thibault sharply increases the number of seals that can be culled to 975,000 seals over the next three years, with the maximum catch in any one year set at 350,000 animals.
April	Almost 11 years after DFO imposes a moratorium on cod fishing, Fisheries Minister Thibault announces the outright closure of what remains of the cod fishery in Newfoundland, the Maritime provinces, and Quebec.
May	Atlantic cod is officially listed as endangered. COSEWIC estimates a 97 per cent decline in cod off the northeast coast of Newfoundland and Labrador over the last 30 years.

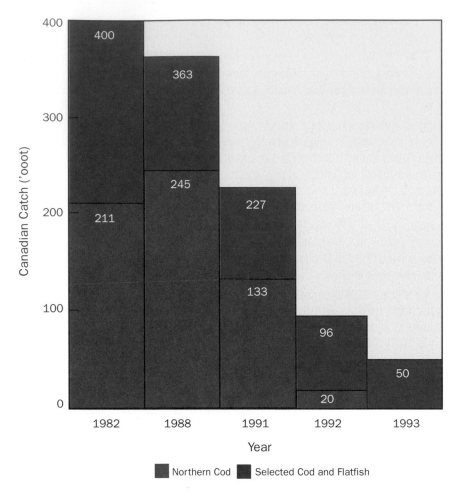

Canadian Catch ('ooot)

400

400

363

300

245

211

200

227

133

100

96

50

20

0

1982   1988   1991   1992   1993

Year

■ Northern Cod   ■ Selected Cod and Flatfish

Figure 8.9 Collapse of Newfoundland's groundfish base: decline in catch from major stocks. Note: Canadian catch of six major groundfish stock: 2J3KL cod, 4RS, 3Pn cod, 3Ps cod, 3NO cod, 3LNO American plaice, 3LNO yellowtail flounder; 1993 catch is an estimate based on catch (31,527t) as of 15 September 1993. SOURCE: Task Force on Incomes and Adjustments (1993: 30).

commercial extinction' (Steele et al., 1992: 37). The politicians and bureaucrats running Canada's Atlantic fisheries created opportunities for overfishing through the provision of inappropriate incentives for processing plants and lucrative subsidies (unemployment insurance) to all fishers and plant workers involved in the fishery.

Foreign Overfishing

Once Canada established the 200-mile fishing limit in 1977, foreign fleets were required to fish outside that boundary or to fish inside the boundary for that portion of the domestic quota that was not taken by Canadian vessels. Foreign fishing fleets were monitored by the Northwest Atlantic Fisheries Organization (NAFO), and during the 1970s and early 1980s foreign fleets for the most part followed the rules for harvesting. However, in 1986 Spain and Portugal entered the European Community (EC) and that year the EC unilaterally estab-

were very small. Local fishers had identified the first signs of serious problems. Unfortunately, the models used by the fishery scientists indicated that stocks were still abundant, so these early warnings were not heeded.

Some Reasons for the Collapse

At the time of the collapse a single cause to account for the depletion of the groundfish stocks seemed unlikely. It was suggested that changing environmental conditions creating colder, less hospitable water temperatures had driven the cod away, while increasing seal populations had devoured entire stocks of both cod and capelin, the favorite food source for cod. However, ongoing research has made it abundantly clear that such environmental factors played only minor roles in the disappearance of the fish. With study after study, it has become increasingly indisputable that 'the northern cod stock had been overfished to

lished quotas considerably higher than those set by NAFO (Figure 8.11). Furthermore, the EC boats harvested fish well beyond the EC limits. The EC then raised the quota the following year; the quota was again exceeded by the actual catch. In 1988, just half of the EC target was achieved, even though it was 4.5 times higher than the target recommended by NAFO. The EC, known now as the European Union (EU), later rejected a NAFO northern cod moratorium. In 1993, however, the EU finally accepted all NAFO quotas, after having set its own quotas at a much higher level since the mid-1980s.

Thus, strong evidence exists that foreign vessels, especially those from Spain and Portugal, were overfishing at least during the mid- and late 1980s. Since cod migrate towards the coast in summer and then move offshore in winter to spawn in deeper waters, the fish are more vulnerable to foreign fishing.

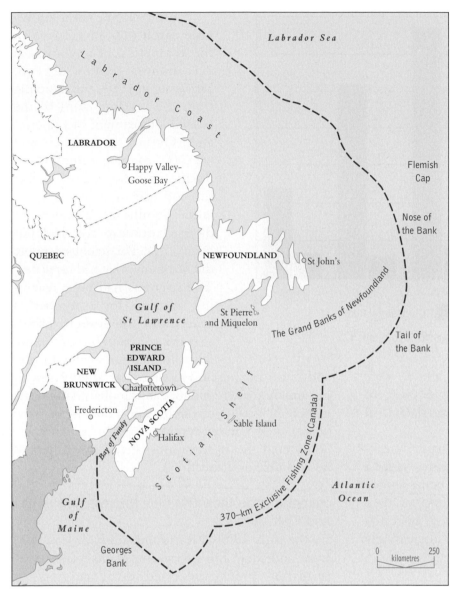

Figure 8.10 Major fishing areas in Atlantic Canada. SOURCE: Cameron (1990: 30).

another 30,000 to 50,000 tonnes each year. Such harvesting did not appear to affect adversely the then estimated breeding stock of 1.6 million tonnes in the North Atlantic.

In the mid-1950s, the introduction of large offshore trawlers that operated year-round in the North Atlantic was a significant change to this pattern. Initially, catches were very high, but the spawning stocks were placed under great pressure because cod require seven years to reach maturity. For example, in the 1970s yields reached a high of 800,000 tonnes per year before they started to drop. Until 1977, foreign trawlers did most of the offshore fishing. Following the establishment of the 200-mile limit, the Canadian offshore fleet expanded and Canadian-based offshore trawlers became the main harvesters of northern cod. By the time the moratorium was placed on the fishery in the summer of 1992, Newfoundland was the base for some 55 large and 30 medium-sized offshore trawlers. Thus, Canadian offshore draggers, operating on a year-round basis, placed considerable pressure on groundfish stocks.

Domestic Overfishing

Despite the pressure placed on the stocks by foreign fishing vessels, most of the principal fishing grounds have been under Canadian control since the 200-mile limit was created in 1977. Two fisheries—both inshore and offshore—must be managed, which has been, and continues to be, a challenge.

For hundreds of years, the inshore fishery consisted of many fishers (particularly from Newfoundland) relying on small wooden boats, lines, traps, and nets to catch cod during the spring and summer months when the cod move close to shore. Until the mid-1950s, the inshore fishery, combined with limited offshore fishing by Canadian boats, resulted in annual landings of 200,000 tonnes or more. Foreign fishers were harvesting

Critical in this regard are the ecology and behaviour of the northern cod. Initially, the harvesters caught a mix of ages and sizes of fish. However, market demand and net mesh sizes led to a focus on larger, older fish. Cod swim in groups or schools of similar ages primarily because the larger cod will eat the smaller and younger cod. The emphasis on larger fish had two consequences. First, the northern cod normally do not begin to spawn until they reach seven years of age. Second, older fish produce more eggs. As the larger fish became scarce, the fishery then concentrated on fish in the five-to-seven-year age range. The result was that, by the early 1990s, most of the older fish had been overharvested and attention

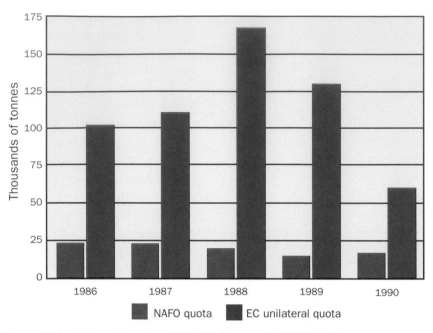

Figure 8.11 NAFO and EC quotas, 1986–90. Source: Hall (1990: 47).

Until 1991, the **total allowable catch (TAC)** was based on the assumption of a biomass of 1.1 million tonnes of cod. However, in 1991, the sampling from the research ships of the federal Department of Fisheries and Oceans indicated only 600,000 tonnes. Sampling in distant areas did not reveal that the cod had migrated to other areas. Significant numbers of diseased or dead fish had not been found. The scientists simply did not know what had happened. The scientists had been receiving warnings for a number of years from the inshore fishers that the fish being caught were fewer in number, smaller in size, and lower in weight. However, the scientists, who put much greater credence in the quantitative data gathered by DFO research vessels and from what were likely under-reported offshore landings, considered the observations of inshore fishers as less credible and anecdotal.

had shifted to pre-adolescent fish, which were being caught before they were old enough to spawn. The consequence was a dramatic decline in the fish stock.

A further complication was that many domestic fishers overharvested. Some estimates suggest that up to 50 per cent more fish were being landed than were being reported (Cameron, 1990). Often, this cheating was driven by a need to catch enough fish to pay for the costs of increasingly sophisticated vessels.

Imperfect Science and Management

Fishery scientists did not anticipate the collapse of the Atlantic fishery, especially the northern cod stocks. One reason is that sampling procedures do not provide sufficient ecological information about fish stocks. An American scientist outlined the difficulties by drawing an analogy with estimating the number of cattle on a ranch. If you prepared such an estimate by dragging a large bag hung from a helicopter across the ranch at night, and then counted how many cows were in the bag by morning, you would be making your estimate in the same way that fishery scientists make their estimates about fish stocks. Few people would put much weight on an estimate of cattle derived from such a sampling method, and yet that is basically the type of information from which fishery scientists work.

Inappropriate Incentives for Processing Plants and Fish Workers

By the early 1990s, Newfoundland had about 100 large and small fish processing plants, some two-thirds of which processed northern cod. In Atlantic Canada as a whole, the number of plants increased from about 500 in 1977 to nearly 900 in 1988, and employment grew from approximately 25,000 full-time jobs to about 33,000. The provincial

The state of our ignorance is appalling. We know almost nothing of value with respect to behaviour of fish. We don't even know if there's one northern cod stock, or many, or how they might be distinguished. We don't know anything about migration patterns or their causes, or feeding habits, or relationships in the food chain. I could go on listing what we don't know…. Our technology has outstripped our science. We have under-estimated our own capacity to find, to pursue, and to kill.

— Leslie Harris, Chairman of the Northern Cod Review Panel, 1990, quoted in Cameron (1990: 35, 29)

governments, which license on-shore fish-processing plants, provided incentives for creation of new processing plants as a way to create new jobs for small communities. This placed political pressure on DFO to keep increasing the total allowable catch.

Another incentive for people to enter or stay in the fishing industry was the federal unemployment insurance program. After working for 10 weeks, fish-plant workers were permitted to collect unemployment insurance for the other 42 weeks of the year. This arrangement resulted in several individuals in a community sharing one job but all qualifying for separate benefits. For fishers, the unemployment benefits were based on the sale value of fish caught during the May–November season (26 weeks), which created the potential to collect unemployment benefits for the other 26 weeks of the year.

This program, which was intended to provide a social safety net, encouraged more people to become involved in the Atlantic fishery than could be justified economically. There was little incentive to consider other types of work, and the program also helped to reinforce an outlook in which little value was placed on education. By the time the fishery was closed, 50 per cent of Newfoundland's 19-year-olds were already on unemployment insurance, and 80 per cent of the fishery workers did not have a high school diploma. Thus, the fishery involved more people than realistically could be supported over the long term, yet unemployment insurance programs provided little incentive for individuals to consider alternatives. This made the trauma of the 1992 moratorium even greater than it might otherwise have been.

Changing Environmental Conditions

One theory for the depletion of groundfish in the North Atlantic fishery is based on the idea of environmental change. Records show that in 1991 the ocean temperatures off Newfoundland were the coldest ever measured. It has furthermore been

It is difficult for urban dwellers to imagine the close relationship that built up over the centuries between the people in the outports of Newfoundland and the sea. Virtually every family would be involved in some way with fishing. When the fish were exposed to such fishing pressure that they could no longer be caught in any numbers, it was not just the economy that suffered but a whole way of life (*Tourism Newfoundland and Labrador/W. Sturge*).

suggested that there had been several years of very cold water temperatures before that. The water warmed slightly in 1992, then cooled again in 1993. It is possible that colder waters, combined with overfishing from the previous two decades, have been preventing or inhibiting the shrunken stocks from regenerating. However, since relatively little is known about the migratory patterns of the northern cod, it is difficult to determine what the specific implications of changing water temperatures may have been. Furthermore, water temperatures began to heat up and stabilize around 1998, restoring a theoretically favourable habitat for groundfish once again. But despite this warmer trend, there has not been any significant recovery observed within the affected groundfish stocks.

Predators

At the time of the fisheries collapse it was also popular to blame predation. Seals, in particular, were identified because of their 'voracious appetites' and their growing numbers due to the closure of the

> *We reject hypotheses that attribute the collapse of the northern cod to environmental change.... We conclude that the collapse of the northern cod can be attributed solely to overexploitation [by humans]....*
> — *Hutchings and Myers (1994)*

seal hunt in the early 1980s (Box 8.5). There is no scientific evidence to support this view.

However, there is some evidence to suggest that seal predation may be a factor in the slow recovery of the east coast groundfish stocks. The seal population has more than doubled in the last three decades and although cod represents only a small percentage of seals' diet, they are consuming more northern cod than fishermen are catching. In fact, the total allowable catch for northern cod in 1999 was only 9,000 tonnes, which is less than 20 per cent of corresponding predation by harp seals in that year (Doubleday, 2000)

The science of seals and cod is inconclusive and will remain inconclusive for the foreseeable future. Seals are a significant source of mortality for northern cod in the Northwest Atlantic and the Gulf of St Lawrence. Nevertheless, reducing the abundance of harp seals may or may not lead to more rapid recovery of depleted cod stocks. Even if a reduced seal population resulted in an increased number of fish in the ocean, there are other predators in marine ecosystems. Any increase in the size of a commercially important fish stock could well be eaten by these other predators before being caught by fishers. Furthermore, seals eat predators of commercially important fish, and so fewer seals could actually mean fewer fish for fishers.

Fishing Down the Web

Fishing down the food chain has been discussed earlier and is exemplified in Atlantic Canada with the replacement of cod fisheries by shrimp and crab fisheries. Recent observations made of the poor physiological condition of many predator fish in the area also suggest an overall lack of prey, forcing predator species such as the Atlantic cod to the same alternative as human fishers, 'fishing down the web'. After the cod were fished down, fishing pressure increased on shrimp. 'Cod feed on shrimp', says Daniel Pauly (1998) from the University of British Columbia Fisheries Centre. 'If you remove the shrimp, how will the cod ever recover?'

Lessons

The collapse of the Atlantic groundfish fishery highlights how some contemporary resource management practices may encourage resource liquidation. It illustrates, too, that fisheries management requires scientific understanding of the biophysical resource system, a greater appreciation of traditional or local ecological knowledge, and parallel understandings of the history, culture, economy, and politics of the region, as well as federal and provincial fisheries and regional development policies. The Atlantic fishery also provides an excellent example of how inexact science often is and the extent to which complexity and uncertainty are paramount. It demonstrates conflict among different values and interests (Chapter 5), and the manner in which conditions can change dramatically over a relatively short time period.

The situation is readily comparable with the framework introduced in Chapter 1. Not only were our simplified models of the complexity of the biophysical system inadequate for a proper understanding of the east coast fisheries, but our resulting attempts to assess the status of the system were inadequate and there was a lack of clarity regarding societal expectations and management directions. Some of the approaches suggested in Chapter 6, relating to identification of stakeholders, resolving conflict, and taking more ecosystem-based and adaptive approaches, are applicable to the resolution of these problems.

ABORIGINAL USE OF MARINE RESOURCES

One of the most challenging aspects of fisheries management in Canada is allocation of catch, especially when catches are declining. This is particularly difficult when allocation involves Aboriginal communities. There is a patchwork of treaties with First Nations in different regions of Canada, and the rights to sustenance from fishing were often written into these treaties. However, it has never been clear which regulations First Nations should be subjected to and how broad a range sustenance might cover. Over the last decade, many important court cases have helped to clarify some of these issues. Nonetheless, high-profile conflicts still occur on both coasts, as detailed in the two examples below.

In the fall of 1999, a violent and complex dispute erupted between Native and non-Native fishermen in Miramichi Bay in northeastern New Brunswick. Both groups immediately hit the water determined to do whatever they deemed necessary to protect their livelihoods. Lobster traps were cut and damaged, threats were exchanged, boats were rammed, and multiple shots were fired.

> *The disaster in the cod fishery is now worse than anyone expected.... It may be a generation before we see a recovery of the cod. That a five-hundred-year-old industry could be destroyed in fifteen years by a bureaucracy is a tragedy of epic proportions.*
>
> — *Ransom Myers, quoted in Harris (1998: 332–3)*

The crux of the dispute lay in the Supreme Court's decision of 17 September 1999 in the case of a Nova Scotia Mi'kmaq, Donald Marshall Jr, who had caught and sold eels out of season and claimed protection under a 1760 treaty. The Court's ruling upheld the 1760 treaty, which effectively gave Mi'kmaq, Maliseet, and Passamaquoddy bands the right to earn a 'moderate livelihood' from year-round fishing, hunting, and gathering. As news of this decision spread, it spurred an immediate reaction among the Natives of the Burnt Church band of Miramichi Bay to resume catching and selling species like lobster, even though the season was formally over. Within days the local Native bands had over 4,000 traps back in the water. Not surprisingly, this infuriated the non-Native fishers who were confined to the short summer season and who believed that this continued harvest would lead to the destruction of the lobster fishery.

Despite the rising tension and conflict between the two groups, the Department of Fisheries and Oceans was hesitant to interfere, stating that Natives now had a right to fish that had been denied for over two centuries. However, DFO did promise a peaceful solution to resolve the ambiguity of the *Marshall* decision. The *Marshall* case specifically indicated that the federal government still retained the right to regulate the fishery, but was equally firm that Ottawa's authority to regulate treaty rights was limited to those actions that could be 'justified'. This led to an examination of the term 'justification', and more specifically, whether or not the DFO's limit on the number of lobster traps and length of fishing season was reasonable and 'justifiable' according to the rights of the Mi'kmaq in their 1760 treaty.

After several years of research into the affected lobster stocks, commissioned in order to 'justify' the decision scientifically, Robert Thibault, Minister of Fisheries and Oceans, announced agreement with the Burnt Church band. The $20-million agreement included enhanced commercial fishery access for Native fishermen, including additional lobster licences and extra boats and gear. However, a quota was set for the Native fall fishery of 25,000 pounds of lobster for food and 5,000 pounds of lobster for ceremonial use. Furthermore, the fishery would be limited to six weeks or until the quota was filled, and the sale of lobster would be strictly prohibited at all times.

Since the federal decision was made, there has been a significant decrease in violent conflicts on the water and an improved relationship has emerged among all parties. The non-Native lobster fishers and many fisheries scientists still maintain that any kind of fall fishery needs to end because, they believe, it exploits the lobster at their most vulnerable stage. Yet, they recognize that while the negotiated agreement does not represent a definitive solution for the conflict, it is a vital step towards the government providing a more appropriate degree of oversight and accountability in the management of the lobster fishery.

On the west coast, Native communities have unemployment rates of between 65 and 85 per cent and the fishery plays a critical role in the communities. There is an Aboriginal right, established by the courts, to fish 24 hours a day, seven days a week, wherever Aboriginal people wish to, for food and ceremonial purposes. In the past this right was sometimes abused, with fish being caught for commercial use. This prompted DFO to establish a special Natives-only commercial salmon fishery for some areas. In 2003, the BC Appeal Court, following complaints by non-Native fishers, struck down the Natives-only commercial salmon fishery. The judge said the opening amounted to 'legislated racial discrimination' and was against the Charter of Rights. DFO then cancelled the program, but Native fishers have vowed that they will continue to catch and sell the salmon as they always have. This was the second court setback for the Natives-only fishery. A few months earlier another judge had given absolute discharges to 40 non-Native fishers who had cast their nets during a Natives-only fishery on another part of Vancouver Island.

In many ways, the story of the BC coastal Aboriginal cultures is a story of the sea in general, and of salmon in particular (Box 8.6). Nowhere else on Earth did hunter-gatherer societies develop such complex social structures, rigid hierarchies, and dense populations in permanent winter villages. From these villages, the people developed complex and effective hunting practices for whales, sea lions, seals, sharks, tuna, wolf eels, sole, oolichan, greenlings, herring, halibut, crabs, clams, mussels, skate, sturgeon, and, above all, salmon. The salmon fishery was managed effectively; no stocks crashed. And the salmon was venerated through myth and legend among the coastal peoples.

Examples, such as the two discussed above, will continue to arise as Canada strives to achieve equitable solutions to fish resource allocation problems involving First Nation peoples. The clocks cannot be rolled back to pre-treaty times, yet there must be some recognition of the central life that fish and fishing played in the societies of many First Nations people in Canada.

HYDROCARBON DEVELOPMENT

Energy consumption has reached new heights, particularly in Western countries, and many countries have begun to investigate the potential of offshore areas to meet their needs for petroleum resources. In Canada, the offshore oil and gas industry has been the focus of heated debates, with various stakeholder groups drawing attention to the costs and benefits of exploratory and extractive activities. The industry could certainly help to secure local access to energy resources and potentially provide valuable jobs for coastal communities, now faced with the decline of traditional industries such as fisheries and forestry. These benefits, however, must be weighed against envi-

> *We're going to continue, we're going to continue to exercise our fishing rights, continue to barter and sell and trade our salmon; it's a way of life for us.*
> — *Bob Hall, Sto:lo Nation fisheries spokesman, in* Victoria Times Colonist, *30 July 2003*

ronmental concerns, specifically the discharge of organic and inorganic wastes into marine waters and the potential effects of these wastes on flora and fauna.

Offshore oil and gas projects are now at various stages of development along Canada's east, north, and west coasts. Canada's earliest offshore venture—the Cohasset-Panuke oil field located just south of Newfoundland—produced petroleum between 1992 and 1999. In the last decade, new production centres have been established in the White Rose, Hibernia, and Terra Nova oil fields off the Grand Banks of Newfoundland, as well as near Sable Island, off Nova Scotia. In 2002, the oil industry contributed $1.3 billion to Newfoundland's economy and directly employed nearly 3,000 people. Furthermore, significant oil and gas reserves have been discovered in the Beaufort Sea, although their exploitation will likely be on hold until the establishment of a pipeline to connect supply to the market.

Although interest has certainly been mounting, British Columbia has yet to develop a viable offshore oil and gas industry. From the 1950s to the 1970s, the search for oil and gas resources focused on the area between the Queen Charlotte Islands and Barkley Sound on the west coast of Vancouver Island. Disputes over ownership of the seabed and mounting concerns about environmental issues soon led to the suspension of all drilling and exploration. A federal moratorium on all new and already existing exploration permits for Canada's west coast was officially issued in 1972. During the next 30 years, the federal and provincial governments continued to negotiate jurisdictional issues and established a formal environmental review process to assess

Native fishers use dip nets to intercept salmon heading up the Fraser River (*Philip Dearden*).

BOX 8.6 SALMON: THE STORIES THEY TELL

From the shores of Japan to almost 2,500 kilometres up the Yukon River, a tangible thread exists—the Pacific salmon. Every year millions of salmon make their way back from the other side of the Pacific Ocean to the streams of their birth. The five species of Pacific salmon—chum, coho, chinook, pink, and sockeye—are *anadromous,* that is, they spend part of their lives in fresh water and part in salt water. They depend on a wide range of conditions that link the mountains to the seas: the amount of snowpack to feed the streams, the lack of floods to wash away spawning gravel, unpolluted rivers and estuaries, the right temperature for entry into the marine environment, avoidance of predators, and avoidance of fishing nets. If any of these myriad factors go awry, then higher mortality rates can drastically reduce the numbers of fish returning to spawn in subsequent years. These factors are the links in a chain reflecting the limiting factor discussed in Chapter 2. It also means that salmon are good indicators of the overall health of our environment and our resource management practices.

What have these *indicator* species been telling us? The story is not a good one. Salmon in their millions sustained populations of coastal Aboriginal peoples. Early descriptions of the Fraser River by explorers talk about a river that could be crossed on the backs of the salmon. But early logging and mining practices soon made considerable dents in these numbers, along with wasteful fishing practices. Habitat destruction and overfishing led to the virtual closing of the fishery in many areas by the 1990s. Scientists estimate that the salmon biomass has been diminished by half from precommercial fishing levels. Some stocks have been declared extinct while others are now on the official endangered species list in both Canada and the US. The value of salmon landings has fallen 73 per cent since 1992.

The salmon have also been trying to tell us something else that scientists are only now starting to realize. Salmon spend anywhere between two and seven years in the ocean environment before returning to spawn and die. When they die, the nutrients they have collected over this sojourn do not disappear (law of conservation of matter), but are released into the surrounding environment. The salmon provide food and nourishment not only for the plankton and insects that feed the next generation of fish and propel their journey to the sea, but also the terrestrial riverine environment. When the fish die they provide a feast for many other species, including eagles, raccoons, and bears. As the fish are digested their nutrients are distributed throughout the forest as feces, providing the rich fertilizer upon which some of the tallest trees in the world depend. Professor Tom Reimchen from the University of Victoria estimates that BC's bears could be transferring 60 million kilograms of salmon tissue into coastal forests each year. When we take away the fish, we take away this fertilizer, and forest growth suffers. Scientists have found that up to 40 per cent of the nitrates in the old-growth forests in coastal BC originate from marine environments. Rivers with barriers to salmon, such as waterfalls, have noticeably poorer forest growth.

Unfortunately, scientists have now discovered another aspect of this remarkable linkage. Not only do salmon collect nutrients from the ocean, they also collect pollutants and concentrate them within their bodies (Chapter 9). When the fish die in their millions after spawning, scientists in Alaska have found a sevenfold increase in the concentration of PCBs in these remote, pristine, freshwater lakes. Lakes with the highest numbers of spawning salmon also have the highest concentrations of PCBs. The salmon have an eloquent and tragic story to tell about how we are treating their environment. All we have to do is listen.

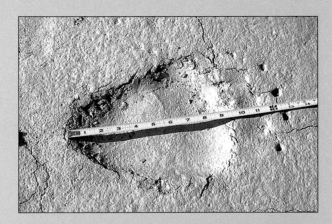

A grizzly bear track from the Tatshenshini Valley in northern British Columbia. Research now shows that the nutrients bears carry back from the ocean and the rivers are key in promoting rich forest growth along many coastal streams (*Philip Dearden*).

the potential effects of offshore exploration. In 2001, a provincially appointed scientific panel reported that new advances in technology no longer justified a full moratorium. On the heels of this finding, the BC government, led by Premier Gordon Campbell, announced that a thriving offshore petroleum industry would be developed by 2010. Still, industrial support for the exploration phase of the project has remained low, with many industry analysts remaining skeptical about the potential for petroleum production in BC's offshore waters. Despite such reservations, both the provincial government and industry representatives have continued to push for the moratorium on exploration to be lifted. In the meantime, Canada's east coast remains the most promising location for oil and gas extraction.

POLLUTION

Approximately 23,000 existing substances currently approved for use in Canada are being reviewed to determine if they are toxic or capable of becoming toxic. As of 4 January 2003, 68 different substances were defined as toxic by the Canadian Environmental Protection Act (CEPA). A substance is considered to be toxic if it enters the environment in a quantity that has, or may have, a harmful effect on the environment or human health.

The main sources of marine toxic pollution in Canada come from the deposition of airborne pollutants from fossil fuel combustion, agricultural runoff, inadequately treated sewage, and by-products or waste materials from refining processes (e.g., the effluent from pulp and paper mills). Some chemicals, known as POPs (persistent organic pollutants), including PCBs (polychlorinated biphenyls) and DDE (the breakdown product of the now-banned pesticide, DDT, discussed in Chapter 10), can take decades or even centuries to degrade, and tend to bioaccumulate in the fatty tissues of organisms over time. The concentrated contaminants are then passed through the food chain (biomagnification—see Chapter 10) and can reach very high concentrations in the tissues of animals in the top trophic levels (such as polar bears, whales, and humans). A recent study by the federal Department of Fisheries and Oceans concluded that the killer whales of the Strait of Georgia are among the most contaminated mammals on the planet

(Ross et al., 2000). This issue is further exacerbated by the long-range, polar transport of toxins in the atmosphere so that many Aboriginal people in Canada's North have bioaccumulated extraordinarily high levels of toxins in their bodies due to their dietary reliance on marine mammals (Dewailly et al., 1993).

Some of these persistent toxins (including dioxins and furans from pulp and paper effluent, PCBs, and pesticide residues) are also *endocrine disrupters*, which have been linked to severe growth, development, and reproductive problems in wildlife populations, as discussed earlier in this chapter. However, even toxic substances that are not persistent or bioaccumulative (such as benzene) can have significant harmful effects on the health of the marine environment.

In recent years, Canada has made progress in reducing emissions from a number of marine toxic pollution sources. Agricultural industries have been forced to develop and use more environmentally friendly pesticides and fertilizers, and to increase conservation tillage to reduce runoff pollution. There has also been a significant decrease in the amount of toxic pollutants coming from other industries, such as pulp and paper, petroleum refinement, and aluminum. For example, discharges of dioxins and furans from Canada's forest product industry have decreased by 99 per cent since 1988 (Environment Canada, 2003b) due to the implementation of new regulations under federal and provincial legislation.

Although the concentration of toxins in the Canadian environment has recently declined, they have certainly not disappeared. Existing toxic residues will be recycled and dispersed throughout ecosystems for some time. In addition, toxic substances from sources outside Canada continue to enter our ecosystems through oceanic and atmospheric transport. One study of contaminant concentrations in the eggs of double-crested cormorants shows a significant decrease (Figure 8.12) in the last 30 years (Canadian Wildlife Service, 2003). However, the lack of further declines, despite the banning of these chemicals in Canada, leads scientists to believe that this may be the result of long-range transport of POPs used outside of Canada, as well as the slow release of contaminant residues from bottom sediments and dump facilities.

In Canada, federal law and policy have

Agriculture and aquaculture, shown here in Prince Edward Island, do not always make good companions due to the pollution coming from the land (*Philip Dearden*).

declared the management and reduction of toxic substances in the environment 'a matter of national priority'. Under the Canadian Environmental Protection Act 1999, the Minister of Environment is mandated to virtually eliminate the production of POPs and manage the discharge of other pollutants and wastes into the environment. Canada has also been active on an international scale and was the first nation to ratify an international treaty, known as the Stockholm Convention on Persistent Organic Pollutants, which aims to identify problematic substances for which comprehensive global action is required.

Organic pollution is also of concern in some areas. The overall process of organic decay has been outlined in Chapter 4. Given the immense volume of water in the ocean it might be thought that, by and large, an infinite adsorption capacity exists for receiving and breaking down organic wastes. However, where there are dense populations and waste is deposited in a site with low adsorptive capacity, even the ocean can become polluted. On the east coast, for example, 52 per cent of all towns and cities lack any sewage treatment. The problem became quite obvious in Halifax, where sewage has been deposited directly into the harbour since 1749 and toilet paper, tampon applicators, and used condoms have become a familiar sight. In late 2003, Halifax made a historic move after 30 years of

delay, announcing it would install a network of treatment plants for the 181 million litres of raw sewage pumped out every day. The total cost is estimated to be $310 million, of which the city has less than half that amount at the moment.

At the other end of the country, Victoria has taken a different approach. The city pumps out 100 million litres of raw sewage into Juan de Fuca Strait every day. The Victoria situation is rather different from Halifax due to the large volume of fast-moving water in the Strait that breaks down and disperses the sewage very quickly. Marine biologists for years have monitored the situation and been able to detect virtually no negative impacts. More sophisticated monitoring devices have now been installed with specific limits established that, if passed, would compel Victoria to treat its sewage before disposal. Most biologists are satisfied that the dilution of organic waste is acceptable. There is, however, much greater concern over the non-organic wastes that are disposed of, illegally, through the sewage system. Over a two-year period, estimates suggest that these include 2,920 kilograms of oil and grease, 17,400 kg of zinc, 9,000 kg of copper, 2,560 kg of cyanide and 1,360 kg of lead. The regional government has introduced educational programs on waste disposal, since the most effective way to deal with these substances is to prevent their entry into the sewage system rather than trying to treat them once they are there.

SOME CANADIAN RESPONSES

Canada's Ocean Strategy

The Oceans Act was passed in 1998 as an attempt to provide a more comprehensive and co-ordinated approach to marine resource management in Canada. One of the main requirements of the Act was for the Minister of Fisheries and Oceans to develop a national Oceans Strategy. This Strategy establishes three principles that guide *all* ocean management decision-making:

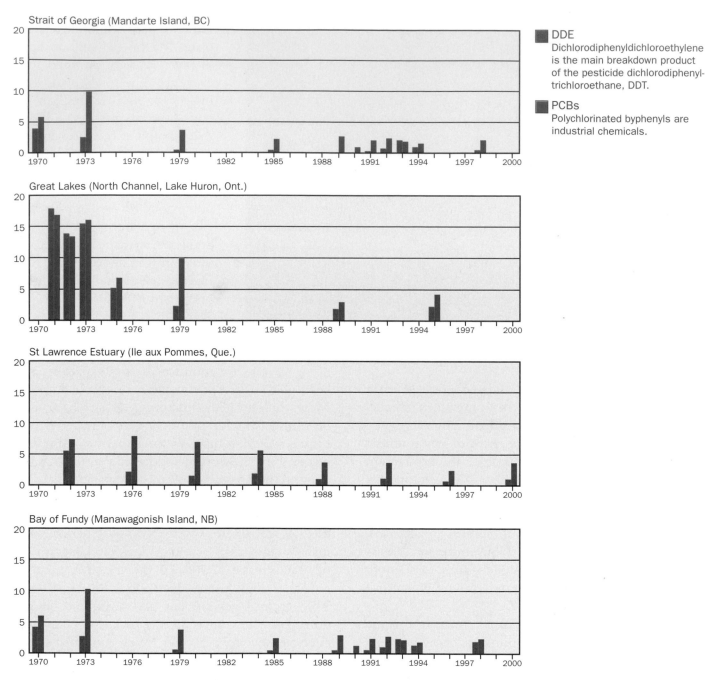

Figure 8.12 Contaminant levels in double-crested cormorant eggs (parts per million). SOURCE: Environment Canada (2003a: 10). Reproduced with the permission of the Minister of Public Works and Government Services, 2005

1. *Sustainable development* 'recognizes the need for integration of social, economic and environmental aspects of decision-making, and that any current and future ocean resource development must be carefully undertaken without compromising the ability of future generations of Canadians to meet their needs.'

2. *Integrated management* 'is a commitment to planning and managing human activities in a comprehensive manner while considering all factors necessary for the conservation and sustainable use of marine resources and shared use of ocean spaces.'

3. The *precautionary approach* is defined in the Oceans Act as 'erring on the side of caution'.

As underlying principles, these provide the essential litmus test against which all ocean management decisions should be judged and to which the government is accountable. Unfortunately, this has not proven to be the case. Globally important glass-sponge reefs off the coast of BC have been heavily damaged due to a lack of protection from fishing—a failure to apply the precautionary principle and poor consideration of values other than economic ones.

The Oceans Strategy also identifies three desired policy objectives or outcomes:

- understanding and protecting the marine environment;
- supporting sustainable economic opportunities;
- providing international leadership.

Again, these are laudable objectives, but they are meaningless unless progress can be shown and resources devoted to them. Seven years after the passing of the Oceans Act, the Oceans Directorate of DFO has virtually no resources and very little progress has been made on realizing any of the lofty objectives of the Act or the subsequent Strategy. Canadians think that progress is being made because of periodic policy announcements, but in fact it is largely business as usual. A public poll undertaken in Atlantic Canada and New England on behalf of a consortium of conservation organizations found that most respondents think that between 20 and 23 per cent of their offshore waters are protected (Seaweb, 2003). In reality, the figure is less than 1 per cent.

Marine Protected Areas in Canada

Canada, like the rest of the world, has paid little attention to protecting the marine environment through marine protected areas (MPAs), especially when compared to the attention and protection given the terrestrial environment. Although Canada is estimated to have some 155 protected areas with a marine component, virtually all of these are terrestrial parks bordering the sea with little or no jurisdiction over marine environments. The Saguenay–St Lawrence Marine Park, created by separate legislation, Fathom Five National Marine Park in Lake Huron, and the Endeavour Hydrothermal Vents MPA off the coast of BC are currently the only established sites at the national level. Under half of 1 per cent of Canada's marine environment is protected, compared with over 7 per cent in the terrestrial environment.

In response to this situation, three programs have been created at the federal level to establish MPAs (Table 8.2). The Oceans Act of 1997 established an MPA program within the Department of Fisheries and Oceans. The purpose of these MPAs is to conserve commercial and non-commercial fishes, protect species at risk, and conserve unique habitats, i.e., areas of high biodiversity or biological productivity. At present, eight such areas have been identified as potential MPAs (Box 8.7). Through the Canadian Wildlife Service (CWS), Environment Canada has several programs that may include designation of marine sanctuaries such as National Wildlife Areas and migratory bird sanctuaries. There is also a new designation, called Protected Marine Areas. At present, the Scott Islands at the north end of Vancouver Island are the only site being considered for protection under this designation.

The third program is for National Marine Conservation Areas (NMCAs), developed by Parks Canada. These areas differ from terrestrial national parks and other Canadian marine initiatives in that they are managed for sustainable use. NMCAs will generally be larger than the MPAs established by DFO, will be selected to represent Canada's ocean heritage, and will also contain an explicit mandate for recreation and education. Although some of the CWS sanctuaries are large, especially in the Arctic, they are designed primarily to protect specific species (particularly seabirds) rather than ecosystems, and they have no minimum standards to control extractive activities.

NMCAs will contain zones where special protection measures such as fishing 'no-take' zones are implemented. In many jurisdictions, such areas are often called marine reserves or sanctuaries. The location and size of these zones will be decided through consultation among fishers, scientists, conservationists, government agencies, and other stakeholders. These decisions will be crucial. Setting aside small fragments in unproductive areas will produce few benefits.

Apart from no-take zones, commercial and recreational fishing will continue in NMCAs, although additional conservation measures may be stipulated. Some activities, such as exploration or exploitation of hydrocarbons, minerals, aggregates, or any other inorganic material, are prohibited. Dumping is not allowed. Conservation interests also sought to exclude bottom trawling, dragging, and fin-fish aquaculture due to their destructive impacts on ocean ecosystems. These prohibitions

Guest Statement

Public Awareness and Ocean Management
Sabine Jessen

In May 2003, scientists at Dalhousie University reported that only 10 per cent of the large ocean fish still remain in the world's oceans. In their 10-year study of global fisheries they found that 90 per cent of tuna, sharks, swordfish, and cod have been fished out. Other scientists point out that this is having huge impacts on marine ecosystems, which are often regulated from the top by these large predators. This study followed on previous ones that showed how we are fishing down marine food webs on a systematic basis, moving from these large predators to the smaller fish further down on the food chain.

Scientists point out that there is a simple solution to these problems—reduce fish mortality by reducing quotas, reducing overall effort, cutting subsidies, reducing bycatch, and creating networks of marine reserves. They acknowledge that it is not an easy task. But why has it been so difficult for most countries to make these needed changes in the face of the overwhelming evidence of the impacts of overfishing?

Part of the answer to this question lies in the serious disconnect between what the scientists are telling us and what the public believes is happening in the oceans. When the public is asked what the greatest threat to ocean ecosystems is, the answer is usually pollution. But these recent scientific studies clearly show that the greatest threat is overfishing—the most serious problem is in what we are taking out of the oceans, not what we are putting in. Couple this with the results of another recent poll of Americans, asking them what the most important decision was that they make every day. You might be shocked to hear that deciding what to wear was considered the day's most difficult decision for one in every 10 people.

Why have I highlighted these points in talking about ocean management? Because without public understanding of the issue, and public demand for

Sabine Jessen

change, it is difficult to persuade politicians to make the difficult decisions that will lead to the much-needed fundamental change in how our oceans are managed. Political support determines the priorities, and the resources allocated to address these issues. While massive amounts of money are spent on fishery allocation decisions, broader ocean management programs and marine protected areas are receiving very few resources.

Over the last few years, the plight of our oceans has been receiving increasing attention. In 2002 Canada released an Oceans Strategy, and more recently the government has committed to the development of an oceans action plan. In the US, two national-level commissions have examined oceans policy and highlighted the need for concerted attention and reform of ocean management. These are reasons for hope. But until concrete steps are taken, we will continue to witness the ongoing destruction of the blue frontier. And by the time the public really understands and pushes for change, it could be too late.

Sabine Jessen is the Conservation Director of the Canadian Parks and Wilderness Society, British Columbia chapter, and one of the foremost activists pushing for increased conservation of Canada's ocean environment.

were not included in the National Marine Conservation Areas Act.

The goals of NMCAs are to conserve areas representative of the ocean environment and the

Great Lakes and to foster public awareness, appreciation, and understanding of our marine heritage. The interpretation aspect of these areas may be their greatest contribution and a unique Parks

Canada has not done very well in protecting the marine environment, with Fathom Five, at the tip of the Bruce Peninsula in Lake Huron, being the only legislated Parks Canada site so far. However, the federal government has committed to the establishment of five new National Marine Conservation Areas by 2010 (*Philip Dearden*).

analogous to its terrestrial system plan (Chapter 11), with 29 marine regions. Their goal is to have representation within each of these regions. The five new NMCAs will get the ball rolling in this direction. Prime Minister Chrétien also committed the government to completing the system within 10 years.

AQUACULTURE

One response to the declining catch in wild fisheries is to produce more seafood through farming or **aquaculture**. Aquaculture is the fastest-growing food production sector in the world, and is expected to surpass beef production by the year 2010. Aquaculture accounts for almost 30 per cent of the volume and 39 per cent of the value of global fish landings. By the year 2030, it is expected that it will be the dominant source of fish and seafood. Aquaculture could play a critical role in reducing world hunger.

Canada mission. Canadians are poorly informed about the marine environment. Creating ocean literacy in Canadians, who must support public policies for the sustainable use and protection of Canada's marine environment, will provide unique challenges to interpreters.

One of the main challenges with all these programs is to actually designate some areas. It is essential that local communities support these conservation measures, and gaining support can be time-consuming. Prime Minister Jean Chrétien made a commitment at the World Summit on Sustainable Development in Johannesburg in 2002 to create five NMCAs within the next five years. These sites will probably be in Lake Superior, in Haida Gwaii (the Queen Charlotte Islands) and in the southern Strait of Georgia in BC, in the Arctic, and in the Magdalen Islands in the Gulf of St Lawrence. Parks Canada has a marine system plan

Canada has been part of this growth, and currently ranks twenty-second in the world in aquaculture production. Between 1986 and 2001 the output grew at an annual rate of 19 per cent and was valued at $711 million, roughly 13 per cent of the total fishery by 2002. Over 12,000 people are employed in aquaculture and it is predicted that employment levels will quadruple over the next 15 years. BC has Canada's largest output, worth $329.6 million in 2002 (Figure 8.13). Most of this comes from salmon. A typical farm consists of 10 to 30 cages, each 12 or 15 metres square, and contains on average 20,000 fish. The cages are made of open nets that allow water to flow through and antibiotics, uneaten food, feces, and chemicals used to prevent excessive marine growth on the cages to flow out.

There is considerable conflict regarding aquaculture development. For example, 5 November 2003 was declared by the Coastal Alliance for

Table 8.2	Federal Statutory Powers for Protecting Marine Areas		
Agency	**Legislative Tools**	**Designations**	**Mandate**
Fisheries and Oceans Canada	Oceans Act	Marine Protected Areas (MPAs)	To protect and conserve: • Fisheries resources, including marine mammals and their habitats; • Endangered or threatened species and their habitats; • Unique habitats; • Areas of high biodiversity or biological productivity; • Areas for scientific and research purposes.
	Fisheries Act	Fisheries closures	Conservation mandate to manage and regulate fisheries, conserve and protect fish, protect fish habitat, and prevent pollution of waters frequented by fish.
Environment Canada	Canada Wildlife Act	National Wildlife Areas	To protect and conserve marine areas that are nationally or internationally significant for all wildlife but focusing on migratory birds.
		Marine Wildlife Areas	To protect coastal and marine habitats that are heavily used by birds for breeding, feeding, migration, and over-wintering.
Parks Canada	National Parks Act; proposed Marine Conservation Areas Act	National Parks; National Marine Conservation Areas (NMCAS)	To protect and conserve for all time marine conservation areas of Canadian significance that are representative of the five natural marine regions identified on the Pacific coast of Canada, and to encourage public understanding, appreciation, and enjoyment.

Aquaculture Reform an international day of protest against fish farms. Demonstrations were held in Campbell River (Vancouver Island), Vancouver, Los Angeles, San Francisco, and Austin, Texas. This followed an advertisement in the *New York Times* against BC fish farms, paid for by the Coastal Alliance. Fish farms felt the effects of consumer pressure after a US buyer subsequently cancelled an order of 68 tonnes from Tofino's Creative Salmon Company after finding out that the fish are fed colourants to turn their flesh red to look like wild salmon.

There are currently over 80 fish farms on the BC coast, mainly concentrated in three small areas. In addition, many applications await approval to expand operations to other areas along the coast. Environmentalists have raised several concerns.

Escapement. Salmon farms in BC raise mainly Atlantic, not Pacific, salmon, primarily because the Norwegian-dominated industry had more experience with and well-developed markets for Atlantic salmon. The Atlantic salmon are also more efficient in converting feed into flesh, are less

aggressive, and tolerate crowded conditions. The farming of Atlantic salmon is an environmental concern because escapement from farm fish cages is a regular occurrence, and escapes often occur in high numbers. Over 1 million Atlantic salmon have been estimated to have escaped. There is irrefutable proof that some Atlantic salmon are now spawning wild in Pacific rivers. DFO maintained for many years that this was impossible until scientists proved otherwise (Volpe et al., 2000). There are concerns that these escapees, an invasive species (Chapter 3), may displace the native salmon. Atlantic salmon have been found in over 80 rivers on the BC coast.

If salmon farming was confined to only native Pacific species, it might not help matters. At the moment there is no evidence that Atlantic and Pacific salmon interbreed. However, with the same species of fish on both sides of the cage (i.e., Pacific salmon), there would undoubtedly be interbreeding. This genetic introgression could have a devastating effect on wild stocks, as determined by McGinnity et al. (2003) on the east

BOX 8.7 THE ENDEAVOUR HYDROTHERMAL VENTS MARINE PROTECTED AREA

The Endeavour Hydrothermal Vents Marine Protected Area (MPA) lies in water 2,250 metres deep, 250 kilometres southwest of Vancouver Island. As part of the Juan de Fuca Ridge system, the Endeavour segment is an active sea-floor-spreading zone where tectonic plates diverge and new oceanic crust is extruded onto the sea floor. In this zone, cold sea water percolates downward through the crust, where it is heated by the underlying molten lava, eventually emerging through the sea floor as buoyant plumes of particle-rich, superheated fluid. The five known vent fields on the Endeavour Segment are separated along the ridge from one another by about two kilometres. Their associated plumes rise rapidly about 300 metres into the overlying water column.

Hydrothermal vents in the Endeavour area consist of large, hot, black smokers (chimney-like structures) and surrounding lower temperature sites. The fields span a wide range of hydrothermal venting conditions characterized by different water temperatures and salt content, sulphide structure morphologies, and animal abundance. Temperatures associated with black smokers are typically in excess of 300° C. Formation of the large polymetallic sulphide chimneys takes place when dissolved minerals and metallic ions carried upward by the smokers precipitate on contact with the cold sea water. Cooler waters below 115° C on the sea floor and along the flanks of the chimneys support an abundance of flora and fauna. This rich ecosystem is supported by microbes whose life processes are fuelled by the chemical energy from the emerging fluids in the hydrothermal vents in the process of chemosynthesis (Chapter 2).

Hydrothermal venting systems host one of the highest levels of microbial diversity and animal abundance on Earth. The deep ocean near the Endeavour area normally only supports sparse animal abundance of about 20 worms and brittlestars per square metre. In the diffuse vent flows around the sulphide structures, these abundances can range up to half a million animals per square metre. An amazing abundance of life is in concentrated areas around the vents, surrounded by a veritable desert in the deep oceans.

Globally, hydrothermal venting systems foster numerous unique species of animals. There are some 60 distinct species native to the Juan de Fuca Ridge. Many of these species are the first in the world to be identified. Hydrothermal vents at Endeavour are home to 12 species that are not known to exist anywhere else in the world.

The Endeavour Hydrothermal Vents Marine Protected Area has been designated to ensure the protection of these hydrothermal vents and the unique ecosystems associated with them. As a marine protected area, the removal, disturbance, damage, or destruction of the venting structures or the marine organisms associated with them is prohibited. The regulation allows for scientific research that will contribute to the understanding of the hydrothermal vents ecosystem.

---

coast. There is a large degree of scientific uncertainty involved with all these issues. The stated policy of DFO is to 'err on the side of caution', as discussed above. Yet the department has not been doing this with respect to the dangers associated with escapement.

Disease. The high stocking levels of fish in nets promote rapid spread of infectious diseases and parasites. Since fish farms are along migration routes for wild salmon, diseases and/or parasites can be passed along easily, with a detrimental impact on wild populations. The transference of sea lice is of particular concern. In Clayoquot Sound on the west coast of Vancouver Island a viral disease, infectious hematopoietic necrosis (IHN), swept through fish farms in the fall of 2002. As a result, the main operator in the region—Pacific National Aquaculture—lost some US$5.7 million in 2002 and closed down its processing plant in Tofino.

Pollution. To combat the diseases mentioned above, farmed salmon are treated with antibiotics. More antibiotic per weight of livestock is used by salmon aquaculture than by any other form of farming. Antibiotics can harm other marine organisms, as they are released directly into the ocean. There is also a large amount of organic pollution created by excess food and feces. The excess builds up on the ocean floor, depleting oxygen levels, releasing noxious gases (as a by-product of decomposition), and smothering benthic organ-

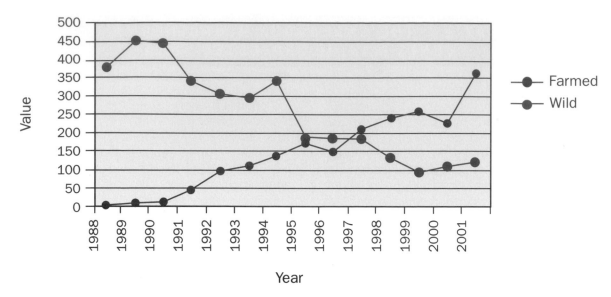

Figure 8.13a Value of BC salmon exports in millions of dollars.

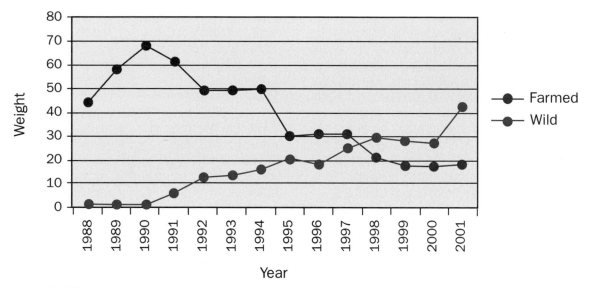

Figure 8.13b Weight of BC salmon exports in millions of kilograms. SOURCE: BC Salmon Market Database, at: <http://www.bcsalmon.ca/database/export/summary/sumwtpd.htm>.

isms. On a daily basis the aquaculture industry in BC dumps the same amount of sewage in the ocean as a city of half a million people.

Predator control. Predators such as seals and sea lions are one of the main problems for the farmers, since they literally eat farmers' profits. Farmers are permitted to shoot animals that rip open nets. In 2001, farmers reported killing over 400 seals and sea lions. Many observers think that this number is grossly under-reported.

Energetics. Unlike the vast majority of fish farms operating around the world that produce herbivorous fish, salmon are carnivorous. As a result, farmed salmon are mainly fed other fish, such as anchovies and mackerel that are caught as far away as South America. As dictated by the second law of thermodynamics (Chapter 2), only 1 kg of farmed salmon is produced for every 3 to 4 kg of feed fish. This is a poor use of fish protein and leads to the reduction of fish stocks elsewhere.

Social dimensions. Most profits from production go to five multinational companies that control 80 per cent of the industry in BC. As a result, a high percentage of the economic benefits attached to salmon aquaculture are exported out of the province. Increased mechanization is leading to

A small fish farm on the BC coast (*M.D. Hauzer*).

The combined impact of hybridization and competition means that, when a large number of farm salmon spawn in a river, the number of adult salmon returning to the river and the potential offspring production in the next generation are reduced.... As repeated escapes are now a common occurrence in some areas, a cumulative effect is produced generation upon generation, which could lead to extinction of endangered wild populations as a result of this 'extinction vortex'.

— McGinnity et al. (2003)

lower employment figures, further limiting the economic benefits accruing to local communities. It is also feared that further growth in the industry will be detrimental to the wild fishery and reduce the health of communities dependent on wild fish. Increased supply of farmed salmon may continue to depress the price of BC's wild salmon.

Human health. To turn the white flesh pink, farmed salmon are fed artificial colouring. The most commonly used colourants are synthetic astaxanthin and canthaxanthin. In 2003, the European Union reduced the amount of canthaxanthin that can be fed to salmon by two-thirds due to concerns over retinal damage caused by ingesting too much of the chemical. A study published in *Science* (Hites et al., 2004) found that farmed salmon contained 11 times the amount of toxic contaminants found in wild salmon.

Most of these problems are not insurmountable. Salmon can be produced in closed, land-based systems that all but eliminate some of the above problems. The main reason why more environmentally and socially sound farming techniques are not being more widely adopted is because of the consumer. If people were willing to pay more for salmon produced using techniques that avoided the problems outlined above, then producers would not be so resistant to adopting these more sustainable systems (Box 8.8). At the moment, however, the environment pays those extra costs. The *2001 Annual Report on Marine Fin Fish Inspections* from the Ministry of Agriculture, Food and Fisheries reported that almost half the fish farms in BC did not meet government requirements for net maintenance, net marking, and net record keeping. There were major discrepancies between approved plans and the on-site management practices.

IMPLICATIONS

The last five years have started to provide proof of the extent of the extreme degradation of the oceans of this planet, something that has been suspected for many years. Scientific efforts have intensified and understanding has increased, yet much remains to be done. Even the largest and best-funded fishery service in the world, the National Marine Fisheries Service of the US, says that it only knows the status of 76 per cent of US fish stocks and cannot complete its job without more funding (http://ens-news.com/ens/feb2003/2003-02-20-10.asp). The very visible collapse of fishing stocks around the world and on the east and west coasts of Canada has helped direct a little more political attention to oceans in general and fisheries in particular. Commitments have been made at both international and national levels to adopt more sustainable ocean practices, including encouraging and enabling sustainable fisheries, limiting pollution, and establishing systems of marine protected areas. At the moment most of these are in the embryo stage. Some plans, such as Canada's Ocean Strategy, have shown little progress. Only time will tell whether international and national commitments will be successful in turning around many of the trends described in this chapter.

BOX 8.8 WHAT YOU CAN DO

1. Fish are an important dietary component for many people, and a healthy one. However, it is important that the fish you eat are not endangered or caught with a method that involves killing other species as bycatch. Use the fish guide produced by the Sierra Club of Canada to inform your consumption: <http://sierraclub.ca/bc/aa-bc_upload/fd5c29a62a2f2c66bdb73ed43cec4361/citizensguidetoseafood_large.pdf>.
2. Only buy certified brands where they are available, such as dolphin-free tuna.
3. If you buy farmed seafood, consider paying a little more for products that have been produced using low-impact methods. For example, Thrifty Foods in BC sells farmed salmon produced using land-based, closed-system methods.
4. Ensure that you dispose of any toxic materials in the correct manner, not down the drain.
5. Use natural cleaners, such as vinegar and water, rather than commercial cleaners.
6. Using less water for your own needs leaves more water in rivers for fish such as salmon.
7. Support NGOs, such as Oceans Blue and the Canadian Parks and Wilderness Society, that are working for ocean conservation and the development of marine protected areas.

SUMMARY

1. Throughout history the resources of the oceans have been thought of as vast and undiminished. The last decade has furnished conclusive proof that this view is far from correct. Over 70 per cent of global fisheries are now at, or over, their maximum exploitation levels.
2. Oceanic ecosystems are controlled by the same general principles that influence terrestrial ecosystems, but their manifestations may be different. There may be up to 5,000 species of fish still awaiting discovery.
3. The carbon balance of the oceans is of great interest due to the relationship with global climatic change.
4. Ocean fisheries supply about 20 per cent of the world's annual animal protein. Catch statistics showed very large increased catches over the last 50 years but these have now levelled off considerably.
5. The oceans are the ultimate sink for many pollutants and about 80 per cent of ocean pollution comes from activities on land.
6. More than 60 per cent of global oil production originates under the oceans. Exploration, drilling, transporting, and processing this oil is a major source of contamination.
7. Almost half the world's population lives within 100 kilometres of the coast and this is expected to increase to 75 per cent by mid-century.
8. Global climatic change will lead to increases in sea level of between 9 and 88 centimetres this century. This will create severe challenges for many coastal communities.
9. There are many international agreements and programs on ocean management. Most have yet to fulfill their true potential in improving oceanic conditions.
10. Marine protected areas have been endorsed by the scientific community as a necessary requirement to improve ocean conservation.
11. Canada has the longest coastline of any country and the second largest continental shelf, equal to 30 per cent of Canada's land mass.
12. Since 1992 commercial fishery landings in Canada have declined by 19 per cent and the number of vessels by 31 per cent. However, the value of the catch has doubled.
13. The east coast fisheries have witnessed profound changes over the last decade with the total collapse of the northern cod stocks.
14. Management of Aboriginal use of marine resources has emerged as an important concern on both coasts.
15. Exploitation of offshore hydrocarbons in Canada has taken place over the last decade, mainly off the east coast. Increased attention is now being given to the Pacific and Arctic coasts.

16. Pollution levels of many substances have declined over recent years in response to government controls. However, a recent study concluded that the killer whales of Georgia Strait in BC are among the most polluted animals on the planet.

17. Canada passed a comprehensive Oceans Act in 1998, but to date it has been ineffectual due to lack of political support and funding.

18. There are three federal programs for establishing marine protected areas (MPAs) in Canada. However, under half of 1 per cent of the area of Canada's marine environment is currently protected.

19. Aquaculture accounts for almost 30 per cent of the volume and 39 per cent of the value of global fish landings. Canada ranks twenty-second in terms of global production, with output increasing at 19 per cent per year between 1986 and 2001.

20. BC has Canada's largest share of the total value of aquaculture production, focusing mainly on salmon.

KEY TERMS

aquaculture	endocrine disruption	marine protected areas	thermocline
biomass pyramid	exclusive economic zones	(MPAs)	thermohaline circulation
bottom trawling	(EEZs)	polynyas	total allowable catch (TAC)
carbon balance	fishing down the food	prey switching	zooxanthellae
coral bleaching	chain	serial depletion	
coral polyps	longline	shifting baseline	

REVIEW QUESTIONS

1. In what ways are oceanic and terrestrial ecosystems the same, and in what ways do they differ?

2. What are the most biologically productive areas of the ocean?

3. What is thermohaline circulation and why is it important?

4. Give an example of a positive feedback loop related to global climate change and the oceans. Are there any negative feedback loops?

5. Explain the concepts of shifting baselines, serial depletion, and fishing down the food chain.

6. Give an example of the destructive effects of bottom trawling.

7. What are the two main forms of chemical pollutants in the oceans and what are their main effects?

8. What are some of the main international conventions concerning ocean management?

9. What are the jurisdictional arrangements for ocean management in Canada?

10. Characterize two of Canada's marine ecozones.

11. Discuss the principal reasons behind the collapse of the Atlantic groundfish stocks and some of the lessons to be learned from this experience.

12. Outline some of the challenges involving the Aboriginal use of marine resources in Canada.

13. Discuss the differing approaches of Halifax and Victoria to ocean pollution resulting from sewage.

14. What are the main principles underlying Canada's Ocean Strategy?

15. Outline the three federal programs for creating marine protected areas in Canada and their similarities and differences.

16. Discuss the positive and negative aspects of aquaculture production.

RELATED WEBSITES

FISHERIES CRISIS: http://www.fisherycrisis.com/

UNITED NATIONS FOOD AND AGRICULTURE
ORGANIZATION, FISHERIES:
http://www.fao.org/fi/default_all.asp

ENDEAVOUR HOT VENTS MARINE PROTECTED
AREA: http://www.pac.dfo-mpo.gc.ca/
oceans/mpa/Endeavour_e.htm

FISHERIES AND AQUACULTURE:
http://www.agf.gov.bc/fisheries/studies_rpts.htm

UNIVERSITY OF BRITISH COLUMBIA FISHERIES
CENTRE: http://wwwl.fisheries.ubc.ca

DAVID SUZUKI FOUNDATION, OCEANS, FISHING,
AQUACULTURE, MPAS:
http://www.davidsuzuki.org/Oceans

SIERRA CLUB, OCEANS, FISHING, AQUACULTURE,
MPAS: http://www.sierraclub.ca/bc/programs/
marine/publications.shtml

WATERSHED WATCH SALMON SOCIETY,
PUBLICATIONS: http://www.watershed-watch.org/
ww/publications.html

VALUE OF MARINE PROTECTED AREAS:
http://www.nceas.ucsb.edu/Consensus

MARINE PROTECTED AREAS, CANADA:
http://www.pac.dfo-mpo.gc.ca/
oceans/mpa/default_e.htm
http://parkscanada.pch.gc.ca/progs/
amnc-nmca/index_E.asp
http://www.cpaws.org

SEAFOOD CONSUMPTION CHOICES:
http://www.seafoodchoices.com
http://www.oceantrust.org
http://environet.policy.net/marine/csb/
http://www.legalseafoods.com
http://www.montereybayaquarium.org

REFERENCES AND SUGGESTED READING

Cameron, S.D. 1990. 'Net losses: the sorry state of our Atlantic fishery', *Canadian Geographic* 110, 2: 28–37.

———. 1998. 'Why aren't heads rolling?', *Globe and Mail*, 20 Jan., 35.

Canadian Environmental Protection Agency (CEPA). *Environmental Registry, Toxic Substances List.* At: <http://www.ec.gc.ca/CEPARegistry/subs_list/Toxic update.cfm>.

Canadian Wildlife Service (CWS), Environment Canada Web site. *Contaminants levels in Double-crested Cormorant eggs, 1970-2000.* At: <http://www.ec.gc.ca/soer-ree/English/indicator_series/techs.cfm?tech_id=9$issue_id=2&supp=4>.

Carr, M.H., T.W. Anderson, and M.A. Hixon. 2002. 'Biodiversity, population regulation, and the stability of coral-reef fish communities', *Proceedings of the National Academy of Sciences* 99, 17: 11241–5.

Clarke, A., and C.M. Harris. 2003. 'Polar marine ecosystems: major threats and future change', *Environmental Conservation* 30: 1–25.

Costanza, R., R. d'Arge, R. deGroot, S. Farber, M. Grasso, B. Hannon, R.G. Raskin, P. Sutton, and M. van den Belt. 1997. 'The value of the world's ecosystem services and natural capital', *Nature* 387: 253–60.

Dearden, P. 2002. 'Marine parks', in P. Dearden and R. Rollins, eds, *Parks and Protected Areas in Canada.* Toronto: Oxford University Press, 345–78.

Dewailly, E., P. Ayotte, S. Bruneau, C. Laliberte, D.C.G. Muir, and R.J. Norstrom. 1993. 'Inuit exposure to organochlorines through the aquatic food chain in Arctic Quebec', *Environmental Health Perspectives* 101, 7: 618–20.

Doubleday, W.G. 2000. 'Seals & Cod', *Isuma* 1, 1. At: <http://www.isuma.net/v01n01/doubleda/doubleda _e.shtml>.

Environment Canada. 1991. *The State of Canada's Environment.* Ottawa: Minister of Supply and Services.

———. 2003a. *Environmental Signals: Canada's National Environmental Indicator Series, 2003.* Ottawa: Environment Canada.

———. 2003b. *NPRI: Substance Information, Dioxins and Furans.* At: <http://www.ec.gc.ca/pdb/npri/npri_ dioxins_e.cfm>.

Fisheries and Oceans Canada. 2003. *Canadian Science Advisory Secretariat Status Report. State of the Eastern Scotian Shelf Ecosystem.* http://www.dfo-mpo.gc.ca/ csas/csas/status/ESR2003_004_E.pdf.

Garshelis, D.L. 1997. 'Sea otter mortality estimated from carcasses collected after the *Exxon Valdez* oil spill', *Conservation Biology* 11, 4: 905–16.

Glavin, T. 2000. *The Last Great Sea: A Voyage through the Human and Natural History of the North Pacific Ocean*. Vancouver: Greystone Books.

Hall, P. 1990. 'Crisis in the Atlantic fishery', *Canadian Business Review* 17, 2: 44–8.

Harris, Michael. 1998. *Lament for an Ocean: The Collapse of the Atlantic Cod Fishery, A True Crime Story*. Toronto: McClelland & Stewart.

Hites, R.A., J.A. Foran, D.O. Carpenter, M.C. Hamilton, B.A. Knuth, and S.J. Schwager. 2004. 'Global assessment of organic contaminants in farmed salmon', *Science* 303: 226–9

Hoegh-Guldberg, O. 1999. 'Climate change, coral bleaching and the future of the world's coral reefs', *Marine Freshwater Research* 50: 839–66.

Hutchings, J.A., and R.A. Myers. 1994. 'What can be learned from the collapse of a renewable resource? Atlantic cod, *Gadus morhua*, of Newfoundland and Labrador', *Canadian Journal of Fisheries and Aquatic Sciences* 51: 2126–46.

Intergovernmental Panel on Climate Change (IPCC). 2001. *Climate Change 2001: The Scientific Basis*. At: <http://www.grida.no/climate/ipcc_tar/wg1/index.htm>.

Jackson, J.B.C., et al. 2001. 'Historical overfishing and the recent collapse of coastal ecosystems', *Science* 293: 629–38.

Leggatt, S.M. 2001. *Clear Choices, Clean Waters: The Leggatt Inquiry into Salmon Farming in British Columbia*. Vancouver: David Suzuki Foundation.

McClanahan, T.R. 2002. 'The near future of reefs', *Environmental Conservation* 29: 460–83.

McGinnity, P., et al. 2003. 'Fitness reduction and potential extinction of wild populations of Atlantic salmon (*Salmo salar*) as a result of interactions with escaped farm salmon', *Proceedings of the Royal Society: Biological Sciences* 270, 1532: 2443–50.

McGowan, J.A., and J.G. Field. 2002. 'Ocean studies', in Field, G. Hempel, and C.P. Summerhayes, eds, *Oceans 2020*. Washington: Island Press, 9–48.

Malcolm, C. 2003. 'The Current State and Future Prospects of Whale-Watching Management with Special Emphasis on Whale Watching in BC, Canada', Ph.D. dissertation, University of Victoria.

Ministry of Agriculture, Food and Fisheries. 2002. *The 2nd Annual Report on Marine Finfish Inspections*. At: <http://www.agf.gov.bc.ca/fisheries/aqua_report/2001/toc.pdf>.

Myers, R.A., and B. Worm. 2003. 'Rapid worldwide depletion of predatory fish communities', *Nature* 423: 280–3.

Naylor, R.L., J. Eagle, and W.L. Smith. 2003. 'Salmon aquaculture in the Pacific Northwest: a global fishery with local impacts', *Environment* 45: 18–39.

Norse, E. 1998. *Environment: Ocean Trawling Worse than Forest Clearcutting*. At: <http://www.sunsonline.org/trade/process/followup/1998/12160698.htm>.

Pauly, D., J. Alder, E. Bennett, V. Christensen, P. Tyedmers, and R. Watson. 2003. 'The future for fisheries', *Science* 302: 1359–61.

———, V. Christensen, J. Dalsgaard, R. Froese, and F. Torres Jr. 1998. 'Fishing down marine food webs', *Science* 279: 860–3.

———, M.L. Palomares, R. Froese, P. Sa-a, M. Vakily, D. Preikshot, and S. Wallace. 2001. 'Fishing down Canadian aquatic food webs', *Canadian Journal of Fisheries and Aquatic Sciences* 58: 51–62.

Peterson, C.H., et al. 2003. 'Long-term ecosystem response to the *Exxon Valdez* oil spill', *Science* 302, 5653: 2082–6.

Roberts, C.M. 2003. 'Our shifting perspectives on the oceans', *Oryx* 37: 166–77.

Rose, G.A., and R.L. O'Driscoll. 2002. 'Capelin are good for cod: can the northern stock rebuild without them?', *ICES Journal of Marine Science* 59: 1018–26.

Ross, P.S., G.M. Ellis, M.G. Ikonomou, L.G. Barrett-Lennard, and R.F. Addison. 2000. 'High PCB concentrations in free-ranging Pacific killer whales, *Orcinus orca*: Effects of age, sex and dietary preference', *Marine Pollution Bulletin* 40: 504–15.

Safina, C. 1997. *Song for the Blue Ocean*. New York: Henry Holt and Co.

———. 2002. *Eye of the Albatross: Visions of Hope and Survival*. New York: Henry Holt and Co.

Schiermeier, Q. 2002. 'How many more fish in the sea?', *Nature* 419: 662–5.

Seaweb. 2003. *Danger at Sea: Our Changing Ocean*. At: <http://www.seaweb.org>.

Smedbol, R.K., and J.S. Wroblewski. 2002. 'Metapopulation theory and northern cod population structure: interdependency of subpopulations in recovery of a groundfish population', *Fisheries Research* 55: 161–74.

Steele, D.H., R. Andersen, and J.M. Green. 1992. 'The managed commercial annihilation of northern cod', *Newfoundland Studies* 8, 1: 34–68.

Task Force on Incomes and Adjustments in the Atlantic Fishery. 1993. *Charting a New Course: Towards the Fishery of the Future*. Ottawa: Minister of Supply and Services Canada.

Thorne-Miller, B. 1999. *The Living Ocean: Understanding and Protecting Marine Biodiversity*. Washington: Island Press.

Volpe, J.P., E.B. Taylor, D.W. Rimmer, and B.W. Glickman. 2000. 'Evidence of natural reproduction of aquaculture escaped Atlantic salmon (*Salmo salar*) in a coastal British Columbia river', *Conservation Biology* 14, 3: 899–903.

Watling, L., and E.A. Norse. 1998. 'Disturbances of the sea bed by mobile fishing gear: a comparison to forest clearcutting', *Conservation Biology* 12: 1180–97.

Watson, R., and D. Pauly. 2001. 'Systematic distortion in world fisheries catch trends', *Nature* 414: 534–6.

Weber, P. 1993. *Abandoned Seas: Reversing the Decline of the Oceans*. Worldwatch Institute Paper 16. Washington: Worldwatch Institute.

Chapter **9**

Forests

Learning Objective

- To understand what the boreal forest is, its significance to Canada, and the main threats it is facing.
- To describe the eight main forest ecozones of Canada.
- To discuss the economic and non-economic values of Canada's forests.
- To appreciate the management arrangements and different approaches for harvesting Canada's forests.
- To understand some of the environmental and social aspects of forest management practices.
- To discuss the theory and practice of 'New Forestry'.
- To describe current directions for forest use in Canada.

THE BOREAL RENDEZVOUS

When the combined effects of climate warming, acid deposition, stratospheric ozone depletion and other human activities are considered, the boreal landscape may be one of the global ecoregions that changes the most in the next few decades. Certainly, our descendants will know a much different boreal landscape than we have today.

— *D.W. Schindler,* 1998

In July 2003, Justin Trudeau, the son of former Prime Minister Pierre Trudeau, broadcast live on CBC from Virginia Falls on the Nahanni River, one of Canada's, and the world's, most spectacular sites. Justin was part of a team canoeing down the Nahanni, just as his father had done before him. The Prime Minister's journey resulted in a directive to establish Nahanni National Park in 1976, later recognized as one of the first World Heritage sites. Justin was on a similar mission, but this time as part of a Canadian Parks and Wilderness Society (CPAWS) campaign to draw greater attention to the

Virginia Falls, Nahanni River, NWT (*T. Parker/Ivy Images*).

great boreal forests that stretch across the roof of Canada.

Justin was not the only celebrity taking part. Others included Tom Cochrane, Silken Laumann, Rick Mercer, Ken Dryden, and David Suzuki, all part of expeditions down 10 of Canada's most

Boreal is a term that literally means 'of the North'. It comes originally from the Greek god of the north wind, 'Boreas'. It is now applied to many northern phenomena, perhaps the most famous being the aurora borealis, or northern lights. Many animal and plant species that live in the North have 'borealis' as part of their Latin name, such as the delicate twin-flower Linnaea borealis, *which is found all across the country. It is also the name used to characterize the great northern forests that stretch not only across Canada, but all across the northern hemisphere.*

spectacular boreal rivers. Participants were united by their recognition of the ecological, recreational, and spiritual values of the boreal forest and the need to create a sustainable development path that recognizes these values, in addition to the economic values that have dominated land-use allocations in the past. All would later meet at the 'Boreal Rendezvous' in Ottawa as part of a national campaign to give voice to these concerns.

Why are these and many other well-known Canadians so concerned about the boreal forests? Stretching 3,800 kilometres from the eastern tip of Newfoundland to the northeastern corner of Alberta, the Boreal Shield, which is coterminous with the geological formation known as the Canadian or Precambrian Shield, is Canada's largest ecozone, covering almost 20 per cent of the country's land mass (over 1.8 million km²), containing 43 per cent of its commercial forest land, and accounting for 22 per cent of the country's freshwater surface area. The forests are home to a wide diversity of terrestrial and aquatic wildlife and many First Nations peoples who depend on the resources of the forests for subsistence.

The boreal forests also support commercial activities such as logging, wood fibre and sawlog production, pulp and paper mills, and fibreboard production. Its wealth of minerals supports prospecting, mining, and smelting activities. There are large-scale hydro-electric developments, and the abundant fish and wildlife resources support subsistence, sport, and commercial harvesting activities, as well as a growing tourism industry. Almost 50 per cent of the boreal forest is allocated to industry and the remaining, intact forests are currently being eyed as the next cutting frontier by multinational logging companies.

Although the Boreal Shield is the largest of Canada's 15 terrestrial ecozones, it has one of the lowest proportions of land (3 per cent) dedicated to protected areas in which all forms of industrial activity are prohibited. Approximately 5 per cent of the ecozone includes additional protected areas where forestry, mining, and other activities may be permitted. As a result, large areas of the Boreal Shield are currently experiencing a number of serious environmental stresses (Table 9.1). These stresses are not different in many ways from those experienced elsewhere in Canada and they epitomize the challenges faced in developing sustainable strategies for the management of Canada's forest ecosystems into the future. This chapter will outline the main challenges and some of the strategies developed to address them.

AN OVERVIEW OF CANADA'S FORESTS

Canada is a forest nation. The symbol on our national flag is a leaf. Along with our northern latitude, the forests have provided the historical context for our national identity. Canada has one-

The use of wood products is an integral part of the livelihood of many Canadians, as illustrated by this local boat building in Newfoundland (*Philip Dearden*).

Table 9.1   Environmental Stresses in the Boreal Shield Ecozone

Logging

- The 1990s witnessed a dramatic northward expansion of forestry-related activities. Approximately 400,000 ha of forest are harvested annually. About 90 per cent of all harvesting is carried out through clear-cutting.

- Technological advances have encouraged the industry to harvest smaller trees and additional trees (poplar, birch) that were formally considered uneconomical. Many areas have been cut over by small northern operators who lack the capacity or the responsibility for replanting after harvest.

- The long-term and cumulative impacts of intensified forestry on the fragile areas of the northern Boreal Shield are entirely unknown—the science behind current forest management practices comes from research in the southern areas of the ecozone, where ecosystems, growth rates, and nutrient dynamics may vary considerably from those in northern portions of the Boreal Shield.

- Logging continues to threaten boreal wildlife, particularly large mammals and migratory birds. A recent report suggested that as many as 85,000 migratory bird nests were destroyed in 2001 by logging in Ontario alone (CPAWS, 2003). There are also a number of socio-economic impacts reported by indigenous communities, including decreased trapping returns, destruction of traditional trails, significant noise pollution, and erosion of cultural values.

- Numerous pulp and paper mills (27) have been major point sources of aquatic pollution by mercury, dioxins and furans, and other organochlorines. The presence of organochlorines such as polychlorinated biphenyls (PCBS) has been linked with liver enzyme dysfunction in fish. Highly toxic leachate from both hardwoods and softwoods stored near the water's edge has been identified. The long-term cumulative effects on lakes and rivers are not well known.

Mining

- Exploration companies hold over 1,200 mineral leases or claims covering more than 2.2 million ha in the boreal forest. The Boreal Shield produces 75 per cent of the total Canadian production of iron, copper, nickel, gold, and silver, generating over $6 billion annually and supporting over 80 communities. Mercury and cadmium are released into the atmosphere through smelting and other industrial processes, then deposited across the landscape. Mine effluents discharged into water systems contain high concentrations of heavy metals as well as cyanide, arsenic, or radioactive compounds.

- Large tracts of land are disturbed by ongoing mineral exploration, which involves road construction, trenching, blasting, and exploratory drilling.

- Metal smelters are point sources of acid and heavy metal pollution that cause widespread aquatic contamination via atmospheric transport. Mercury and cadmium enter the food chain and accumulate in tissues of organisms, reaching toxic levels (Chapter 10). Improper disposal of mine tailings and associated acid mine drainage has caused acute toxicity problems in localized areas and more widespread chronic effects downstream. Such impacts can continue for decades after a mine is abandoned. There are 6,000 abandoned mine sites in Ontario alone.

Hydroelectric Development

- Eighty-five per cent of the drainage basins contained whole or in part in the Boreal Shield have been altered by hydro development in one way or another; 77 per cent of the drainage areas contain major dams, 25 per cent have major reservoirs, and 33 per cent have rivers whose flow has been either augmented or diminished by water transfers.

- Impacts associated with dam construction and reservoir impoundment include soil erosion and scouring of river channels, and loss of terrestrial habitats. Transmission lines reduce the wilderness character of remote regions, especially when they create long, wide swaths through undisturbed terrain. Increased hunting access provided by transmission corridors may be linked to local depletions of wildlife. For example, abrupt declines in moose and other large game animals have been linked to increased access associated with hydroelectric projects.

- Aquatic ecosystems are affected in several ways: increases in water volume caused by the river flooding and reservoirs disperse existing species, thus altering competitive and predatory regimes; lake and brook trout spawning beds are adversely affected by reservoir impoundments, causing abandonment; river diversion can result in high levels of turbidity owing to increased flow rates and bank erosion; and flooding of organic matter by reservoirs results in the release of naturally occurring mercury, with toxic effects on aquatic and terrestrial food chains. Extensive flooding of land also causes the release of significant quantities of carbon dioxide and methane, two contributors to climate warming (Chapter 7).

Continued

Climate Change

- Atmospheric models predict that forests and wetlands within the Boreal Shield are likely to be highly vulnerable to the effects of global warming. Many distinctive characteristics of the Boreal Shield are closely tied to climate, including plant productivity, carbon storage, insects, hydrology, and fish habitat. Changes in climate include: a shift towards shorter periods of ice cover, increases in air temperature, decreases in precipitation, increases in evapotranspiration rates, increases in forest fire frequencies, and increases in wind velocity.

- Possible impacts include: decreased stream flow resulting in decreased transportation of chemicals such as dissolved organic carbon (DOC) and phosphorous (important elements in food webs) (Chapters 2–4); decreased phytoplankton abundance; increased transparency (caused by reduced DOC) resulting in deeper penetration of UV-B radiation; and deeper thermoclines, reducing habitat for cold-adapted invertebrates and fish. Climate change is also affecting the life cycle, behaviour, and range of many boreal insect species—there are numerous examples of insect outbreaks associated with warmer temperatures resulting in changes to tree species composition. The range of most boreal insects is controlled by low threshold temperatures, and current climate trends may remove this climatic control in some areas resulting in more widespread outbreaks of some insects.

Acid Precipitation

- Acid precipitation, caused by human emissions of sulphur and nitrogen oxides to the atmosphere, has caused considerable stress to ecosystems throughout eastern North America (Chapter 4). The poorly buffered lakes of the eastern half of the Boreal Shield are particularly sensitive to acidification.

- Thousands of fish populations and perhaps millions of invertebrate populations from Canada's boreal region have been lost as a result of acid precipitation. Widespread forest decline, growth reductions, and degraded soils have already been documented in southern regions of the Boreal Shield. Increased acidity may contribute to increased mercury contamination in some boreal fish species, limiting human consumption of fish and impairing reproduction in fish-eating birds, particularly the common loon. Low pH in lakes is also linked to high concentrations of cadmium and lead in fish, as well as loss of DOC in surface waters.

- The impacts of acid precipitation reduce the overall ability of forests to cope with climate warming, drought, insect outbreaks, disease, increased UV radiation, and other stressors.

SOURCES: Environment Canada (2000a); CPAWS (2003).

tenth of the world's forests, and these forests cover almost half of the nation's land area (Figure 9.1) and a much higher proportion of southern Canada, where most Canadians live. These forests include one-quarter of the world's temperate rain forests and more than one-third of the world's boreal forests. Furthermore, estimates suggest that more than half of Canada's forest area consists of tracts that are over 50,000 hectares in size and, as yet, undisturbed. Over one-third of Canada, however, is naturally treeless, and most of it occurs in the North. Together, Quebec, the NWT, Ontario, and British Columbia account for almost two-thirds of the country's boreal forest. Canada clearly has a major international role to play in forest conservation and management.

Ecosystem and Species Diversity

Although all the provinces are dominated by forest land, forest types differ significantly. Cana-dian forest ecosystems are a mixture of forest, woodlands, wetlands, lakes, glaciers, and rock, providing habitat for a great variety of plants, animals, insects, fungi, and other organisms. Of the estimated 140,000 species in Canada, approximately 66 per cent occur in forests, although many species currently occupy only a portion of their former range owing to habitat modifications (Chapter 11). There are 15 terrestrial ecozones in Canada (Figure 9.2), although the majority of Canada's forests lie within the eight ecozones discussed below.

The **Boreal Cordillera** ecozone, to the south and west of the Tundra Cordillera, is found in northern BC and southern Yukon. Physiographi-cally, this zone is made up of the mountains in the west and east, separated by intermontane plains. The wetter climate gives rise to more tree growth, hence the boreal designation. Much of the terrain along the western side of the ecozone is covered

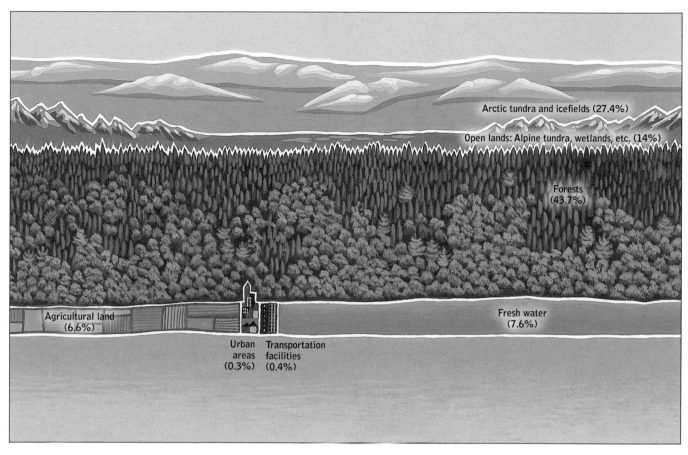

Figure 9.1 Canada's physical makeup.

by permanent ice and snow. In parts of the ecozone located in BC, south-facing slopes are covered with meadows, and north-facing slopes are covered with boreal forest. Vegetative cover ranges from dense to sparse on a large part of the plateaus and valleys. Tree species include trembling aspen, balsam poplar, white birch, black and white spruce, alpine fir, and lodgepole pine. At higher elevations there are large areas of rolling alpine tundra, with sedge meadows and stone fields colonized by lichens. Characteristic mammals include woodland caribou, moose, Dall's sheep, mountain goats, black and grizzly bears, martens, lynx, hoary marmots, and Arctic ground squirrels. Representative bird species include willow, rock, and white-tailed ptarmigan, spruce grouse, and a range of migratory songbirds and waterfowl. The rich resources contained within this ecozone have fostered mining, forestry, and tourism industries, in addition to hydroelectric development and localized agriculture.

Stretching along the entire British Columbia coast, the **Pacific Maritime** ecozone is characterized by the influence of the Pacific Ocean. West-

erly winds sweeping across the ocean pick up considerable amounts of water. Most of this is precipitated on the windward side of the coastal mountains, giving the highest rainfall figures in Canada (up to 3,000 mm). The maritime influence also results in the warmest average temperatures in the country, with mild winters and relatively cool summers. These climatic characteristics provide ideal growth conditions and give rise to the most productive growth conditions in the country in the temperate rain forests at lower elevations. Some of the tree species in the temperate rain forests are long-lived and grow to impressive heights—yellow cedars can live more than 1,000 years, Douglas firs 750 years, and Sitka spruces 500 years, and Douglas fir and Sitka spruce can grow to over 70 metres high. The most common tree species in this ecozone are western red cedar, western hemlock, amabilis fir, Sitka spruce, and yellow cedar. The Pacific Maritime ecozone is also home to a variety of mammals and birds, including black-tailed deer, elk, otters, black and grizzly bears, cougars, fishers, wolves, American black oystercatchers, mountain and California quails,

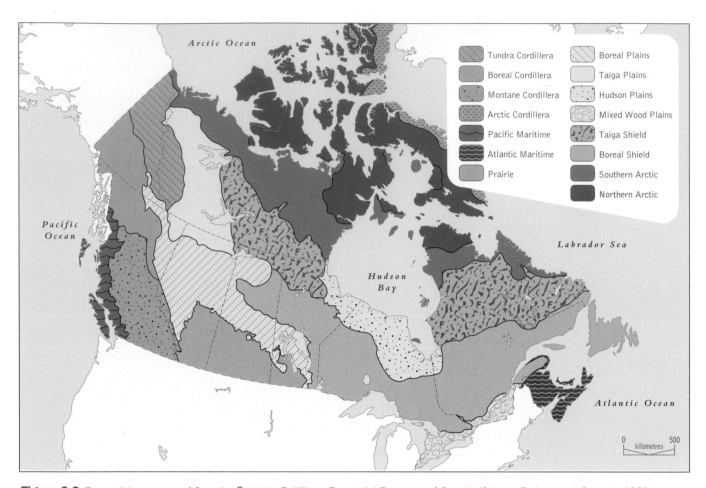

Figure 9.2 Terrestrial ecozones of Canada. SOURCE: E. Wiken, *Terrestrial Ecozones of Canada* (Ottawa: Environment Canada, 1986).

blue grouse, tufted puffins, pygmy owls, Steller's jays, and northwestern crows.

Forestry is the major industry in most of the ecozone. Douglas fir and western red cedar form the backbone of the British Columbia lumber industry, but they are no longer in the infinite supply that they were once assumed to be. Area estimates show that approximately 25 per cent of the coastal temperate rain forest within BC has been logged and reforested, and an additional 3 per cent of the total area has been logged and urbanized. Approximately 50,000 hectares of coastal temperate rain forest are logged annually.

Like the Boreal Cordillera ecozone, the great altitude variation in the **Montane Cordillera** ecozone leads to considerable contrast between the summits of the snow-bound peaks through to high montane valleys, rolling plateaus, and deeply entrenched desert-like conditions in the BC Interior. The climate is similarly varied, but is generally characterized by long, cold winters and short warm summers. Precipitation ranges from highs of over 1,200 mm along the mountain summits to as

little as 205 mm in the valleys in the rain shadow. Vegetation varies according to these conditions, and can be thought of as a series of vertical zones increasing in altitude. At the summits, the vegetation is alpine, characterized by lichens, herbs, and small shrubs. In the lower subalpine environment, trees such as alpine fir, Englemann spruce, and lodgepole pine become more common. Below this zone there is considerable variation depending on local conditions. Ponderosa and lodgepole pines, Douglas fir, and trembling aspen are found towards the north, and in moister conditions in the southeast, western hemlock, red cedar, and Douglas fir may be found. Characteristic mammals include black and grizzly bears, woodland caribou, mule deer, moose, mountain goats, California bighorn sheep, wolverines, and fishers. Common bird species include the pileated woodpecker, Clark's nutcracker, and red crossbill.

Commercial forest operations have been established in many parts of the Montane Cordillera ecozone, particularly in the northern interior. Other significant industries include mining, oil

British Columbia has some of the largest trees in the world. This is Cathedral Grove, a stand of large Douglas fir and western red cedar on Highway 4 en route to Pacific Rim National Park Reserve (*Philip Dearden*).

Many Canadians are familiar with the spectacular mountain landscapes of the Montane Cordillera ecozone, which includes Banff, Canada's first and most visited national park (*Philip Dearden*).

and gas production, tourism, and agriculture. The southern valleys are well known for their orchards and vineyards.

The **Boreal Plains** ecozone extends from the southern part of the Yukon in a wide sweeping band down into southeastern Manitoba. The underlying glacial moraine and lacustrine deposits give a generally flat to undulating surface similar to the Prairie zone to the south. Climatic differences, making it wetter and cooler than the Prairie zone, produce vegetation dominated by trees rather than grasses. Coniferous trees include tamarack, Jack pine, and black and white spruce, although deciduous trees such as trembling aspen, white birch, and balsam poplar are common, especially at the transition zone into the true prairie. Mammals found in the Boreal Plains include woodland caribou, white-tailed deer, bison, wolves, black bears, mule deer, and elk. Characteristic bird species include boreal and great horned owls, blue jays, a number of warblers, grouse, red-tailed hawks, cormorants, gulls, herons, and terns. One of Canada's most famous endangered species, the whooping crane, nests in wetlands in Wood Buffalo National Park, at the extreme north of the ecozone. This ecozone contains a number of other species considered at risk of extinction, including woodland caribou, wolverines, grizzly bears, and wood bison. The southern and northwestern areas of the ecozone have been transformed by agricultural development and timber, mining, and oil production.

'**Taiga**', a Russian word, is used to describe coniferous forests in that country. In North America it describes that portion of the boreal forest lying between the southern boundary of the tundra and the closed-crown coniferous forest to the south. Topography in the **Taiga Plains** is gently rolling with a high proportion of surface water storage, wetlands, and organic soils. The climate is cold and relatively dry, with as little as 200 mm of precipitation in the northern sections. Canada's largest watercourse, the Mackenzie River and its tributaries, is located in the Taiga Plains.

Trembling aspen is characteristic of the Boreal Plains (*Philip Dearden*).

Winters are long and cold, with mean daily January temperatures ranging from −22.5 to −30° C. These conditions, plus the topography, give rise to large areas of wetlands dominated by species such as Labrador tea, willows, dwarf birch, mosses, and sedges. On better-drained localities and uplands are found mixed coniferous-deciduous forests containing white birch, trembling aspen, balsam poplar, lodgepole pine, tamarack, and black and white spruce. A number of large mammals are found in this ecozone, including moose, caribou, black bear, wolf, and wood buffalo. Common bird species include the bald eagle, the peregrine falcon, and the osprey. The Mackenzie Valley forms one of North America's most travelled migratory corridors for waterfowl breeding along the Arctic coast. Economic activity in the area is based on subsistence hunting, trapping, and fishing, in addition to a few industrial activities, such as mining and oil extraction.

The **Boreal Shield** is the largest ecozone in Canada, stretching along the Canadian Shield from Saskatchewan to Newfoundland. It is also part of one of the largest forest belts in the world, the **boreal forest**, extending all across North America and Eurasia, encompassing roughly a third of the Earth's forest land and 14 per cent of the world forest biomass. It is the belt that generally separates the treeless tundra regions to the north from the temperate deciduous forests or grasslands to the south. Winters are cold and summers warm to hot, with moderate precipitation. In Canada the zone is influenced by cold Hudson Bay air masses that yield relatively high precipitation, ranging from 400 mm annually in the east to over 1,000 mm in

The boreal forest in Newfoundland has bedrock outcrops, lakes, and muskeg (*Philip Dearden*).

the west. Average January temperatures range from −10 to −20° C with the July range of 15 to 18° C. The terrain is characteristically rolling with bedrock outcrops, glacial moraine, and many lakes dominated by coniferous forests of black and white spruce, balsam fir, tamarack, and Jack pine with significant cover by aspens and white birch. This forest cover is interrupted by large areas of wetlands, particularly moss-dominated bogs. Overall, the boreal forest is characterized by its lack of diversity of tree species; large areas are covered in just one or two species, particularly the spruces. Balsam fir becomes more dominant in areas of heavier precipitation towards the east. Further south, hardwoods such as white birch, yellow birch, and trembling aspen are more common, along with softwoods such as red pine, Jack pine, and eastern white pine.

Characteristic mammals include the woodland caribou, moose, white-tailed deer, black bear, wolf, marten, snowshoe hare, striped skunk, Canadian lynx, and bobcat. Characteristic bird species include the common loon, boreal owl, great horned owl, and evening grosbeak. The rich natural resource base of the Boreal Shield ecozone supports various industries, including mining, forestry, energy, and tourism, in addition to commercial and subsistence hunting, trapping, and fishing.

The **Mixed Wood Plains** ecozone is the most urbanized in Canada, spreading from the lower Great Lakes north up through the St Lawrence Valley. The topography is gentle, mainly resulting from the lacustrine, marine, and morainic deposits. The climate is continental, with warm, humid summers and cool winters. Precipitation ranges from 720 to 1,000 mm per year. These conditions have produced the most diverse tree coverage in Canada, with over 64 species according to Schueler and McAllister (1991). However, few intact areas of natural vegetation remain. In the northern part of the ecozone, the mixed coniferous-deciduous forest is dominated by red and white pine, oaks, maples, birches, and eastern hemlock. Further south, the warmer zones contain deciduous species such as sugar maple, beech, white elm, basswood, and red and white oaks. The white elm is also native to this zone. Unfortunately, many of them have been devastated by an imported fungus that causes Dutch elm disease. First noticed in 1944, the fungus is spread from tree to tree by beetles. It is a good example of the problems created by alien organisms, discussed in more detail in Chapter 3.

White-tailed deer and black bears were once common in this ecozone, but small mammals such as the raccoon, striped skunk, black squirrel, groundhog, and eastern cottontail rabbit now dominate.

Common bird species include the whip-poor-will, blue jay, red-headed woodpecker, great blue heron, cardinal, and northern oriole.

The Mixed Wood Plains is the most densely populated ecozone in the country. Most of the deciduous forest has been cleared for agriculture and urban development, with small patches of forest cover found scattered throughout the zone. Two tree species, eastern white pine and eastern hemlock, were over-harvested and are now under-represented. Service industries and the manufacturing sector are the largest employment sectors.

The **Atlantic Maritime** ecozone, stretching from the mouth of the St Lawrence River across New Brunswick, Nova Scotia, and Prince Edward Island, is heavily influenced by the Atlantic Ocean, which creates a cool, moist maritime climate. However, conditions vary considerably between the upland masses of hard crystalline rocks, such as the Cape Breton and New Brunswick highlands, through to the coastal lowlands that support most of the population. Mean annual precipitation is as high as 1,425 mm on the coast but falls to less than 1,000 mm further inland. Temperatures are also moderated by the ocean, with mean daily January temperatures of −2.5 to −10° C and a mean July daily temperature of 18° C. Forests are generally mixed stands of deciduous and coniferous species, such as balsam fir, red spruce, yellow birch, sugar maple, eastern hemlock, and red and white pine, mixed in with boreal species such as black and white spruce, white birch, Jack pine, and balsam poplar. Characteristic animal species include the white-tailed deer, moose, black bear, bobcat, snowshoe hare, wolf, eastern chipmunk, mink, whip-poor-will, blue jay, eastern bluebird, and rose-breasted grosbeak.

Major land-oriented activities include forestry, agriculture, and mining. Forest landscapes in the region are undergoing significant change due to over-harvesting on private woodlots and spruce budworm outbreaks, which have increased dramatically in frequency, extent, and severity.

The Mixed Wood Plains ecozone has the highest tree diversity in Canada (*Philip Dearden*).

Cape Breton Highlands (*Philip Dearden*).

Forest Ecosystem Services

Canada's forest ecosystems provide an array of beneficial services arising from ecological functions such as nutrient and water cycling, carbon sequestration, and waste decomposition. For example, plant communities are important in moderating local, regional, and national climate conditions. Biological communities are also of vital importance in protecting watersheds, buffering ecosystems against extremes of flood and drought, and maintaining water quality (Power et al., 1996). Indeed, the contributions of forest lands to the maintenance of ecological processes (Chapters 2 and 4) within Canada are substantial. However, the sheer scale of Canada's forests means that they are significant contributors on a global scale. It is estimated that 20 per cent of the world's water originates in Canada's forests. The forests are also major carbon sinks, with an estimated 224 billion tonnes stored and a yearly accumulation of some 72 million tonnes. It is likely that this accumulation rate will grow in the future as a result of reforestation, improved fire suppression, reduced slash burning, and increased recycling of forest products.

Forests are also places of exceptional scenic beauty, and millions of Canadians travel each year to participate in nature-related recreational activities such as wildlife viewing in parks and protected areas, nature walks, and birdwatching. The monetary value of these activities can be significant. For example, over 20 million Canadians participate in nature-related recreational activities, creating approximately 245,000 jobs and contributing over $12 billion to Canada's GDP (National Roundtable on the Environment and the Economy, 2003).

Forest Ecosystem Products

Non-timber forest products. In addition to the important 'services' that forest ecosystems in Canada provide, forests are also a valuable source of commodities. Wild rice, mushrooms and berries, maple syrup, edible nuts, furs and hides, medicines, ornamental cuttings, and seeds—collectively known as **non-timber forest products (NTFPs)**—currently contribute an estimated $442 million annually to the Canadian economy, with a potential to contribute over $1 billion (Natural Resources Canada, 2003b). Some NTFPs are harvested commercially and are allocated by licence, while others are freely available and contribute significantly to recreational values, including tourism. These commodities are also important in sustaining First Nations communities. With careful management, NTFPs are renewable.

Timber forest products. Despite recognition from federal and provincial governments that Canadian forests provide a broad range of values (wilderness, recreation, wildlife habitat, etc.), over the past century forest management paradigms have focused on the management of Canadian forests to supply wood. The economic benefits arising from timber products are substantial, providing a livelihood for 800,000 Canadians in the forest and associated industries. Over 350 communities are totally dependent on forestry and another 600 are partially dependent. The industry directly employs over 353,000 Canadians, representing 2.3 per cent of total employment. Canada is number one in the world for newsprint production and export, and ranks first in exports of softwood lumber and wood pulp and second in production of softwood lumber and wood pulp (ibid.). In 2001, Canada's forests contributed over $74 billion per year to the Canadian economy and $28.5 billion (2.9 per

Many people find forest lands a source of recreational and spiritual fulfillment (*Philip Dearden*).

cent) to national GDP (Table 9.2). The forestry industry is the largest single contributor to Canada's balance of trade, with exports totalling over $34 billion (or 8.4 per cent of total exports) in 2001. Canada accounts for almost 20 per cent of the global trade in forest products (Box 9.1).

The volume of wood produced per unit area differs across the country, rising to highs in excess of 800 cubic metres per hectare on the most productive sites in coastal British Columbia where mild temperatures, deep soils, and abundant rainfall create some of the most productive growing sites in the world. Volumes harvested also vary by province. For example, in 2001 British Columbia harvested close to double the volume of any other province, while less area was cut in BC than in either Quebec or Ontario (Table 9.2)

The Canadian forestry industry is also a frequent flashpoint of conflict. Names such as Carmanah, Temagami, and Clayoquot became well known across the country in the 1990s as they appeared in newspaper headlines and on national news broadcasts. All these conflicts revolved around questions of whether particular areas should be logged or preserved. These conflicts reflected the increasing appreciation of the many values provided to society by forests besides the usual economic benefits. Few of these non-economic values are easy to calculate in monetary terms and compare against the financial returns of the forest industry. However, these values are gaining increasing recognition by the public and decision-makers as the process of converting old-growth forests across Canada into managed forests continues.

The sheer scale of the industry has also served to attract international attention. Concern over the destruction of the tropical forests (Box 9.2) and resulting impacts on the biosphere and on biodiversity has spread from countries such as Malaysia, Indonesia, and Brazil to temperate forest nations, and Canada in particular, which has been dubbed, rightly or wrongly, 'the Brazil of the North'. Governments in Canada have launched large public relations missions around the world to

Guest Statement

Reforming Forest Tenure
Kevin Hanna

Forest tenure refers to the conditions that govern forest ownership and use. It is an important and fundamental element in determining forest policy and management, and must play a role in creating solutions to the significant challenges in Canadian forestry. Because it is a difficult and thorny subject, governments, industry, labour, and environmental groups have tended to ignore tenure. But this is beginning to change.

About 94 per cent of Canada's forest land is owned by governments, largely the provinces. About 80 per cent of Canada's private forestland is located east of Manitoba, most of it in the Maritimes. British Columbia has the highest level of provincial forest ownership—96 per cent. Compared to other major forest nations, Canada's tenure profile is quite unique. For example, Finland's forests are mainly privately owned. Individuals, usually farmers, hold about 62 per cent of forest land; timber companies about 6 per cent, and the government 31 per cent. There are about 280,000 private forest holdings with an average size of 37 hectares. These small holdings are particularly productive. They supply the majority of the Finnish wood industry's fibre requirements and provide about 80 per cent of the stumpage income and 80 per cent of annual growth and cut. In Sweden, small-scale landowners own about 50 per cent of the forest land, while the state and forest companies each have about 25 per cent. There are about 240,000 private forests in Sweden, and about 30 per cent of these are less than 50 hectares. They also provide a major part of Sweden's timber needs. In Canada, while production on private lands has grown, the great majority of timber comes from provincial forests.

In Sweden and Finland forestry investment levels (regeneration, tending and harvesting techniques, and worker training) tend to be relatively high. The woods also hold a special place in the psyche of each nation. This is not to say that Scandinavian forestry is without problems and controversies, but a strong culture of forest stewardship has certainly developed and their forest sectors have realized a relatively stable employment and fibre supply.

Kevin Hanna

Many who work in Canada's forest industry would say that a culture of stewardship also exists here. Workers care about sustaining the resource on which their livelihoods depend. They might also suggest that it's the short-term vision of companies and governments that limits the potential for such a culture to really flourish. But industry counters that tenures are unstable, the times too short, or conditions too uncertain to allow them to increase investment levels, develop non-timber forest products, or acknowledge the services that forests provide beyond timber.

So how might tenure be changed in Canada? A solution offered at various times has been to emulate the Scandinavian model. Large companies might be more willing to invest in innovative forestry practices if they owned the land. But there might not be much public support for selling provincial forests, certainly not to large companies, nor would many large firms necessarily want to buy forests. After all, the present system has served them well, and some firms have been innovative under existing tenure forms. Another option is to increase the management role of communities through community-held tenures. Or tenure opportunities for small firms and individuals could be provided or expanded, perhaps creating many small private forests. But would small holdings, community forestry, or corporate ownership result in better forestry practices?

Experience from the US and Canada shows that some companies with private forests are not always good forest managers. Some have logged their lands quickly for short-term profit and made few investments in regeneration. But others treat their forests as the foundation of their long-term survival. Community tenures may also result in a more stable, long-term vision of forest management. Alternatively, some communities might support a short-term timber emphasis as they seek immediate employment and prosperity. Small private holdings would require capacity-building and stable investment sources. And just as some large firms have, individuals may also log their lands quickly for the sake of short-term profit. In Scandinavia the governments act to blunt some of these negative effects by regulating private forestry, requiring forest plans, setting cutting rates, and financing forestry, all while supporting the private forest context.

In Canada tenure reform will require a careful consideration of the lessons seen in other places. While we can look to other jurisdictions for information and experience, Canada's forests, geography, and culture are distinct. As part of addressing Canada's forest management problems, tenure reform will require innovative approaches. Outlining the various solutions and evaluating their potential is more than can be done here; each tenure alternative has its promises and pitfalls.

The answer lies in creating a tenure context that provides new opportunities for individual landowners, community-based tenures, and conditions that encourage stable long-term business investments, and that supports a more complex vision of what forests provide. Tenure reform must be part of realizing more sustainable forests and forest communities across Canada.

Kevin Hanna teaches in the Geography and Environmental Studies Department at Wilfrid Laurier University in Waterloo, Ontario, where his main research interests are in community-based resource management.

combat this perception. They have also started to look more closely at the basis for these accusations, and substantial changes are underway in many aspects of the forest industry in Canada.

FOREST MANAGEMENT PRACTICES

The provincial governments are responsible for 71 per cent of the nation's forests, with the federal and

BOX 9.1 THE COMPETITIVENESS OF CANADA'S FOREST INDUSTRY

Canada is the world's largest exporter of forest products. In the international marketplace, Canada has a number of assets, including:

- abundant forest resources;
- high-quality forest resources;
- access to low-cost energy to transform industrial roundwood into secondary forest products;
- proximity to major markets;
- access to well-established export markets;
- rapid expansion of the value-added products sector;
- world-class forestry product research.

Despite these many advantages, the forest industry is vulnerable. Forest products are highly sensitive to economic fluctuations and can suffer severe losses during periods of recession. In addition, the imposition of non-tariff barriers, including codes, standards, eco-labelling requirements, and policies regarding recycled content, can retard growth of the forestry industry. The cost of production commodities such as energy, labour, and transportation also plays a key role in competitiveness. As commodity prices continue to rise, the industry must increase efficiency to maintain market share. However, improved efficiency often entails an investment in technology, leading in turn to increased skill requirements for workers and/or a loss of jobs. Furthermore, rising environmental standards, including international certification schemes, are putting significant pressure on the Canadian government to develop ways of implementing standards at minimal cost. To remain competitive and retain a dominant position in the world marketplace, Canada's forest industry will have to tailor products to consumers' increasing expectations, develop new value-added products, develop new markets, and cultivate an image of an environmentally responsible industry.

SOURCE: Natural Resources Canada (2003b).

Table 9.2	Canada's Forests	
Total land		921.5 million ha
Total forest		417.5 million ha
Commercial forest		234.5 million ha
Managed forest		119.0 million ha
Harvested forest		1.0 million ha
Status of harvested Crown land (1999):		
Stocked (88%)		14.5 million ha
Understocked (12%)		2.0 million ha
Value of international and domestic shipments (1999)		$73.6 billion
Value of exports (2001)		$44.1 billion
Contribution to the balance of trade (2001)		$34.4 billion
Contribution to the GDP (2001)		$28.5 billion
Direct employment (2001)		352,800
Annual allowable cut (1999)		225.3 million m^3
Harvest (1999)		193.2 million m^3
Value of non-timber forest products		$442 million

SOURCE: Natural Resources Canada (2002).

territorial governments responsible for 23 per cent. The remaining 6 per cent is managed by 425,000 private landowners. These figures are likely to change over the next decade as an increased number of land claims by Aboriginal peoples are settled and more land comes under their control. Areas under current land claims account for about one-quarter of large, intact forest landscapes in Canada. Currently, forest management is primarily a provincial responsibility with governments managing forest resources on behalf of the public through agreements with private logging companies. Different forms of tenure exist, but generally all involve the submission of plans by the

The Clayoquot Peace Camp became the centre for the Clayoquot protests (*Philip Dearden*).

BOX 9.2 FORESTS: A GLOBAL PERSPECTIVE

- Some 40 per cent of the land surface of the Earth supports trees or shrubs. Approximately 27 per cent is covered in trees, with the rest in open scrub.
- The world's total forest cover in 2000 was 3.86 billion hectares, of which 44 per cent is found in the developed countries and 56 per cent in the developing countries.
- The largest forested areas are in Europe (1,040 million ha), South America (874 million ha), Africa (650 million ha), Asia (548 million ha), and North America (471 million ha). Four countries—Canada, Russia, the US, and Brazil—contain over 50 per cent of the world's forests.
- The world's forests declined by 90.4 million ha during the 1990s, a net loss of 9 million ha, or 0.2 per cent annually. In the tropical countries, total forest loss is estimated at 12 million ha per year, for a total of 120 million ha during the 1990s. During the 1980s, the area of tropical forests declined by an average of 15.4 million ha annually.
- Forest plantations now cover more than 187 million ha (less than 5 per cent of total forested area), and account for 20 per cent of wood production.
- Estimates suggest that over 80 per cent of the world's terrestrial species are found in forests. The tropical forests are our richest terrestrial biome. Tropical forests occupy only 7 per cent of the world's land area, but they contain over half of the world's species. The volume of global wood consumption has more than doubled since 1960.
- Developing countries consume more than 80 per cent of their wood as fuel; in developed countries only 16 per cent goes to fuel, with the rest being processed as wood products. Approximately 1.5 billion tonnes of wood is harvested for fuel annually worldwide.
- Most wood products (85 per cent) are used domestically.
- Wood products, including wood for fuel, are valued at an estimated US$400 billion, while exports account for about 3 per cent of world trade. Between 1961 and 1990 global trade in forest products increased 3.5 times. Developing countries are responsible for only about 13 per cent of world trade in wood products.
- Global forests provide wage employment and subsistence equivalent to 60 million work years worldwide, 80 per cent of which is in developing countries.
- People in developing countries consume much less wood products (30 cubic metres/1,000 people) and paper (12 tonnes/1,000 people) than people in developed countries (300 cubic metres/1,000 people and 150 tonnes of paper).
- It is estimated that an additional 50–100 million ha of forests will be needed to meet the projected wood demands of developing countries by the year 2010.
- Worldwide, an estimated 290 million ha of forested land are under protection from logging. However, of 200 areas of high biological diversity, illegal logging threatens 65 per cent. Illegal logging is estimated to cost governments approximately $15 billion annually.

SOURCE: FAO (2003).

Can tropical forests, such as these in Sri Lanka, sustain the needs of both the people and animals that depend on them? (*Philip Dearden*).

logging company that outline where it intends to cut, the details of the harvesting process (including the location of roads), and **reclamation** plans. The governments provide regulations and guidelines for these practices and have the authority to ensure that these are followed.

Of the total of 417.5 million hectares of forest land in Canada, some 119 million hectares are currently managed for timber production (Table 9.2). On these lands, forest ecosystems are being transformed from relatively natural systems to control systems, as described in Chapter 1, in which humans, not nature, influence the species that will grow there and the age that they will grow to. Over the last decade, increasing awareness and concern about the environmental impacts of forest harvesting have raised questions about the environmental sustainability of forestry, and about the different kinds of management approaches that might lead to sustainability. Key questions relate to the amount of forest reserved from logging, the amount of fibre harvested over a specific time period, the way in which it is logged, and what happens to the land after harvesting.

Rate of Conversion

The rate of conversion of natural to managed forests is one of the most controversial issues in Canadian forestry. It has increased dramatically over the years (Figure 9.3). Each province establishes an **annual allowable cut (AAC)**, which is based on the theoretical annual increment of merchantable timber, after taking into account factors such as quantity and quality of species, accessibility and growth rates, and amounts of land protected from harvesting because of other use values, such as parks and wildlife habitat (Table 9.3). The AAC should reflect the long-run sustained yield (LRSY) of a given unit of land or what that land should yield in perpetuity. This is ultimately limited by the growth conditions, the biological potential of the site, and how that potential can be augmented by silvicultural practices. It is not sustainable to have an AAC that consistently exceeds this biological potential. Economists, however, often argue for the need to maximize the monetary return of the first cut in order to invest in other wealth-producing programs and to provide social services. The dominance of this line of thought has led to rates of conversion significantly higher than can be supported biologically.

To calculate AACs it is also necessary to know the rotation period for each forest type. This is the age of economic maturity of the tree crop and varies widely, but usually falls within the 60–120 year range in Canada. Foresters call this the **culmination age**.

The AAC will also vary quite substantially depending on the proportion of old-growth to second-growth timber included in the proposed cutting unit. Old-growth forests have very high timber volumes, for example, up to 750 cubic metres/ha in BC's coastal forests. However, at the culmination age for **second growth** on these

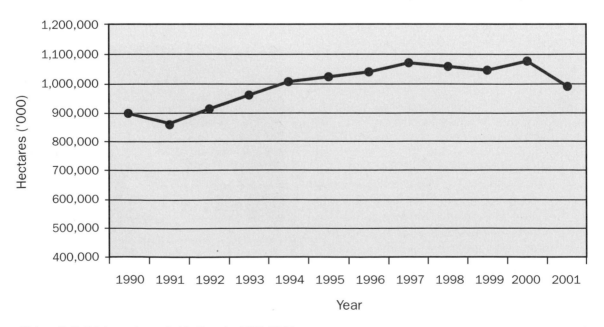

Figure 9.3 Total area harvested in Canada, 1990–2001.

| Table 9.3 | Selected Forest Resource Statistics |

Province/ Territory	Land Area (million ha)	Forest Land (million ha)	AAC (1999)	Harvest Volume, Industrial Roundwood (1999)	Harvest Area (ha), Industrial Roundwood (1999)	Area (ha) Defoliated by Insects (1999)	Area (ha) Burned (2001)
NWT	329.3	61.4	236,500 m^3	71,271 m^3	547	487,556	111,262
Que.	135.7	83.9	58.2 million m^3	43.3 million m^3	384,208	1.1 million	33,068
BC	93.0	60.6	71.1 million m^3	75.0 million m^3	176,312	n.a.	9,668
Ont.	89.1	58.0	0.4 million ha	28.1 million m^3	201,522	13.5 million	10,733
Alta	64.4	38.2	27.3 million m^3	21.9 million m^3	42,210	3.6 million	153,459
Sask.	57.1	28.8	0.7 million m^3	4.5 million m^3	21,169	438,883	183,820
Man.	54.8	26.3	9.6 million m^3	2.2 million m^3	15,509	181,614	86,199
Yukon	48.3	47.9	352,000 m^3	253,326 m^3	1,034	n.a.	17,772
Nfld & Lab.	37.2	22.5	2.7 million m^3	2.7 million m^3	17,415	35,121	1,275
NB	7.2	6.1	11.1 million m^3	11.3 million m^3	111,077	760	604
NS	5.3	3.9	6.7 million m^3	6.2 million m^3	49,680	0	530
PEI	0.57	0.29	0.5 million m^3	0.7 million m^3	5,780	0	27
Canada	921.5	417.6	225.3 million m^3	193.2 million m^3	1.03 million	6.3 million	629,836

SOURCE: Natural Resources Canada (2002).

sites, volumes will be much lower, in the region of 500 cubic metres. This is known as the **falldown effect** and will result in AACs that are up to 30 per cent less as old-growth forests are cleared out.

For Canada, the total AAC is calculated by adding together all the provincial and territorial AACs where these figures are available. In 1999 the AAC stood at 225.3 million cubic metres. In most provinces the AAC is greater than the harvest, but in some regions the AAC is approaching or exceeds the harvest for the most important component, the softwoods. This is leading many provinces to review their rates of conversion. Case studies estimate that the rate of logging in Canada would have to decline by 10 to 25 per cent in the boreal forests of Canada and 30 to 40 per cent on the coast of BC to address broader forestry sustainability objectives (Global Forest Watch, 2000). In the face of increasing world demand (which is expected to rise between 20 and 50 per cent by 2020) and the economic value of timber products, it will be difficult to reduce harvesting rates to a level to ensure that the industry can proceed on a sustainable basis.

Silvicultural Systems

Silviculture is the practice of directing the establishment, composition, growth, and quality of forest stands through a variety of activities, including harvesting, reforestation, and site preparation.

Harvesting Methods

Perhaps no aspect of resource or environmental management has created as much conflict in Canada as the dominant forest harvesting practice of **clear-cutting**. In 2001, more than 85 per cent, or 889,157 hectares, of the forest lands harvested were clear-cut (Figure 9.4) (National Forestry Database, 2003). The size of the clear-cuts varies widely, from approximately 15 hectares to over 250 hectares, and in some cases clear-cuts extend for many thousands of hectares. Although openings of this size have few ecological advantages, it should not be assumed that more, smaller clear-cuts are necessarily superior to fewer, larger clear-cuts. More clear-cuts create more fragmentation and less undisturbed forest area to the detriment of 'interior' forest species.

Not only are clear-cuts aesthetically unappealing to many Canadians, but their environmental

The rate of conversion from natural to managed forests has alarmed many environmentalists, who claim that forest companies have been allowed to extract too much wood too quickly. This will result not only in environmental problems but also in a lack of adequate fibre for industrial use in the future (*Philip Dearden*).

impacts, especially cumulatively as they spread across the landscape, can be substantial. For this reason, Nova Scotia has implemented guidelines for the establishment of wildlife corridors to connect habitat. If clear-cuts are in excess of 50 hectares, the guidelines recommend that at least one corridor be created, with irregular borders and a minimum width of 50 metres. While this is a step in the right direction, it is the implementation of such guidelines, not just their specification, that is important. British Columbia tried to take a tougher approach by legislating forest practices in detail through the Forest Practices Act (Box 9.3). Yet, current government policy has seen considerable weakening of the implementation of the legislation.

Clear-cutting is the preferred means of harvest on vast areas. It is the most economical way for the fibre to be extracted and also allows for easier replanting and tending of the regenerating forest.

A clear-cut area (*Philip Dearden*).

In certain types of forests it may mimic natural processes more closely than selective or partial cutting systems. This is especially true where natural fires have created even-aged stands of species such as lodgepole and Jack pine, black spruce, aspens, and poplars. Researchers working in the eastern boreal forest, however, are starting to question this assumption (Bergeron et al., 2001; Lesieur et al., 2002). They have found that there has been a dramatic decrease in fire frequency in this area over the last 150 years as a result of climatic change, and they suggest that this trend will continue into the future, leading to higher proportion of old-growth forest in the landscape. In turn, this will lead to natural species replacement in these forests, with deciduous and mixed stands replaced by balsam fir and Jack pine by black spruce. Clear-cutting would counteract these changes rather than mimic natural processes, and would lead to a dramatic decrease in stand diversity

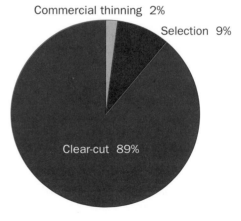

Figure 9.4 Area harvested by harvesting method, 2001. SOURCE: Canadian Council of Forest Ministers (2003).

BOX 9.3 THE BC FOREST PRACTICES ACT

Concern over how forest harvesting was being practised in British Columbia led that province to introduce a Forest Practices Act and Forest Practices Code in 1994 that established a consolidated package of legislation, regulations, standards, and field guides to govern forest practices. The Code sets minimum provincial standards for various practices. For example, clear-cutting will not be allowed on sites:

- with very unstable terrain;
- within critical wildlife habitat areas, for which the forest canopy is essential to maintain wildlife populations;
- in streamside management zones;
- in old-growth management areas;
- where visual quality is a prime consideration;
- on other sensitive sites identified in the Standards documents.

Also, new requirements maintain biodiversity and provide for higher-than-normal standards of forest practice in areas that have high non-timber values. Furthermore, all resource values must be considered in forest operations with full public involvement in all key plans and prescriptions. Although many environmentalists welcomed the Act enthusiastically, this has waned considerably over the years as many loopholes have become apparent where less than rigorous forestry practices can still take place. Considerable discretion, for example, is given to the regional manager for the Ministry of Forests about whether the Code has been applied and whether public involvement is necessary. In a few high-profile cases, the judgement of the manager in making pro-industry decisions has been called into question, even by other government departments.

One specific example of this relates to the case study discussed in Box 9.9 on the endangered northern spotted owl, which is dependent on old growth for its survival. As there is no endangered species legislation in BC, the Forest Practices Act is the main legislation available to protect the dwindling population. A regional forest manager recently approved an application to log in old-growth forests inhabited by the owl despite the objections of conservationists.

Some people, however, question whether it is realistic to have a province-wide set of practices in such a diverse province. The regulations form a stack of paper over one metre high, but even this voluminous documentation cannot account for every diverse situation that might arise. There is also a very real cost to the forest industry and the province through implementing the Code. Even now, BC logging costs are among the highest in the world, with BC pulp having the distinction of being the highest in the world according to the accounting firm Price-Waterhouse. In its annual report on the pulp industry for 1995 Price-Waterhouse determined that the Code is costing companies $10 per cubic metre of wood sold on the coast and $8 in the interior of the province. In 1995, 67 million cubic metres were harvested from Crown land. It cost the companies $600 million to comply with the Code, half as much as their 1995 earnings of $1.28 billion. In addition to these costs are others related to loss of revenue from trees not harvested as a result of the Code, administration, retraining workers, and the research necessary to implement the Code. UBC forest economist David Haley has estimated the total to be in the region of $2.1 billion, or about 40 per cent of the province's education budget.

at the landscape level. The authors suggest that in the future forest management will have to employ mixed harvesting systems (Table 9.4) if forest management is to more closely emulate natural systems.

Reforestation

Until 1985, Canada's forests were considered to be so extensive that little effort was given to reforestation. Sites, once logged, might be burned to facilitate rapid nutrient return to the soil, but then

were abandoned in the hope that they would be satisfactorily recolonized by seeds from the surrounding area. Sometimes this was successful, but often it was not. Thus, with increased harvesting levels, the amount of land that no longer supported trees that could be harvested in the future gradually grew. This land, known as not satisfactorily restocked or NSR land grew from 585,000 ha in 1975 to 2.0 million ha by 1999. In 2001, 1.03 million hectares of land were harvested, a further 629,826 hectares were burned by fire,

Figure 9.5 New forestry. SOURCE: *Victoria Times-Colonist.*

and 6.3 million hectares were defoliated by insects, yet only 538,036 hectares were planted and reseeded, and the success of these plantings cannot be guaranteed. Given this annual deficit between what is cut and that which is replanted, many conservationists have difficulty understanding further allocation of old-growth timber to the forest industry. Nonetheless, the amount of land replanted over the last decade has increased considerably (Figure 9.6), largely due to the intervention of the federal government in subsidizing these efforts.

Site Preparation—Biocide Use

Biocides are used on forest lands in Canada to reduce competition for seedlings on replanted sites and to protect seedlings against insect damage. Sites regenerating from forest harvesting return to an earlier successional phase (Chapter 3), and, under natural conditions, a vigorous and diverse secondary succession takes place. However, for

many years this community will not be dominated by the commercial species desired by foresters. Chemicals are used to suppress early successional species, to compress the successional time span, and to maximize the growth potential of the commercially more desirable species, usually conifers. Chemical use is generally quicker, easier, and more effective than using mechanical alternatives for weed suppression. Three herbicides (2,4-D, glyphosate, hexazinone) are registered for forest management in Canada. Glyphosate (or 'Roundup'), the most widely used, affects a broad spectrum of plants but degrades quickly and is relatively non-toxic to terrestrial animals.

Research shows that early colonizers often compete more effectively for soil nitrogen than conifers (Hangs et al., 2003). Where nitrogen is the most limiting factor, as it is in the boreal forest, then this can inhibit conifer growth over the short term. However, the law of conservation of matter

Table 9.4	Main Characteristics of Common Silvicultural Systems Practised in Canada

Clear-cutting:

The most commonly applied silvicultural system in Canada, clear-cutting involves the removal of all trees in a cutblock, in one operation, regardless of species and size. Some trees are left along riparian zones to protect streams. The objective is to create a new, even-aged stand, which will be regenerated naturally or through replanting.

ADVANTAGES:	DISADVANTAGES:
• Cost-effective. Clear-cut areas are easily accessed for site preparation and tree planting.	• Nutrients stored in the bodies of trees are removed from the ecosystem.
• Stands of even-aged trees are created, producing wood products with more uniform qualities.	• Loss of habitat for some wildlife species. Loss of biodiversity. New vegetation does not maintain the complexity and stability of mature forests.
• Newly planted seedlings quickly take root and grow in the sunlight reaching the ground. This can benefit certain animal species.	• Clear-cutting in sensitive ecosystems can cause soil erosion, landslides, and silting, which can damage watersheds, lead to flooding, and inhibit successful fish reproduction.
• In some respects, clear-cutting simulates natural disturbances such as wildfire and insect disease outbreaks.	• Large gaps are opened up, fragmenting the forest and exposing more area to the edge effect.
• Safest harvesting method with least risk of worker injury.	• Aesthetically unattractive. Clear-cutting can conflict with other forest values.
	• No timber products for a long period of time (e.g., 50–70 years).

Seed tree:

Method of clear-cutting in which all trees are removed from an area in a single cut, except for a small number of seed-bearing trees, which are intended to be the main source of seed for natural regeneration after harvest. Tree species that have been managed under this system in Canada include western larch, Jack pine, eastern white pine, and yellow birch.

ADVANTAGES:	DISADVANTAGES:
In addition to the above	In addition to the above
• Next to clear-cutting, this system is the least expensive to implement.	• Regeneration can be delayed if seed production and/or distribution are inadequate.
• This system can result in improved distribution of seedlings and a more desirable species mix, since the seed source for natural regeneration is not limited to adjacent stands.	

Shelterwood:

Mature trees are removed in a series of two or more partial cuts. Residual trees are left to supply seed for natural regeneration and to supply shelter for the establishment of new or advanced regeneration. The remaining mature cover is removed once the desired regeneration has been established. Thirty to 50 per cent canopy removal on the first cut is common. In Canada, this system has mainly been applied to conifers (e.g., red spruce in the East, white pine in Ontario, interior Douglas fir in the West).

Continued

Advantages:	Disadvantages:
• Trees left after the first cut grow faster and increase in value.	• Complex and costly system to plan and implement.
• For eastern white pine, this system can be an effective management tool against white pine weevils, which are more attracted to pine shoots under full light exposure than under shade.	• Young trees can be damaged during removal of mature trees.
• Visually more appealing than clear-cutting.	• Windthrow is a serious concern. Uprooted stems can displace significant amounts of soil.

Selection:

Involves the periodic harvest of selected trees of various ages in a stand. Trees are harvested singly or in groups as they reach maturity. Valuable, mature trees, along with poorly shaped, unhealthy, crooked and leaning trees, and broken or damaged trees, are selected for removal. The objective of this method is to create and maintain an uneven-aged stand. Small gaps created by harvesting leave room for natural seeding.

Advantages:	Disadvantages:
• This method is often favoured in areas where recreation or scenic values are important, since the harvested area is less visually offensive.	• This system can only be successfully applied to stands containing shade-tolerant tree species (e.g., sugar maple, western red cedar, red spruce, balsam fir, eastern/western hemlock).
• This method results in a continuous, regular supply of mature trees over time. Overall stand quality should improve after each harvest cycle.	• Requires skilled workers to implement successfully.
• Biodiversity loss is minimized.	• This method can require the maintenance of more roads and skid trails per unit area.
	• Complex and costly system to plan and implement.
	• In some instances, landowners take the best trees in a forest, leaving poor shaped and unhealthy trees to provide seed for the next generation, resulting in forest degradation over the long-term.

Source: Natural Resources Canada (1995); Alberta Food and Rural Development (2001); Canadian Institute of Forestry (2003).

(Chapter 4) tells us that these nutrients have not disappeared—they are simply being held by different species. As these species die and decay the nutrients will once more be returned to the soil and become available for uptake. Furthermore, by holding nutrients in this way, early colonizers often slow down the loss of nutrients from the site that might otherwise occur as a result of leaching. They act as a biological sponge over the short term. Herbicide application may also eliminate species that are ecologically advantageous, such as nitrogen fixers like red alder, exacerbating nutrient loss from logged sites.

The balance between these effects needs to be evaluated over long time periods and probably differs from site to site. Lautenschlager and Sullivan

(2002) reviewed the literature on the effects of forest herbicide applications on major biotic components of regenerating northern forests. They conclude that there is currently very little evidence to support any long-term negative consequences. Indeed, they suggest that, at the landscape scale, application of herbicides might help the return to a more natural species composition in northern forests. This is because, since colonization, hardwoods have expanded in these areas as coniferous species have been harvested. Use of herbicides may help to reverse this process. Nonetheless, Lautenschlager and Sullivan caution that there are still large gaps in knowledge before we can truly understand all the biotic implications of herbicide applications.

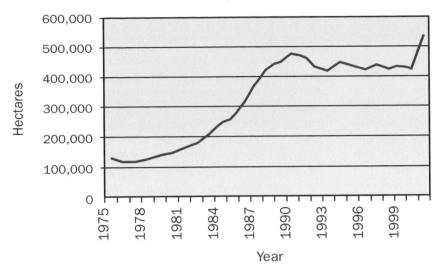

Figure 9.6 Estimated forest area planted in Canada (ha), 1975–2001.

Insecticides are used to attack pests such as the spruce budworm, Jack pine budworm, hemlock looper, mountain pine beetle, gypsy moth, and forest tent caterpillar. The amount sprayed varies, depending on the population dynamics of these insects, which changes in response to environmental factors. Between 1980 and 1996, consecutive years of defoliation by the spruce budworm affected a total of more than 69 million hectares of forest (Global Forest Watch, 2000). Most spraying has occurred in eastern Canada; the spruce budworm spraying program in the Maritime provinces is the best-known incidence of such spraying and has caused considerable controversy (Box 9.4). The total area treated with forest insecticides in Canada, however, has declined from a high of 1,312,915 ha in 1990 to 211,511 ha in 2000 (Canadian Council of Forest Ministers, 2003).

The health and ecological concerns arising from widespread application of chemicals (discussed further in Chapter 10) have led to several high-profile confrontations. Attention is being increasingly directed towards the replacement of synthetic insecticides with biological control agents such as *Bacillus thuringiensis* (Bt). Bt now accounts for some 60 per cent of the insecticides used in Canadian forests and is non-toxic to humans and most wildlife, although it does affect moth and butterfly larvae of some non-target species. Plants also manufacture many chemicals themselves as protection against insects. Several of these appear to be good prospects for the development of insecticides for forestry use in the future (Helson, 1992).

Intensive Forest Management

After a new crop of trees has been established and reaches a free-growing condition, future timber resource values can be further enhanced by intensive silvicultural practices. Activities undertaken by foresters to improve stand growth include:

- pre-commercial thinning (homogenizes the stand, increases mean tree size, and lowers the age at which the stand can be harvested);
- commercial thinning (attempts to recover lost volume production in a stand due to competition-induced mortality);
- scarification (physical disturbance of the forest floor to create improved seedbeds for natural regeneration);
- prescribed burning (removes slash and woody debris, sets back competing vegetation, provides ash as fertilizer, and increases nutrient mobilization and availability through increased soil temperatures);
- pruning and shearing (increases the value of individual trees by prematurely removing the lower branches so that clear wood, free of knots, is laid down around an unpruned knotty core);
- timber stand improvement (cutting down or poisoning all deformed and unwanted trees

Alder is a common pioneer of secondary succession on logged-over sites. The smaller species in the foreground is broom, a non-native plant from Europe. Both these species biologically fix nitrogen into the soil and help fertility levels to recover (*Philip Dearden*).

BOX 9.4 THE SPRUCE BUDWORM CONTROVERSY

Every year since 1952, aircraft have doused the forests of eastern Canada, particularly New Brunswick and Nova Scotia, with an array of chemicals in the competition to see who would harvest the area's lumber—humans or the eastern spruce budworm. A native in Canada, the budworm feeds primarily on balsam fir, but will also eat white spruce, red spruce, and, to a lesser extent, black spruce. Its range extends wherever balsam fir and spruce are found, from Atlantic Canada across to the Yukon. Damage varies considerably from one stand to another. Mortality rates are related to stand composition and age, as well as site quality (soil, water, climate). Ordinarily, the influence of the insect goes unnoticed in a forest, but its impact occasionally reaches epidemic levels with massive damage to the host trees. Outbreaks last between six and 10 years and have been documented for the last two centuries as the product of a long-term budworm-fir ecological cyclic succession. However, spruce budworm epidemics are occurring with increasing frequency as a result of human intervention related to forest harvesting. Human intervention has reduced the natural diversity of the forest through removal of preferred species such as white pine, creating a less diverse forest composed of large areas of mature balsam fir, the budworm's preferred food. Extensive mortality occurs in stands that have suffered defoliation for several years. In 1975, at the height of infestation, 54 million hectares of forest were defoliated, resulting in serious economic losses for the forestry industry.

The long-term ecological stability of the budworm-fir system does not match the shorter-term demands of the economic system dependent on the forests for products. As a result, the forests have been extensively sprayed to limit defoliation and mortality. When spray programs started in 1952, DDT was the chemical of choice, and some 5.75 million kg were sprayed in New Brunswick alone between 1952 and 1968, when use was suspended. Other chemicals replaced DDT, such as phosphamidon, aminocarb, and fenitrothion, until questions were raised about their ecological and health impacts. Phosphamidon, for example, is very toxic to birds. Up to 1985, 118.5 million ha (mostly in New Brunswick) were sprayed, some areas on an annual basis. Continual spraying appears necessary once natural controls are disrupted.

The organophosphate, fenitrothion, became the most popular chemical. It also became the source of heated controversy as to its health and ecological impacts. In particular, there was concern over the link between the chemical and Reye's syndrome, a rare and fatal children's disease. As a result, and due to unfavourable reviews of the ecological impacts of the chemical, particularly on songbirds, the use of fenitrothion for aerial applications was cancelled by 1998. The biological control *Bacillus thuringiensis* (Bt) is now being used more extensively. In some areas this is supplemented by use of an insecticide called Mimic that kills the budworm through interruption of the moulting process, starving the larvae to death. Although Mimic is not a broad-spectrum biocide, it also has the same lethal effect on the larvae of butterflies and moths. Nonetheless, it has been used in some areas, such as Manitoba, since 1998.

Attention is also being devoted to other less toxic approaches, such as species and landscape diversification, so that large areas of mature balsam fir do not dominate the landscape. Biological control is also being investigated with a parasitic wasp that attacks the larvae of the budworm.

Management of spruce budworm outbreaks provides a graphic example of the challenges presented by the conflict between longer-term ecological cycles and shorter-term economic dependencies. An interesting paper by Miller and Rusnock (1993) uses the case of the spruce budworm to analyze the role of science in policy-making. The authors conclude that the scientific uncertainties involved render the advice of scientists of little use to policy-makers.

within older stands) (Natural Resources Canada, 1995).

The long-term impacts of these activities are, in general, not well known. Thompson et al. (2003) reviewed the published information on some of the likely impacts on vertebrate wildlife on boreal forests in Ontario. One important impact identified was the structural simplification that occurs with increased tending. However, overall they conclude that intensive forest management will benefit some species and have negative effects on others and that currently there is little evidence suggesting that these techniques will have any greater impact than more traditional silvicultural activities.

New forestry would devote much more attention to stand tending on commercial sites, such as this one, where stand limbing and thinning enhances growth (*Philip Dearden*).

Fire Suppression

In certain areas, fire is a frequent occurrence and necessary to the reproduction of forest tree species. Fire is part of the long-term dynamics of these ecosystems. Fire initiates secondary successions, renewing vegetation through: regeneration involving a complete change of species; regeneration of the same species; or diversification of the species. Jack pine, birch, and trembling aspen are common in areas where fires have recently occurred. Fire suppression, viewed as essential to protect lives, property, and commercially valuable timber, has resulted in ecological changes not characteristic of fire-dominated ecosystems. For example, it has contributed to the very dense regeneration of almost pure Douglas fir in old-growth ponderosa pine stands in the interior of BC. Old-growth ponderosa pine stands are maintained by low-intensity, naturally occurring surface fires that burn brush and prevent maturation of the more shade-tolerant Douglas fir.

In the absence of recurring fires, ground fuel may accumulate over time, increasing the risk of a major wildfire event (Box 9.5). The area affected by wildfires in Canada each year is immense: over the decade of the 1990s, an average of 8,248 fires burned 3.2 million hectares annually, including more than 700,000 hectares of commercial forest

land, or 74 per cent of the annual area harvested (Neave et al., 2002). In addition to fire suppression policies, other reasons for the gradual increase in area burned over the past 30 years include higher temperatures and dry and hot summers.

ENVIRONMENTAL AND SOCIAL IMPACTS OF FOREST MANAGEMENT PRACTICES

Change in species and age distributions arising from forest management practices has a major impact on ecological processes such as energy flows, biogeochemical cycles and the hydrological cycle, and the habitat for other species. We have only a rudimentary knowledge of how forest ecosystems function. It is therefore difficult to be precise about the possible impacts of wholesale conversion from complex natural to more simple human-controlled systems. In addition, there are important differences among forest ecosystems. Some, such as the boreal forest, have naturally evolved with periodic disturbances such as fire or insect attack that stimulate forest renewal. Others, such as the rain forests of the west coast, have little history of disturbance. The difference between disturbance through forestry and natural processes (Box 9.6) must be considered against this background. One essential difference is that natural disturbances such as fire or insect attack do not result in the physical removal of the biomass from the site; it is merely converted from one form to another at that site, consistent with the law of conservation of matter (Chapter 2). In contrast, logging results in the physical translocation of nutrients from the site. The closer that forest harvesting approximates the conditions of natural perturbations, the less disturbing it will be to ecosystem processes.

Chapter 2 described how energy flows through ecosystems. It is also stored in various compartments. Trees, for example, are energy stored in the autotrophic component of the ecosystem; deer represent storage in the herbivorous component. Thus, a forester may wish to maximize the energy storage in trees and minimize energy losses to herbivores, whereas a wildlife or recreation manager may wish to move the energy storage further up the food chain to support more wildlife. Conflicts therefore arise about where energy should be stored in ecosystems to optimize societal values. This section describes some of the environmental and social implications of forestry activities.

BOX 9.5 BRITISH COLUMBIA'S BURNING

'Imagine a landscape in only black and white ... a place between a beginning and an end where the only sound is the whisper of flames and the only colour is fire-red.'
— McLure, BC, firefighter, 2003

This was the sinister scene as catastrophic flames surged across the BC Interior, consuming over 175,000 hectares, demolishing more than 350 homes and businesses, and forcing the evacuation of over 50,000 people from their homes. By midsummer of 2003, a state of emergency was declared. Highways and parks were closed as thousands of firefighters, Canadian military, emergency response workers, and community volunteers battled over 800 forest fires burning throughout the province. The areas that suffered the greatest hardships and adversity from the disaster were the smaller communities, such as Barriere, McLure, Louis Creek, and Kelowna. The total economic impact on the province as a whole was estimated at more than $500 million.

Although many factors contributed to the violent summer fires, including several years of drought, global warming, and carelessly discarded cigarette butts, perhaps the greatest culprit was decades of poor forest management practices that focused on timber production and fire suppression. For example, the 2003 fires may well have been aggravated by the fact that naturally occurring, low-intensity ground fires, which normally release nutrients into the soil, kill insect infestations, and clear the forest floor of debris and small trees, have been suppressed for generations. When a fire finally started, instead of being a restorative force, the excessive debris caused an environmental disaster burning the entire forest area to the ground.

Catastrophic fires like those that swept through the BC Interior do not just move along the ground—they race through the treetops and the root networks, killing even those tree species, like ponderosa pine, western larch, and Douglas fir, that have evolved thick bark to survive fire. Furthermore, such huge fires do not release nutrients into the soil but vaporize them into the air, generating enough heat to create their own weather system with conditions that tend to perpetuate even more fires.

In future, we need to recognize that fire is a natural and essential part of the forest ecosystem. Rather than trying to stop all fires, or letting every fire burn, we need to use selection, cutting, and controlled burns as a way to thin out fuel and restore vigour and diversity to our over-managed forests. In short, we need to start managing forests for overall ecosystem health rather than for timber production alone.

Firefighting in Kitimat, British Columbia (*Andrew Farquar/Valan Photos*).

BOX 9.6 FOREST DISTURBANCE: NATURAL VERSUS CLEAR-CUT

Some forests are more susceptible to disturbance than others, and different disturbances have differing impacts. There are similarities between some disturbances and clear-cutting, but there are also important differences. Some of the differences between the effects of fire and clear-cutting are summarized below.

- Openings created by fire are generally irregular in shape with high perimeter-to-edge ratios that facilitate natural reseeding. Boundaries tend to be gradual rather than the abrupt edge of a clear-cut.
- Fires will leave standing vegetation in wet areas that continue to provide habitat for wildlife and act as a natural seed source. Clear-cuts remove all trees.
- Fires tend to kill pathogens; clear-cutting allows many pathogens to survive.
- Fire releases nutrients to the soil; clear-cutting removes nutrients in the bodies of the trees.
- Fire helps break up rock that aids in soil formation; clear-cutting tends to physically disturb the site leading to compaction and erosion.
- Fire stimulates growth of nitrogen-fixing plants that help maintain soil fertility; clear-cutting does not.
- Fire encourages the continued growth of coniferous species in many areas through stimulating cone opening; clear-cutting often leads to dominance by shade-intolerant hardwoods. Ultimately, this changes species composition, as was found in studies in Ontario where in 1,000 sampled boreal sites regenerating poplar and birch had increased by 216 per cent and spruce had fallen by 77 per cent (Hearnden et al., 1992).

Forestry and Biodiversity

Most natural forest land is dominated by forests with old-growth characteristics, although the age of the trees and degree of structural and compositional attributes described for old growth vary greatly across Canada. Old-growth forests have ecological attributes that tend to be absent from forests that have been harvested (Box 9.7). Logged forests undergo a number of changes, including modifications of the physical structure of the ecosystem and changes in biomass, plant species mixtures, and productivity. These changes affect biodiversity directly and indirectly by changing the nature of the habitat. As more than 90,000 species depend on forest habitats in Canada, this is obviously of concern.

Direct changes arising from forestry practices include the effects on the genetic and species richness of a biotic community. Most species contain a wide range of genetic variability that helps them adapt to changes in the environment. As natural forests are replaced by plantations, this natural variability is reduced since most plantation-grown trees are specially selected from the same genetic base to have desirable characteristics. This makes them more susceptible to pest infestations and disease, and less able to adapt to future environmental changes (Box 9.8). For example, it has been suggested that the emergence of the spruce forest moth in central and eastern Canada since 1980 is at least partially due to the establishment of white spruce plantations (Neilson, 1985). Over 85 per cent of the forest harvesting in Canada is done by clear-cutting in which large blocks of forest are removed at the same time. This does not destroy the ecosystem per se in that an ecosystem still exists on that unit of land, but it does dramatically alter the attributes of that ecosystem. For vegetation, changes include removal of the previously dominant trees and their ecological influence, followed by the vigorous growth of other assemblages of plants (where herbicides have not been applied) as the successional process starts again.

Early successional species dominate in the immediate post-harvesting phase, but are much reduced over time as the canopy closes. Tree species such as alder, birch, cherry, Jack pine, poplar, and aspen often fit into this category, along with semi-woody shrubs such as elderberry and blackberry, annual and short-lived perennial herb species such as members of the aster family, and various grasses and sedges. Other species that exist in low abundance in the original forest may survive the harvest and, freed from the competitive suppression of the harvested trees, may dominate the community for some time. Trees such as red maple, yellow birch, and white pine often fall into this category in central and eastern Canada. As the canopy closes, most of these species will eventually be out-competed. Some species that may survive

BOX 9.7 OLD-GROWTH FORESTS

Various definitions of old-growth forests have emerged. This one was suggested by the Forest Land Use Liaison Committee of British Columbia: 'Old growth forests include climax forests, but do not exclude sub-climax or even mid-seral forests.' The age structure of old growth varies significantly by forest type and from one biogeoclimatic zone to another. The age at which old-growth forests develop their characteristic structural attributes will vary according to forest type, climate, site characteristics, and disturbance regime. However, old growth is typically distinguished from younger stands by several of the following attributes:

- large trees for species and site;
- wide variation in tree sizes and spacing;
- accumulations of large dead, fallen, and standing trees;

- multiple canopy layers;
- canopy gaps and understorey patchiness;
- decadence in the form of broken tops or boles and root decay.

Old-growth forests typically contain trees that span several centuries. Based on age, approximately 18 per cent of Canada's forests can be classified as old growth. In Canada, the longest-lived tree species are yellow cedar on the west coast and eastern white cedar in central Canada. Trees of both these species may live more than 1,000 years.

Old-growth forests supply high-value timber, contain large amounts of carbon, contain a large reservoir of genetic diversity, provide habitat for many species, regulate hydrologic regimes, protect soils and conserve nutrients, and have substantial recreational and aesthetic values.

the harvest or invade from surrounding areas may be found through all stages of succession. They have low light compensation thresholds, allowing survival under heavy shading. Balsam fir, hemlock, sugar maple, and beech fall into this category.

Once clear-cuts have had the opportunity to start regenerating, species diversity increases rapidly and usually results in a plant community that is more species-rich and diverse than the harvested community. The exception is where replanting takes place, which involves few species and where steps are taken to reduce competition for plantation trees. Herbicides such as Roundup are commonly applied to achieve this, and may be applied consistently until the planted trees become established. Under these circumstances, an artificial lack of diversity is created, just as a farmer creates a similar system to maximize the amount of energy stored in the particular component that he or she wishes to harvest (Chapter 10).

Naturally regenerating clear-cuts may also produce a higher biomass of herbivorous species such as white-tailed and mule deer that require brushy habitats for at least part of the year. Both the quantity and quality of browse is greater in regenerating clear-cuts up to the time that the canopy starts to close, and is usually optimal in the first 8–13 years. In many parts of their range,

white-tailed deer are more abundant than they were before European colonization when the landscape was covered mostly in mature forest. However, even these species benefit most from a pattern of small clear-cuts as they prefer edge habitats where the protective cover of the forest is not too far distant. Optimal clear-cut size for deer in southern Ontario, for example, was recommended by Euler (1978) as two hectares.

Unharvested forests in most areas across Canada comprise a patchwork of forest stands of different ages and different diversities regenerating from the effects of various natural disturbances such as fire and insect attack. Clear-cutting at a certain rate and scale may not be inappropriate in some of these ecosystems. However, some forests, such as the coastal forests of BC and the mixed deciduous forests of southern Ontario and Quebec, are heavily influenced by the pattern of death of individual canopy trees. For example, more than half of the coastal rain forest is over 250 years in age, and much smaller interventions are required to mimic natural processes.

Some animal and bird species require forests with old-growth characteristics, such as ample lichen growth for woodland caribou or the presence of dead trees to provide nesting cavities for birds such as woodpeckers (Table 9.5). If forest

BOX 9.8 THE WAR AGAINST THE MOUNTAIN PINE BEETLE—WHO'S WINNING?

A large outbreak of mountain pine beetles in the Interior forests of BC is fast becoming one of the highest-profile forest issues to hit the province in the past few years. The mountain pine beetle epidemic is spreading throughout BC's range of lodgepole pine forests due to a combination of natural beetle population cycles, continuous mild winters, and an abundance of uniformly mature pine stands. Since 1997, mountain pine beetles have infested more than 300,000 hectares of pine forests in the central Interior. In previous outbreaks, the beetles have killed as many as 80 million trees over 450,000 hectares. Bark beetles are small, cylindrical insects that attack and kill mature trees by boring through the bark and mining the phloem (the layer between the bark and wood of a tree). The mountain pine beetle and other bark beetles are native species and natural and important agents of renewal and succession in Interior forests. However, when the beetles reach epidemic levels, natural predators like woodpeckers cannot reproduce quickly enough to maintain the insect population at manageable levels.

The beetles have damaged and/or killed an estimated $9 billion worth of lodgepole pine, enough to build 3.3 million homes. The dead and dying trees, visible from the air as huge swaths of red and grey forest throughout central BC, are the source of a hotbed of controversy as government agencies try to combat the rate of infestation through increases in the annual allowable cuts for some areas, reductions in environmental regulations and planning requirements on treatment units, and reduced stumpage (fees paid for logging on public lands). Beetle-infested trees retain economic value and can be salvage harvested, but since wood quality and value decline with time since attack, there is significant pressure to cut beetle-infested timber promptly to maximize the economic value of salvaged trees. The recommended size of clear-cut areas is set at 60 hectares in the BC Interior, but exemptions for salvage loggers have resulted in cutblocks as large as 1,300 hectares, over 20 times the recommended maximum. The salvage clear-cuts can be placed adjacent to previously logged areas that have not yet regenerated, creating potentially very large contiguous openings.

Accelerated cutting of large attacked areas can prevent the decline of timber value in the short term, but a number of social and environmental costs are associated with this approach. Increases in AACs in response to insect infesta-

tions can disrupt forest plans, oversupply markets and cause short-term decreases in timber value, and affect employment levels when workers are no longer needed to support the temporary increase in harvest levels. Research in forest ecology has increasingly recognized the essential role that natural disturbances play in shaping forest structure and maintaining ecosystem processes. Small infested areas create patchiness at the stand level, increasing local diversity, and create habitat and food for a large number of species. In addition, outbreaks accelerate forest succession (Chapter 3) to other tree species in some areas, as well as create the conditions for stand-replacing fires that maintain the dominance of lodgepole pine.

The approach of the Ministry of Forests has received considerable attention from environmental groups, which view the response of government agencies as drastic, particularly in light of a recent government announcement that BC parks and protected areas would not be exempt from the war against pine beetles. Environmentalists generally advocate natural control methods. Cold winters or fires will kill mountain beetle larvae, reducing the size of beetle populations. Some environmental groups, such as the David Suzuki Foundation, are calling for an **ecosystem-based management (EBM)** approach for outbreaks of the pine beetle. An EBM approach (Chapter 6) would seek to decrease the size of the affected area as well as the duration of outbreaks, while not compromising forest biodiversity and ecosystem integrity.

Despite nearly a century of actively managing the mountain pine beetle, efforts to suppress the outbreaks across BC and other parts of North America have been largely unsuccessful. Why do you think this is? What approach would you recommend, given the value of lodgepole timber stands? Do you think provincial parks should be open to clear-cutting as a response to the beetle infestation?

SOURCE: Hughes and Drever (2001).

The mountain pine beetle (*Ron Long/invasive org/IVY IMAGES*).

A view of mountain pine beetle damaged landscape (*Al Harvey/ www.slidefarm.com*).

| Table 9.5 | Forest-dwelling Species at Risk | | |

Mammals	Birds	Plants	Reptiles
Endangered			
American marten	Acadian flycatcher	American ginseng	Blue racer (snake)
Vancouver Island	Kirkland's warbler	Bashful bulrush	Night snake
marmot	Northern spotted owl	Blunt-lobed woodsia	Rocky Mountain tailed frog
Wolverine	Prothonotary warbler	Cucumber tree	
(eastern population)	White-headed woodpecker	Deltoid balsamroot	
Woodland caribou	Western yellow-	Drooping trillium	
	breasted chat	Heart-leaved plantain	
		Large whorled pogonia	
		Nodding pogonia	
		Prairie lupine	
		Purple twayblade	
		Red mulberry	
		Seaside centipede (lichen)	
		Small whorled pogonia	
		Spotted wintergreen	
		Tall bugbane	
		Wood-poppy	
Threatened			
Ermine haidarum	Hooded warbler	American chestnut	Black rat snake
subspecies	Marbled murrelet	Deerberry	Blanding's turtle
Pallid bat	Queen Charlotte goshawk	Goldenseal	Eastern Massasauga
Wood Bison		Kentucky coffee tree	rattlesnake
Woodland caribou		Lyall's mariposa lily	Jefferson salamander
		Phantom orchid	Pacific giant salamander
		Purple sanicle	
		Round-leaved greenbrier	
		Scouler's corydalis	
		White wood aster	
		White-top aster	
		Yellow montane violet	

SOURCE: Natural Resources Canada (1995); Alberta Food and Rural Development (2001); Canadian Institute of Forestry (2003).

harvesting takes place at a rate and scale that eliminates stands with these characteristics from the landscape, these species will decline in number and may become extirpated. Bunnell and Kremsater (1990) suggest that approximately 65 per cent of the bird species found on Vancouver Island, for example, require habitat with snags or logs. The case study presented in Box 9.9, the spotted owl, is a good example of the difficulties associated with maintaining populations of species dependent on old-growth habitat. Other species that have suffered in this regard include the woodland caribou, American pine marten, and marbled murrelets (Neave et al., 2002).

The pressure on high-profile species such as the grizzly bear (Box 9.10) is not the only concern in considering the destruction of old-growth forests; less well-known species, even unknown ones, and the kinds of ecological functions undertaken by these species are also at risk as the old-growth forests are cut away. Some 85 per cent of the species in some forests are arthropods such as insects and spiders. Only recently is the richness of this fauna being realized. There may be over 1,000 species of invertebrates within a single forest stand (Franklin, 1988). Many of these species are new to science and we have little idea of their ecological role in maintaining healthy ecosystems. However, research has indicated a clear pattern of different spider species associated with different stages of the successional process, with at least 30 years required in even the fastest-growing forests for the spider community to recover from clear-cutting (McIver et al., 1990). Studies of the upper branches of coastal forests in BC are revealing very complex predator-prey relationships among many different species. These relationships are not replicated in managed forests, suggesting that continued elimination of old-growth habitat will lead to species extinctions, a decrease in genetic diversity in these communities, and removal of natural controls on forest pests.

One kind of impact on biodiversity that is often overlooked is that resulting from human intrusion to undertake silvicultural and other activities. For example, in looking at the impacts of forestry on grizzly bears in the Selkirk Mountains of BC, Wielgus and Vernier (2003) reported that the main impact seemed to be disturbance from roads rather than avoidance of clear-cuts or young forests.

From this brief review of the impacts of forest-harvesting activities on biodiversity it can be concluded that substantial changes can occur. These changes are beneficial to some species and detrimental to others. It is critical, therefore, to understand the complex ecological relationships involving forests, if these changes are to be fully evaluated and taken into account in designing harvesting activities. At a regional scale, it is essential to maintain some areas in their original state to maintain landscape biodiversity. The size, number, location, and shape of these remnants are critical and will be discussed more thoroughly in Chapter 11.

Forestry and Site Fertility

Forest harvesting removes nutrients from the harvested site (Figure 9.7). The amount of nutrients removed depends on the kind and extent of harvesting. Selective tree whole-stem harvesting will remove relatively few nutrients compared with large clear-cuts of complete-tree (above and below ground biomass) harvesting. The latter will maximize the short-term yield of biomass from the forest, but may compromise the potential of that site over the long term to produce further harvests. More than half of the harvesting in Canada is done with full-tree systems (Figure 9.4).

To judge the potential effects of forest harvesting on site fertility it is necessary to consider the size of the soil nutrient pool, the amount of nutrients being removed, the net accretions and depletions of nutrients in the forests, and the ways in which these variables interact. The process of site impoverishment over time as a result of harvesting is shown in Figure 9.8. On some sites, the proportion of nutrient capital removed in the biomass will be relatively minor, while on others it may be substantial. This depends, to a large degree, on the existing nutrient capital of the site. Areas with high soil fertility will be less affected. Some sites will recover quickly from harvesting and can sustain relatively short rotations. Other sites will not recover adequately between rotations and site nutrient capital will fall, making tree regeneration difficult and in some cases impossible. Thus, nutrient-deficient sites should have long rotations with just stem harvesting in order to maintain productivity.

The amount of nutrients removed by harvesting is influenced by tree species, age, harvesting method, season of harvesting, and other factors.

BOX 9.9 CASE STUDY: THE NORTHERN SPOTTED OWL

In the latter part of the 1980s, the northern spotted owl in the western United States became the focal point of high-profile conflicts between conservationists and logging interests. The owl was accorded threatened status throughout its entire range in the US under the US Endangered Species Act. A 'threatened status' designation means that the owl is likely to become an endangered species within the foreseeable future throughout all or a significant part of its range. This designation requires that critical habitat be identified and a recovery plan implemented. Due to the dependence of the owl on old-growth forests, this decision led to severe conflicts with forest harvesting activities. In 1994 the Clinton administration established reserves on over 4 million hectares where harvesting would be severely restricted.

The northern spotted owl (*John and Karen Hollingsworth/US Fish and Wildlife Service*).

The spotted owl also occurs in Canada in the old-growth forests of southwestern British Columbia. This northern extension of the range is important since individuals at the extremes of a species' range are often of most significance for protection because they may have the genetic diversity best suited for future adaptability. Furthermore, just as in the US, the old-growth forest upon which the owl depends was allocated for harvest. The spotted owl therefore provides a good case study of the impacts of forest harvesting on biodiversity.

Biology and Range

The northern spotted owl is found from northern California to southwestern BC. Between 1985 and 1993, some 39 active spotted owl sites were recorded in BC, totalling a minimum of 71 adult owls. By 2003, government research reported only 25 breeding pairs left in the province. The historic range of the species in the province is probably not that much different from the current range, although the distribution of the owl within that range has changed significantly, due primarily to destruction of its prime habitat, old-growth forests.

Superior habitat for the owls has old-growth characteristics, such as 'an uneven-aged, multi-layered multi-species canopy with numerous large trees with broken tops, deformed limbs and large cavities; numerous large snags, large accumulations of logs and downed woody debris; and canopies that are open enough to allow owls to fly within and beneath them' (Dunbar and Blackburn, 1994: 19). Main prey items for northern spotted owls are small mammals such as the northern flying squirrel and dusky-footed and bushy-tailed wood rats. Both prey species are abundant in old-growth forests.

Threats

The single greatest threat to the spotted owl is the continued logging of old-growth forests, leading to loss of suitable habitat. Estimates suggest that probably less than 50 per cent of the old-growth habitat that once covered the Lower Mainland of BC is suitable habitat, and much of this is highly fragmented. At the present rate of decline, northern spotted owl survey results suggest extirpation will occur within the next decade (wcwc, 2002).

Some protection is provided by provincial and regional parks in two main blocks, covering some 110,000 hectares in total. Unfortunately, these blocks are about 85 kilometres apart, with little interconnecting habitat. It is extremely unlikely that these two populations will be able to interbreed. Provincial forests managed for timber production contain much larger amounts of suitable habitat, but much of this land is scheduled to be harvested over the next 100 years. In 2001, the provincial government gave the logging company Cattermole Timber preliminary permission to log 88 hectares of old-growth forest near Anderson River, north of Hope, a decision that was challenged and later upheld in the BC Supreme Court of Appeal. The northern spotted owl is listed as endangered by the Committee on the Status of Endangered Wildlife in Canada (COSEWIC), but nothing in BC law 'precludes approval of a forest development plan if there is any element of risk to a forest resource, even where that forest resource is an endangered species' (*Vancouver Sun*, 2003).

Management Options

Concern over the future status of the spotted owl resulted in the formation of a Spotted Owl Recovery Team (SORT) in 1995 to develop a recovery plan for the species. The primary goal was to outline a course of action to stabilize the current population and lead to an improvement of the status of the species, allowing it to be removed from the Endangered Species category. Sixteen different management options, which correspond with the different categories of abundance status outlined by COSEWIC, are discussed in detail in Chapter 11. The options by which these changes in category could occur range from banning all timber harvesting that could degrade suitable owl habitat within all forested habitats within the entire range of the owl, to no owl management beyond existing parks and protected areas. The costs associated with withdrawing this amount of land from the operable forest have yet to be assessed, but they are likely to be significant. This, however, is the price that society must pay in terms of forgone timber revenues if an old-growth-dependent species such as the spotted owl is going to survive.

Older trees contain larger amounts of nutrients—such as nitrogen, phosphorus, potassium, calcium, and manganese—than younger trees. There is also considerable variation among species in the amount of nutrients organically bound and the nutrients preferentially held by different species. The differences between whole-tree and stem-only harvests can also be significant. In a study of black spruce in Nova Scotia, Freedman et al. (1981) found almost a 35 per cent increase in biomass take for whole-tree harvesting. This resulted in an increased loss of 99 per cent nitrogen, 93 per cent phosphorus, 74 per cent potassium, 54 per cent calcium, and 81 per cent magnesium derived mainly from the nutrient-rich foliage and small branches. When deciduous trees are harvested, the loss of nutrients can be reduced significantly by cutting in the dormant period when leaves are not present.

Forest harvesting can also lead to dramatically increased rates of nutrient loss through leaching (the downward movement of dissolved nutrients) to the hydrological system. The amount of loss varies according to the intensity and scale of the

harvest and the particular ecosystem under consideration. Loss of nitrate is of most concern since not only is it often a dominant limiting factor (see Chapter 2), but it is also the nutrient that is lost most often in large quantities. One reason for this is the disturbance in the nitrogen cycle (Chapter 4) by logging, which results in an increase in the bacterial process of nitrification turning ammo-

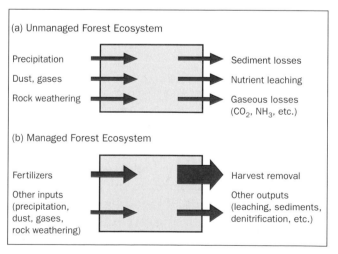

Figure 9.7 Nutrient inputs and outputs from managed and unmanaged forest ecosytems.

BOX 9.10   FORESTRY AND GRIZZLIES

The grizzly bear once extended across most of North America. Unlike its smaller cousin, the black bear, however, it is not able to tolerate human disturbance, and as human populations grew the grizzly populations shrank. Some of the densest remaining populations are in the lowlands of the coastal valleys in BC, where the bears are attracted not only to the annual salmon runs but also to the abundant berry-producing shrubs and other nutritious vegetation found on the flood plains. These nutrient-rich sites also constitute some of the most productive forestry sites.

Studies over the last 10 years have documented declines in grizzly populations following logging activities. One of the main problems was with the changes in vegetation that occurred as a result of logging. The favourite forage foods of the grizzly are those that compete with the re-establishment of trees as secondary succession takes place (Chapter 5), and because the re-establishment of trees has meant profit for logging companies and governments down the road the mixed vegetation has often been suppressed using chemicals. In turn, this helps to create dense stands of conifers, which have the long-term effect of shading out preferred forage for the grizzly. The result is a critical shortfall in grizzly food over an extended period of time.

As a result of these concerns, new approaches are being tried. For example, lower replanting densities for conifers and more open space between tree clusters are being used in an effort to mimic the natural environment. The use of chemicals is restricted to the tree stands. A new adaptive management approach has been formulated for silviculturists dealing with these grizzly bear habitats. This involves not only the measures outlined above for new logging activities, but also revisiting past sites to undertake remedial activities. The approach is not only adaptive but also exemplifies an ecosystem approach, as discussed in Chapter 8, which takes a more holistic view of resource management by accounting for the limits of tolerance (Chapter 4) of one of the most spectacular animal species in Canada. Although it is too early to assess the success of these efforts, in combination with the absolute protection of some of these sites, such as a grizzly reserve established in the Khutzeymateen Valley, they should help to ensure that grizzly bear populations survive in the coastal valleys of BC.

The grizzly bear is not very tolerant of human disturbance and has suffered large declines in number and distribution since European colonization (*World Wildlife Fund Canada/Karl Sommerer*).

nium to nitrate. Nitrate is highly soluble, resulting in significant losses of nitrogen site capital in some ecosystems, particularly if the soil is not too acidic. Other factors, such as warmer soil temperatures, decreased uptake by vegetation, and abundant decaying organic matter on the forest floor, also contribute to increased losses of soluble nutrients following logging. In addition, younger stands (under 145 years old) support a smaller biomass of lichens with nitrogen–fixing abilities, compared to old–growth trees on the west coast of BC.

It is also important to consider nutrient inputs. Substantial amounts of nutrients will be added over time by precipitation. For a maple-birch stand in Nova Scotia, for example, Freedman et al. (1986) calculated that it would take 96 years of

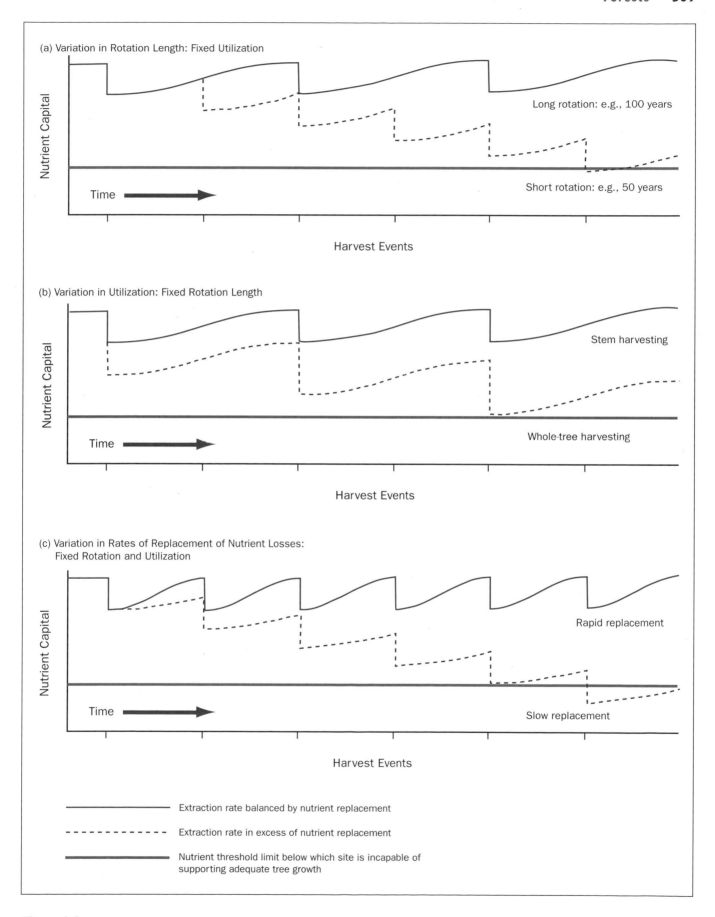

Figure 9.8 Site impoverishment as a result of forest harvesting. SOURCE: Adapted from J.P. Kimmins, 'Evaluation of the Consequences for the Future Tree Productivity of the Loss of Nutrients in Whole-Tree Harvesting', *Forest Ecology and Management* 1 (1977): 169–83.

precipitation to replace the nitrogen lost through whole-tree removal, 83 years for potassium, 166 years for calcium, and 41 years for magnesium. Other nutrient inputs occur through dry deposition of gases and particulate matter, the weathering of minerals, and the fixation of atmospheric dinitrogen. Soil mycorrhizae are particularly important for the fixation of atmospheric dinitrogen (Chapter 4), but land treatments subsequent to clear-cutting, such as slash burning and pesticide use, may adversely affect fixation rates.

Forestry and Soil Erosion

Besides the direct influence on nutrients described above, forest harvesting can also have substantial impacts on soil through erosion, especially on steep slopes in areas of heavy precipitation. Such losses also contribute to losses of site fertility, remove substrate for further regrowth, and contribute to flooding and the destruction of fish habitat. Poor design and maintenance of roads are often key factors behind accelerated soil erosion. Cutting of roads across steep terrain exposes large banks of unprotected top soil. Compacted road surfaces encourage overland flow with high erosive power. Much stricter regulations are now being put in place in many jurisdictions regarding road construction.

Forestry and Hydrological Change

Large-scale forest harvesting can have a significant impact on hydrology. Under natural conditions, large amounts of water are evapotranspired into the atmosphere by the trees. Removal of the trees significantly reduces this evapotranspiration mechanism and other storage capacities, releasing large amounts of water into stream flow. Flooding in high discharge periods can often be the result. Furthermore, without the delaying mechanism of the trees, and with soils that have suffered compaction through harvesting, the speed of flow is often increased, again raising the potential for flooding as well as for erosion. In turn, sediment from erosion can damage spawning beds used by fish.

Forestry and Non-Timber Forest Products and Services

Historically, non-timber uses have received little attention in Canada. However, as timber harvest levels have increased and the public has become more aware and vocal about declines in these

When an old-growth tree such as this giant Sitka spruce on the west coast of Vancouver Island dies, the nutrients it contains recycle to fuel new tree growth. When logging removes the whole tree these nutrients are lost to the ecosystem (*Philip Dearden*).

other forest values, forestry companies are being increasingly required to take these values into account in their cutting plans. In other words, they are having to take a more ecosystem-based approach, as described in Chapter 6. Non-timber forest products (NTFPs) are also seen as bringing diversification to rural economies and can yield valuable economic returns. Canada has a long way to catch up to most other countries in bringing this kind of product diversification to the forest land base.

NTFPs may be wild or managed and may come from both natural and managed forests. It is important to understand these differences if NTFPs are to play a fuller role in forest valuations and decision-making. For example, the harvesting of some wild stocks, such as mushrooms, from managed forests may conflict with timber production activities. Harvesting of some managed NTFPs can be encouraged alongside timber production and raise the overall level of return from the land. This has been promoted in some areas of Quebec with blueberries. This is an example of symbiotic resource use between the different resource uses, where both kinds of resource use can benefit. This is common in agroforestry ecosystems commonly developed in the tropics but rarely used in Canada.

Other kinds of relationships are complementary, competitive, and independent resource use. Complementary relationships occur where NTFPs and timber are extracted from the same land base in non-conflicting ways. Craftspeople, for example, may get improved access to their raw materials (e.g., tree bark, boughs) through the development of logging roads. In contrast, competitive relationships often involve mutually exclusive uses. Logging old-growth forests in western Canada, for example, would devastate the lucrative pine mushroom industry. Finally, independent systems develop where the two uses operate on different units of land, for example, from commercial and non-commercial forests.

Besides the tangible non-timber forest products, a host of less tangible values relate to cultural and spiritual fulfillment and knowledge and understanding. Such values are more difficult to assess, let alone manage. Aesthetics in particular has received growing attention over the last decade. Our forest lands include some of our most scenic lands. Not only a main attraction for tourists, they are also a main source of recreational and spiritual satisfaction for residents. Most provinces now have some procedures for the inclusion of assessments of aesthetic quality into harvesting plans. Unfortunately, many procedures still rely on those in charge of timber extraction to make the judgements as to what viewpoints will be used for making scenic assessments and what harvesting prescriptions might follow. As a result, modifications to cutting plans to take aesthetics into consideration often tend to be minimal.

Given the large range of values involved with forest landscapes and forestry in Canada, judging the overall health of forest ecosystems is not an easy task. Box 9.11 describes Environment Canada's approach to this challenge.

NEW FORESTRY

Concerns over the impacts and sustainability of forest practices have given rise to calls for what has been termed **new forestry**, which involves new

Logging has resulted in severe changes in the morphology of many coastal streams in BC, with large amounts of logging debris accumulating following flooding (*Philip Dearden*).

ways of looking at the management of forest ecosystems. Current approaches usually emphasize economic maximization over the short term through intensive forest management subsidized by auxiliary energy flows, leading to a simplification of forest biology. This entails genetic simplification through the exclusion of non-commercial species from regrowth areas and genetic manipulations to homogenize the species grown. Intensive forestry emphasizes production of a young, closed-canopy, single-species forest, usually the least diverse of all successional stages. Furthermore, the strength and reliability of the wood produced from such forests have been questioned, and these plantations are susceptible to **windthrow**, insect infestations, and gradual nutrient depletion. Structural simplification also takes place as the range of tree sizes and growth forms is reduced, snags and fallen trees are removed, and trees are regularly spaced to optimize growth. At the landscape scale, simplification occurs as old growth is removed and the irregularity of wind- and fire-created openings is replaced by the regularity of planned clear-cuts. Successional simplification also takes place, as intensive management seeks to eliminate early and late successional stages from the landscape.

New forestry embraces an approach that mimics natural processes more closely through emphasizing long-term site productivity by maintaining ecological diversity. This includes rotation periods sometimes longer than the minimum economic periods, reinvesting organic matter and nutrients in the site through snag retention and

BOX 9.11 CANADA'S NATIONAL ENVIRONMENTAL INDICATORS SERIES: FORESTS

The health of Canada's forests is one of many 'headline' indicators used to measure the state of Canada's environment. The health of Canada's forests is assessed by Environment Canada using four indicators:

- percentage of ecozone with strictly protected forest area;
- population status of forest bird species in selected forested ecozones;
- total area harvested;
- number of forest fires in Canada and area burned.

The collective amount of strictly protected area in the Boreal Shield, Atlantic Maritime, Pacific Maritime, and Montane Cordillera ecozones increased from 5.3 per cent in 1992 to 6.8 per cent in 2001, accounting for approximately 18 per cent of the total strictly protected area in Canada. In these ecozones, the populations of most forest bird species are showing little change. However, forest harvest levels have steadily increased over the past decade—since 1994, an area almost twice the size of Prince Edward Island has been cut each year. According to the data, there has been no obvious increase in forest fires since 1990.

Environment Canada chose the trend in strictly protected area for all four ecozones from 1992 to 2001 as the indicator that best summarizes the health of Canada's forests. According to the meter, the health of Canada's forests are 'improving'. However, the federal agency also recognizes that challenges remain. Two have been identified: (1) protecting representative forests; and (2) developing methodologies to quantify the value of ecosystem services provided by forests.

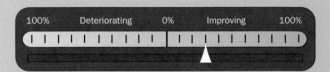

Do you think the indicators used to assess the state of Canada's forests are adequate? Are there other indicators (ecological, social, economic) that would provide a more accurate representation of ecosystem health?

SOURCE: Environment Canada (2003).

---

stem-only harvesting, minimizing chemical inputs, and diversifying the range of tree species and other forest products. Growth of traditionally non-commercial species such as alder and other early successional species, particularly nitrogen fixers, is permitted, and all stages of the successional process are accommodated. Old-growth big-leaf maple, for example, provides excellent growth sites for many epiphytes (plants that use other plants for physical support but not nourishment), which provide valuable nutrient accumulation and water retention. Riparian or riverbank habitats receive special attention; litter from streamside vegetation provides the primary energy base for the aquatic community, and management of coarse woody debris is particularly important for the structure of smaller streams. Needless to say, large woody debris cannot be produced by a forest that no longer contains large trees.

New forestry also emphasizes the maintenance of non-timber parts of the forest community. Special attention is directed towards the impacts on wildlife of the size, shape, and location of forest patches and how these may be connected to sustain populations. The ecological complexities of the forests are only just starting to be revealed. Recent research, for example, suggests that the younger the forests, the lesser the conifer seed production. Species that rely on these seeds, such as crossbills, will also experience a decline. In western Canada, five species of crossbill have evolved, each of which specializes on a different species of conifer and even different varieties of the same species. Protection of this diversity of crossbills will require protection of old-growth stands and an increase in rotation ages throughout the range of each conifer. Similar considerations must be given to the entire range of forest biodiversity if it is to be maintained into the future.

BOX 9.12 NEW FORESTRY IN ACTION

The ideas contained in new forestry must be put in motion if change is to occur. Several examples exist where alternatives to the dominant way of managing our forests are already in operation, both in regard to individual woodlots and for management of more extensive areas by communities.

At the individual scale, one well-known example is that of Merv Wilkinson and his 55-hectare woodlot on southern Vancouver Island. Since 1936, Wilkinson has been practising **sustained yield** forestry, and despite over 4,000 cubic metres of timber having been removed, his woodlot still contains as much wood as it did when it was first assessed in 1945. He removes forest products by cutting in five-year rotations. The straightest, most vigorous trees with good foliage and abundant cone production are left as seed trees, including some estimated to be as old as 1,800 years. There is no clear-cutting, slash burning, or use of chemicals. The canopy is left intact to shield seedlings but thinned a little to promote good growth. Sheep are used for brush control. Wilkinson's model may not apply everywhere, but it has worked for him and is one demonstrable example of how a forest can be maintained while still retaining its essential ecological characteristics.

At a regional scale, attention recently has focused on 'community forests'. Decisions on forest use are often made in boardrooms at the dictates of international capital flow. Such decisions may not be to the benefit of the local communities dependent on the forests for their livelihoods. Concern over this situation has prompted interest in how to manage forests to maximize the benefits to local communities. In an overview of the situation in Canada, Duinker et al. (1994: 712) defined a community forest as 'a tree-dominated ecosystem managed for multiple community values and benefits by the community'.

There are many different manifestations of community forests in Canada with different types of land tenure and administrative arrangements. However, they all seek to achieve local benefits for the community and encourage local involvement in decision-making. In BC, for example, the provincial government amended the Forest Act to create 'Community Forest Agreements', a new form of tenure designed to allow more communities and First Nations to participate directly in the management of local forests. In 1998 the Ministry of Forests launched a pilot project in which it issued a special form of the tenure, called a community forest pilot agreement (CFPA). To date, eight agreements have been issued to a range of communities and First Nations, and several more agreements are pending. The pilot agreements are located throughout the province and currently range in size from about 400 hectares to more than 60,000 hectares. The pilot agreements are limited to a term of five years, during which the tenure is evaluated. Holders who successfully operate the pilot agreements may be offered a long-term community forest agreement with a term of 25–99 years.

The BC government has a number of objectives for the program:

- provide long-term opportunities for achieving a range of community objectives, including employment, forest-related education and skills training, and other social, environmental, and economic benefits;
- balance the use of forest resources;
- meet the objectives of government with respect to environmental stewardship, including the management of timber, water, fisheries, wildlife, and cultural heritage resources;
- enhance the use of and benefits derived from the community forest agreement area;
- encourage co-operation among stakeholders;
- provide social and economic benefits to British Columbia.

Similarly, community forests are typically managed to reflect community goals:

- supporting the local economy by hiring and buying supplies locally and selling timber to local timber-processing facilities;
- diversifying the local economy by making small volumes of wood fibre available to new and existing niche markets such as small-scale, value-added manufacturers and artisans;
- maintaining and enhancing local recreational opportunities;
- protecting drinking water, viewscapes, wildlife, and other environmental attributes;
- providing a source of income to support local community initiatives.

Continued

Although the issuance of pilot agreements is complete, under the Forestry Revitalization Plan announced in March 2003 the BC government committed to increase significantly the volume of timber allocated to community-based forest tenures.

It is clear that alternative approaches to forestry, besides those that currently dominate operations, do exist. It is essential, however, that the goals of the forestry activities are specified before the most appropriate approaches can be chosen, as emphasized by the framework in Chapter 1. Current models have evolved to maximize economic returns over the short term; the alternatives described above have different goals, which are more consistent with the demands of today.

Source: BC Ministry of Forests (2003).

The kinds of changes suggested by new forestry make it unlikely that the dominant practices of today, such as extensive clear-cuts, can continue, and indicate that other harvesting systems (Table 9.4) will play a larger role as an ecosystem-based perspective becomes more widespread. New forestry may not suit the needs for short-term economic return of the large corporations that now dominate the industry, and it is likely that more, smaller, and community-based companies will become involved (Box 9.12). Monetary returns over the short term will probably fall as the amount of wood fibre extracted from the forest declines. Proponents of new forestry argue, however, that these changes will have to occur anyway. Continuation of the old approaches will simply lead to an abrupt decline in the amount of timber available and consequently will provide little in the way of future prospects. This way of thinking is now gaining wider acceptance. The development of Canada's National Forest Strategies, described in the next section, are one result.

CANADA'S NATIONAL FOREST STRATEGIES

During the 1980s it became increasingly clear that forestry in Canada could not continue as it had in the past and new ways had to be found to develop more sustainable management practices. This realization led the Canadian Council of Forest Ministers (CCFM), a body composed of all the provincial and territorial forest ministers and the Canadian Minister of Natural Resources, to sponsor a series of public forums across the country. These resulted in the formation of a national forest strategy, Sustainable Forests: A Canadian Commitment (1992–7). Canada was the first country in the world to develop such a national strategy. This was revised and extended for a second strategy covering 1998–2003. The National Forest Strategy Coalition (NFSC), composed of 52 governmental and non-governmental agencies, was formed to oversee implementation of the strategy. The strategy made 121 commitments under nine strategic directions to move Canada along the road to a more sustainable use of forest ecosystems.

Such an approach is only useful, however, if the commitments are meaningful and some progress is made to achieve them. An independent evaluation concluded that there had been substantial progress on 37 commitments, some progress on 76 commitments, little progress on six commitments, and no progress on two commitments. The NFSC commended Canada for showing international leadership in this area but suggested a simplification of any future strategy, as well as the inclusion of clear targets and timetables.

In 2003, at the Ninth National Forest Congress held in Quebec, a further five-year strategy was adopted, along with a commitment of the NFSC partners to work towards its completion. Eight themes were outlined, each of which specifies an objective and action items to be undertaken (Table 9.6). Two themes of noteworthy attention are the urban forest, emphasizing the need to actively engage more of Canadian society in forest questions, and the recognition of the importance of private woodlots to sustainability. This is particularly important in the Maritime provinces, where over half of the forests are in private hands and not subject to provincial forestry regulations. Over-cutting and neglect of these lands have been problems in the past

Table 9.6   Canada's National Forest Strategy: Objectives and Actions

Objective	Action Examples
Ecosystem-based Management Manage Canada's natural forest using an ecosystem-based approach that maintains forest health, structure, functions, composition, and biodiversity, and includes, but is not limited to: using integrated land-use planning; maintaining natural forested ecosystems; completing a system of representative protected areas; maintaining carbon reservoirs and managing the forest to be a net carbon sink over the long term; and conserving old-growth forests and threatened forest ecosystems.	• Develop guidelines for integrating watershed-based management and wildlife habitat conservation into forest management across Canada and measures for evaluating implementation. • Establish a process involving forest-based communities leading to the implementation of land-use management plans, which include all forest benefits. • Implement systems and decision-making that set resource-use levels as an output of the planning process.
Sustainable Forest Communities Develop legislation and policies to improve the sustainability (social, environmental, and economic) of forest-based communities by: fostering participation and involvement in forest management decision-making; improving access to resources; sharing benefits; enhancing multiple benefits; and supporting community resilience and adaptive capacity.	• Develop and adapt forest legislation and policies to provide involvement of forest-based communities in sustainable forest management decision-making and implementation. • Expand the area and use of community-based tenure systems and resource allocation models in remote, rural regions of Canada to increase benefits to Aboriginal peoples and forest-based communities. • Support capacity-building in local communities so that they can effectively participate in processes that lead to community sustainability.
Rights and Participation of Aboriginal Peoples Accommodate Aboriginal and treaty rights in the sustainable use of the forest, recognizing the historical and legal position of Aboriginal peoples and their fundamental connection to ecosystems.	• Initiate processes with Aboriginal peoples and appropriate levels of government for establishing: a shared and grounded understanding of Aboriginal rights, Aboriginal title, and treaty rights; the roles and responsibilities of Aboriginal peoples, governments, and forest stakeholders; and measures to fulfill government fiduciary responsibilities and the legal duty to consult. • Implement institutional arrangements between Aboriginal peoples and governments that reflect a spirit of sharing responsibilities and benefits for the management, conservation, and sustainable use of forest lands and resources; and give effect to land claim settlements, treaties, and formal agreements on forest resource use and management. • Incorporate traditional knowledge in managing forest lands and resources in accordance with the Convention on Biological Diversity.

Continued

Objective	Action Examples

Forest Products Benefits

Stimulate the diversification of markets, forest products and services, and benefits (both timber and non-timber) by: understanding current and emerging markets and developing new domestic and international markets; promoting value-added and best end-use through expanded research and design; and attracting manufacturers of finished products and promoting markets for forest environmental services.

- Create and maintain policies and programs that encourage human capacity, investment, productivity, innovation, and competitiveness in: existing and potential primary and value-added timber industries; non-timber and service-based industries, such as tourism and recreation, hunting and fishing, trapping, and wild foods; and specialty forest products and services (e.g., medicinal plants, carbon sinks, etc.)
- Create and maintain policies and programs that encourage, develop, and maintain access to markets for primary and value-added timber and non-timber-based industries.
- Develop strategies for increasing domestic and export markets.

Knowledge and Innovation for Competitiveness and Sustainability

Maintain and enhance the skills and knowledge of forest practitioners and mobilize the broader Canadian knowledge community to establish a new forest innovation agenda for Canada by: developing 'clusters' of forest sector science and technology (S&T) co-operation, both nationally and regionally, to use available S&T resources more efficiently and effectively; supporting innovative post-secondary education institutions, continuing education, and technology transfer to ensure that the principles of adaptive management improve the management of our resources; improving the processes for bringing new and traditional knowledge and ideas to policy evolution, decision-making, and field practices; and informing investors about opportunities for innovative uses of renewable forest materials and services in relation to the Kyoto Protocol.

- Integrate individual S&T efforts into innovation networks.
- Develop a framework to use traditional knowledge along with current scientific knowledge and to protect the intellectual property rights of Aboriginal peoples.
- Develop strategies for improving the forest sector's success in competing for funding to support leading-edge S&T programs, using research and development, tax credits, research extension, and education.

The Urban Forest and Public Engagement in Sustainability

Actively engage Canadians in sustaining the diversity of benefits underlying the importance of Canada's forest by: establishing mechanisms to advance the planning, maintenance, and management of urban forests based on an ecosystem-based approach; and enhancing communication and outreach programs.

- Develop and implement a national urban forestry strategy.
- Develop guidelines and support tools to help municipalities maintain and enhance their urban forest.
- Develop guidelines and support tools to protect the surrounding forest and watersheds from urban pollution.

Objective	Action Examples
Contribution of Private Woodlots to Sustainability Increase the economic, social, and environmental contribution by Canadian woodlot owners to Canadian society through a concerted effort by stakeholders to strengthen policies and services that encourage and support viable woodlot businesses.	• Identify and remove obstacles hindering sustainable development with particular attention to market incentives, silvicultural programs, and tax policies. • Increase the capacity of private woodlot owners by expanding extension programs. • Develop and implement incentives for the provision of environmental services from private woodlots.
Reporting and Accountability Create a comprehensive national forest reporting system that consolidates data, information, and knowledge for all valued features of the forest, both urban and rural.	• Assess socio-economic and environmental impact analysis of policy and management options. • Institute forest data standards for forest inventories, including monitoring protocols, to create a publicly accessible forest information system that provides high-quality information on the status of the forests. • Enhance programs to monitor and inform the public about invasive species.

SOURCE: *A Sustainable Forest: The Canadian Commitment* (2003). The complete report can be found at: <http://nfsc.forest.ca/strategy.html>.

and the action items in the strategy attempt to provide more incentives and support for landowners to manage their woodlots on a sustainable basis.

The Model Forest Program

One commitment from Canada's 1992–7 National Forest Strategy that has born fruit is to develop a system of model forests in the major forest regions. Proposals were solicited for areas between 100,000 and 250,000 hectares where partners would develop a management structure to facilitate co-operation and include a vision and objectives to balance a variety of values, as well as actions to demonstrate sustainable forest management. Key attributes of model forests include:

- a partnership that includes principal land users and other stakeholders from the area;
- a commitment to sustainable forest management, based on an ecosystem-based approach;
- operations at the landscape or watershed level;

- activities that reflect stakeholder needs and values;
- a transparent and accountable governance structure;
- commitment to networking and capacity-building.

Eleven model forest agreements in six major forest regions across the country are now underway, each with a unique management structure designed to address the particular situation. Core issues relate to ecosystem management, Aboriginal participation, public participation, science and innovation, and the integration of non-market values into decision-making.

Model forests have a deliberate strategy of intra-site and inter-site demonstration and networking. This has now been expanded to the international context with sites in Mexico, Chile, Argentina, China, the US, Japan, Indonesia, Thailand, Myanmar, the Philippines, and Russia as part of an International Model Forest Program. These initiatives have been supported by Canadian aid programs totalling over $11 million, with additional support from other donors of over $7.5 million. The programs have received this support as they constitute a tangible demonstration of co-operatively working together towards sustainability and illustrate many of the approaches outlined earlier in Part C.

IMPLICATIONS

Forestry is at a watershed in Canada in terms of how forests, their values, and their management are viewed. The next decade will be crucial in determining whether we as Canadians will still consider ourselves a forest nation in 50 years. Although society in general and government and industry in particular now have a much greater appreciation of the changes that need to be made in the industry to move towards more sustainable practices, actually making these changes will take some time. However, the world is watching and the power of consumers to effect change is well in evidence

(Box 9.13). Large companies such as Ikea and Home Depot have reacted to consumer pressure by agreeing not to sell products from forestry operations that have not been certified as being sustainable (Box 9.14). Meanwhile, Canada continues to fight against the softwood tariffs implemented by the US, our largest market, in response to criticisms that the Canadian industry is unfairly subsidized by taxpayer money.

The boreal forest continues to get eaten away. Timoney (2003: 512), in an overview of the changing disturbance regime of the boreal forest, found that average forest age and biomass are declining, the forest is increasingly fragmented, and alien plants are becoming more prevalent. He concludes that 'without a rapid decrease in the rate of disturbances, the establishment of a more complete protected areas network, and the adoption of ecosystem-based management the subhumid boreal ecosystem will continue to be degraded.' Given the importance of the boreal ecosystem to Canada and the world, we cannot afford to let this happen.

BOX 9.13 WHAT YOU CAN DO

A number of environmental stresses threaten the health of Canada's forests, but there are ways in which individuals can help:

1. Reduce your consumption of paper products. There are several ways to reduce paper consumption: use all available space on your paper (write in the margins and on both sides); carry along linen to replace tissues, paper towels, and napkins; resist the temptation to print materials from the Internet—read on-line instead; insist that no unsolicited flyers be delivered to your home.
2. Recycle paper products and purchase recycled paper products.
3. Choose unbleached paper products where possible. Unbleached paper is less harmful to the environment, since the toxic chemicals used to whiten paper are not used.
4. Purchase certified wood products. Make an

effort to purchase wood that has been certified by at least one certification agency. This will help ensure that the wood you purchase comes from a logging company that has introduced measures to promote long-term ecological, social, and economic sustainability.
5. Reduce the risk of human-caused fires. Many forest fires are started by human carelessness. Obey fire restrictions when visiting parks and protected areas, and do not discard fire accelerants (e.g., cigarettes) along highways.
6. Join one of the NGO groups that support sustainable use of forest resources.
7. Write or phone your MP or MLA, and/or e-mail a letter to the editor of your local newspaper. Canadians own Canadian forests. If you are unhappy about how forests in your province/territory are being managed, voice and/or publicize your discontent.

BOX 9.14 THE SEAL OF APPROVAL

Increasingly, customers of wood products around the globe are asking for guarantees that the products they buy (kitchen chairs, 2 x 4s, paper products, etc.) come from forests that are managed and logged according to ecologically responsible standards. The trend towards responsible consumerism is supported by independent certification and labelling, a process by which an independent audit is conducted of a forestry company to assess whether it meets internationally and/or nationally recognized guidelines of responsible forest management (ForestEthics, 2003). Certification is designed to allow consumers and participants to measure forest management practices against approved standards, and also provides forest owners with an incentive to maintain and improve forest management practices.

In Canada, there are four main certification systems currently in use:

- Forest Stewardship Council (FSC)
- Canadian Standards Association (CSA)
- Organization for International Standardization (ISO)
- Sustainable Forestry Initiative (SFI)

There are important differences between these certification systems that relate to standards, policies, procedures, and on-the-ground results, but in general, forest certification systems are designed to link environmentally and socially conscious consumers with like-minded producers, retailers, and distributors. Certification systems typically involve:

- independent, third-party auditing;
- chain of custody procedures (verification of compliance from the forest through to the final product);
- on-the-ground inspections of forested areas to determine whether they are managed according to established sets of environmental and social standards;
- certified product labelling;
- multi-stakeholder involvement.

The range of issues considered in defining responsible forest management include: wildlife habitat protection; the identification and mainte-nance of endangered forests; riparian and water utility protection; indigenous people's rights; and the equitable sharing of benefits with forestry-dependent workers and communities. For example, the FSC advocates that all functions of a forest ecosystem remain intact after an area is logged. This requires that a mix of different tree species of different ages still remain standing after the forest is logged, and that the functions of tree and other plant species also remain intact.

Forest products given the seal of approval should give consumers confidence that the products they purchase are derived from responsibly managed forests. As of April 2001, roughly 44 million hectares or 37 per cent of Canada's managed forest land had been certified under one or more of the main certification systems listed above (CSA, 5 million ha; FSC, 36,000 ha; ISO, 44 million ha; SFI, 4 million ha).

To date it has been difficult for Canadian forest companies to attain certification from the FSC, since the current dominant harvesting system in Canada is clear-cutting. However, one company operating in the boreal forest of northeastern Ontario has earned the Forest Stewardship Council logo for voluntarily meeting the FSC's high standards for forest management. Clear-cutting once dominated the 2-million-hectare forest managed by Tembec Inc., but today considerable patches of trees are left standing, large tracts of old-growth forest are being protected, and selections of all forest types are being set aside to serve as wildlife habitat. Tembec Inc. pledges to certify all 13 million hectares of its woodlands in Canada to FSC standards by 2005.

Tembec Inc. has demonstrated that it is possible to dramatically improve forest management practices to attain the coveted FSC logo. In a climate where consumers are increasingly concerned about environmental and social issues surrounding primary resource extraction, securing a seal of approval from one or more of the certification programs will become increasingly important.

Sources: BC Ministry of Forests (2004); ForestEthics et al. (2003); World Wildlife Fund Canada (2003).

SUMMARY

1. Canada is a forest nation. Not only do we have 10 per cent of the world's forests, we are also the largest exporter of forest products in the world. Some 800,000 Canadians and almost 1,000 communities rely on forestry as a main source of income. The forests, along with the North, are dominant elements in the history and culture of the nation.

2. There are 15 terrestrial ecozones in Canada. Most of Canada is within eight of these ecozones: Boreal Cordillera, Pacific Maritime, Montane Cordillera, Boreal Plains, Taiga Plains, Boreal Shield, Mixed Wood Plains, and Atlantic Maritime. Each differs with respect to climate, geology, genetic diversity, and level of human development. The Boreal Shield, Canada's largest ecozone, is threatened by a number of environmental stresses, including logging, stand replacement, mining, hydroelectric development, acid precipitation, and climate change.

3. Ecosystems provide an array of beneficial services arising from ecological functions such as nutrient and water cycling, carbon sequestration, and waste decomposition. Forests are also places of exceptional scenic beauty, and millions of Canadians travel each year to participate in nature-related recreational activities.

4. Forests are also a valuable source of commodities. Non-timber forest products contribute millions of dollars to the Canadian economy and are also an important aspect of First Nations subsistence economies. Despite government recognition that Canadian forests provide a broad range of values, forest management paradigms have traditionally focused on the management of Canadian forests to supply wood.

5. The provincial and territorial governments are responsible for 71 per cent of the nation's forests and the federal government for 23 per cent on behalf of the owners, the people of Canada. The remaining 6 per cent is owned privately. Governments enter into contract arrangements with private companies, in which they can specify the forest management practices to be followed.

6. Approximately 119 million hectares are currently managed for timber production. On these lands, forest ecosystems are being transformed from relatively natural systems to control systems in which humans, not nature, influence the species that will grow there and the age to which they will grow.

7. The rate of conversion from natural to managed forest is controlled by provincially established annual allowable cuts (AACs). In theory, the AAC should approximate what the land should yield in perpetuity. It is not sustainable to have an AAC that consistently exceeds this biological potential. In most provinces, the AAC is greater than the harvest, but in some regions the AAC is approaching or exceeds the harvest for softwoods.

8. Silviculture is the practice of directing the establishment, composition, growth, and quality of forest stands through harvesting, reforestation, and site preparation.

9. Clear-cutting is the dominant harvesting system used in Canada. It is the most economical way for the fibre to be extracted and also allows for easier replanting and tending of the regenerating forest. In certain types of forests it may mimic natural processes more closely than selective or partial cutting systems. However, clear-cutting may not be the most appropriate way to harvest timber in some areas. Clear-cuts are aesthetically unappealing to many Canadians, and their environmental impacts can be substantial.

10. Biocides are used in forestry to control populations of vegetation and insect species that compete with, or eat, commercial species. Several high-profile conflicts have resulted over application of chemicals. Concern over the spraying of the spruce budworm in the Maritime provinces is one of the most significant. Increasing use is now being made of a biological control agent, *Bacillus thuringiensis*, against insect attacks in Canada.

11. Intensive forest management techniques are used to further enhance future timber resource values. Intensive silvicultural practices include pre-commercial thinning, commercial thinning, scarification, prescribed burning, pruning and shearing, and timber stand improvement. The long-term impacts of these activities are not well understood.

12. Fire suppression has resulted in ecological changes not characteristic of fire-dominated ecosystems, as well as a gradual increase in the area burned over the past 30 years.

13. Various environmental impacts are associated with current forestry management systems, including: changes to ecosystem, species, and genetic diversity; changes to biogeochemical and hydrological cycles; and soil erosion.

14. Timber harvesting can significantly alter species composition and abundance, as the proportion of forest with old-growth characteristics is reduced. Species such as the woodland caribou and marten that depend on old-growth characteristics decline in abundance. Other species, such as deer, may increase as regenerating cut areas produce more forage for them.

15. The spotted owl is perhaps the best-known example of the potential impacts of logging on biodiversity. The spotted owl requires old-growth forests to maintain populations, but logging in BC's old-growth forests continues to threaten this endangered species. Management options to maintain populations are currently being considered.

16. Forest harvesting removes nutrients from the site. The significance of this for future growth varies, depending on the nutrient capital of the site and type of harvesting system used. Sites with abundant capital and/or selective harvesting systems that leave branches behind will suffer less chance of growth impairment of future generations than nutrient-poor sites or sites that are clear-cut with complete tree removal.

17. Forest harvesting may also contribute to increased soil erosion and water flows.

18. Forests produce many values for Canadians. In the past, attention has focused almost exclusively on the monetary returns from forest harvesting. However, as the amount of forest brought under management has increased and as the public becomes increasingly aware of the changes occurring in Canadian forests, more attention is being devoted to the assessment and management of other values besides timber production. An ecosystem perspective is being adopted.

19. Concern over the impacts and sustainability of forest practices has given rise to calls for what has been termed 'new forestry'. Such an approach embraces an ecosystem and adaptive management perspective that seeks to mimic natural processes more closely and give greater attention to the full range of values from the forests.

20. Management of Canada's forests is directed by a National Forest Strategy, developed by provincial and territorial forest ministers and the Canadian Minister of Natural Resources. The aim is to develop and implement more sustainable management practices.

21. Canada's Model Forest Program is one commitment arising from Canada's 1992–7 National Forest Strategy. Eleven model forest agreements in six major forest regions across the country are now underway. Core issues relate to ecosystem management, Aboriginal participation, public participation, science and innovation, and the integration of non-market values into decision-making.

KEY TERMS

annual allowable cut (AAC)	clear-cutting	Mixed Wood Plains	reclamation
Atlantic Maritime	culmination age	Montane Cordillera	second growth
biocides	DDT	new forestry	silviculture
Boreal Cordillera	ecosystem-based manage-	non-timber forest prod-	sustained yield
boreal forest	ment (EBM)	ucts (NTFPs)	taiga
Boreal Plains	falldown effect	old-growth forests	Taiga Plains
Boreal Shield	forest tenure	Pacific Maritime	windthrow

REVIEW QUESTIONS

1. Outline some of the ways in which forests are important to Canada.
2. Compare and contrast the physical and biological attributes of the Boreal Shield and Pacific Maritime ecozones. What major economic activities affect the health of these ecozones?
3. How is forestry an ecological process?
4. What is an AAC?
5. Outline some of the advantages and disadvantages to clear-cutting.
6. Is Canada reforesting all lands that are harvested? What are some of the issues associated with current replanting schemes?
7. Outline some of the pros and cons of using chemical sprays to control insect infestations in Canada's forests.
8. List all the different values that society realizes from forests. What do you think the priorities should be between these different and sometimes conflicting uses?
9. Name some species that might increase in abundance as a result of forest harvesting and others that might decline. What are the characteristics of these species that would encourage this response?
10. What are the impacts of forest harvesting on site fertility and how do they differ between sites?
11. What attributes of old-growth forests appear to explain their use by spotted owls?
12. How is forest management administered in Canada? What are the main strengths and weaknesses of this approach? What alternatives might you suggest? Do examples of such alternatives exist in you region?
13. What is 'new forestry'?
14. What tools are used to evaluate the sustainability of Canadian forests? Are these tools adequate to the task?

RELATED WEBSITES

BC MINISTRY OF FORESTS:
http://www.for.gov.bc.ca/hfp/forsite/Forest_Health.htm

CANADIAN INSTITUTE OF FORESTRY:
http://www.cif-ifc.org/

CANADIAN FOREST SERVICE:
http://www.pfc.forestry.ca

CANADIAN FOREST SERVICE, CRITERIA AND INDICATORS REPORT:
http://nrcan.gc.ca/cfs/proj/ppiab/ci/indica_e.html

CANADIAN MODEL FOREST NETWORK:
http://www.modelforest.net

CANADIAN PARKS AND WILDERNESS SOCIETY (CPAWS): http://www.cpaws.org

ENVIRONMENT CANADA: WHAT YOU CAN DO:
http://www.ec.gc.ca/eco/main_e.htm

DAVID SUZUKI FOUNDATION, FORESTS AND LANDS:
http://www.davidsuzuki.org/Forests/

NATIONAL FORESTRY DATABASE PROGRAM:
http://nfdp.ccfm.org

NATURAL RESOURCES CANADA, SILVICULTURAL TERMS IN CANADA:
http://nfdp.ccfm.org/silviterm

NATURAL RESOURCES CANADA, ECOZONE REPORTS:
http://www.health.cfs.nrcan.gc.ca

REFERENCES AND SUGGESTED READING

Alberta Food and Rural Development. 2001. *Woodlot Harvesting*. At: <http://www.agric.gov.ab.ca/sustain/woodlot>.

Allan, K., and D. Frank. 1994. 'Community forests in British Columbia: Models that work', *Forestry Chronicle* 70: 721–4.

Benkman, C.W. 1993. 'Logging, conifers, and the conservation of crossbills', *Conservation Biology* 7: 473–9.

Bergeron, Y., S. Gauthier, V. Kafka, P. Lefort, and D. Lesieur. 2001. 'Natural fire frequency for the eastern Canadian boreal forest: Consequences for sustainable forestry', *Canadian Journal of Forest Research* 31: 384–91.

Blue Ribbon Panel for the National Forest Strategy Coalition. 1994. *Mid-Term Evaluation Report*. Ottawa. National Forest Strategy Coalition.

British Columbia Ministry of Forests. 2003. *Community Forest Pilot Project*. At: <www.for.gov.bc.ca/hth/community/>.

———. 2004. *Forest Management Certification*. At: <www.for.gov.bc.ca/het/certification/>.

Buchert, G.P., O.P. Rajora, J.V. Hood, and B.P. Dancik. 1997. 'Effects of harvesting on genetic diversity in old-growth eastern white pine in Ontario, Canada', *Conservation Biology* 11, 3: 747–58.

Bunnel, F., and L.L. Kremsater. 1990. 'Sustaining wildlife in managed forests', *Northwest Environment Journal* 6: 243–69.

Busby, D.G., L.M. White, and P.A. Pearce. 1990. 'Effects of aerial spraying of fenitrothion on breeding white-throated sparrows', *Journal of Applied Ecology* 27: 743–55.

Canadian Council of Forest Ministers. 2003. *Compendium of Canadian Forestry Statistics.* At: <http://www.nfdp.ccfm.org>.

Canadian Institute of Forestry (CIF). 2003. *Silviculture Systems.* At: <http://www.cif-ifc.org/practices>.

Canadian International Development Agency (CIDA). 2003. *Forestry Profiles: The International Model Forest Network Program.* At: <http://www.rcfa-cfan.org/english/profile>.

Canadian Parks and Wilderness Society (CPAWS). 2003. 'Boreal forests: For the birds', *Boreal Forest Factsheet Series.* At: <www.cpaws.org/boreal/english/about boreal/boreal-bird-factsheet.pdf>.

Cumming, H.G., and D.B. Beange. 1993. 'Survival of woodland caribou in commercial forests of northern Ontario', *Forestry Chronicle* 69: 579–87.

Dufour, J. 2004. 'Towards sustainable development of Canada's forests', in B. Mitchell, ed., *Resource and Environmental Management in Canada: Addressing Conflict and Uncertainty*, 3rd edn. Toronto: Oxford University Press, 265–86.

Duinker, P.N., P.W. Matakala, F. Chege, and L. Bouthillier. 1994. 'Community forests in Canada: An overview', *Forestry Chronicle* 70: 711–20.

Dunbar, D., and I. Blackburn. 1994. *Management Options for the Northern Spotted Owl in British Columbia.* Victoria: Ministry of Environment.

Environment Canada. 2000a. *Ecological Assessment of the Boreal Shield Ecozone.* At: <http://manitobawildlands.org/pdfs/BorealShield_Ecozone.pdf>.

———. 2000b. *The Importance of Nature to Canadians: The Economic Significance of Nature-Related Activities.* Ottawa: Environment Canada.

———. 2003. *Environmental Signals: Canada's National Environmental Indicator Series, 2003.* Ottawa: Environment Canada.

Euler, D. 1978. *Vegetation Management for Wildlife in Ontario.* Toronto: Ministry of Natural Resources.

Food and Agriculture Organization (FAO). 2003. *State of the World's Forests, 2003.* Rome: FAO.

ForestEthics, Greenpeace, and Sierra Club of Canada BC Chapter. 2003. *On the Ground: Forest Certification, Green Stamp of Approval or Rubber Stamp of Destruction?* At: <www.goodwoodwatch.org>.

Freedman, B. 1981. *Intensive Forest Harvest—A Review of Nutrient Budget Considerations.* Information Report M-X-121. Fredericton, NB: Maritimes Forest Research Centre.

———. 1995. *Environmental Ecology: The Ecological Effects of Pollution, Disturbance, and Other Stresses.* Toronto: Academic Press.

———, P.N. Duinker, and R. Morash. 1986. 'Biomass and nutrients in Nova Scotia forests, and implications of intensive harvesting for future site productivity', *Forest Ecology and Management* 15: 103–27.

Hangs, R.D., J.D. Knight, and K.C.J. Van Rees. 2003. 'Nitrogen uptake characteristics for roots of conifer seedlings and common boreal forest competitor species', *Canadian Journal of Forest Research* 33: 156–63.

Hartmann, G.F., and J.C. Scrivener. 1990. *Impacts of Forestry Practices on a Coastal Stream Ecosystem: Carnation Creek, British Columbia.* Ottawa: Department of Fisheries and Oceans.

Hearnden, K.W., S.V. Millson, and W.C. Wilson. 1992. *A Report on the Status of Forest Regeneration.* Toronto: Ontario Independent Forest Audit Committee.

Hudson, A.J. 1993. 'The importance of mountain alder on the growth, nutrition, and survival of black spruce and Sitka spruce in an afforested heathland near Mobile, Newfoundland', *Canadian Journal of Forestry Research* 23: 743–8.

Hughes, J., and R. Drever. 2001. *Salvaging Solutions: Science-based Management of BC's Pine Beetle Outbreak.* Report commissioned by the David Suzuki Foundation, Forest Watch of British Columbia, and Canadian Parks and Wilderness Society, BC Chapter. At: <http://www.davidsuzuki.org/files/salvaging_solutions.pdf>.

Kimmins, J.P. 1995. 'Sustainable development in Canadian forestry in the face of changing paradigms', *Forestry Chronicle* 71: 33–40.

Lautenschlager, R.A., and T.P. Sullivan. 2002. 'Effects of herbicide treatments on biotic components in regenerating northern forests', *Forestry Chronicle* 78, 5: 695–731.

Lee, P., D. Aksenov, L. Laestadius, R. Nogueron, and W. Smith. 2003. *Canada's Large Intact Forest Landscapes: A Report by Global Forest Watch Canada.* Edmonton: Global Forest Watch Canada.

Lesieur, D., S. Gauthier, and Y. Bergeron. 2002. 'Fire frequency and vegetation dynamics for the south-central boreal forest of Quebec, Canada', *Canadian Journal of Forest Research* 32: 1996–2009.

McIver, J.D., A.R. Moldenke, and G.L. Parsons. 1990. 'Litter spiders as bio-indicators of recovery after clear-cutting in a western coniferous forest', *Northwest Environmental Journal* 6: 410–12.

McLachlan, S.M., and D.R. Bazley. 2003. 'Outcomes of long-term deciduous forest restoration in southwestern Ontario, Canada', *Biological Conservation* 113: 159–69.

McRae, D.J., L.C. Duchesne, B. Freedman, T.J. Lynham, and S. Woodley. 2001. 'Comparisons between wildfire and forest harvesting and their implications in forest management', *Environmental Reviews* 9, 4: 223–60.

Miller, A., and P. Rusnock. 1993. 'The ironical role of science in policymaking: The case of the spruce budworm', *International Journal of Environmental Studies* 43: 239–51.

Ministry of Natural Resources, Province of Ontario. 2002. *State of the Forest Report, 2001.* Toronto: Ministry of Natural Resources.

National Forestry Database Program (NFDP). 2002. *Compendium of Canadian Forestry Statistics.* At: <http://nfdp.ccfm.org/framescontents_e.htm>.

National Forest Strategy Coalition. 1992. *Sustainable Forests: A Canadian Commitment.* Ottawa: Canadian Council of Forest Ministers.

———. 2003. *National Forest Strategy (2003–2008), A Sustainable Forest: A Canadian Commitment.* Ottawa: Canadian Council of Forest Ministers.

National Roundtable on the Environment and the Economy. 2003. *Securing Canada's Natural Capital: A Vision for Nature Conservation in the 21st Century.* At: <http://www.nrtee-tree.ca>.

Natural Resources Canada. 1995. *Silvicultural Terms in Canada.* At: <http://nfdp.ccfm.org/silviterm>.

———. 2001. *The State of Canada's Forest, 2000–2001: Sustainable Forestry, A Reality in Canada.* Ottawa: Canadian Forest Service, Natural Resources Canada.

———. 2002. *The State of Canada's Forest, 2001–2002: Reflections of a Decade.* Ottawa: Canadian Forest Service, Natural Resources Canada.

———. 2003a. *Old-Growth Forests in Canada.* At: <http://www.atl.cfs.nrcan.gc.ca>.

———. 2003b. *Competitiveness of the Forest Industry.* At: <http://www.cfl.scf.rncan.gc.ca>.

———. 2003c. *Forest Pests: Eastern Spruce Budworm (Choristoneura fumiferana Clem.).* At: <http://www.nrcan.gc.ca/cfs-scf/science/prodserv/pests>.

———. 2003d. *The State of Canada's Forests, 2002–2003: Looking Ahead.* Ottawa: Canadian Forest Service, Natural Resources Canada.

———. 2003e. *Canada's Model Forest Program.* At: <http://www.nrcan.gc.ca/cfs-scf/national>.

Neave, D., E. Neave, T. Rotherham, and B. McAfee. 2002. *Canada's Forest Biodiversity: A Decade of Progress in Sustainable Management.* Ottawa: Canadian Forest Service, Natural Resources Canada.

Ontario Forest Policy Panel. 1993. *Diversity: Forests, People, Communities—A Comprehensive Forest Policy Framework for Ontario.* Toronto: Queen's Printer for Ontario.

Reynolds, P.E., J.C. Scrivener, L.B. Holtby, and P.D. Kingsbury. 1993. 'Review and synthesis of Carnation Creek herbicide research', *Forestry Chronicle* 69: 323–30.

Schueler, F.W., and D.E. McAllister. 1991. 'Maps of the number of tree species in Canada: A pilot project GIS study of tree biodiversity', *Canadian Biodiversity* 1: 22–9.

Sierra Club of Canada. 2002. *The State of Ontario's Forests: A Cause for Concern.* At: <http://www.sierra-club.ca/national/forests/state-of-the-forests/scc-ontario-forests.pdf>.

Simard, J.R., and J.M. Fryxell. 2003. 'Effects of selective logging on terrestrial small mammals and arthropods', *Canadian Journal of Zoology* 81, 8: 1318–26.

Swanson, F.J., and J.F. Franklin. 1992. 'New forestry principles from ecosystem analysis of Pacific Northwest forests', *Ecological Applications* 2: 262–74.

Thompson, I.D., J.A. Baker, and M. Ter-Mikaelian. 2003. 'A review of the long-term effects of post-harvest silviculture on vertebrate wildlife, and predictive models, with an emphasis on boreal forests in Ontario, Canada', *Forest Ecology and Management* 177: 441–69.

Timoney, K.P. 2003. 'The changing disturbance regime of the boreal forest of the Canadian prairie provinces', *Forestry Chronicle* 79: 502–616.

Vancouver Sun. 2003. 'Spotted owl's survival threatened', 11 July.

Western Canada Wilderness Committee (WCWC), Sierra Legal Defence Fund, and Forest Watch of British Columbia. 2002. *Logging to Extinction: The Last Stand of the Spotted Owl in Canada.* At: <www.wildernesscommittee.org/campaigns/endangered-species/spotted_owl>.

Westfall, J. 2003. *2002 Summary of Forest Health Conditions in British Columbia.* Victoria: Ministry of Forests, Forest Practices Branch.

Wielgus, R.B., and P.R. Vernier. 2003. 'Grizzly bear selection of managed and unmanaged forests in the Selkirk Mountains', *Canadian Journal of Forest Research* 33: 822–9.

Wildlands League. n.d. *Forest Diversity-Community Survival Project.* Toronto: Canadian Parks and Wilderness Society.

World Resources Institute (WRI). 2000. *Canada's Forests at a Crossroads: An Assessment in the Year 2000.* At: <http://www.wri.org>.

World Wildlife Fund Canada. 2003. *The Nature Audit: Setting Canada's Conservation Agenda in the 21st Century.* At: <www.wwf.ca>.

Chapter 10

Agriculture

Learning Objectives

- To understand some of the environmental and social impacts associated with the growth of agriculture.
- To appreciate the global food situation and some of the factors that influence it.
- To gain an understanding of the role of energy inputs in agriculture and the Green Revolution.
- To realize the main trends in Canadian agriculture and Canada's contribution to the global food supply.
- To know some of the main environmental implications of agriculture in Canada.
- To understand some of the main problems arising from the use of agricultural chemicals.
- To analyze the implications of a diet with a high level of meat consumption.
- To discover some of the changes that have to be made to move towards more sustainable modes of agricultural production.

INTRODUCTION

The origins of agriculture have been traced to at least 9,000 and perhaps 11,000 years ago in a few regions in which societies domesticated both plant and animal species. Through domestication, such desired traits as increased seed/fruit concentration and fleshiness, reduced or increased seed size, controlled seed dispersal, and improved taste were possible. Various agricultural practices such as seedbeds, improved animal nutrition, and water management were also devised. In turn, the increased availability of food, feed, and fibre provided the impetus for societies to prosper and support a larger non-farming population. Societies around the globe flourished by improving their capacity to expand agricultural production (Box 10.1).

The domestication of plants and animals continues today, but under a much different set of social, economic, and environmental conditions than existed even a century ago. According to the United Nations Food and Agriculture Organization (FAO, 2002a), agriculture is now a dominant influence on the global landscape outside the major urban centres, if not *the* dominant influence.

Over much of history, agricultural output has been increased through bringing more land into production. For example, the global extent of cropland increased from around 265 million ha in 1700 to around 1.2 billion ha in 1950, predominately at the expense of forest habitats, and now stands at around 1.5 billion ha (Wood et al., 2000). Today, the opportunity for further geographic expansion of cropland is greatly reduced due to the comparatively limited amount of land well-suited for crop production, the increasingly concentrated patterns of human settlement, and the growing competition from other uses for land (FAO, 2003a). Between the early 1960s and the mid-1990s, for example, the total amount of agricultural land in Western Europe and North America showed a decline of approximately 39 million ha, the first time in modern history that the extent of agricultural land in these two regions had diminished (Wood et al., 2000). This was mainly a result of urban encroachment (Box 10.2).

BOX 10.1 SOCIAL IMPLICATIONS OF THE DEVELOPMENT OF AGRICULTURE

Agriculture has had a profound influence on society, which in turn has further implications for ecosystems.

• More reliable food supplies permitted growth in populations.
• A sedentary life became more possible as a result of these food supplies and the ability to store food; this allowed larger, permanent settlements to become established.
• Permanent settlements allowed for greater accumulations of material goods than was possible under a nomadic lifestyle.
• Agriculture allowed food surpluses to be generated so that not all individuals or families had to be involved in the food-generating process, and specialization of tasks became more clearly defined. The end result of this situation is that now only some 4 per cent of Canada's population is directly involved in food production, allowing the rest of the population to direct their energies to other tasks, usually the processing of raw materials into manufactured goods and thereby increasing the speed of flow-through of matter and energy in society. As indicated in Chapter 4, this high rate of throughput is at the core of many current environmental problems.
• The creation of food surpluses and increased material goods promoted increasing trade between the now sedentary settlements. This led to the development of road and later rail connections to facilitate the rapid transport of materials, involving the consumption of large amounts of energy.
• Land and water resources became more important, leading to increased conflict between societies regarding control over agricultural lands.
• Aggregation of large numbers of people together in sedentary settlements also served to concentrate waste products in quantities over and above those that could be readily assimilated by the natural environment. Today we call this pollution.

Huge monuments, such as Angkor Wat in Cambodia, could not have been constructed, nor the large cities that surrounded them, without the development of agriculture (*Philip Dearden*).

BOX 10.2 URBANIZATION OF AGRICULTURAL LAND

Despite the fact that Canada is the second largest country in the world and one of the biggest exporters of foodstuffs worldwide, only 673,000 km² or 7 per cent of Canada's overall land mass is used for agricultural production. The amount of arable land free from severe constraints on crop production is even smaller, totalling less than 5 per cent of the land base. Limitations such as climate and soil quality reduce the amount of land that can be relied on for agricultural use; consequently, about 40 per cent of agricultural activities occur on marginal or poorer quality land, which may not be dependable for long-term agricultural activity (Statistics Canada, 2001b).

Of the total of dependable agricultural land (Classes 1, 2, and 3) in Canada, Saskatchewan has the largest amount (40 per cent), followed by Alberta (24 per cent), Ontario (16 per cent), and Manitoba (11 per cent). Most of the prime agricultural land is in southern Canada, where 90 per cent of Canadians live. In Ontario, for example, over 18 per cent of Class 1 farmland is now being used for urban purposes. This juxtaposition of prime agricultural land and the main urban centres has meant that suburban expansion invariably leads to losses in agricultural land.

Concern over the loss of agricultural land from urbanization led to the undertaking of a major monitoring program on 70 Canadian cities with populations over 25,000 between 1966 and 1986. During that period, 301,440 hectares of rural land were converted to urban use. Furthermore, some 58 per cent of the converted land was of prime agricultural capability. To replace the productivity of this land would require twice as much land to be brought under cultivation on the agricultural margins. Whether these losses should be of concern is open to debate. However, in some regions, urbanization of agricultural land affects specialty crops that have a limited ability to flourish in Canada. These crops often represent an important resource to the local economy (e.g., the fruit belts in the Niagara and Okanagan regions) (ibid.). Cities also impact the use of surrounding lands in indirect ways—golf courses, gravel pits, and recreational areas are often located on agricultural land in areas adjacent to urban areas, and as a result the effects of urban areas extend beyond their physical boundaries (ibid.).

The Canada Land Inventory (CLI) indicates that only 18 per cent of the renewable resource base of the country has prime capability for crops, and less than 2 per cent is categorized as having the highest capability (Class 1). To try to slow down the rates of conversion, several provinces have enacted legislation regarding the protection of agricultural lands. In 1972, for example, British Columbia enacted the Agricultural Land Reserve (ALR). At the time of its inception some 6,000 hectares of prime agricultural land were lost to urbanization in the province each year. This has now fallen to less than one-seventh of this figure. The ALR initially covered 4.7 million hectares (5 per cent of BC); despite boundary changes over the decades, its area remains approximately the same. A similar program exists in Quebec. Quebec's territory is vast, but less than 2 per cent of its land is suited to agriculture. Farmland was constantly disappearing in the face of expansion and rampant urbanization. In 1978, the Quebec government introduced the Agricultural Land Preservation Act, which now protects over 63,000 square kilometres of prime agricultural land.

Most Canadian cities of any size are surrounded by good agricultural lands. As the cities, such as Montreal, increase in size, they invariably encroach on the surrounding lands (*Philip Dearden*).

Intensification of production—obtaining more output from a given area of agricultural land—has become a key development strategy in most parts of the world. World grain production has more than doubled since 1961, mainly because farmers are harvesting more grain from each hectare—

only slightly more land today is planted in grain: 671 million ha in 2002 compared with 648 million ha in 1967 (Worldwatch Institute, 2003a). Worldwide, intensification of production, dubbed the 'Green Revolution', has doubled the average harvest of grain from a given hectare from 1.24 tons in 1961 to 2.82 tons in 2002 (ibid.). However, the future capacity to deliver agricultural outputs also depends on the continuing ecological viability of agro-ecosystems, and the Green Revolution carries with it a number of significant long-term environmental impacts that will ultimately lead to long-term productivity losses.

The overriding purpose of agriculture is to provide an adequate and stable supply of food. Yet, agriculture faces an enormous challenge to meet future food needs (Box 10.3). Despite past successes, affordably feeding the current world population—and the more than 70 million people per annum by which that population will continue to grow over the next 20 years—remains a formidable challenge. An analysis by the International

Intensification has resulted in a doubling of grain production over the last 40 years (*Philip Dearden*).

Food Policy Research Institute (IFPRI) suggests that between 1995 and 2020, global demand for cereals will increase by 40 per cent, while meat demand is projected to grow by 63 per cent and the demand for roots and tubers by 40 per cent (Wood et al., 2000). Demand for fruits, vegetables, and seasonings, as well as for non-food farm products, will also rise. All regions will face difficulties in meeting the growing demand for agricultural products while also preserving the productive capacity of their agro-ecosystems. In 2002, global grain production declined for the third time in four years, due mainly to drought in North America and Australia. At 1,833 million tonnes, the harvest was the smallest since 1995. Furthermore, total cereal use exceeded world production for the second year in a row. World cereal stocks by the close of the seasons ending in 2002 were forecast to reach 587 million tonnes, down 8 per cent from the previous season's level. The very large subsidies involved with global agriculture are a major confounding factor in examining food production capabilities. Industrial nations collectively pay their farmers over $300 billion each year in subsidies, and these go to the largest farmers in the richest countries, promote chemical dependencies, inhibit change, and discriminate against the produce from lesser-developed countries.

As the world becomes more crowded and as pressures on biological systems and global biogeochemical cycles mount (Chapter 3), it is no longer sufficient to ask whether we can feed the planet. We need to ask ourselves more difficult questions. Can the world's agro-ecosystems feed today's planet and remain sufficiently resilient to feed tomorrow's hungrier planet? Will intensive production systems cause some agro-ecosystems to break down irreversibly? Presuming we can maintain current food production, are we paying too high a price in terms of the broader environmental effects of agriculture?

Agriculture is fundamentally an ecological process, as solar radiation is converted through one or more transformations into human food supplies. Rapid growth in human population has entailed increasing disruption of natural systems in order to feed burgeoning populations and, particularly in developed countries, growing appetites. This chapter will provide some context for this ecological process before considering some of the environmental challenges facing Canadian agriculture.

BOX 10.3 HUNGER

Agriculture provides approximately 94 per cent of the protein and 99 per cent of the calories consumed by humans. People need, on average, about 2,500 calories per day, although this will vary from person to person depending on weight, age, level of activity, and other factors. These calories come mainly from carbohydrates, such as potatoes and rice, but to remain healthy, humans also need protein, vitamins, fatty acids, and minerals. Fatty acids are essential for hormonal control and minerals are necessary for bone formation and growth.

People are classified as *undernourished* if their caloric intake is under 90 per cent of the recommended level for their size and occupation. When only 80 per cent of recommended caloric intake requirements are met, people are considered *severely undernourished*. *Malnourished* people have adequate caloric intake, but are deficient in other requirements. Protein deficiencies are particularly noticeable in children who appear pot-bellied, skinny limbed, and undersized. Such images have become well known from news coverage of famines, largely in Africa. Vitamin deficiencies lead to many different diseases depending on the particular vitamin that is lacking. Effects can be fatal, such as the vitamin B deficiency that causes beri-beri disease. Other effects include blindness, loss of hair and teeth, and bow-leggedness. Malnourishment also increases susceptibility to other diseases. Vitamin A deficiency, for example, impairs the immune system and is a significant contributor to the nearly 2.2 million annual child deaths worldwide caused by diarrhea and 1 million annual deaths from measles.

The effects of malnutrition also cross generations. The infants of women who are themselves malnourished and underweight are likely to be small at birth and more susceptible to death and disease. Overall, 60 per cent of women of child-bearing age in South Asia, where half of all children are underweight, are themselves underweight. In Southeast Asia the proportion of underweight women is 45 per cent, in sub-Saharan Africa, 20 per cent.

Most Canadians cannot appreciate what it is like to go hungry, but for almost 20 per cent of the world's population, hunger seldom goes away. Although global food supply could provide adequate nutrition for the entire global population, over 1.5 billion people are currently malnourished. The United Nations Food and Agriculture Organization estimates that there are at least

Hunger is a daily reality for many of the world's children (*Philip Dearden*).

815 million chronically hungry people in the world, a modest decline from the 956 million estimated in 1970. There are 777 million undernourished people in the developing countries, 27 million in the countries in transition, and 11 million in the developed market economies. More than half of the undernourished people (61 per cent) are found in Asia, while sub-Saharan Africa accounts for almost a quarter (24 per cent). Hunger is most pronounced where there are natural disasters and, more commonly, where warfare is ongoing. Every year, over three million children die as a result of starvation. In stark contrast, the World Health Organization (WHO) reports that over 300 million people are clinically classified as obese, and about half a million people die from obesity-related diseases every year. Obesity rates have tripled since 1980.

The 1974 World Food Summit promised to eradicate hunger within the next decade. The 1996 Summit was a little more realistic and promised to cut the number of hungry people in half by the year 2015. This would have required a reduction in the number of hungry people by 22 million each year; the actual number has been less than six million. The time frame to meet this goal has now been extended to 2060.

There will be significant challenges facing global food production systems in the future, as the global population increases by an estimated 70 million people each year until 2020. No single solution addresses the challenge of increasing the quantity and quality of *affordable* foods. The solution will involve a variety of approaches, including: exploration of new marine and terrestrial food sources; continued research to increase yields of existing crops; improvements in the efficiency of natural resource use; family planning programs that aim to reduce population growth rates; the breakdown of global agricultural tariffs; more efficient food distribution systems to address chronic hunger; and a moderation in demands from the already overfed countries.

SOURCES: FAO (2002b); UNDP et al. (2000).

AGRICULTURE AS AN ECOLOGICAL PROCESS

Agriculture can be considered a food chain, with humans as the ultimate consumers. Energy flows through this food chain in a manner similar to natural food chains. As such, the second law of thermodynamics is also important to agricultural food chains—the longer the food chain, the greater the energy loss (Chapter 2). This constitutes one part of the argument for a vegetarian diet. By eating at the lowest level on the food chain, as herbivores, humans will maximize the amount of usable energy in the food system. There are, however, other aspects of food production higher up the food chain that should also be taken into account.

Food is more than energy requirements. Important protein and mineral demands must also be considered. Animal products, by and large, are the main suppliers of these proteins and minerals, including such elements as calcium and phosphorus. Areas currently used as rangeland to support animal production often cannot be used as cropland. It may be too dry or otherwise ecologically marginal, and severe problems, such as excessive soil erosion, have arisen in the past when humans have tried to bring such lands under cultivation. Thus, it is not always correct to presume that rangelands can produce more food under tillage; grazing animals in many parts of the world provide not only food supplies but also other products/services, such as their energy as draft animals and use of hides and other animal parts for clothing.

These points do not mean that there are no valid arguments for reducing meat consumption in favour of a diet with a higher vegetable content, especially in industrialized nations where consumers annually eat more than 80 kg of meat per person, compared with just 28 kg for people in developing countries (Worldwatch Institute,

2003a). However, these arguments include not only ecological aspects, but also health, moral, and religious elements.

The domestication of plants and animals thousands of years ago led to profound changes to the global land base. Complex natural systems that once dominated the landscape have been replaced by relatively simple control systems in which humans are in command of the species and numbers that exist in a given area. However, unlike the Industrial Revolution, these changes took place over an extended time period, allowing greater potential for adaptation. Furthermore, until the last 150 years or so, energy inputs were largely limited to photosynthetic energy from the sun, the energies of domesticated draft animals such as oxen, camels, and horses, and human energy input. It was not until the Industrial Revolution unlocked past deposits of photosynthetic energy in the form of coal and later oil that industrial agriculture started and energy inputs and environmental impacts

Horse power is one form of auxiliary energy used in agricultural production, but the environmental impacts of this system of cutting hay by hand and collecting it by horse in Newfoundland are much less than those resulting from the fossil-fuelled, mechanical processes favoured by most Canadian farmers (*Philip Dearden*).

BOX 10.4 THE GREEN REVOLUTION— CONTRIBUTOR TO SUSTAINABILITY?

Starting in the 1940s, research stations in various parts of the world were able to increase dramatically the yields of crops such as wheat and rice through selective breeding and the application of auxiliary energy inputs, as described in the text. Yields of rice, for example, tripled in some locations. This was known as the Green Revolution. At first glance, these boosts in yield appeared to address any questions that may have been raised regarding the ability of agricultural systems to produce enough food to satisfy the growing world population. A second glance, however, revealed some problems.

Rice yields, where sufficient water was available, tripled in some locations (*Philip Dearden*).

Most of the new hybrid seeds grew better than their native counterparts only if fertilizers and biocides were applied with sufficient frequency and in sufficient quantity. As these chemicals became more expensive, many poorer farmers were unable to take advantage of the high-yield seeds. The Green Revolution also encouraged a narrowing of the genetic base of the crop, and with each farmer growing exactly the same strain, any disease or pest that managed to adapt to the strain had an almost unlimited food supply.

Today, critics of the Green Revolution point to stagnation in scientific progress—the gains that can be made through technologies have already been made, and the damage done to agro-ecosystems through modern production systems will limit the ability of farmers to increase yields significantly. For example, macro-nutrients in the soil, such as nitrogen, phosphorus, and potassium, can be replenished by fertilizers, but many of the micro-nutrients required in trace amounts cannot be replaced. As we take out more crops from the soil these nutrients may become exhausted, leading to greatly diminished returns in the future. There are already signs of such declines in some areas.

However, advances are still being made. In 2001, for example, a new rice strain, called Nerica, was produced from hybridization of African and Asian rice varieties. According to the United Nations Development Program (UNDP), the project's main supporter, Nerica produces 50 per cent more yield, uses less fertilizer, is richer in protein, and is more resistant to drought, disease, and pests. It was also developed in full consultation with farmers and consumers. By bringing down world rice prices and helping to nourish poor people, promoters are confident that this rice strain will be a success.

increased dramatically. Indeed, only during the latter half of the twentieth century did some of the most damaging impacts of agriculture begin.

MODERN FARMING SYSTEMS IN THE INDUSTRIALIZED WORLD

The Green Revolution

Dramatic changes in food production systems have occurred through a variety of technological advances that were in turn influenced by changes in demographics (e.g., increases in population densities), social structure (e.g., urbanization, social stratification, etc.), and economic conditions (e.g., global trade). Early food production systems were characteristically small in scale since they were largely dictated by fixed environmental conditions—the quantity and quality of food produced depended heavily on existing local climatic conditions (e.g., annual rainfall), native vegetation (i.e., plants indigenous to the region), and availability of human and/or animal labour. Soil fertility was maintained or enhanced using locally available, natural elements such as manure, bones, and ashes.

BOX 10.5 GENETICALLY MODIFIED ORGANISMS

Yields of many crops in the developed world will not increase significantly using conventional agricultural techniques, even those relating to the Green Revolution. Instead, farmers are turning to biotechnology or the genetic modification of crops to increase production. The technology already dominates the production of a few crops in several countries, yet in other countries its use has been banned. Some nations require labelling of foods that have been genetically modified, while others such as Canada have rejected such openness. Why are there such differences in approach towards this new 'genetic revolution'? Before answering that question it is necessary to explain what GMOs entail.

Advances in understanding of DNA have shown that the basic building blocks of life are all very similar. Traits controlled by single genes can be transplanted from one species to another. Transgenic crops are produced when a single species contains pieces of DNA from at least one other species. One of the most common examples is the implantation of genes containing the toxin from *Bacillus thuringiensis* (Bt). For years, farmers have sprayed this naturally occurring toxin on their fields as an insecticide. Geneticists were able to isolate the toxic gene in the bacteria and have now inserted it into crops such as corn and soybeans. The crops now produce their own toxin. The other common genetic modification is to produce crops with genes that make them resistant to a particular herbicide. When the farmer sprays the crop, all competing plants are killed.

The growth in use of such transgenic crops has been very rapid. At the time of the Earth Summit in Rio (1992), transgenic crops were not in use. The total area cultivated with genetically modified crops now stands at over 40 million hectares. About 75 per cent of this area is in industrialized countries. Substantial plantings largely concern four crops: soybeans, maize, cotton, and canola. About 16 per cent of the total area planted to these crops is now under GM varieties, and two traits—insect resistance and herbicide tolerance—dominate. Throughout the world, several thousand GMO field tests have been conducted or are underway, and many more crop-trait combinations are being investigated with greater focus on virus resistance, quality, and in some cases, tolerance to abiotic stresses.

In Canada, scientists at the University of Victoria in British Columbia have developed a potato that resists bacteria and fungi, allowing it to be stored for 10 times longer than 'normal' potatoes. In Guelph, Ontario, scientists have engineered a pig that produces manure 20–50 times lower in phosphorus content than that of 'regular' pigs. As phosphorus is a main cause of eutrophication (Chapter 4), the numbers of pigs that can be kept in a given area is often limited. The new pig can be stocked in higher densities.

There is no doubt that GMOs hold great potential for the future. It is estimated that they will generate an industry worth US$350 billion by 2015. However, development has been very rapid and there are several areas of uncertainty regarding their effects:

- *Pleiotropic effects.* These are unexpected side effects that might be suffered by the target organism as a result of incorporation of the new gene. For example, there may be a change in the toxins produced or the nutrient content.
- *Environmental effects.* There will be impacts on natural processes, such as pollination and biogeochemical cycles, as a result of creating new crops. It is also feared that there may be unanticipated gene flow to other organisms and perhaps interbreeding with wild relatives. This has the potential to produce 'superweeds'. This has happened in Canada with canola, where the biocide-resistant superweeds are now growing in wheat fields. The only way to get rid of them is to resort to broad-spectrum herbicides, ones that will kill everything.
- *Unintentional spread.* Pollen and seeds from transgenic crops may spread onto lands where they are not intended to grow. This is already a problem for organic canola and honey producers in the Prairies, and they can no longer guarantee that their products are transgenic free.

There is little consensus among governments, consumers, farmers, and scientists concerning the benefits and risks associated with biotechnology or genetic engineering. A lack of perceived benefits for consumers and uncertainty about their safety have limited their adoption in many countries, particularly in Europe. The European Union, Japan, Australia, New Zealand, South Korea, Thailand, and other nations are developing or have implemented legislation requiring mandatory labelling on genetically modified foods. Canada has chosen not to do so. The government voted down a bill that would have required mandatory labelling for all foods containing more than 1 per cent genetically engineered ingredients, even though polls suggested that a large majority of Canadians would prefer to have a choice.

Today, local conditions are manipulated to improve both the quantity and quality of outputs. The amount of food produced per unit of land has increased dramatically through a variety of technological advances, including genetic engineering, greater mechanization (tractors for plowing and seed sowing, mechanized food processing, etc.), and the creation of auxiliary energy flows. The package of inputs or agricultural techniques, which together are referred to as the **Green Revolution**, includes the introduction of higher-yield seeds (e.g., shorter maturation, drought resistance, etc.) and a reliance on auxiliary energy flows (Box 10.4).

The development and commercialization of higher-yielding seeds through **hybridization** led to significant gains in grain yields throughout the world. In the 1940s, scientists developed a 'miracle wheat seed' that matured faster, was shorter and stiffer than traditional breeds, and was less sensitive to variation of daylight. India more than doubled its wheat production in five years with the new technology, and other Asian and Latin American countries recorded similar productivity increases. Miracle rice seeds and high-yield maize (corn) were developed by scientists soon after, and their use has diffused rapidly around the world.

The development of these 'miracle seeds' relied only on genetic combinations found in nature. The development of **genetically modified organisms (GMOs)**, on the other hand, combines genes from different and often totally unrelated species. Such genetic manipulation is now a multi-billion dollar industry, and Canada is one of the world's leading participants (Box 10.5)

The reliance on large auxiliary energy flows in modern industrialized agricultural systems is one

Irrigation water is critical for producing crops on much of Canada's landscape. This is near Kamloops in the BC Interior (*Philip Dearden*).

of the main differences between natural and agro-ecosystems. Auxiliary energy flows include fertilizers, biocides (insecticides, herbicides, fungicides), fossil fuels, and irrigation systems. These energy subsidies have significantly increased crop yields over the last century, particularly over the last two decades. However, in the most energy-intensive food systems, such as in Canada and the United States, on average 10 times as much energy needs to be invested through auxiliary energy flows for every unit of food energy produced. In contrast, **subsistence farming**, where the production of food is intended to satisfy the needs of the farm household, relies on natural energy supplies and may produce 10 food units for every unit of energy invested.

Improved water management, a key component in the Green Revolution technologies, helped boost worldwide productivity or output of 'crops per drop' by an estimated 100 per cent since 1960 (FAO, 2003b). The water needs of humans and animals are relatively small—the average human drinks about four litres a day. But producing the same person's daily food can take up to 5,000 litres of water, which is why the production of food and fibre crops claims the biggest share (70 per cent) of fresh water withdrawn from natural sources for human use. For example, it takes 1,000 tonnes of water to produce one tonne of grain.

Although agriculture is not Canada's largest user of water, it is its largest consumer. Agriculture removes significant quantities of water from the landscape, tying it up in agricultural products or evaporating it back into the air rather than returning it directly to streams or groundwater. Agriculture relies on a reliable supply of good-quality water to grow crops, raise livestock, and clean farm buildings. Approximately 75 per cent of all agri-

The agricultural system of these hill tribe people in northern Thailand, based on dry rice cultivation with livestock such as pigs and chickens, may seem primitive but yields a much higher return than modern farming systems in terms of the energy budget (*Philip Dearden*).

BOX 10.6 AGRICULTURAL IMPACTS ON CANADA'S WATER RESOURCES AND AQUATIC ECOSYSTEMS

Technological advances over the past 50 years have allowed Canadian farms to become more productive, but some farming practices have contributed to environmental degradation, including the decline of water quality. The main pollutants of water coming from farmland are: nutrients (particularly nitrogen and phosphorus), pesticides, sediment, and bacteria. Agriculture has also changed the physical presence of water across the Canadian landscape through the construction of dams and reservoirs, distribution of irrigation water, drainage of wetlands, and sedimentation of streams and lakes.

Nutrient losses. In certain parts of Canada, the use of additional nutrients in the form of mineral fertilizer, manure, compost, and sewer sludge to increase crop productivity has led to a nutrient surplus in the soils with the potential loss of nutrients to surface and groundwater. Nutrient addition to aquatic ecosystems promotes eutrophication (Chapter 4). Other indirect consequences include changes in the abundance and diversity of higher trophic levels (e.g., benthic invertebrates and fish), increased abundance of toxic algae, and fish kills caused by loss of oxygen from the water. One factor contributing to the decline in population numbers for 17 of Canada's 45 frog, toad, and salamander species is nitrate concentrations in agricultural streams and runoff water that exceed thresholds for chronic and acute toxicity of amphibians.

Pesticide losses. The use of pesticides for disease, weed, and insect control has resulted in pesticide losses to the atmosphere and subsequent deposition from the atmosphere to surface water and non-agricultural lands, as well as runoff and leaching to surface and groundwater. Depending on the compound and the concentrations involved, pesticides introduced into surface waters can: kill fish and other aquatic organisms; cause sub-lethal effects on reproduction, respiration, growth, and development; cause cancer, mutations, and fetal deformities in aquatic organisms; inhibit photosynthesis of aquatic plants; and bioaccumulate in an organism's tissues and be biomagnified through the food chain. Pesticide concentrations in surface waters sometimes exceed Canadian water-quality guidelines for irrigation water or protection of aquatic life.

Sedimentation. Agricultural practices such as tillage and allowing livestock access to streams increase erosion and the movement of soil from farmland to adjacent waters. Alterations to soil conditions caused by tillage and cropping patterns may cause soil degradation, which can lead to less infiltration of water into soil and increased runoff and movement of nutrients and pesticides to surface waters. The introduction of soil into aquatic ecosystems can increase turbidity and thereby reduce plant photosynthesis, interfere with animal behaviours dependent on sight, impede respiration (by gill abrasion) and feeding, degrade spawning habitat, and suffocate eggs. Over the past 40 years, tillage practices have moved towards greater use of conservation practices, including no-till. These tend to encourage infiltration rather than runoff.

Pathogens. Pathogens from farm livestock can migrate into groundwater or be carried in runoff into surface waters. Pathogen contamination of irrigation waters, drinking waters, or shellfish areas can pose significant risks to human food supplies, and can also pose a threat to aquatic ecosystems and biodiversity. Regions of the country with high livestock densities are reporting fecal coliform counts that exceed Canadian water-quality guidelines for both drinking and irrigation water. For example, in Ontario the number of wells with E. coli counts exceeding the guideline almost doubled over a 45-year period, from 15 per cent in 1950–4 to 25 per cent in 1991–2.

Wetland drainage. Drainage of wetlands has increased the area of agriculturally productive soils, but it has also modified local ecosystems and changed the pattern of water partitioning between evaporation, stream flow, and infiltration. Wetland drainage destroys the habitat of a number of species of arthropods, reptiles, amphibians, birds, mammals, and fish. More than 85 per cent of the decline in Canada's original wetland area has been attributed to drainage for agriculture.

SOURCE: Chambers et al. (2001).

cultural water withdrawals in Canada occur in the semi-arid prairie region. Over 780,000 ha of cropland in Canada are under irrigation (Agriculture and Agri-Food Canada, 2000). Agricultural practices consume large quantities of water, but they also have an impact on water quality. Box 10.6 details some of agriculture's impact on water resources and aquatic ecosystems.

Fertilizers and biocides are also important inputs to modern farming systems. New strains of crops, such as wheat, will only produce superior yields if fertilized adequately. Global fertilizer consumption stands at about 137,729,730 metric tons per year (FAOSTAT, 2001). In Canada, over 24 million ha were treated with commercial fertilizers (mainly nitrogen, phosphorus, and potash) in 2000. Large increases in fertilizer inputs, particularly nitrogen, have occurred over the last few decades in Canada. Western Canada, for example, saw a fivefold increase in fertilizer applications between 1970 and 2000, and the three Prairie provinces now account for 69 per cent of Canada's total fertilizer sales, up from 29 per cent in 1970.

Fertilizer sales have been on the increase for over 30 years. In 1970, Canadian farmers spent $1.37 million on fertilizers, compared with $1.98 billion in 2000. Canadian farmers are spending more on fertilizers, but the trend since the mid-1990s has been towards an overall decline in the total area treated with commercial fertilizers, with a decrease of 3.7 per cent between 1995 and 2000; Prince Edward Island, Ontario, and Saskatchewan reported the highest declines in crop area treated. However, these statistics mask an important trend.

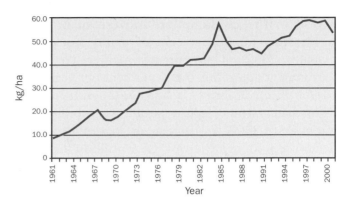

Figure 10.1 Canadian fertilizer use intensity (kg/ha). SOURCE: World Resources Institute Earth Trends (2003).

The *amount* of fertilizers applied *per hectare* has increased considerably since the early 1970s, although the rate of increase has slowed since the mid-1980s (Figure 10.1). In 1970, 1980, 1990, and 2000, fertilizers were applied at a rate of 18.4 kg/ha, 42.4 kg/ha, 45.1 kg/ha, and 54.2 kg/ha, respectively. These rates are not high by international standards, with the United States, Australia, and Japan applying 103.4, 151.7, and 301.0 kg/ha, respectively (Worldwatch Institute, 2003b). Nonetheless, fertilizer application is of environmental concern because fertilizers are a main contributor to the speed of the eutrophication process (Chapter 4) and are also a main contributor to groundwater pollution in some areas.

A central strategy in improving agricultural output is to limit losses from the effects of pests and diseases and from weed competition. Since the mid-1990s, the approach to crop protection has relied increasingly on the use of biocides (insecticides, nematocides, fungicides, and herbicides). Pesticides should be referred to as biocides, since their application often affects more than just the target species. Biocides are applied to boost yields. Yields are improved by reducing the amount of energy flowing to the next trophic layer of the food chain through the respiration of heterotrophs (often insects) and by eliminating non-food plants that compete with the crop plants for available growth resources.

Biocide use continues to increase dramatically, indicating that farmers find biocides cost-effective from a production perspective, particularly where alternative forms of crop protection are labour-intensive and labour costs are high, as they are in Canada. Canada and the United States lead the

Not all crops require the same level of chemical inputs. Tomatoes are particularly demanding compared with, for example, cereal crops (*Philip Dearden*).

world in the consumption and use of biocides, accounting for 36 per cent of world biocide use (UNEP, 2002). The biocide market has been growing at approximately 6 per cent per year since 1990 (ibid.). Statistics Canada (2003) estimates that Canadian farmers spent about $1.5 billion on biocides in 2000. In that year, of 36.4 million ha of land used to grow crops, 25.9 million ha were treated with herbicides, 2.2 million ha were treated with insecticides, and 2.6 million ha were treated with fungicides (Table 10.1). In total, over 30 million ha were treated with at least one biocide in 2000. In 1970, less than 10 million ha were treated with biocides. These large increases in the application of agricultural biocides over the last 20 years, with almost a threefold increase in value of application per hectare in Canada, have profound environmental implications, as will be discussed in more detail later.

Modern, industrial cropping systems that rely on auxiliary energy flows and other inputs are responsible for the production of a large percentage of the cereals, pulses, oil crops, roots and tubers, fruits, and sugar crops produced and consumed worldwide. However, these crops represent only 60 per cent of the total value of output from the

world's agro-ecosystems. Livestock production is responsible for the remaining 40 per cent.

The Livestock Revolution

Hundreds of years ago, livestock (e.g., cattle, sheep, goats, etc.) raised for local consumption were permitted to graze on surrounding natural vegetation. Stocking densities were dictated by surrounding environmental conditions, i.e., the availability of water and food supplies on a given unit of land, and the ability of the local environment to assimilate animal wastes. As a result, farms were typically small, with less than 100 animals. However, as the industrialized world's appetite for meat continues to grow exponentially, traditional livestock production systems are being replaced. Worldwide, meat consumption has doubled since 1977, and over the last half-century it has increased fivefold. Production of beef, poultry, pork, and other meats has risen to nearly 40 kilograms per person per year, more than twice as much as was available in 1950 (Worldwatch Institute, 2003a).

The **Livestock Revolution** has led to a number of changes with respect to how animals are brought to market. In industrialized countries like Canada, the desire to supply the phenomenal

| **Table 10.1** | **Area Treated with Auxiliary Inputs in Canada, 2000 (hectares)** |

Province	Irrigation	Fertilizer	Commercial Herbicides	Insecticides	Fungicides	Total Area Treated with Pesticides
British Columbia	111,181	346,521	149,256	27,787	24,395	201,438
Alberta	499,240	6,700,045	6,623,945	342,903	541,789	7,508,637
Saskatchewan	68,490	9,908,558	12,326,980	930,481	913,961	14,171,422
Manitoba	28,145	3,531,175	3,556,024	382,328	741,376	4,679,728
Ontario	49,271	2,231,823	2,208,982	360,821	194,149	2,763,952
Quebec	22,578	1,001,733	848,220	90,415	73,323	1,011,958
Newfoundland & Labrador	188	6,415	1,067	641	310	2,018
Prince Edward Island	739	110,102	97,732	45,260	44,548	187,540
New Brunswick	1,144	90,067	54,018	29,118	26,504	109,640
Nova Scotia	3,491	88,374	29,686	16,183	12,034	57,903
Canada	784,469	24,014,813	25,900,910	2,225,937	2,572,388	30,699,235

SOURCES: Statistics Canada (2001a); Commissioner of the Environment and Sustainable Development (2003).

growth in demand for protein has led to a reliance on industrial feedlots, which now produce more than half of the world's pork and poultry and 43 per cent of the world's beef (ibid.). Industrial systems of livestock production depend on outside supplies of feed, energy, and other inputs. Technology, capital, and infrastructure requirements are based on large economies of scale, and production efficiency is high in terms of output per unit of feed. As the world's main providers of eggs, poultry, beef, and pork at competitive prices, intensive farm operations (or 'factory farms') meet most of the escalating demands for low-cost animal products.

Agricultural production of livestock has grown across Canada, while the number of farmers has declined and the size of the average farm has increased. In Ontario, for example, one-quarter of farms account for three-quarters of total farm revenues. New farms are often capital-intensive operations, with very large numbers of livestock. Farms with 3,000 or more pigs or 1,200 cattle are increasingly common (Miller, 2000), and some farms in Ontario and Quebec house in excess of 10,000 animals. As the livestock industry expands and becomes more intensive, health and environmental concerns over livestock manure are growing, particularly when livestock are produced in large numbers under confined conditions such as beef feedlots and intensive hog and poultry barns. The social and environmental impacts associated with intensive farming operations will be discussed in greater detail later.

Agriculture's Impact on the Global Landscape

A variety of impacts are associated with the development of agriculture and, more specifically, the development and spread of modern farming systems:

- Natural selection determines the plants and animals to be found in an area, but following the development of agriculture, humans became the major influence on species distributions. In modern agriculture this has led to large areas of **monoculture cropping**, often made up not only of identical crop species, but with each individual having exactly the same genetic code. New species have been created for the purpose of maximizing the output of food for humans, while native species have

Agriculture has had a profound impact on the distribution of species. Large areas of the agricultural landscape are dominated by monocultures—in this case sunflowers—in which each plant has the same genetic makeup (*Philip Dearden*).

been displaced. Humans have influenced not only distributions but also the numbers of each species, with domesticated plants and animals greatly increasing in number and range.

- Energy flows are directed increasingly into agricultural as opposed to natural systems. It is now estimated that humans appropriate some 40 per cent of the net primary productivity of the planet.

- Biogeochemical cycles are interrupted as natural vegetation is replaced by domesticates that are harvested on a regular basis. Auxiliary energy flows in the form of fertilizers are used to try to replenish some of the nutrients extracted through harvesting.

- Auxiliary energy flows used in modern agricultural systems to supplement the natural energy flow from the sun are often in excess of those derived from natural sources.

- In many areas, agriculture involves supplementing rainfall with irrigation to provide adequate water supplies. This has led to large-scale water diversions, changes in groundwater, soil characteristics, precipitation patterns, and water quality.

- Soils are altered not only chemically but also physically through plowing. There is no natural process that mimics the disturbance created by plowing.

- Natural food chains are truncated as humans destroy and replace natural consumers and predators at higher trophic levels.

- Natural successional processes are altered to keep the agricultural systems in an early seral stage; auxiliary energy flows in the form of herbicides and mechanical weeding are often used to accomplish this.

BOX 10.7   CONSERVING CANADA'S BREEDS AT RISK

Modern livestock production systems rely on a few specialized breeds that produce high yields. Consequently, many less-productive but genetically valuable traditional breeds are threatened or have disappeared. According to the Canadian Farm Animal Genetic Resources Foundation:

- The genetic base for future breeding of dairy cattle is threatened by relying on a single breed. Within the Holstein breed, 80 per cent of the cows are bred to 20 sires or their sons.
- The poultry breeding industry is limited to a few large breeding organizations, all of which appear to have similar genetic material. About a dozen such breeding companies provide stock for the world's poultry industry. The farm birds of the 1950s are now seen only as 'fancy' flocks.
- The Canadian swine industry is based predominantly on crossbreds of four breeds.

Loss of farm animal diversity is not just a Canadian problem. Of the estimated 6,500 breeds of domesticated mammals and birds worldwide, 1,350 are at risk of extinction, 119 are officially confirmed as extinct, and 620 are reported to be extinct (FAO, 2000). According to the FAO, the number of mammalian breeds at risk of extinction has risen from 23 per cent to 35 per cent since 1995, while the total percentage of avian breeds at risk of being lost has increased from 51 per cent in 1995 to 63 per cent in 1999 (ibid.). The loss of genetic diversity in our farm animals may represent a serious food security risk. Modern production systems are very efficient, but they are not well prepared for unexpected challenges, such as new diseases, feed crop failures, or sustained changes in consumer demand (e.g., the desire for low-fat foods and issues of animal welfare and environmental sustainability). Efforts are now underway to raise awareness about the state of the world's animal genetic resources and to encourage improved technical documentation of each country's domestic animal diversity, and some livestock conservancies seek to maintain this diversity. However, at present, there is no agricultural research program to further research activity in the field of farm animal genetic resource conservation in Canada.

- The stocking densities of domesticated herbivores are often much higher than that of natural herbivores, leading to a reduction in standing biomass and changes in the structure and composition of the primary production system.
- The industrial system of livestock production acts directly on land, water, air, and biodiversity through the emission of animal waste, use of fossil fuels, and substitution of animal genetic resources. It also affects the global land base indirectly, through its effect on the arable land needed to satisfy its feed concentrate requirements. The industrial system requires the use of uniform animals of similar genetic composition, contributing to within-breed erosion of domestic animal diversity, as discussed in Box 10.7 (de Haan et al., 1997).

TRENDS IN CANADIAN AGRICULTURE

Approximately 7 per cent of Canada's total land area (68 million ha) is agricultural land, of which 46 million hectares are cropland, pasture, or summer fallow. The total area of farmland in Canada has remained relatively constant for the last 50 years. The Prairie provinces contain 83 per cent of the agricultural land base, while Quebec and Ontario together contain approximately 13 per cent. The agriculture and agri-food sector is a $95 billion industry exporting more than $21 billion in products annually (Statistics Canada, 2001a). Agriculture is responsible for about 8.5 per cent of Canada's GDP (29 per cent of GDP in the

South Saskatchewan from the Cypress Hills. The Prairie provinces contain 83 per cent of the agricultural land base of the country (*Philip Dearden*).

primary production sector), 10 per cent of employment (third largest employer), and about 6.1 per cent of total merchandise exports. There are approximately 247,000 farms in Canada. The number of farms has declined since the early 1970s, while average farm size has been increasing.

Wheat is still the dominant crop in Canada, with over 8.4 million hectares dedicated to its production. Saskatchewan grows 51 per cent of Canada's wheat. Wheat still has the largest field crop area (30 per cent of the total), but its share has decreased since 1996. Other traditional grains, such as barley and oats, are also declining. The production of pulses (e.g., dry field peas, lentils, field beans, soybeans) has increased dramatically since the late 1970s and now accounts for almost 8 per cent of national crop area. Soybeans, the second largest oilseed crop grown in Canada after canola, arc a major field crop in eastern Canada, along with grain corn.

Western Canada produces a majority (76 per cent) of Canada's second largest crop, hay. Hay and other fodder crops have increased by 36 per cent since 1986, to keep up with the increasing numbers of livestock that require feed. The livestock industry is an important part of Canadian agriculture, accounting for $18.9 billion in farm cash receipts in 2001, or 57.9 per cent of the total farm cash receipts (Statistics Canada, 2003a). The number of cattle and calves on Canadian farms rose over 4 per cent between 1996 and 2001, to a record 15.6 million head. Alberta accounts for 43 per cent of the national herd. The poultry sector has also experienced significant growth since 1996. Most production is in eastern Canada, with Ontario and Quebec producing 58 per cent of the 126 million hens and chickens brought to market in 2001. Abundant supplies of low-cost feed have pushed hog numbers to an all-time high—in a five-year period (1996 to 2001) the total number of pigs in Canada increased by 26 per cent to 14 million pigs. Quebec and Ontario house more than half of all the pigs in Canada.

Innovations and technological advances ensure that many Canadians have access to a ready supply of fresh produce year-round. Innovations include a thriving greenhouse subsector as well as some of the most advanced storage technologies in the world. Other technologies have increased the availability of fresh food through production techniques that improve yields. Yield-enhancing technologies include mechanization, fertilizers and biocides, and genetic research. Canada is a main producer of food for the global market (Box 10.8). Unfortunately, many of the innovations and technologies currently employed by Canadian farmers have implications for ecosystem health. These will be discussed in more detail in the next section.

ENVIRONMENTAL CHALLENGES FOR CANADIAN AGRICULTURE

Land Degradation

Land degradation includes a number of processes that reduce the capability of agricultural lands to produce food. As agricultural activities have intensified with increased cultivation and addition of agricultural chemicals to produce better yields, so has pressure on the soil resource.

Soil Erosion

Soil erosion is a serious land degradation problem in Canada and is estimated to cause up to $707 million damage per year in terms of reduced yields and higher costs. In some parts of southwestern Ontario, erosion has caused a loss in corn yields of 30 to 40 per cent. Further costs are incurred off the farm when sedimentation blocks waterways, impairs fish habitat, lowers water quality, increases the costs of water treatment, and contributes to flooding. One study in Ontario estimated these costs to be in excess of $100 million annually.

Soil erosion is a natural process whereby soil is removed from its place of formation by gravitational, water, and wind processes. Under natural conditions in most ecozones, soil erosion is relatively minimal as the natural vegetation tends to bind the soil together and keep it in place. Agricultural activities may totally remove this natural vegetation and replace it with intermittent crop plantings, thereby exposing the bare soil to erosive processes, or keep the land under full vegetation for grazing purposes. The latter approach provides much better protection for the soil, but may still result in erosion, particularly under conditions of high livestock density.

The rate of soil formation varies as a function of different environmental factors. Due to the latitude of Canada, soil formation is slow and an annual rate of 0.5 to 1.0 tonne/ha may be considered average. Any soil erosion above this amount will result in some loss of productive capacity.

BOX 10.8 FOOD PRODUCTION AND CONSUMPTION IN CANADA

Food Production

- In Canada's early years, agriculture employed more than 80 per cent of the population. Today, only 3 per cent of Canadians are directly occupied in farming.
- Canada exports a wide range of products to more than 200 trading partners around the world. Value-added and processed goods, together with prime-quality meats, live animals, bulk grains, oilseeds, and vegetables, are Canada's top agricultural exports. Other important export foods include milk products, fish and seafood, maple syrup and honey, organic, natural, and health foods, and confectionaries and beverages.
- Between 1990 and 2000, canola production in Canada more than doubled. Saskatchewan and Alberta are the largest producers.
- There are about 16,500 vegetable growers in Canada, producing close to seven million tonnes of vegetables worth approximately $2 billion annually.
- During the past five years, the volume of exports of fresh vegetables (excluding potatoes) has grown at a rate of 31 per cent annually, to reach 347,000 tonnes. Greenhouse vegetables and mushrooms make up the bulk of vegetable exports.
- Potatoes, along with sweet corn and green peas, are the most extensively grown vegetables in Canada. Approximately 4.5 million tonnes of potatoes (valued at over $700 million) are grown each year. More than half of them are processed, mostly into french fries.
- Apples are Canada's largest fruit crop, with about 542,859 tonnes grown in 2000. Commercial apple production is worth more than $185 million annually.
- There are over 12,000 maple syrup producers in Canada. Canada accounts for more than 85 per cent of world production.
- Canada's red meat and meat products industry includes beef, pork, lamb, venison, and bison. From 1990 to 2003, red meat and live animal exports increased in value from $1.9 billion to $5.3 billion. However, in 2003, after bovine spongiform encephalopathy (BSE) was detected in one cow on a farm in Alberta, Canadian beef exports to the US were immediately banned, which had a significant impact on the Canadian beef cattle industry. The US government is considering lifting the ban in 2005.

- In 2000, there were: 12.8 million cattle and calves on 121,375 farms and ranches earning farm cash receipts of $6.6 billion; 12.2 million hogs on 13,500 farms, earning farm cash receipts of $3.4 billion; 978,500 sheep and lambs on approximately 10,000 farms, earning farm cash receipts of $79 million; 122,000 elk, fallow deer, white-tailed deer, and other venison species on about 2,000 farms, with a total industry value of approximately $207 million.
- Canada's commercial chicken and turkey meat production totals 1,030 million kg. In 2000, Canada exported almost 12 million chicks, domestic fowl, turkeys, ducks, geese, and guinea fowl worth $36.4 million.
- Poultry production and processing are among the most highly mechanized sectors in agriculture. One person can operate a unit of 50,000 broiler chickens. Poultry processing plants in Canada are effectively mechanized, which allows for the slaughter and preparation of 25,000 broiler chickens for market *per hour*.

Food Consumption

- In Canada, food costs about 14 per cent of the average person's disposable income, making our food among the best and least expensive in the world.
- The level of food energy consumed per Canadian, which remained relatively stable from the mid-1970s to the early 1990s, increased by 17 per cent between 1991 and 2001. Total fat consumed from the food supply has climbed by 23 per cent per person since 1991.
- The levels of protein per person provided by the Canadian food supply have also increased since the early 1990s. Overall, meat is the major source of protein.
- Red meat consumption totalled 27.1 kg per person in 2002.
- Consumption of poultry surpassed 13 kg per person in 2002, up 23 per cent over the last decade. In 2002, egg consumption stood at 12.8 dozen eggs per person.

SOURCES: Agriculture and Agri-Food Canada (2003); Statistics Canada, Agricultural Division (2003); Statistics Canada (2003a).

Large areas of tilled soil are particularly vulnerable to erosion (*Philip Dearden*).

Losses in excess of 5–10 tonnes/ha per year may lead to serious long-term problems. These figures have often been exceeded, and 30 tonnes/ha have been recorded in the Fraser River Valley of BC under row crops, while 20 tonnes/ha is not uncommon in Prince Edward Island. Wind erosion is more difficult to measure, but is estimated as a significant problem in the Prairie provinces in particular with high wind speeds, dry soils, and cropping practices that often leave the soil unprotected. One study in Saskatchewan detected a net output of soil of 1.5 tonnes/ha for a near-level field due to wind erosion, whereas a field with a greater incline (three degrees) was found to lose 6.6 tonnes/ha, as water and wind erosion combined.

To illustrate the complexities of dealing with soil erosion, the results from a study in the Avon River catchments in Perth County in southwestern Ontario are revealing (McNairn and Mitchell, 1991). Water erosion of agricultural land has been the main contributor to soil degradation and water quality impairment in that river basin. A study of 75 farms revealed that 17 per cent of the farms experienced high soil erosion (more than 10 tonnes of soil loss per hectare per year), 67 per cent had medium erosion problems (2–10 tonnes of soil loss per hectare per year), and 16 per cent had low erosion problems (less than 2 tonnes of soil loss per hectare per year).

Interviews with the farmers indicated that 79 per cent underestimated the soil erosion situation on their farms, 3 per cent overestimated the problem, and 18 per cent had an accurate perception about erosion occurring on their farms. Furthermore, the findings showed that farmers with the most serious erosion problem had the poorest understanding of the problem. Finally, farmers had a better understanding of the overall soil erosion problem in their township than they did for their own property. Until farmers have a realistic appreciation of the nature of soil erosion on their land, it is unlikely that they will be inclined to adopt soil conservation methods. These findings also emphasize the importance of understanding the social dimensions of resource management.

Soil Compaction

Soil compaction occurs through frequent use of heavy machinery with wet soils or from overstocking with cattle. Compaction serves to break down the soil structure and inhibit the throughflow of water. Crop yields can be reduced by up to 60 per cent in such conditions. Soil compaction is a problem mainly in the lower Fraser River Valley in BC and in parts of central and eastern Canada. One estimate puts the annual cost of compaction at between $68 million and $200 million in Canada.

Soil Acidification and Salinization

Acidity in soils can occur naturally, but can also be augmented by fallout from acid precipitation (see Chapter 4) and the use of fertilizers. Nitrogen fertilizers undergo chemical changes in the soil that result in production of H+ ions, causing greater acidity. In the Maritime provinces, where significant declines in soil pH have been measured, it is estimated that 60 per cent of the change can be attributed to fertilizer use and 40 per cent to acid precipitation. In the Prairies, concern over acidity is relatively recent due to the generally alkaline nature of the substrate. However, increased use of fertilizer has now led to acidification in some areas. Excess acidity reduces crop yields and also leads to nutrient deficiencies and the export of soluble elements such as iron and aluminum into waterways. The yields of crops such as barley and alfalfa fall sharply at soil pH of less than 6. Liming is a common agricultural practice to combat the effects of acidity.

Salinization is a major problem in many areas of the world where irrigation is common. As water evaporates, it leaves behind dissolved salts. Over time these can accumulate in sufficient quantities to render the land unusable. Ancient civilizations that designed complex irrigation systems were unable to counter the effects of salinization, contributing to their eventual decline. Estimates suggest that 50 to 65 per cent of irrigated croplands worldwide are now less productive as a result of salinization.

Alkaline soils occur naturally in areas of western Canada that have high sodium content and shallow water tables. Salinization can also be exacerbated through cropping practices in which natural vegetation is removed and surface evaporation increases to concentrate salts at the surface. Summer fallow has this effect. **Summer fallow** is a practice common on the Prairies in which land is kept bare to minimize moisture losses through evapotranspiration. On the Prairies, crop yields have been reduced by 10 to 75 per cent. Despite the increased use of fertilizers, it is estimated that in some regions salinization is increasing in area by 10 per cent every year.

Organic Matter and Nutrient Losses

Cultivation involves a continuous process of removing plant matter from a field. In so doing both the organic content and nutrient content of the soil are reduced. Organic matter is critical for maintaining the structure of the soil, influencing water filtration, facilitating aeration, and providing the capacity to support machinery. It also helps to maintain water and nutrient levels.

On the Prairies, current organic matter levels are estimated to be 50–60 per cent of original levels, representing a probable annual loss of about 112,000 tonnes of nitrogen (Figure 10.2). This nitrogen is replaced by the addition of synthetic fertilizers, which, in turn, contribute to the problem of acidification. An alternative way to replace the nitrogen is through the growth of leguminous crops to enhance biological nitrogen

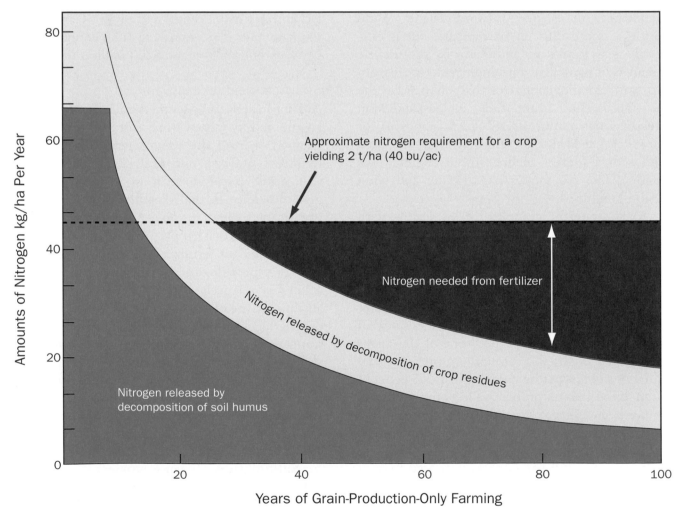

Figure 10.2 Diagrammatic illustration of approximate sources of nitrogen needed to maintain grain yields of about 2 tonnes per ha (40 bu/ac) of barley under a system of continuous grain production in the Prairie region. Note that this diagram illustrates plant requirements, not supply, i.e., the amount of fertilizer nitrogen applied would normally be greater than the plant requirements because of losses due to denitrification and/or leaching. SOURCE: C.F. Bentley and L.A. Leskiw, *Sustainability of Farmed Lands: Current Trends and Thinking* (Ottawa: Canadian Environmental Advisory Council, 1958), 22.

fixation (BNF), as discussed in Chapter 4. Before the increased use of fertilizers in the 1960s, it is estimated that nitrogen exports from prairie grain exceeded fertilizer applications by more than tenfold, and phosphate removals exceeded inputs by threefold. Current estimates still show a depletion of the soil, but with nitrogen now reduced to double the exports over inputs and phosphorus inputs about 50–60 per cent of the export.

Soil degradation, including erosion and nutrient depletion, is undermining the long-term capacity of many agricultural systems worldwide. The FAO (2002a) estimates that soil degradation has already had a significant impact on the productivity of approximately 16 per cent of the globe's agricultural land. According to scientists at the International Food Policy Research Institute (IFPRI) who have carried out the most comprehensive mapping to date of global agriculture, the situation looks worse—nearly 40 per cent of the world's agricultural land is seriously degraded (WRI, 2003). The global significance of this degradation remains unclear, but it will likely have serious implications for future generations, since the production of food in sufficient quantity and quality requires a healthy natural resource base.

Biocides

Biocide use affects almost all Canadians—they are used to produce and preserve the food we eat, and homeowners use biocides to control weeds in their lawns, insects in their gardens and homes, and parasites on their pets. Since the publication of Rachel Carson's classic book, *Silent Spring*, some 40 years ago, which outlined some of the environmental problems associated with the use of biocides, there has been considerable controversy over the risks of using chemicals to control pests. Much of the controversy stems from the fact that biocides are designed to be toxic and are deliberately released into the environment. Biocides have helped to boost yields throughout the world to meet the food demands of rising populations (many more people would be starving without the use of biocides), and biocides have also saved countless lives throughout the world by assisting in the control of various diseases by attacking vectors, such as malaria-carrying mosquitoes. But scientific evidence indicates that many chemicals have profound negative impacts on ecosystems. The possible environmental and health impacts may be

delayed, in some cases, for decades. Despite the risk, biocide use continues to grow on a global scale.

Most biocides now fall into five main groups (Tables 10.2 and 10.3). Various environmental effects can be attributed to these chemicals, but it should be noted that there is considerable variation among them in terms of these effects.

> *It was a spring without voices. On the mornings that had once throbbed with the dawn chorus of robins, catbirds, doves, jays, wrens and scores of other bird voices there was now no sound; only silence lay over the fields and woods and marsh.*
> — *Rachel Carson,* Silent Spring *(1962)*

Resistance

Part of the scientific debate on crop protection relates to the ability of pests, weeds, and viruses to develop resistance to biocides. When a population of insects, for example, is sprayed with a chemical, individuals within the population will react in various ways. If the biocide is effective, most of the population will be killed, but it is likely that a small number of individuals will have a higher natural resistance and survive the chemical onslaught. This remnant resistant population may then grow rapidly in numbers, a result of the lack of competition from all the dead insects. Seeing a resurgence of the pest insect, the farmer sprays again and is again successful in killing a proportion of the population, but not as high a proportion as before since the natural resistance has been passed on to a larger proportion of the population. As this process repeats itself, the use of the chemical creates a population that will ultimately be quite resistant to it. This results in a constant need to develop new biocide products (or pest-resistant plant varieties) to keep one step ahead of biological adaptation.

The 'biocide treadmill' has led to biological adaptations resistant to most commercially available biocides. In Canada and the United States, one estimate suggests that over 900 major agricultural pests are now immune to biocides, including some 500 insects and mites, 270 weed species, and 150 plant diseases (UNEP, 2002). As a result, more frequent applications are needed today to accomplish the same level of control as in the early 1970s (ibid.). Across Canada, farmers have experienced increasing difficulty in controlling pests through spraying. In New Brunswick, Colorado potato

Table 10.2	Major Types of Pesticides

Type	Examples
Insecticides	
Chlorinated hydrocarbons	aldrin, chlordane, DDT, dieldrin, endrin, heptachlor, mirex, toxaphene, kepane, methoxychlor
Organophosphates	malathion, parathion, dizainon, TEPP, DDVP
Carbamates	aldicarb, carbaryl (Sevin), carbofuran, propoxur, maneb, zineb
Botanicals	rotenone, nicotine, pyrethrum, camphor extracted from plants
Microbotanicals	bacteria (e.g., Bt), fungi, protozoans
Fungicides	
Various chemicals	captan, pentachorphenol, methyl bromide, carbon bisulphide
Fumigants	
Various chemicals	carbon tetrachloride, ethylene dibromide (EDB), methyl bromide (MIC)
Herbicides	
Contact chemicals	atrazine, paraquat, simazine
Systemic chemicals	2,4-D, 2,4,5-T, daminozide (Alar), alachlor (Lasso), glyphosate (Roundup)

Table 10.3	Characteristics of Common Chlorinated Hydrocarbons

Pesticide	Characteristics
Aldrin	A pesticide applied to soils to kill termites, grasshoppers, corn rootworm, and other insect pests; aldrin can also kill birds, fish, and humans.
Chlordane	Used extensively to control termites and as a broad-spectrum insecticide on a range of agricultural crops; chlordane remains in the soil for a long time.
DDT	An insecticide controlled in many nations, but still widely used in tropical countries, particularly to combat malaria; DDT builds up in the food chain and has caused harm to wildlife, including thinning of eggshells.
Dieldrin	Used principally to control termites and textile pests, dieldrin persists in soil; it is highly toxic to fish, aquatic animals, and predatory birds.
Endrin	A chemical used against insects and rodents; endrin is highly toxic to fish.
Heptachlore	Primarily used to kill soil insects and termites, this chemical is believed responsible for the decline of several wild bird populations; heptachlore also kills or causes reproductive problems in test animals.
Mirex	An extremely persistent chemical that was used as a fire retardant and a pesticide; mirex is toxic to a number of plants and animals and is classified as a possible human carcinogen.
Toxaphene	This insecticide is chemically similar to DDT; toxaphene is highly toxic to fish and is classified as a possible human carcinogen.

beetles that used to be killed with one spraying a season must now be sprayed five or six times, and they still cause substantial damage. In British Columbia, the pear psylla, an aphid-like insect that feeds on pears, has become resistant to the main biocide registered for use against it.

One of the main beneficial impacts that biocides have had is in their use to control disease-bearing organisms, such as mosquitoes. Some of the world's deadliest diseases are spread through mosquito bites, including malaria. Over one million people are infected every year, and one person dies from malaria every 30 seconds. The numbers of infections and deaths are rising. Fifty years ago medical experts predicted that malaria would be eliminated due to the control of mosquitoes with DDT. Unfortunately, the mosquitoes soon developed resistance to DDT and since that time have become resistant to virtually all control mechanisms. Given global warming trends it is possible that hitherto tropical diseases, such as malaria, could invade Canada. The advent of West Nile virus may well be a forerunner of what is to come in the future.

Non-selective

Many biocides are popular because they are broad-spectrum poisons. In other words, there is no need to identify the specific pest because a broad-spectrum poison will kill most insects. Unfortunately, they tend to eliminate not only the pest species but also other valuable species, including some that may act to control the population of the pest. This may result in a population explosion of the resis-

Malaria kills many people a year but receives little funding because it has little impact on the richer countries of the North. This is an educational sign in Sri Lanka showing villagers how to minimize the risks (*Philip Dearden*).

tant members of the pest population after spraying and also the development of new pests previously kept in check by natural predators.

Many biocides are also extremely toxic to species other than those directly targeted, such as soil micro-organisms, insects, plants, mammals, birds, and fish (Box 10.9). For example, in PEI more than 20 instances of fish kills since 1994 have been attributed to pesticides, with up to 35,000 dead fish collected in each incident (Commissioner of the Environment and Sustainable Development, 2003). Non-target organisms poisoned by biocide use may be beneficial to agriculture or other human economic activities, and they are part of a biodiversity valued by society for recreational, cultural, ethical, or other reasons.

Mobility

The purpose of biocides is to reduce the impact of a particular pest species (or several species) on a particular crop in a particular area. However, the effects of the chemical application are often felt over a much wider area, sometimes spanning thousands of kilometres, due to the mobility of the chemicals in the Earth's natural cycles, particularly the hydrological cycle, and the manner in which chemicals are applied (Figure 10.3). The US Department of Agriculture, for example, estimates that aerial spraying of insecticide results in less than 2 per cent reaching the target, and for herbicide applications, less than 5 per cent. The remainder finds its way into the ecosystem, where it may contaminate local water supplies or be transported by atmospheric processes to more distant sites (Box 10.10).

Places with well-developed agricultural sectors and frequent use of chemicals might find that the entire environment is becoming contaminated. This seems to be the case in PEI according to studies undertaken by government agencies that were only made available to the public under the Access to Information Act. The province uses more than eight kilograms of biocides per person every year, compared with the average of 1.3 kg used in the US. The tests on airborne pollution found every sample to be contaminated. Even tests taken at the end of a wharf, far from any farms, showed chemicals. One of the most heavily used chemicals on the island, chlorothalonil, has been identified by the US government as a potential carcinogen.

BOX 10.9 CARBOFURAN AND BIRDS

Carbofuran is a carbamate (Table 10.2) registered for agricultural use in Canada. It is available either as a liquid or in granular form on particles of grit. It is often applied in the latter form during seeding to protect recently germinated seedlings from insects. Considerable amounts of the chemical (up to 30 per cent of that applied) are often exposed on the soil surface following application. These granules are highly attractive to many birds that ingest grit for use in their gizzards for grinding seed. The chemical is highly toxic to birds, and the consumption of just one grain can be fatal to small seed-eaters. Carbofuran also contaminates invertebrates, such as earthworms, which in turn will poison organisms higher on the food chain. Flooded fields are also highly dangerous, as the chemical goes in to solution. This is particularly serious in acidic fields, where the breakdown of carbofuran is very slow.

There have been many documented bird kills as a result of the use of carbofuran since its introduction. Some Canadian examples include:

- December 1973: 50–60 pintails and mallards killed in flooded turnip fields in BC.
- November 1974 to January 1975: 80 ducks killed in flooded turnip fields in BC.
- October to December 1975: 1,000 green-winged teal killed in flooded turnip fields in BC.
- May 1984: over 2,000 Lapland longspurs were killed in a rapeseed field in Saskatchewan.
- September 1986: an estimated 500–1,200 seed-eating birds were killed in turnip and radish fields in BC.
- June 1986: 45 gulls died after eating carbofuran-contaminated grasshoppers in Saskatchewan.

Of particular concern is the impact of carbofuran on one of Canada's designated threatened species, the burrowing owl. The owl nests on the Prairies in abandoned mammal burrows, feeding on small mammals and insects. Two-thirds of the Canadian breeding population nests in Saskatchewan, where studies suggest that breeding numbers declined by 50 per cent in the south-central region between 1976 and 1987. Carbofuran spraying for grasshopper infestations is suspected to be the main cause. There are significant declines in nesting success and brood size with increasing proximity of carbofuran spraying to the nests. A high percentage of adult owls also disappeared after spraying. There were

The burrowing owl is an endangered species that has been particularly threatened by the use of agricultural biocides (*World Wildlife Fund Canada/T. Muir*).

particularly severe infestations of grasshoppers in the early 1980s, and it is estimated that in Saskatchewan alone over three million hectares were sprayed in 1985, with 40 per cent of the area being sprayed with carbofuran.

The registration of carbofuran was finally cancelled in 1996. Meanwhile, a 1995 report on the status of the burrowing owl recommended a change in its listing from threatened to endan-

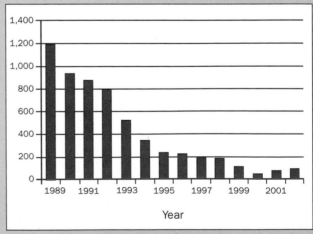

Estimated Pairs of Burrowing Owls Reported by Alberta and Saskatchewan Landowners. SOURCE: <www.pnr-rpn.ec.gc.ca/nature/endspecies/burrowing/db04s05.en.html>.

gered (Chapter 11), and concluded that 'Unless the population trend is reversed, all indications are that the Burrowing Owl will be extirpated from Manitoba within a few years and from the entire country within a couple of decades' (Wellicome and Haug, 1995: 1). The burrowing owl remains on Canada's endangered species list, but a recent evaluation suggests that the voluntary habitat stewardship program (see Chapter 11), Operation Burrowing Owl, is having some success in protecting the remaining habitat (Warnock and Skeel, 2004).

Persistence

Not only do biocides spread over vast areas, they also contaminate through time, as many of them are very persistent. The time taken for some common organochlorine insecticides to decay is shown in Table 10.4. DDT is one of the best-known insecticides. First synthesized over a hundred years ago, it was not until the 1940s that the chemical became widely used, first in health programs in World War II to control disease vectors and then later as an agricultural chemical. Production peaked by 1970 when 175 million kg were manufactured. By that time the environmental effects of DDT were becoming better understood, and its use, but not manufacture, was banned in the US in 1972. It was not until 1985 that registration of all DDT products was discontinued in Canada. More than seven million kg of DDT were sprayed on forests in New Brunswick and Quebec between the early 1950s and the late 1960s. DDT is extremely persistent, as can be seen in Table 10.4. Even now there are still considerable residues of DDT and its main breakdown product, DDE, in the environment. Because it is soluble in fat, DDT may also gradually accumulate over time in the tissues of organisms. This is known as **bioaccumulation** (Figure 10.4).

Organisms with long lifespans are particularly susceptible to bioaccumulation. In British Columbia, for example, an insecticide banned in the late 1970s called hexachlorocyclohexane (HCH), once used as a timber preservative and agricultural spray, has now been found in geoducks off the west coast of Vancouver Island and in Puget Sound. These large clams are filter feeders that may live as long as 140 years. They are therefore very susceptible to

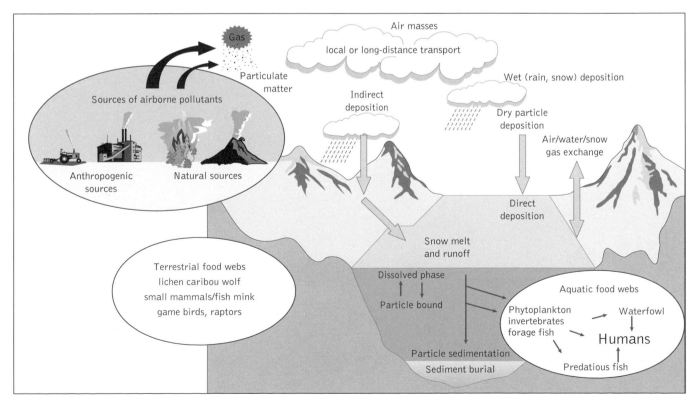

Figure 10.3 Pesticide transportation in the environment. SOURCE: Adapted from Indian and Northern Affairs (1997a). Reproduced with the permission of the Minister of Public Works and Government Services, 2004.

Table 10.4	Persistence of Some Organochlorine Insecticides in Soil		
Chemical	Typical Annual Dose (kg/ha)	Half-life (years)	Average Time for 95% Disappearance (years)
Aldrin	1.1–3.4	0.3	3
Isobenzan	0.3–1.1	0.4	4
Heptachlor	1.1–3.4	0.8	3.5
Chlordane	1.1–2.2	1.0	4
Lindane	1.1–2.8	1.2	6.5
Endrin	1.1–3.4	2.2	7
Dieldrin	1.1–3.4	2.5	8
DDT	1.1–2.8	2.8	10

SOURCE: Edwards (1975: 231).

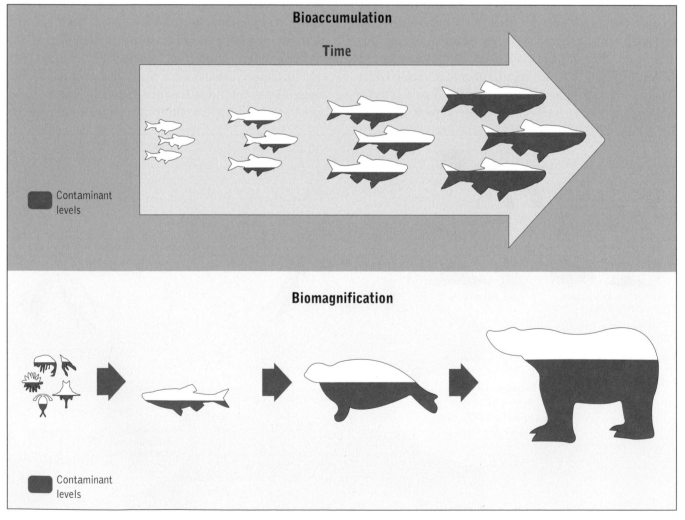

Figure 10.4 Bioaccumulation and biomagnification. SOURCE: Adapted from Indian and Northern Affairs (1997b). Reproduced with the permission of the Minister of Public Works and Government Services, 2004.

Although Prince Edward Island is known for its beautiful rural landscape, the intensity of agricultural production has left large biocide residues, even in ocean sediments (*Philip Dearden*).

organochlorine biocides have been detected in top predators of the Arctic food chain, including indigenous peoples. The ability of biocides and other toxic chemicals to traverse long distances, combined with the ability of biocides to concentrate and accumulate in the lipids of pelagic marine organisms (e.g., Atlantic cod, whales, seals), is to blame for the introduction of contaminants into the Arctic food web (Box 10.10). Concentrations of POPs multiply five- to tenfold with every step in the food chain. This process is illustrated in Figure 10.5, showing the concentration of DDT and its derivatives along a food chain in the North Pacific Ocean. The relatively low concentrations at the lower end of the food chain are magnified many times by the time they reach top fish-eating predators. Most of the visible effects of POPs on animals are related to the ability to conceive and raise young. Malformations in reproductive organs, fewer young, and even complete failures to reproduce are some of the detrimental signs of high contaminant levels. Animals with a high load of organic contaminants are also more susceptible to infections, and POPs are suspected of being responsible for increased rates of malignant tumors in wildlife.

Some species build up very large concentrations. The beluga whales of the St Lawrence estuary, for example, show concentrations of 70,000–100,000 parts per billion (ppb) of DDT. The fact that their population has fallen to less than 10 per cent of the original population in the area, and that individual lifespans are about half of the normal lifespan for the species, indicates that this level of toxic burden exceeds their levels of tolerance. Scientists examined 73 of 175 carcasses that washed ashore between 1983 and 1994 and found that 20 per cent of them had intestinal cancer. These whales account for more than half the cancer cases among all dead whales in the world. Of the 1,800 whales washed ashore and examined in the US, scientists found cancer in only one. Cancer has never been reported in Arctic belugas.

bioaccumulation, and concentrations have been sufficient to have the clams refused by processing plants. In another example of the persistence of biocides in aquatic environments, researchers have detected high levels of toxaphene in trout in Bow Lake in Banff National Park, Alberta. The toxaphene had been applied in 1959 to rid the lake of what were then seen as undesirable fish species. DDT and other chemicals were also detected in trout in many of the lakes in Waterton, Banff, Jasper, and Yoho National Parks.

Biomagnification

The Arctic has been viewed, generally, as one of the few unpolluted regions of the world. But a closer look reveals a different story. High concentrations of persistent organic pollutants (POPs) such as

3,700 ppb PCBs
5,200 ppb total DDT[a]
in dolphin blubber

800–9,800 ppb PCBs
600–2,800 ppb total DDT
in storm-petrel eggs

68 ppb PCBs
22 ppb total DDT
in squid

48 ppb PCBs
43 ppb total DDT
in fish

1.8 ppb PCBs
1.7 ppb total DDT
in zooplankton (copepods)

0.0003 ppb PCBs
0.0001 ppb total DDT
in seawater

[a] Total DDT=DDD+DDE+DDT.

Figure 10.5 Organchlorines in a North Pacific food chain. SOURCE: P.G. Noble, *Contaminants in Canadian Seabirds*, SOE Report no. 90–2 (Ottawa: Environment Canada, 1990).

BOX 10.10 THE GRASSHOPPER EFFECT

Imagine for a moment, if you will, the emotions we now feel: shock, panic, grief—as we discover that the food which for generations nourished us and keeps us whole physically and spiritually, is now poisoning us. You go to the supermarket for food. We go out on the land to hunt, fish, trap and gather. The environment is our supermarket.... As we put our babies to our breasts, we feed them a noxious chemical cocktail that foreshadows neurological disorders, cancers, kidney failure, reproductive dysfunction. That Inuit mothers—far from areas where POPs are manufactured and used—have to think twice before breastfeeding their infants is surely a wake-up call to the world.

—Sheila Watt-Cloutier, President, Inuit Circumpolar Conference (Canada)

We tend to think of the Arctic as 'pristine wilderness'. Recently, however, research has indicated that this is far from the truth. At Ice Island, for example, a floating ice-research station 1,900 km above the Arctic Circle, concentrations of a family of pesticides called hexachlorocyclohexanes (HCHs) have been measured that are twice as high as those in agricultural southern Ontario. Yet, there is not a single pesticide-dependent product grown in the North! How did the chemicals get there?

The so-called **grasshopper effect** is one reason behind Arctic pollution. Indeed, atmospheric transport and deposition is a major pathway of contamination. After chemicals are introduced into an environment they are absorbed into the soils and/or plant tissues, or deposited into rivers, lakes, and wetlands. Persistent and volatile pollutants evaporate into the air in warmer climates and travel in the atmosphere towards cooler areas, condensing out again when the temperature drops. The cycle then repeats itself in a series of 'hops' until the pollutants reach climates where they can no longer evaporate. Chemicals released in southern Canada, for example, may go through the grasshopper cycle several times and take 10 years to reach the Arctic. Extremely volatile chemicals will travel further and recondense in greater concentrations. For example, biocides such as lindane (an HCH are quite volatile compared with DDT, and therefore usually reach the North in greater quantities.

The implications of this long-distance transport are serious. Arctic ecosystems are more vulnerable to toxic chemicals because they last longer in the North. Degradation processes are inhibited due to low temperatures and reduced

People have the perception of the Arctic and Rockies as pristine wilderness areas. Marketers use this image to sell products such as water. In reality, these cold environments can have high toxic burdens due to the grasshopper effect (*Philip Dearden*).

ultraviolet radiation from the sun. The cold also condenses the toxins, keeping them locked up and slowing evaporation rates. There are measurable concentrations of DDT, toxaphene, chlordane, and PCBs found in the Arctic, and when fish and other species ingest these chemicals, they travel up the food chain, accumulating in the fatty tissue of animals at the top of the food chain.

Concentrations of several persistent organochlorine pesticides (POPs) remain high in many aquatic food webs in Canada. This has serious implications for the Inuit in particular because of their high consumption levels of wildlife. Over 80 per cent of Inuit consume caribou, almost 60 per cent consume fish, and almost 40 per cent consume marine mammals. Due to bioconcentration, the consumption of traditional foods places the Inuit at greater risk for the development of several ailments related to toxic chemical exposure, including endocrine disruption, reproductive impairment, and cancer. These concerns are not restricted to the Arctic, as large concentrations of toxic chemicals have also been found in the mountains of British Columbia. Fish in alpine lakes have chemical levels that make them toxic to eat in large quantities.

Since POPS travel great distances, a global approach to tackle the issue is required. Various initiatives have arisen to control or eliminate POPS, including the Global POPS Protocol, signed in Johannesburg in December 2000. Signatories of the Protocol, including Canada, agreed to a global ban of 12 chemicals, including the follow-ing pesticides: aldrin/dieldrin, endrin, DDT/DDE, HCH/lindane, chlordane, heptachlor, chlordecone, mirex, and toxaphene. Canada is also a signatory to the United Nations Economic Commission for Europe (UNECE) POPS Protocol, which lists 16 chemicals for phase-out. But while governments around the globe have a significant role to play in reducing Arctic pollution, consumers also share in the responsibility. Chemicals used in consumers' everyday environment can end up polluting some of the most 'pristine' environ-ments on Earth. Consumers must make an effort to phase out domestic use of toxic chemicals.

A major component of the toxic burden for the belugas is mirex and its by-products. Mirex is a biocide, now banned, which was never produced along the St Lawrence. Biologists think that the source is Lake Ontario, where American eels accu-mulate the chemical. During their downstream migration, these eels constitute a significant part of the whales' food supply.

Biomagnification is largely responsible for the drastic population reductions of many birds during the 1960s and 1970s. Birds of prey such as ospreys, peregrine falcons, and bald and golden eagles and fish eaters such as double-crested cormorants, gannets, and grebes were particularly affected by the widespread use of insecticides. Some birds were killed directly through bioaccu-

BOX 10.11 BIOCIDES AND YOU

Although most biocides in Canada are used by commercial producers, large amounts are also used domestically in homes and on gardens and lawns to control unwanted organisms. How people use, store, and dispose of these chemi-cals is very important in terms of minimizing envi-ronmental damage. Here are a few tips.

• Use chemicals only as a last resort. Ask your-self why you need to kill the organism. If it's just for aesthetic reasons, such as dandelions on your lawn, then maybe you need to change your perceptions rather than automatically reach for a chemical solution. For each pest or weed, there are usually several other approaches you can take as part of your own integrated pest management strategy. These are too numerous to document here, but you can find out more specific strategies from government ministries such as the Ministry of Agriculture in your area.
• Use the safest chemicals available in the minimum quantities. Many plant nurseries now sell products that are less toxic than traditional biocides. Often they need more skill in applica-tion, but they are less environmentally damag-ing than regular chemicals. Examples of these include the 'Safer Soap' line of products.

• Apply all chemicals in strict accord with the manufacturer's instructions.
• Store unused chemicals so that they do not leak and cannot be accidentally upset.
• Dispose of chemicals and containers in a safe manner. Contact your local Ministry of Environ-ment to see what programs are in place in your province for safe disposal. Some provinces, for example, have specific sites where biocides and other toxic chemicals can be disposed of. If your province does not have such a program or acceptable alternative, start lobbying for one!

You should also protect yourself against the risk of ingesting chemicals that have been applied to food:

• Grow your own food; don't use chemicals.
• Buy organically grown produce wherever possible.
• Fruit and vegetables that look perfect often do so because they have had heavier applications of fertilizers and pesticides. Choose products that show more natural blemishes; this is a sign that chemical use has not been as high.
• Carefully wash all produce in soapy water.
• Remove the outer leaves of vegetables such as cabbage and lettuce, and peel all fruit.

The beluga whale population has declined rapidly in the St Lawrence, and the toxic burden from biocides appears to be one cause (*Fred Bruemmer*/Valan Photos).

mulation and biomagnification, while many others were unsuccessful in breeding. DDT affects the calcium metabolism of these species, resulting in thinner eggshells and leading to breakage and chick mortality. The banning of DDT and similar chemicals has led to a recovery of many of these species in temperate countries (see Chapter 11).

However, the continued use of the chemicals in some tropical countries still affects populations in these areas and also the populations of migratory species such as peregrines.

Together, biomagnification and bioaccumulation are often known as **bioconcentration**. Humans are also exposed to the harmful effects of biocides through bioconcentration. They may come into contact with chemicals through contaminated water supplies, ingestion of food products, their occupation, and/or domestic use (Box 10.11). Researchers found, for example, that women involved in farming in Ontario's Windsor and Essex counties have nine times the risk of developing breast cancer as opposed to non-farm women.

A greater dependence on fish or game birds for food may elevate the risk associated with biocide exposure, since fish and game birds may have already concentrated significant amounts of toxic matter in their fat deposits. Many First Nations communities are highly dependent on marine fish and mammals and may be particularly at risk. For example, levels of the biocide chlordane are significantly higher in the breast milk of Inuit women in the North than in women in southern Canada (UNEP, 2002).

BOX 10.12 COSMETIC USE OF PESTICIDES

The Federation of Canadian Municipalities estimates that two-thirds of Canadian households use pesticides. Half of the homeowners apply the products themselves, often with no protection, an average of three times a year (Standing Committee on Environment and Sustainable Development, 2002). In addition to this private use, municipalities also use substantial amounts of pesticides. Figures are difficult to obtain, but in Ontario in 1993 this amount was estimated at 1.3 million kilograms, about one-quarter of the amount used in agriculture.

Due to potential negative impacts on human health, the Standing Committee on Environment and Sustainable Development recommended a ban on the use of biocides for cosmetic purposes, i.e., lawn care, but the federal government refused to endorse the recommendation (Figure 10.6). The government preferred to take a voluntary, educational approach to reduce the cosmetic use of pesticides and launched a 'Healthy Lawns Strategy' to address the issue.

However, the government's educational approach did not satisfy all Canadians, particularly residents of a small community in Hudson, Quebec. Residents were worried about the health consequences of lawn and park applications of herbicides and insecticides, particularly on children. The community wanted the municipal government to enact a bylaw that would ban the cosmetic use of pesticides, but chemical companies won the first battle, arguing that municipal governments lacked the authority to introduce such bylaws. The community took the matter to court, and in June 2001 Canada's Supreme Court unanimously ruled that towns and cities have the right to enact bylaws banning the purely cosmetic use of pesticides. The Supreme Court ruling grants municipalities across the country the right to impose similar pesticide restrictions. More than 30 communities in Canada now have bylaws banning the use of chemicals for cosmetic lawn purposes. Toronto's Board of Health has endorsed a similar bylaw, making it the first large city to begin steps to phase out the use of pesticides on lawns for cosmetic purposes.

Figure 10.6 Biocides and human health. Used by permission CAM. SOURCE: *Ottawa Citizen.*

Synergism

When chemicals are tested for their harmful effects they are tested individually in controlled situations. When applied on farmers' fields, however, the chemicals are free to interact with each other and the environment in myriad ways. A single biocide may contain up to 2,000 chemicals and as the chemicals break down, new ones are created that may again react with each other in unpredicted ways. The combined effects are often greater than the sum of their individual effects. This is called synergism, and can result in many unanticipated effects.

Biocide Regulation

The concerns tied to the use of biocides—resistance, non-selection, mobility, persistence, bioaccumulation, biomagnification, and synergism—are significant. However, the economic value tied to the use of biocides is also very significant. Because the use of biocides will likely continue to grow in the foreseeable future, there is a pressing need to strengthen the regulatory and enforcement mechanisms governing biocide import, production, and use.

> *We have been waiting years for the federal government to get its priorities straight when it comes to the registration of pesticides. We need to ensure that Health Canada puts the health and protection of the Canadian public and workers ahead of that of the profits of pesticide manufacturers.*
> — Dr Warren Bell, Executive Director of the Canadian Association of Physicians for the Environment
> <http://www.cela.ca/media/mr020326.htm>

In Canada, an estimated 1,000 new chemicals are introduced annually, in addition to the over 20,000 already in industrial, agricultural, and commercial use. New chemicals cannot be introduced onto the market without undergoing scientific tests regarding their capacity to cause cancer, birth defects, and mutations. The Pest Management Regulatory Agency (PMRA), a branch of Health Canada created in 1995, has the primary responsibility for regulating biocides. Other Health Canada branches and other federal departments and agencies that play important roles in biocide management include Agriculture and Agri-Food Canada, the Canadian Food Inspection Agency, Environment Canada, Fisheries and Oceans Canada, and Natural Resources Canada. The federal government shares the responsibility for managing biocides with provincial, territorial, and in some cases municipal governments (Box 10.12).

To predict the effectiveness of biocides and their risks to human health and the environment, the PMRA relies on the expertise and judgement of the Agency's scientists and managers. Evaluation of a new biocide ends with the approval of the biocide label, which describes the biocide's hazards and its proper use. This process, however, has not been foolproof.

The House of Commons Standing Committee on Environment and Sustainable Development reviewed biocide management in Canada and published a report, *Pesticides: Making the Right Choice*, in 2002. The report expressed strong concern over current practices and urged a much stronger regulatory approach based on the precautionary principle. This was followed in 2003 by a report titled *Managing the Safety and Accessibility of Pesticides* released by the Commissioner of the Environment and Sustainable Development (Box 10.13). This report also documents the adequacy (or lack thereof) of the current regulatory practices used in Canada to approve biocide use, as well as the health and environmental standards relating to compliance, the government's commitment to research, and monitoring. The report identifies several instances of poor overall compliance. To predict occupational exposure to pesticides and pesticide residues on food, PMRA evaluators assume that agricultural users will follow good practices, i.e., users will follow label instructions. But available evidence suggests otherwise. For example, a 2001 survey of farmers by Statistics Canada concluded that only 14 per cent

BOX 10.13 A REPORT CARD ON CANADA'S USE OF PESTICIDES

In 2003, the Commissioner of the Environment and Sustainable Development released a report card on Canada's use of pesticides. The document, titled *Managing the Safety and Accessibility of Pesticides*, details the shortcomings of the Pest Management Regulatory Agency (PMRA), a branch of Health Canada responsible for the evaluation of new pesticides, monitoring and compliance of existing pesticides, and the re-evaluation of old pesticides to ensure compliance with modern standards. Overall, the Commissioner concluded, 'the federal government is not managing pesticides effectively.' A few of the main findings taken from the report are highlighted here.

Some pesticides are approved based on inadequate information. The temporary use of a pesticide may be granted pending the submission of further studies. Of new pesticide registrations in 2001–2, 58 per cent were temporary. For some temporary registrations, the missing information was to have been included with the original submission. Examples of information gaps at the time of temporary registration include: what happens to the pesticide after it is released into the environment, what impact it is likely to have on children's central nervous systems, and how toxic it is to invertebrates and non-target plants. As a result, many pesticides are used before they have been evaluated fully against current health and environmental standards.

Key assumptions are not tested and some are not correct. Agency evaluators must make a series of assumptions to link the laboratory studies they receive to the possible impacts of the pesticide's use. Such assumptions include how large a crop area will be treated, how much treated food Canadians will eat, and how the pesticide will be applied. Despite the uncertainties in all of the different assumptions evaluators make, they have not determined how reliable their predictions of the risks are. For example, evaluators have not tried altering their assumptions slightly to see if that would reverse the decision to approve a pesticide. In addition, Agency staff have unrealistic assumptions about user behaviour. They assume that pesticide users will follow label instructions, even though the Agency's own compliance reports show that they may not. An unrealistic assumption of full compliance means that evaluators are underestimating the risks of pesticide use.

The Agency manages a legacy of older pesticides. Many pesticides have been registered in Canada based on evaluations that did not apply the more stringent methods and standards used today. Some of the changes in requirements include considering the impacts on bystanders, the reproductive impacts on later generations, and the greater susceptibility of children. To ensure that older pesticides meet today's standards, the Agency has implemented re-evaluation programs. However, many re-evaluations do not consider new information about the pesticide's effectiveness resulting from new research, and as a consequence, opportunities may be missed to reduce the rate of application. In addition, progress on re-evaluations has been very slow. By 2005, the Agency plans to re-evaluate all products registered by 1994, which will require reviewing 405 active ingredients or about 75 per cent of all active ingredients currently registered in Canada. As of March 2003, only 1.5 per cent of the 405 active ingredients had been fully re-evaluated. All pesticides re-evaluated to date were found to pose unacceptable risks and had to be restricted or removed from the market.

The Agency does not know to what extent users are complying with pesticide labels. The Pest Control Products Act influences how pesticides are produced, distributed, and used. Lack of compliance could cause serious environmental impacts and expose users and bystanders to unnecessary risks. The Food and Drugs Act deals with residues in food; lack of compliance would affect those who eat treated or contaminated food. The Agency conducts inspections to determine whether pesticide registrants, distributors, and users are complying with the Pest Control Products Act. In 2002–3, the Agency conducted only 510 inspections of users, although in agriculture alone roughly 216,000 farms across Canada could have used pesticides. Inspection programs check which pesticides are being used, but do not determine systematically whether the label requirements are being met. Problems with unclear labels also make it difficult to ensure compliance. Ambiguous labels mean that enforcement action cannot be taken for some possible violations of the Act.

Methods for measuring pesticide residues on food are not up to date. The Canadian Food Inspection Agency (CFIA) conducts an extensive chemical sampling program that includes testing each year for pesticide residues in food. However, methods for measuring pesticide residues are not entirely inclusive or thorough.

The CFIA uses a risk-based, multi-residue testing method that screens for 269 different pesticides in various commodities, but the Agency has identified more than 190 additional pesticides used in Canada or in other countries that export food to Canada. In addition, small samples tested for any given pesticide on one type of food may prevent meaningful conclusions about compliance with the limits; consequently, the CFIA can provide only limited assurance that pesticide residues on food comply with the Food and Drugs Act. Furthermore, some of the residue limits set by the PMRA are based on old assessments and are inconsistent with current standards.

Critical data on pesticide use and exposure are still missing. The federal government has not yet set up a pesticide sales database, despite a commitment to do so in 1994. In addition, the adverse effects of pesticides on human health and the environment are tracked and reported only on an ad hoc basis. Reports of pesticide problems by registrants, doctors, provincial agencies, university researchers, and pesticide users could help the Agency understand the impacts of pesticides. In 1994 and again in 2000, the federal government committed to developing a program of mandatory reporting on adverse effects of pesticides, but it has yet to do so. The lack of reliable information on pesticide use, exposure, and impacts is a major hurdle that continues to interfere with the Agency's ability to regulate pesticides.

Federal research on the health impacts of pesticides has not been a priority. Health Canada has very limited dedicated funding for research on human exposure to pesticides or the resulting health effects. Three researchers are working on current pesticides, and they rely primarily on outside funding.

---

calibrate their pesticide spraying equipment between applications of different pesticides; as a result, the amounts they actually use on their crops may be higher than the levels specified on the label. In 2001, evaluators also collected soil samples from 20 onion growers in Ontario. Of those, 14 (or 70 per cent) had violated the Act by using pesticides not registered in Canada, while four other growers were using pesticides not registered for use on onions.

Lack of compliance is partly due to problems with pesticide labels, as some agricultural pesticides may have 30 or more pages of directions in fine print, while other label instructions are difficult to follow. For example, labels are often ambiguous, and application is therefore left to the applicator's interpretation. Ambiguous, vague terms used on pesticide labels include:

- '*Appropriate* buffer zones should be established between: treatment areas and aquatic systems, and treatment areas and *significant* habitat.'
- 'Do not apply in areas where soils are *highly* permeable and ground water is *near* the surface.'
- 'Do not apply *near* buildings inhabited by humans or livestock.'
- 'Do not apply where fish and crustaceans are *important* resources.'

Failure to follow label instructions could increase the risks to consumers and the environment, but it is difficult to apply pesticides appropriately when the directions are unclear. Many poisonings have been attributed to inappropriate application, and this is one reason why farmers in Ontario have welcomed mandatory biocide safety courses dealing with the use, mixing, handling, transportation, and laws governing biocide use. The certificate from these courses must be renewed every five years, and it must be presented in order to buy agricultural chemicals.

Many pesticides have been registered in Canada for decades, but they have not been evaluated against current standards. The PMRA has implemented re-evaluation programs, but the progress has been slow. All pesticides re-evaluated to date have been found to pose unacceptable risks for some uses, leading to restrictions on their use or removal from the market. For example, the re-evaluation of *phorate*, the active ingredient in an organophosphate insecticide first registered in 1969, concluded that the pesticide poses extremely high environmental

Canada has no system in place to track pesticide sales and use, to report their adverse health and environmental effects or to monitor the quality of products on the market; this must be amended.
— *Edith Smeesters of the Coalition for Alternatives to Pesticides*
<http://www.cela.ca/media/mr020326.htm>

risks that may not have been assessed when it was first registered—one granule can kill a small bird or mammal. This chemical is used to control insects on corn, lettuce, beans, rutabagas, and potatoes. As of 31 December 2004, use of this pesticide was no longer permitted in Canada.

Intensive Livestock Operations

Another aspect of modern agriculture that can have significant and far-reaching environmental implications is intensive livestock farming. The production of livestock manure has both environmental benefits and drawbacks. Although manure is a valuable fertilizer for crop production, it can also become a source of pollution if not managed properly.

Manure consists of a variety of substances, including nitrogen, phosphorus, potassium, calcium, sodium, sulfur, lead, chloride, and carbon (Statistics Canada, 2003b). Manure also contains countless micro-organisms, including bacteria, viruses, and parasites. Some of these micro-organisms are pathogenic, and therefore direct consumption or recreational use of water containing these organisms can lead to a variety of illnesses and even death. The contamination of drinking water with E. coli that killed seven residents of Walkerton, Ontario, in May 2000 was identified as having been related to livestock manure (Chapter 12). Pathogens from manure that have reached watercourses also have the potential to spread disease to livestock. The recent spread of bovine spongiform encephalopathy (BSE), commonly referred to as 'mad cow disease', is an example of an inter-species disease transmission. BSE is thought to cause Creutzfeldt-Jakob disease among humans.

Other risks to human and ecosystem health arise from air pollution. Odour and air pollution are identified as serious environmental and human health concerns related to **intensive livestock operations (ILOs)**. High concentrations of noxious gases such as methane, hydrogen sulphide, carbon dioxide, and ammonia are often found in manure pits and confinement barns. Pigs and poultry, for example, excrete some 65 and 70 per cent of their nitrogen and phosphorus intake, respectively. Nitrogen, under aerobic conditions, can evaporate in the form of ammonia (Chapter 4). Ammonium nitrate and ammonium sulphate emitted to the air from animal housing can be harmful to human

Beef feedlot in southern Alberta (*Philip Dearden*).

and animal health. Foul odours emitted by ILOs are a significant problem for neighbours, and studies have shown an increase in chronic respiratory diseases reported by people who live in close proximity to a large animal farm. Ammonia can also have toxic and acidifying effects on ecosystems. Ammonia in high concentrations in the air can have a direct effect on plant growth by damaging leaf absorption capacities, but its indirect effect on soil chemistry is even more important—ammonia acidifies the soil, interfering with the absorption of other essential plant elements (de Haan et al., 1997).

While air pollution from intensive livestock operations is a significant issue, larger issues arise from the storage and use of livestock manure. There are two basic manure storage types in operation in Canada: liquid manure storage (concrete enclosures; steel tanks, either open or covered; earthen basins; lined/unlined lagoons) and solid manure storage (e.g., manure stored indoors with bedding or as a pack in the barn; manure stored outside as a pile on the ground). Problems arise when storage systems are inadequately built or when they are sited too close (less than 30 metres) to water supplies. Liquid manure stored in lagoons, for example, may overflow during periods of heavy rainfall, or the lagoons can fail to prevent the leaching of organic and inorganic materials into the surrounding environment. The health and environmental impacts associated with manure spills are similar to those arising from the improper use of manure as fertilizer.

Raw, well-rotted manure is often spread onto farm fields as fertilizer, a reasonable environmental practice as long as farmers have sufficient cropland

Guest Statement

Canada Feeding the World?

Peter Schroeder

There is a myth woven into agricultural policies and subliminally embedded in all agricultural thinking throughout the developed countries that agricultural output must continually increase, at all costs, or humanity will face the consequences of global starvation. This myth formed the base for the Green Revolution and is the primary reason stated for the global rush to accept and adopt genetically modified organisms in all aspects of modern commercial farming systems.

The 'need to feed the world' myth would not be a bad thing if in some form modern agriculture's hyper-productivity actually helped reduce global hunger, but there is little if any evidence of that. There is ample evidence that maximizing production from every hectare has caused more problems than it has solved by ignoring some of history's most important lessons.

Since its inception, agricultural production has been based on the 'natural fertility' of the land on which it is practised. As farmers became more adept at farming they began to understand what depletes as well as enriches the inherent fertility of the soil. By paying close attention to these factors they began to manipulate their environment to increase yields and harvest reliability. In a relatively short time they had a thorough understanding of the direct linkage between sustainable yield and sustainable fertility. Left to themselves farmers are loath to destabilize this critical balance because they have nothing but the land to sustain them.

The great civilizations throughout antiquity rose and fell according to how well their farming and herding people were able to balance the land's sustainable fertility with the harvest output. The fundamentals of that technology and science have not changed because, now as then, agriculture is all about repackaging solar energy into calories that can be digested by humans. This includes the use of all food animals that convert plant material not digestible by humans. Considering that so much of this planet is more pasture land than 'farm' land, food animals form an important link in the solar energy reprocessing system that is agriculture.

Modern agriculture, which began in the 1950s and 1960s, is based on cheap petroleum energy and a

Peter Schroeder

surplus of nitrogen production left over from World War II explosives manufacturing. These industrial surpluses were all redirected at agricultural production and marketing. Without fossil fuel in all its forms to power modern agricultural equipment and provide the fertilizer as well as transportation, contemporary commercial agricultural models cannot exist.

It was at this point the Canadian government turned prairie agriculture into a full-blown export-earning component of the Canadian economy. Grain, specifically wheat, was produced beyond any possible local demand and shipped around the world. Everybody involved in growing grain on the Prairies had a fuzzy warm feeling—they were working hard, making a decent living, and feeding the world. What could be better?

The long-term truth is the wheat was sold to the highest bidder, often disrupting local markets and bankrupting farmers as far away as India and other developing countries. The cheap Canadian wheat was often used to lower local grain prices, forcing massive disruptions in the local agricultural economies. Canadian wheat was cheap in part because Canadians had learned to use modern farming systems but also because it was subsidized by cheap fuel and transportation.

Intensive livestock production can only happen when the supply of grain in the world is too large for the human population to consume. In the overstimulated production environment of North America and

Western Europe this artificially created surplus of grain is used as livestock feed for everything from poultry to pigs, cattle, and sheep. In addition to producing the cheapest food the world has ever seen, this farming model brings with it environmental and biological challenges that have their own costs.

For almost 100 years government subsidies have been directed at agricultural production in western Canada. The best example is how for almost a century the Canadian taxpayer paid the bulk of the freight costs to get western wheat and barley from the Canadian Prairies to deep-water ports on Canada's west coast or the Great Lakes. The same story is now being repeated with canola, minus the transportation subsidy. Today's incentives come in the form of tax breaks and risk management programs. A similar scenario is also being played out in Manitoba with its massive export-dependent hog industry and across the Prairies where the beef industry expanded to supply a growing demand in the US.

Modern agriculture in Canada and most developed nations is all about producing exportable commodities cheap enough to displace local production. To do this farmers are using all the tools they have available to them as they compete for market share. Their goal in maximizing production at the lowest possible costs is the perfect formula to supply an abundance of raw material to the food industry's processing companies.

From a farmer's perspective modern agriculture has nothing to do with feeding the world's hungry and has everything to do with simply working hard and not going broke. The world's hungry can't afford to pay the freight for the food it would take to solve their problem, never mind cover the production costs. At the same time as the world's starving poor are pictured on TV every other night, the developed nations are coping with an epidemic of obesity. This can only happen when in one part of the world food is so cheap that eating becomes a recreational pastime while in another the same food is so expensive people die of malnutrition-related diseases and outright starvation.

The Green Revolution has been underway for more than 40 years and one has to wonder how it has changed the world for the better. There are fewer farmers on the land in Canada's Prairies every year. Those who are left earn less per acre and animal unit than their parents did before the Green Revolution started. In their struggle they've abandoned many of the fundamentals that sustained the generations on the land before them. Many of today's farming practices, instead of keeping the land productive and healthy, now contribute to ecological damage with chemical and nutrient runoff contributing to contaminating waterways and lakes.

There are farmers in North America who have opted to step off the industrial farming treadmill. They distinguish themselves as grass farmers, holistic ranchers, natural and organic farmers. The common link between all of them is that they are striving to achieve a balance between the 'natural' fertility, a sustainable yield, and their own economic needs without resorting to the farming stimulants: chemical fertilizers, herbicides, pesticides, and genetically modified seeds. These people are a minority but their numbers are growing.

In my own farming experience covering 25 years of sheep ranching, we had our best economic return after we adopted and applied the grass farming principles. Simply put, grass farming is all about using livestock to repackage the solar energy in plants for human consumption. Grass farming requires the rancher to put the needs of pasture ahead of the livestock. It also requires the rancher to forget about maximizing production in favour of profits earned through sustainable yields. The credo for many grass farmers is the 80:20 rule: 'Take the easy 80 per cent nature is offering for free and forget about the last 20 per cent.' The net return to our ranching enterprise increased substantially once we began to work with nature and learned to use the livestock's strengths in their natural environment. In addition to an increased financial return there was a marked improvement in lifestyles, with less labour and lower stress levels along with a marked improvement in soil fertility, which resulted in still better pastures.

The biggest threat facing food production today is that by well-intentioned regulatory means the traditional subsistence and small-scale commercial farms are eliminated in the drive to accommodate intensive agri-business operations as they continue to perpetuate the myth that they are feeding the world's hungry.

Peter Schroeder farmed the Canadian Prairies for many years and now shares his experience as a columnist with the Winnipeg Free Press.

to absorb the manure of their livestock. However, new large-scale farms produce vast quantities of manure and often do not have correspondingly large areas of farmland. In 1996, Canadian livestock produced an estimated 361 million kg of manure *daily*, or over 132 billion kg of manure annually (Statistics Canada, 2003b). Ontario currently has over 3.4 million hogs and they produce as much raw sewage as the province's 13 million people (Miller, 2000).

If manure and commercial fertilizers are misused, spilled, or applied in excessive quantities, the result is contamination of soil and water by nitrogen, phosphorus, and bacteria. Although crops take up the bulk of added nutrients, a portion—the nutrient surplus—remains in the field. For all agricultural land in Canada, annual inputs of nitrogen and phosphorus from commercial fertilizers and livestock manure exceed annual outputs. There is a national surplus of approximately 0.3 million tonnes of nitrogen and 56,000 tonnes of phosphorus, or 8.4 kg/ha of nitrogen and 1.6 kg/ha of phosphorus. Fertilizers, whether organic or synthetic, are viewed as essential to maintain crop yield and soil health, but their application in excess of what crops can utilize can have significant implications for human and environmental health.

Nutrient loading can result in runoff to streams, rivers, lakes, and wetlands, spurring additional growth of algae and other aquatic plants. Accelerated eutrophication results in loss of habitat and changes in biodiversity (Chapter 4). For example, long-term exposure to elevated nitrate concentrations has contributed to the recent decline in frogs and salamanders in Canada. Concentrations of nitrate greater than 60 mg/litre in water kill the larvae of many amphibians.

Excess aquatic plant growth associated with nutrient loading also reduces oxygen levels in the water, leading to fish kill. Between 1987 and 1997, 353 fish kills were reported to Environment Canada's Environmental Emergencies Centre. Of these, 22 were caused by discharges of nutrient-containing materials, and almost all of these contained nitrogen compounds. Most of the reported fish kills were associated with agricultural activities, particularly the release of manure through storm or flood runoff, underground tank leaks, overflowing storage facilities, or spraying of fields. From 1988 to 1998, a total of 274 manure spills

were reported in Ontario, of which 53 resulted in fish kills, primarily due to the ammonia in liquid manure. Reporting on fish kills from accidental spills/discharges of nutrient-related compounds is currently on a voluntary basis only, and so they are believed to be widely under-reported.

The application of fertilizers can also affect the quality of drinking water. Twenty-six per cent of Canadians—approximately eight million people—rely on groundwater for domestic water supply. In most provinces, 5 to 20 per cent of these wells are contaminated with unsafe levels of nitrate, and up to 60 per cent of wells are contaminated in regions with high-demand crops or intensive livestock operations. An increase in the frequency and spatial extent of drinking-water advisories due to nitrate concentrations in excess of Canadian standards has been reported (Chambers, 2001).

Factors influencing the effect of manure and commercial fertilizers on the environment include soil type, climate, precipitation, topography, and the quantities of manure produced. A majority of these factors are beyond the farmer's control. However, manure management practices also influence the magnitude and extent of ecological impacts. Unfortunately, the environmental laws governing manure management were created when small operations were the norm and, therefore, fail to address the environmental and health risks that come with more intensive livestock operations (Miller, 2000). For example, in Ontario there are no legally binding standards for constructing manure storage facilities or for the application of manure. There are also no monitoring mechanisms to ensure that farmers use best practices for managing manure.

SUSTAINABLE FOOD PRODUCTION SYSTEMS

While Canadians continue to enjoy a wide variety of high-quality foodstuffs year-round, concern over the vulnerability of the productive capacity of our agro-ecosystems mounts (Box 10.14). The future capacity to deliver agricultural outputs depends on the continuing ecological viability of agro-ecosystems, yet the stresses imposed on them by intensification are significant. The challenge is to foster agro-ecosystem management practices that will meet growing

BOX 10.14 SUSTAINABLE FOOD PRODUCTION SYSTEMS

Some of the environmental challenges confronting Canadian agriculture have been highlighted in the text. However, socio-economic challenges such as the decline of family farming, product price volatility, rising input costs, and macroeconomic policies are also cause for concern. Allied with the environmental challenges outlined earlier, these have given rise to increasing interest in the concept of *sustainable food production systems.* Brklacich, Bryant, and Smit (1991) suggest that six key concepts underpin sustainability in agricultural systems.

1. *Environmental accounting* concentrates on the biophysical limitations for agricultural production. It does not, however, develop explicit linkages between the quality of the environment and productive capabilities.

2. For agriculture, the concept of *sustained yield* refers to output per unit area and helps to describe the conditions that will maintain yields from year to year. As a result, there is a more explicit link between environmental degradation and productivity than in environmental accounting.

3. *Carrying capacity* seeks to explore the relationship between the number of organisms that can be supported indefinitely on a given unit area. The concept was initially devised to determine the number of grazing animals that could be supported without impairing the quality of the range supporting them. It has since been extended into many different areas. It can, for example, be used to try to estimate the population carrying capacity of different nations in terms of food production, although many different variables besides strictly agricultural ones must be taken into account. Carrying capacity is also determined by considerations such as what constitutes adequate nutrition and what kinds and amounts of environmental change are acceptable. In most cases no absolute number can be calculated to represent carrying capacity, as many management variables must be considered.

4. Sustainability requires that farming is not only biophysically but also economically viable. In other words, *production unit viability* is necessary. Increasing costs of inputs and volatile markets for produce have created increasing stress for many farmers. The decay of rural communities has also meant a decline in the social support for farming. Several authors have suggested that a vigorous rural community must underlie sustainable agriculture. Thought must also be given to the relationship between the short-term costs of achieving long-term environmental sustainability and how these costs can be met by society. Investigations into resilience—the ability of different kinds of farms to withstand stress—are important here.

5. *Product supply and security* are required because a sustainable food production system must be able to provide food on a secure and continuous basis to meet the demands of the population. Although food productivity has increased dramatically over the last few decades, increasing signs indicate that this trend is abating and that population growth is outstripping food production in several areas of the world, especially sub-Saharan Africa. In 1970, Africa was virtually self-sufficient in food supply. Now, about 30 per cent of the people south of the Sahara suffer from chronic hunger and malnutrition.

6. *Equity*, both across the globe and intergenerationally, is important for sustainable food production systems. Although enough food is produced to feed all people on the planet, millions of people suffer from malnutrition due to the inequitable distribution of food. Not only must food be distributed more equitably, but we should not inhibit the abilities of future generations to produce food. Food production systems that cause environmental deterioration and pass the costs of this on to future generations are not sustainable.

On the basis of these concepts, Brklacich et al. suggest that a sustainable food production system should have three qualities:

1. It should maintain or enhance environmental quality.
2. It should produce adequate economic and social rewards to all individuals and firms in the production system.
3. It should produce a sufficient and accessible food supply.

Achieving sustainable food production systems is complicated. Brklacich et al. note that their definition is generic and does not prescribe particular cultural or economic structures to achieve sustainability. There may be a number of different ways to proceed, depending on the situation.

food, feed, and fibre needs, while providing more environmental amenities.

Improving agro-ecosystem management so that all levels of agricultural production can be associated with better environmental performance requires new knowledge and better skills, which can be achieved through improvements in technology, natural resource management systems, and landscape planning, as well as in the policies and institutional arrangements that help integrate environmental values into agricultural investment and management decisions (Wood et al., 2000). Examples include the spread of integrated pest management, integrated plant nutrient systems, and no-till and conservation agriculture. These approaches seek to meet the dual goals of increased productivity and reduced environmental impact.

Integrated Pest Management

The goal of **integrated pest management (IPM)** is to avoid or reduce yield losses caused by diseases, weeds, insects, mites, nematodes, and other pests, while minimizing the negative impacts of pest control (resistance, non-selection, mobility, persistence, bioconcentration, etc.). Originally used to reduce excessive use of pesticides while achieving *zero* pest incidence, the concept has broadened over time. The presence and density of pests and their predators and the degree of pest damage are monitored, and no action is taken as long as the level of pest population is expected to remain within specified limits.

IPM considers the crop and pest as part of a wider agro-ecosystem, promoting biological, cultural, and physical pest management techniques over chemical solutions to pest control. Combinations of approaches are used, including:

- bacteria, viruses, and fungi (pathogens);
- insects such as predators and parasites (biological management);
- disease- and insect-resistant plant varieties;
- synthetic hormones that inhibit the normal growth process;
- behaviour-modifying chemicals and chemical ecology products (such as pheromones, kairomones, and allomones).

If pesticide use is deemed essential to pest control, only pesticides with the lowest toxicity to humans and non-target organisms are applied.

The adoption of IPM practices has economic and other benefits for farmers, but its use does require more expertise than simply applying chemicals. For this reason, Ontario has established a formal system of IPM for 21 of the province's 95 agricultural commodities. Producers can obtain expert and current advice on these products, their pests, and optimal courses of action by phone. An IPM program in Ontario cut the use of pesticides by 53 per cent since its introduction in 1987. Overall pesticide use in fruit and vegetable crops decreased by 20 per cent in the last five years alone. Specifically, insecticide use in fruit growing declined by 57 per cent, while fungicide use for both fruit and vegetables has been reduced by 54 per cent since 1998. Increased adoption of IPM and alternative pest control strategies such as border sprays for migratory pests, mating disruption, alternate row spraying, and pest monitoring are major reasons for these large declines. Internationally, countries such as Indonesia have developed an aggressive IPM program, managing to cut pesticide use by as much as 90 per cent. Sweden has adopted a similar aggressive approach, reducing pesticide use by 50 per cent.

Integrated Plant Nutrient Systems

As discussed earlier in the chapter, agricultural production removes plant nutrients from the soil, reducing the organic and nutrient content of the soil. Imbalances in nutrient availability can lead to excessive depletion of nutrients that are in short supply, with corresponding reductions in crop yield. The goal of **integrated plant nutrient systems (IPNSs)** is to maximize nutrient use efficiency by recycling all plant nutrient sources within the farm and by using nitrogen fixation by legumes (Chapter 4) to the extent possible. Soil productivity is enhanced through a balanced use of local and external nutrient sources, including manufactured fertilizers. Plant nutrient losses arising from nutrients supplied in excess can pollute soils and waters, so IPNSs also seek to minimize the loss of nutrients through the judicious use of external fertilizers. IPNSs aim to optimize the productivity of the flows of nutrients passing through the farming system during a crop rotation (Scialabba, 2003). The quantities of nutrients applied are based on estimates of crop nutrient requirements, i.e., knowledge of the quantities of nutrients removed by crops at the desired yield level.

No-Till/Conservation Agriculture

To destroy weeds and loosen topsoil to facilitate water infiltration and crop establishment, agricultural land is plowed, harrowed, or hoed before every planting. Topsoil disturbance of this magnitude and frequency destabilizes the soil structure, leading to soil erosion and soil compaction, negatively affecting productivity and sustainability. The economic and ecological costs associated with conventional tillage systems are becoming more apparent, leading farmers to search for alternative land preparation techniques, such as **no-till/conservation agriculture (NT/CA)**. NT/CA (or zero, minimum, or low tillage) protects and stimulates the biological functioning of the soil while maintaining and improving crop yields.

According to the FAO (2001), essential features of NT/CA include: minimal soil disturbance restricted to planting and drilling (farmers use special equipment to drill seeds directly into the soil instead of plowing); direct sowing; maintenance of a permanent cover of live or dead plant material on the soil surface; and crop rotation, combining different plant families (e.g., cereals and legumes). Crops are seeded or planted through soil cover with special equipment or in narrow cleared strips. Soil cover inhibits the germination of many weed seeds, minimizing weed competition and reducing reliance on herbicides. Soil cover also reduces soil mineralization, erosion, and water loss,

Strip farming in Saskatchewan (*V. Last, Geographical Visual Aid.*)

builds up organic matter, and protects soil micro-organisms. Crop sequences are planned over several seasons to minimize the buildup of pests or diseases and to optimize plant nutrient use by synergy among different crop types, and involve alternating shallow-rooting crops with deep-rooting ones to utilize nutrients throughout various layers of the soil. Other advantages associated with NT/CA include: increased yields in the order of 20–50 per cent higher than with conventional tillage practices; reductions in the variability of yields from season to season; significant reductions in labour costs; and lower input costs, particularly for machinery (e.g., smaller tractors can be used, reducing fuel costs).

Despite these advantages, conventional tillage-based agricultural systems continue to dominate worldwide. There are several reasons for this—a reluctance to change (conventional approaches have been working for decades), a lack of knowledge of the damage to soil systems associated with plow-based agriculture, and the complex management skills required for a successful transition. The short-term economic costs associated with the transition from conventional tillage to NT/CA are also a deterrent. During the transition years, there are extra costs for tools and equipment, higher weed incidence may increase herbicide costs initially, and yields will improve only gradually.

In Canada, farmers appear to recognize the advantages of NT/CA, with more than 63 per cent employing some or all of the elements of NT/CA. Figure 10.7 shows the percentage of farms in

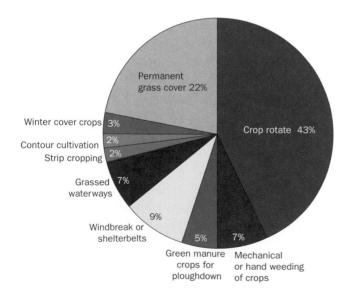

Figure 10.7 Percentage of farms in Canada using different soil conservation practices. SOURCE: Statistics Canada (2001a).

Canada using various soil conservation methods. **Crop rotation** is the most common, and is an important means of recharging soil nitrogen through use of legumes such as alfalfa and clover. Grassed waterways are used to control overland flow of runoff, thereby controlling the formation of gullies on exposed soil surfaces. **Contour cultivation** involves cultivating the soil parallel to the contour of the slope, which serves to reduce the speed of runoff by catching soil particles in the plow furrows. Over 16 per cent of PEI and Saskatchewan cropland is protected in this way. **Strip cropping** is a similar technique in which different crops may be planted in strips parallel to the slope. While one crop may be harvested, leaving bare soil, the other crop serves to provide some protection. This technique is commonly used against wind erosion, and is most prevalent in western Canada. The soil surface can also be protected in winter through growth of a winter cover crop. This is effective not only for wind erosion in winter, but also to protect the soil from intense rainfall in the spring.

There is considerable regional variation in the use of these practices. Newfoundland, for example, has the smallest proportion of farmers using erosion control, due largely to the small amount of land devoted to seed crops in the province. In Saskatchewan, however, almost 72 per cent of farmers use some form of erosion control. Overall, 63 per cent of farms in Canada use one or more erosion control practices. During 2000, 24 per cent of farms reported using conservation tillage techniques, and a further 7 per cent used no-tillage, equal to over 4 million hectares of land. Comparatively, the United States, Brazil, Australia, and Mexico have approximately 19,750,000 ha, 13,470,000 ha, 8,640,000 ha, and 650,000 ha, respectively, under zero tillage.

SUSTAINABLE AGRICULTURE IN ACTION: ORGANIC FARMING

Current approaches to sustainable agriculture such as IPM, IPNS, and conservation tillage consider only one aspect of the farming system components—pest ecology, plant ecology, and soil ecology, respectively. Organic agriculture, however, combines these and other management strategies into a single approach, focusing on food web relations

and element cycling to maximize agro-ecosystem stability. Organic agriculture is a production management *system* that aims to promote and enhance ecosystem health. It is based on minimizing the use of external inputs and represents a deliberate attempt to make the best use of local natural resources while minimizing air, soil, and water pollution. Synthetic pesticides, mineral fertilizers, synthetic preservatives, pharmaceuticals, genetically modified organisms (GMOs), sewage sludge, and irradiation are prohibited in all organic standards (FAO, 2001; Canadian General Standards Board, 2003).

Organic agriculture encompasses a range of land, crop, and animal management procedures. According to the World Health Organization (WHO) and the United Nations FAO (1999), organic production systems are designed to:

- enhance biological diversity within the whole system;
- increase soil biological activity;
- maintain long-term soil fertility;
- recycle wastes of plant and animal origin in order to return nutrients to the land, thus minimizing the use of non-renewable resources;
- rely on renewable resources in locally organized agricultural systems;
- promote healthy use of soil, water, and air as well as minimize all forms of pollution that may result from agricultural practices;
- handle agricultural products with emphasis on careful processing methods in order to maintain the organic integrity and vital qualities of the product at all stages;
- become established on any existing farm through a period of conversion, the appropriate length of which is determined by site-specific factors such as the history of the land and the type of crops and livestock to be produced.

For example, organic practices that encourage soil biological activity and nutrient cycling include: manipulation of crop rotations and strip cropping; the use of **green manure** and organic fertilizer (animal manure, compost, crop residues); minimum tillage or zero tillage; and avoidance of pesticide and herbicide use. Research indicates that organic agriculture significantly increases the density of beneficial invertebrates, earthworms, root symbionts, and other micro-organisms essential to maintaining soil health (Scialabba, 2003). For example, the biomass of earthworms in organic systems is 30–40 per cent higher than in conventional systems, and earthworm density is 50–80 per cent higher.

Enhanced soil fertility is the cornerstone of organic agriculture, but there are other benefits associated with organic production systems. For example, organic fields in excess of 15 hectares contain a greater diversity and abundance of flora and fauna than conventional fields, including endangered varieties (e.g., organic grasslands contain 25 per cent more herb species than conventional grasslands). Organic farming systems are also more energy efficient per unit crop than conventional farming techniques, as organic systems resemble closed or semi-closed nutrient cycles. Organic land management permits the development of a rich weed flora, and a versatile flora attracts more kinds of beneficial insects. Organic farming systems are also better at controlling erosion, since organic soil management techniques improve soil structure. Organically grown foods are also beneficial to human health, as they contain fewer pesticide residues than foodstuffs grown under intensive farming methods.

Like most industrial countries, Canada has national organic standards, regulations, and inspection and certification systems that govern the production and sale of foods labelled as 'organic'. However, in true Canadian fashion the Canadian National Organic Agriculture Standards are voluntary. There are no federal policies aimed specifically at the organic industry, and in 2002 the House of Commons Standing Committee on Environment and Sustainable Development recommended that the government move beyond this voluntary standard.

Worldwide, an estimated 20 million hectares of land are under certified organic agriculture, with Australia, Argentina, Italy, Canada, and the United States with the largest areas of organic farmland. In 2000, agricultural land under certified organic management averaged 0.25 per cent of total agricultural land in Canada, compared with 2.4 per cent of total agricultural land in Western Europe, 1.7 per cent in Australia, and 0.22 per cent in the United States (Scialabba, 2003). In most developing countries, agricultural land reported under certified organic production is less than 0.5 per cent,

although the extent of non-market, non-certified organic agriculture may be considerable. For example, an estimated one-third of West African agricultural produce is produced organically (ibid.).

There are currently more than 3,000 organic farmers in Canada with an estimated total certified cropland of 340,000 hectares (Agriculture and Agri-Food Canada, 2002). Field crops such as buckwheat, rye, and caraway dominate Canadian organic production. Canada is among the top five world producers of organic grains and oilseeds. Farm cash receipts from organic production in Canada reached about $1 billion in 2002. The number of certified producers increased 34 per cent between 1999 and 2000. The domestic organic market is strongest in British Columbia, Alberta, Quebec, and Ontario.

Consumer health and food quality concerns (e.g., concerns about growth-stimulating substances, genetically modified [GM] food, dioxin-contaminated food, and livestock epidemics such as bovine spongiform encephalopathy and foot-and-mouth disease) continue to drive demand for organic products in Canada and around the globe. Organic food production systems are considerably more respectful of the environment than intensive farming practices, but intensive farming systems still dominate. In the absence of governmental support for the expansion of organic production, farmers may be reluctant to convert to organic farming for several reasons. Conversion from conventional, intensive systems to organic production causes a loss in yields, the extent of which varies depending on the biological attributes of the farm, farmer expertise, the extent to which synthetic inputs were used under previous management, and the state of natural resources. Yields can be 10 to 30 per cent lower in organic systems, and it may take several years (e.g., three to five) to restore the ecosystem to the point where organic production becomes economically viable. In addition, production costs per unit of production (e.g., labour, certification and inspection fees, etc.) and marketing expenses can be higher with organic produce, but once produce qualifies as *certified* organic, some of these costs can be offset by price premiums. In developed countries, retail organic products can command 10 to 50 per cent above conventional prices for the same commodity. In Canada, demand for most organic products continues to exceed supply, despite the higher

BOX 10.15 ENVIRONMENT CANADA'S NATIONAL ENVIRONMENTAL INDICATORS SERIES: AGRICULTURE

The health of Canada's agricultural soils is one of many 'headline' indicators used by Environment Canada to measure the state of Canada's environment. The health of Canada's agricultural soils is assessed using five indicators:

- reduction in number of bare-soil days on agricultural land between 1981 and 1996;
- changes in residual nitrogen levels between 1981 and 1996;
- agricultural land subject to unsustainable water erosion;
- prairie agricultural land subject to unsustainable wind erosion;
- prairie agricultural land subject to unsustainable salinization.

Results

Soil cover by crop or crop residue is one way of offsetting the impacts of erosion. Between 1981 and 1996, the average number of bare-soil days in Canada's agricultural regions dropped from 98 to 78, almost 20 per cent. Residual nitrogen refers to the difference between the amount of nitrogen available to the growing crop and the amount removed by the harvested crop. Residual nitrogen often leaches out of soils into ground and surface water where it negatively affects water quality. Residual nitrogen levels in agricultural soils increased markedly between 1981 and 1996 in all provinces except British Columbia. Provincially, the share of farmland showing an increase of at least five kilograms of residual nitrogen per hectare during this period ranged from 27 per cent in BC to 80 per cent in Manitoba. The reasons behind this increase are currently unknown.

Environment Canada uses the term 'sustainable erosion' to describe a condition where rates of erosion are compensated for by natural soil-forming processes and agricultural practices that result in the addition of soil organic matter. The percentage of agricultural land in Canada experiencing unsustainable water erosion decreased from 1981 to 1996, largely due to decreases achieved in the Prairies. During the same period, the percentage of prairie agricultural land at risk from unsustainable levels of wind erosion decreased from 59 per cent to 36 per cent. The percentage at risk from unsustainable levels of salinization has not changed. Elevated concentrations of salts make it difficult for plants to absorb water from the soil, and are also toxic to plants at extreme levels.

Environment Canada chose the trend 'reduction in number of bare-soil days on agricultural land between 1981 and 1996' as the indicator that best summarizes the condition of Canadian agricultural soils. According to the meter, the health of Canada's agricultural soils is 'improving'. However, the federal agency also recognizes that challenges remain. For example, indicator calculations are based on generalized census, landscape, and climate data, and consequently, small land areas where soil degradation is a concern may be overlooked. In addition, there are problems with estimates of nitrogen input into the soil, and erosion indicators need to consider a broader range of concerns, including erosion control practices, severe weather events, and small land areas that are susceptible to soil degradation.

Do you think the indicators used to assess the state of Canada's agricultural soils are adequate? Would other indicators provide a more accurate representation of ecosystem health?

SOURCE: Environment Canada (2003).

prices charged for certified organic produce. Consumers in industrialized countries are willing to pay a premium for organic food because they perceive environmental, health, or other benefits.

In Canada and elsewhere, cases are emerging where organic production is constrained or no longer feasible due to the advent of GM crops. In North America, the contamination of organic fields by GM crops has been estimated to result in an economic loss of over US$90 million a year (Scialabba and Hattam, 2002). Organic farmers in Canada can no longer grow organic canola (i.e.,

BOX 10.16 WHAT YOU CAN DO

Although the challenges faced by agriculture at the global and national levels are immense, there are still some ways in which individuals can help.

1. Eat less. This entails finding out about good nutritional habits so that we consume only the food that we really need.
2. Eat lower on the food chain. Most North Americans eat far too much meat. Eating more vegetables will benefit not only the global food situation, but also your own health.
3. Feed your pet lower on the food chain. Dogs and cats will also be healthier if fed on balanced grain pet foods rather than meat.
4. Waste less food. Studies indicate that as much 25 per cent of food produced in North America is wasted.
5. Grow at least some of your own food. If Canadians were to devote a fraction of the time and resources on growing food that they spend on their lawns, this would allow more food for others elsewhere.
6. Support local food growers and food co-ops. This helps protect agricultural land in Canada from being transformed to other uses.
7. Join one of the NGO groups that specializes in rural development in lesser-developed countries.

oilseed rape) because of GM canola contamination in Saskatchewan. In 2002, two organic farmers from Saskatchewan filed a class action suit against Monsanto Canada and Aventis Canada in response to the commencement of field trials of genetically engineered (GE) wheat for introduction as a commercial crop in 2004. Wheat is a self-pollinated crop, but there is danger of spread of pollen from the wheat, and the resultant genetic contamination of organic wheat would devastate Saskatchewan's organic sector. The suit calls for compensation for damage caused to certified organic farmers resulting from the introduction of GE canola into the rural environment and an injunction to prevent the introduction of GE-wheat61. The outcome of this precedent-setting legal case will be of major importance to future policies regarding the impact of biotechnology on organic agriculture.

IMPLICATIONS

Agricultural modification is arguably the main impact that humans have had on natural ecosystems. It is, however, also one of the oldest and one that is basically a modification of ecological systems to benefit humans. Over centuries, natural and human-modified agricultural landscapes have existed and transformed from one state to the other with little lasting damage to planetary life support systems. However, as additional auxiliary energy flows were applied to boost the productivity of agriculture, the differences between these two ecological systems became more distinct, and the impacts of agriculture on natural ecosystems increased. Agricultural production (certainly in Canada's commercial agricultural sector) is now more similar to industrial production than to the natural ecosystems from which agriculture was derived.

This industrialization has led to many environmental challenges for agriculture. Yields are declining in some areas as crops become less responsive to fertilizer input; biocides continue to eliminate many natural enemies of pests; and soils are eroded, salinized, and compacted. In response, researchers are suggesting that a fundamental restructuring is required in how agriculture is undertaken, with the emphasis changing from maximizing productivity to ensuring sustainability.

Sustainability will require greater attention to the agro-ecosystem and particularly to the soil base that sustains agriculture (Box 10.15). In addition, the socio-economic and regulatory dimensions will need to be integrated into systemic change. The Canadian government repeatedly fails to follow advice from its own commissioned reports in terms of regulations. In contrast, the Danish government, for example, established policy, an action plan, and funding for organic farming that aimed for 10 per cent of its farmers to be organic by 2003. Organic farming will not be successful, however, unless customers are willing to pay for the produce, not only as a benefit to themselves but also to sustain and nourish healthier ecosystems overall. In addition to buying organic produce, we can do a number of things as individuals to ensure that we at least do not exacerbate the challenges now facing agricultural systems (Box 10.16).

SUMMARY

1. Agriculture originated at least 9,000 years ago, when societies domesticated both plant and animal species. Societies around the globe flourished by improving their capacity to expand agricultural production.

2. Over much of history, agricultural output has been increased through bringing more land into production. Today, intensification of production—obtaining more output from a given area of agricultural land—has become a key development strategy in most parts of the world to meet the increased demand for foodstuffs. The future capacity to deliver agricultural outputs depends on the continuing ecological viability of agro-ecosystems.

3. Agriculture is a food chain with humans as the ultimate consumers. The second law of thermodynamics dictates that the shorter the food chain, the more efficient it will be.

4. The Green Revolution and the Livestock Revolution have led to profound changes to the global land base. Complex natural systems have been replaced by relatively simple control systems in which humans are in command of the species and numbers that exist in a given area.

5. The Green Revolution relies on auxiliary energy flows, such as fertilizers, biocides, fossil fuels, and irrigation systems, to increase yields. Industrial systems of livestock production also depend on outside supplies of feed, energy, and other inputs to satisfy growing worldwide demand for meat.

6. Agriculture is an important industry in Canada, accounting for 8.5 per cent of gross domestic production, 10 per cent of employment, and about 6 per cent of total merchandise export earnings. Many of the innovations and technologies currently employed by Canadian farmers to produce agricultural products have extraordinary implications for ecosystems.

7. Large increases in fertilizer inputs have occurred in Canada over the last two decades. Western Canada, for example, saw a fivefold increase in fertilizer applications between 1970 and 2000, and the three Prairie provinces now account for 69 per cent of Canada's total fertilizer sales, up from 29 per cent in 1970.

8. Between 1970 and 2000, the value of biocides applied to agricultural crops in Canada increased threefold. Over 30 million hectares were treated with at least one biocide in 2000, compared with less than 10 million hectares in 1970.

9. Land degradation includes a number of processes to reduce the capability of agricultural lands to produce food. One study suggests that such processes cost Canadian farmers over $1 billion per year.

10. Soil erosion is estimated to cause up to $707 million damage per year in terms of reduced yields and higher costs. Soil formation in Canada is slow. A rate of 0.5 to 1.0 t/ha may be considered average. Losses of 5–10 t/ha are common in Canada and figures of 30t/ha have been recorded in the Fraser River Valley.

11. Increasing acidity as a result of application of nitrogen fertilizers and acid deposition is also a problem that reduces crop yields. Salinization occurs where there are high sodium levels in the soils and shallow water tables, such as in the Prairies. One estimate suggests that salinization causes economic losses four to five times as great as losses due to erosion, acidification, and loss of nitrogen.

12. Cultivation involves a continual process of removing plant matter from a field. In so doing, both the organic content and nutrient content of the soil are reduced. On the Prairies, current organic matter levels are estimated to be 50–60 per cent of the original levels.

13. Biocides are applied to crops to kill unwanted plants and insects that may hinder the growth of the crop. They have boosted yields throughout the world and helped feed many hungry mouths. There is also clear scientific evidence that they have serious negative impacts on ecosystem health.

14. Biocides have been shown to promote the development of resistance among target organisms. Over the last 40 years, more than 1,000 insects have developed such resistant populations. They are non-selective, and tend to kill non-target as well as target organisms. They may also be highly mobile and move great distances from their place of application. In addition, they may persist for a long time in the environment and accumulate along food chains. Such biomagnification has resulted in drastic reductions in the populations of some species at higher trophic levels, such as ospreys and bald eagles.

15. Chemicals, and their constituents as they break down, may interact synergistically.

16. The government must register chemicals before they can be used. Registration involves the chemical company providing evidence that the chemical is not carcinogenic or oncogenic. Many chemicals have been registered but subsequently found to fail these tests.

17. Attention is now being devoted to sustainable food production systems that maintain or enhance environmental quality, produce adequate economic and social rewards to all individuals/firms in the production system, and produce a sufficient and accessible food supply. Integrated pest management is now becoming more popular, and several provinces have such programs.

KEY TERMS

bioaccumulation	green manure	intensive livestock	soil compaction
bioconcentration	Green Revolution	operations (ILOs)	soil erosion
biomagnification	hybridization	Livestock Revolution	strip cropping
contour cultivation	integrated pest manage-	monoculture cropping	subsistence farming
crop rotation	ment (IPM)	no-till/conservation	summer fallow
genetically modified	integrated plant nutrient	agriculture (NT/CA)	
organisms (GMOs)	systems (IPNSs)	salinization	
grasshopper effect		*Silent Spring*	

REVIEW QUESTIONS

1. How do the laws of thermodynamics apply to agriculture?
2. Indicate how the law of tolerance and law of the minimum, discussed in Chapter 4, relate to agriculture.
3. Considerable interest is being directed towards ecosystem management. What is the relevance, if any, of this concept for agriculture?
4. In this chapter, mention was made of the resilience of farm systems. In Chapter 6, in relation to adaptive management, attention was also given to resilience. What value do the concepts of resilience and adaptive management have regarding agriculture?
5. How would you go about identifying and assessing the impacts (environmental and social) of agricultural policies and practices in Canada? What ideas from Chapter 6 might be helpful in this?
6. In Chapter 6 attention focused on the ideas of partnerships and stakeholders. What relevance do such concepts have regarding agriculture if partnerships are to be formed among the federal and provincial governments, agribusiness, and family farms?
7. It has been suggested here that conflict or tension can exist between the needs of urban areas and adjacent rural areas. To what extent do you think alternative dispute resolution ideas might be applied to deal with such tensions?
8. If you were a commercial farmer in Canada, to what extent would it be important for you to consider the implications of climate change for your farming operations?

RELATED WEBSITES

AGRICULTURE AND AGRI-FOOD CANADA:
http://www.agr.gc.ca/site_e.phtml#env

CANADIANS AGAINST PESTICIDES:
http://www.caps.20m.com

CONSULTATIVE GROUP ON INTERNATIONAL AGRICULTURAL RESEARCH (CGIAR):
http://www.cgiar.org/

ENVIRONMENT CANADA:
http://www.ec.gc.ca/envhome.html

FOOD AND AGRICULTURE ORGANIZATION OF THE UNITED NATIONS (FAO):
http://www.fao.org/

GOVERNMENT OF CANADA, AGRICULTURE AND THE ENVIRONMENT:
http://www.agr.gc.ca/policy/environment/pubs_sds_e.phtml

PESTICIDES AND WILD BIRDS:
http://www.hww.ca/hww2.asp?cid=4&id=230

WORLD HUNGER: WE FEED PEOPLE:
http://sustag.wri.org

WORLD RESOURCES INSTITUTE, EARTH TRENDS:
http://earthtrends.wri.org/

WORLD WILDLIFE FUND, CHEMICALS AND WILDLIFE:
http://panda.org/downloads/toxics/causesforconcern.pdf

REFERENCES AND SUGGESTED READING

Agriculture and Agri-Food Canada. 1997. *Biodiversity in Agriculture: Agriculture and Agri-Food Canada's Action Plan*. Ottawa: Environment Bureau, Agriculture and Agri-Food Canada.

———. 2000. *The Health of Our Water: Toward Sustainable Agriculture in Canada*. Publication Number 2020/E. At: <http://www.res2.agr.ca/research>.

———. 2001a. *Agriculture in Harmony with Nature: Agriculture and Agri-Food Canada's Sustainable Development Strategy, 2001–2004*. Ottawa: Agriculture and Agri-Food Canada.

———. 2001b. *Canadian Fertilizer Consumption, Shipments and Trade, 1999/2000*. At: <http://www.agr.gc.ca/policy/cdnfert/text00-00.pdf>.

———. 2002. *Canada's Agriculture, Food and Beverage Industry: Organic Industry*. At: <http://ats-sea.agr.ca/supply/factsheet-e.htm>.

———. 2003. 'Did you know?' At: <http://www.agr.gc.ca/fact>.

Brklacich, M., C.R. Bryant, and B. Smit. 1990. 'Review and appraisal of the concept of sustainable food production systems', *Environmental Management* 15: 1–14.

Canadian Biodiversity Strategy. 1995. *Canada's Response to the Convention on Biological Diversity*. Ottawa: Ministry of Supply and Services.

Canadian Farm Animal Genetic Resources Foundation. n.d. Publicity brochure. At: <http://www.cfagrf.com/English_Publicity_Brochure.htm>.

Canadian General Standards Board. 2003. *Draft National Standard: Organic Agriculture*. Ottawa: Canadian General Standards Board, Committee on Organic Agriculture.

Carson, R. 1962. *Silent Spring*. Boston: Houghton Mifflin.

Chambers, P., C. DeKimpe, N. Foster, M. Goss, J. Miller, and E. Prepas. 2001. 'Agricultural and forestry land use impacts', in *Threats to Sources of Drinking Water and Aquatic Ecosystem Health in Canada*. Ottawa: Environment Canada, 57–62.

———, M. Guy, E.S. Roberts, M.N. Charlton, R. Kent, C. Gagnon, G. Grove, and N. Foster. 2001. *Nutrients and Their Impact on the Canadian Environment*. Ottawa: Agriculture and Agri-Food Canada, Environment Canada, Fisheries and Oceans Canada, Health Canada, and Natural Resources Canada.

Cassman, K.G., A. Dobermann, and D.T. Walters. 2002. 'Agroecosystems, nitrogen-use efficiency, and nitrogen management', *Ambio* 31, 2: 132–40.

Commissioner of the Environment and Sustainable Development. 2003. *Managing the Safety and Accessibility of Pesticides*. At: <www.oag-bvg.gc.ca/domino/

reports.nsf/html/c20031001ce.html/$file/c20031001ce.pdf>.

de Haan, C.H. Steinfeld, and H. Blackburn. 1997. 'Livestock and the environment: finding a balance', Commission of the European Communities, the World Bank and Food and Agriculture Organization (FAO). At: <www.fao.org/docrep>.

Döös, B.R. 2003. 'The problem of predicting global food production', *Ambio* 31, 5: 417–24.

Environment Canada. 2001. *Nutrients in the Canadian Environment: Reporting on the State of Canada's Environment*. Ottawa: Environment Canada.

———. 2003. *Environmental Signals: Canada's National Environmental Indicator Series, 2003*. Ottawa: Environment Canada.

——— and United States Environmental Protection Agency. 2002. *State of the Great Lakes 2001*. At: <www.binational.net/sogl2001/download.html>.

Food and Agriculture Organization (FAO). 2000. 'Domestic and animal diversity at risk'. At: <http://www.fao.org/ag/magazine/0011sp2.htm>.

———. 2001. 'Conservation Agriculture'. FAO Magazine Spotlight. At: <www.fao.org/ag/magazine>.

———. 2002a. *The State of Food and Agriculture 2002*. At: <www.fao.org>.

———. 2002b. *Food Insecurity: When People Must Live with Hunger and Fear Starvation*. Rome: FAO.

———. 2003a. *World Agriculture: Towards 2015/2030. An FAO Perspective*. Rome: FAO.

———. 2003b. *Water Management: Towards 2030*. At: <www.fao.org/ag/magazine>.

——— and World Health Organization (WHO). 1999. *Guidelines for the Production, Processing, Labelling and Marketing of Organically Produced Foods*. At: <www.fao.org/organicag/doc/glorganicfinal.doc>.

FAOSTAT. 2001. *Agricultural Data: World Fertilizer Consumption*. At: <http://apps.fao.org/faostat/collections?version=ext&hasbulk=0&subset=agriculture>.

Indian and Northern Affairs Canada. 1997a. 'Pathways of Transportation of Persistent Organics and Metals to Arctic Freshwater and Marine Ecosystems' in *Canadian Arctic Contaminants Assessment Report*.

———. 1997b. *Highlights of the Canadian Arctic Contaminants Assessment Report: A community Reference Manual*.

McNairn, H.E., and B. Mitchell. 1991. 'Farmers' perceptions of soil erosion and economic incentives for conservation tillage', *Canadian Water Resources Journal* 16: 307–16.

Maguire, R.J., P.K. Sibley, K.R. Solomon, and P. Delorme. 2001. 'Pesticides', in *Threats to Sources of*

Drinking Water and Aquatic Ecosystem Health in Canada. Ottawa: Environment Canada, 9–11.

Miller, G. 2000. 'Intensive farming', in *The Protection of Ontario's Groundwater and Intensive Farming: Special Report to the Legislative Assembly of Ontario.* At: <http://www.eco.on.ca>.

Muir, D., M. Alaee, D. Dube, L. Lockhart, and T. Bidleman. 2001. 'Persistent organic pollutants and mercury', in *Threats to Sources of Drinking Water and Aquatic Ecosystem Health in Canada.* Ottawa: Environment Canada, 13–25.

Roy, R.N., R.V. Misra, and A. Montanez. 2002. 'Decreasing reliance on mineral nitrogen—yet more food', *Ambio* 31, 2: 177–83.

Scialabba, N.E. 2003. *Organic Agriculture: The Challenge of Sustaining Food Production while Enhancing Biodiversity.* Rome: Food and Agriculture Organization.

——— and C. Hattam, eds. 2002. *Organic Agriculture, Environment and Food Security.* Rome: Food and Agriculture Organization.

Standing Committee on Environment and Sustainable Development. 2000. *Pesticides: Making the Right Choice for the Protection of Health and the Environment.* Ottawa: House of Commons, May.

Statistics Canada. 2001a. *2001 Census of Agriculture.* At: <www.statcan.ca/english/Pgdb/census.htm#far>.

———. 2001b. 'Urban consumption of agricultural land', *Rural and Small Town Canada Analysis Bulletin* 3, 2. Catalogue no. 21–006–XIE.

———. 2003a. *Farming Facts 2002.* Ottawa: Ministry of Industry.

———. 2003b. 'Manure storage in Canada', *Farm Environmental Management in Canada* 1, 1. Catalogue no. 21–021–MIE.

———, Agriculture Division. 2003. *Food Statistics, 2003* 2, 2. At: <www.statcan.ca/english/freepub/21-020-XIE/21-020-XIE02002.pdf>.

United Nations Development Program (UNDP), United Nations Environment Program (UNEP), World Bank, World Resources Institute (WRI). 2000. *World Resources 2000–2001: People and Ecosystems: The Fraying Web of Life.* At: <http://sustag.wri.org/pubs_description.cfm?PubID=3027>.

United Nations Environment Program (UNEP). 2002. *Global Environmental Outlook 3 (GEO-3).* At: <http://www.unep.org/geo/geo3/>.

Warnock, R.G., and M.A. Skeel. 2004. 'Effectiveness of voluntary habitat stewardship in conserving grassland: case of Operation Burrowing Owl in Saskatchewan', *Environmental Management* (25 Mar.). At: <http://wwwspringerlink.com>.

Watt-Cloutier, S. 2000. *Wake-up Call.* At: <www.ourplanet.com/imgversn/124/watt.html>.

Wellicome, T.I., and E.A. Haug. 1995. *Updated Report on the Status of the Burrowing Owl (Speotyto cunicularia) in Canada.* Ottawa: Report to COSEWIC, Canadian Wildlife Service.

Wood, S., K. Sebastian, and S.J. Scherr. 2000. *Pilot Analysis of Global Ecosystems: Agroecosystems.* Worldwatch Institute. New York: W.W. Norton.

World Resources Institute (WRI). 2003. 'Global study reveals new warning signals: Degraded agricultural lands threaten world's food production capacity'. At: <http://wri.igc.org/press/goodsoil.html>.

Worldwatch Institute. 1992. *Vital Signs: The Trends That Are Shaping Our Future.* New York: W.W. Norton.

———. 2003a. *Vital Signs.* New York: W.W. Norton.

———. 2003b. *Earth Trends: Agricultural Inputs 2003.* At: <http://earthtrends.wri.org/datatables/index.cfm?theme=8&CFID=504446&CFTOKEN=74873335>.

Yussefi, M., and H. Willer, eds. 2003. *The World of Organic Agriculture: Statistics and Future Prospects.* At: <http://www.soel.de/oekolandbau/weltweit.html>.

Chapter 11

Endangered Species and Protected Areas

Learning Objectives

- To understand why endangered species are important and the factors leading to endangerment.
- To become aware of the extrinsic and intrinsic values of nature.
- To learn why some species are more vulnerable to extinction than others.
- To be able to discuss the main responses to endangerment at the international and national levels.
- To appreciate the many roles played by protected areas.
- To gain an international and a Canadian perspective on protected areas.
- To know some of the main management challenges faced by protected areas in Canada.

INTRODUCTION

Most people are aware that many more species are becoming endangered than is natural, that we have a 'biodiversity crisis'. A common perception, however, is that this is a problem more for the tropics than for countries such as Canada. Although it is true that threat levels and the numbers of endangered species are higher in the tropics, Canada also has plenty of challenges in this regard.

Although Canada has a long-established national parks system designed to protect species and their habitats, parks do not necessarily afford adequate protection to all species that need it. The decline in turtle populations in Point Pelee National Park, Ontario, provides a good example of some of the problems in our national parks and the challenges faced by park managers.

Turtles have evolved for hundreds of millions of years. Historically, their adaptations—terrestrial nesting, low adult mortality, late maturation, and longevity—have served them well. But these attributes are no longer adequate, and throughout the world many turtle species are experiencing

dramatic declines. Historical records show that Point Pelee, the southern tip of Canada's mainland, at one time had seven different species of indigenous turtles. Research by Browne and Hecnar (2003) from Lakehead University revealed that

Point Pelee, Ontario (*Bill Ivy/Ivy Images*).

several species, including the stinkpot, map, and Blanding's turtles, now exist only in small populations, while other species may be headed towards extirpation—no spotted turtles were found in the park, and only one individual of the threatened spiny softshell turtle was recorded. Reasonably large populations exist only for two species—painted and snapping turtles. The authors examined the age structures of the populations and found a preponderance of older animals for most species, indicating aging populations especially for Blanding's and snapping turtles.

Although the populations are 'protected' within the park, at least three problems continue to threaten the long-term viability of turtle populations. Roads are implicated in two of the three and illustrate the need to minimize development within parks (discussed later in the chapter). Roads are a source of direct mortality. Some species are attracted by the soft shoulders of roads for nesting, and while migrating across roads in search of suitable sites many turtles are killed. In addition, roadside nesting sites were found to be very vulnerable to nest predation (100 per cent loss), compared with the 62 to 64 per cent loss in more remote areas of the park. Roadsides are a favourite scavenging area for predators such as raccoons. These and other predators, such as striped skunks and opossums, are reportedly at higher levels in the park than previously.

Contaminants are also implicated in turtle population declines. There are still elevated levels of DDT and DDE (see Chapter 10) in areas of the park from past agricultural practices, illustrating the vulnerability of parks to threats from surrounding land uses.

This example illustrates the plight of many species to which we give relatively little attention. Plant and animal populations are often assumed to be 'healthy', especially if they are within the boundaries of a national park. However, *no* national park—anywhere on the planet—is big or remote enough to exclude the impacts of modern society. Endangered species and the factors behind endangerment are the focus of this chapter. One of the main responses to endangerment is to try to set lands aside in park systems. The designation and management of park systems comprise the second main topic discussed here.

Extinction is a natural process that has been occurring since life first evolved on Earth. However, it is the speed of current extinction rates across many different forms of life that concerns scientists (*Philip Dearden*).

But why should we be concerned about endangerment? Extinction is a natural process that has been taking place since life first evolved on this planet over four billion years ago (Chapter 3). Consequently, concern for the **extinction** of species does not focus on the process itself, but on the increasing rates of extinction over time, i.e., what humans are doing to speed up the process. Before looking at some of the pressures responsible for this increase, we need to have a clear understanding of why high extinction rates are undesirable.

VALUING BIODIVERSITY

Extrinsic and Intrinsic Values

Extrinsic values are those values that humans derive from other species. These values can be *consumptive* (i.e., the organism is harvested) or *nonconsumptive* (i.e., the organism is not harvested or the resource is not destroyed). There is no universally accepted framework for assigning value to biological diversity, but various approaches have been proposed. For example, the value of biodiversity can be calculated by examining import and export statistics for products bought and sold in markets. However, it is often difficult to assign an economic value to biodiversity. How do you put a price tag on environmental services provided by biological communities, such as photosynthesis, protection of watersheds, or climate regulation?

These services are not directly consumed by humans, but they are vital for survival.

While important, the extrinsic reasons for species protection should not be allowed to dominate our thinking. Such thinking could lead to the protection of a selection of species believed to be of higher value, while species with less use value are afforded little or no protection. Thus, arguments for biological conservation focus on the **intrinsic value** of nature—nature has value in and of itself, apart from its value to humanity. Although there are many reasons why we should be concerned about the rapid destruction of biodiversity worldwide, the following discussion will focus on some of the key ecological, economic, and ethical reasons for conservation.

> *If the land mechanism as a whole is good, then every part is good, whether we understand it or not. If the biota, in the course of aeons, has built something we like but do not understand, then who but a fool would discard seemingly useless parts? To keep every cog and wheel is the first precaution of intelligent tinkering.*
>
> — *Aldo Leopold,* Round River *(1953)*

Ecological Values

The elimination of species affects ecosystem functioning, which can lead to unfortunate consequences, such as the impact on coastal marine ecosystems on the Pacific coast when the sea otter was extirpated (discussed in Chapter 3). Species become **extirpated** when they have been eliminated from one part of their range but still exist somewhere else (Table 11.1). A species is considered **ecologically extinct** in an area when it exists in such low numbers that it can no longer fulfill its ecological role in the ecosystem. For example, the eastern mountain lion may still exist in the Maritime provinces and has not yet been declared extinct, but if this species does persist, it exists in such low numbers that it no longer acts as a significant control for species in the preceding trophic level. The eastern mountain lion may therefore be considered ecologically extinct in this region.

All species in a community combine to maintain the vital ecosystem processes that make human life possible on this planet—oxygen to breathe, water to drink, and food to eat. Humans are part of this web of life, but if we continue to eliminate components of the web its strength will be compromised, which will, in turn, have a significant impact on the ability of humans to survive.

Table 11.1	Definitions Used by the Committee on the Status of Endangered Wildlife in Canada (COSEWIC)
Not at Risk (NAR)	A species that has been evaluated and found to be not at risk.
Data Deficient (DD)	A species for which there is insufficient scientific information to support status designation.
Delisted	A species not at risk (no designation required).
Special Concern (SC)	A species of special concern because of characteristics that make it particularly sensitive to human activities or natural events.
Vulnerable (V)	A species particularly at risk because of low or declining numbers, small range, or for some other reason, but not a threatened species.
Threatened (T)	A species likely to become endangered in Canada if factors threatening its vulnerability are not reversed.
Endangered (E)	A species threatened with imminent extinction or extirpation throughout all or a significant portion of its range in Canada.
Extirpated (XT)	A species no longer existing in the wild in Canada, but occurring elsewhere.
Extinct (X)	A species that no longer exists.

The great hornbill is found throughout the tropical forests of Southeast Asia where it has been extirpated from many areas by hunting. Even where it exists in small numbers, it is often considered ecologically extinct as there are no longer sufficient numbers to crack and distribute the seeds of many of the tree species. Ultimately this will also cause changes in the tree species composition of the forests (*Philip Dearden*).

The elimination of species from an ecosystem has been likened to the removal of rivets from an airplane—the system may continue to function after losing a few components, but sooner or later the system will crash. One of the roles of science is to understand how these systems work, but it is difficult to achieve this understanding if components are missing due to extinction.

In addition to the value of species in ecosystem functioning, species should be protected for their evolutionary value, their value to future generations. Species evolve over time, as discussed in Chapter 3, and as more species become extinct, genetic variation in the ecosphere on which to base future adaptability is reduced. Fewer species mean a more impoverished biosphere on which to base evolutionary adaptability for future generations. In short, the need for preservation of species exemplifies the precautionary principle on a grand scale. In a fanciful manner, the movie *Star Trek: The Voyage Home* showed Captain Kirk, Spock, and their colleagues returning back in time to the present to rescue a humpback whale and transport it to the future, where it had become extinct, in order to save the planet. This is science fiction, to be sure, but it makes the point well—we cannot presume to know what importance or value a particular species might have for future generations.

Economic Values

The preservation of ecosystem components and functions is also economically beneficial. Countless products used in agriculture and industry originate in the natural world. Naturally occurring plants in the tropics, for example, are the source of 90 per cent of the world's food supply. Corn, or maize, feeds millions of people and is now estimated to be worth at least $50 billion annually worldwide. Corn was first domesticated by indigenous people in Central America some 7,000 years ago.

Over 99.8 per cent of the world's plants have never been tested for human food potential. Some of these plants may become important food staples in the future, and so preservation of their habitat is important. Furthermore, wild animals still provide an important source of food for millions of people worldwide, particularly indigenous peoples, including Canada's First Nations communities.

Whether a modern or traditional pharmacy, many of our medicinal products are based on products found in nature (*Philip Dearden*).

Other products besides food are also of economic importance. Many plant and animal products are used extensively in a variety of industries. Rubber, for example, is an important commodity in the automotive sector. It is just one example of a chemical produced by tropical plants in order to prevent insect damage. Many other chemicals produced by plants are used in the pharmaceutical industry. Fifty-six per cent of the top 150 prescribed drugs in the United Sates contain ingredients from wild species, with an economic value of US$80 billion according to the UN. Yet less than 1 per cent of the world's tropical plants have been screened for potential pharmaceutical applications. Next time you take an aspirin tablet, thank the white willow, the species wherein the active ingredient was first discovered. Taxol, found in the bark of the western yew—a small understorey tree found in the forests of the Pacific Northwest—was recently discovered as a treatment for cancer. Prior to the discovery, the species was of little commercial value, and it was routinely cut down in clearcuts and left to rot.

It is difficult to assess the economic value of many of the products we derive from nature. For example, natural gene pools provide a source of material to aid in the development of new genetic strains of crops needed to feed the world's burgeoning human population. Wheat, the mainstay of the western Canadian agricultural economy, originated in Mediterranean countries, where most of its wild forebears have disappeared. But preservation of such wild strains is necessary to allow selective breeding based upon the widest range of genetic material to continue.

Ecosystems also provide humans with a wide array of economically important services. Natural pollinators, for example, provide an essential service for commercial crops. Environment Canada (2003a) has conservatively estimated the value of pollination services for crops in Canada at $1.2 billion annually. When New Brunswick switched from spraying its forests with DDT to fenitrothion in 1970 to control spruce budworm (see Chapter 9), there was a devastating impact on pollination of the blueberry crop due to the high toxicity of fenitrothion for bees. The commercial crop fell by 665 tonnes per year and the growers successfully sued the government.

Natural predator/prey relationships also aid in food production. It has been shown, for example,

It is impossible to fully assess the value of natural processes such as pollination (*Philip Dearden*).

that woodpeckers provide an economically important service in the control of pests such as coddling moths in the orchards of Nova Scotia. Such predator/prey relationships can significantly reduce the need to apply biocides to control pests. Similarly, the natural toxicity found in some species can occasionally be refined into a natural biocide for use in agriculture. For example, a powerful insect repellent—trans-pulegol—was recently discovered in an endangered member of the mint family.

Other Extrinsic Values

The ecological and economic reasons mentioned above do not encompass all the values associated with protecting other species. How many of us will be permanently enriched and emotionally uplifted by a wildlife encounter at some point in our lives? Wildlife contributes to the joy of life, but this joy often translates into a contribution to the economic values attached to wildlife. It is estimated that over 20 million Canadians spend over $11 billion participating in nature-related activities, creating approximately 245,000 jobs (National Roundtable on the Environment and the Economy, 2003). In fact, viewing wildlife is a major reason for travel to some areas, both in Canada and elsewhere. The economic value attached to nature-based tourism can provide a significant impetus to encourage conservation when managed so as to enhance biodiversity and educational values. This form of tourism can also provide a sustainable livelihood for local residents.

How much is it worth to see Pacific white-sided dolphins jump around your boat? (*Philip Dearden*).

Ethical Values

Ethical arguments can be made for preserving all species, regardless of their use value to humans. Arguments based on the intrinsic value (value unrelated to human needs or desires) of nature suggest that humans have no right to destroy any species. In fact, humans have a moral responsibility to actively protect species from going extinct as a result of our activities. This philosophy reflects an *ecocentric view*—humans are part of the larger biotic community in which all species' rights to exist are respected. In stark contrast, an *anthropocentric view* of life favours protection of species with an economic and/or ecological use to humans (Chapter 5).

Beazley (2001) has provided an overview of the reasons why we should be concerned about extinction. In the past, extinction has been viewed simply as a biological problem. The points raised above emphasize the necessity of making links among the biological process of extinction and the ethical and economic reasons why extinction is undesirable. However, decisions to protect species and communities often come down to arguments over money—how much will it cost, and how much is it worth? All too often, governments demonstrate a willingness to protect biodiversity only when its loss is perceived to cost money. Unfortunately, standard economic systems tend to undervalue natural resources, and as a consequence, the cost of environmental damage has been ignored and the depletion of natural resources has been disregarded. Ecosystems are being destroyed and species are being driven to extinction at a rate that is greater now than at any time in the past. An economic system that undervalues natural resources is one of the main underlying causes of extinction. But what are the direct causes of biodiversity loss?

MAIN PRESSURES CAUSING EXTINCTION

Chapter 3 described some human activities that are contributing to the increasing rates of extinction witnessed over the last 200 years. This section discusses some of the main pressures on biodiversity in greater detail. Rarely do these pressures act alone; they must therefore be seen as part of the overall stress that human demands are placing on the biosphere.

Humans currently appropriate about 40 per cent of the net primary productivity (NPP) of the planet (Chapter 2). With global populations predicted to increase by 70 million people each year for the next 20 years, this figure will only increase. As the amount of NPP increases to support one species—*Homo sapiens*—the amount available to support all other species decreases. The extinction vortex (Figure 11.1), therefore, is ultimately driven by human pressures.

Most attention regarding extinction has been devoted to the tropical countries, mainly because they are 'hotspots' for biodiversity and because they are experiencing many of the pressures responsible for increasing rates of extinction. Of the 10–15 million terrestrial species thought to be on Earth, up to 90 per cent are estimated to live in the tropics, particularly in the tropical rain forests. Tropical ecosystems are among the most diverse on Earth for a variety of reasons:

THE CATASTROPHOZOIC

Soon a millennium will end. With it will pass four billion years of evolutionary exuberance. Yes, some species will survive, particularly the smaller, tenacious ones living in places too dry and cold for us to farm or graze. Yet we must face the fact that the Cenozoic, the Age of Mammals, which has been in retreat since the catastrophic extinctions of the late Pleistocene, is over, and that the 'Anthropozoic' or 'Catastrophozoic' has begun. Our task now is to salvage some samples of the megafauna and protect enough habitat to give future human beings an opportunity to restore a semblance of evolutionary integrity in the 22nd century.

— *Soule (1996: 25)*

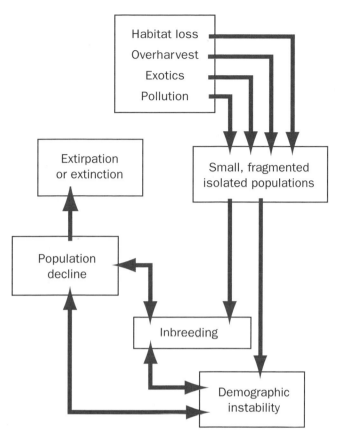

Figure 11.1 Extinction vortex.

- *Long evolutionary time period.* This has allowed for a greater degree of specialization and local adaptation to occur. For example, the greater diversity of plant species in the tropics has promoted the evolution of large numbers of insect species that feed on particular plant species.
- *Warm temperatures and high humidity.* Tropical climates provide favourable conditions for the growth and survival of many species.
- *Tropical regions receive more solar energy than temperate regions.* Many tropical regions have a higher rate of productivity in terms in the number of kilograms of living material produced each year per hectare of habitat. High productivity results in a greater resource base that can support a wider range of species.

Unfortunately, tropical ecosystems are being degraded and/or destroyed at alarming rates (Box 11.1). Estimates suggest that perhaps 50 per cent of tropical rain forests have already disappeared, and at current rates of destruction only a few forest fragments will remain in 30 years. About 7.4 million hectares of tropical forest are destroyed every year. Causes of tropical deforestation include:

Although habitat destruction is a main cause of biodiversity declines in tropical countries, so is hunting. Species are shot to put on display, as in this hotel foyer in Vietnam, and are also killed for various superstitious and so-called 'medicinal' purposes, such as these bear parts offered for sale in a hill tribe village visited by Taiwanese tourists in northern Thailand (*Philip Dearden*).

BOX 11.1 THE GLOBAL TOLL

Through its Species Survival Commission, the International Union for the Conservation of Nature and Natural Resources (IUCN), commonly called the World Conservation Union, evaluates and categorizes species according to their degree of endangerment:

- Extinct
- Extinct in the Wild
- Critically Endangered
- Endangered
- Vulnerable
- Near Threatened
- Least Concern
- Data Deficient
- Not Evaluated

This so-called **Red List** is produced by thousands of scientific experts and is the best source of knowledge on the status of global biodiversity.

Some highlights from the 2003 list:

- Since the release of the 2002 Red List, over 2,000 new entries have been added and 380 taxa reassessed. The IUCN Red List now includes 12,259 species threatened with extinction, falling into the Critically Endangered (CR), Endangered (EN), or Vulnerable (VU) categories.
- Almost one-third of all primate species are in danger of becoming extinct over the next decade. In 2000, 120 of the world's 683 species of primates were at risk of extinction; in 2002 and 2003 respectively, an additional 75 and 3 primate species were found to be at risk.
- 7,158 plants have now been assessed as threatened (1,046 CR, 1,291 EN, 3,377 VU), an increase of 1,444 since 2002. However, with only approximately 4 per cent of the world's described plants evaluated, the true percentage of threatened plant species is much higher. Most of the plant species listed are trees, since these have been relatively thoroughly assessed.
- There are now 762 plant and animal species recorded as Extinct, with a further 58 known only in cultivation or captivity (820 in total). In 2000, the number of species assessed as Extinct and Extinct in the Wild was 811.
- In 2003, new kingdoms—seaweeds and lichens—entered the Red List. The boreal felt lichen (*Erioderma pedicellatum*) is Critically Endangered; previously found in Canada, Norway, and Sweden, it has suffered a major population decline over the last 100 years and is thought to be extinct in the latter two countries.

- Indonesia, India, Brazil, China, and Peru are among the countries with the most threatened mammals and birds, while plant species are declining rapidly in Ecuador, Malaysia, Indonesia, Brazil, and Sri Lanka.
- The main areas where mammals, birds, and plants (trees) seem to require the most conservation effort are in the Neotropics (Brazil, Colombia, Ecuador, and Mexico), East Africa (Tanzania), and Southeast Asia (China, India, Indonesia, and Malaysia). For mammals alone, the situation appears most critical in Africa, while for the birds, Argentina and the Southeast Asian block of Myanmar, Vietnam, Cambodia, and Lao PDR emerge as important areas.
- Birds and mammals have shown an increasing move towards the higher threat categories (i.e., more bird and mammal species are entering the Critically Endangered and Endangered categories).
- It is more difficult to assess the trends for reptiles, amphibians, fish, and invertebrates because there are still many information gaps. Within these taxa, however, declines appear to be occurring. For example, a number of amphibian species have shown rapid and unexplained disappearances in Australia, Costa Rica, Panama, and Puerto Rico. There are similar indications that freshwater fish species may be suffering a serious deterioration, particularly river-dwelling species.
- Habitat loss and degradation affect 89 per cent of all threatened birds, 83 per cent of mammals, and 91 per cent of threatened plants assessed. Habitats with the highest number of threatened mammals and birds are lowland and mountain tropical rain forest. Freshwater habitats are extremely vulnerable with many threatened fish, reptile, amphibian, and invertebrate species.
- Invasive alien species—a major threat to global biodiversity—are responsible for the loss of island populations of native plants and animals. Hundreds of plant assessments in 2003 from Hawaii, the Falkland Islands (Malvinas), British Virgin Islands, the Seychelles, Tristan da Cunha, St Helena, and Ascension reveal a bleak outlook. Hawaii's plants are particularly under serious threat from invasive species. Of the 125 endemic Hawaiian plant species added to the Red List this year, 85 are threatened and the number is expected to increase.

SOURCES: World Conservation Union (2002, 2003).

Table 11.2	Number of Threatened Species in Each Major Group of Organisms in Canada, North America, and Worldwide		
	Canada	**North America**	**Worldwide**
Mammals	14	51	1,137
Birds	8	64	1,192
Reptiles	2	29	293
Amphibians	1	26	157
Fishes	16	146	742
Molluscs	0	256	939
Other Invertebrates	10	311	993
Plants	1	170	5,714
Total	52	1,053	11,167

SOURCE: World Conservation Union (2002).

- rapidly growing population levels—more people equals less biological diversity, since people use natural resources;
- overconsumption of resources—the rise of industrial capitalism and materialistic modern societies has greatly accelerated demands for natural resources, particularly in developing countries;
- inequality in the distribution of wealth—poor rural people with no land or resources of their own are forced to destroy biological communities and hunt endangered species just to stay alive.

Extinctions also occur in developed and temperate countries. Since the European colonization of North America, more than 500 species and subspecies of native plants and animals have become extinct. Some comparative figures among Canada, North America and the rest of the world are shown in Table 11.2. The Committee on the Status of Endangered Wildlife in Canada (COSEWIC) has listed 12 extinctions and 21 extirpations in Canada since the arrival of Europeans (Table 11.3). Most of the following examples have been chosen to illustrate extinction pressures in Canada.

Overharvesting

There are many examples in Canada of species under pressure due to overharvesting. A couple of historical examples that took place at least partly in Canadian territory are well known.

The great auk. The great auk, a large flightless bird, inhabited the rocky islets of the North Atlantic. For many years fishers in these waters used the great auk as a source of meat, eggs, and oil. It was reasonably easy to catch and club these birds to death, and once the feathers became an important commodity for stuffing mattresses in the mid-1700s, extinction soon followed. On Funk Island off the east coast of Newfoundland, the species was extirpated by the early 1800s. The last two great auks were clubbed to death off the shores of Iceland in 1844.

The passenger pigeon. There are claims that the passenger pigeon was the most abundant land bird on Earth, totalling up to five billion birds. These great flocks used to migrate annually from their breeding grounds in southeastern Canada and the American Northeast to their wintering grounds in the southeastern US. So great were their numbers that tree limbs would break under the pigeons' weight, and trees would die due to the amount of guano (bird excrement) deposited. Many eyewitness accounts tell of flocks so huge that they blotted out the sun for hours and sometimes days. Flocks of passenger pigeons provided an easy target for hunters, and they were slaughtered in great numbers for food to feed growing urban

Table 11.3	Summary of COSEWIC's Assessment Results for the Risk Categories				
	Extinct	**Extirpated**	**Endangered**	**Threatened**	**Special Concern**	**Total**
Mammals	2	4	22	13	23	64
Birds	3	2	21	8	22	56
Reptiles		4	5	11	11	31
Amphibians		1	5	5	8	19
Fishes	5	2	19	22	32	80
Lepidopterans		3	5	3	2	13
Molluscs	1	2	11	2	2	18
Plants		2	58	37	39	136
Mosses	1	1	5	1		8
Lichens			2		4	6
Total	12	21	153	102	143	431

Note: Results include the May 2003 COSEWIC meeting.
SOURCE: COSEWIC (2003).

Great auks, painted by John James Audubon (*Metropolitan Toronto Reference Library*).

A pair of passenger pigeons, painted by John James Audubon (*Metropolitan Toronto Reference Library*).

populations. It is estimated that over one billion birds were killed in Michigan alone in 1869. They were shot, netted, and clubbed into extinction. Hunting took place concurrently with a reduction in their breeding grounds as habitat was converted into agricultural lands. The last passenger pigeon sighted in Canada was at Penetanguishene, Ontario, in 1902; the last pigeon died in a zoo in Cincinnati in 1914. The world will never again experience the sound and sight of millions of passenger pigeons darkening the heavens.

This astonishing tale of a species going from such abundance to extinction is not that unusual. Three fish species of the Great Lakes, the blue walleye, deepwater cisco, and longjaw cisco, were at one time all very abundant and millions of kilograms of the fish were harvested commercially. By 1950, they were fished into extinction. Similarly, the northern cod, once one of the most abundant fish on the planet, was fished into commercial extinction by the early 1990s. Two populations of Atlantic cod were designated as threatened and endangered by COSEWIC in 2003 (see Chapter 8).

Perhaps extinction of the passenger pigeon and the great auk are sufficiently in the past that we can excuse their demise due to a lack of knowledge. Such a case cannot, however, be made for these recent extinctions. They stand as the ultimate symbols of the failure of resource managers and decision-makers to understand the natural dynamics of the species they are supposed to be managing.

THE PASSENGER PIGEON

The noise they made, even though still distant, reminded me of a hard gale at sea, passing through the rigging of a close-reefed vessel. As the birds arrived and passed over me, I felt a current of air that surprised me. Thousands of the Pigeons were soon knocked down by the pole-men, while more continued to pour in.... The Pigeons, arriving by the thousands, alighted everywhere, one above another, until solid masses were formed on the branches all around. Here and there the perches gave way with a crash under the weight, and fell to the ground, destroying hundreds of birds, beneath, and forcing down the loaded. The scene was one of uproar and confusion. I found it quite useless to speak, or even to shout, to those persons nearest me. Even the gun reports were seldom heard, and I was made aware of the firing only by seeing the shooters reloading.
— *John James Audubon*

One aspect of overharvesting generally given little consideration is the demand for captive species. In the past, the zeal of zoo collectors to exhibit various species, particularly rare species such as pandas and orangutangs that visitors would pay to see, was of serious concern. International regulations, such as the Convention on International Trade in Endangered Species of Wild Fauna and Flora (CITES), of which Canada is a signatory, now make it difficult for this kind of trade to occur. Despite international regulations, trade in rare and endangered species still occurs through demand for private collections. Exotic species such as tigers, monkeys, parrots, and tropical fish belong in their native habitats, not in people's homes. Some of the most sought-after Canadian species are falcons, particularly the gyr and peregrine falcons, and they can fetch thousands of dollars each on the international market. Although there is an allowable harvest of wild falcons in some provinces and territories, poaching is a problem due to the high price tags attached to these birds.

Hunting/fishing and the capture of live specimens for captivity can have significant impacts on populations. There are, however, more subtle cases where 'non-consumptive' activities can have detrimental impacts on species by causing displacement from valuable habitat or even death. For example, research is underway on both the Atlantic and Pacific coasts of Canada to assess the potential impacts of whale-watching vessels on the well-being of whales. Such research requires detailed knowledge of a species' natural distribution and behaviour before it can be ascertained whether changes have occurred as a result of disturbance. Impacts associated with the non-consumptive use of natural resources are usually much more difficult to document than the more direct effects of consumptive use.

Predator Control

Several species have been targeted for elimination by humans due to direct competition for consumption of the same resource. One North American example is the Carolina parakeet, the only member of the parrot family native to North America. It was exterminated in the early part of the century because of its fondness for fruit crops. Another example is the prairie dog, which was extensively poisoned due to the mortality suffered

Whales were hunted to the point of extinction around the world because of their commercial value. Baleen, such as what is shown here, was used for such 'indispensable' things as umbrellas, buggy whips, and ladies' corsets (*Philip Dearden*).

by horses and cattle after stepping into prairie dog burrows and breaking their legs. This extermination program has been highly successful—prairie

Sometimes the pressures on species that may cause them to decline are subtle and difficult to understand. For example, researchers have been studying the impacts of whale-watching on whales to see if it causes whales to alter their behaviour in ways that may be detrimental to their overall well-being. Results suggest that, by and large, this is not the case (*Dave Duffus*).

dog populations have declined by 99 per cent. Once such a decline occurs, repercussions occur elsewhere in the food chain. In this case, the drastic decline in prairie dogs led to the collapse of their main predator, the black-footed ferret, for which prairie dogs made up more than 90 per cent of their diet. Fortunately, the ferrets have been successfully bred in captivity, an example of ex situ conservation. They have been reintroduced to the wild in places where prairie dogs are protected. The swift fox provides another example of predator control and ex situ conservation (Box 11.2).

Predator control is not a thing of the past. For example, deer are sometimes culled to prevent or stop overbrowsing of vegetation. Across parts of their North American range, white-tailed deer populations are literally eating themselves out of house and home—deer populations are in excess of what can be supported by the natural resource base. This is due in large part to reductions in range and population declines of their predators, particularly wolves. In stark contrast, other parts of the country, such as Vancouver Island, do not have enough deer, and provincial governments have considered shooting wolves in order to increase the number of deer that humans can kill for sport (Box 11.3). A similar debate is also underway in Newfoundland where coyotes, which first invaded the island in the 1980s, are now taking significant numbers of caribou.

The seal hunt off Canada's eastern seaboard is also regaining momentum as a predator control program to protect cod, in hopes of reviving the fishing industry (Chapter 8). Recently, there have been calls in Prince Edward Island and Ontario to kill more double-crested cormorants, accused of taking too many fish. As the human population grows and as our demands increase, conflicts over who will consume another organism will escalate. So far, other species appear to be losing the battle.

Habitat Change

Habitat change is *the* most important factor causing biodiversity loss at national and international scales. For **endangered** species in Canada, habitat degradation is responsible for: 100 per cent of the listed reptiles, amphibians, invertebrates, and lichens; 99 per cent of the listed plants; 90 per cent of the listed birds; 85 per cent of the listed fish; and 67 per cent of the listed mammals.

BOX 11.2 RECOVERY OF THE SWIFT FOX

Not all species subject to heavy pressures are pushed to extinction. Some, such as the beaver and the swift fox, may recover in numbers and start to repopulate their old range. The beaver was able to repopulate with relatively little help. The swift fox, on the other hand, is the target of a 20-year, $20-million reintroduction program, which emphasizes the difficulties and costs associated with trying to reverse extinction trends.

The swift fox is so called because of its ability to run down rabbits and other prey in its home terrain, the dry, short-grass prairie. Speeds of over 60 km/hour have been recorded. The swift fox is a small fox (about half the size of a red fox) that at one time roamed all the way from Central America through to the southern Prairies of Canada. Unlike most other members of the dog family, the swift fox uses dens throughout the year, which are preferably located on well-drained slopes close to a permanent water body. This may be for protection, due to their small size. Natural enemies include coyotes and birds of prey such as eagles and red-tailed and rough-legged hawks.

The last swift fox in the wild was spotted in Alberta in 1938. A combination of factors led to its demise, including habitat degradation, over-hunting, and predator control programs. The short-grass prairie came under heavy pressure from cultivation, leading to a loss of habitat for the swift fox and many other species. In addition, the fox was heavily trapped in the mid- and late 1800s for its soft, attractive pelt. The Hudson's Bay Company sold an average of 4,681 pelts per year between 1853 and 1877; by the 1920s the take declined to just 500 pelts per year. However, predator control programs against the coyote and wolf finally removed the swift fox from the Canadian Prairies. Predator control programs often relied on extermination methods that were non-species specific, such as poisoned bait and leg traps. As with many species, more than one factor typically drives a species to extinction, and these factors often interact synergistically.

Since 1978, efforts have been made to return the fox to the prairie. Foxes were bred in captivity, and wild populations from the United States were relocated to Alberta. The captive breeding program was initiated by two private citizens, which illustrates the positive impacts that individuals can have on environmental issues. The first reintroductions took place 20 years ago, and there are now over 200 swift foxes living and breeding in the wild on the Canadian Prairies.

The best strategy for the long-term protection of wildlife species is the preservation of populations in the wild—only in natural communities are species able to continue their process of evolutionary adaptation to a changing environment. Conservation strategies that focus on the organism within its natural habitat are referred to as **in situ preservation**. The Thelon Game Sanctuary in the Northwest Territories, for example, was established in 1927 to help protect the remaining population of muskoxen. Since that time, much of the mainland habitat of the animal has been recolonized by out-migration from this sanctuary. However, in situ preservation may not be a viable option for many rare species, including the swift fox. If remnant populations are too small, on-site preservation strategies will be ineffective. In such cases, the only way to prevent extirpation or extinction is to temporarily maintain individuals in artificial conditions under human supervision. This strategy is known as **ex situ preservation**. Zoos, game farms, aquaria, private breeders, and botanical gardens are all examples of ex situ facilities. The swift fox is an example of ex situ conservation.

The swift fox feeds mostly on small mammals, but also on insects (© *Wayne Lynch*).

Hunting of coyotes is allowed in most parts of Canada as part of predator control programs. However, numbers continue to increase in most areas (*Philip Dearden*).

The Prairies have been virtually completely transformed into an agricultural landscape, with the result that many species native to this habitat are now endangered (*Philip Dearden*).

Human demands are causing both physical and chemical changes to the environment. Physical changes such as deforestation *remove* important habitat components, while chemical pollution may *degrade* habitats to the point where they are no longer able to support wildlife, even if the physical structure of the habitat remains. Humans place further pressure on species by the introduction of alien species (Chapter 3).

Physical Changes

Some impacts arising from physical changes in the natural environment have already been discussed within the context of forestry and agricultural practices (Chapters 9 and 10). It is difficult historically to separate these influences from the more general impacts of colonization in North America. Large areas of forests in central and eastern Canada were cleared to make way for agriculture. Species dependent on such forests, such as the eastern cougar and wolverine, suffered accordingly. Forests are still being replaced by agriculture in some areas. In Saskatchewan, for example, Hobson et al. (2002) calculated that conversion of the boreal forest between 1966 and 1994 occurred at an annual rate of 0.89 per cent, three times the global average.

Before the mid-1850s, there were some 101 million hectares of long-grass prairie in central North America; now less than 1 per cent remains. Other prairie ecozones have not fared much better, with only 13 per cent short-grass prairie, 19 per cent mixed-grass prairie, and 16 per cent of aspen

parkland remaining. Millions of bison and antelope once grazed these regions, and the land trembled with their migrations. The bison and the antelope have now been replaced by cattle. Not surprisingly, one-half of Canada's endangered and **threatened** mammals and birds are from the Prairies.

Accompanying the transformation of the natural prairie grassland for agricultural purposes, thousands of hectares of wetlands have been drained to create more agricultural land. Until the early 1990s, the Canadian Wheat Board Act made it financially attractive for farmers to expand cropland instead of managing their land more effectively, and as a result, much marginal land was brought under the plow. It is now estimated that over 70 per cent of prairie wetlands have been drained. Of the remaining wetlands, 60 to 80 per cent of the habitat surrounding the basins is affected by farming practices. Such changes have been a major factor behind declines in the numbers of waterfowl breeding on the Prairies, a trend that is only just being reversed by wildlife management practices (Box 11.4).

Draining of wetlands is not restricted to the Prairies. Eighty per cent of the wetlands of the Fraser River delta have been converted to other uses, as have 68 per cent of the wetlands in southern Ontario and 65 per cent of the Atlantic coastal marshes. Drainage not only has a negative impact on marsh-dwelling species, but also serves to increase pollutant loads and sediment inputs accumulated in drainage water. High pesticide,

BOX 11.3 THE WOLF KILL

The wolf was vilified as a rapacious killer and enemy of humans for centuries. It was shot, poisoned, and extirpated throughout large areas of its range, particularly in parts of the United States where it was listed as an endangered species. Canada has some of the healthiest wolf populations in the world, numbering around 58,000, and these are being used for reintroductions, such as the recent transfer of wolves from Alberta to Yellowstone National Park. However, wolves are still being shot and poisoned in Canada for predator control programs. For example, in 1994 at least 29 wolves were killed by government agencies and local ranchers in Alberta. Over 400,000 cattle graze in the area of the east slopes of southern Alberta, and in that year 14 cattle were killed by wolves. Biologists estimate that it could take decades for wolf numbers to recover in the area. The Alberta government had a program that would compensate ranchers by as much as $1,000 for every cow that was killed by a predator, but in 1993 this program was discontinued due to budget restrictions. Fortunately, the program has been reintroduced by a non-governmental organization (NGO). It remains to be seen what kind of relationship will evolve between the wolves and ranchers over the next few years in this area.

Despite the estimated number of wolves in Canada, local populations are clearly at risk. In Banff National Park, for example, the wolf population is estimated at about 25 animals, half of what it was in 1998. Since 1981, 52 wolves have been killed on roads in Banff, Kootenay, and Yoho national parks while many others have been shot or trapped legally outside park boundaries. At least 13 wolves were shot or trapped legally in close proximity to the boundary of Banff National Park in 1999–2000. A new provincial regulation now requires all hunters and trappers north of the Bow Valley to register their kills. A 10-km buffer zone around the park, where wolves cannot be killed, has been suggested. This is also the approach taken in Ontario, where in the spring of 2004 the government announced that there would be a permanent moratorium on wolf hunting and trapping in the 39 townships that surround Algonquin Provincial Park. A moratorium was first enacted in 2001, designed to protect the largest remaining population of the eastern wolf. However, a loophole allowed traps to be set if coyotes were the intended target species, which led to the death of several wolves.

Wolf-kill programs are also generating considerable tension in Yukon, where the conflict is over who should harvest various ungulate species such as moose and caribou—humans or wolves. The conflict has been a long one, with the Yukon Game Branch conducting a systematic wolf poisoning program throughout the territory during the 1950s and 1960s in response to a reported decline in ungulate numbers. Private aerial hunting was initiated in the 1980s in the Coast Mountains, although subsequent research indicated that grizzly bears were the primary predators on moose calves. Between 1983 and 1985, wolf numbers were reduced from 161 to 47. The wolf kill resulted in a decline of only 20–30 per cent in the amount of prey biomass killed during the winter. Biologists subsequently concluded that short-term removal of wolves would have no positive effects on prey populations.

Nonetheless, in 1993 another wolf-kill program was initiated in the Aishihik region east of Kluane National Park Reserve, a World Heritage site. The three-year program planned to remove some 140 wolves from the area in order to boost caribou and moose populations in response to requests from Native peoples. Although a 2–12-km-wide buffer zone was established between the killing area and the park, by 1994 no intact packs remained inside the park. In late 1994, the government decided to extend the wolf kill for

Wolves have not fared well wherever humans dominate the landscape and have been extirpated from many areas of the world. Even in lightly populated regions in Yukon there are calls for their control (*World Wildlife Fund of Canada/Frank Parhizgar*).

another five years, and by 1995, 127 wolves had been removed. After favouring aerial shooting in previous years, the government switched to snares in 1995. One moose, five wolverines (listed as a vulnerable species by COSEWIC), and 12 coyotes were caught in addition to the wolves.

Predator control programs still continue today. In 2003 the Minister of Water, Land and Air Protection in BC approved the culling of predators around two Vancouver Island marmot populations. Despite government estimates that point to a decline in wolf and cougar populations on the island, a cull of up to 30 wolves (at least three wolf packs) and 20 cougars was approved in an effort to protect one of Canada's most endangered species. In a study conducted on marmots in 2002, six of 18 fitted with radio-transmitter collars were killed by predators—wolves killed four, an eagle killed one, and a cougar killed the other. Fewer than 30 Vancouver marmots live in the wild; another 70 have been bred in captivity. The cull has stirred up a backlash from several environmental groups that claim that if the government was truly concerned about recovering marmot populations, it would advocate an end to clear-cut logging.

fertilizer, sediment, and salt levels may negatively impact organisms further downstream.

Another ecozone that has been particularly hard hit by habitat destruction is the Carolinian forests of southwestern Ontario. These southern deciduous forests support a greater variety of wildlife than any other ecosystem in the country, including 40 per cent of the breeding birds. More than 90 per cent of this habitat has now been transformed by forestry, agriculture, and urbanization; less than 5 per cent of the original woodland remains. An estimated 40 per cent of Canada's species at risk are in this zone. Most of the remaining forests are in tracts belonging to regional conservation authorities or in privately owned woodlots that are potentially open to logging. Landscapes such as these are particularly suited to stewardship initiatives, discussed in greater detail later in the chapter.

Concern regarding the loss of habitat for Canadian species extends beyond the Canadian border. The harsh winters and productive summers that characterize much of Canada mean that many species in Canada, especially birds, are migratory. Over the last few decades, significant population declines have occurred for species that spend most of the year in tropical habitats but migrate to Canada to breed. In BC, for example, statistically significant declines have occurred among northern flickers, Swainson's thrushes, chipping sparrows, yellow warblers, and dark-eyed juncos. The reasons for these declines probably involve several factors, including loss of winter range through tropical deforestation and increased fragmentation within their northern breeding habitat. More long-term data and detailed studies are required to sort out the complexities of these changes.

Sometimes the impacts of physical habitat change can be indirect. A good example is provided by the parasitic habit of the brown-headed cowbird. This species lays its eggs in the nests of other species. The unsuspecting parents often lavish more attention on this large interloper and neglect their own young, leading to their death. The cowbird is an indigenous grassland species, but its distribution has expanded dramatically with human disturbance, as it prefers fragmented habitats and is adept at interloping on the forest-edge nesting sites of other birds. Some of its favourite targets are endangered species, such as Kirtland's warbler, where rates of up to 70 per cent parasitism have been recorded.

Chemical Changes

As the number of chemicals introduced into the environment continues to increase, concern over chemical degradation of habitats intensifies. Over 20,000 chemicals are in use in Canada, and more than 1,000 new chemicals are introduced every year. The effects of chemical pollution are often more difficult to assess than those of physical destruction. Unless there is a catastrophic chemical spill, the signs of declining populations often go unnoticed for several years and even decades. Even after population declines have been documented, it may take many years of careful analysis before a conclusive link to chemical pollution can be made. This was the case with the decline in the numbers of birds at the top trophic level (Box 11.5), where pesticide biomagnification (Chapter 10) led to thinner eggshells and ultimately lower breeding success. Bald eagles, for example, had been persecuted for a long time around the Great Lakes, but it was the total breeding failure due to high chem-

BOX 11.4 IS CANADA'S 'DUCK FACTORY' DISAPPEARING?

Millions of ducks, geese, and swans darken the skies every year as they migrate across the length of the continent and back again. This annual migration evokes a sense of wonder and mystery in the over 60 million North Americans who watch migratory birds each year. But for some, wonder and mystery is accompanied by anxiety over the future status of the 35 species of waterfowl that spend part of each year in Canada. Waterfowl depend on a complex and increasingly vulnerable chain of habitats that extend across international borders. Accelerated conversion and degradation of habitat caused by human activities have led to a series of record low populations of most duck species. Between 1998 and 2002, the number of ponds across the Prairies decreased by 12 per cent, while the total ducks on the Canadian Prairies declined by 12.6 per cent.

Many of the most productive wetlands in Canada have been drained in an effort to bring more land under cultivation. Wetlands found in the Prairie provinces are particularly productive. The retreat of glaciers that at one time covered all of Canada left behind significant nutrient deposits, which have formed the basis of highly productive ecosystems. Waterfowl such as mallards and pintails feed on the plants and invertebrates that feed on the nutrients. But as farm intensification has increased over the last 50–60 years, prairie wetlands have reduced in size and extent, making it difficult for ducks to secure adequate food supplies and nesting sites along their long migratory routes.

Unfortunately, habitat loss and degradation are not the only pressures on migrating waterfowl. Duck mortality rates also vary in response to weather, climate, competition for resources, environmental contamination, and hunting. In the Canadian and US Prairies, weather has a particularly strong influence on waterfowl breeding habitat conditions and consequently on the abundance of waterfowl populations. According to Environment Canada (2002), drought in the late 1980s and early 1990s created particularly difficult breeding conditions for ducks. Spring habitat conditions, measured by the number of ponds in May, improved into the late 1990s from the low levels during the drought of the 1980s, but have been in decline since 1997. Severe drought in 2002 in the Canadian Prairies resulted in a 48 per cent decline in the estimated number of May ponds compared with 2001. In 2002, pond numbers were 58 per cent below the 10-year and the long-term (1961–2002) averages. As a result of the recent drought, total duck populations for the Canadian Prairies declined by 33 per cent (to 7.2 million ducks), illustrating the dramatic impact that weather can have on the reproductive potential of waterfowl.

With over three million Canadians and Americans shooting migratory waterfowl each year, hunting also has a significant impact on waterfowl populations. For example, in 2001 over 5.5 million mallard ducks were killed by hunters (see figure).

Hunting regulations were introduced decades ago by the Canadian and American governments to protect waterfowl populations, but the governments have been slow to recognize the impact of land-use practices on waterfowl habitat and therefore abundance. The issue was not formally addressed until 1986, when Canada and the US signed the North American Waterfowl Management Plan (NAWMP) (Mexico joined in 1994). The goal of the NAWMP is to conserve 3.4 million hectares (8.2 million acres) of wetland and associated upland habitat for the benefit of waterfowl and other wetland-related species. To date, Canada has spent $787 million protecting 1.8 million hectares. Conservation efforts under the

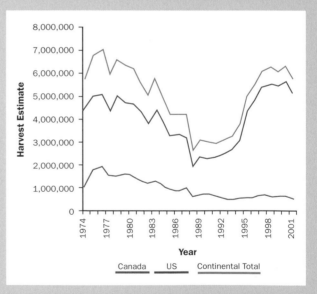

Harvest estimates of mallards in Canada and the United States.

NAWMP include involvement from various stakeholder groups—government agencies at all levels, industry, conservation groups, hunters, farmers, and other landowners. Duck conservation practices include maintaining nesting areas on land close to shallow water for land breeders such as mallards, pintails, teal, gadwalls, wigeons, and shovellors, and ensuring water levels are managed for diving ducks such as redheads and canvasbacks.

Success of the Management Plan remains to be seen. In the interim, duck populations across the Prairie provinces continue to decline, leaving some conservationists wondering if Canada's 'duck factory' is disappearing.

Another form of habitat degradation caused the demise of this pilot whale off the east coast of Newfoundland. It drowned after getting caught in a fish trap (*Philip Dearden*).

ical levels that led to their extirpation from the Ontario side of Lake Erie in 1980. By 1980, there were only seven nests along the entire Canadian shoreline of the Great Lakes, including Lake Superior, and not one healthy chick was produced.

Bald eagles have now recolonized many areas in the Great Lakes region where they were extirpated as a result of chemical use. There are now 136 nests and the birds are producing strong, healthy chicks. But the birds are dying young—at 13 to 15 years old, less than half their natural lifespan. Autopsies completed on dead birds show high levels of lead and mercury contamination. The former is likely persisting in the environment from when lead was used in the manufacture of bullets and fishing lures; it should decrease over time. The source of mercury, which is highly toxic, is a mystery. It has now been eliminated from most products in which it was once used, and discharges from human sources are down about 80 per cent. However, the metal is emitted as a by-product of burning coal to produce electricity. Mercury tends also to build up in fish that live in the reservoirs behind power dams. Scientists are now investigating these sources and their links to bald eagles.

Alien Species

Introduced species can also have a significant impact by out-competing native species for necessary resources or by direct predation on native species. Examples of successful alien species competing for resources were described in Chapter 3. The introduction of new species to insular habitats provides some graphic examples of the destruction that can be wrought. On the Queen Charlotte Islands, for example, the introduction of both raccoons and Norway rats is having a catastrophic impact on ground-nesting seabirds. The breeding population of ancient murrelets on Langara Island off the north coast of the Queen Charlottes declined by approximately 40 per cent between 1988 and 1993, to less than 10 per cent of the original size of the population. Further south in the new national park reserve of Gwaii Hanaas, the main ancient murrelet colony on Kunghit Island decreased in size by approximately one-third between 1986 and 1993. In both cases predation by Norway rats appears to be mainly responsible for the declines.

Aside from direct predation, alien species can impact native species in other ways. For example, populations of the Newfoundland crossbill have declined significantly; competition for food (pine cones) with the introduced red squirrel is believed responsible.

VULNERABILITY TO EXTINCTION

The effects of overhunting and habitat degradation differ among species, as not all species are equally vulnerable to extinction. Ecologists have identified a set of extinction-prone characteristics that help to identify which species are most vulnerable to extinction. Through the identification of such characteristics, conservationists are better able to anticipate the need for protection, and as a result, governments are able to allocate limited resources

BOX 11.5 RAPTORS AS INDICATORS OF CHEMICAL DEGRADATION OF HABITAT

Sitting as they do at the top of the food chain, birds of prey or raptors are recognized as powerful indicator species of ecosystem health. Raptors were discovered to be useful indicators of environmental health during the 1960s, when research into the drastic population declines in bird- and fish-eating species revealed that eggshell-thinning and reproductive failure were caused by organochlorine pesticides (Kennedy, 2003). The decline of peregrine falcon populations is particularly well documented.

In Canada there are three subspecies of peregrine falcon: the continental *anatum* subspecies that breeds south of the treeline from the Atlantic to the Pacific oceans; the northern *tundrius* subspecies that nests along Arctic rivers, lakes, coastline, and inland escarpments; and the western *pealei* subspecies that occupies coastal islands and areas of adjacent mainland BC. Peregrine falcons are extremely powerful birds of prey, catching other birds in flight while attaining speeds as high as 300km/hour. Prey are killed by a direct blow of the closed fist delivered at great speed. Favourite prey include songbirds, water-

fowl, pigeons, shorebirds, and seabirds, and, especially among the Arctic peregrines, small mammals such as lemmings. Falcons nest on cliffs or trees where they can overlook water bodies. Tall buildings may act as a substitute where urban pigeons are the main prey. Falcons are quite territorial during the breeding season, with nests seldom closer than one km apart.

The peregrine falcon once bred all across Canada. Populations appeared remarkably stable up until the 1940s, when populations started to crash. Over 300 known nesting sites in the eastern US became vacant. The falcon's precipitous decline was directly linked to the bioaccumulation of pesticides (Chapter 10) such as DDT, BHC, dieldrin, and heptachlor epoxide (Rowell et al., 2003). Surveys in the 1970s documented the continued downfall of the peregrine, and by this time the species had been extirpated from large areas of its previous range. In 1978, COSEWIC classified *anatum* peregrines as endangered, *tundrius* as threatened, and *pealei* as rare.

In the late 1980s, urban populations were established in southern Canada through the reintroduction of captive-raised young (an example of ex situ conservation), a program initiated by the Canadian Wildlife Service. The program was expanded, and to date over 700 birds have been released to the wild at over 20 sites from the Bay of Fundy to southern Alberta. Over a dozen released birds are now breeding in the wild.

Peregrine falcons appear to be on the road to recovery. However, recent bird surveys suggest that these raptors may still be threatened by environmental contaminants. For example, peregrine falcon nesting populations in Yukon are surveyed every five years; by the year 2000, data suggested a possible collapse in productivity, with about 20 per cent of the sites showing no evidence of breeding adults at all. The production of young is cause for further concern—about 60 per cent of all sites had no young, with only 50 per cent of the sites occupied by breeding pairs being productive (Mossop, 2003). According to the survey, sites with the longest history in the population are showing productivity problems, which could imply a building concentration of chemical contaminants (ibid.). Some of the chemicals that have been banned in Canada due to their harmful effects on wildlife are still in use in other regions (e.g., DDT is still used in South America), and so they are still affecting the ability of falcons to reproduce successfully.

Chemical habitat degradation is not the only stress threatening the long-term viability of peregrine falcons across Canada. In BC, for example,

The peregrine falcon suffered badly as a result of the biomagnification of agricultural chemicals, but is now recovering in numbers in many areas (*World Wildlife Fund Canada/Edgar T. Jones*).

falcon populations are threatened by declines in the raptor's supply of colonial seabirds due to habitat loss and competition from alien predators.

In 1999 COSEWIC downlisted *anatum* peregrines because extinction was no longer a threat, but the subspecies was retained in the threatened category due to the slow growth and uncertain status of populations across their range. The *pealei* and *tundrius* subspecies remain on COSEWIC's special concern list.

more effectively. Species with one or more of the following characteristics are more **vulnerable** to extinction.

- *Species with specialized habitats for feeding or breeding.* Once a habitat is altered, the environment may no longer be suitable for specialized species. A good example is the northern spotted owl, discussed in Chapter 9. Northern spotted owls require old-growth habitat for survival and reproduction.
- *Migratory species.* Many songbirds are experiencing population declines in Canada as a result of their long and hazardous migrations to South and Central America. Species that migrate seasonally depend on two or more distinct habitat types, and if either one of these habitats is damaged, the species may be unable to persist.
- *Species with insular and local distributions.* Dawson's caribou, endemic to the Queen Charlotte Islands, became extinct due to the ease with which it could be hunted in such a restricted habitat, with no hope of an emigrating population for replacement.
- *Valuable species.* Many organisms are overharvested to the point of extinction due to their high economic value. The American ginseng was once abundant in forests of eastern North America but is now rare due to the demand in Asian countries for dried ginseng roots for medicinal purposes. Similarly, populations of Asian bears have been all but eliminated across their range. Their gall bladders are highly valued in Asian markets, and North American bears are now coming under pressure from the same markets. A single gall bladder can be worth over $5,000. One illegal dealer in BC was found with 1,125 gall bladders in his possession. BC has now passed a law making possession of endangered animal contraband an offence.
- *Animal species with large body size.* Large animals tend to have large home ranges, require more food, and are more easily hunted by humans. Top carnivores, for example, depend on the abundance of many different species lower in the food chain. If the numbers of prey species are disrupted, the impacts are felt at the top of the food chain. The proposed bison slaughter in Wood Buffalo National Park, for example, would have a catastrophic effect on the wolf population of the park (discussed in greater detail in Box 11.12). Furthermore, animals higher up the food chain are more vulnerable to the concentration of toxic materials (Chapter 10).
- *Species that need a large home range.* Species that need to forage over a wide area are prone to extinction when part of their range is damaged or fragmented.
- *Species with only one or a few populations and species in which population size is small.* Any one population may 'blink out' as a result of chance factors (e.g., earthquakes, fire, disease), increasing the species' vulnerability to extinction. Small populations are also more likely to become extinct locally due to their greater vulnerability to demographic and environmental variation.
- *Species that are not effective dispersers.* Species unable to adapt to changing environments must migrate to more suitable habitat or face extinction. Species that cannot migrate quickly have a greater chance of extinction.
- *Species with behavioural traits that leave them susceptible.* Some species have behavioural traits that make them particularly vulnerable to clashes with human activities. For example, the red-headed woodpecker flies in front of cars, and the Florida manatee appears to be attracted by motorboats, which are a main cause of death for the animal. Other species, although not attracted by human activities, may be too slow to get out of the way. Tragically, this is the case with the few remaining right whales in the Gulf of St Lawrence area. The whales do not appear to be able to avoid the heavy marine traffic in the area (Box 11.6).

The extinction of species may now be occurring at roughly 100 to 1,000 times faster than the natural rate of extinction. This rate of extinction is much faster than the evolution of new species, and so we are in a period in which the world's biological diversity is in decline. What are nations doing to arrest this decline in biodiversity? What are the best strategies for the long-term preservation of biological diversity? In the sections that follow, the international and Canadian responses to our biodiversity crisis will be discussed.

RESPONSES TO THE LOSS OF BIODIVERSITY

The International Response

Awareness of the fact that we are living in an unprecedented period of mass extinction has led to the formulation of several international conventions and programs. Some programs have a regional orientation, such as the North American Waterfowl Management Plan with the United States and Mexico (see Box 11.4), while others include many different nations. Some conventions have a fairly long history. Established in 1973, the **Convention on International Trade in Endangered Species of Wild Fauna and Flora (CITES)** is one of the longest-standing treaties. Currently ratified by more than 120 countries (including Canada), this treaty establishes lists of species for which international trade is to be controlled or monitored (e.g., orchids, cacti, parrots, large cat species, sea turtles, rhinos, primates, etc.). International treaties such as CITES are implemented when a country passes laws to enforce them. Once CITES laws are passed within a country, police, customs inspectors, wildlife officers, and other government agents can arrest and prosecute individuals possessing or trading in CITES-listed species and seize the products or organisms involved. Countries that ratify CITES also vote on the species that should be protected by the treaty.

Several well-known Canadian species are listed under Appendix II of the Convention, such as the lynx, bobcat, cougar, polar bear, river otter, and burrowing owl. These species may only be traded with a valid permit from the Canadian Wildlife Service. When travelling through inter-

Species that have high economic values are targeted by hunters and poachers. Here a leopard has fallen victim to a poacher's snare in a park in Sri Lanka, and these cobras in Laos are now part of a local drink (*Philip Dearden*).

national airports in Canada, you will commonly see information on CITES to warn travellers about trying to import listed plants or animals.

CITES has been instrumental in restricting trade in certain endangered wildlife species. Its most notable success has been a global ban on the ivory trade instituted in 1989. Without this ban, it is unlikely that any elephants would be left in East Africa today. Despite the treaty's success in protecting some species, Canada's support has been disappointing. According to Le Prestre and Stoett (2001: 197), the Canadian voting record on whether species should be listed has reflected 'domestic economic and/or cultural interests at every turn', and Canada has consistently failed to pay its dues to support the Convention.

Other international treaties with a focus on conserving biodiversity include:

BOX 11.6 NORTHERN ATLANTIC RIGHT WHALES: WRONG PLACE AT THE WRONG TIME

OTTAWA—Transport Minister David Collenette and Fisheries and Oceans Minister Robert Thibault, together with Dr Moira Brown of the Canadian Whale Institute, announced that new shipping lanes in the Bay of Fundy, designed to protect the endangered North Atlantic Right Whale population from ship strikes, will officially be put into operation on July 1, 2003.

The new lanes will help to protect the Right Whale by organizing the ship traffic flow in and around an area where the whale densities are the greatest.

'The Government of Canada has created new safe and effective shipping lanes in order to protect Right Whales in the Bay of Fundy', said Mr Collenette. Canada has taken a leading role internationally through the creation of the world's first shipping lanes designed to protect the whale population.

Amendments to the navigational charts and vessel traffic control procedures, as well as distribution and notification procedures, have been completed. As a result, the additional protection for the Right Whales provided by the lane changes is now in place prior to the expected seasonal return to Fundy waters later this summer.

Since there are only about 350 of these whales in existence, they are one of the world's most endangered large whales, said Minister Thibault. There are reports of 18 calves born this year. Most of them will be in the Bay of Fundy throughout the summer, with the main concentration of Right Whales. The vessel traffic lane change is a positive step on the part of the marine transportation and fishing sectors to help move the species towards recovery.

The new lanes are part of the Government of Canada's North Atlantic Right Whale Recovery Plan led by Fisheries and Oceans Canada. A marine industry working group, co-chaired by Transport Canada and the Canadian Whale Institute, determined that this approach would be the most effective in reducing strikes and maintaining safe commercial marine operations. The working group included representatives from the shipping, fishing and whale-watching sectors. 'This initiative will greatly lessen the ship strike threat to this highly vulnerable population', said Dr Brown. 'Canadians can be proud of the role their government played in making it happen.'

The new shipping lanes are based on considerable scientific whale research and were reviewed by several marine industry stakeholders and experts to ensure safety would be maintained.

SOURCE: Transport Canada press release, 26 June 2003. At: <http://www.tc.gc.ca/mediaroom/releases/atl/2003/03-a007e.htm>.

ADDENDUM: 'We might have to consider putting some mandatory speed conditions in the Bay of Fundy, or having a lookout posted on the bow of a ship' (Jerry Conway, a marine adviser, commenting on the death of an endangered right whale in a collision with a ship *after* new routes were established).

SOURCE: Canadian Press, 9 Oct. 2003.

- Convention on Conservation of Migratory Species of Wild Animals (the Bonn Convention, 1979);
- Convention on Conservation of Antarctic Marine Living Resources (1982);
- International Convention for the Regulation of Whaling, which established the International Whaling Commission (1946);
- International Convention for the Protection of Birds (1950);
- Benelux Convention on the Hunting and Protection of Birds (1970).

Unfortunately, participation in all of these treaties is voluntary, and countries can withdraw at any time to pursue their own interests when they find the conditions of compliance too arduous. Canada, for example, withdrew from the whaling convention in order unilaterally to permit indigenous whaling in the Arctic.

One of the most important international agreements to protect biodiversity—the **Convention on Biological Diversity (CBD)**—emerged from the World Summit on Sustainable Development, held in Rio de Janeiro in 1992. Canada was the first indus-

Alive or dead, exotic species do not belong in your home. Do not be tempted to buy such souvenirs because by doing so, you only create a market for their death. These dead hawksbill turtles are hanging outside a souvenir ship in Hanoi. Other common tourist offerings in many locations are seashells. Although some may have washed up on beaches, many species are killed just to supply the tourist trade and will be forbidden, through CITES regulations, from entering another country (*Philip Dearden*).

trialized nation to sign the Convention and provided extra funding to house the secretariat in Montreal. The CBD requires signatories to develop biodiversity strategies, identify and monitor important components of biodiversity, develop endangered species legislation/protected areas systems, and promote environmentally sound and sustainable development in areas adjacent to protected areas. Canada has been implementing many of the strategies required by the Convention over the last decade. The *Canadian Biodiversity Strategy* (Government of Canada, 1995) was developed as a direct response to the requirements of the CBD, and other developments, such as the new Species at Risk Act, described below, are also consistent with these requirements.

The Canadian Response

In 1973, the United States became the first country to pass endangered species legislation. Australia followed some 20 years later, and a host of other countries, including the European Union and Japan, have also developed similar legislation. In Canada, six provinces have endangered species legislation (Ontario: 1973, New Brunswick: 1974, Quebec: 1989, Manitoba: 1990, Nova Scotia: 1998, and Newfoundland and Labrador: 2001). However, this provincial legislation is fairly weak in that many of the provisions to protect endangered species are discretionary. It has taken many years for the federal government to respond to the legislative challenge of protecting endangered

species. The **Species at Risk Act (SARA)** was finally passed in 2002. The federal government had little choice; as a signatory to the CBD, there was a requirement to enact such legislation.

One of the reasons the government was so reluctant to introduce and pass federal endangered species legislation was related to the Canadian Constitution. Unlike most other countries, a majority of the land in Canada is publicly owned, with 71 per cent held by the provinces and 23 per cent held by the federal government (see Chapter 1). Most responsibilities are shared by these two levels of government, with the federal government responsible for oceans and freshwater ecosystems, migratory birds, and the management of federal lands including the Northwest Territories and Nunavut. Yukon now has responsibilities for its own land base and a similar devolution is underway for the NWT and Nunavut. The federal government also has responsibility for Aboriginal lands south of the sixtieth parallel, although this will change as land claim negotiations are settled. Therefore, most of Canada's public land, and the resources contained within, are under provincial jurisdiction.

The provinces have numerous individual programs related to nature conservation and they also participate in joint programs with the federal government. The Canadian Endangered Species Conservation Council (CESCC) consists of the three federal ministers responsible for Environment, Canadian Heritage, and Fisheries and

Oceans, as well as provincial and territorial government ministers responsible for the conservation and management of wildlife. The Council co-ordinates federal, provincial, and territorial government activities relating to the protection of species at risk and provides general direction on the activities of the Committee on the Status of Endangered Wildlife in Canada (COSEWIC) and the preparation of recovery strategies and action plans.

Since 1976, COSEWIC has been responsible for determining the status of endangered species. The Committee—which includes representatives from relevant federal agencies, provincial and territorial wildlife agencies, and the Aboriginal community, as well eight scientific subcommittees that are species specialist groups—meets annually to consider status reports on candidate species and to assign them to various categories (Table 11.1). By 2003, COSEWIC had assigned 431 species to five risk categories (Table 11.3). Species are reassessed every 10 years to see if their status has changed. During the period 1985–2002, the status of half of the reassessed species remained unchanged, a third deteriorated, and 16 per cent improved (Figure 11.2).

The Committee's assessment is the first step in the process for protecting a proposed species at risk under SARA (Box 11.7). However, even if COSEWIC lists a species, this does not guarantee that the species will receive any form of protection. The ultimate decision is in the hands of the politicians who make up the CESCC, and for this reason SARA has been strongly criticized by many who feel this should be a scientific and not a political process. The danger of allowing politicians to make such decisions was illustrated in Alberta in 2003. According to the accepted international criterion, a species having less than 1,000 mature breeding individuals should be listed as threatened. Following this criterion, members of the Endangered Species Conservation Committee in Alberta (appointed by the government) recommended that Alberta's grizzly bear population be listed as a threatened population. Nonetheless, the responsible minister, Mike Cardinal, refused to act on the recommendation despite the fact that a breeding population of only 300 to 650 grizzlies remains.

The influence of politics was clear even by the time of the first new species listings under SARA, in April 2004, when the Fisheries Minister decided to delay by nine months a decision on whether to list 12 aquatic species recommended by COSEWIC. The Atlantic cod, for example, is estimated to have a population of less than 1 per cent of historic levels and

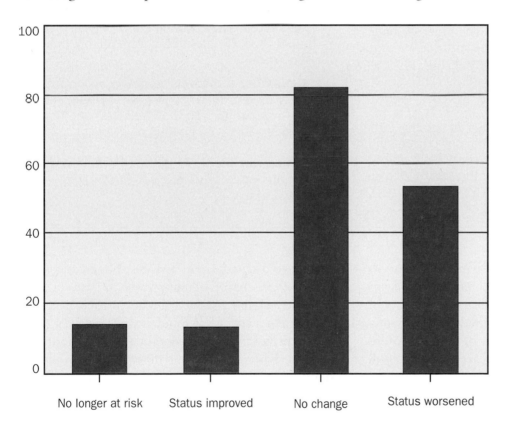

Figure 11.2 Change in status of reassessed species at risk, 1985–2002 (number of species reassessed). SOURCE: Environment Canada (2003). Reproduced by the permission of the Minister of Public Works and Government Services, 2005.

NOTES: 1. The data are based on status reassessments conducted by COSEWIC. Reassessments based on existing status reports only were not included. These were re-evaluated using new IUCN criteria and not based on any new information.

2. Some downlistings and delistings were as a result of new information gathered rather than a change in the status of the species.

3. Species reassessments that result in splitting a species into smaller units (i.e., populations) are considered new assessments.

BOX 11.7 THE PROCESS FOR PROTECTING A SPECIES AT RISK

1. COSEWIC assesses and classifies a wildlife species as extinct, extirpated, endangered, threatened, special concern, data deficient, or not at risk (see Table 11.1). COSEWIC provides its report to the Minister of the Environment and the Canadian Endangered Species Conservation Council, and a copy is included in the Public Registry.

2. The Minister of the Environment indicates how he or she intends to respond to a COSEWIC assessment within 90 days. Within nine months, the government makes a decision about whether or not to add the species to the List of Wildlife Species at Risk. If no government action is taken, the species is automatically added.

3. When a species is on or added to the List of Wildlife Species at Risk, extirpated, endangered, or threatened species and their residences have:
   - immediate protection on federal lands (except for those species in the territories that go through the safety net process described below);
   - immediate protection if it is an aquatic species;
   - immediate protection if it is a migratory bird;
   - protection through a safety net process if it is any other species in a province or territory.

4. For all species included on the List of Wildlife Species at Risk on 5 June 2003:

- a recovery strategy must be prepared within three years for endangered species and within four years for threatened species or extirpated species;
- a management plan must be prepared within five years for a special concern species.

Or, for all species added to the List of Wildlife Species at Risk after 5 June 2003:
- a recovery strategy must be prepared within one year for endangered species and within two years for threatened or extirpated species;
- a management plan must be prepared within three years for a special concern species.

5. Recovery strategies and action plans, which must include the identification of critical habitat for the species, if possible, and management plans are published in the Public Registry. The public has 60 days to comment on these documents.

Five years after a recovery strategy, action plan, or management plan comes into effect, the minister must report on the implementation and the progress towards meeting objectives.

SOURCE: Government of Canada. 2004. Species at Risk Act. At: <www.sararegistry.gc.ca/background/process_e.cfm>.

---

shows no signs of recovery (Chapter 8), but its listing has not been confirmed. Similar concerns occur over the bottlenose whales off the coast of Nova Scotia. Although these animals have a population of only 130, the inclusion of the bottlenose whale has been delayed for fear inclusion will inhibit oil and gas exploration.

For species listed under SARA, recovery and management plans must be developed and implemented unless the minister responsible feels that recovery is not 'feasible', a caveat that provides another political opportunity to block action. The initial listing placed 233 species on the List of Wildlife Species at Risk (as of 2003, COSEWIC had 255 species listed as endangered or threatened and an additional 143 listed as special concern). 'Recovery strategies' are to be devised for these species, complete with 'action plans' that will be

subject to a cost-benefit analysis. Nineteen recovery strategies have been approved to date.

The Act has received further criticism in that even when a species is listed, it only receives automatic protection on federal lands. In southern Canada where many endangered species live, a significant proportion of the federal lands are national parks where the species are already protected. Therefore, there is no incremental gain in protection unless provincial jurisdictions agree

The one process ongoing in the 1990s that will take millions of years to correct is the loss of genetic and species diversity by the destruction of natural habitats. This is the folly that our descendants are least likely to forgive us.

— *E.O. Wilson, Harvard University*

BOX 11.8 WHAT YOU CAN DO FOR ENDANGERED SPECIES

Although the challenges created by endangered species can seem quite daunting to the individual, there are several things that you can do.

1. If you own land, even your own backyard, try to encourage the growth of native species and promote high diversity among these species. Provide the three staples—food, water, and shelter. Plant perennials such as fruit and nut trees, nectar-producing flowers, and berry bushes. Don't use chemicals!
2. Write letters to politicians at all levels encouraging them to adopt specific measures. For example, write to local politicians urging protection for a natural habitat in your area.
3. Join and support an environmental group that takes a special interest in endangered species.
4. Take part in an active biodiversity monitoring project such as the Christmas Bird Count, the Canadian Lakes Loon Survey, Frogwatch, Project FeederWatch, or one of the other host of organized activities that take place across

the country. Again, details on these projects will be available from local NGOs and university and college departments.
5. Don't keep exotic pets.
6. If you have a pet, try to make sure that it does not injure or harass wildlife. Put a bell on your cat. Domestic cats kill large numbers of songbirds every year.
7. Don't buy products made of endangered animals or plants.
8. Vote for political candidates who share your views on conservation matters.
9. Keep informed of biodiversity issues by watching nature programs on television, reading books, attending public lectures, and having discussions with local conservationists.
10. Actively try to learn more about wildlife, not just by reading and watching television, but also by becoming more aware of the wildlife in your region through field observation. Encourage others, especially children, to do likewise.

to provide it. There is a so-called 'safety net' whereby the federal government can invoke powers to act if a provincial government refuses to do so and if the case is seen as critical, but many observers feel that the federal government is unlikely to take this course of action, given the history of federal-provincial relations.

The Species at Risk Act is quite different from the American approach to endangered species protection in that it lays out a framework for co-operation towards the protection of endangered species and relies primarily on volunteerism. This same approach is also seen in relationship to habitat protection, whereby the federal government established the Habitat Stewardship Program to provide information to landowners on how best to manage their lands to protect endangered species. Compensation may be paid for any economic losses incurred by landowners. Warnock and Skeel (2004), in their evaluation of Operation Burrowing Owl in Saskatchewan, present evidence that some voluntary habitat stewardship programs can be effective.

Only time will tell whether the Canadian approach will be successful in protecting endangered species in Canada. If there is political inter-

ference at any of the numerous opportunities provided, if the provinces disagree with the federal government, or if cost-benefit analysis indicates that it is less expensive to let a species become extinct, we will have failed in this experiment. The costs of such failure will be extremely high. Fortunately, endangered species legislation is not the only means of protecting biodiversity (Box 11.8).

PROTECTED AREAS

Protected areas have emerged as one of the key strategies to combat the erosion of biodiversity both internationally and in Canada (Box 11.9). **Protected areas** play different roles in society (Box 11.10), and in many cases their conservation role has been recognized only recently as the dominant one. This is especially true in Canada, where Banff, our first national park (1885), was set aside mainly to promote tourism and generate income rather than to protect species and ecosystems. However, since that time the crucial role in species and ecosystem protection played by protected areas in Canada has led to both legislation and policy directives that make it clear that biodiversity protection is the prime mandate for the national park system.

The 'playground' role is one of the most controversial issues for park managers. This Parks Canada golf course sign in Pacific Rim National Park, for example, is advertising a facility that is not actually located on park lands. Nonetheless, by advertising it, a link is created in the public mind between national parks and certain recreational activities. What about traditional activities in some parks, such as horseback riding, which can have a significant impact on the environment, or are national parks places where we should rely on our own feet to explore the beauty of the landscape? (*Philip Dearden*)

Overall, Canada's national park system is central to the protection of rare and endangered species (Figure 11.3). Although the 39 parks cover only 2.6 per cent of the land base, they contain 70.6 per cent of the native terrestrial and freshwater vascular plants and 80.9 per cent of the native vertebrate species (Table 11.4). This high degree of coverage is a result of the explicit systems plan of Parks Canada to establish parks in the 39 representative natural regions of the country. Over 50 per cent of endangered vascular plant species and almost 50 per cent of the endangered vertebrate species are found in national parks. Canada's national parks have also played a critical and increasingly important role as sites for reintroduction of endangered species (Table 11.5).

Protected Areas: A Global Perspective

The World Conservation Union (IUCN) is an international body that draws together governments, non-governmental organizations (NGOs),

and scientists who are concerned with nature conservation. The IUCN helps to provide leadership and set global standards for conservation. A protected area, as defined by the IUCN, is 'an area of land and/or sea especially dedicated to the protection and maintenance of biological diversity and of natural and associated cultural resources, and managed through legal or other effective means'. There are many different kinds of protected areas, such as national and provincial parks, wilderness areas, and biosphere reserves. They all offer some form of protection, but often with differing degrees of stringency.

To help bring some order and understanding to the different types of protected areas, the IUCN has developed a system of classification that ranges from minimal to more intensive use of the habitat by humans (Table 11.6). Given the level and type of use permitted under some classifications, not all conservationists believe that a 'protected area' status is appropriate. For example, Category VI allows for the extraction of natural resources such as timber and wildlife. Category VI, then, should not be considered a true protected area, since habitat is not necessarily managed primarily for biological diversity. Internationally, this type of 'protected area' is the most rapidly growing, and is now the dominant one in terms of area covered. In Canada, when the World Wildlife Fund (WWF) was assessing the progress of each of the provincial governments in protecting natural ecosystems for the Endangered Spaces Campaign (Hummel, 2000), areas that fell into IUCN categories V and VI were excluded.

The most recent international listing of protected areas (Chape et al., 2003) shows very strong growth. In 1994, approximately 8,600 protected areas had been designated worldwide, covering some 8 million km^2; by 2003, 102,102 protected areas had been designated covering more than 18.8 million km^2 (Figure 11.4). This is equivalent to the combined area of Southeast and South Asia and China, and represents 12.65 per cent of the Earth's land surface. Figure 11.5 shows the area and percentage distribution of categorized and non-categorized protected areas. The rapid growth in protected area establishment is explained by several factors.

- *Increased realization of the rate of biodiversity loss and the severity of the issue.* In 1990, for example, statistics on endangered wildlife in Canada

BOX 11.9 PROTECTED AREAS: A GLOBAL PERSPECTIVE

From the origins of the conservation movement in America, the idea of having areas protected by the government for conservation and public benefit, education, and enjoyment has spread throughout the world. Implementation has differed to reflect local conditions, but one ubiquitous concern to managers from Canada to England to Southeast Asia is the relationship between protected areas and local populations. As human populations grow, so do pressures for increased use of protected areas. In the UK this overuse might be mainly recreational, and significant biophysical impacts may result just from the sheer numbers of people enjoying the parks. In many tropical areas, such as Thailand, conflicts arise as local people, often driven by poverty and land-use pressures, encroach on the parks in large numbers, hunting wild animals and cutting down trees to make way for agriculture and/or to sell on international markets. Estimates suggest that over half a million people are illegally occupying national parks and wildlife sanctuaries in Thailand, with almost half the area of some parks suffering environmental degradation.

Needless to say, these kinds of management problems are challenging. There is little point in emphasizing the area of land officially designated as a protected area if, in fact, it is not protected from resource use. In the past, in Thailand, management activities have focused on a preventative approach, with the use of armed guards to patrol boundaries. Since most large remaining areas of forest and most wild animal populations are within the protected area system, there have been some benefits to this approach. On the other hand, it is clear that large-scale poaching continues in many areas, and shootouts between poachers and park guards are not an ideal management tool. Attention, therefore, has also spread to trying to address some of the underlying motives behind poaching, such as poverty, though here, too, there are substantial challenges. Economic development programs initiated in some villages have caused land prices to rise, leading some villagers to sell their lands and encroach further into the park lands. Unscrupulous local leaders may also use the villagers in this way so that they can gain control over more land area.

As with many environmental management problems, the answers do not lie in one single solution. Each case is different and an adaptive management approach (see Chapter 6) to the protected area ecosystem is essential. In the long term, education must play a lead role. Many people are unaware of the vital functions played by protected areas. It is better to achieve voluntary compliance with more flexible management regimes than have armed standoffs and mass non-compliance, as has often occurred in the past.

Park wardens in Thailand receive little pay and risk their lives to protect what remains of the wildlife. Every year lives are lost in battles with poachers (*Philip Dearden*).

Table 11.4 The Role of National Parks in the Protection of Designated Vascular and Vertebrate Species at Risk in Canada

	Native Vascular Plant Species	Vascular Plant Species at Risk*	Native Vertebrate Species	Vertebrate Species at Risk*	Area (km²)
Canada	4,521	109	1,061	190	9,900,000
National parks	3,192	62	858	92	256,385
% of total protected in national parks	70.6	56.0	80.9	48.4	2.59

*Species at risk in Canada (vulnerable, threatened, endangered) as designated by COSEWIC.
SOURCE: Parks Canada.

BOX 11.10 THE MANY ROLES OF PROTECTED AREAS

Art Gallery: Many parks were designated for their scenic beauty, and this is still a major reason why people visit parks.

Zoo: As one component of the art gallery, parks are usually easy places to view wildlife in relatively natural surroundings. Due to protection from hunting in most parks, the wildlife is not as shy of humans as wildlife outside parks.

Playground: Parks provide excellent recreational settings for many outdoor pursuits.

Movie Theatre: Just like seeing a movie, parks are able to lift us into a different setting from our everyday life.

Cathedral: Many people derive spiritual fulfillment from communing with nature, just as others go to human-built places of worship—churches, temples, and mosques.

Factory: The first national parks in Canada were designated with the idea of generating income through tourism. Since these early beginnings, the economic role of parks has been recognized, although it is a controversial one due to the potential conflict with most of the other roles.

Museum: In the absence of development, parks act as museums reminding us of how landscapes might have looked to early settlers. These museums also provide a valu-

able ecological function, as they provide important areas against which to measure ecological change in the rest of the landscape.

Bank: Parks are places in which we store and protect our ecological capital, including threatened and endangered species. From these accounts we can use the 'interest' to repopulate areas with species that have disappeared.

Hospital: Ecosystems are not static and isolated phenomena, but are linked to support processes all over the planet. Protected areas constitute one of the few places that such processes still operate in a relatively natural manner. As such they may be considered ecosystem 'hospitals', where air is purified, carbon stored, oxygen produced, and ecosystems 'recreated.'

Laboratory: As relatively natural landscapes, parks provide outside laboratories for scientists to unravel the mysteries of nature. Killarney Provincial Park in Ontario, for example, provided an important laboratory for the early research on acidic precipitation in Canada.

Schoolroom: Parks can play a major role in education as outdoor classrooms.

Source: Dearden (1995).

---

listed 194 species at risk, compared with 431 in 2003 (COSEWIC, 2003).

- *Growing awareness at the political level of the links between environmental and societal health.* When ecosystems collapse, livelihoods and economies collapse, too. This interdependency between ecosystem protection and poverty is now recognized by many international development agencies and has helped spur the worldwide interest in protected area establishment. The Global Environmental Facility of the World Bank, for example, has funded protected area projects worth almost US$1 billion since 1991, and an equal investment is planned over the next four years (GEF, 2002, 2003). Increased methodological sophistication has also allowed monetary values to be placed on ecosystem values. One team of researchers put an average price tag of US$33 trillion a year on fundamental ecosystem

services like nutrient cycling, soil formation, and climate regulation. This figure is nearly twice the annual global GNP of US$18 trillion and demonstrates, in economically understandable terms, the value of the so-called 'free' services of functioning environments (Costanza, 1997).

- *Growing evidence of the effectiveness of protected areas in helping combat environmental degradation.* A study by Bruner et al. (2001: 126) based on a survey of tropical countries found that parks were 'surprisingly effective at protecting ecosystems and species within their borders in the context of chronic under-funding and significant land use pressures'.

- *International treaties have furthered the development of protected area systems.* The Convention on Biological Diversity, for example, which entered into force in 1993, requires signatories to develop protected areas systems.

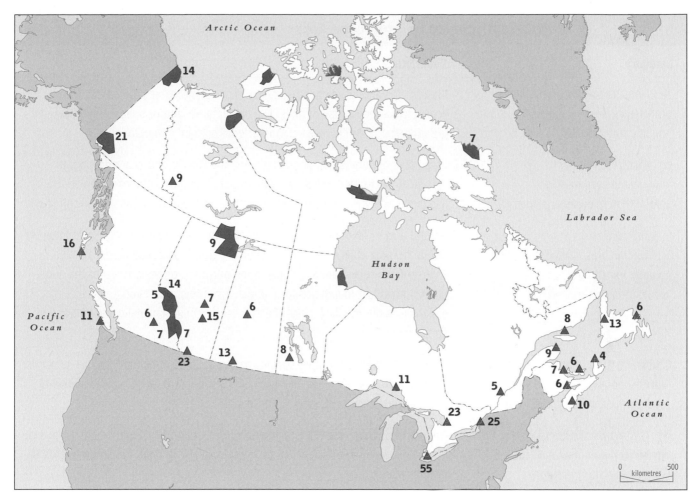

Figure 11.3 Number of COSEWIC species in national parks. SOURCE: Dearden (2001: 78).

Table 11.5	Examples of Reintroductions of Endangered Species into Canadian National Parks
Species	**Park**
American Beaver	Cape Breton Highlands National Park, Prince Edward Island National Park
American Bison	Prince Albert National Park, Riding Mountain National Park
Plains Bison	Elk Island National Park
Wood Bison	Nahanni National Park Reserve, Jasper National Park, Waterton Lakes National Park, Elk Island National Park
Fisher	Georgian Bay Islands National Park, Riding Mountain National Park, Elk Island National Park
American Marten	Fundy National Park, Kejimkujik National Park, Terra Nova National Park, Riding Mountain National Park
Moose	Cape Breton Highlands National Park
Muskox	Ivvavik National Park
Trumpeter Swan	Elk Island National Park
Caribou	Cape Breton Highlands National Park
Swift Fox	Grassland National Park

SOURCE: Dearden (2001).

Table 11.6   IUCN Classification of Protected Areas

Category	Description
1a: Strict Nature Reserve (protected area managed mainly for science)	Area of land and/or sea possessing some outstanding or representative ecosystems, geological or physiological features, and/or species, available primarily for scientific research and/or environmental monitoring.
1b: Wilderness Area (protected area managed mainly for wilderness protection)	Large area of unmodified or slightly modified land and/or sea, retaining its natural character and influence, without permanent or significant human habitation, which is protected and managed so as to preserve its natural condition.
II: National Park (protected area managed mainly for ecosystem protection and recreation)	Natural area of land and/or sea, designated to (a) protect the ecological integrity of one or more ecosystems for present and future generations, (b) exclude exploitation or occupation inimical to the purposes of designation of the area, and (c) provide a foundation for spiritual, scientific, educational, recreational, and visitor opportunities, all of which must be environmentally and culturally compatible.
III: Natural Monument (protected area managed mainly for conservation of specific natural features)	Area containing one or more specific natural or natural/cultural feature, which is of outstanding or unique value because of its inherent rarity, representative or aesthetic qualities, or cultural significance.
IV: Habitat/Species Management Area (protected area managed mainly for conservation through management intervention)	Area of land and/or sea subject to active intervention for management purposes so as to ensure the maintenance of habitats and/or to meet the requirements of specific species.
V: Protected Landscape/Seascape (protected area managed primarily for landscape/seascape conservation and recreation)	Areas of land, with coast and sea as appropriate, where the interaction of people and nature over time has produced an area of distinct character with significant aesthetic, ecological, and/or cultural value, and often with high biological diversity. Safeguarding the integrity of this traditional interaction is vital to the protection, maintenance and evolution of such an area.
VI: Managed Resource Protected Area (protected area managed mainly for the sustainable use of natural ecosystems)	Area containing predominately unmodified natural systems, managed to ensure long-term protection and maintenance of biological diversity, while providing at the same time a sustainable flow of natural products and services to meet community needs.

SOURCE: World Conservation Union (2001).

Although major gains have been seen in terms of terrestrial park systems, there is still a lot of progress required in the marine realm, where less than 0.5 per cent of freshwater and oceanic ecosystems enjoy any effective protection.

Protected Areas: A Canadian Perspective

Canada has a large variety of protected areas ranging from small ecological reserves to vast multiple-use areas. They are protected by a wide range of authorities, from municipal to federal and even international agencies. The most important protected areas are those in our national and provincial park systems. The amount of land protected in these systems varies from province to province. The location of national parks is strongly

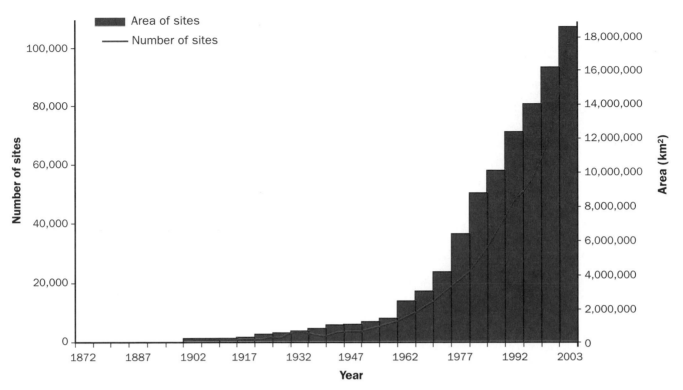

Figure 11.4 Cumulative growth in protected areas, 1872–2003. SOURCE: Chape et al. (2003). Courtesy UNEP World Conservation Monitoring Centre (UNEP–WCMC).

influenced by Parks Canada's national park system plan, which divides the country into 39 physiographic regions that are representative of Canada's natural heritage. The goal of the national park system is to have one or more national parks in each of these regions (Figure 11.6). At present, only 25 of these regions are represented. However, in 2002 Prime Minister Chrétien announced a five-year plan that would see the establishment of 10 new parks, increasing representation to 35 of the 39 regions. Representation of the remaining four regions is uncertain due to jurisdictional challenges with Quebec.

In addition to a national park system plan administered by the federal government, each province has a protected area system plan. British Columbia, for example, has 110 natural regions, Nova Scotia 27, and Alberta 20. Across Canada, there are 486 ecoregions in total at the provincial level. Only 27 per cent of these ecoregions are adequately or moderately protected according to World Wildlife Fund Canada (2000). For Canada as a whole, an estimated 6.84 per cent of ecoregions are now protected, compared to 2.95 per cent in 1989 (McNamee, 2002). Canada has endorsed the international goal of protecting a minimum of 12 per cent of the land base within

protected area systems, a goal first suggested in 1987 by the World Commission on Environment and Development. To date, BC is the only province to attain this goal.

Although Canada still has a way to go to attain the 12 per cent protection goal, the quality of Canadian protected areas is very high. Canada has 13 World Heritage sites, among them the four mountain parks (Banff, Jasper, Kootenay, and Yoho) as well as Nahanni, Kluane, and Gros Morne (Table 11.7). These sites are judged to be of national and global significance. High international accolades mean that park management practices have to be first rate. The next section identifies some of the management challenges facing protected areas in Canada. Most of the section is devoted to national parks, but many of the challenges also apply to provincial parks.

Park Management Challenges
Development within the Parks
Park management is ultimately guided by legislation. In the case of national parks, the National Parks Act guides park management. Although Canada's first national park was created in Banff, Alberta, in 1885 (Rocky Mountain National Park, as it was then known), the first National Parks Act

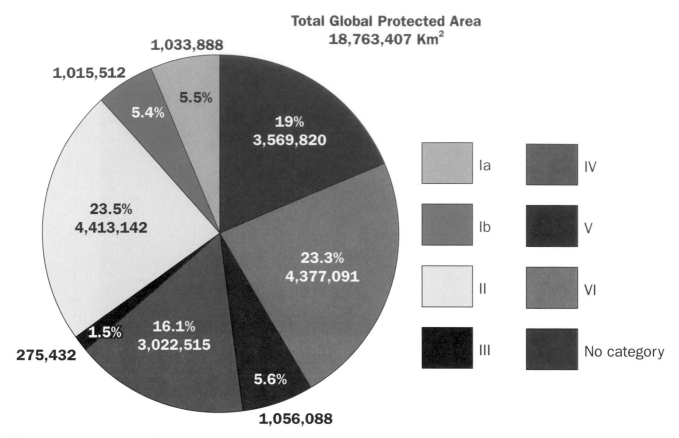

Figure 11.5 Global area (km²) and percentage distribution of categorized and non-categorized protected areas. SOURCE: Chape et al. (2003).

was not passed until 1930. Both this Act and earlier legislation dedicated the parks to 'the people of Canada for their benefit, education and enjoyment ... such Parks shall be maintained and made use of so as to leave them unimpaired for the enjoyment of future generations' (Canada, 1930: s. 4). Reconciling the balance between 'making use of' and maintaining the parks 'unimpaired' has been a major topic of debate ever since.

Without doubt, tourism and income generation were the main reasons behind the early establishment of many parks, including Banff. As such, catering to the demands of tourists was the most important management priority, and those involved with providing services for tourists were very influential. This is shown as the 'entrepreneurial' period in Figure 11.7. However, by the early 1920s, development voices were not the only ones being heard. James Harkin, the Chief Commissioner for National Parks, was well aware of the economic value of tourism, but he pressed management for a greater sense of nature protection. In 1930 the National Parks Association was formed; this NGO was the first to focus on national parks and lobby for conservation.

By the 1960s the number of visitors to Canada's national parks had increased tremendously, as had developments to serve them, including ski hills, golf courses, roads, hotels, and similar infrastructure. A massive proposal to expand the Lake Louise ski area was rejected in the early 1960s, signifying that the environmental movement was at last starting to be heard in the parks (Figure 11.7). The next 40 years witnessed many debates on the issue of controlling development in protected areas. Even through the Lake Louise expansion was thwarted, smaller developments permeated the parks. In 1988, the balance between development and protection was clarified in amendments to the National Parks Act. Protection of ecological integrity became the prime mandate of Canada's national parks.

Despite this legislative mandate, development pressures continued throughout the parks. For example, the level of development at Banff townsite escalated rapidly throughout the 1980s. Between 1985 and 1992, building permits worth over $360 million were issued in Banff. Shopping space almost doubled between 1986 and 1994, compared to increases of between 10 and 15 per

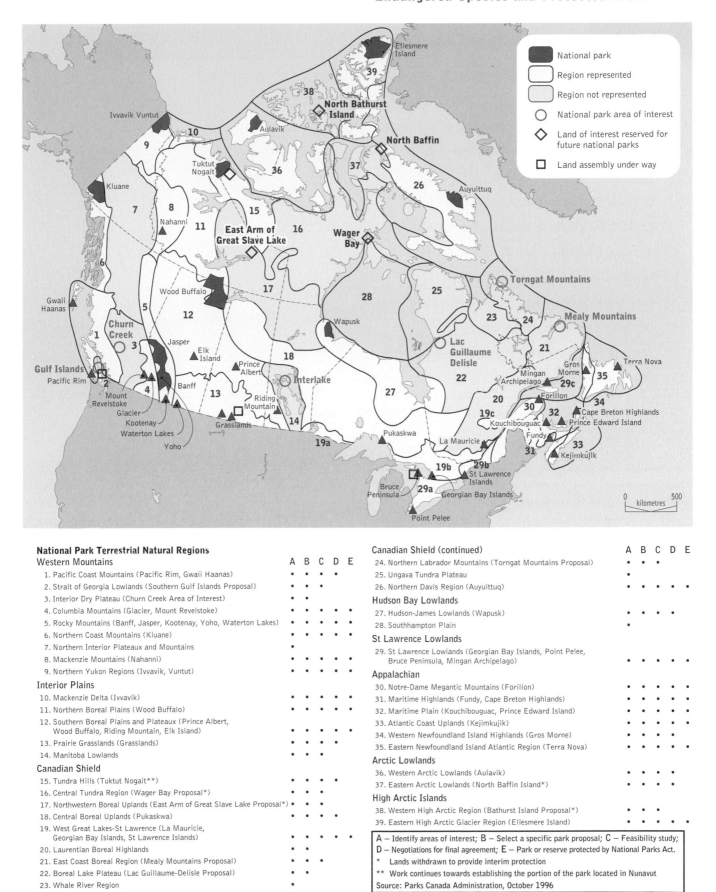

Legend	
■	National park
▢	Region represented
▢	Region not represented
○	National park area of interest
◇	Land of interest reserved for future national parks
☐	Land assembly under way

National Park Terrestrial Natural Regions

Western Mountains	A	B	C	D	E
1. Pacific Coast Mountains (Pacific Rim, Gwaii Haanas)	•	•	•	•	
2. Strait of Georgia Lowlands (Southern Gulf Islands Proposal)	•	•	•		
3. Interior Dry Plateau (Churn Creek Area of Interest)	•	•			
4. Columbia Mountains (Glacier, Mount Revelstoke)	•	•	•	•	•
5. Rocky Mountains (Banff, Jasper, Kootenay, Yoho, Waterton Lakes)	•	•	•	•	•
6. Northern Coast Mountains (Kluane)	•	•	•	•	•
7. Northern Interior Plateaux and Mountains	•				
8. Mackenzie Mountains (Nahanni)	•	•	•	•	•
9. Northern Yukon Regions (Ivvavik, Vuntut)	•	•	•	•	•

Interior Plains					
10. Mackenzie Delta (Ivvavik)	•	•	•	•	•
11. Northern Boreal Plains (Wood Buffalo)	•	•	•	•	•
12. Southern Boreal Plains and Plateaux (Prince Albert, Wood Buffalo, Riding Mountain, Elk Island)	•	•	•	•	•
13. Prairie Grasslands (Grasslands)	•	•	•	•	•
14. Manitoba Lowlands	•	•	•		

Canadian Shield					
15. Tundra Hills (Tuktut Nogait**)	•	•	•	•	
16. Central Tundra Region (Wager Bay Proposal*)	•	•	•		
17. Northwestern Boreal Uplands (East Arm of Great Slave Lake Proposal*)	•	•	•		
18. Central Boreal Uplands (Pukaskwa)	•	•	•	•	
19. West Great Lakes-St Lawrence (La Mauricie, Georgian Bay Islands, St Lawrence Islands)	•	•	•	•	•
20. Laurentian Boreal Highlands	•	•			
21. East Coast Boreal Region (Mealy Mountains Proposal)	•	•	•		
22. Boreal Lake Plateau (Lac Guillaume-Delisle Proposal)	•	•			
23. Whale River Region	•				

Canadian Shield (continued)	A	B	C	D	E
24. Northern Labrador Mountains (Torngat Mountains Proposal)	•	•	•		
25. Ungava Tundra Plateau	•				
26. Northern Davis Region (Auyuittuq)	•	•	•	•	•

Hudson Bay Lowlands					
27. Hudson-James Lowlands (Wapusk)	•	•	•	•	•
28. Southhampton Plain	•				

St Lawrence Lowlands					
29. St Lawrence Lowlands (Georgian Bay Islands, Point Pelee, Bruce Peninsula, Mingan Archipelago)	•	•	•	•	•

Appalachian					
30. Notre-Dame Megantic Mountains (Forillon)	•	•	•	•	•
31. Maritime Highlands (Fundy, Cape Breton Highlands)	•	•	•	•	•
32. Maritime Plain (Kouchibouguac, Prince Edward Island)	•	•	•	•	•
33. Atlantic Coast Uplands (Kejimkujik)	•	•	•	•	•
34. Western Newfoundland Island Highlands (Gros Morne)	•	•	•	•	•
35. Eastern Newfoundland Island Atlantic Region (Terra Nova)	•	•	•	•	•

Arctic Lowlands					
36. Western Arctic Lowlands (Aulavik)	•	•	•	•	•
37. Eastern Arctic Lowlands (North Baffin Island*)	•	•	•		

High Arctic Islands					
38. Western High Arctic Region (Bathurst Island Proposal*)	•	•	•	•	
39. Eastern High Arctic Glacier Region (Ellesmere Island)	•	•	•	•	•

A – Identify areas of interest; B – Select a specific park proposal; C – Feasibility study; D – Negotiations for final agreement; E – Park or reserve protected by National Parks Act.

*   Lands withdrawn to provide interim protection

** Work continues towards establishing the portion of the park located in Nunavut

Source: Parks Canada Administration, October 1996

Figure 11.6 Status of planning for natural regions. SOURCE: Parks Canada Agency (n.d.).

Table 11.7   World Heritage Sites in Canada

1. Kluane/Wrangell–St Elias/Glacier Bay/Tatshenshini-Alsek (Yukon and British Columbia)

2. SGaang Gwaii (Anthony Island, British Columbia)

3. Nahanni National Park Reserve (Northwest Territories)

4. L'Anse aux Meadows National Historic Site (Newfoundland and Labrador)

5. Head-Smashed-In Buffalo Jump (Alberta)

6. Dinosaur Provincial Park (Alberta)

7. Wood Buffalo National Park (Alberta and Northwest Territories)

8. Canadian Rocky Mountain Parks (Alberta and British Columbia)

9. Historic District of Québec (Quebec)

10. Gros Morne National Park (Newfoundland and Labrador)

11. Old Town Lunenburg (Nova Scotia)

12. Waterton Glacier International Peace Park (Alberta)

13. Miguasha Park (Quebec)

SOURCE: <http://parkscanada.pch.gc.ca/progs/spm-whs/itm2-/index_e.asp>.

cent for the rest of the country. The amount of shopping space per person in Banff is three times larger than it is in Toronto. The development in Banff is greater than all the development in Yellowstone, Grand Canyon, Yosemite, and Great Smoky Mountain National Parks in the US, combined. Banff National Park is also the core of the four mountain parks (Banff, Yoho, Jasper, and Kootenay) that were accorded World Heritage status in 1985 due to their importance for wildlife protection on a continental basis.

Gros Morne National Park in western Newfoundland is a World Heritage site (*Philip Dearden*).

Although the park itself is large, most of it is rock and ice; the town of Banff is situated in one of the most productive and rare habitats throughout the Rocky Mountains. The 'montane' zone, which covers between 2 and 5 per cent of the park, is characterized by frequent chinook winds, low snow accumulation, and warm winter temperatures. These conditions permit a dense concentration of wildlife and critical winter refuge areas.

The montane zone is also very attractive for development purposes. Over 70 per cent of the zone has already been developed into highways, golf courses, towns, and resorts, and more develop-

I do not suppose in any portion of the world there can be found a spot, taken all together, which combines so many attractions and which promises in as great a degree not only large pecuniary advantage to the Dominion, but much prestige to the whole country by attracting the population, not only on this continent, but of Europe to this place. It has all the qualifications necessary to make it a place of great resort.... There is beautiful scenery, there are curative properties of the water, there is a genial climate, there is prairie sport, and there is mountain sport; and I have no doubt that it will become a great watering-place.

— *Sir John A. Macdonald on Banff, 1887*

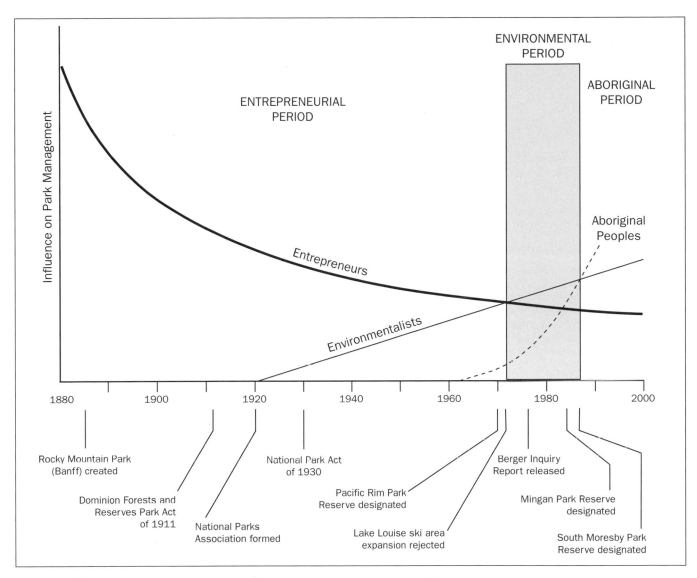

Figure 11.7 Administrative penetration model. SOURCE: Dearden and Berg (1993: 198).

ment is planned. Within Banff National Park, the four-kilometre-wide Bow Valley contains the Trans-Canada Highway, the 1A highway, a national railway, an airstrip, a 27-hole golf course, three ski resorts, the village of Lake Louise, and the town of Banff with a population 7,500! Between 1989 and 1994, 176 animals were killed in the park on the railway tracks alone and the annual toll on the roads was over 200 dead. These roadkill figures were obviously unacceptable in a national park and management responded by erecting fences along 44 km of road and directing animals to a series of underpasses and overpasses. The deer and elk adapted to these immediately; it took the carnivores a little longer. There now have been more than 3,000 passes by wolves, nearly a thousand crossings by cougars, and 150 by grizzly bears. Overall, it has been estimated that roadkill has been

reduced by more than 80 per cent through these measures (Clevenger and Waltho, 2000).

Banff is also of major economic importance. Over three and a half million people visit every year, and Banff tourism operators employ thousands of workers and generate hundreds of millions of dollars in annual revenue. Tour operators argue that they have to compete on an international basis and that if further developments are not permitted, tourism will dry up.

In response to these pressures, the Minister of Canadian Heritage created the Banff-Bow Valley Task Force in 1994 to advise on the management of Banff and the surrounding areas of the Bow Valley. Swinnerton (2002) provides a detailed overview of the Task Force, including the stakeholder participation process discussed in Chapter 6. The Task Force found that grizzly bear populations

Although the environmental lobby was successful in its fight against the construction of an upper and lower village at Lake Louise funded by Imperial Oil and supported by Parks Canada about 20 years ago, many would argue that the subsequent incremental developments have achieved almost the same result. This photograph shows the enlarged Château Lake Louise in front of what is advertised as the largest ski hill in Canada. Is this a national park landscape? (*Philip Dearden*)

were declining rapidly and that aquatic ecosystems were compromised due to exotic introductions and dams. The Task Force predicted that current rates of growth would cause 'serious, and irreversible, harm to Banff National Park's ecological integrity' (Banff-Bow Valley Study, 1996: 4).

The federal minister committed to implement the recommendations of the Task Force and in 1998 created another inquiry—the Ecological Integrity Panel—to look at similar issues in all of Canada's national parks. The Ecological Integrity Panel's report concurred with the earlier study on Banff and made strong recommendations for a more adaptive approach to park management, with greater attention to ecosystem-based management and greater consultation with stakeholders (Chapter 6). The Panel prepared 127 recommendations and delivered one central message: *ecological integrity in all the national parks is in peril*. The minister accepted the Panel's findings and started to implement the recommendations. The proclamation of a new National Parks Act in 2000, which further strengthens the ecological mandate of the parks, was one response to the Panel's recommendations.

Over time, there has been a progressive change regarding attitudes towards development in Canada's parks. Initially, development for tourism and recreation was encouraged. Parks were created for the enjoyment of Canadians and protection of wildlife was secondary—Canada still had large

intact wilderness areas, and so there seemed little need to protect them. There was, however, a need to generate revenue, and parks were regarded as playing an important role in generating income out of wilderness. But as the Canadian landscape became increasingly developed, this perception changed. Particularly in southern Ontario, parks became one of few areas that were not developed.

Research has quite clearly shown that development within parks is detrimental to many species (Box 11.11). Fortunately, the federal government has acted accordingly. Over the last 15 years, the government has revised the National Parks Act twice in favour of a mandate that favours wildlife protection over recreational opportunities. But while these changes are positive, there has been a growing realization that development is not the only threat to ecological integrity within our parks. Many management challenges arise from threats originating beyond park boundaries.

Wildlife overpass built over the Trans-Canada Highway in Banff National Park (*Philip Dearden*).

External Threats

Parks do not exist in isolation—they are intimately linked to surrounding and global ecosystems. It is therefore necessary to be aware of any influences from outside park boundaries that may have a detrimental impact on wildlife resources within park boundaries (Box 11.12). This awareness is relatively recent (Figure 11.8). In the early days of parks, boundaries were easily penetrated as society adjusted to the idea of preserving nature. But as time passed, park boundaries became less permeable and protection of wildlife resources was more assured. However, as development surrounding parks intensified, park managers began to realize that development pressures outside park boundaries were affecting resources within parks. In response, park managers began to develop integrated management plans that took external threats into account.

Although the creation of integrated management plans is a step in the right direction, external threats to parks are often difficult to eliminate or even control. For example, external threats may originate from private landowners surrounding park boundaries, and it may be difficult or impossible to restrict activities on private land. External threats also come in all shapes and sizes—to tackle them all would require significant human and economic resources. Examples of external threats include mining, logging, agriculture, urbanization, water projects, hunting, exotic species, tourism, acid precipitation, and chemical pollution.

The 1997 *State of the Parks Report* indicated that many Canadian parks are showing signs of stress from these external sources. The stress survey showed that 53 per cent of parks (or 19 out of 36) reported stresses originating from park management practices, while 72 per cent reported stresses from visitor/tourism facilities. Other external stresses included forestry (50 per cent), agriculture (47 per cent), mining (42 per cent), and sport hunting (31 per cent).

Exotic vegetation (Chapter 3) was also a significant concern identified by the *Report*. According to Woodley (2002), invasive species may now represent 50 per cent of the flora in some parks. Purple loosestrife, for example, has invaded Point Pelee National Park, and knapweed has invaded Elk Island National Park. Utility corridors, which cross 19 of 36 parks, can also have significant impacts where major transportation

Building greater awareness among visitors of the role of national parks in society is a central facet of sound management. Parks Canada has developed some of the best interpretive facilities in the world. This is the visitor centre at Greenwich in PEI National Park (*Philip Dearden*).

routes cross animal migration routes, as the Banff example illustrates.

Sometimes external threats can be readily identified and even managed, such as timber cutting along a park boundary; other times, the influences are too distant and diffuse for park managers to influence, such as the effects of global climate change (see Chapter 7). Suffling and Scott (2002) have used climate change models to indicate some of the possible effects on the national park system over the next century. The researchers conclude that all regions and parks will be dramatically affected, and suggest that Parks Canada is going to have to interfere increasingly in park ecosystems to maximize the capacities of ecosystems and species to adapt to climate change.

The idea of park managers actively interfering in park ecosystems rather than leaving change to the vagaries of nature is known as **active management**. It recognizes that the human forces of change are now so ubiquitous throughout the landscape that even parks are affected. Active management activities include habitat restoration, creation of wildlife corridors, reintroduction of extirpated species (Table 11.5), prescribed burning, and management of hyper-abundant species, such as culling white-tailed deer populations in Point Pelee National Park in Ontario. However, due to a lack of knowledge of ecosystem processes there is

BOX 11.11 THE BANFF LONGNOSE DACE: EXTINCTION IN A NATIONAL PARK

The Banff longnose dace was a small minnow found in only one location on the planet, a marsh downstream from the famous Cave and Basin hot springs in Banff National Park. Discovered in 1892, the minnow was declared extinct by COSEWIC in 1987. As has been the case with many extinctions, a combination of factors likely led to the animal's demise. Tropical fish were introduced into the warm waters of the marsh, where they flourished and were able to out-compete the dace. The dace may also have interbred with other species of dace, and collecting for scientific purposes may have also reduced numbers. However, it seems that the most important factor in its extinction was the decision to allow a hotel to chlorinate hot spring water and discharge it into the marsh. The chlorine reacted to form chlorinated hydrocarbons, which are very toxic to fish, even at low concentrations. Shortly thereafter, the Banff longnose dace disappeared forever, and so it was pressure from tourism that ultimately led to Banff's first extinction.

tion, leading to genetic inbreeding and a higher susceptibility to extinction. This raises two questions. (1) How many individuals are necessary to ensure the long-term survival of a species? (2) How large an area of habitat is required to sustain the population? The first question, related to the **minimum viable population (MVP)** of a species (i.e., the smallest population size that can be predicted to have a very high chance of persisting for the foreseeable future), can be estimated using genetic and demographic models. Estimates of MVP are then multiplied by the area required to support each animal. In western Canada, for example, calculations suggest that 15,000 km^2 would be required to support a viable wolf population. A study by Landry et al. (2001: 19) concludes that 'the small size, high visitation rates, and ecological isolation of southern parks mitigate against minimum viable populations of wolves, black bears and grizzly bears being maintained there ... most of Canada's national parks cannot indefinitely sustain [MVPs] of these large carnivorous mammals.' Similar conclusions have also been reached regarding large herbivores in many parks. Flanagan and Rasheed (2002), for example, used two different models and both models indicated that caribou in Jasper National Park will go extinct within the next 40 years if conditions do not change.

These concerns have made it obvious that most of our parks are too small, too few, and too far apart to be able to sustain populations of many species throughout the next century. More attention is now being directed towards ways of linking the parks through corridors of natural habitat. One such scheme, for example, seeks to link American parks such as Yellowstone north through the Canadian Rockies up into Yukon and Alaska. In fact, one wolf that was marked for tracking in Montana was actually shot along this corridor on the Alaska Highway. These schemes explicitly acknowledge the limitations of our park systems and encourage a more integrated perspective of resource management on lands outside the parks that involves other actors such as private landowners and private foundations.

This approach is often called **stewardship** and refers, in general, to many different activities that can be undertaken towards caring for the Earth. In the context of protected areas it generally means encouraging landowners to modify their activities in ways that will help protect ecosystems. In practice, stewardship takes many forms. It includes:

often considerable debate among scientists as to how such programs should be implemented.

Effective influence over threats that originate from outside parks requires an ecosystem-based management approach (Chapter 6), combined with methods that protect wildlife resources along ecosystem rather than legal/political boundaries. Such approaches attempt to mitigate external threats while also counteracting the forces of **fragmentation**.

Fragmentation

Parks are becoming islands of natural vegetation totally surrounded by human-modified landscapes, as illustrated by the Riding Mountain case discussed in Box 11.13. Studies of Fundy National Park in New Brunswick showed that only 20 per cent of the surrounding area remained in forest patches large enough to be 500 metres from disturbed areas. This situation creates several problems since many animal species and some bird species cannot cross modified landscapes. As a result, they become an isolated breeding popula-

BOX 11.12 THE ROLE OF PARKS IN ENDANGERED SPECIES PROTECTION: WOOD BUFFALO NATIONAL PARK

Straddling the Alberta/Northwest Territories boundary, Wood Buffalo covers 44,807 km². The park is recognized both as a World Heritage site and as a Ramsar site. Ramsar sites are wetlands of global significance, and the Ramsar Convention on Wetlands is named after Ramsar, Iran, where it was signed in 1971. Wood Buffalo contains critical habitat for two endangered species: North America's largest terrestrial mammal, the bison, and the tallest bird, the whooping crane.

We came to places where, as far as the eye could see, untold thousands were in sight; the country being fairly black with them...these immense herds were moving north and there seemed no end to them.
— *Cecil Denny in Saskatchewan, 1874*

Bison: The image of vast herds of bison ranging back and forth along the Great Plains of North America is one that will never be seen again. With up to 60 million animals, bison herds probably constituted the greatest large-mammal congregations that ever existed on Earth and were important in the subsistence lifestyles of many Aboriginal peoples in western Canada. But by the 1860s, the bison had been extirpated from the plains of Manitoba. Trainloads of meat and hides were sent back east from the slaughter. As American Indians flooded into Canada to seek protection of the 'Great White Mother' (Queen Victoria), the pressure on the remaining herds increased dramatically, and the wild bison herds were extirpated from the Canadian Prairies.

All through today's journey, piled up at the leading stations along the road, were vast heaps of bones of the earliest owners of the prairie—the buffalo. Giant heads and ribs and thigh bones, without one pick of meat on them, clean as a well washed plate, white as driven snow, there they lay, a giant sacrifice on the altar of trade and civilization.
— *Traveller on CP Rail, 1888*

Several remnants remained, however. A small number had been protected by the earlier establishment of Yellowstone National Park. Yellowstone is the only place where wild, free-ranging plains bison have survived since colonial

times. Banff also had a growing population kept as a tourist attraction in an animal compound. In addition, two remnants had been brought together by an American rancher. The herd was bought by the Canadian government, and the 703 animals were transported to a national park (created for that purpose) adjacent to the railroads near Wainwright, Alberta. In the mid-1920s the herd, then numbering 6,673, was relocated to Wood Buffalo National Park.

These were the plains bison (*Bison bison*). Less well known are their non-migratory, taller, and darker cousins, the wood bison (*Bison bison athabascae*). These bison were at one time widely distributed through the aspen parklands of Saskatchewan and Alberta to the eastern slopes of the Rockies and British Columbia, and north to the coniferous forests of the Mackenzie Valley. The wood bison was never as abundant as the plains bison, and by 1891 fewer than 300 of them survived. Wood Buffalo National Park was established at least partially to protect this remnant of 300, and by 1922 the herd had grown to 1,500–2,000 animals. Shortly thereafter the herd of plains bison from Wainwright was imported and interbreeding led to the disappearance of the distinctive wood bison characteristics. Wood bison were believed to have become extinct.

In 1957, however, an isolated group of wood bison was located in a remote area of the vast park. This herd was relocated to guard against further interbreeding within Wood Buffalo. Some animals were removed to the Mackenzie Bison Sanctuary in the North; the herd now numbers over 200, and individuals have expanded their range outside the Sanctuary. Other animals were removed to Elk Island National Park near Edmonton, where, due to the small area available, their numbers have to be closely controlled. This herd has provided animals for satellite herds in Yukon, the Northwest Territories, northwestern Alberta, and Manitoba. In total, there are now 3,100 wood bison in the wild and 700 in captivity. In 1988 the wood bison was downlisted from endangered to threatened by COSEWIC. In May 2000, the species was reassessed; its status did not change.

The dangers for the bison are not yet over, however. Recovery efforts have re-established free-ranging wild herds in Canada along with four captive-bred herds, but only one population (Mackenzie bison herd) exceeds what is consid-

The vast herds of plains bison had been extirpated from Canada until efforts were made to reintroduce them from the US and eventually transport them to Wood Buffalo National Park, where they mixed with the wood bison population (*Philip Dearden*).

ered to be the minimum viable population size for wood bison (400 animals) (Environment Canada, 2003b). In addition, only six of the free-ranging herds (or 2,400 bison) are disease-free. When the plains bison were imported from Wainwright, they brought with them bovine diseases such as brucellosis and tuberculosis. The diseases have already taken a toll on bison populations—from highs of over 12,000 animals, by the early 1990s they had dropped to a quarter of this number. Diseases afflicting the bison have raised concerns from the agricultural sector. Bison represent the last focus for both diseases in Canada, and as agriculture has impinged upon the western boundary of the park, farmers are concerned that domestic stock will become infected. This has led to calls for elimination of the herd by the agricultural lobby, a move that was supported by a federal environmental assessment review in 1990. Environmentalists strenuously opposed this course of action.

In April 1995, the Canadian Heritage Minister announced that the herd would not be slaughtered. In addition, the minister introduced a management plan calling for more research on the impact of habitat change on bison ecology, a buffer zone between the diseased animals and the Mackenzie Bison Sanctuary to the north, and more investigation of the impact of the diseases on the park's ecosystem. These directions were followed but there is still considerable uncertainty, even among conservation scientists, about whether to proceed with 'depopulation' (slaughter) and then repopulate with disease-free calves.

Several factors besides the diseases are also threatening bison populations. The Peace-Athabasca delta, for example, supported the highest concentrations of bison. However, since the construction of the W.A.C. Bennett Dam upstream in British Columbia, water levels on the delta have fallen considerably, causing habitat changes that have negatively affected many animal species, including the bison. This impact from outside the park again emphasizes the need to take an ecosystem-based perspective on park management (Chapter 6).

Whooping Crane: Unlike the bison, whooping cranes (*Grus americana*) were never very numerous. Historical accounts suggest a population of 1,500. What they lacked in numbers they made up for in presence. Over 1.5 metres high and pure white except for black wing tips, black legs, and a red crown, with wing spans in excess of two metres, these majestic birds migrate annually from wintering grounds on the Gulf of Mexico coast of Texas to the Northwest Territories. These wintering grounds are all that remain of a breeding summer range that at one time stretched from New Jersey in the east to Salt Lake City in the west and as far north as the Mackenzie delta, and a winter range that included marshes from southern Louisiana into central Mexico (see figure). Requiring undisturbed breeding habitat, the cranes soon declined under the expansion of agriculture. Unrestricted hunting along their long migration routes also

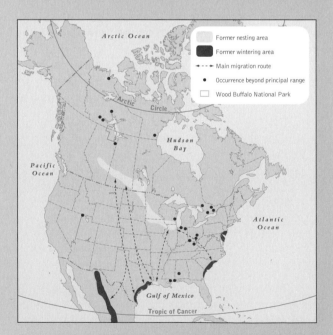

The original range of the whooping crane in recent times.

contributed. By 1941 there were only 22 whooping cranes left.

The governments of the US and Canada agreed to a joint program to try to save the species from extinction. The 1916 Migratory Bird Treaty between the US and Canada was used to stop legal hunting. In 1937, the US government bought the Aransas National Wildlife Refuge to protect the wintering habitat on the Gulf coast. In 1954, the only known nesting area was discovered, in the northern part of another protected area, Wood Buffalo. Finding the breeding grounds also allowed for direct human interventions, such as artificial incubation of eggs, to be attempted. Whooping cranes generally lay two eggs, but only one chick usually survives. A captive propagation program in the 1960s and 1970s moved one of the eggs for incubation. By the late 1990s, the world population of whooping cranes had risen to over 250, including captive populations, with over 40 breeding pairs in Wood Buffalo National

The whooping crane, the tallest North American bird and one of the rarest, makes its habitat in muskeg, prairie pools, and marshes (*Parks Canada/R.D. Muir*).

Park. The species is currently listed as endangered by COSEWIC.

- landowners voluntarily restricting damaging land use, planting native species rather than exotic ones, and placing protective covenants on their land;
- community members contributing to wildlife monitoring programs, providing passive education for tourists and visitors, and participating in collective restoration;
- park visitors voluntarily choosing to avoid hikes along sensitive trails, or participating in park host programs;
- corporations introducing sustainable land practices that reduce damage to wildlife habitat.

Stewardship initiatives also include the activities of NGOs that acquire land for conservation purposes outside the government park systems. The Nature Conservancy of Canada (NCC)—one of many conservation organizations—has protected approximately 1.6 million hectares since 1962. In 1999, Ducks Unlimited Canada secured almost 80,000 hectares of wetland and enhanced over 80,000 hectares. In total, Ducks Unlimited has been responsible for protecting over seven million hectares of Canadian wetlands since 1938. Smaller, provincially based and local land trusts are also increasing in number. In the southern Vancouver Island–Gulf Islands area, where natural heritage is fading quickly and land prices are increasing, at

least eight land trusts are operating to protect endangered spaces, most of which have been established since 1998. Government agencies like Environment Canada and various provincial ministries are also embracing stewardship. For example, numerous funding programs exist for community stewardship, and some ministries publish guides or maintain websites (e.g., see Environment Canada's 'A Guide to Conservation Programs and Funding Sources for Agro-Manitoba') to educate and to support local initiatives. Legislative changes, particularly tax deductions, have also recently been passed, resulting in incentives and encouragement for ecological gifts and donations.

These initiatives are critically important for the future of conservation in Canada. They will never replace the role played by strictly protected areas, but they do play an essential role in 'gluing together' the larger wilderness areas set aside in government parks. The international Biosphere Reserve Program is one of the best-known initiatives promoting greater stewardship surrounding protected areas (Box 11.13).

Many visitors to our park system are unaware of the challenges faced by park managers. One of the reasons is that although most park systems in Canada have nature interpretation services (e.g., visitor centres and guided walks), park administrators are generally uncomfortable with interpreta-

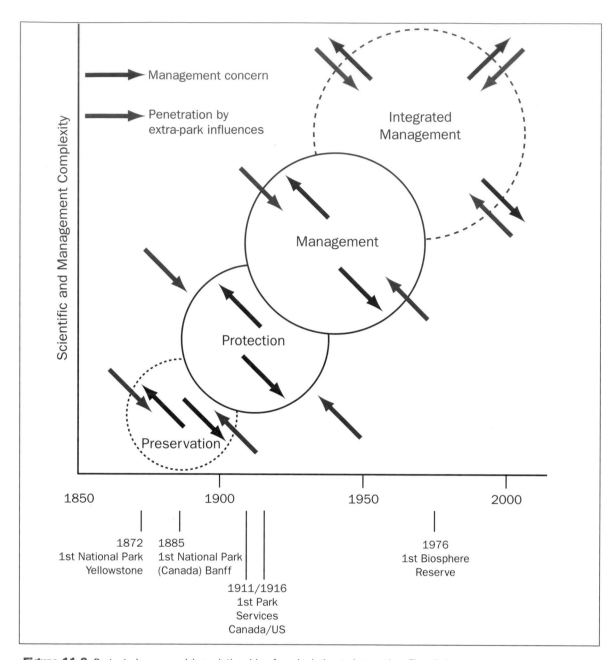

Figure 11.8 Protected area: evolving relationships from isolation to integration. The circles represent the growing size of the protected area system over time. Boundaries (circle circumferences) were initially of little importance, but assumed greater significance in the protection and management phases. It is now realized that for park management (arrows) to be effective, it must also pay equal attention to environmental changes outside park boundaries. SOURCE: Dearden and Rollins (1993: 10).

tion programs that focus on problems rather than on successes. Hence visitors learn only about the latter. Although this is understandable, it does not create a realistic picture for the Canadian public, nor does it emphasize the need for greater stewardship activities outside the parks (see Box 11.14).

Stakeholder Interests

Balancing the interests of the variety of stakeholders that may be affected by the establishment and/or management of parks is a formidable chal-

lenge. Private landowners, local communities, Aboriginal peoples, industry, tourists, conservation organizations, and government agencies all have an influence over how parks are managed.

Aboriginal peoples have a particularly powerful role in the designation and management of many parks (Figure 11.7). An amendment to the National Parks Act in 1972 created a special category of park, the *national park reserve*, which does not prejudice future land claim negotiations. Several large parks have been created in the Arctic

Logging on the border of Pacific Rim National Park Reserve shows both the impacts of activities outside park boundaries as well as the fragmentation of habitat that follows. Many parks are now islands of natural habitat with no connection to natural habitat elsewhere (*Philip Dearden*).

The totem poles of Ninstints, an abandoned Haida village on Gwaii Hanaas (Moresby Island in the Queen Charlotte Islands), give some impression of the Haida's spiritual connection with the environment. The village is now part of Gwaii Hanaas National Park Reserve and is co-managed by Parks Canada and the Haida (*Philip Dearden*).

as a result. In the south, on the other hand, where the provincial governments have jurisdiction over the land base, progress has been slow. It was not until after several court cases that the legitimacy of Native claims over land and resources in southern Canada were taken seriously. Now more than 50 per cent of the land area in Canada's national park system has been protected as a result of Aboriginal peoples' support for conservation of their lands, and this proportion will only increase in the future as Native land claims are settled.

Caribou from the porcupine caribou herd at Vuntut National Park in Yukon (*World Wildlife Fund Canada/Douglas Harvey*).

BOX 11.13   RIDING MOUNTAIN NATIONAL PARK AND BIOSPHERE RESERVES

Riding Mountain National Park is a good example of the kinds of external pressures that threaten the ecological integrity of many of our national parks. The park is located on the Manitoba Escarpment and is an isolated boreal forest area completely surrounded by agricultural land. Nonetheless, the 3,000 km^2 park provides habitat for 5,000 elk, 4,000 moose, over 1,000 black bears, and populations of cougars and wolves.

Large mammal populations have become increasingly threatened by the intensification of agricultural activities surrounding the park. Between 1971 and 1986 the amount of land under agriculture within 10 km of the boundary increased from 77 to 93 per cent. Not only is the amount of farmland increasing, so is the intensity of use, with a 42 per cent increase in cropland area in the same zone over the same time period. During the same 15-year period, the area of woodland declined by 63 per cent within a 70-km radius of the boundary, the volume of agricultural fertilizers used quintupled, and pesticide expenditures indicate an increase of 744 per cent in pesticide applications.

Agricultural expansion is not the only challenge confronting park wildlife. Until recently, bear-baiting was permitted directly on the park boundary. Farmers conditioned bears to feed from barrels full of meat. In the hunting season, the bears formed easy targets for 'sportsmen'. Some 70 'bear-feeding' stations existed around the park, causing unnatural bear distributions, very large bears, irregularities in breeding behaviour, and death. On average, 122 bears are killed each year in this manner. Scientists suggest that these mortality levels cannot be maintained if the bear population of the park is to survive. Farmers are now required to move bait barrels back from the park boundary itself. However, bears are highly mobile animals and will have little difficulty in locating these barrels.

These types of external pressures can have significant impacts on biodiversity within park boundaries. The Biosphere Reserve Program of the United Nations Educational, Scientific and Cultural Organization (UNESCO) is one of the most highly touted means of trying to deal with similar kinds of external threats to protected areas. The concept behind this program is sound. Biosphere reserves exist to represent global natural regions and should consist of a protected core area, such as a national park, surrounded by a zone of co-operation where socio-economic activities may take place, but which are modified to help protect the integrity of the core area. The reserves also have important educational and scientific roles. Unfortunately, there is no legislation to ensure co-operation on the privately owned lands in the zone of co-operation. Continued hunting around the park boundary at Riding Mountain National Park, designated as a biosphere reserve in 1986, is a graphic illustration of the dependency on local landowners to co-operate. Similar challenges face other biosphere reserves, such as Georgian Bay Islands National Park in Ontario, the Niagara Escarpment Biosphere Reserve in Ontario, the Greater Fundy Ecosystem in Nova Scotia, and Waterton Lakes National Park in southern Alberta, Canada's oldest biosphere reserve.

The goals of Aboriginal peoples and conservationists are not always identical, however. Aboriginal peoples, for example, retain the right to hunt in many parks. At the moment there are few restrictions on the size or means of harvest, a source of concern for some conservation scientists. In Pacific Rim National Park Reserve several proposals have been put forward by First Nations that are not consistent with the National Parks Act. One Native band, for example, opened a burger stand in the middle of the supposedly 'wilderness' West Coast Trail section of the park. Several parks in both northern and southern Canada have some form of co-management arrangement between Parks Canada and Aboriginal peoples. In general these have worked to the satisfaction of both groups, but the issue of shared responsibility remains an ongoing challenge for both sides.

IMPLICATIONS

There is no debate about whether rates of extinction have increased as a result of human activities over recent times. There is still some uncertainty, however, regarding the impact of this reduction. Is there ecological redundancy so that the Earth can

Guest Statement

Canada's Great Bear Rainforest
Ian McAllister

Environmental campaigns do not originate in boardrooms or around coffee tables. They begin when one person or a small group of people fall head-over-heels in love with a mountain, a river, an animal, or some combination of all three. You can trace the roots of any successful environmental campaign back to the simple emotion of love, and that's why there is such passion in the environmental movement today.

I fell in love with the central coast of British Columbia about 15 years ago. My father had chartered a sailboat to visit the Koeye River Valley, a mysterious estuary halfway up the BC coast where grizzly bears were rumoured to walk white-sand beaches and to fish in waters teeming with salmon. I had managed to squeeze aboard that boat and by the end of the trip I was hooked. It turned out the rumours had been true.

At that time, none of the river valleys on the coast had been extensively explored or protected by legislation, yet the majority had already been licensed out to logging companies. In other words, British Columbia didn't know what it was about to lose! Six months after that fateful voyage, my wife Karen and I helped co-found the Raincoast Conservation Society.

Now, 15 years later, the campaign to save the Great Bear Rainforest (GBR) is one of the largest environmental movements in Canadian history. It has had a huge impact on the international marketplace for timber and has become a keystone example of how a group of normal individuals can combine their strengths to bring about change. From scientists to celebrities, from letter-writers to lawyers, the Raincoast family includes people from all walks of life joined together by a mutual love of nature and the commitment to help preserve it. I guess it makes sense that we are such a diverse group. After all, diversity is what we strive to protect.

This year, 22 per cent of the GBR is poised for legislated protection with an additional 10 per cent designated off limits to logging. Scientists continue to warn that more needs to be protected to ensure survival over time of all species, such as Spirit bears (an endemic 'white' black bear), grizzly bears, salmon, and wolves.

Ian McAllister

However, this is a huge departure from the 12 per cent cap that previous governments have put on environmental protection in British Columbia—and maybe the greatest achievement of the campaign to protect this wildlife-rich ecosystem to date.

There's no question it's been a long time coming, but there's also no doubt it's been worth it. Over the years we have been called heroes and we've been called traitors (we've even been called colonialists for renaming the coast the Great Bear Rainforest!). But now, governments and businesses respect our voice and recognize the strength in numbers behind it. And these numbers continue to grow.

Environmental work is one of the most exciting and creative occupations I can think of. It is a constant challenge filled with unrivalled spiritual and physical rewards. Perseverance is a prerequisite, as is a sensitivity to other people's points of view, and the positive result of this is that environmentalism attracts fun-loving, hard-working, passionate individuals who believe very strongly in the value of teamwork.

I didn't know what I was getting into when I hopped aboard that boat 15 years ago. But that's the best thing about environmental work: you can learn everything you need to know along the way and I wouldn't trade it for anything.

Ian McAllister is a co-founder of the Raincoast Conservation Society: <www.raincoast.org>.

afford to lose some species without major impacts? Is the loss of any species as a result of human activities ethically and morally unacceptable? There are many unanswered questions regarding the implications of reduced biodiversity.

Extinction is commonly viewed as simply a biological problem. Yet, there is a need to link the biological process of extinction with the economic and ethical reasons why extinction is undesirable, the reasons why human-caused extinctions are increasing, and the kinds of measures that have to be taken to prevent this from happening. In other words, extinction is not just the domain of biologists but involves consideration from a broad range of perspectives, including all the social sciences, geography, law, and ethics. These interactions, from a Canadian perspective, are emphasized in Beazley and Boardman's *Politics of the Wild: Canada and Endangered Species* (2001) and in Coward, Ommer, and Pitcher's *Just Fish: Ethics and Canadian Marine Fisheries* (2000).

A Biosphere Reserve surrounding Waterton Lakes National Park has been set aside to allow for joint decision-making between park authorities and local landowners. The agreement has been particularly useful in reducing conflicts between grizzly bears and local ranchers (*Philip Dearden*).

BOX 11.14 WHAT YOU CAN DO FOR PROTECTED AREAS

1. Visit parks and other protected areas often throughout the year. Enjoy yourself. Tell others that you have enjoyed yourself, and encourage them to visit.
2. Always follow park regulations regarding use. Feeding wildlife, for example, may seem kind or harmless, but it can lead to the death of the animal.
3. If you have questions regarding the park's management or features, do not be afraid to ask. A questioning public is a concerned public.
4. Many park agencies have public consultation strategies relating to topics ranging from park

policy to the management of individual parks. Let them know your interests so you can be placed on the mailing list to receive more information.
5. Join a non-governmental organization, such as the Canadian Parks and Wilderness Society or the Canadian Nature Federation, which has a strong interest in parks issues.
6. Many parks now have co-operating associations in which volunteers can help with various tasks. Find out if a park near you has such an organization.
7. Write to politicians to let them know of your park-related concerns.

SUMMARY

1. Extinction levels have reached unprecedented levels. There are several reasons why we should be concerned. Life-supporting ecosystem processes depend on ecosystem components. As we lose components through extinction, these processes become more impaired. We also derive many useful and valuable products from natural biota, including medicines. In addition to these utilitarian reasons, there are ethical and moral reasons why we should be concerned about species extinction.

2. Many factors are behind current declines. The underlying factor is human demand as population and consumption levels grow. Much attention has concentrated on the tropics due to the high biodiversity levels and high rates of destruction there. However, Canada has also experienced 12 extinctions and 21 extirpations since European colonization.

3. Some of the main pressures causing extinction include overharvesting, predator control, and habitat change. Habitat change includes not only physical changes (e.g., conversion of habitat into agricultural land) but also those caused by chemicals and the introduction of alien species.

4. Not all species are equally vulnerable to extinction. Species that have specialized habitat requirements, migratory species, species with insular and local distributions, valuable species, animal species with a large body size, species that need a large home range, species that are not effective dispersers, and species with low reproductive potential tend to be the most vulnerable.

5. Canada is party to several international treaties for the protection of wildlife and was the first industrialized nation to sign the Biodiversity Convention at the Earth Summit in 1992. As a signatory to the Convention, Canada was required to introduce legislation to protect endangered species. In 2002, the federal government passed the Species at Risk Act (SARA).

6. The Committee on the Status of Endangered Wildlife in Canada (COSEWIC) is responsible for determining the status of rare species and categorizing them as extinct, extirpated, endangered, threatened, and vulnerable. As of May 2003, 431 species have been classified. The Committee's assessment is the first step in the process for protecting a proposed species at risk under SARA.

7. For species listed under SARA, recovery and management plans must be developed and implemented, unless the minister responsible feels that recovery is not 'feasible'.

8. Protected areas have emerged as one of the key strategies to combat the erosion of biodiversity both internationally and in Canada. Protected areas fulfill many roles in society, including species and ecosystem protection, maintenance of ecological processes, and as places for recreation and spiritual renewal, aesthetic appreciation, tourism, and science and education in natural outdoor settings.

9. There are many different kinds of protected areas in Canada, including national and provincial parks, wilderness areas, First Nations parks, wildlife refuges, ecological reserves, and regional and municipal parks. The amount of protection given to ecosystem components varies among these different types.

10. National parks are outstanding natural areas protected by the federal government because of their ecological importance and aesthetic significance. There are 39 national parks in Canada. The goal is to have at least one national park in each of the 39 regions of the national system plan. At present, only 25 of these regions are represented. In 2002, the Prime Minister announced a five-year plan that would establish 10 new parks, increasing representation to 35 of the 39 regions.

11. Banff, the first national park in Canada, was protected in 1885. Since that time the national parks have fulfilled a dual mandate that required protection of park resources in an unimpaired state, but also required their use. This conflicting mandate was first clarified in a policy statement in 1979, then enshrined in an amendment to the National Parks Act in 1988 and further clarified in the current policy document of 2000. All of these make it clear that the first and overriding priority of Canada's national parks is to protect the natural environment.

12. Current management challenges to the national parks system include external threats and fragmentation, which require more of an ecosystem approach. The involvement of First Nations peoples also represents a broader view of the stakeholders for many national parks. Banff National Park demonstrates some of the difficulties encountered in trying to control tourism development.

KEY TERMS

active management	Convention on Biological	extirpated	protected areas
Convention on	Diversity	extrinsic values	Red List
International Trade in	ecologically extinct	fragmentation	Species at Risk Act
Endangered Species of	endangered	in situ preservation	(SARA)
Wild Fauna and Flora	ex situ preservation	intrinsic value	stewardship
(CITES)	extinction	minimum viable	threatened
		population (MVP)	vulnerable

REVIEW QUESTIONS

1. What are the main reasons we should be concerned about species extinctions?
2. Why are some species more vulnerable to extinction than others?
3. What is being done to protect endangered species in your province?
4. What are some of the strengths and weaknesses of Canada's Species at Risk Act?
5. What do you think should be the relative importance of the various roles played by protected areas?

6. What different classifications of protected areas exist in your province and what kinds of protection are offered by these different systems?
7. What is your province doing to achieve the 12 per cent protected area that all jurisdictions in Canada have committed to establishing?
8. What ecozone is Banff townsite situated in, and what is the importance of this zone?

RELATED WEBSITES

BIODIVERSITY PROJECT:
http://www.biodiversityproject.org

CANADIAN PARKS AND WILDERNESS SOCIETY:
http://www.cpaws.ca

CANADIAN NATURE FEDERATION:
http://www.cnf.ca

NATURE SERVE:
http://www.natureserve.ca

NATURE WATCH:
http://www.naturewatch.ca/english

SIERRA YOUTH COALITION:
http://www.syc-cjs.grg

WESTERN CANADA WILDERNESS COMMITTEE:
http://www.wildernesscommittee.org

WILD CANADA:
http://www.wildcanada.net

WORLD COMMISSION ON PROTECTED AREAS (IUCN):
http://www.iucn.org/themes/wcpa

WORLD WILDLIFE FUND CANADA:
http://www.wwf.ca

REFERENCES AND SUGGESTED READING

Abbey, E. 1968. *Desert Solitaire: A Season in the Wilderness.* New York: Simon & Schuster.

Banff-Bow Valley Study. 1996. *Banff-Bow Valley: At the Crossroads. Summary Report of the Banff-Bow Valley Task Force,* eds R. Page, S. Bayley, J.D. Cook, J.E. Green, and J.R.B. Ritchie. Prepared for the Honourable Sheila Copps, Minister of Canadian Heritage. Ottawa: Ministry of Canadian Heritage.

Beazley, K. 2001. 'Why should we protect endangered species? Philosophical and ecological rationale', in Beazley and Boardman (2001: 11–25).

———— and R. Boardman, eds. 2001. *Politics of the Wild: Canada and Endangered Species.* Toronto: Oxford University Press.

Bondrup-Nielsen, S., N.W.P. Munro, G. Nelson, J.H.M. Willison, T.B. Herman, and P. Eagles, eds. 2002. *Managing Protected Areas in a Changing World: Proceedings of the Fourth International Conference on Science and Management of Protected Areas.* Wolfville, NS: Acadia University, SAMPAA.

Browne, C.L., and S.J. Hecnar. 2003. 'Dwindling turtle populations', *National Park International Bulletin* no. 9 (May).

Bruner, A.G., R.E. Gullison, R.E. Price, and G.A.B. da Fonseca. 2001. 'Effectiveness of parks in protecting tropical biodiversity', *Science* 91, 5501: 125–8.

Canada. 2000. 'National Park Act'. At: <http://laws.justice.gc.ca/en/N-14.01/>.

Canadian Wildlife Service Waterfowl Committee. 2002. *Population Status of Migratory Game Birds in Canada.* CWS Migratory Birds Regulatory Report Number 7. At: <http://www.cws-scf.ec.gc.ca/publications/status/nov03/nov03_e.pdf>.

Chape, S., S. Blyth, L. Fish, P. Fox, and M. Spalding, compilers. 2003. *2003 United Nations List of Protected Areas.* Gland, Switzerland, and Cambridge: IUCN and UNEP-WCMC.

Clevenger, A.P., and N. Waltho. 2000. 'Factors influencing the effectiveness of wildlife underpasses in Banff National Park, Alberta, Canada', *Conservation Biology* 14: 47–56.

Committee on the Status of Endangered Wildlife in Canada (COSEWIC). 2003. *Canadian Species at Risk.* At: <http://www.cosewic.gc.ca>.

Coward, H., R. Ommer, and T. Pitcher, eds. 2000. *Just Fish: Ethics and Canadian Marine Fisheries.* St John's: ISER Books.

Dearden, P. 1995. 'Park literacy and conservation', *Conservation Biology* 9: 1654–6.

————. 2001. 'Endangered species and terrestrial national parks', in Beazley and Boardman (2001: 75–93).

————. 2004. 'Parks and protected areas', in B. Mitchell, ed., *Resource and Environmental Management in Canada: Addressing Conflict and Uncertainty,* 3rd edn. Toronto: Oxford University Press, 314–41.

———— and L. Berg. 1993. 'Canada's national parks: a model of administrative penetration', *Canadian Geographer* 37: 194–211.

———— and R. Rollins, eds. 1993. *Parks and Protected Areas in Canada: Planning and Management.* Toronto: Oxford University Press.

———— and ————, eds. 2002. *Parks and Protected Areas in Canada: Planning and Management,* 2nd edn. Toronto: Oxford University Press.

Dempsey, J., P. Dearden, and J.G. Nelson. 2002. 'Stewardship: expanding ecosystem protection', in Dearden and Rollins (2002: 379–400).

Duffus, D.A., and P. Dearden. 1993. 'Recreational use, valuation and management of killer whales (*Orcinus orca*) on Canada's Pacific Coast', *Environmental Conservation* 20: 149–56.

Edwards, R., S. Brechtel, R. Bromley, D. Hjertaas, B. Johns, E. Kuyt, J. Lewis, N. Manners, R. Stardom, and G. Tarry. 1994. *National Recovery Plan for the Whooping Crane.* Ottawa: Recovery of Nationally Endangered Wildlife Committee.

Environment Canada. 2002. *Population Status of Migratory Game Birds in Canada.* CWS Migratory Birds Regulatory Report No. 7. Ottawa: Canadian Wildlife Service Waterfowl Committee.

————. 2003a. 'Protecting Plant Pollinators', *EnviroZine: Environment Canada's On-line Newsmagazine* 33 (26 June). At: <http://www.ec.gc.ca/EnviroZine>.

————. 2003b. 'Species at Risk: Wood Bison (*Bison bison athabascae*)'. At: <http://www.speciesatrisk.gc.ca/publications/renew>.

Environment Canada. 2003. *Environment Signals: Canada's National Environmental Indicator Series 2003.* Ottawa, Canada. Cat. En40-775/2002E. At: <http://www.ec.gc.ca/soer-ree/English/Indicator_series/default.cfm>.

Flanagan, K., and S. Rasheed. 2002. 'Population viability analysis applied to woodland caribou in Jasper National Park', *Research Links* 10: 16–18.

Global Environment Facility. 2002. 'Donor countries agree to the highest replenishment ever for the Global Environment Facility', news release, Washington, DC.

————. 2003. *Strategic Business Planning: Directions and Targets.* Document prepared for the meeting of the GEF Council, 14–16 May.

Government of Canada. 1995. *Canadian Biodiversity Strategy: Canada's Response to the Convention on Biological Diversity.* Ottawa: Minister of Supply and Services Canada.

Harfenist, A. 1994. *Effects of Introduced Rats on Nesting Seabirds of Haida Gwaii.* Technical Report Series no. 218. Vancouver: Canadian Wildlife Service, Pacific and Yukon Region.

Hayes, R.D., A.M. Baer, and D.G. Larsen. 1991. *Population Dynamics and Prey Relationships of an Exploited and Recovering Wolf Population in the Southern Yukon.* Whitehorse: Yukon Department of Renewable Resources.

Hobson, K.A., E.M. Bayne, and S.L. Van Wilgenburg. 2002. 'Large-scale conversion of forest to agriculture in the boreal plains of Saskatchewan', *Conservation Biology* 16, 6: 1530–41.

Hummel, M., ed. 2000. *Protecting Canada's Endangered Spaces: An Owner's Manual.* Toronto: Key Porter Books.

Kirk, D.A. 2003. 'Overview of raptor status and conservation in Canada', in J. Kennedy, ed., *Bird Trends: A Report on Results of National Ornithological Surveys in Canada.* Ottawa: Canadian Wildlife Service, 1–9.

Landry, M., V.G. Thomas, and T.D. Nudds. 2001. 'Sizes of Canadian National Parks and the viability of large mammal populations: Policy implications', *The George Wright Forum* 18, 1: 13–23.

Le Prestre, P.G., and P. Stoett. 2001. 'International initiatives, commitments, and disappointments: Canada, CITES, and the CBD', in Beazley and Boardman (2001: 190–216).

McNamee, K. 2002. 'From wild places to endangered spaces', in Dearden and Rollins (2002: 21–50).

Parks Canada Agency. 1994. *Guiding Principles and Operational Policies*. Ottawa: Minister of Supply and Services.

———. 1998. *State of the Parks, 1997 Report*. Ottawa: Minister of Supply and Services.

———. 2000. *Unimpaired for Future Generations? Protecting Ecological Integrity with Canada's National Parks*, vol. 2, *Setting a New Direction for Canada's National Parks*, Report of the Panel on the Ecological Integrity of Canada's National Parks. Ottawa: Minister of Public Works and Government Services.

———. 2003. *State of Protected Heritage Areas, 2002 Report*. Ottawa: Minister of Supply and Services.

———. n.d. *National Park System Plan*. Ottawa: Minister of Supply and Services.

Peepre, J., and P. Dearden. 2002. 'The role of Aboriginal peoples', in Dearden and Rollins (2002: 323–53).

Rivard, D.H., J. Poitevin, D. Plasse, M. Carleton, and D.J. Currie. 2000. 'Changing species richness and composition in Canadian national parks', *Conservation Biology* 14: 1099–1109.

Rowell, P., G.L. Holroyd, and U. Banasch. 2003. 'Summary of the 2000 Canadian Peregrine Falcon survey', in J. Kennedy, ed., *Bird Trends: A Report on Results of National Ornithological Surveys in Canada*. Ottawa: Canadian Wildlife Service, 52–6.

Searle, R. 2000. *Phantom Parks: The Struggle to Save Canada's National Parks*. Toronto: Key Porter Books.

Slocombe, D.S. 1998. 'Defining goals and criteria for ecosystem-based management', *Environmental Management* 22: 483–93.

Soule, M. 1996. 'The end of evolution?', *World Conservation* (Apr.): 8–9.

Statistics Canada. 2000. *The Importance of Nature to Canadians: The Economic Significance of Nature-Related Activities*. Ottawa.

Stewart, A., A. Harries, and C. Stewart. 2000. 'Waterton Biosphere Reserve landscape change study', in *Canada MAB 2000 Landscape Changes at Canada's Biosphere Reserves*. Toronto: Environment Canada, 13–20.

Suffling, R., and D. Scott. 2002. 'Assessment of climate change effects on Canada's National Park system', *Environmental Monitoring and Assessment* 74: 117–39.

Swinnerton, G.S. 2002. 'The Banff-Bow Valley: Balancing human use and ecological integrity within Banff National Park', in Dearden and Rollins (2002: 240–64).

Theberge, J.B., and D.A. Gauthier. 1985. 'Models of wolf-ungulate relationships: When is wolf control justified?', *Wildlife Society Bulletin* 13: 449–58.

Trant, D. 1993. 'Land use change around Riding Mountain National Park', in *Environmental Perspectives 1993: Studies and Statistics*. Ottawa: Statistics Canada, 33–46.

Warnock, R.G., and M.A. Skeel. 2004. 'Effectiveness of voluntary habitat stewardship in conserving grassland: case of Operation Burrowing Owl in Saskatchewan', *Environmental Management* (25 Mar.). At: <http://www.springerlink.com>.

Willcox, L., and P. Aengst. 1999. 'Yellowstone to Yukon: Romantic dream or realistic vision of the future?', *Parks* 9: 17–24.

Woodley, S. 2002. 'Planning and managing for ecological integrity', in Dearden and Rollins (2002: 97–114).

World Conservation Union (IUCN—International Union for the Conservation of Nature and Natural Resources). 2001. *Protected Area Management Categories*. At: <http://www.unep-wcmc.org/protected_areas/categories>.

———. 2002. *2002 IUCN Red List of Threatened Species*. At: <http://www.redlist.org>.

———. 2003. 'News Release: Release of the 2003 IUCN Red List of Threatened Species'. At: <http://www.iucn.org>.

Chapter 12

Water

Learning Objectives

- To appreciate the need for credible science to inform decision-making.
- To realize the water endowment in Canada.
- To understand the hydrological cycle.
- To know the environmental and social impacts associated with water diversions.
- To understand the significance of point and non-point sources of pollution.
- To learn the concept of 'water security'.
- To gain an understanding of the concept of a 'multiple barrier approach' to drinking water protection.
- To realize that water is both resource and hazard.
- To appreciate the difference between structural and non-structural approaches to flood damage reduction.
- To understand the significance of droughts.
- To appreciate the significance of heritage related to protection of aquatic systems.

INTRODUCTION

Statistics Canada (2003) reports that, in terms of water, Canada has just 0.5 per cent of the world's population, yet Canadians have access to almost 20 per cent of the global stock of fresh water, and Canada has 7 per cent of the total flow of renewable water. This apparent natural bounty is often taken for granted, it seems, as Canadians are among the highest consumers of water in terms of per capita water use, second only to citizens of the United States. This high use was characterized by Foster and Sewell (1981: 7) as due to a 'myth of superabundance'. With or without a myth of superabundance, Canada has a relatively and absolutely generous endowment of water. Indeed, as O'Neill (2004: xi) has observed, 'There can be no question that Canada's freshwater supply is an immensely valuable national resource. Recent estimates of water's measurable contribution to the

WATER USE BY CANADIANS

Canada's per capita water demands on water resources are the second highest in the world. At about 326 litres per person per day at home, Canadians use twice as much water as the average European. <www.sdinfo.gc.ca/reports/en/mongraph6.wateruse.cfm>

Canadian economy range from $7.5 to $23 billion annually, values comparable to the gross figures for agricultural production and other major economic components.' In contrast, countries in the Middle East and Sahelian Africa usually experience significant water deficits that are a major impediment to overcoming poverty and facilitating development. However, even in Canada, frequently water is not available in the right place or at the right time.

WATER AVAILABILITY IN 2025

By 2025, the greater part of the earth's population will likely live under conditions of low and catastrophically low water supply. Approximately 30–35 per cent of the world population will have catastrophically low fresh water supply (less than 1,000 m³ per year per capita). At the same time…high water availability can be found in Northern Europe, Canada and Alaska, almost all of South America, Central Africa, Siberia, the Far East, and Oceania.

— *Shiklomanov (2000: 28)*

Hydrological Cycle

Water or aquatic resources are one component of a system that includes the atmosphere, cryosphere, biosphere, and terrestrial components. Evaporation from surface water (rivers, lakes, wetlands) and transpiration from plants release water vapour into the atmosphere that condenses and forms clouds while moving upward. The tiny droplets of water in clouds eventually fall to the earth as rain, fog, hail, or snow. After reaching the surface, the water evaporates back into the atmosphere, moves into rivers, lakes, or oceans, or percolates into the soil to become groundwater. More detailed discussion of the hydrological cycle was provided in Chapter 4.

GROUNDWATER

Over 6 million Canadians, about one-fifth of the population, rely on groundwater for daily water needs. Groundwater accounts for more than 50 per cent of the total available fresh water in Canada. The province of Prince Edward Island is totally reliant on groundwater.

— *Environment Canada: <www.sdinfo.gc.ca/ reports/en/monograph6/wateruse.cfm>*

About 12 per cent of Canada (1.2 million km²) is covered by *lakes* and *rivers*, with only 3 per cent of that area located in inhabited regions. There are more than 2 million lakes, with the largest being the Great Lakes shared between Canada and the United States. Other large lakes are Great Bear Lake and Great Slave Lake in the Northwest Territories and Lake Winnipeg in Manitoba. Lake water represents about 98 per cent of the surface water available for human use. Canada has over 8,500 named rivers, and the Mackenzie River, with an average surface flow of 8,968 m³ per second, has the highest volume. There are over 1,000 named *glaciers*, and these are an important source of freshwater for rivers and lakes. Various types of **wetlands** exist, all being hybrid aquatic and terrestrial systems. They are a key habitat for waterfowl, and also store and gradually release water, thus serving as an important 'sponge' to aid in reducing flooding. Wetlands are found in the greatest number and extent in the Prairie provinces and in northern Ontario. Canada has about 25 per cent of the wetlands in the world, the largest amount of any country. *Groundwater* is a key source of water for rivers and lakes, and is created by surface water passing into the ground and becoming contained in sand and gravel, as well as in pores and cracks in bedrock. During dry periods, many rivers receive much of their water via base flow from groundwater aquifers.

HUMAN INTERVENTIONS IN THE HYDROLOGICAL CYCLE: WATER DIVERSIONS

Given that water is often not in the right place at the right time, humans modify aquatic systems to store, divert, or modify flows. There are some 600 large dams in Canada, and about 60 large interbasin diversions. Diversions are completed for one or more of the following reasons:

- to increase water supplies for a community or in a region, as illustrated by the St Mary Irrigation District in Alberta. While diversions for irrigation are important in the southern Prairies, this type of diversion is not as typical of the Canadian experience as it is for countries such as India and the United States.
- to deflect watercourses away from or around areas to be protected, such as the Portage Diversion in Manitoba. Here, the purpose is not to move water to a place of need, but to protect a community from flood damages. Other reasons are to drain land to allow agricultural production or to drain a mine site.
- to enhance the capacity of a river so that it can be used to support activities such as floating logs or allowing passage of ships, disposing of

and scale of diversions in Canada (Day and Quinn, 1992: 10–11).

While diversions have the positive capacity for increasing power production, expanding irrigation, enhancing water-borne commerce, and reducing flood damages, they also can cause negative environmental impacts and impose costs on people or regions not benefiting directly from them. The James Bay Cree and their homeland in northern Quebec are a case in point.

wastes, or sustaining fish. For example, dams on the Ottawa River were partially designed to facilitate moving logs down river to sawmills.

- to combine or consolidate water flows from several sources into one channel or route to facilitate hydroelectric generation, such as the James Bay hydroelectric project in northern Quebec. Canada is a global leader regarding water diversions for hydroelectricity generation, and diversions for hydropower purposes dominate overwhelmingly in both number

The James Bay Hydroelectric Project

Governments and private corporations have pursued many **megaprojects** in Canada to meet energy demands, and virtually every region in the country has experienced such megaprojects (Figure 12.1). Perhaps the most massive, and one that has garnered a great deal of national and international attention, has been the **James Bay Project** in Quebec. Other huge hydroelectric developments have been Churchill Falls in Labrador, the Nelson-Churchill in Manitoba, and

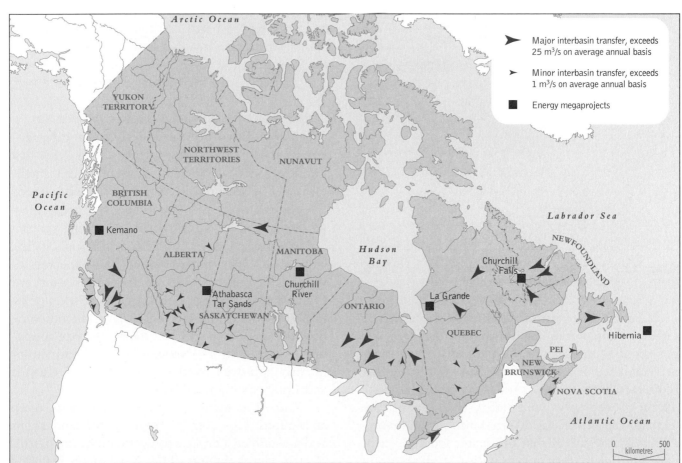

Figure 12.1 Energy megaprojects in Canada. SOURCE: Adapted from Day and Quinn (1992: 16).

the Columbia and Nechako rivers in British Columbia, while nuclear power plants in Ontario, the development of the Hibernia oil fields off the coast of Newfoundland, the Sable Island natural gas exploration off of Nova Scotia, and the exploitation of the tar sands in northern Alberta are among other major Canadian energy projects. Selected aspects of the James Bay Project are considered below.

Background

In 1971, Premier Robert Bourassa proposed hydroelectric development using the rivers on the eastern side of James Bay. The purpose was to satisfy future electricity needs in Quebec. The estimated cost was to be $2 billion. After alternative projects were examined, the decision was taken to develop La Grande River basin to double the flow in that river by diverting water from adjacent catchments (Figure 12.2). Other river systems north and south of La Grande were to be developed in later phases.

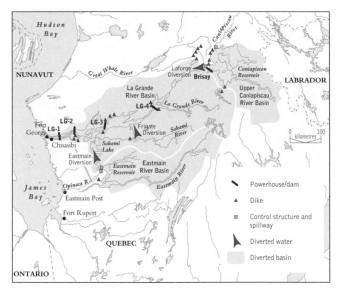

Figure 12.2 La Grande River hydroelectric development project, phase I. SOURCE: Day and Quinn (1992: 134).

Two major diversions channelled water into La Grande basin. From the south, 845 cubic metres per second (m³/sec) or 87 per cent of the flow measured at the mouth of the Eastmain River was redirected into the LG2 reservoir on La Grande. Twenty-seven per cent (790 m³/sec) of the Caniapiscau River was diverted into the LG4 reservoir, with the intent that this water would in turn flow through all four power stations to be built on La Grande. These diversions together add an

average of 1,635 m³/sec to La Grande, almost doubling the natural flow in that river. Over a 15-year period the cost increased to $14.6 billion, compared to the $2 billion announced by Premier Bourassa in 1971.

In Stage 1 of the development, three hydroelectric plants (LG 2, 3, 4) with a combined 10,283 megawatt (MW) capacity were built. LG2, with 5,328 MW capacity, became the most powerful underground generating station in the world. The first electricity was generated from LG2 in 1979, and LG4 was completed in 1986. LG1 and other dam construction were deferred to Stage 2.

The scope and magnitude of the James Bay development has been described as 'breathtaking'. It produces electricity from rivers flowing in a 350,000 km² area of Quebec, more than one-fifth of the province or an area equivalent to France. The provincial government and Hydro-Québec justified the James Bay development for the jobs to be created, the industrial growth to be attracted to the province, and the stability to be created.

In the enthusiasm for the perceived benefits from hydroelectricity in James Bay, an area remote from the settled part of the province, little regard was given to the fact that this was the homeland for about 10,000 Cree and Inuit, who had lived and hunted in this region for centuries.

James Bay and Northern Quebec Agreement

The **James Bay and Northern Quebec Agreement** is considered by many to be the first 'modern' Native land claims agreement in Canada. However, when Premier Bourassa first announced the construction of the hydroelectric megaproject no systematic environmental or social impact assessments had been completed, although such a large-scale development could be expected to have major impacts. The Cree people in northern Quebec, who had not been consulted, soon organized themselves to fight the project. The outcome was the James Bay and Northern Quebec Agreement, signed on 11 November 1975 and subsequently approved by the government of Canada and Quebec's National Assembly.

The Agreement is complex, and also often ambiguous. However, it provided for land rights and guaranteed a process to deal with future hydroelectric developments. The Agreement included provisions for environmental and social impact

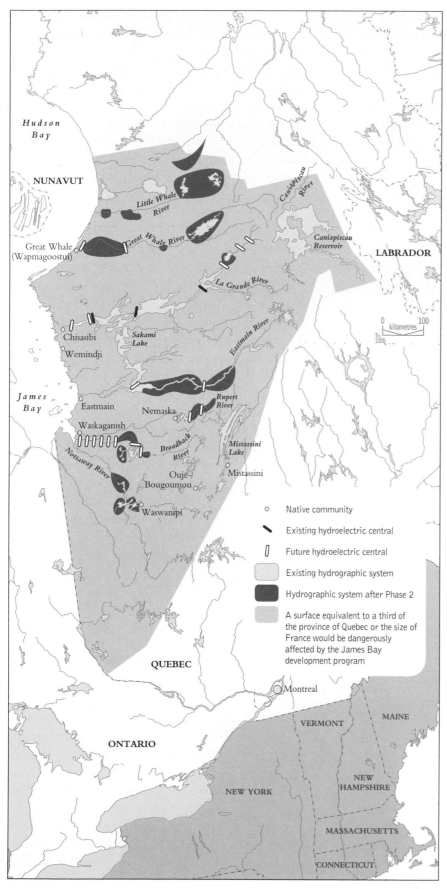

Figure 12.3 The Great Whale project. SOURCE: Diamond (1990: 32).

Native community

Existing hydroelectric central

Future hydroelectric central

Existing hydrographic system

Hydrographic system after Phase 2

A surface equivalent to a third of the province of Quebec or the size of France would be dangerously affected by the James Bay development program

assessment for future developments, monetary compensation, economic and social development, and income security for Cree hunters and trappers.

James Bay II

When Premier Bourassa announced Phase II in 1985, he explained that the development would (1) generate revenue for Quebec through exports of electricity to the United States under long-term contracts, and (2) attract energy-intensive industries (such as aluminum and magnesium smelters) as a result of competitively priced electricity. James Bay II involved completion of development in La Grande basin, particularly the building of LG1, as well as new hydroelectric development in the Great Whale and Nottaway-Broadback-Rupert river systems. During 1986, the Cree agreed to the completion of the development in La Grande basin but opposed the projects to be started in the basins to the north and south of La Grande.

The projects in the Great Whale basin would provide just under 3,000 MW of new power by diverting several adjacent rivers. One outcome would be a reduction by 85 per cent in the flow of the Great Whale River at the community of Whapmagoostui (Figure 12.3). The Nottaway-Broadback-Rupert development would produce 8,000 MW of additional power and, as with development on La Grande, would involve inundation of land due to dam construction.

During the 10-year construction period from 1974 to 1984, various concerns had emerged. These included the relocation of Fort George to the new site at Chisasibi, the quality of the drinking water in the new community, the problems in maintaining traditional hunting activity in areas that had become accessible from the new roads built for the construction of dams, and, because of the altered patterns of ice breakup on the lower river and estuary due to release of relatively warmer water from the reservoirs in winter and early spring, the difficulty for hunters in travelling to the northern coastal area across the river from Chisasibi.

At the community level, other concerns emerged. For example, increased erosion along the banks of La Grande, due to the fluctuating water levels in the river caused by releases from the upstream reservoirs, at one point threatened the site of the new community. The newly built road exposed the community to other people and values, contributing to problems such as alcohol abuse for some individuals.

Following completion of the first three dams on La Grande, the major problem became the very high levels of mercury in fish caught in the reservoirs or connecting rivers. As a result, by the end of 1985, the Cree completely stopped fishing in the LG2 area. Another problem was that hunters from Chisasibi purchased vans to travel to distant inland hunting grounds. However, after construction of LG4 was completed, maintenance of the road network east of LG4 was stopped and the vans no longer could be used.

Against this changing mix of issues and concerns, the challenges in estimating impacts related to some specific matters are considered below. These changing issues and concerns also reinforce the arguments for an adaptive management approach, discussed in Chapter 6, and highlight the presence of uncertainty, complexity, and change related to both science and management.

1. *Mercury in reservoirs.* None of the environmental impact assessment studies had predicted the appearance of mercury in reservoir fish. This is a puzzling aspect of the impact assessment work, since evidence about elevated mercury levels in fish was available from earlier hydroelectric projects at the Smallwood Reservoir in Labrador and from Southern Indian Lake in Manitoba. Such impacts apparently were dismissed by investigators as being of only short duration and not significant for La Grande.

   As Gorrie (1990: 27–8) explained, mercury

is common in rocks throughout the North. It exists in an insoluble form that does not affect air or water. However, when such rocks are inundated by a reservoir, bacteria associated with the decomposition of organic material in the reservoir water transform the insoluble mercury into methyl mercury that vaporizes, is released to the atmosphere, and returns to the water. Once in the water, the mercury enters the food chain and through biomagnification reaches the highest trophic levels in fish species that prey on other fish, as reviewed in Chapter 10. Such predator fish—walleye, pike, lake trout—had been an important source of high-quality protein food for the local people. Berkes (1988) indicated that in most years about one-quarter of the total community wild food harvest came from fishing, averaging about 60 kg per year for every man, woman, and child.

In new reservoirs, a burst of decomposition often accelerates the release of mercury. In the La Grande river system, few trees were removed prior to the flooding of the reservoir area, so there was lots of organic matter to decompose. Downstream from the dams on La Grande, levels of mercury in fish climbed to six times their normal levels within months of completion of the dams. By the sixth year following the impoundment, concentrations of mercury were four to five times higher in all species sampled. A survey in 1984 of the Cree at Chisasibi showed that 64 per cent of the villagers had unsafe levels of mercury in their bodies. In terms of presence in food sources, the highest levels of mercury were in walleye and northern pike, which also ruined any potential for a recreational fishery in the reservoirs behind the three dams.

It was expected that, as time passed and the drowned vegetation completely decomposed, the release of mercury would return to normal (and safe) levels. Monitoring, as reported on by Chevalier et al. (1997) and Dumont et al. (1998), focused on the species most often consumed by the Cree people, including whitefish (*Coregonus clupeaformis*), Cisco (*Coregonus artedii*), lake trout (*Salvelinus namaycush*), northern pike (*Esox luncius*), walleye (*Stizostedion vitreum*), and longnose sucker (*Catostomus catostomus*). The results indicate that 15 years after the impounding of the LG2 reservoir, the concentrations of methyl mercury were higher than in natural lakes but also that they were decreasing in both predatory (northern pike) and non-predatory (whitefish) species.

Monitoring revealed that mercury concentrations in the hair of the Cree people had decreased and were stabilized, and that, for most of the population, these concentrations 'do not present a health risk' (Chevalier et al., 1997: 79). Dumont et al. (1998) reached a more qualified conclusion. While mercury levels had dropped in all communities among both sexes and in all age groups between 1988 and 1993–4, in their view, 'the significance of these mercury levels for the health of the Cree population remains unknown.' Their reasoning was that the decrease in mercury concentrations in humans could be due to the Cree having changed the type of fish eaten, from contaminated to less contaminated fish, or having decreased their total fish consumption, or both. A reduction in other types of bush food (because of game cycles) or new roads providing access to contaminated sites could lead to increases in future fish consumption. As a result, Dumont et al. (1998: 1444) concluded that 'Communities with historically low [mercury] levels can thus experience increases rapidly. The present low mercury levels in the Cree communities may not be permanent and must not lull health authorities into believing that the mercury problem has been definitively resolved.'

2. *La Grande estuary fish.* The pre-construction impact assessments indicated that the estuarine fishery in the La Grande was unlikely to survive the development of the dams and reservoirs. The impact study predicted that, as a result of interruption of the water flow in the river during the construction and when the reservoirs began to fill behind the dams, and the resulting absence of ice cover to dampen the impact of ebb and flow of the tidal water from James Bay, salt water would move further up into the river. The consequence would be the elimination of the freshwater overwintering fish habitat for species important for the local fishery. On the other hand, if the river water flow were reduced *after* the formation of

ice cover, then saltwater intrusion would be impeded and a critically important pocket of fresh water could be maintained in the key habitat area.

Partially as a result of pressure exerted by the local fishers, the river flow was not cut off until after an ice cover had formed on the river. Monitoring revealed that this action did result in the creation of the necessary freshwater pocket, which remained in place throughout the winter. The freshwater pocket was created even though no minimum flow was provided other than what entered the river from very minor tributaries. The outcome was that the predicted fish kill did not occur, and subsequent fish populations were about the same as in the pre-construction period. In this situation, the impact prediction was incorrect, but the outcome was positive for the fishery.

LIMITING FACTOR PRINCIPLE

In Chapter 2, we discussed the limiting factor principle, which indicates that all factors necessary for growth must be available in certain quantities if an organism is to survive. We also noted that the weakest link is known as the dominant limiting factor. Are these ideas helpful in understanding the impact from interrupting river flow, and changing patterns of ice cover, on the overwintering fish habitat in La Grande estuary?

Improving Impact Assessment

The James Bay experience reinforces the viewpoint that surprises should be expected in impact assessment and other aspects of environmental management, a point made in Chapters 1 and 6. Berkes (1988) concluded that little mystery surrounds some sources of uncertainty. Impacts do not occur simultaneously but over an extended time period and are the result of decisions taken—filling schedules for reservoirs, release of 'excess' water through spillways, and including more or fewer turbines and powerhouses. Furthermore, impacts do not happen in isolation; in many instances they combine to form *cumulative impacts*. Such a situation is particularly important in instances such as James Bay, for which development is contemplated over decades for adjacent catchments on the eastern side of James Bay. The state of the art in impact assessment is still very weak regarding how to deal with such cumulative impacts and, therefore, how to prepare mitigation and compensation strategies.

WATER QUALITY

Humans can adversely affect water quality in numerous ways. The most important is pollution from various sources, but especially from industrial and other urban wastes, and from agricultural runoff. The former two are easier to identify because they usually are associated with *point sources*, such as end-of-pipe sources at manufacturing plants or sewage treatment plants. Agricultural runoff is more challenging, as it is usually diffuse pollution from *non-point sources*, such as fertilizers, herbicides, and pesticides from farm fields. Some types of urban runoff, such as oil and salt from road surfaces, are also characterized as non-point, since they cannot be identified with specific places.

POLLUTION CHALLENGES IN CANADA

The Great Lakes and the St Lawrence River basins continue to suffer from industrial and municipal pollution, urban and agricultural runoff, and deposition of airborne pollutants. The Red River and other prairie rivers are being degraded by agricultural runoff and inadequately treated sewage. The Fraser River is under pressure from industrial effluents, landfill pollutants, wood treatment chemicals, and forestry and agricultural runoff.

— *Environment Canada:<www.sdinfo.gc.ca/ reports/en/monograph6/watqulty.cfm>*

Point Sources

Urban wastewater from households, industries, other commercial operations, and institutions can be treated to varying degrees. Up to three levels of treatment exist: (1) *primary*, which removes only insoluble material; (2) *secondary*, which removes bacterial impurities from water previously having received primary treatment; and, (3) *tertiary*, which removes chemical and nutrient contaminants, following secondary treatment. In 1999, 78 per cent of Canadians living in municipalities serviced by sewers had secondary and/or tertiary treatment, an increase from the 58 per cent with such service in 1983. Nine per cent of all Canadians had

their wastes go to waste stabilization ponds (also known as sewage lagoons), simple treatment systems that provide the equivalent of secondary treatment. In 1998, 3.46 million Canadians living in larger municipalities were not connected to wastewater systems and either used septic tanks or other arrangements to handle their wastes. Examples of ocean disposal and the different approaches taken by Victoria and Halifax are discussed in Chapter 8. Another 4.88 million people lived in smaller municipalities, and almost 57 per cent of them had no wastewater treatment facilities.

A growing concern is that many wastewater treatment facilities are old, and need expensive maintenance or upgrading or replacement. The Canadian Water and Wastewater Association calculated that $5.4 billion in new investment is required *each year* between 1997 and 2012 to modernize and upgrade all existing water and wastewater treatment facilities, as well as to provide such facilities to those presently without them.

Industry is an important source of wastes deposited into water bodies. Tables 12.1 and 12.2 identify the most prevalent chemicals and the water bodies receiving over 1,000 tonnes of pollutants in 2001. These data are from the National Pollutant Release Inventory (NPRI), which tracks the release of some 200 chemicals into the environment. Ammonia and nitrogen represented more than 94 per cent of the total releases to water. Other chemicals, such as mercury, are released in much smaller amounts but have serious negative impacts on human and aquatic system health. Mercury bioaccumulates and biomagnifies (see Chapter 10) in the liver, kidneys, and muscles of affected organisms, and chronic exposure can result in brain and kidney damage. Monitoring

---

MUNICIPAL WASTEWATER

Municipal wastewater can result in increased nutrient levels, often leading to algal blooms; depleted dissolved oxygen, sometimes resulting in fish kills; destruction of aquatic habitats with sedimentation, debris, and increased water flow; and acute and chronic toxicity to aquatic life from chemical contaminants, as well as bioaccumulation and biomagnification of chemicals in the food chain.

— *Environment Canada (2003a: 34)*

Table 12.1 Top Releases of Chemicals to Water, 2001

Chemical	Releases (tonnes)
Ammonia (total)*	26,106
Nitrate ion in solution at pH equal to or greater than 6.0	22,450
Manganese (and its compounds)	1,157
Methanol	697
Zinc (and its compounds)	308

*Total includes both ammonia (NH_3) and ammonium ion (NH_4^+) in solution.
SOURCE: Statistics Canada (2003: 18).

Table 12.2 Water Bodies Receiving over 1,000 Tonnes of Pollutants, 2001

Water Body	Total Release (tonnes)	Dominant Release	Share of Total Release (%)
Fraser River	9,168	Ammonia*	49.2
Lake Ontario	8,877	Ammonia*	41.6
Bow River	8,264	Nitrate ion	90.8
Ottawa River	3,066	Ammonia*	76.6
North Saskatchewan River	2,953	Nitrate ion	61.3
Red River	2,766	Ammonia*	72.7
Hamilton Harbour	1,516	Ammonia*	70.6
South Saskatchewan River	1,275	Nitrate ion	62.4
St Lawrence River	1,086	Nitrate ion	43.6

*Total includes both ammonia (NH_3) and ammonium ion (NH_4+) in solution.
SOURCE: Statistics Canada (2003: 18).

shows that mercury levels in the Canadian environment continue to rise and that the main sources are metal mining and smelting, waste incineration, and coal-fired power plants.

Runoff from urban areas flows either directly into water bodies from roads and other non-point sources or can be channelled by stormwater systems. Storm water can contain various contaminants, such as suspended solids, sediment, and grit; nutrients, including different forms of phosphorus and nitrogen; toxic metals, including copper, lead, and zinc; hydrocarbons, including oil, grease, and polycyclic aromatic hydrocarbons; trace organic contaminants including pesticides, herbicides, and industrial chemicals; and fecal bacteria. As a result, it is desirable that storm water be treated in municipal wastewater plants, along with waste from homes, industries, and other urban institutions. Unfortunately, such treatment does not always occur.

A third important source of wastes into water bodies is from agricultural activity, but that is more appropriately discussed under the category of non-point sources.

Non-point Sources

As discussed in Chapter 10, crop and livestock production has increased significantly as a result of more effective farm machinery, new genetic crops, agrochemicals, and irrigation. However, the latter two also contribute to environmental impacts, especially through fertilizers, pesticides, and herbicides being carried in runoff from farm fields, which end up in streams, rivers, and lakes (see Table 10.6). In this section, the experience with diffuse pollution in the Great Lakes basin is examined.

Diffuse pollution has been recognized as a policy issue in the Great Lakes basin. Since the early 1960s, interest has evolved from concern about sedimentation from soil erosion and eutrophication from phosphorus and nitrate loading to persistent toxic chemicals. As the definition of the problem has evolved, so also have ideas regarding appropriate responses.

What has been learned about the strategic implications of how the problem has been defined? One of the key lessons is that diffuse pollution represents a 'layered' problem. That is, it is much like an onion, with many layers. Too often with such problems, attention does not go beyond the first layer, although there is a need to go much deeper. To elaborate, at the first layer, concern about diffuse pollution focused on *environmental degradation* and the *economic costs* imposed on downstream users. The motivation for defining the problem in this manner appears to be that people will see the connection between diffuse pollution and loss of economic production or increased costs for economic production. A second layer, of increasing focus, is to link diffuse pollution to negative impacts on *ecosystem health* or *integrity*, and especially on *human health*. It has been believed that making the connection to human health should create a powerful image in the minds of both policy-makers and residents in any area affected by diffuse pollution. A third layer is to interpret diffuse pollution as a problem regarding human *values, beliefs, attitudes, and behaviour*. In other words, from this perspective, the fundamental dilemma is behaviour by individuals and groups, driven by inappropriate values, beliefs, and attitudes. If attention is focused at this third level, then the prescription to resolve diffuse pollution is certainly different from what it is if attention is maintained at the first level.

What have been effective mechanisms to achieve recognition of diffuse pollution as a policy issue? From experience in Canada in general, and

> ### PHOSPHORUS LOADINGS IN CANADA
>
> *For Canada as a whole, estimated yearly loadings of phosphorus fell by 44 per cent between 1983 and 1999, despite the 24 per cent increase in urban population.*
>
> — *Environment Canada (2003a: 35)*

the Great Lakes basin in particular, an evident need is for *credible science* to document the nature of the problem. Furthermore, a high-profile *advocate* or *champion*, either an institution or an individual, is very important if the results from the science are to be shared with, and understood and accepted by, the general community and key decision-makers.

Credible Science and Institutional Commitment
Appreciation of diffuse pollution as a policy issue has been helped in Canada and the Great Lakes basin by a combination of science, institutions, and individuals. Several initiatives by the **International Joint Commission**, a bilateral institution created to manage interjurisdictional water issues between Canada and the United States, have been significant.

During the 1960s, the media declared that 'Lake Erie is dying', a reference to the highly eutrophic state of that lake (see Chapter 4). In 1972, the governments of Canada and the United States entered into an agreement to restore and enhance water quality in the Great Lakes. Initial attention focused on reducing phosphorus loading from municipal sewage treatment plants and other point sources. Initiatives in that regard were effective, but it was suspected that non-point sources might also be significant. However, data were not available to indicate the importance of non-point sources.

Under the 1972 agreement between Canada and the United States, the International Joint Commission was requested 'to conduct a study of pollution of the boundary waters of the Great Lakes System from agricultural, forestry and other land use activities'. Subsequently, an intensive inquiry was completed by the International Reference Group on Great Lakes Pollution from Land Use Activities, otherwise known as PLUARG. PLUARG examined two major pollution problems in the basin: eutrophication from elevated nutrient inputs, especially in the lower lakes (Erie and Ontario), and increasing contamination by toxic substances. PLUARG studied the pollution potential from various land uses, including agriculture, urbanization, forestry, transportation, and waste disposal, as well as natural processes such as lakeshore and riverbank erosion.

In its final report, PLUARG concluded that, depending on the magnitude of point source loads, the 'combined land drainage and atmospheric [non-point] inputs to individual Great Lakes ranged from 32 per cent (Lake Ontario) to 90 per cent (Lake Superior) of the total phosphorus loads (excluding shoreline erosion). Phosphorus loads in 1976 exceeded the recommended target loads in all lakes' (International Reference Group on Great Lakes Pollution from Land Use Activities, 1978: 4–5). The PLUARG study was the first credible science to document the important contribution of non-point sources to phosphorus loading. Produced by a group of international scientists and government officials for a respected bilateral institution, these results were difficult to ignore. The PLUARG report also stated that toxic substances such as PCBs were entering the Great Lakes system 'from diffuse sources, especially through atmospheric deposition. Through land drainage, residues of previously used organochlorine pesticides (e.g., DDT) are still entering the boundary waters in substantial quantities.' In terms of the sources of these loadings, it was reported that 'intensive agricultural operations have been identified as the major diffuse source contributor of phosphorus.' In addition, 'Erosion from crop production on fine-textured soils and from urbanizing areas, where large scale land developments have removed natural ground cover, were found to be the main sources of sediment. Urban runoff and atmospheric deposition were identified as the major contributors of toxic substances from non-point sources' (ibid., 6).

Regarding necessary actions, PLUARG concluded that:

remedying non-point source pollution will be neither simple nor inexpensive. Non-point sources of water pollution are characterized by their wide variety and large numbers of sources, the seemingly insignificant nature of their individual contributions, the damaging effect of their cumulative impact, the intermittent nature of their inputs, the complex set of natural processes acting to modify them and the variety of social and economic interactions which affect them.

...The level of awareness about pollution from non-point sources among Great Lakes Basin residents is inadequate at present. Control of non-point sources will require all basin residents to become involved in reducing the generation of pollutants through conservation practices. Improved planning and technical assistance are prerequisites to long term solutions of land drainage problems. (Ibid., 10–11)

The PLUARG report, and other analysis, led to renewal of the Great Lakes Water Quality Agreement in 1978 and to the signing in 1987 of a protocol amending the 1978 agreement. The 1987 amendments made the agreement broader in scope than the previous agreements, but diffuse pollution was still recognized as one of the priority problems. For example, in commenting on the 1987 protocol, the International Joint Commission remarked that, through it, 'Specific improvements seek to control air-borne toxics, to deal with contaminated sediments, to designate and remedy particular areas of concern, to reduce pollution from non-point sources, to protect groundwater, and to develop water quality objectives for each of the Great Lakes' (Government of Canada and the Government of the Province of Ontario, 1988: 5). More specifically, Annex 13 of the protocol focused exclusively on 'Pollution from Non-point Sources' and identified 'programs and measures for abatement and reduction of non-point sources of pollution from land-use activities' in order 'to further reduce non-point source inputs of phosphorus, sediments, toxic substances and microbiological contaminants contained in drainage from urban and rural land, including waste disposal sites, in the Great Lakes System' (ibid., 55).

Many observers would agree that the PLUARG study commissioned by the International Joint Commission, along with the Commission's prestige and watchdog role, was significant in helping both elected government officials and the public to understand the severity of the diffuse pollution problem in the Great Lakes. Without such a credible voice to draw attention to the issue of non-point source pollution, it is likely that action would not have been forthcoming when it did.

The Role of an Individual Advocate

A second important factor for change was the work of Senator Herbert Sparrow, a farmer from Saskatchewan, who chaired a standing committee of the Canadian Senate that produced a report in 1984 entitled *Soil at Risk: Canada's Eroding Future.* For many Senate committee reports, their publication is less than an auspicious event in Canada, but this report was an exception. The Sparrow Report, based on public meetings held across the country, documented the significant environmental and economic implications of soil erosion, including the costs of diffuse pollution. More importantly, Senator Sparrow did not stop when the report was published. He continued to speak out and to lobby, and the efforts by him and others led to the establishment of the National Soil Conservation Program in December 1987 as a three-year, $150 million cost-shared program between the federal and provincial governments. The lesson from Sparrow's activities is that a committed, energetic, determined, enthusiastic, and high-profile individual can make a difference. Indeed, given the impact of comparable other prominent people in Canada over the years, such as David Crombie's role in leading the study of the future of the Toronto waterfront (Royal Commission on the Future of the Toronto Waterfront, 1992) and Thomas Berger's (1977) role in leading an examination of environmental impacts from development on the land and people in the Canadian North, it appears that a key consideration for any group interested in dealing with diffuse pollution should be to recruit a respected individual in a position of prominence and influence to become the spokesperson and advocate for necessary change.

Agricultural Non-point Source Pollution

After the PLUARG studies were completed, the governments of Canada and the United States agreed to deal with the issue of high phosphorus loadings from rural non-point sources. In Canada, the federal and Ontario governments in 1987 created a five-year, cost-shared program—the Soil and Water Environmental Enhancement Program,

or SWEEP. The purpose was to meet the target reduction of 200 metric tonnes per year for Canada for phosphorus loading in Lake Erie by 1990 from non-point sources.

SWEEP consisted of various programs. The first focused on technology evaluation and development, and was intended to stimulate adoption of soil management and cropping practices needed to improve water quality and to reduce soil erosion and degradation. A second thrust focused on pilot watershed programs, local demonstrations, and technical assistance at the farm level. The pilot watershed program was particularly innovative, with emphasis on testing the effectiveness of state-of-the-art conservation practices on working farms. The effects of using different practices on all farms in three experimental watersheds were compared with conventional practices in three control watersheds. Control and treatment watersheds were evaluated for impacts on water quality, hydrology, soil quality, crop production, and economics at the farm level. A third component involved information services, with attention to informing the public about the nature and consequences of soil and water quality problems and of the SWEEP objectives. Cressman (1994: 421), whose consulting firm was actively involved in the pilot watershed studies, later remarked that 'It was during this time that interest in, and the practice of, conservation tillage grew significantly among Ontario farmers.'

In parallel with SWEEP, the Ontario Land Stewardship Program was introduced by the Ontario Ministry of Agriculture and Food. The Land Stewardship Program was a three-year, $40 million program that provided financial incentives for first-time adoption of conservation measures on farmland. Financial assistance was provided for practices to protect soil structure, to build structures to ameliorate soil erosion, to purchase conservation equipment, and to obtain technical training. Funds also were dedicated for research projects related to stewardship practices.

The five-year SWEEP program (1987–91) overlapped with the National Soil Conservation Program, a three-year, $150 million cost-shared program. There were many delays in getting activities underway in the different provinces for the latter program. As Cressman (1994: 422–5) explained, lengthy negotiations were required by the federal government with each of the provincial governments. Ottawa expected the provinces to provide 'new' money for this program, but several provinces wanted to count funds in existing programs as part of their share of the costs. There were other disagreements, as the federal government was most interested in removal of fragile lands from production and in on-farm soil conservation. In contrast, the provinces preferred financial and technical support for on-farm conservation and education programs. In Cressman's view, 'For many conservation interests, these delays were incomprehensible considering the gravity of the soil degradation problems that needed attention.'

When SWEEP and National Soil Conservation Program both finished in the early 1990s, a newly elected federal government introduced another program—the Green Plan—which provided funds for soil conservation and diffuse pollution control, but by that time attention had shifted from soil erosion, sedimentation, and eutrophication to toxic substances. One observer remarked to one of the authors in 1999 that, in his view, these three programs were developed with little consultation among key agencies, resulting in duplication and overlap and the time frame not being long enough to allow measurement of program impacts. Indeed, the monitoring for soil erosion, sedimentation, and eutrophication had been cursory rather than systematic—*perceived* results and the provision of support to farmers for production improvements dominated over concern for ameliorating environmental degradation problems. Cressman (1994: 429) evaluated the effectiveness of the National Soil Conservation Program and summarized the general views expressed by farmers:

A concern expressed by many farmers is the confusion created by the plethora of programs and agencies, the redundancy among certain efforts, and the competition among agencies for recognition, power, and control. In some cases, the system's need to control was viewed as so overbearing that farmers were reluctant to get involved or to stay involved in conservation programs. Many innovative farmers who had invested in conservation practices before the programs started were cut off from financial support directed only at 'new' practices. Mean-

while, many of these farmers were asked to be leaders in farmer-to-farmer extension efforts. For some of the strongest farm leaders in soil conservation, justice and equity seem remote.

Nonetheless, some positive initiatives were undertaken and results achieved. In the Lake Erie and Lake Ontario watersheds, the loads from phosphorus have been reduced significantly due to initiatives dealing with both point and non-point sources. The main initiatives for non-point sources helped farmers modify how their land was cropped, especially by encouraging use of conservation tillage to reduce erosion and thereby reduce sediment and toxics placed into aquatic systems. The main actions regarding point sources required upgrading of municipal sewage treatment plants, and regulations were established to reduce phosphorus in laundry detergents. The latter action is consistent with the principles of anticipation and prevention, that is, to minimize a problem in the first place rather than to pursue remedial measures after the pollutants enter the environment.

Toxic and Persistent Toxic Substances

The International Joint Commission (1998: 3) concluded that the Great Lakes Water Quality Agreement 'is sound, effective and flexible. Review and renegotiation are not necessary. Rather, the Parties need to renew and fulfill their commitments and focus on implementation, enforcement and other actions, including review of institutional arrangements, to achieve the Agreement's purpose.' This is a timely reminder that it is not necessary to continuously introduce new programs and projects. The IJC indicated that it had identified core principles on which action should be based, and those principles included involvement of all stakeholders, anticipation and prevention, precaution, application to all sources and pathways, and consideration of all places where contaminants reside in the ecosystem, including water, land, sediment, air, and biota.

The IJC also assessed the changing approaches to pollution from non-point sources and considered their effectiveness. For example, the IJC (1998: 15) observed that, 'Increasingly, the emphasis has shifted to embrace not only control of discharges and emissions but also prevention (including bans on production and use of DDT, mirex and other chemicals) and to blend regulation with voluntary, beyond-compliance measures, partnerships and greater flexibility.' Furthermore, the IJC asked whether appropriate responses should be regulatory reform, greater use of existing authority, more flexible approaches to problem-solving, reorganization or better integration, and better communication regarding issues. Not surprisingly, it concluded that all of these factors provide opportunities to improve performance.

The IJC concluded that the governance of the Great Lakes 'is in the midst of a profound and continuing evolution characterized by a shift from an exclusive command-and-control emphasis to voluntary measures, and a move from top-down management to environmental partnerships' (ibid., 16), and argued that some of the profound and significant changes included the following.

Voluntary measures and partnerships. Voluntary measures and partnerships create opportunities for governments to move away from bureaucratic command-and-control management and for participants to work co-operatively.

Science and research. The IJC believes that restoration and protection of the Great Lakes ecosystem requires strong science-policy links, models, and surveillance and monitoring. While the IJC advocates the need for 'good science', it distinguishes between 'good' science and 'certain' science. The IJC has stated that science is always evolving, and that its conclusions are not necessarily invalid simply because they are provisional or uncertain. Indeed, in the words of the IJC, 'It may be difficult or impossible to distinguish a strong correlation between levels of exposure and resulting health damage, but that does *not* mean there is *no* link, only that the mechanisms remain obscure…. The call for unequivocal evidence of injury to humans is inappropriate and must not preclude corrective and preventive action.'

Communication of scientific information—linking science and policy. Science and research should provide information in a manner that is understandable and defensible to those who fund restoration and protection, a point also highlighted in the discussion about communication in Chapters 6 and 7. The IJC concluded that an unfortunate gap remains between research undertaken by the scientific community and the information required by decision-makers to strengthen and

underpin actions to evaluate, restore, and maintain ecosystems in the Great Lakes basin. Furthermore, 'scientists are often not connected to the regulatory, policy and jurisdictional arenas.'

Surveillance and monitoring. Surveillance and monitoring are needed to provide information about the status of the environment and progress towards achieving desired objectives. They are 'the basic tool that informs decision-making'. The IJC identified nine specific desired outcomes for the Great Lakes: (1) fishability, (2) swimability, (3) drinkability, (4) healthy human populations, (5) economic viability, (6) biological community integrity and diversity, (7) virtual elimination of inputs of persistent toxic substances, (8) absence of excess phosphorus, and (9) physical environmental integrity.

Communication and public participation. The IJC believes that achieving a desired future requires leading, rather than being led by change. Such leadership requires communication of information, active public participation, and changes in governance (see also Chapter 6). And, in the view of the IJC, 'public awareness of environmental issues is the first step. The key is information. Information and education provide understanding. Informed discussion allows the public to take ownership at the community level and to develop consensus and support for action.'

Key players. The IJC concluded that progress towards the desired future has resulted from combined efforts and participation from governments, industry, environmental NGOs, Native Americans and First Nations, labour, the public, and the IJC itself.

Governance. In the opinion of the IJC, federal, state, and provincial governments must provide strong leadership to protect the public good, including human and ecosystem health. In particular, the IJC believes that the governments 'must reduce bureaucracy, apply the precautionary principle, and foster co-operation and partnership among basin stakeholders.'

Relative to the observations above, the accompanying box provides a summary of the environmental status in 2003 of the Great Lakes, the St Lawrence River, and the St Clair River–Lake St Clair–Detroit River ecosystem. The overall conclusions by Environment Canada and the US Environmental Protection Agency (2003: 5) was

BOX 12.1 STATE OF THE GREAT LAKES IN 2003

Positive signs of recovery include:
1. Lake trout stocks in Lake Superior have remained self-sustaining.
2. Reproduction of lake trout in Lake Ontario is now evident.
3. Bald eagles nesting and fledging along the shoreline are recovering.
4. Persistent toxic substances are continuing to decline.
5. Phosphorus targets have been met in all the lakes except Lake Erie.

Negative signs of degradation include:
1. Phosphorus levels are increasing in Lake Erie.
2. Long-range atmospheric transport is a continuing source of contaminants to the basin.
3. Non-native species are a significant threat to the ecosystem and continue to enter the Great Lakes.
4. Scud (*Diporeia*) are continuing to decline in Lakes Ontario and Michigan.
5. Type E Botulism outbreaks, resulting in the deaths of fish and aquatic birds, are continuing in Lake Erie.
6. Native mussel species are being lost throughout Lake Erie and Lake St Clair as a result of invasive zebra mussels.
7. Land-use changes in favour of urbanization continue to threaten natural habitats in the ecosystems of Lake Ontario, Lake Erie, St Clair River–Lake St Clair–Detroit River, and Lake Huron.

SOURCE: Environment Canada and US Environmental Protection Agency (2003: 5).

that 'the status of the chemical, physical, and biological integrity of the Great Lakes basin ecosystem is mixed, based on Lake by Lake and basin-wide assessments of 43 indicators.'

Implications

Stewart (1993: 98) has observed that '*What is* can be a great barrier to *what could be*. Those who want to move forward through bold and effective change, should begin at the end—with where they want to be.' In his view, it is essential to determine

what should be done before addressing *how* to do it. The intent here has been to identify a process to define a desired future condition, to identify and assess the present condition, and to develop actions that facilitate progress towards the desired future. Such an approach has started to be used within the Great Lakes ecosystem, shared by Canada and the United States, with regard to non-point or diffuse pollution. A vision is being articulated, and complementary means are being combined to move forward. No one individual or group has the necessary information, understanding, authority, or resources to overcome diffuse pollution. As a result, partnerships and alliances are being formed, and will continue to be needed, to ensure solutions fit local and regional conditions and needs. While command-and-control measures are no longer considered to be the only or best means to deal with diffuse pollution, such measures have a continuing role, along with voluntary and cross-compliance measures.

HIGH WATER USE BY CANADIANS

One of the key factors explaining the high residential consumption rates is the lack of financial incentive to Canadian households to use less water. For instance, in 1999, unmetered households, which pay a flat rate for water, used 50 per cent more water than metered households, which pay for water by volume used. About 57 per cent of Canada's municipal population had water meters in 1999, showing a gradual increase since 1991.
— *Environment Canada (2003a: 30)*

WATER SECURITY: PROTECTING QUANTITY AND QUALITY

A central concern in water management is to ensure a sufficient quantity of water of adequate quality for human use. At the start of the twenty-first century, it was estimated that 1.1 billion people did not have access to safe water supplies, and two out of five people did not have access to adequate sanitation, notwithstanding substantive efforts during the United Nations' International Drinking Water Supply and Sanitation Decade throughout the 1980s to improve conditions. In terms of current per capita water use, the range extends from as little as 20 litres to over 500 litres each day. Only 4 per cent of the world's population use water in the range of 300 to 400 litres per person per day, with people in the United States, Canada, and Switzerland being the highest per capita users. In contrast, about two-thirds of the global population get by on less than 50 litres for each person daily.

To place the above information in perspective, most humans become thirsty after losing only 1 per cent of their bodily fluid and are in danger of death once the loss approaches 10 per cent. The minimum water requirement to replace loss of fluid for a normal healthy adult in an average temperate climate is about three litres each day. In tropical or subtropical conditions, the minimum amount becomes about 5 litres per person per day.

GROUNDWATER QUALITY

Groundwater sources generally provide water that is safe to drink. This is especially true if the well field is protected from pollutants. Aquifers that are close to the surface are more prone to contamination by pollution, which partially explains the poor water quality that is characteristic of many shallow wells in Canada.

— *Statistics Canada (2003: 24)*

Most Canadians receive their drinking water from the 4,000 municipal water treatment plants across the country, but a significant number depend on private wells or other arrangements. About 9 million Canadians, most living in small towns or rural areas, draw on groundwater for their drinking water.

The significant abundance of water in Canada, the high levels of water use, and the myth of superabundance, all referred to at the beginning of this chapter, made most Canadians complacent about the adequacy and safety of their water supplies. For many, this all changed in mid-May 2000 when the small town of Walkerton in southwestern Ontario, with population of about 5,000, experienced contamination of its water supply system by deadly bacteria, *Escherichia coli* O157:H7, or **E. coli** as it is often called, and *Campylobacter jejuni*. These bacteria can cause bloody diarrhea and sometimes extreme abdominal pain. Children under five years of age and the elderly are at greatest risk. Seven people died and more than 2,300 became ill. It is not known for sure, but is anticipated that some individuals who became sick in Walkerton, especially children, may have effects for the rest of their lives.

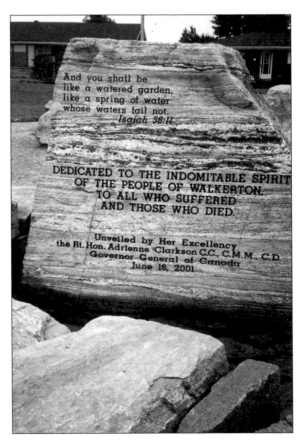

Walkerton water memorial (*Bruce Mitchell*).

The concern generated by the Walkerton experience was reinforced during March 2001 in North Battleford, Saskatchewan, a community of 14,000, where thousands of residents suffered from vomiting, diarrhea, and high fever due to contamination of the municipal water system by the parasite *cryptosporidium*. The parasite got into the water supply system over a period of three weeks following routine maintenance of a chemical filter at the treatment plant. Residents of the city were under a boil-water order for three months. An inquiry ordered by the provincial Premier concluded that the Saskatchewan government had not been effective in safeguarding drinking water in the province.

The Walkerton Inquiry

A public inquiry into the Walkerton tragedy by Justice Dennis O'Connor (2002a, 2002b) established that:

1. The E. coli, contained in manure spread on a farm near one well of the Walkerton water supply system, entered the system through that well. The well was very shallow—its casing went only five metres below the surface, and all of its water was taken from a shallow area 5–8 metres below the surface. In addition, the groundwater was drawn from an area of highly fractured bedrock. Due to the fracturing and the shallow overburden of soil, surface bacteria could rapidly enter the fractured rock and move directly into the well. This was exacerbated by an extraordinary rainfall between 8–12 May 2000 following the spreading of manure on the adjacent field in the last week of April.
2. The farmer who spread the manure followed proper practices, and was not at fault.
3. The outbreak caused by the contamination would not have occurred if the water had been treated by continuous chlorine residual and turbidity monitors at the well. The water was not treated because the chlorination equipment was being repaired and thus was not operating.
4. The operators of the well did not have sufficient education and training to appreciate the vulnerability of the well to contamination from non-point surface contamination, and the provincial government's approvals and monitoring programs were inadequate.

FROM THE WALKERTON INQUIRY

The story of the outbreak involves much more than a description of the clinical symptoms of the illnesses, the medical treatment, and the numbers of people who became ill and died. Most important are the stories of the suffering endured by those who were infected; the anxiety of their families, friends and neighbours; the losses experienced by those whose loved ones died; and the uncertainty and worry about why this happened and what the future would bring.

— O'Connor (2002a: 5)

5. In addition to lack of training, the operators of the well system had a history of improper operating practices, ranging from failing to monitor wells daily to making false entries about water quality in daily operating records.
6. When people began to fall ill, the general manager of the water system withheld from the public health unit critical information about adverse water quality test results. This resulted in delay in a boil-water advisory and in 300–400 more people becoming ill than would have been the case if water had been boiled.
7. Budget reductions by the provincial government had led to closure of government laboratory testing services for municipalities, and private laboratories were not required to submit adverse test results to the Ministry of the Environment or to the Medical Officer of Health.

Walkerton: Lessons and Recommendations

In Part Two of his report, Justice O'Connor offered recommendations to ensure the safety of drinking water across the province. Some of these are provided here, as they are relevant to all regions of Canada. Overall, he recommended a **multi-barrier approach** to drinking water safety. In his words, 'Putting in place a series of measures, each independently acting as a barrier to passing waterborne contaminants through the system to consumers achieves a greater overall level of protection than does relying exclusively on a single barrier (e.g., treatment alone or source protection alone). A failure in any given barrier will not cause a failure of the entire system' (O'Connor, 2002b: 5). He argued that in a multiple-barrier approach, the first barrier involves selecting and protecting reliable, high-quality drinking water sources. As a result, he recommended 'a source protection system that includes a strong planning component on an ecologically meaningful scale—that is, at the watershed level', and that 'the Province adopt a watershed-based planning process' (ibid., 6, 3). Within a watershed-based approach, he recommended the following:

1. A comprehensive approach for managing all aspects of watersheds is needed and should be adopted by the province.

2. The provincial Ministry of Environment should be responsible for establishing a framework for developing watershed-based source protection plans, assisting in funding and participating in their development, and approving completed plans. A Watershed Management Branch should be created to take on this role. In his view, the establishment of a centralized branch responsible for watershed management 'should provide consistency in planning across the province and provide the expertise and support necessary for ensuring that good plans are developed' (ibid., 14).

3. To ensure that local considerations are taken into account fully, and to create goodwill in and acceptance by local communities, source protection planning should be undertaken as much as possible at a local (watershed) level, by those who will be most affected (municipalities and other affected local groups).

4. Regarding policy, the provincial government should prepare a comprehensive, source-to-tap, government-wide drinking water policy and enact a Safe Drinking Act containing the key components of such a policy.

The multi-barrier approach was subsequently endorsed by other governments. For example, in 2002, at a national level, Environment Canada, Health Canada, the Canadian Council of Ministers of the Environment, and the Committee on Environmental and Occupational Health collaborated to prepare what was labelled a multi-barrier approach. Details about this initiative can be seen at the website of the Canadian Council of Ministers of the Environment (http://www.ccme.ca/sourcetotap/), under 'Source to Tap—Protecting Our Water Quality'.

WATERSHED-BASED SOURCE PROTECTION PLANS

Watershed management plans usually take a comprehensive ecosystem approach to water, dealing with all water-related natural features, terrestrial resources, fisheries, water linkages, and green space planning.

— *Ontario Advisory Committee on Watershed-based Source Protection Planning (2003: 2)*

In April 2003, an Ontario Advisory Committee on Watershed-based Source Protection Planning released a response to the O'Connor recommendations and provided a set of guidelines for the Ontario government. The Advisory Committee (2003: iii–iv) recommended the following principles as the basis for development of source protection plans to reflect a multi-barrier approach:

- *Sustainability.* Water is essential for human health and ecosystem viability and must be valued as finite. Source protection plans should consider historical, existing, new, and future land uses when considering how to ensure clean sources of drinking water now and in the future.

- *Comprehensiveness.* All watershed-based source protection plans must take a precautionary approach that uses the best available science and is subject to continuous improvement as knowledge increases. The plan must be defensible and have the flexibility to accommodate the province's diverse watersheds.

- *Shared responsibility and stewardship.* While the Ministry of Environment has ultimate responsibility for ensuring source water protection, responsibility for specific outcomes is shared among all water managers, users, and landowners.

- *Public participation and transparency.* There must be open discussion and communication of the source protection planning process and its results, from development to implementation. Stakeholders and the public will have opportunities for meaningful input.

- *Cost-effectiveness and fairness.* The costs and impacts on individuals, landowners, businesses, industries, and governments must be clear, fair, and economically sustainable. Source protection planning must be based on all practical and reasonable information, and must use technologies and risk management practices to maximize the protection of public health.

- *Continuous improvement.* Source protection planning is built on a commitment to continuous improvement, including peer review, which requires ongoing support of stakeholders to ensure successful implementation based on assessment, monitoring, evaluation, and reporting, followed by appropriate modifications.

To highlight that the Advisory Committee's views reflect ideas about best practice, they can be compared with the principles used by the Saskatchewan Watershed Authority, which was established in October 2002. Its vision emphasizes 'excellence in watershed management that promotes safe, sustainable water supplies within healthy ecosystems', and the following principles guide its activities: source water protection; water as a provincial resource; shared responsibility; stewardship; partnerships; a watershed/ecosystem approach; sustainable development; beneficial practices; empowerment; and accountability (Saskatchewan Watershed Authority, 2003: 5).

The approaches of both Saskatchewan and Ontario reflect the view of the Ontario Advisory Committee on Watershed-based Source Protection Plans (2003: 50) that 'If Ontario is to have a safe, reliable source-to-tap drinking water system, the province must ensure that all the separate parts come together and create an integrated whole.'

WATER AS HAZARD

Flooding

Humans settle adjacent to rivers and lakes for good reasons. Proximity to them provides access to potable water, a place to dispose wastes, and, sometimes, a source of power and a means for transportation. The relatively flat land beside many rivers and lakes facilitates construction of roads, homes, and places of business. The aesthetic quality—serenity, beauty, close awareness of nature—of a river or lake view also often means that waterfront lots command a premium price. However, the term **flood plain** exists for a reason. From time to time, rivers and lakes extend beyond their normal limits to cover adjacent areas or flood plains. Flooding is a normal hydrological function. Indeed, many species of flora and fauna depend on flooding to survive and flourish. In addition, humans often benefit from flooding, as when flood plains are enriched by the deposit of silt that then supports agriculture.

When flooding occurs as a result of a river or lake 'reclaiming' adjacent land, the result often is only minor inconvenience, and this usually was the case when settlements were relatively small. However, as population concentrations on flood plains increase, the potential flood damages go up. Examples of major floods and serious associated damages in Canada include the floods in the lower Fraser River Valley in BC (1948), Winnipeg (1950, 1997 [estimated $300 million damages]), Toronto (1954 [Hurricane Hazel, with over 80 deaths and millions of dollars in damages; see Box 12.2]), Fredericton (1973), Cambridge, Ontario, Maniwaki, Quebec, and Montreal (1974), and the Saguenay River Valley, Quebec (1996 [10 deaths and $800 million damages]).

Humans have various ways to reduce flood damage potential. One relies on *structural approaches*, in which the behaviour of the natural system is modified by delaying or redirecting flood waters. Common methods are upstream dams and storage reservoirs, protective dykes or levees, and deepening or straightening of river channels to increase their capacity. All these measures provide protection. However, because they are designed and built with regard to a standard, such as the magnitude of flood that occurs once in a hundred years, eventually a flood of greater magnitude will occur (such as a flood that occurs once every 200 or 500 years). If people perceive the structural measures as having 'protected' the flood plain, thereby resulting in more development there, then when the inevitable flood event greater than the design capacity of the structural measures does happen, the potential and actual flood damages will be increased.

A second approach, referred to as *non-structural*, focuses on modifying the behaviour of people. Methods include land-use zoning to restrict or prohibit development in flood-prone areas, relocation of existing flood-prone structures, information and education programs to alert people about the hazard of occupying flood plains, and insur-

BOX 12.2 THE AFTERMATH OF HURRICANE HAZEL

The valleys, ravines, and trees of Toronto are some of the city's defining natural features—and we have a hurricane to thank for them.

Half a century ago ... Hurricane Hazel swept across southern Ontario, unleashing floods that took at least 81 lives and caused millions of dollars in damage. In the years afterward, efforts to prevent a repeat of Hazel resulted in a collection of protected green spaces sprinkled throughout the city and region.

Islands of green amid the concrete and asphalt, places like Riverdale Park, the Toronto Brick Works, the G. Ross Lord Reservoir, and Étienne Brûlé Park were all created or developed in the aftermath of the hurricane.

'If that (hurricane) hadn't happened 50 years ago, I think we probably would have lost the vast majority of the green space in our valleys', says Gregor Beck, director of conservation science with Ontario Nature (the former Federation of Ontario Naturalists)....

The key event in Hazel's immediate wake, Beck says, was the establishment three years later of what is now known as the Toronto and Region Conservation Authority. Its initial role was to mitigate the damage that any future Hazels could cause. But the authority evolved into the most important player in the protection and revitalization of the GTA's natural environment.

In October 1954, when Hazel dropped up to 28 centimetres of rain over just two days on to the already sodden city, most of the deaths and property damage after the hurricane occurred near rivers and creeks. Hazel made it clear that encroaching development in these settings had to be halted, says Brian Denney, chief administrative officer of the conservation authority.

'There was a recognition that the valley systems really belong to the rivers', Denney says.

People, Beck adds, like to live near rivers and lakes, but rarely consider the water's latent power and destructive potential....

The authority began to aggressively acquire land, buying or expropriating homes in watershed areas and taking over open lands and forests abutting rivers and creeks. Places like Raymore Dr. in Etobicoke, where a dozen homes were washed away in the Hazel flooding, were transformed from residential streets into riverside parks.

Today, the authority owns more than 14,164 hectares of land, mainly along the ... Humber, Don, and Rouge rivers and their tributaries, between the Oak Ridges Moraine and Lake Ontario....

SOURCE: Hall (2004: B1–B2).

ance programs to help people deal with the costs from flood damages. The best strategies to reduce flood damages use a mix of structural and non-structural approaches, since each specific method has limitations. In the following discussion of the 1997 flood of the Red River in Manitoba, it is possible to see how both structural and non-structural approaches are used.

Red River Flood, 1997

In late April and early May of 1997, the Red River 'experienced a catastrophic regional flood that far surpassed any previous flood in the historical record' (Todhunter, 2001: 1263). As a result, areas in Minnesota, North Dakota, and Manitoba had what the media called the 'Flood of the Century'. Total damages in the United States part of the basin have been calculated to be US$4 billion, nearly US$3.6 billion within the Grand Forks (North Dakota) and East Grand Forks (Minnesota) metropolitan area. Per capita, this event was the most costly flood for a major metropolitan area in the United States. In the Grand Forks metropolitan area, almost 13,000 structures, or 85 per cent of all structures, were damaged by the flood, and about 55,000 residents had to evacuate their homes for periods ranging from days to months. Some 1,300 homes were destroyed. Downstream, southern Manitoba experienced Cdn$500 million in damages. The Winnipeg area just avoided a catastrophe, with only 54 homes being flooded. While there were no deaths, 28,000 rural residents had to evacuate their homes.

The Red River originates in the United States, in southern North Dakota/Minnesota, and flows northward, draining into Lake Winnipeg (Figure 12.4). It is 880 km long, and the watershed is 290,000 km², including the Assiniboine River basin (163,000 km²), which joins the Red River at the 'Forks' in Winnipeg (Brooks and Nielson, 2000). The valley is underlain by Mesozoic bedrock, which has a moderate relief. In the

THE RED RIVER VALLEY

The topography of the Red River Valley is very flat resulting in slow movement of floodwaters and considerable areal extent of flooding when a flood occurs.

— Burn and Goel (2001: 356)

central valley, the bedrock is covered by late Pleistocene glacial sediments, which in turn are capped by glaciolacustrine clay-rich deposits that accumulated within glacial Lake Agassiz, which at one time covered most of Manitoba, the northern part of North Dakota and Minnesota, northwestern Ontario, and east-central Saskatchewan in the late Pleistocene and early Holocene. In Manitoba, the modern Red River is on the bed of glacial Lake Agassiz, in an eroded valley up to 15 metres deep and 2,500 metres wide.

The river has a long history of flooding. The largest recorded flood occurred in 1826. Records also document major floods in 1852 and 1861. More recently, significant floods happened in 1950, 1979, 1996, and 1997. Various factors combined to cause the major flood of 1997: (1) high precipitation in the autumn of 1996, resulting in the soil being saturated by the time of the winter freeze-up; (2) near-record levels of precipitation during the winter, with twice or more the normal amounts of snowfall; (3) a long and unusually cold winter, leading to a high water content of

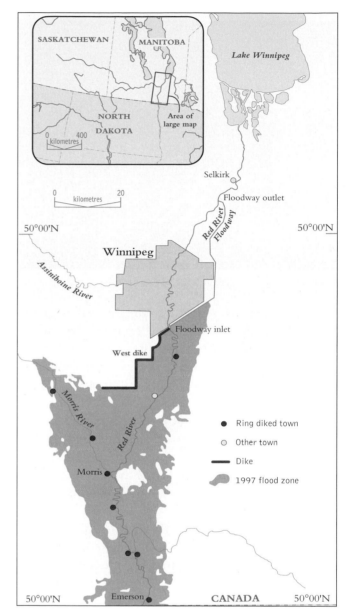

Figure 12.4 Extent of Red River flooding, 1997. SOURCE: Brooks and Nielson (2000: 307).

the snow pack into the spring; and (4) after spring melting had started, a major blizzard over much of the basin on 5 April, with snow accumulations up to 50 cm. These factors combined to produce ideal conditions for flooding.

The flooding of the Red River was typical for that river system, but unusual compared to most Canadian rivers. During the flood, an area of about 2,000 km^2 and up to 40 km wide was covered by water from the Canada–US border to the southern edge of Winnipeg (Figure 12.4). For most rivers, flood waters are confined to the river valley, even when they spread out onto the flood plain. Extensive flooding occurs in the Red River Valley because of the low capacity of the river to accom-

Flooding on the Red River in 1979 (*Philip Dearden*).

Flooding on the Red River in 1997 (*Al Harvey/www.slidefarm.com*).

Control gates for the Red River Floodway (*Bruce Mitchell*).

modate extreme flows, due to the low valley gradient and shallowness of the valley itself. This situation is exacerbated because the areas adjacent to the river are broad and flat, allowing the flood waters to spread out over many kilometres on each side of the river. The flood peaked at Emerson, Manitoba, the most southern community along the river in Manitoba, on 27–8 April, and then at the Floodway inlet in Winnipeg on 3–4 May. This pattern was typical of the Red River, where floods usually rise and fall slowly over a period of several weeks.

After the devastating flood in 1950, when a large part of Winnipeg was inundated, structural measures were used to protect the city, smaller towns to the south, and transportation routes. These measures include:

1. *Red River Floodway*. Completed in 1968, the **Floodway** is an excavated channel (48 km long, 210 to 305 m wide, and 9.1 m average depth) that can divert up to 1,700 m³/second of the flow of the Red River around the eastern side of Winnipeg, to rejoin the Red River downstream of the city. The capacity is able to divert the river stage to a level 0.6 m below the top of the primary dykes in the city. During the 1997 flood, use of the Floodway resulted in only minor flooding in Winnipeg, compared to major damage from the 1950 flood. However, the Floodway operated at maximum capacity (a once-in-160-years flood) during the 1997 flood, indicating that a flood of larger magnitude would have caused serious damage.

2. *Portage diversion*. This **diversion** is located at Portage La Prairie, 84 km west of Winnipeg, and is an excavated channel, 29 km long and 54 to 366 m wide, that can divert up to 700 m³/second from the Assiniboine River to Lake Manitoba. Completed in 1970, the diversion is designed to reduce the flow of the Assiniboine River into Winnipeg during high water in the Red River.

3. *Shellmouth Dam and Reservoir*. Completed in 1972, this storage reservoir is able to hold water for later release at times of high flows

4. *Earth dykes*. A system of **dykes** has been built along the Red, Assiniboine, and Seine rivers within Winnipeg. One hundred twenty kilometres of primary dykes have been built within the city, plus the 32-km-long West Dyke that extends west and south from the Floodway.

5. *Ring dykes*. South of Winnipeg, large dykes encircle various small towns on the Red River flood plain. Smaller ring dykes surround individual buildings, such as homes and barns. Dugouts also often remained after the earth dykes were constructed, and have become small perennial ponds in the valley.

6. *Elevated roads and railroad beds*. Roads and railroad beds are raised 1–2 metres above the valley. This elevation protects transportation infrastructure from flooding and also can serve as dykes.

There were both positive and negative outcomes from the 1997 Red River flood (Burn and Goel, 2001; Shrubsole, 2001; Todhunter, 2001). First, the Winnipeg Floodway prevented an estimated $6–$10 million in damages in Winnipeg. Second, flood fighting by local and provincial people was supplemented by over 8,000 military personnel. Over 8 million sandbags were filled and delivered in Winnipeg, over 600,000 m³ of clay were excavated to provide material for new, temporary dykes, and 50 temporary dykes were modified to become permanent structures. Third, in addition to filling sandbags, thousands of volunteers looked after children and provided food and other support to their neighbours. Fourth, non-government organizations provided critical assistance, such as the $10,000 grants from the Red Cross to residents whose homes were damaged beyond repair.

Balancing these positive outcomes were some negative aspects. First, the operation of the gates at the Winnipeg Floodway caused the upstream water level to increase, exacerbating flooding by 0.64 metres in some smaller upstream communities (south of Winnipeg). This led some rural upstream residents to conclude that their well-being was sacrificed to protect the residents of Winnipeg. Second, although there was considerable forewarning of the flood, the peak flow at the Floodway was underestimated by 1.5 to 1.7 metres. Furthermore, the flooding of the upstream community of Ste Agathe was a surprise to many, since the town had never been flooded before and thus a ring dyke had not been built. Eight other small communities on the flood plain, but within ring dykes, were not flooded. Third, some rural municipalities hesitated to spend their own funds on fighting the flood until arrangements were clarified with the provincial government, since provincial law does not allow municipalities to have operating deficits. Fourth, some Aboriginal communities experienced problems because of confusion regarding which agencies were responsible for them.

The International Joint Commission (2000) conducted an inquiry on the 1997 flood, in which it offered the following conclusions and recommendations: (1) although the 1997 flood was natural but rare, floods of the same or greater magnitude are possible; (2) both people and property in the Red River Valley will remain at risk to flooding until comprehensive, integrated, and binational solutions are developed; (3) a mix of structural and non-structural approaches is needed, since no one approach is adequate; (4) specific communities in both Canada and the US need to take flood damage initiatives; and (5) aspects of the ecosystem need more explicit attention, and, in that regard, hazardous material needs to be more carefully controlled and banned substances need to be removed from flood-prone locations.

Since the 1997 flood, the Canada and Manitoba governments together have spent over $130 million on structural measures. Much of this amount ($110 million) has been for protective adjustments for 16 rural communities. In December 2003, the two governments announced an agreement to allocate $240 million for the joint funding of expansion of the Red River Floodway around Winnipeg. This will allow an increase in the capacity of the existing 49-km channel to meet a flood event comparable to the one in 1826, calculated to have been a once-in-280-years occurrence. This will be realized by increasing the width of the diversion channel by about 110 metres and the depth by up to 2 metres. When the entire expansion is completed, at an estimated cost of $700 million, it will provide protection against a once-in-700-years flood.

The expansion will be complicated and challenging, as it will involve upgrading and improvements to 12 bridge crossings, the Floodway outlet, the West Dyke, utilities, and drainage services. There also is potential to disrupt the groundwater table through the excavation, which will move more than 30 million cubic metres of earth. An environmental impact assessment was started in 2004 to determine what adjustments would need to be made to the design. This project, which may take up to a decade to complete, is being managed by the Manitoba Floodway Expansion Authority, and up-to-date details can be found at its website (http://www.floodwayauthority.mb.ca/flood-facts.html).

Canadians will continue to live and work on flood plains, for the reasons identified at the beginning of this section. Thus, there will continue to be damage from floods. In that context, the comments of Shrubsole (2001: 462) deserve consideration:

The present practice of flood management in Canada is characterized by at least three realities. First, it is impossible to provide absolute protection to people and communities. Second, a mix of structural and non-structural adjustments that cover the entire range of protection, warning, response and recovery is needed to effectively protect lives and property. Third, implementation of flood adjustments requires the effective participation of all levels of government and the public.

Droughts

If flooding represents situations in which there is too much water, droughts represent the opposite problem—insufficient water. While flooding is immediate and apparent, the beginning or end of a drought is more difficult to determine, as droughts are a function of a lack of precipitation, temperature, evaporation, evapotranspiration, capacity of soil to retain moisture, and resilience of flora and fauna in dry conditions.

Consequently, as Gabriel and Kreutzwiser (1993) have noted, a significant challenge when studying droughts or seeking to identify 'drought-prone' areas is to define what is meant by a **drought**. As they noted, interpretations are based on different factors: causes and effects. Regarding those based on *causes*, an example is a meteorological drought due to a prolonged deficiency of precipitation, which reduces soil moisture. This type of drought can trigger a second type (hydrological drought, an effect that then becomes a cause), manifested by reduced stream flows and lowered **water table** and/or lake levels. In terms of *effects*, an example is an agricultural drought, which results in reduced crop yields due to lack of moisture. An urban drought happens when there is insufficient water, because of lower stream flows or water tables, to support all demands in the community. The point here is the importance of understanding the definition for a 'drought' when preparing or examining results or recommendations pertinent to management decisions.

Droughts are of concern because, as Gabriel and Kreutzwiser (1993: 119) explain, they reduce the amount of water for use through depleting soil moisture and groundwater reserves, as well as by lowering stream flows and lake levels. These reductions can start a 'depletion cycle': less than normal rainfall leading to low soil moisture, triggering

demand for irrigation development, in turn depleting non-recharging surface and groundwater supplies. The high evapotranspiration associated with hot, dry periods also contributes to depleting soil moisture and surface water supplies, which are not restored to normal levels without unusually high rainfall.

What has been the experience with drought in Canada, where southern Ontario, interior British Columbia, and the Prairies are most vulnerable? In Ontario, drought can occur in any season, but is most likely in the summer when demand for water is usually the highest. Southwestern Ontario is most vulnerable, especially in the summer and early autumn. Extended dry periods for more than a month are unusual, but shorter droughts are not uncommon. For example, dry periods of at least seven consecutive days occur at least once a month during the agricultural growing season in southern Ontario, and short-term (10–20 days) dry spells occur every year. Longer droughts (more than four weeks) happen once in three years. One of the most severe and extensive droughts occurred during 1966 when, during a 41-day period between mid-June and the end of July, most of southern Ontario received less than 25 mm of precipitation, or less than a quarter of normal rainfall. Areas from Lake Huron and Georgian Bay to Lake Ontario received less than 15 per cent of normal rainfall in this period. Another significant drought occurred in the summer of 1988, when southwestern Ontario received less than 40 per cent of the average precipitation from early May to mid-July.

Lake levels are affected by dry periods. The Great Lakes illustrate this effect. The variation between minimum and maximum lake levels is 1.2 metres on Lake Superior, 1.8 metres on Lakes Huron and Erie, and 2 metres on Lake Ontario. Low levels affect shipping, especially regarding cargo tonnages, which have to be reduced so that ships don't run aground. To reduce the draft by only 2.5 cm requires a reduction of up to 90 tonnes on most ships and over 180 tonnes on ships between 244 and 305 metres in length. During low levels in the St Clair River in 1964, ships carrying iron ore and grain from Lake Superior to the lower Great Lakes had to reduce their loads by between 725 to 1,360 tonnes to pass through the river.

In Ontario, many streams are almost totally supplied by groundwater discharges during low rainfall periods. In average conditions, groundwater discharge provides 20 per cent of the water for streams or rivers in most of Ontario. For some rivers or streams, the contribution can be up to 60 per cent, extending up to 100 per cent in the summer months. Thus, depletion of groundwater reserves resulting from drought can have a serious impact on surface flows, especially for smaller streams. The lowering of water tables due to drought can also lead to the drying up of wells dependent on shallow aquifers.

However, droughts are most usually associated with the Prairie provinces, especially in that area of southern Alberta and Saskatchewan and extreme southwest Manitoba known as **Palliser's Triangle** (Figure 12.5). This area is named after Captain John Palliser, sent by the British government and the Royal Geographical Society to explore the territory between the Laurentian Shield and the Rocky Mountains between 1857 and 1860 to determine the nature of the soil, its capacity for agriculture, the quantity of its timber, and the presence of coal or other minerals. Based on his surveys, Palliser divided the area into two sections—a fertile belt and a semi-arid area. He considered the southern or semi-arid area to be unfit for settlement. Palliser commented in his journal that this area 'has even early in the season a dry patched look.... The grass is very short on these plains, and forms no turf, merely consisting of little wiry tufts. Much of the arid country is occupied by tracts of loose sand, which is constantly on the move before the prevailing winds' (Mackintosh, 1934: 11). Palliser concluded that 'There is no doubt that the prevalence of a hard clay soil derived from the cretaceous strata which bakes under the heat of the sun, has a great deal to do with the aridity of these plains, but it is primarily due more to want of moisture in the early spring' (ibid., 34). Thus, more than 145 years ago, Palliser identified the drought-prone nature of the southern Prairies. His 'heads up' was reinforced during both 2001 and 2002, when large areas of the Prairie provinces experienced drought conditions.

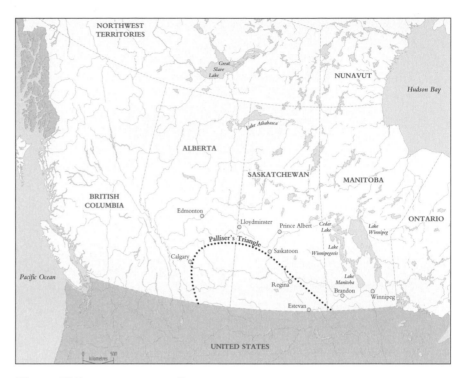

Figure 12.5 Palliser's Triangle. SOURCE: Adapted from Bone (2005: 410).

Figure 12.6 illustrates that 'in 2002, drought-stricken areas covered over three-quarters of the Prairies (including the north-eastern part of British Columbia)' (Statistics Canada, 2003: 13). While there were many impacts from these dry conditions, the most pronounced was inadequate water to support agriculture. As the data in Table 12.3 show, yield of spring wheat, barley, and canola fell significantly during 2002 relative to the average yields between 1991 and 2000, which were non-drought years.

Livestock were also affected negatively due to the drought. The greatest impact was in Alberta, where the inventory dropped by 605,000 cattle, a decrease of 10.4 per cent between January 2002 and January 2003. At the same time, declining supplies of cattle feed, a result of the drought conditions, pushed up feed prices and led many ranchers to reduce herds.

Another indicator of the drought conditions was the drying up of many dugouts (small, human-made ponds), potholes (small natural ponds), and sloughs. By September 2002, 80 per cent of Prairie farms were in regions in which dugouts were half-empty, and 20 per cent reported their dugouts were completely dry. The drying up of potholes and sloughs not only affected agriculture. These are also critically important habitat for migratory wildfowl, which also were adversely affected

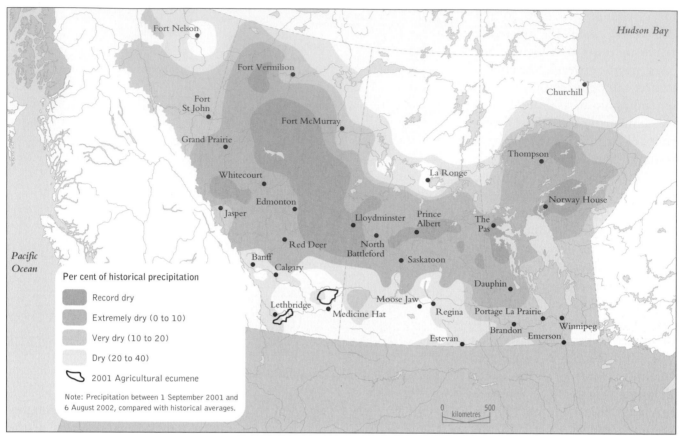

Figure 12.6 Precipitation below historical averages, 2002. SOURCE: Statistics Canada (2003: 13).

(Statistics Canada, 2003: 13; see also Table 11.4 regarding the affect of drought on wetlands).

HERITAGE RIVERS

The mission of the Canadian Heritage Rivers Program (CHRP) is to 'develop a river conservation program that is nationally valued, internationally recognized and reflects the significance of rivers in the identity and history of Canada: and ensure that the natural, cultural and recreational values for which rivers are designated are managed in a sustainable manner.' Started in the early 1980s, the CHRP is overseen by a Board, with representatives from the federal, provincial, and territorial governments, under the oversight of appropriate ministers from the three levels of government.

The intent is to achieve the mission by 2006. By the end of 2003, 39 rivers had been nominated under this program and 30 had been formally designated (Figure 12.7 and

Table 12.3	Crop Yields and Insurance Payments, 2002 Percentage Variation from 1991–2000 Average, Prairie Provinces			
Province	Spring Wheat	Barley	Canola	Crop Insurance Payment
Alberta	−29.4	−26.8	−13.0	399.1
Saskatchewan	−32.0	−34.1	−21.4	224.1
Manitoba	2.6	7.6	3.9	69.9

SOURCE: Adapted from Statistics Canada (2003: 13).

RIVERS AS HERITAGE

Rivers teach valuable lessons about renewal. It is said, and it is true, that you cannot enter a river at the same spot twice; because, of course, the river 'rolls along'. This very character of rivers generates health and well-being. It also connects one part of the waterway to another.

— *Harry Collins, Chairperson, Canadian Heritage Rivers Board, 2001–2*

Table 12.4). The process leading to designation has several steps, as shown in Box 12.3.

As revealed in Table 12.4, for almost a decade after the beginning of the program, designated rivers were located in federal or provincial parks, the territories, or in areas within provinces with relatively few people. Designation of such places involved rivers that were primarily on Crown

land, avoiding the complication of having to deal with private landowners and municipalities, who often were suspicious of the Heritage Rivers Program, viewing it as possible government intrusion related to property or municipal rights.

However, in 1994, the entire Grand River basin, located in southern Ontario and with most of its land in private ownership, was designated. The designation of this river was a bold and imaginative move, as not only did it involve working with municipalities, private landowners, and Aboriginal people, but it also addressed the cultural criteria more systematically than for the earlier designations, which had emphasized the natural environment criterion. As a result, considerable research was conducted to document the cultural and natural values in the river basin. The designation of the Grand River had to deal with concerns both about perceived curtailment of landowner rights and freedoms and about restrict-

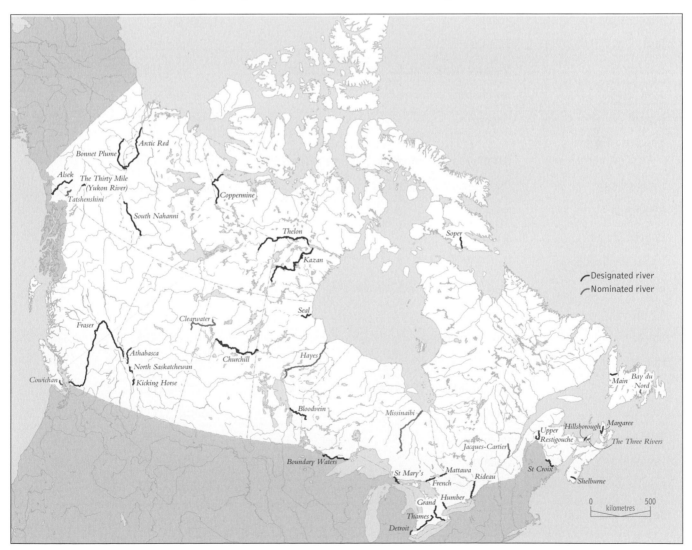

Figure 12.7 Heritage rivers in Canada. SOURCE: Canadian Heritage Rivers Board (2003: 5).

Table 12.4	Designated and Nominated Heritage Rivers

DESIGNATED RIVERS

River	Location Province/Territory (Park[1])	Date of Designation	Length (in km)
French	Ontario (French River PP)	February 1986	110
Alsek	Yukon (Kluane NP)	February 1986	90
South Nahanni	Northwest Territories (Nahanni NP Reserve)	January 1987	300
Clearwater[2]	Saskatchewan (Clearwater River PWP)	June 1987	187
Mattawa[3]	Ontario (Mattawa PP and Samuel de Champlain PP)	January 1988	76
Athabasca[7]	Alberta (Jasper NP)	January 1989	168
North Saskatchewan[7]	Alberta (Banff NP)	January 1989	49
Kicking Horse[7]	British Columbia (Yoho NP)	January 1990	67
Kazan	Nunavut	July 1990	615
Thelon	Nunavut	July 1990	545
St Croix	New Brunswick	January 1991	185
Yukon – The Thirty Mile[7]	Yukon	January 1992	48
Seal	Manitoba	June 1992	260
Soper[5]	Nunavut (Katannilik Territorial Park Reserve)	June 1992	248
Arctic Red	Northwest Territories	September 1993	450
Grand[6]	Ontario	January 1994	627
Boundary Waters/ Voyageur Waterway	Ontario (La Verendrye/Quetico/Pigeon River PPs)	September 1996	250
Hillsborough	Prince Edward Island	January 1997	45
Shelburne	Nova Scotia	June 1997	53
Bonnet Plume[6]	Yukon	February 1998	350
Upper Restigouche	New Brunswick	February 1998	55
Bloodvein[4]	Manitoba (Atikaki PP), Ontario (Woodland Caribou PP)	June 1998	306
Margaree[5]	Nova Scotia	June 1998	120
Fraser	British Columbia	June 1998	1,375
Humber[6]	Ontario	February 1999	100
Rideau	Ontario (Rideau Waterway – Parks Canada)	February 2000	202
Thames[6]	Ontario	February 2000	273
St Mary's	Ontario	February 2000	125
Detroit[6]	Ontario	February 2001	51
Main	Newfoundland	February 2001	57
Total			**7,387**

NOMINATED RIVERS

River	Location Province/Territory (Park)	Anticipated Date	Length (in km)
Missinaibi	Ontario (Missinaibi PP)	May 2004	501
Clearwater[2]	Alberta	October 2004	139
Cowichan	British Columbia	October 2004	47
Bay du Nord	Newfoundland (Bay du Nord Wilderness Park Reserve)	May 2005	75
Jacques-Cartier	Quebec (Jacques-Cartier PP)	May 2004	128
Tatshenshini	Yukon	May 2004	45
Hayes	Manitoba	May 2005	590
The Three Rivers	Prince Edward Island	May 2004	73
Churchill	Saskatchewan	May 2004	487
Coppermine	Nunavut	May 2005	450
Total			**2,535**
Total km of nominated and designated rivers			**9,922**

1. PP denotes provincial parks;
   NP denotes national parks;
   PWP denotes provincial wilderness park.
2. Clearwater River has been nominated in two sections by Saskatchewan and Alberta. The Saskatchewan section has been designated.
3. Extensions to the Mattawa were added in 2001, including the 11 km historic La Vase Portages connecting the headwater to Lake Nipissing, as well as a downstream extension to its confluence with the Ottawa River.
4. Bloodvein River (Manitoba section) was designated in 1987.
5. Includes mainstream and major tributaries.
6. Includes entire watershed.
7. Includes a segment of the river.

SOURCE: Canadian Heritage Rivers Board (2003: 6).

Grand River, Cambridge, Ontario. A former factory has become a park setting adjacent to the river, making the river accessible to the public while retaining a sense of heritage (*Bruce Mitchell*).

ing development, costing taxpayers money, and attracting too many people who would trespass and vandalize property. Through extensive discussion with stakeholders in the river basin, these issues were all handled, resulting in designation of the Grand River in January 1994 (Nelson et al., 2004). The guest statement by Barbara Veale elaborates on what was involved in achieving heritage designation for the Grand River. Other rivers in highly settled areas have since been designated, such as the Humber River in Toronto (February 1999), the Thames in

BOX 12.3 PROCEDURE FOR DESIGNATION OF A HERITAGE RIVER

1. *Submissions and public involvement.* The first step is a community-based initiative to recognize and protect a local river. The initiative can be taken by one or more of landowners, recreational, historical, or tourism organizations, Aboriginal peoples, environmental groups, local businesses, and interested members of the public. This step requires a submission to the Board.
2. *Selection.* The governments with responsibility for the river identified in a submission assess the case relative to the natural, cultural, and recreational values of the river and the level of public support for the nomination. The decision is also based on consideration of the number and complexity of conflicting land uses and the capacity for effective management of the river.
3. *Nomination.* If the government(s) determine the river is a credible candidate, then the relevant government agencies work collaboratively with local stakeholders to prepare a formal nomination to the Board. The nomination must demonstrate that the river is of outstanding Canadian value relative to the CHRP criteria (natural, cultural, and recreational significance), and that sufficient measures will be implemented to ensure

maintenance of those values. Once the Board has reviewed the nomination, it forwards a recommendation to the federal minister responsible for Parks Canada and the relevant provincial or territorial minister.

4. *Management plan.* The formal designation process starts when the nominating government submits a management plan to the Board. The management plan identifies the policies and practices that will protect the outstanding values of the river.
5. *Designation.* Once the management plan has been approved, the responsible federal and provincial or territorial governments formally designate the river, and a plaque is unveiled at a key location along the river. Designation is intended to ensure that the river will be managed so that the outstanding heritage resources will be protected, and the recreational potential will be realized.
6. *Monitoring.* To ensure that designated rivers are managed to protect the outstanding natural, cultural, and recreational values, at least every 10 years the Board requires the managing jurisdictions to conduct a systematic assessment to verify the heritage status of each designated river.

southwestern Ontario (February 2000) and the Detroit River in southwestern Ontario (February 2001).

Various initiatives, besides those for the Grand River described by Barbara Veale, have been taken related to designated heritage rivers. For example, the Rideau River Biodiversity Project has been started in the Rideau River system in eastern Ontario. Involving researchers from the Canadian

Museum of Nature and the Canadian Wildlife Service, plus local partners, the purpose is to assess the status of plants, wildlife, and water conditions. The ultimate intent is to evaluate the ecological health of the river system and to recommend local uses of the river and adjacent land that will be consistent with protecting biodiversity.

In the Fraser River basin in British Columbia, a Charter for Sustainability, which outlines what is

BOX 12.4 WHAT YOU CAN DO

1. Use a bucket, sponge, and trigger nozzle on a hose to wash your vehicle. This will save about 300 litres of water each time you wash the vehicle.
2. Install water-saving devices in your home, including low-flow showerheads and toilets, or use toilet dams.
3. When purchasing new appliances (dishwashers, washing machines), aim to buy water- and energy-efficient equipment.
4. When watering your garden or lawn in the summer, do so early in the morning or in the evening to reduce the amount of water that will evaporate in the heat of the day.
5. Become a voice in your community for water conservation policies and practices, such as water rates that provide an incentive to conserve water or incentives that encourage people to replace high-water-use appliances with water-efficient ones.

Guest Statement

How Becoming a Heritage River Can Influence Water Management

Barbara Veale

The Grand River is located in the heart of southern Ontario. Its rich diversity illustrates key elements in the history and the development of Canada, and many of the river-related heritage resources remain intact today. These resources include the historical buildings and settlements in the watershed, the land use associated with different ethnic groups, such as Mennonite farmers still using horse-drawn ploughs, and the Six Nations area near Brantford. In addition, the river and its tributaries provide a broad range of excellent recreational opportunities.

The designation of the Grand River as a Canadian heritage river marked the beginning of a second generation of heritage rivers. Prior to 1990, almost all nominated rivers either were within protected areas or were short sections of larger rivers. In contrast, the Grand River is located in one of the most densely populated and fastest-growing parts of Canada, where almost all lands are privately owned and managed within a complex multi-agency, multi-jurisdictional setting. The designation included the entire Grand River and its four major tributaries, the Nith, Conestogo, Speed, and Eramosa rivers.

In 1987, the Grand River Conservation Authority spearheaded a participatory process to have the Grand River and its major tributaries declared a Canadian heritage river. This was achieved for the Grand River and its major tributaries in 1994.

The management plan presented to the Canadian Heritage Rivers Board—*The Grand Strategy*—deviated from past management plans for heritage rivers in that it provided a framework for an ongoing, community-based watershed approach sustained by consensus, co-operation, and commitment. It was based on a common vision of beliefs, values, principles, and goals, and the designation of the Grand River set a precedent for the Canadian Heritage Rivers Board to accept other rivers in highly settled areas of Canada where river management is complex and shared among all levels of government, First Nations, and non-government entities.

Since 1994, *The Grand Strategy* has evolved into a shared management approach for integrated watershed management. Management partners, including

Barbara Veale

federal and provincial governments, watershed municipalities, First Nations, non-government groups and organizations, and educational institutions, jointly identify critical resource issues, develop creative solutions, pool resources, implement actions, monitor results, and evaluate progress on an ongoing basis.

Under the umbrella of *The Grand Strategy*, the Grand River Conservation Authority works with its partners to address a wide array of existing and emerging resource issues and to identify priorities for action. Progress is documented and celebrated through monthly newsletters, special events, an annual Registry of Accomplishments and Commitments, and an annual Watershed Report. Opportunities to directly participate in *The Grand Strategy* are offered through various working groups and public forums. The philosophy is that everyone who shares the resources of the Grand River watershed is encouraged to be part of a collective effort to address key watershed issues, including:

- reducing non-point sources of pollution in rural areas;
- pursuing excellence in wastewater treatment;
- slowing increases in water use and advocating the wise use of water;
- protecting groundwater resources;
- developing long-term water quality and water budget/water supply plans;

- maintaining the water control system;
- implementing a watershed-wide fisheries management plan;
- developing community-based plans that advance forest management, wildlife management, and natural heritage management;
- developing community riverfront plans;
- developing the watershed's potential for outdoor recreation, cultural education, and eco-tourism;
- building a sense of community around the river and celebrating successes.

Building on the Grand River Basin Water Management Study, a multi-agency initiative completed in 1982 to review the issues of flood protection, water supply, and water quality within the Grand River watershed, *The Grand Strategy* relies on good science and strong technical tools. Within the context of water management, partners are currently evaluating the cumulative impacts of municipal wastewater on river water quality, determining what needs to be done to improve water quality, and monitoring and reporting on water quality conditions and long-term trends in the river. Partners are also assessing the cumulative demand for water, identifying where water shortages may occur, and putting drought contingency plans in place. Groundwater sensitivities have been mapped, priority areas for water protection programs are being identified, and water protection policies will be incorporated into municipal planning documents.

The merits of a collaborative watershed management approach for improving river health were recognized in 2000 when the Grand River Conservation Authority was awarded the international Thiess River Prize for excellence in river management in Brisbane, Australia.

In keeping with Canadian Heritage Rivers System requirements, the first Ten-Year Monitoring Report for the Grand River was completed in 2004. This assessment was undertaken within the participatory structure created by *The Grand Strategy* and provided participants with the opportunity to revisit and reaffirm the vision, values, principles, and goals of *The Grand Strategy*, to acknowledge successes, and to identify additional actions required.

The heritage river designation has resulted in increased awareness of the river as a community asset. Residents increasingly explore and enjoy its many natural and cultural resources in growing numbers. The general level of concern for the way the river is treated has increased, resulting in more stewardship activities taking place.

The Grand Strategy illustrates that strong science and technical tools must be combined with an ongoing participatory approach that allows people to regard water management within the broadest societal context. This integration is central for motivating collective actions to resolve current water issues and address potential water issues before they are cause for concern.

Barbara Veale is Co-ordinator of Policy Planning and Partnerships, Grand River Conservation Authority, Cambridge, Ontario.

needed to achieve sustainability for the Fraser River and its watershed, has been signed by federal, provincial, and municipal governments, as well as by First Nations and other organizations. A Fraser Basin Council, a not-for-profit organization, was created in 1997 and provides oversight for activities to achieve the future identified by the Charter.

These examples illustrate the types of initiatives being taken as part of the management strategies for designated **heritage rivers**. All are oriented to protecting the integrity and health of the ecosystems, ranging from biophysical to cultural components.

IMPLICATIONS

Canada is fortunate to have an abundance of high-

quality water. As a society, however, we have placed stress on aquatic systems, sometimes degrading them significantly. As the discussion in this chapter has shown, considerable scientific understanding can be drawn upon to assist in the management of water systems. And there also have been some significant improvements, confirming that individuals, groups, communities, and societies can reverse degradation and deterioration.

Each of us has an opportunity to modify our basic values, and change behaviour, to help protect our water resources and to maintain the integrity and health of aquatic ecosystems. Box 12.4 provides some instances of action that individuals can take to make further contributions to improve water resources in Canada.

SUMMARY

1. While comprising only 0.5 per cent of the world's population, Canadians have access to almost 20 per cent of the global stock of fresh water and to 7 per cent of the total flow of renewable water. About one-fifth of the Canadian population relies on groundwater for daily water needs.

2. Canada's per capita demands on water resources are the second highest in the world and have been calculated to be about 326 litres per person per day at home.

3. Canada is a global leader regarding water diversions for hydroelectric generation, and diversions for hydro power purposes dominate overwhelmingly in both number and scale of diversions.

4. Canada has used many megaprojects to meet energy demands, and virtually every region in the country has energy megaprojects. One of the most significant has been James Bay in Quebec, while others have been Churchill Falls in Labrador, the Nelson-Churchill in Manitoba, and the Columbia and Nechako Rivers in British Columbia.

5. Pollution involves point and non-point sources, with the latter being the most challenging to manage.

6. Progress regarding diffuse pollution has been associated with *credible science* to document the nature of the problem and the involvement of a prominent *advocate* or *champion*.

7. The governance of the Great Lakes is evolving, and key aspects are a shift from an exclusive command-and-control emphasis to voluntary measures, and from top-down management to environmental partnerships.

8. At the start of the twenty-first century, it was estimated that 1.1 billion people worldwide did not have access to safe water supplies and that two out of five people did not have access to adequate sanitation.

9. Many Canadians have been complacent about the adequacy and safety of their water supplies. For many, this changed in mid-May 2000 when the small town of Walkerton, Ontario, experienced contamination of its water supply system by deadly bacteria. Seven people died and more than 2,300 became ill.

10. The judge who conducted a public inquiry into the Walkerton tragedy recommended that a 'multi-barrier approach' to drinking water safety be used because a series of measures, each independently acting as a barrier to passing water-borne contaminants through a system, provides more robust protection of drinking water supplies.

11. Floods are a normal component of the behaviour of rivers and lakes. Humans adapt to flooding through some mix of structural and non-structural approaches that aim to modify behaviour of natural systems and humans.

12. Droughts are a function of precipitation (or lack thereof), temperature, evaporation, evapotranspiration, capacity of soil to retain moisture, and resilience of flora and fauna in dry conditions.

13. The most drought-prone areas in Canada are south-central British Columbia, the southern Prairie provinces, and southern Ontario.

14. The Canadian Heritage Rivers Program is a conservation program to recognize the significance of rivers in the identity and history of Canada, and to ensure that their natural, cultural, and recreational values are protected.

KEY TERMS

diversion	floodway	James Bay and Northern	multi-barrier approach
drought	heritage rivers	Quebec Agreement	Palliser's Triangle
dykes	International Joint	James Bay Project	water table
E. coli	Commission	megaprojects	wetlands

REVIEW QUESTIONS

1. Does the 'myth of superabundance' adequately account for the high per capita water use by Canadians?

2. What are the main reasons for water diversions in Canada?

3. Why is the presence of mercury in freshwater a serious concern?

4. Explain the significance of the 'limiting factor principle'.

5. Why are there different levels of treatment for municipal wastewater?

6. Explain the significance of bioaccumulation and biomagnification of chemicals in the food chain and what role water plays in this.

REVIEW QUESTIONS

7. Why is diffuse pollution from non-point sources a challenge for managers?
8. What is meant by 'water security'?
9. What are the implications of the experience with contaminated drinking water in Walkerton, Ontario, and North Battleford, Saskatchewan?
10. What is the significance of structural and non-structural approaches for reducing flood damages?
11. What lessons were learned from major floods in Canada in the past five years?
12. How do we know when a drought begins and ends?
13. Where are the most significant flood- and drought-prone areas in Canada?
14. What is Palliser's Triangle?
15. What criteria and indicators should be used to identify the natural, cultural, and recreational values of rivers?
16. Are there river systems in your region that could be candidates to be nominated as heritage rivers?

RELATED WEBSITES

ATLANTIC COASTAL ACTION PROGRAM:
http://www.ns.es.gc.ca.acap

CANADA'S WATER QUALITY GUIDELINES:
http://www.ec.gc.ca/cwqg.english/default.html

CANADIAN COUNCIL OF MINISTERS OF THE ENVIRONMENT:
http://www.mbnet.mb.ca/cmme

CANADIAN GLOBAL CHANGE PROGRAM:
http://www.cgcp.rsc.ca

CANADIAN POLLUTION PREVENTION INFORMATION CLEARINGHOUSE:
http://www.ec.gc.ca/cppic

CANADIAN WATER AND WASTEWATER ASSOCIATION:
http://www.cwwa.ca

CANADIAN WATER RESOURCES ASSOCIATION:
http://www.cwra.org

CLEAN ANNAPOLIS RIVER PROJECT:
http://fox.nstn.ca/~carp

ENVIRONMENT CANADA'S WATER INFORMATION COLLECTION:
http://www.ec.gc.ca/water.index.html

EXPERIMENTAL LAKES AREA:
http://www.umanitoba.ca/institutes/fisheries

FISHERIES AND OCEANS CANADA—FRESHWATER INSTITUTE:
http://www.cisti.nrc.ca/programs/indcan/fedlabs/text.111.html

FRASER RIVER ACTION PLAN:
http://www.pwc.bc.doa.ca/ec/frap/fr_sel.html

GREAT LAKES INFORMATION MANAGEMENT RESOURCE:
http://www.cciw.ca/glimr/intro.html

GREAT LAKES INFORMATION NETWORK:
http://www.greatlakes.net

INTERNATIONAL JOINT COMMISSION:
http://www.ijc.org

NATIONAL WATER RESEARCH INSTITUTE:
http://www.cciw.ca/nwri_e/intro.html

NORTHERN RIVER BASINS STUDY:
http://www.gov.ab.ca/~env.nrs/nrbs/nrbs.html

ST LAWRENCE VISION 2000 ACTION PLAN:
http://www.siv2000.qc.ec.gc.ca/slv2000/english/indeng.html

WATERCAN:
http://www.watercan.com

WETNET: THE WETLANDS NETWORK:
http://www.wetlands.ca

REFERENCES AND SUGGESTED READING

Berger, T.R. 1977. *Northern Frontier; Northern Homeland*, vol. 1. Ottawa: Minister of Supply and Services.

Berkes, F. 1988. 'The intrinsic difficulty of predicting impacts: lessons from the James Bay hydro project', *Environmental Impact Assessment Review* 8: 201–20.

Bone, R.M. 2005. *The Regional Geography of Canada*, 3rd edn. Toronto: Oxford University Press.

Bourassa, R. 1985. *Power from the North*. Scarborough, Ont.: Prentice-Hall Canada.

Brooks, G.R., and E. Nielson. 2000. 'Red River, Red River Valley, Manitoba', *Canadian Geographer* 44: 306–11.

Burn, D.H., and N.K. Goel. 2001. 'Flood frequency analysis for the Red River at Winnipeg', *Canadian Journal of Civil Engineering* 28: 355–62.

Canadian Heritage Rivers Board. 2002. *The Canadian Heritage Rivers System: Annual Report 2001–2002*. Ottawa: Minister of Public Works and Government Services Canada.

———. 2003. *The Canadian Heritage Rivers System: Annual Report 2002–2003*. Ottawa: Minister of Public Works and Government Services Canada.

Chevalier, G., C. Dumont, C. Langlois, and A. Penn. 1997. 'Mercury in Northern Québec: role of the Mercury Agreement and the status of research and monitoring', *Water, Air and Soil Pollution* 97: 75–84.

Cressman, D.R. 1994. 'Remedial action programs for soil and water degradation problems in Canada', in T.L. Napier, S.M. Camboni, and S.A. El-Swaify, eds, *Adopting Conservation on the Farm*. Ankeny, Iowa: Soil and Water Conservation Society, 413–34.

Day, J.C., and F. Quinn. 1992. *Water Diversion and Export: Learning from Canadian Experience*. Department of Geography Publication Series No. 36. Waterloo, Ont.: University of Waterloo.

Diamond, B. 1985. 'Aboriginal rights: the James Bay experience', in M. Boldt and J.A. Long, eds, *The Quest for Justice: Aboriginal People and Aboriginal Rights*. Toronto: University of Toronto Press, 265–85.

———. 1990. 'Villages of the dammed', *Arctic Circle* (Nov.–Dec.): 24–34.

Dumont, C., M. Girard, F. Bellavance, and F. Noël. 1998. 'Mercury levels in the Cree population of James Bay, Quebec, from 1988 to 1993/94', *Canadian Medical Association Journal* 158: 1439–45.

Environment Canada. 2003a. *Environmental Signals: Canada's National Environmental Indicator Series 2003*. Ottawa: Environment Canada.

———. 2003b. *Performance Report for the Period Ending March 31, 2003*. Ottawa: Environment Canada.

———. 2004. *Threats to Water Availability*. NWRI Scientific Assessment Report Series No. 3 and ACSD Science Assessment Series No. 1. Ottawa and Burlington, Ont.: Environment Canada and National Water Research Institute.

——— and United States Environmental Protection Agency. 2003. *State of the Great Lakes 2003*. Ottawa: Environment Canada.

Foster, H.D., and W.R.D. Sewell. 1981. *Water: The Emerging Crisis in Canada*. Ottawa: Canadian Institute for Economic Policy.

Gabriel, A.O., and R.D. Kreutzwiser. 1993. 'Drought hazard in Ontario: a review of impacts, 1960–1989, and management implications', *Canadian Water Resources Journal* 18: 117–32.

Gorrie, P. 1990. 'The James Bay Power Project: the environmental cost of reshaping the geography of northern Quebec', *Canadian Geographic* 110, 1: 21–31.

Government of Canada and the Government of the Province of Ontario. 1988. *First Report of Canada under the 1987 Protocol to the 1978 Great Lakes Water Quality Agreement*. Toronto: Environment Canada, Communications Directorate, Dec.

Hall, Joseph. 2004. 'Hazel's gift wrapped in green', *Toronto Star*, 15 Oct., B1–B2.

International Joint Commission (IJC). 1998. *Ninth Biennial Report on Great Lakes Water Quality*. Windsor, Ont.: International Joint Commission.

———. 2000. *Living with the Red*. Ottawa and Washington: International Joint Commission.

International Reference Group on Great Lakes Pollution from Land Use Activities. 1978. *Environmental Management Strategy for the Great Lakes System: Final Report to the International Joint Commission*. Windsor, Ont.: International Joint Commission, July.

McCutcheon, S. 1991. *Electric Rivers: The Story of the James Bay Project*. Montreal: Black Rose Books.

Mackintosh, W.A. 1934. *Prairie Settlement: The Geographic Background*. Toronto: Macmillan.

Mainville, R. 1992. 'The James Bay and Northern Quebec Agreement', in M. Ross and J.O. Saunders, eds, *Growing Demands on a Shrinking Heritage: Managing Resource Conflicts*. Calgary: Canadian Environmental Law Association, 176–86.

Mitchell, B., and D. Shrubsole. 1994. *Canadian Water Management: Visions for Sustainability*. Cambridge, Ont.: Canadian Water Resources Association.

Nelson, J.G., B. Veale, B. Dempster, et al. 2004. *Towards a Grand Sense of Place*. Waterloo, Ont.: University of Waterloo, Heritage Resources Centre.

O'Connor, D.R. 2002a. *Report of the Walkerton Inquiry: Part One, The Events of May 2000 and Related Issues*. Toronto: Ontario Ministry of the Attorney General, Queen's Printer for Ontario.

———. 2002b. *Report of the Walkerton Inquiry: Part Two, A Strategy for Safe Drinking Water*. Toronto: Ontario Ministry of the Attorney General, Queen's Printer for Ontario.

O'Neill, D. 2004. 'Threats to water availability in Canada—a perspective', *Threats to Water Availability in Canada*, NWRI Scientific Assessment Report Series No. 3 and ACSD Science Assessment Series No. 1. Ottawa and Burlington, Ont.: Environment Canada and National Water Research Institute, xi–xvi.

Ontario Advisory Committee on Watershed-based Source Protection Planning. 2003. *Protecting Ontario's Drinking Water: Toward a Watershed-based Source Protection Planning Framework: Final Report*. Toronto: Ministry of the Environment.

Perkel, C.N. 2002. *Well of Lies: The Walkerton Water Tragedy*. Toronto: McClelland & Stewart.

Quinn, F., J.C. Day, M. Healey, R. Kellow, D. Rosenberg, and J.O. Saunders. 2004. 'Water allocation, diversion and export', *Threats to Water Availability in Canada*, NWRI Scientific Assessment Report Series No. 3 and ACSD Science Assessment Series No. 1. Ottawa and Burlington, Ont.: Environment Canada and National Water Research Institute, 1–8.

Royal Commission on the Future of the Toronto Waterfront. 1992. *Regeneration: Toronto's Waterfront and the Sustainable City, Final Report*. Ottawa and Toronto: Minister of Supply and Services Canada and Queen's Printer of Ontario.

Saskatchewan Watershed Authority. 2003. *Saskatchewan Watershed Authority: Annual Report*. Regina: Saskatchewan Watershed Authority.

Shiklomanov, I.A. 2000. 'Appraisal and assessment of world water resources', *Water International* 25: 11–32.

Shrubsole, D. 2001. 'The cultures of flood management in Canada: insights from the 1997 Red River experience', *Canadian Water Resources Journal* 26: 461–79.

Standing Committee on Agriculture, Fisheries and Forestry. 1984. *Soil at Risk: Canada's Eroding Future*. Ottawa: Senate of Canada, Committees and Private Legislation Branch, Standing Committee on Agriculture, Fisheries and Forestry.

Statistics Canada. 2003. *Human Activity and the Environment: Annual Statistics 2003*. Cat. 16–201, 3 Dec. 2003. Ottawa: Statistics Canada.

Stewart, J.M. 1993. 'Future state visioning—a powerful leadership process', *Long Range Planning* 26, 6: 89–98.

Todhunter, P.E. 2001. 'A hydroclimatological analysis of the Red River of the North snowmelt flood catastrophe of 1997', *Journal of the American Water Resources Association* 37: 1263–78.

Chapter 13

Minerals and Energy

Learning Objectives

- To understand the characteristics of non-renewable resources relative to the renewable resources discussed in previous chapters.
- To appreciate the significance of minerals and energy for Canada.
- To understand the management issues associated with non-renewable resources in general and minerals and energy in particular.
- To realize the significance of the Whitehorse Mining Initiative.
- To identify the relative importance of different minerals for the Canadian mining industry.
- To know how science can be used as a foundation for remediation strategies for mining landscapes, as illustrated by ongoing work in Sudbury, Ontario.
- To discover how science is used in environmental assessments, illustrated by the Ekati diamond mine in the Canadian North.
- To appreciate that energy resources can be both renewable and non-renewable.
- To learn of the potential of alternative, renewable energy sources, particularly wind power.
- To know what you can do to have a lighter 'footprint' related to use of minerals and energy.

INTRODUCTION

Previous chapters have focused on **renewable (flow) resources**, usually defined as those that are renewed naturally within a relatively short period of time, such as water, air, animals, and plants (Rees, 1985: 14). Other renewable resources are solar radiation, wind power, and tidal energy. Given this mix, a distinction is often made between renewable resources not dependent on human activity (e.g., solar radiation) and those that renew themselves as long as human use allows reproduction or regeneration (e.g., fish).

Figure 13.1 highlights that flow or renewable resources can exist in critical or non–critical zones. Those in the critical zone can be harvested or exploited to exhaustion. The most vulnerable depend on biological reproduction for renewal. Whether through overhunting, overfishing, polluting, or destroying habitats, humans can create conditions so that renewable resources cannot replace or replenish themselves. Indeed, the 'collapse' of the northern cod in the Northwest Atlantic (see Chapter 8) is often noted as a classic case of overharvesting of a renewable resource that led to its depletion.

In this chapter, emphasis is mostly on **non-renewable (stock) resources**, which take millions of years to form. As a result, from a human viewpoint, such resources are for practical purposes fixed in supply and therefore not renewable. However, as with renewable or flow resources,

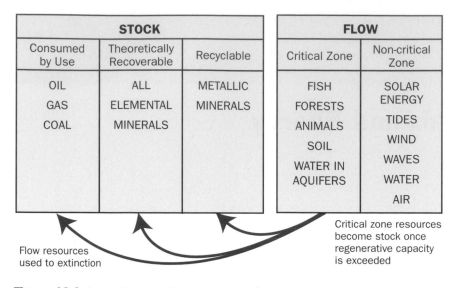

STOCK			FLOW	
Consumed by Use	Theoretically Recoverable	Recyclable	Critical Zone	Non-critical Zone
OIL GAS COAL	ALL ELEMENTAL MINERALS	METALLIC MINERALS	FISH FORESTS ANIMALS SOIL WATER IN AQUIFERS	SOLAR ENERGY TIDES WIND WAVES WATER AIR

Flow resources used to extinction

Critical zone resources become stock once regenerative capacity is exceeded

Figure 13.1 A classification of resource types. SOURCE: Rees (1985: 13). Reprinted by permission of Thompson Publishing Services.

Figure 13.1 indicates that non-renewable or stock resources are not homogeneous. Some are consumed through use, whereas others can be recycled. Those consumed by use are best illustrated by fossil fuels (coal, oil, natural gas). Once used, they are effectively not available to humans, even though you know from earlier chapters that they do not really disappear but are changed into another form, often polluting material. In contrast, stock resources such as metals can be recycled many times, so the stock of resources in the ground is not the only source. However, recycling often requires significant amounts of energy, so the recycling of one type of stock resource (aluminum) may hasten the depletion of another (coal, oil, or natural gas) stock resource.

Our attention here focuses on both minerals and energy. And, to emphasize that non-renewable resources are not homogeneous, the discussion of energy focuses on wind power, usually viewed as one type of renewable or flow resource. Whatever categories are considered, it is important to appreciate that mineral and energy resources are important in and for Canada. For instance, Canada is ranked as the sixth largest user of primary energy in the world. Such a high level of use is attributed to a large country with long travel distances, as well as a cold climate, an energy-intensive industrial base, relatively low energy prices, and a high standard of living. Regarding minerals, Canada is one of the major global exporters. More details about energy and minerals will be provided later

in the chapter, but the key message here is that both are important for regional and national economies in this country and that, in their extraction and use, they can have significant environmental impacts.

FRAMING ISSUES AND QUESTIONS

The challenge for renewable resources is to manage them so that they remain renewable or sustainable. For non-renewable resources, however, their extraction usually results in absolute depletion in any time frame other than a geological one. Given the characteristics of non-renewable resources, the management issues are usually different from those associated with renewable resources (Sanchez, 1998: 522; Hilson, 2000: 202–3). Specifically, concerns focus on:

1. how to use the proceeds from resource extraction to generate new wealth, which, as new capital, can be used to benefit generations today and in the future;
2. how to conserve mineral or fossil fuel assets to extend the longevity of reserves, and how to identify substitutes for use in the long run;
3. how to minimize negative environmental impacts at each stage in the life cycle of use: exploration, extraction, transformation, consumption, recycling, and final disposal;
4. how to create improved socio-economic relationships with stakeholders, especially the local communities located in the mining area, after the resource has been exhausted;
5. how to manage recyclable non-renewable resources, i.e., many metals and minerals, as a renewable or flow resource.

What is the motivation for mining and fossil fuel firms to engage in environmental management, given that their priority is to maximize profits and remain competitive in international markets? As Hilson (2000: 203) noted, if done systematically and correctly, enhanced 'environmental management practices and extended social responsibility almost always generates some kind of economic return on investment for business, although usually over the long term. A documented reduction in effluent discharges, for example, leads to a reduction in costly government inspections and auditing practices.'

As Sanchez (1998: 522–3) observed, the main environmental issues for the mining and energy sectors include **acid mine drainage**, **sulphur dioxide emissions**, and **metal toxicity**.

1. *Acid mine drainage*. Most non-ferrous metals exist as sulphides and usually are accompanied by iron sulphides. When ore minerals are separated from minerals without economic value, significant quantities of waste rock and tailings are created, and these contain iron sulphides that can readily oxidize to become sulphuric acid. When exposed to precipitation (rain or snow), sulphuric acid can dissolve residual metals, leading to acidic drainage, which can continue for centuries. It has been estimated that liabilities in the Canadian mining industry related to acidic drainage may range between $2 billion and $5 billion.

2. *Sulphur dioxide emissions*. One outcome of smelting sulphide ores is the release of huge quantities of sulphur, mainly in the form of dioxides, into the atmosphere. The Canadian mining sector has been estimated to be the main contributor to sulphur dioxide emissions in Canada. The burning of fossil fuels is also a major source of atmospheric emissions of sulphur dioxides, creating pressure for alterna-

tive sources of energy that are less damaging to the environment.

3. *Metal toxicity*. As understanding increases through research, the mining industry is being challenged about the toxic effects of metals on human and ecosystem health. For example, many uses of asbestos are now not acceptable because of connections established between it and cancer. Lead is also considered to be toxic. Emissions from smelting and steelmaking processes also can threaten health, as illustrated by the case study of the Sydney Tar Ponds in Chapter 1.

To these three issues can be added the challenges noted in the above box related to energy:

- disruption of remote ecosystems due to exploration, test drilling, and operation of oil fields or gas wells, ranging from habitat degradation to alteration of nesting, denning, and migration patterns of birds and animals;
- disturbance to aquatic ecosystems from escape of waste heat produced from nuclear energy production;
- threat to human and ecosystem health from radioactive waste associated with nuclear energy production over a period of thousands of years;

BOX 13.1 WHITEHORSE MINING INITIATIVE

During September 1992, at the annual meeting of the mining sector in Whitehorse, the Mining Association of Canada proposed a multi-stakeholder process to examine challenges and opportunities for mining (McAllister and Alexander, 1997). The challenges included growing competition for mineral investment from other mining countries and growing uncertainty within Canada due to questions related to land access, environmental regulations, and taxation. Four groups were established to examine issues related to: land access; environment; workforce, workplace, and community; and finance and taxation.

Two years later, in September 1994, at the annual meeting of federal and provincial mines ministers, leaders from the mining industry, governments, labour, Aboriginal groups, and environmental organizations signed the Whitehorse Mining Initiative Accord. Sixteen principles and 70 goals are in the Accord, designed to support an overall vision of a 'socially, economically, and environmentally sustainable and prosperous mining industry, underpinned by political and community consensus'. The five principles and selected goals related to the environment, which are equally relevant to the fossil energy sector, are:

Environmental Protection
Principle: Environmentally responsible mining exploration, development, operations, and public policies are predicated on maintaining a healthy environment and, on closure, returning mine sites and affected areas to viable and, wherever practicable, self-sustaining ecosystems compatible with a healthy environment and with human activities.

Goals include: (1) voluntary and regulatory means, including appropriate environmental effects monitoring, to ensure minimal environmental impact during mining exploration, development, operations, and closure; (2) preparation of comprehensive reclamation plans to return all mine sites to viable and, when practicable, self-sustaining ecosystems; (3) establishment in each jurisdiction of an acceptable way to identify responsible parties to undertake reclamation of old mine sites that pose health, safety, or environmental problems.

Planning and Environmental Assessment
Principle: Environmental assessment is an essential tool for identifying potential environ-

mental impacts of proposed projects, determining their acceptability, and evaluating potential mitigation and remediation measures, thereby enabling economic activity to proceed while safeguarding the health of the environment.

Goals include: (1) effective, efficient, and well-defined project-specific environmental assessments that are conducted in the broader context of an integrated land-use planning process as well as government policies and programs; (2) terms of reference and scope of environmental assessments that are ecologically relevant and determined early in the process; (3) environmental assessment processes that are formally structured, credible, balanced, and fair; (4) efficient and effective monitoring programs to provide adequate feedback to stakeholders.

Use of Information and Science in Environmental Decision-making
Principle: For sound environmental decisions to be taken during the life cycle of a mine, (1) all stakeholders need access to high-quality, relevant, and unbiased information grounded in sound science, but (2) complete scientific certainty is not a prerequisite to appropriate action to protect the environment where risk of serious adverse impacts to the ecosystem is evident.

Goals include: (1) broadening and improving the information base on the environmental effects of mining, and ensuring that all information is accurate, unbiased, and developed in a manner consistent with professional standards and scientific methods; (2) meaningful participation by Aboriginal peoples and the use of traditional and local knowledge; (3) decisions that could lead to serious adverse impacts on ecosystems are made cautiously, are based on the best available information, and address the limitations of science.

Land Use and Land Access
Principle: Access to land for exploration and development is a fundamental requirement for the mining industry.

Goals include: (1) accessibility of land-use and land-access policy and decision-making processes to all stakeholders whose interests are affected; (2) decision-making processes that consider the requirements of the mining industry and other stakeholders for land access and use;

(3) collaborative mechanisms to allow stakeholders to address and resolve contentious issues on an ongoing basis, both in the context of specific projects and for broader policy matters.

Protected Areas
Principle: Protected area networks are essential contributors to environmental health, biological diversity, and ecological processes, as well as being a fundamental part of the sustainable balance of society, the economy, and the environment.

Goals include: (1) establishing and setting aside from industrial development protected areas representative of Canada's land-based natural regions; (2) government policies clearly stating that, subject to complying with all applicable legislation and regulatory requirements, mining is an acceptable and permitted activity in non-protected areas; (3) the involvement of Aboriginal peoples in the selection and management of protected areas in such a manner that they can benefit from economic opportunities related to development and operation of these protected areas and have access consistent with management plans for traditional economies and ceremonial, cultural, subsistence, and social practices; (4) co-ordination in the selection of protected areas across jurisdictions to achieve representation without unnecessary duplication.

- alteration to ecosystems from building of hydroelectric dams and generating stations.

These issues, individually and collectively, provide a strong rationale for increased attention to environmental aspects and management in the mining and fossil fuel sectors.

'Best practice' related to environmental management for mining and fossil fuel firms in Canada should include a combination of basic scientific research to ensure understanding of natural and social systems that can be affected by operations, and design of appropriate mitigation measures, environmental impact assessments and reporting, environmental audits, corporate policies that explicitly include environmental aspects, environmental management systems (Hilson and Nayee, 2002), and life-cycle assessments.

As noted in Chapter 5, whichever mix of approaches is used, it can be expected that increased emphasis will be placed on partnerships, alliances, and stakeholder involvement. In that regard, the Canadian mining sector took an innovative step, known as the **Whitehorse Mining Initiative**, which is outlined in Box 12.1.

The outcomes of the Whitehorse Mining Initiative Accord have been numerous and varied. Mining company representatives have become more engaged in thinking about, promoting, and incorporating environmental considerations in private-sector activities. The federal government, in 1996, created a Minerals and Metals Policy that built directly on the Accord and was viewed around the world as the first initiative at a global scale seeking to balance economic, environmental, and social aspects. Various provincial governments, such as those of Nova Scotia and Manitoba, created new policies to guide the mining industry towards more sustainable practices. As Hilson (2000: 206) observed, six years after the Accord was signed, it 'has been instrumental in shifting the goals of the mining industry, its regulators, and its stakeholders from a "business as usual" mode to a sustainable development mode.'

The Whitehorse Accord helps to give credibility to the concepts of consultation, collaboration, and partnerships. However, it is important not to overdramatize this initiative. As Whiteman and Mamen (2001: 30) commented in their assessment of experience related to community consultation in mining, throughout the world the history of mining operations and their relationships with indigenous peoples 'has been grim, pervaded by many examples of human rights abuses and severe social and environmental impacts. . . . While the fundamental rights of indigenous peoples to make decisions concerning their lands and resources has been widely acknowledged, significant gaps exist between the rhetoric and reality.' Further comments from them are highlighted in the following box.

NON-RENEWABLE RESOURCES IN CANADA: BASIC INFORMATION

According to Natural Resources Canada (2003), Canada is one of the largest exporters of minerals

CONSULTATION NOT ALWAYS ENOUGH

In the search for ways to promote indigenous peoples' perspectives, community consultation processes are often proposed...because they can empower local communities to influence decisions on mining projects. Yet despite such potential, many mining consultation processes serve as a smokescreen, lacking substance and creating the illusion of democratic process. Public relations ploys justified under the name of consultation abound. The issues facing indigenous peoples and the international mining sector cannot be solved solely through the development of guidelines for community consultation or progressive corporate policies on indigenous peoples. At the crux of the problem are power inequities and the lack of institutional will to share decision-making power with local communities.

— Whiteman and Mamen (2001: 31)

and mineral products in the world. During 2001, minerals and mineral products accounted for almost 13 per cent of total exports for the nation and contributed to its trade surplus. About 80 per cent of Canadian mineral and metal production is exported. As an example, Canada is the seventh largest exporter of coal, in 2001 exporting about 30 million tonnes, which was 43 per cent of total coal production in Canada, valued at $1.9 billion. Furthermore, as documented by Lemieux (2002), Canada has more large mineral exploration companies than any other country. Canada accounted for 33 per cent of such companies, and the closest rival is Australia, accounting for 20 per cent.

Mining and mineral processing industries are important to the national and local economies. In 2002, they added $36.1 billion to the Canadian economy, which represented 3.7 per cent of gross domestic project. Furthermore, they directly employed 361,000 Canadians, with 47,000 working in mining, 52,000 in smelting and refining, and 262,000 in manufacturing of mineral and metal products. At the beginning of 2003, about 190 metal, non-metal, and coal mines existed, not including thousands of stone quarries and sand/gravel pits, as well as about 50 non-ferrous smelters, refineries, and steel mills.

In 2002, the value of mineral products was about $19.6 billion. The top five products in terms of production value in 2002 were gold ($2.3

billion), nickel ($1.9 billion), potash ($1.6 billion), copper ($1.4 billion), and cement ($1. 4 billion). When coal, petroleum, and natural gas are included, the total for non-renewable resources becomes $77 billion, which highlights the importance of the fossil fuel industry. Canada ranks first at a global scale as a producer of potash and uranium, and fourth for primary aluminum. It is ranked in the top five for nickel, asbestos, zinc, cadmium, titanium concentrate, platinum group metals, salt, gold, molybdenum, copper, gypsum, cobalt, and lead.

Potash is the most important product in the non-metals group. Canadian potash exports totalled 8 million tonnes in 2002. The potash industry began in Saskatchewan during the early 1960s and expanded steadily during the 1970s and 1980s so that now Canada is the largest producer and exporter of potash in the world. Furthermore, Saskatchewan's potash industry has been judged to be the most productive.

In 1991, following more than a decade of geological detective work, two Canadian geologists, Chuck Fipke and Stewart Blusson, discovered minable diamonds underneath Lac de Gras in the Northwest Territories, leading to the opening in 1998 of the first diamond mine (BHP's Ekati mine) in Canada, after an investment of $700 million. Ekati, located 200 kilometres south of the Arctic Circle, supplies about 6 per cent of the world's annual $8 billion in rough diamonds. A second diamond mine, Diavik, began producing in 2003 after a $1.3 billion investment by Rio Tinto/Aber Resources. Canada produced almost 15 per cent of the world's diamonds in 2003, making it the third largest producer after Botswana and Russia. The output in 2003 was 11.2 million carats, with a value of $1.7 billion, more than twice the value of production in 2002.

In 2003, the Nunavut Impact Review Board conditionally approved what could become Canada's third diamond mine and the first in Nunavut. The mine would be operated by Vancouver-based Tahera Corporation, and its Jericho mine, located 420 km northeast of Yellowknife, near Contwoyto Lake, could begin operation by the end of 2005. And, in 2006, De Beers is expected to begin producing diamonds from a fourth mine at Snap Lake, NWT. In addition to the diamond production and exploration in the

RECYCLING GROWING IN IMPORTANCE

Canada not only is a producer of metal and non-metal products. Natural Resources Canada (2002) has estimated that about 10 million tonnes of metals are recycled each year, and in 2002 this had a value of $3 billion. Canada imports and exports recycled metals, and 85 per cent of trade in recycled metals is with the United States. Over 3,000 metal recycling companies operate in Canada, employing some 15,000 people.

NWT and Nunavut, major exploration is underway in Alberta, Saskatchewan, Manitoba, Ontario, and Quebec, and one of the largest bodies of diamond-bearing ore in the world has been found some 50 km northeast of Prince Albert, Saskatchewan. Some predictions indicate that in 20 years Canada could be the global leader in terms of value production of diamonds. More details are provided in a later section discussing the development of Ekati.

Canadian mineral production is not only exported. It is also used within the country. For example, coal and uranium are the basis for one-third of electricity production. Alberta, which produces nearly half of the coal mined in Canada, depends on coal for almost 85 per cent of its electrical power.

In the following sections, attention is given to how science has been incorporated into initiatives to remediate landscape degradation associated with mineral extraction and to how it has been used in understanding and mitigating environmental impacts when a new mining venture is being designed. Finally, we consider the role of science in exploring alternatives to fossil fuel energy sources.

REMEDIATING MINED LANDSCAPES: SUDBURY, ONTARIO

Located about 400 km north of Toronto, Sudbury at one time was 'notorious across the country for the air pollution and the barren, blackened landscape created by its smelters' (Richardson et al., 1989: 4). However, starting in the 1970s, some remarkable initiatives were taken to rehabilitate the landscape and restructure the economy of Sudbury. The Sudbury story of remediation and rehabilitation of an industrialized landscape illustrates that challenging circumstances can lead to opportunities and positive changes.

Sudbury was established in the early 1880s as a construction camp for the Canadian Pacific Railway. During building of the railway, copper and nickel deposits were discovered only a few kilometres north of the construction camp. The result was that the construction camp evolved into a mining community, and Sudbury became the second largest world producer of nickel. In addition to Sudbury, other mining communities such as Falconbridge were established in the 30 x 60 km Sudbury basin (Figure 13.2).

As with most mining-based communities, Sudbury experienced boom-and-bust cycles. Peak production occurred in 1970, when nickel output from the Sudbury basin reached an annual high of 200,000 tonnes at the same time as prices for nickel had climbed to their highest levels. At that time, Sudbury had one of the highest average household incomes in Canada. In addition, the economy had begun to diversify, as cultural and advanced educational facilities were built and good transportation links to southern Ontario were in place. The economic prospects of the Sudbury basin at the start of the 1970s appeared to be bright.

On the negative side, however, two aspects were of concern. First, the community had some serious social divisions as a result of bitter and prolonged labour-management and inter-union conflicts within the mining industry. As a result,

RECLAMATION IN SUDBURY

At times the environmental devastation humankind has wrought on our planet seems overwhelming. But when individuals come together and bring their time, energy, money, quick minds, strong backs and collective will to a task they not only can move mountains, they can recreate a landscape. Sudbury is living, thriving proof.

. . . The Sudbury story is a beacon of environmental hope. If our experience can teach the world anything it is that we are all responsible for our natural environment, we can learn from our mistakes and nature is forgiving. Each of us has a role to play and collectively we can make a difference.

— Ross (2001: 8)

Sudbury was a deeply divided community. Second, the strong economic growth, based on the mining industry, was offset by a dramatically degraded physical environment due to previous mining activities. Ten thousand hectares of the Sudbury area were devoid of significant vegetation due to air pollution and past mine practices (Figure 13.3). Many lakes in the area had become highly acidic and degraded by metal contaminants, partially as a result of acid deposition associated with the smelter, as outlined in Chapter 4. The city periodically experienced episodes of choking air pollution from the mining smelters. Sudbury's notoriety was highlighted when the international media mistakenly reported that American astronauts went to Sudbury for training purposes since its stark landscape most closely resembled the terrain on the moon.

While Sudbury's economic growth had been strong and impressive, the reversal of its economic fortunes during the 1970s was equally dramatic. Increasing pressure from international and domestic nickel-producing competitors caused a steady drop in nickel prices, which in turn led to a significant drop in mining employment. For example, in 1970 the International Nickel Company (Inco) and the smaller Falconbridge Nickel Mines together employed nearly 30,000 people. By the late 1970s, that number was fewer than 17,000 employees. A short-term recovery slowed the layoffs, but subsequent extensive shutdowns reduced employment at both companies to below 14,000 by 1983. Thus, over a 12-year period (1971 to 1983), mining employment in the Sudbury basin fell from 47 per cent to 11 per cent of the labour force. This resulted in significant out-migration, but even with that, the unemployment rate in 1983 had climbed to 17 per cent. By 1986, the mining employment had fallen further to 7,700, and between 1971 and 1986 the population

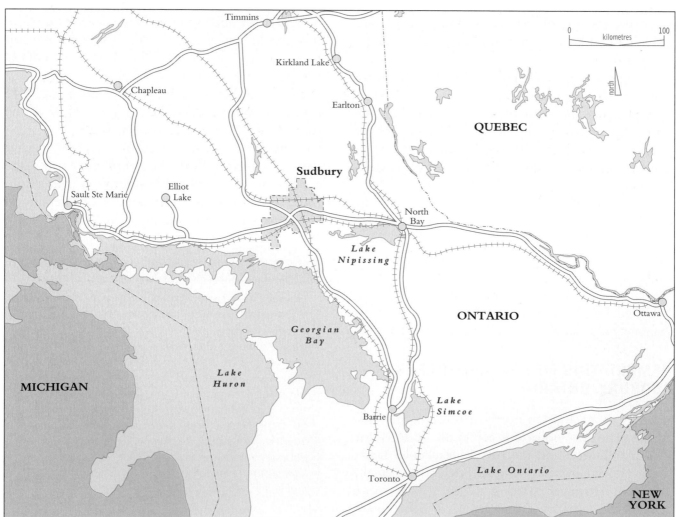

Figure 13.2 Location of Sudbury.

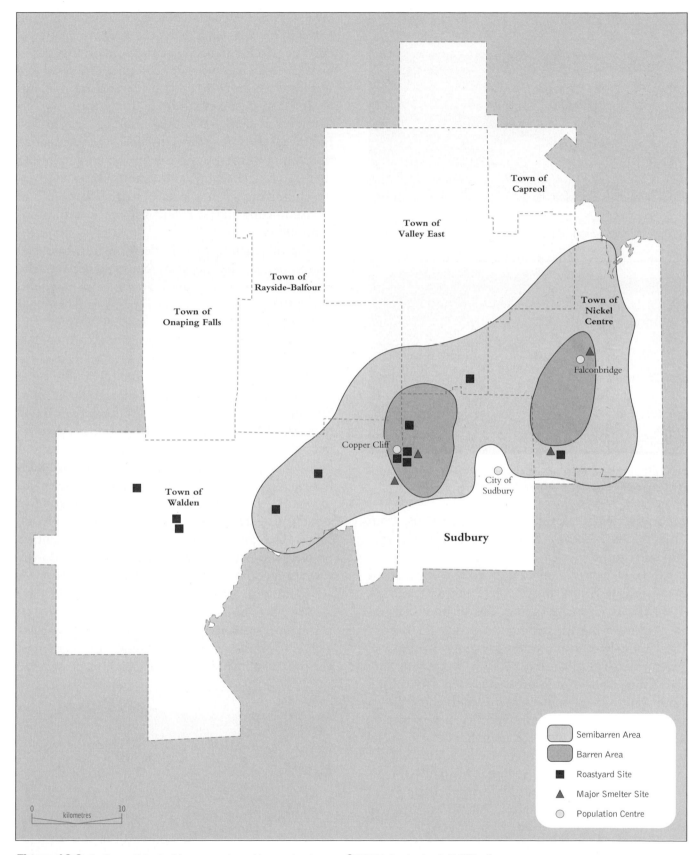

Figure 13.3 Sudbury: Extent of barren and semi-barren landscape. SOURCE: Lautenbach (1985: 4).

A view from the Trans-Canada Highway in 1975 prior to reclamation (© *Regional Municipality of Sudbury*).

A view from the Trans-Canada Highway following reclamation (© *Regional Municipality of Sudbury*).

of Sudbury had dropped by 10 per cent, from 169,200 to 152,000. The prospects for the area appeared to be bleak.

Several factors contributed to the subsequent economic restructuring in the Sudbury basin. In 1973 the Regional Municipality of Sudbury was formed, creating for the first time a federation of the 15 smaller communities that allowed a regional perspective on planning and development. The daunting economic prospects led to much improved labour-management and inter-union relationships, as people realized that it was imperative for them to work together. The provincial and federal governments relocated some of their offices to Sudbury, and regional hospital services were established. Machinery manufacturing also started to grow. In addition, significant growth occurred in service-oriented businesses.

Another important initiative was the development of a community-led strategy called Sudbury 2001 to help rejuvenate and diversify the local economy, with the ultimate goal being to create a sustainable community. A trilateral strategy focused

on economic, social, and physical aspects (Katary, 1982). The *economic goal* was to help Sudbury become a self-sustaining metropolitan area based on a diversified economic structure. The *social goal* was to allow individuals to discover their role and potential in the community, and one thrust was to develop self-help and mutual assistance programs. For the *physical dimension*, the goal was to enhance the quality of the physical environment, including the ecosphere. Particular attention was to be given to minimizing adverse environmental impacts and to maximizing the aesthetic ambience within the community.

On the environmental side, some initiatives had begun before the negative consequences of the economic downturn had become so evident. During the 1970s, Inco and Falconbridge Limited had shut down part of their smelter capacity and had significantly reduced emissions of sulphur dioxide. Inco also built its 381-metre 'Superstack' to disperse the emissions further afield. The result was that the average level of sulphur dioxide in the air at Sudbury, in parts per million, dropped from 0.042 to 0.007 between 1970 and 1988, a decrease of 83 per cent. The outcomes were tangible. The city was no longer periodically subjected to air pollution fumigations. Vegetation could grow and began to return. At the same time, some of those positive aspects were offset by the fact that the Superstack simply had shifted a portion of the air pollution problem from the Sudbury basin to areas downwind, sometimes as far away as parts of Quebec.

With local air quality having been significantly improved, opportunities for revegetation or regreening presented themselves. Beginning in 1971, scientists established small plots to test various combinations of soils and plant species.

Impacted terrain and water environments as a result of past mining activity (© *Regional Municipality of Sudbury*).

Based on this experience, techniques were developed at Inco and in the Department of Biology at Laurentian University to grow grass, clover, and then tree seedlings on formerly degraded land. There were setbacks with the initial plantings, but experience through trial and error led to procedures that resulted in healthy vegetation. The main lesson was that applying crushed limestone resulted in neutralization of acidic soil, inhibition

of the uptake of metals, and enhanced bacterial activity, which allowed some vegetation types to do well. The context for this work has been highlighted by Lautenbach and his colleagues (1995: 112), who commented that 'There was little scientific information to guide the design and implementation of an effective reclamation program for this type of landscape. Restoration ecology was and still is a very new and developing field of study.

Experimental vegetation test plots demonstrated that extensive land reclamation was possible (© *Regional Municipality of Sudbury*).

Therefore, testing and monitoring were essential for achieving the objectives of the reclamation program.' The objectives and outcomes of this work are highlighted in Box 13.1.

The Regional Municipality became involved in the reclamation program during 1974 with the establishment of VETAC—a Vegetation Enhancement Technical Advisory Committee to Regional Council. This committee consists of botanists, ecologists, landscape architects, horticulturalists, agriculturalists, planners, fisheries experts, gardeners, and interested citizens from the mining companies, university and college, provincial ministries, Ontario Hydro (later to become Hydro One), municipalities, and the general public. Together, this group addressed Sudbury's vegetation restoration. After surveying the degraded areas, they identified a target to rehabilitate about 30,000 hectares of barren and semi-barren land.

During the summer of 1978, 174 students worked to apply crushed limestone, fertilizer, and grass seed on an area adjacent to the highway close to the airport and on another area beside the Trans-Canada Highway. At the end of the summer, these students had planted grass on 115 hectares of barren land, removed debris from a further 206 hectares, planted 6,000 trees, shrubs, and

An experimental vegetation test plot successfully reclaimed (© *Regional Municipality of Sudbury*).

Students liming a barren site as part of the region's grassing program (© *Regional Municipality of Sudbury*).

Students planting trees in a previously reclaimed area (© *Regional Municipality of Sudbury*).

plants, collected some 30,000 samples for pH and nutrient testing, created 122 new test plots, and gathered 365 kg of native seed. The next summer, 1979, the number of students increased to 325, and from that time on 150 to 300 students were employed each summer on the reclamation project.

By 1982, records showed nearly 400,000 trees had been planted, and by 1984 more than $12 million had been spent, with 2,636 hectares planted. In 1982, the Regional Municipality began to hire laid-off workers from Inco and Falconbridge to plant trees in the spring before summer students could work. The work continued during the 1980s and 1990s, and in the 1990s, due to

BOX 13.1 OBJECTIVES AND EVALUATION OF THE GRASSING AND REFORESTATION ACTIVITIES IN SUDBURY

The objectives were to:
- create a self-sustaining ecosystem with minimal maintenance;
- use plant species tolerant of acidic soils and low nutrient concentrations;
- use seed application rates that allow for natural colonization and thus increase species diversity;
- give preference to the use of native species;
- restore nutrient cycles and pools by the use of species that fix nitrogen (legumes);
- use species that attract and provide cover for wildlife;
- undertake initiatives that speed up natural successional changes.

Evaluation of the initial reclamation activities provided the following results:
- Vegetation cover in limed areas remained in the range of 10 per cent to 25 per cent.
- There was a rapid, spontaneous colonization of treated sites by herbaceous and woody species.
- Over time, the percentage of grasses tended to decrease and the percentage cover of woody species increased.
- The cover and vigour of legumes increased, relative to grasses.
- Red pine and Jack pine were the most vigorous of the planted trees.
- Surface soil pH increased from 3.5–4.5 before treatment to 4.0–5.5 following treatment and remained in that range.
- Vegetation in reclaimed areas contained elevated aluminum levels in grasses and trees relative to 'normal' sites, but levels of copper or nickel were low.
- The number of insects, birds, and some mammals had increased in reclaimed areas.

SOURCE: Lautenbach (1987, 230; 1995: 112)

BOX 13.2 THE REGREENING RECIPE IN SUDBURY

Amount of lime added:

    10 tonnes per hectare of crushed dolomite limestone

Amount of fertilizer added:

    6–24–24 (NPK) at 400 kg/ha

Seed mixture, at 40 kg/ha:

Grasses (75 per cent)

    1. red top (Agrostis gigantea)
    2. red fescue (Festuca rubra)
    3. timothy (Phleum pratense)
    4. Canada bluegrass (Poa compressa)
    5. Kentucky bluegrass (Poa pratensis)

Legumes (25 per cent)

    1. bird's-foot trefoil (Lots corniculatus)
    2. alsike clover (Trifolium hybridum)

Trees planted (major species):

    1. Jack pine (Pinus banksiana)
    2. red pine (Pinus resinosa)
    3. white pine (Pinus strobus)
    4. white spruce (Picea glauca)
    5. black spruce (Picea mariana)
    6. white cedar (Thuja occidentalis)
    7. larch (Larix laricina)
    8. northern red oak (Quercis rubra)
    9. black locust (Robinia pseudoacacia)

SOURCE: Ross (2001: 117).

improved air quality, some tree species were colonizing previously barren areas. As Ross (2001: 60) reported, one Vegetation Enhancement Technical Advisory Committee member commented on the amazement felt by team members regarding the natural regeneration: 'One day we wandered around up behind the smelter, and our mouths dropped open. We saw all these birch seedlings coming back, and we hadn't planted them.'

Over time, emphasis shifted to planting trees rather than grass, as trees were capable of growing in some places without needing limestone applied first. Thus, while at the outset it was usual for hundreds or more of hectares to be limed, fertilized, and seeded, this was reduced to fewer than 50 hectares annually during the late 1980s and 1990s. In contrast, tree plantings grew to over 100,000 annually for red and Jack pine, red oak, and tamarack between 1983 and 2000. The one millionth tree was planted in 1990, and in 1998 alone 985,574 trees were planted.

The results have been impressive. While slag heaps and bare and blackened rock are still present, by the end of 2002, 3,314 hectares had been limed, 3,135 had been fertilized, and 3,015 had been seeded, and 7.7 million trees had been planted, at a cost of just over $22 million. In summers this formerly barren land is now green with grass and trees (Boxes 13.1 and 13.2). All of this reclamation activity provided short-term employment for students and unemployed individuals, much of the labour cost being covered through employment

programs funded by the federal government. In addition, the mining companies (Inco and Falconbridge) have restored degraded company-owned land and between them have planted 3.5 million trees on their properties. Overall, about 50 per cent of the barren and semi-barren lands have been reclaimed. It will take an estimated 10–15 years and planting of another 5 million trees to achieve the targets established by the Vegetation Enhancement Technical Advisory Committee. A 1,300-hectare

RECOGNITION FOR SUDBURY'S RECLAMATION INITIATIVES

1986: Ontario Horticultural Association's Community Improvement Award

1990: Arboricultural Award of Merit from the International Society of Arboriculture Ontario Inc.; the Lieutenant Governor's Conservation Award from the Council of Ontario; a Government of Canada Environmental Achievement Award.

1992: United Nations Local Government Honours Award; the United States Chevron Award.

1993: Chatelaine magazine stated that Sudbury was one of the 10 best Canadian cities in which to live.

1995: Model Project Award from the Society for Ecological Restoration.

1997: Elaine Burke Memorial Award, from Active Living.

untreated site of blackened rock without any vegetation has been left just south of one smelter to provide a reminder of the conditions before the work started and to serve as a benchmark for measuring future progress.

Other benefits have emerged. Almost a century after elk disappeared from the local landscape, over 150 of these animals have been reintroduced and appear to be thriving in traditional habitat south of Sudbury. Following an absence of 30 years, pairs of peregrine falcons now nest in Sudbury, and during the spring of 2000 a pair of trumpeter swans hatched cygnets.

An intriguing question, posed by Richardson et al. (1989), is the extent to which the environmental enhancement contributed to the economic restructuring and recovery in Sudbury. They concluded that a definitive answer cannot be provided. However, they did conclude that the changes in environmental quality and economic structure had a common cause, modernization of the production facilities at Inco and Falconbridge. The technological changes in the smelting process allowed the companies to become more efficient and capital-intensive, significantly reducing the number of employees. Simultaneously, the modernization allowed the two companies to reduce substantially the emissions of sulphur dioxide into the atmosphere, and thereby to meet increasingly more demanding environmental regulations. And once air quality started to improve, other improvements became possible relating to revegetation, reduction of acidity in lakes, and re-establishment of fisheries.

Richardson and his colleagues (1989: 24–6) drew the following conclusion:

> environmental improvement was probably not a direct cause of economic diversification, but rather a necessary condition without which the economy might have languished. In other words, while government policies and other factors certainly

---

KIMBERLITE

Diamonds are a crystalline type of carbon, stable at depths of 150 km or more beneath the Earth's surface. Kimberlite is a rare igneous rock found at the same or greater depth. Eruptions of kimberlite can transport diamonds to the surface of the Earth. Diamond content can be highly variable in a carrot-shaped kimberlite pipe, ranging from nothing to economic concentrations. Because of glaciation in Canada's North, the top part of the diamond-yielding kimberlite pipes often have been scoured out, leaving a circular depression. Glacial alluvium and water then fill the depression, creating a circular lake over the kimberlite pipe, as was the case at what has become the Ekati mine. There are 136 known kimberlite pipes in the NWT.

— Rylatt and Popplewill (1999: 39); Voynick (1999); Couch (2002: 267)

---

changed Sudbury's economic structure, we suggest that these policies might not have achieved their intended result without the improvements in environmental quality that took place....

In other words, while environmental improvement may not have generated economic benefits directly, it could have provided the conditions necessary for the positive economic changes in Sudbury to take root and flourish. In ecological terms, environmental quality could be a factor which does not directly *produce* economic benefits but which, when lacking, can *limit* economic development.

Improvement of the environment by itself was unlikely to stimulate economic renewal, but as has been pointed out, it could have become a serious constraint on renewal initiatives. As a result, the experience in Sudbury reinforces the argument from sustainable development that economic development and environmental management should be viewed as complementary, that each can support the other, and that new decision-making processes that consider a broader mix of factors and include the public are necessary.

DEVELOPING A DIAMOND MINE: EKATI, NWT

The story of the opening of Ekati is now part of Canadian mining lore. In 1980, exploration geologists Chuck Fipke and Stewart Blusson noticed allu-

---

EKATI: A FIRST FOR CANADA

Ekati is not just another mine. It is a diamond mine, Canada's first, and it is one of the world's better ones, with roughly 4 per cent of the world production by weight and 6 per cent of world production by value.

— Werniuk (1998: 9)

vial traces of pyrope garnet, ilmenite, and chrome diopside—all indicator minerals associated with diamonds—while working near the border of the Yukon and Northwest Territories. These indicator minerals had been dispersed during the last ice age (10,000 BP), and Fipke and Blusson realized that their presence did not mean diamonds were close by. Analysis of paleoglaciation and drainage patterns led them to believe that they would have originated in **kimberlite pipes**, somewhere in a 65 million km^2 area of tundra to the east.

They began an extensive program of exploration. Usually travelling by helicopter, they collected thousands of alluvial samples and examined each one for traces of indicator minerals. By 1983, they observed a promising pattern. As they moved east, the concentration of indicator minerals increased, and, more significantly, the crystals had less alluvial wear. By 1989, the trail of indicator minerals had led them some 640 km to the east, and other geologists also were searching. In that year, although no diamonds had yet been found, Fipke and Blusson staked claims on 1,800 km^2 of tundra. Later that year, travelling again by helicopter, they noticed that Point Lake was circular and much deeper compared to other lakes in the area. They sampled along its shoreline, and discovered a chrome diopside crystal with no alluvial wear, suggesting that a kimberlite pipe was close by, maybe even underneath the lake. Verification required very expensive core drilling. A partnership was arranged with a major Australian mining company, BHP World Minerals, and in 1991 core samples from Point Lake yielded the first diamonds found in Canada. The subsequent public announcement, required under Canadian law, triggered a huge mineral rush, with 260 companies from eight nations attracted to the Canadian North staking

---

PROCESS OF DISCOVERY

Economically viable gem-bearing diamond pipes are very rare. There were only 15 such areas known in Africa, Siberia, Australia, and Brazil. In a seemingly impossible task, Canadian geologists Charles (Chuck) E. Fipke and Dr Stewart Blusson crisscrossed the Arctic by foot and light aircraft for a decade looking for diamonds until they found indicator minerals near Lac de Gras in 1989.

— *Couch (2002: 266)*

MEANING OF 'EKATI'

'Ekati' is a Dene word, meaning 'fat lake', but it has been said that it actually refers in this context to the white granite rock outcrops that resemble caribou fat.
<www.ccrs.nrcan.gc.ca/ccrs/learn/tour/43/43nwt_e.html>

---

claims totalling 194,000 km^2. Following an environmental impact assessment review during 1994 to 1996, in October 1998 Canada became a major producer of diamonds, when the Ekati Diamond Mine started up. Today, it produces $1.7 million of diamonds each day.

The Environmental Context

A claim was staked by BHP Minerals for an area of 3,400 km^2 situated some 300 km northeast of Yellowknife (Figure 13.4). The mining activity is located mainly in the Koala River watershed, which drains into Lac de Gras and then northward into the Coppermine River and on to the Arctic Ocean. The mine is found in the Low Arctic Ecoclimate region, in which the average annual temperature is −11.8° Celsius. The temperature range is large, with daily temperatures in summer reaching 25° C, and winter temperatures often falling below −30° C. Precipitation is low, averaging only 300 mm, most of which falls as snow.

The BHP claim area is in the tundra region, 100 km north of the tree line. About one-third of it is covered by some 8,000 lakes, and the landscape has continuous permafrost, with up to 250 metres deep of permanently frozen subsoil and rock, with an over layer of about one metre that thaws during summer. The main vegetation includes stunted shrubs and grass tussocks, with willows and scrub bush appearing in low areas. Wetlands include a combination of water sedges and sedge-willow communities.

Two important mammals exist in this environment: the Bathurst caribou herd and grizzly bears. The caribou herd has approximately 350,000 animals and moves around a range of about 250,000 km^2. The caribou spend the winter south of the tree line, then in the spring start their northward migration to calving grounds near Bathurst Inlet on the Arctic Ocean. The grizzly bear, due to low numbers, density, and reproduction rates, has been designated as vulnerable (see Chapter 11).

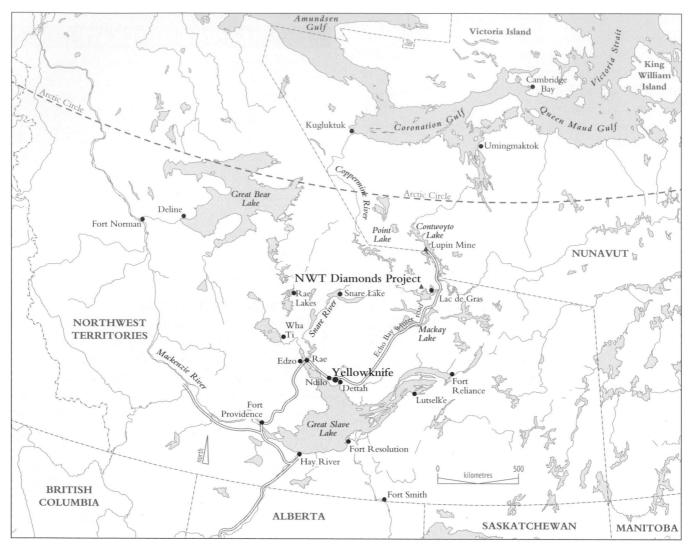

Figure 13.4 Location of NWT Diamonds Project. SOURCE: Canadian Environmental Assessment Agency (1996: 6). Reproduced with the permission of the Minister of Public Works and Government Services Canada, 2005.

Economic and Social Context

The economic aspects of the Ekati Diamond Mine are significant. It has been estimated that the total project capital cost will be $1.2 billion (Canadian), the contribution to the Canadian gross national product will be $6.2 billion, and the direct, indirect, and induced benefits to the NWT will be $2.5 billion (60 per cent of that would be wages and benefits). In terms of employment, the Ekati mine will employ twice as many people as any other mine in the NWT, a total of 830 over the estimated 25-year lifespan. The mining company's policy has been to first hire NWT Aboriginal people, then non-Aboriginal NWT residents, and finally other Canadians. When Aboriginal people do not have necessary skills, the mining company provides appropriate education and training, related, for example, to literacy and technical skills.

The mining company committed to give preference for contracting to businesses owned by Aboriginal people, to establish scholarship programs, on-the-job training programs for Aboriginal students, and cross-cultural training in the workplace, and to draw upon elders, youth, and organizations to ensure traditional knowledge is included when dealing with community issues (see the guest statement in Chapter 7 regarding traditional knowledge). Furthermore, the workplace is drug- and alcohol-free, and offenders are fired without exception.

Numerous challenges were encountered by the company. The area staked for mining had been used for traditional hunting and fishing by Aboriginal people representing four groups: (1) Treaty 11 Dogrib Dene, (2) Treaty 8 Akaitcho Dene (Chipewyan Dene), (3) North Slave Métis Alliance, and (4) Kitkmeot Inuit Association. Members of these groups speak one or more of English, several Athapaskan languages, or Inuinnaqtun (a dialect of Inuktitut spoken in Canada's western Arctic), creating communication problems. The situation was complicated further because two separate land claims agreements were under negotiation and a third was being considered. Another issue was the role of *traditional knowledge*. Such knowledge is transmitted verbally from generation to generation, and, as a result, there is usually no written record of it. Western scientists have tended to view traditional knowledge as anecdotal and unreliable, and therefore inappropriate to use in impact studies. A further difficulty was that, because of the land claims negotiations, the Aboriginal people were reluctant to share their traditional knowledge publicly.

The Diamond Mine and Its Impacts

Five kimberlite pipes were initially identified for mining. One set consists of a group of four, while the fifth is about 30 km away, on the edge of Lac de Gras. Mining would cause at least seven lakes covering 890 hectares to be lost—five due to mining, one for tailings disposal, and one as a source of aggregate material for construction (Figure 13.5). The extracted ore is being processed based on a plan aimed at handling 9,000 tonnes each day; after 10 years, this will increase to 18,000 tonnes daily. The ore is crushed, and the diamonds are separated by physical rather than chemical processes.

Long Lake was chosen to hold tailings. Dams have been built at the outlet of the lake and around its perimeter to increase is holding capacity. Rock dykes have been used to divide the catchment area into five cells. As the tailings held in each cell settle, consolidate, and evolve to permafrost, rocks and soil will be spread over the surface. Revegetation will be started, with the goal of having the entire holding area become a wetland once the mining is completed.

A permanent work camp has been built to accommodate the 400-person workforce, which rotates on the basis of two weeks in/two weeks out with 70-hour workweeks.

No all-weather road has been built. Over the winter months, a 476-km winter ice road over frozen lakes and rivers is used, including by trucks carrying fuel. Air transport is used to transport workers in and out, as well as to bring food in and take the diamonds out.

Assembling Data Related to Environmental Impacts

The mining company, BHP, collected some baseline data during 1992, with systematic and intensive field sampling begun in 1993. The sampling program addressed biological, cultural, and socioeconomic issues. Through the environmental assessment process, various federal government departments provided comments, in the fall of 1993, about the field sampling design. The most significant comment focused on the need for BHP to make significant additional effort to incorporate traditional knowledge into the collection of conventional scientific data.

In response, BHP observed that it had faced serious challenges in its efforts to include traditional knowledge into its research program. First, the Treaty 8 and Treaty 11 Dene groups were in the midst of land claims negotiations, and as a result were reluctant to release traditional knowledge into the public domain because the traditional knowledge was important for their negotiation strategy. Second, concern was being expressed by Aboriginal people about using traditional knowledge outside of the context of the cultures and broader system of knowledge that gave it meaning and value. Third, there was not one set of traditional knowledge, as the Inuit, Métis, and Dene each have their own traditional knowledge, which do not always coincide. Fourth, traditional knowledge was viewed by Aboriginal people as their intellectual property, and therefore its use and management had to remain within their control. And fifth, there was no documented baseline of traditional knowledge, nor were there any generally accepted standards or methods to guide traditional knowledge research.

Mining Tailings

As observed in the report from the federal Environmental Impact Assessment Review Panel (1996: 24), the management of mining tailings is critically important because of the potential impact on downstream water quality. During the mining operations, 35–40 million tonnes of waste rock will be excavated each year. The ore is being crushed, and diamonds are separated through physical means. The crushed rock or tailings will

Figure 13.5 Development plan area, NWT Diamonds Project. SOURCE: Canadian Environmental Assessment Agency (1996: 7). Reproduced with the permission of the Minister of Public Works and Government Services Canada, 2005.

BHP APPROACH TO ENVIRONMENTAL MANAGEMENT AT EKATI

The basis of BHP's environmental management approach is an adaptive environmental strategy which involves the establishment of criteria or indicators used to indicate change so that appropriate management actions can be implemented. A sensitive and effective monitoring program is key to the success of this strategy. BHP told the Panel that the environmental management plan is intended to be flexible so that it can be modified in response to changes in the mine development plan, natural environmental or technological advances, research results, and improved understanding of traditional knowledge.

— *Canadian Environmental Assessment Agency (1996: 24)*

be placed in the Long Lake tailings impoundment basin for the first 20 years and then in one of the mined-out pits for the final five years.

The capacity of Long Lake has been increased through building three perimeter dams. The dams have a frozen core design, meaning that each has a central core of frozen soil saturated with ice and bonded to the natural permafrost. The core is surrounded with granular fill to ensure both stability and thermal protection. The use of a frozen core and permafrost foundation is intended to ensure that no water can escape from the dams as long as the soil remains saturated with ice.

The basin has been divided into five cells, and the cells will be filled sequentially, starting with the cell furthest upstream, leaving time for the tailings to settle and consolidate (Figure 13.5). The last cell, the farthest downstream, will not receive any tailings. It will act as a final settling pond.

The design intends that the tailings gradually will consolidate and become permafrost. Once a frozen crust has formed over a tailings cell, it will be covered with waste rock and then topped with fine granular soil. Such a covering will be thick and moist enough to facilitate creation of a new active layer in the new permafrost system. Subsequently, the soil will be revegetated, with the ultimate purpose being to create a wetland. After the mine is closed, the spillway dam at Long Lake will be breached, with water flowing naturally from the fifth cell into a nearby lake and from there into the Coppermine River system.

The **frozen core dam** design was chosen because of lack of impervious fill needed for construction of a conventional dam, because the climate is conducive to a frozen core design, and because previous experience with frozen core designs could be drawn upon from both Canada and Russia. As the Environmental Impact Assessment Panel observed, the critical feature for a frozen core dam is ensuring that the core stays frozen. The design for the dam is based on a criterion that the long-term average temperature of the core cannot be higher than $-2°$ C when the mine is operational. Modelling indicated that this criterion would be satisfied, even in future conditions suggested by global warming scenarios (see Chapter 7). The frozen core will be monitored systematically, and various contingency plans are in place in case of an unexpected increase in temperature.

Migratory Caribou

BHP conducted and supported research related to the Bathurst caribou herd, the largest one in the NWT (Figure 13.6). Baseline data were collected in 1994 and 1995 to determine the numbers of animals using the Lac de Gras area during the migrations, the location of migration corridors, and the use of habitat. The migration patterns were different in the two years. This variability was confirmed by the Environmental Assessment Panel, which commented in its report that 'This [variability] was consistent with the observations of several that there is a considerable natural variability in caribou migration and habitat use. For example, the GNWT [Government of the Northwest Territories] agreed that the ability to predict, on an annual basis, the timing and numbers of caribou in the vicinity of the proposed mine was low. A representative of the NWT Barren Ground Caribou Outfitters Association said that over his 18 years of experience, the exact migration route varied from year to year' (Canadian Environmental Assessment Agency, 1996: 39).

Since the caribou herd does not follow the same migration route each year and the areas affected by the mining represented less than 0.01 per cent of the range of the herd, it was believed that the mining activity would have a very small impact. Attention was also given to the possible effects of roads and the new airport landing strip, either through collisions of caribou with vehicles or as a barrier to migration. It was concluded that

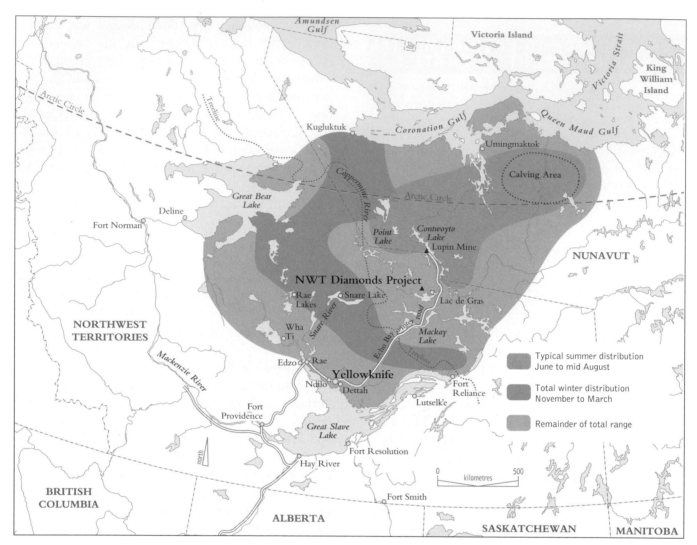

Figure 13.6 Distribution of Bathurst caribou herd. SOURCE: Canadian Environmental Assessment Agency (1996: 40).

these would not cause problems. Indeed, caribou had been noticed using the new habitats adjacent to the roads and landing strip during the exploration period. Finally, modelling indicated that the water in the tailings impoundment would be within federal guidelines for protection of livestock (and, therefore, of wildlife).

BATHURST CARIBOU HERD

Among environmental issues, the potential effect of the Project on the health, numbers and migratory patterns of the Bathurst caribou was the most important concern raised.... The GNWT told the Panel that the cultural value of the herd could not be estimated but that the dollar value of the harvest, based on meat replacement costs, was $11.2 million annually.

— *Canadian Environmental Assessment Agency (1996: 39)*

Water Issues

BHP began to collect baseline data on hydrology in the autumn of 1992. Data were collected for lake bathymetry, lake water levels, and surface hydrology. During 1993, preliminary investigations were conducted related to eight lakes, followed in 1994 with sampling at 25 lakes and in 1995 with sampling in four lakes and seven stream sites. The water samples were sent to an independent laboratory for analysis.

It was concluded that water flow changes would be caused by draining of the lakes to facilitate open-pit mining, as well as by diversion of flows around the pits and by the infilling of Long Lake with the tailings. In total, 15 lakes would be affected by the mining activity. It was decided that the drainage of the lakes prior to open-pit mining would be managed so that flows would not be greater than 50 per cent of the mean annual flood levels in any downstream water system containing fish. As a result, the main consequence of draining the lakes would be to extend the peak spring flows for a longer period of time. And, because the connecting channels between the lakes are both wide and braided, the affects of the extended period of higher flow were judged to be negligible.

The potential impact of mining operations on water quality attracted considerable attention. Primary concerns were that contaminants from the mining operation could affect downstream users of fish and drinking water in the Coppermine River watershed. The primary concern was whether the tailings impoundment in Long Lake would ultimately release water of acceptable quality. Analysis focused on three water quality variables (suspended solids, total nickel, and total aluminum). The design of the holding cells in Long Lake is intended to ensure that water would be held there until the water to be released met regulatory standards. During the impact assessment process, it was agreed that the design should meet all regulatory standards for water quality.

Concern also arose about possible contamination from toxicity of kimberlites, acid generation from waste rock, and nitrogen from blasting. Analysis of the toxicity of kimberlites and waste rock, and the design of the holding pond, led to agreement that such contamination would be controlled satisfactorily. Effects on groundwater also were addressed. Baseline data were collected and a long-term monitoring program was established so that affects on hydrogeology could be tracked.

Fish

Twelve of the 15 lakes that would be affected by the mining through draining of lakes prior to open-pit mining, by filling lakes with tailings, or by covering lakes with waste rock contained fish. In addition, 43 connecting streams, outflow streams, and inflow streams would be affected. The main species of fish in the 12 lakes are lake trout, followed by round whitefish, Arctic grayling, and burbot. The federal Department of Fisheries and Oceans (DFO) has a policy of 'no net loss' of productive capacity of fish habitat. The outcome is that whenever fish habitat is degraded or lost, DFO expects there to be a counterbalancing habitat replacement.

BHP has compensated for the lost fish habitat in streams by creating a diversion channel between two of the key lakes in a way that made the channel a quality fish habitat. The cost of this initiative was $1.5 million. The diversion channel is 2.25 times longer than the natural connecting channels that were lost, and DFO stated that this approach was an acceptable way to offset the loss of natural stream habitat.

Costs

Couch (2002: 274) reported that the scientific research funded by BHP cost more than $10 million. In addition, the environmental assessment review process cost the Canadian government about $1 million, plus $255,000 provided for participant funding. These amounts do not include the costs incurred by various federal departments, such as Environment, Fisheries and Oceans, and Indian Affairs and Northern Development, as well as the government of the Northwest Territories. In Couch's view, 'in comparison with the Project's capital cost, the anticipated profits to BHP Diamonds Inc. and the tax revenue to governments, this outlay was very small' (ibid.).

Environmental Assessment Process

Beginning in 1992, BHP began research to understand the impacts of the proposed mining activity and to develop mitigation measures. In addition, the company visited all communities in the project area at least twice. BHP made public presentations, organized field trips, held community meetings and open houses, facilitated cultural exchanges and workshops, and sent a group of Aboriginal people to its mines in New Mexico, where 76 per cent of its employees were Native Americans.

In July 1994, the Minister of Indian Affairs and Northern Development referred the mining project for an environmental assessment, and in July 1995 BHP submitted its environmental impact statement. From late January to late February 1996, an Environmental Assessment Panel appointed by the Minister of the Environment held public meetings, and its report was submitted to the federal government in June 1996. In February 1997, the federal government gave its formal approval and construction started in May 1997. In January 1999, the first diamonds from Ekati were sold in Antwerp, Holland.

Agreements and Arrangements

Emerging from the process outlined above were a number of agreements and arrangements.

Announcement from the Minister of Indian Affairs and Northern Development. In August 1996, the minister announced his acceptance of the Environmental Assessment Panel's report and gave participants only 60 days to work out detailed agreements. Given that land claims negotiations were also underway, there was the potential for much debate and disagreement. The tight time frame put all participants under much pressure to work out detailed agreements.

Environmental Agreement. The Environmental Agreement is legally binding, and requires BHP Diamonds to (1) prepare a plan for environmental management during the construction and operation of the diamond mine; (2) submit annual reports related to the environmental management plan; (3) prepare an impact report every three years related to the impacts of the project; (4) establish a monitoring program for air and water quality and for wildlife; (5) submit a reclamation plan for approval; (6) establish a security deposit ($11+ million) for potential land impacts and a guarantee of $20 million for potential water impacts; and (7) incorporate traditional ecological knowledge into all environmental plans and programs.

In addition, an Independent Environmental Monitoring Agency (IEMA) was to be established as a public watchdog. The IEMA is to (1) prepare annual reports on the project's environmental implications, (2) review impact reports, and (3) provide a public document repository at its Yellowknife office. This was an innovative feature, as it had not been recommended by the Assessment Panel.

Socio-economic Agreement. This agreement was reached between BHP and the government of the Northwest Territories, and focuses on commitments beyond existing statutory requirements. The concern was economic benefits and social impacts related to all NWT residents, not just for those who were traditional users of the project area. The agreement covered matters such as preferential hiring of NWT residents, criteria to guide recruitment, employment targets, employment of local contractors, training programs, and employment support. Targets were specified for awarding contracts to and purchases from northern businesses, as well as for employment of Aboriginal and northern residents.

Although not part of the agreement, a noteworthy initiative has been the establishment of diamond-cutting and -polishing businesses in Yellowknife, with a population of about 18,000. The traditional centres for polishing are Antwerp, Belgium, Tel Aviv, New York City, and India. In 1999, a small Vancouver-based diamond-polishing company opened a facility in Yellowknife, recruiting a South African diamond cutter from Antwerp. Shortly after, another company opened, with cutters recruited from Armenia. Other firms also have opened facilities, and local people are learning the trade under the guidance of cutters brought in from Europe, Israel, and Africa.

Impact and Benefit Agreements. In 1994, BHP began negotiations with the four Aboriginal groups. Each group was involved with land claim negotiations, and BHP did not want to get entangled in those processes. **Impact and benefit agreements (IBAs)** are one tool to address community and industry relations in mining or other extractive resource activities. They are voluntary agreements, beyond formal impact assessment requirements, and are intended to facilitate extraction of resources in a way that contributes to the economic and social well-being of local people and communities. IBAs create opportunities for communities to realize direct economic benefits from natural resource development projects and also to participate in the management, monitoring, and mitigation of impacts. All of these aspects were addressed in the IBA between BHP and the four Aboriginal groups. In the accompanying guest statement, Michael Hitch provides further insight about IBAs.

ENERGY RESOURCES

Energy resources are classified as renewable and non-renewable. As noted at the beginning of this chapter, renewable resources are those that can be replenished in a relatively short time period. Figure 13.7 identifies three renewable energy sources, and one of these, gravity, is ongoing and widespread but remains as potential unless associated with significant motion, such as tides or river flow. Geothermal heat, created by nuclear energy, also is persistent and widespread, but at great depths below the surface. Manifestations of geothermal heat at the surface or shallow depths are much more limited and usually are associated with the heat being carried by water or steam, so the renewability for geothermal heat depends on a reliable and ongoing supply of water. Solar supplies come from continuous emission of radiation from the sun, but this arrives discontinuously on the surface of the Earth due to diurnal and seasonal variation, as well as cloud cover. As a result, renewable, solar-based energy supplies are intermittent and often cyclic, meaning that they usually must be supplemented by other sources.

Biomass energy sources are frequently used in rural areas in developing countries, and can take the form of millions of people and their draft animals doing subsistence work. Metabolic energy (muscle power) is supplemented by heat created from burning firewood, from crop and animal wastes in basic biogas converters, and from direct sunlight used to dry and preserve agricultural or marine products (e.g., dried fish). Biomass energy is renewable as along as the rate of use and capacity to produce biomass are balanced.

The non-renewable sources cannot be replenished in a period of time to support humans. These sources result from geological processes over millions of years, which lead to solid (coal) and liquid (oil) fuels, natural gas, and nuclear fuels. While they all share the characteristic of offering high energy content per unit or weight or volume, they also differ. Solid fuels are mined, thereby being labour-intensive and requiring expensive infrastructure. For efficient transport, they must be carried in bulk or batch containers, such as rail cars or ships. When used (burned), solid fuels release gaseous and particulate matter in large quantities. In contrast, oil and natural gas can be produced with facilities requiring relatively little labour but capital-intensive refineries or processing plants. Furthermore, once processed, the product can be transported continuously through pipelines or in batches (trains, ships, trucks). Nuclear fuels contain the highest content per unit of weight, but require sophisticated facilities and highly skilled human resources. They are only used to generate electricity, and demand careful handling in processing and waste disposal. Given these different attributes, it is important to appreciate that

the most appropriate sources of energy will vary depending on circumstances. Box 13.3 highlights the different variables that need to be considered when making that choice.

Energy Use and Issues in Canada

As noted at the beginning of this chapter, Canada is ranked as the sixth largest user of primary energy in the world. According to Environment Canada (2002: 57), in the year 2000, energy consumption in Canada was 9.9 exajoules, an increase of 10 per cent relative to 1990. In the same period, however, energy use per capita had decreased, indicating that initiatives to improve energy efficiency may be starting to show results. In terms of fossil fuel, use increased by 20 per cent between 1990 and 2000 (Figure 13.9). In 1999 Canada's fossil fuel energy use was 2.5 per cent of total global fossil fuel use, as was its total energy use relative to total global energy consumption.

Fossil fuels are the main type of energy consumed by Canadians, and one of the largest consumers of energy is the transportation sector, which depends overwhelmingly on fossil fuels. Automobile travel increased by 9 per cent between 1990 and 2000, and in 2000 the automobile accounted for 74 km of every 100 km travelled by Canadians. Fossil fuel use by automobiles increased by 20 per cent between 1990 and 2000. Some fuel efficiencies in automobiles occurred between the early 1970s and the early 1980s, but since then there have not been significant

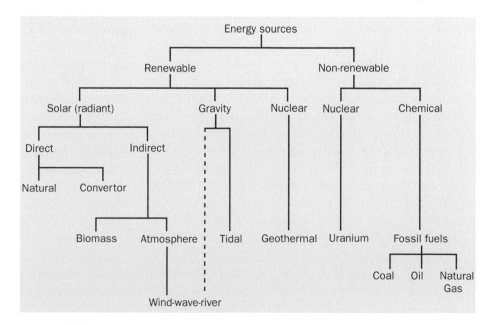

Figure 13.7 Energy sources. SOURCE: Chapman (1989: 4).

Guest Statement

Minerals, Mining, and Impact and Benefit Agreements
Michael Hitch

Interest in mining and mineral exploration has grown globally over the past two decades. This phenomenon, for the most part, is due to two factors: (1) improved technology used in mineral exploration, and (2) the desire of nations to participate in the new wealth generated from the extraction of these minerals. As technology advances and more accessible deposits are being exploited, mining companies increasingly penetrate into remote areas. These areas often encroach upon forests, watersheds, and mountainous regions. Many of these areas are also indigenous peoples' lands, whether officially recognized or claimed.

Mining has been a major economic activity in the Canadian North for the last century. It has made a valuable contribution to the development of this fragile economy and to the living standards of the inhabitants. The benefits include jobs and income, tax revenues and the social programs they finance, foreign exchange earnings and all that they purchase, frontier development, support for local infrastructure, and economic diversification into a broad range of activities beyond the life of the mine. All of these elements constitute the one-third (i.e., economic sufficiency) of the modern-day definition of sustainable economic development.

At the same time, mining continues to be controversial because it has generated costs of a biophysical and social nature (i.e., relative to biophysical integrity and social well-being), especially at the level of local communities. The benefits, which could accrue to the local communities, have the potential to be more substantial than they have sometimes been in the past. Indeed, the principal objective of the new generation of mine enterprise and community relationships is to improve the net benefits of mining activity for the local communities, which were often overlooked in the past. Such new relationships have been a work in process during the last decade.

Impact and benefit agreements (IBAs) are one approach to solving conflict in the early stages of a

Michael Hitch

project's development, and form a negotiated bond or agreement between industry and the local communities that can provide a foundation for mutual understanding. Items covered in the more comprehensive of these agreements include: preferential employment and business contracting opportunities, training and education (including apprenticeships and scholarships), equity participation, revenue-sharing, cash compensation, social and environmental monitoring and/or mitigation measures, archaeological site preservation, access to facilities and infrastructure, information exchange, agreement management, and dispute resolution mechanisms.

An IBA's primary purposes are: (1) to address the adverse effects of mining activity on local communities and their environment, and (2) to ensure that Aboriginal peoples receive benefits from the development of their mineral resources. To date, four IBAs have been negotiated and a further two are in progress in Nunavut. Their ultimate success in promoting a more sustainable pathway for northern communities remains to be seen. The success of the IBA mechanism depends very much on the configuration of the dynamic power relationships between stakeholders and their biophysical, social, and economic environments.

Michael Hitch, a geologist, has worked for 22 years in the mining industry. He has worked on mining projects in 141 countries. He particularly has experience with mining operations in the Canadian North and has participated in the negotiations to create the first IBAS. Currently, he is the Mining and Metals analyst with Clarus Securities in Toronto.

BOX 13.3   CHOOSING AMONG ENERGY SOURCES

1. *Occurrence.* Many energy sources are confined to specific environments and locations, and are only available at other locations when transport systems exist. Even physically present sources may not actually be available because of technical, economic, or other constraints.
2. *Transferability.* The distance over which an energy source may be transported is a function of its physical form, energy content, and transport technology.
3. *Energy content.* This is the amount of usable energy by weight or volume of a given source. Low energy content sources are inadequate when demand is large and spatially concentrated.
4. *Reliability.* Uninterrupted availability gives one source an advantage over one that is intermittent.

5. *Storability.* To meet interruptions of supply or peaks of demand, a source that can be stored has an advantage over one that cannot.
6. *Flexibility.* The greater the variety of end uses to which a given source or form may be put, the more desirable it is.
7. *Safety and impact.* Sources that may be produced or used with low risk to human health and the environment will be preferred over less benign sources.
8. *Cleanliness and convenience.* The cleaner and more convenient source will be preferred over the dirty and the cumbersome.
9. *Price.* The less expensive source or form will be preferred over the more expensive.

Source: Chapman (1989: 5).

improvements. Instead, there has been an increased use of less fuel-efficient vehicles, such as light-duty trucks and sport utility vehicles. For example, the percentage of automobile passenger-kilometres travelled in light-duty trucks almost tripled in 25 years (10 per cent in 1976 to 27 per cent in 2000). Fifty per cent growth occurred in air travel in the same period. These increases in use of automobiles and planes occurred at the same time that there was a drop in use of buses and trains.

A clear message is that Canada is an intensive energy-using nation, and that, despite growing efficiencies, energy use is climbing. Given our dependency on fossil fuels, it is appropriate to examine what are often referred to as 'alternative energy supplies', and this is done in the next section. However, it is important to recognize that alternative sources, usually viewed as renewable sources (solar, geothermal, hydro, tides, wind), are not problem-free from an environmental perspective. For example, the impacts of the hydro power development at James Bay, examined in Chapter 12, included increased levels of mercury in aquatic systems, with consequences for both humans and other living species in that area, and changes in fish populations. In the following section, attention turns to wind power.

Wind Power
Wind power is the fastest-growing sector in the world's energy market, with annual growth rates between 1995 and 2000 averaging 25 per cent. In the United States, Europe, and other countries, progressive policies have been introduced to stimulate wind power. For example, in the United States, up to 15 states require that all energy suppliers provide a portion (1–5 per cent) of their total output from renewable sources such as wind, and in 2000 the US had over 2,000 MW of wind energy. Other examples of production in 2000 are Denmark, with 1,800 MW of wind energy, meeting more than 8 per cent of its energy needs, Germany, with 3,900 MW of wind energy, Spain, with 1,200 MW, and India, with 1,100 MW. In contrast, Canada does not have such a stipulation, and by 2000 the country had only 124 MW of wind power generating capacity, and most of that was installed in 1998 and 1999 in Quebec. At that time, Ontario had only one utility-scale wind turbine, generating 600 kW. To illustrate how quickly wind energy has been growing, at the start of 2003 the comparable numbers were 4,645 MW for the US, 2,889 MW for Denmark, 12,001 MW for Germany, 4,830 for Spain, 1,702 MW for India, and 236 MW for Canada.

The Toronto Wind Turbine Initiative
In 2000, approval was given for construction of a wind turbine in the downtown waterfront area on the Canadian National Exhibition grounds. This

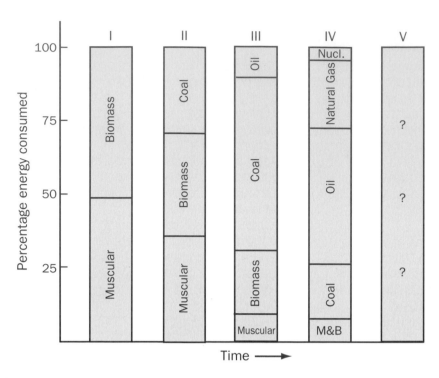

Figure 13.8 The evolution of energy consumption mixes by source. SOURCE: Chapman (1989: 6).

ENVIRONMENTAL IMPACTS OF ENERGY CONSUMPTION

When the nine factors in Box 13.3 are considered, it can be understood why the use of energy sources varies around the world and even within nations. Figure 13.8 illustrates how over time and around the world the major sources of primary energy have evolved from muscular and biomass energy, to coal, to oil and natural gas, and to nuclear fuels. Some countries, such as Japan, have moved rapidly from Phase I to Phase IV. In contrast, many developing countries struggle to move beyond Phase I except in their major cities. Canada is now well into Phase IV. A challenging question is to determine what a Phase V might look like.

The combustion of fossil fuels emits greenhouse gases, such as carbon dioxide and nitrous oxide, which accumulate in the atmosphere and contribute to climate change. Pollutants such as sulphur dioxide and nitrogen oxides are also by-products of fossil fuel combustion and are primary contributors to acid rain and poor air quality.... Fossil fuel spills, waste heat, and habitat destruction associated with mining and damming pose a risk to wildlife and contribute to changes in biodiversity.

— Environment Canada (2002: 57)

was the first urban-area wind turbine to be built in North America. At least one more wind turbine is scheduled to be built on the Toronto waterfront.

This is a joint venture between Toronto Hydro and WindShare. The latter was established by the Toronto Renewable Energy Cooperative (TREC), founded in 1997 by the North Toronto Green Community as one response to growing concern related to smog and climate change. TREC's goals are to promote green energy initiatives in Ontario and to educate the public about green energy alternatives.

The wind turbine installed at Exhibition Place is 94 metres high (30 stories), has blades 29 metres long, and produces 1,800 megawatt hours (MW hours) of electricity each year, enough to provide electricity for 250 homes. WindShare has an agreement to sell its wind-generated power to Toronto Hydro Energy Services. The capital cost for the turbine was $800,000, the projected 'payback' period is 12 years, and the life expectancy of the wind turbine is 20 to 25 years. Furthermore, the wind turbine will displace about 1.8 million kilograms of CO_2 (the leading greenhouse gas) emissions each year, as well as between 5,000 to 8,000 kilograms of NOx and SO_2, the leading contributors to smog and acid rain, respectively. However, at the moment the electricity from the wind turbine is not competitive with conventional sources because the Ontario provincial government has fixed the market price of electricity at $0.043/kWh, and the cost of electricity from the wind turbine ranges between $0.08 to $0.10 kWh. As a result, the main purpose of the wind turbine is educational and consciousness-raising. More specifically, the intent is to highlight the pollution reduction and employment creation possibilities of wind energy. For these reasons, a deliberate decision was taken to locate the first wind turbine at a highly visible site in Canada's largest and most visited city.

A basic question was whether there was sufficient wind to drive the wind turbine. TREC

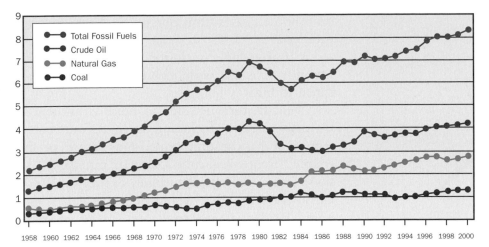

Figure 13.9 Fossil fuel consumption in Canada (exajoules). SOURCE: Environment Canada (2003). Adapted in part from Statistics Canada Energy Division. Reproduced with the permission of the Minister of Public Works and Government Services, 2005.

installed two anemometers on towers for one year on the Etobiocoke lakeshore and at the foot of Leslie Street, the two proposed sites. Data from these measurement devices, along with 20 years of wind data from federal government anemometers at the Toronto City Centre Airport and at the Leslie Street spit, indicated that the wind regime was sufficient to generate an average annual electricity output of 1,800 MW hours.

However, as with all technologies created by humans to provide power, whether from renewable or non-renewable sources, potential exists for environmental affects. Consequently, research was conducted by Dillon Consulting Limited to examine the possible impacts of the proposed wind turbine. Many aspects were considered, but here attention focuses on possible impacts related to wildlife (especially birds), noise, safety considerations, and aesthetic aspects.

Wildlife. One of the main worries during the operation of a wind turbine is the possible adverse impact on wildlife, especially birds. However, monitoring programs in North America and Europe have shown that bird mortality from such turbines is very low. Indeed, for sites with a small number of turbines, it is not unusual to have no recorded fatalities. For sites with many turbines, it is common for there to be not more than one bird fatality per wind turbine per year, and the maximum recorded mortality in North America has been 1.9 birds/turbine/year. Furthermore, most studies have been conducted for variable-speed turbines, which are more dangerous, and the turbine in Toronto is a fixed-speed type. Even for variable-speed turbines, studies have shown that up to 80 per cent of birds can fly unharmed through rapidly moving blades.

Most birds travel during the day, have good vision, and can readily see and avoid obstacles.

WIND POWER IN CANADA

Canada has utility wind turbines installed in Alberta, Saskatchewan, Ontario, Quebec, Prince Edward Island, Nova Scotia, and Yukon. Wind farms have been established in Alberta, Saskatchewan, Ontario, Quebec, and Prince Edward Island.

Canada's largest wind power operation is in the Gaspé region of Quebec. Le Nordais project has 133,750-kilowatt turbines producing 100 megawatts of electricity at two locations, Cap Chat and Matane.

In October 2004, Quebec announced a massive wind power project. Two private companies will establish eight wind turbine farms in eastern Quebec, and these will produce 1,000 megawatts of power by 2012. The result will be that Quebec will generate more than the present 371 megawatts of wind power produced in 2004 across all of Canada.

Pincher Creek in Alberta has a number of large-scale wind projects.

WIND POWER

The energy of a moving mass of air has been used for centuries to propel vessels and lift water.... Wind speed varies considerably from place to place on the Earth's surface and at any one place varies seasonally, diurnally, hourly and over shorter time periods. Consequently, before any installations of a commercial scale can be considered, the wind regime must be assessed in terms of the persistency of various wind speeds. Even in the most windy locations the short-term variations in wind speed result in an uneven flow of primary energy and thus irregularity of delivered energy. To compensate for this, storage systems of various kinds must be provided, back-up energy systems must be available or consumers must adjust their use patterns to such irregularities.

— *Chapman (1989: 95)*

Toronto wind turbine on Canadian National Exhibition grounds. (*Toronto Hydro Energy Services Inc.*)

each home causes from one to 10 deaths to birds each year.

A reasonable conclusion from research and monitoring is that wind turbines pose a negligible threat to birds.

Noise. Disturbance due to noise is influenced by many variables, including distance from source, nature of background noise, and nature of the source (frequency, time pattern, intensity). All noise levels from wind turbines during operation are lower than what would be experienced in a quiet, residential area, and are similar to what would be experienced inside an average home. Dillon Consulting concluded that, given the normal background noise in an average suburban residence, the noise from a wind turbine would be inaudible at a distance of 260 metres. In this context, the City of Toronto adopted the following standards: (1) a 200-metre separation between a wind turbine and residential low-rise buildings; (2) a 300-metre separation between a wind turbine and high-rise residential buildings; and (3) a 50-metre separation between a wind turbine and sensitive natural areas or sensitive park use areas.

Safety. Ice thrown from turbine blades or falling off the tower is the main safety concern. There have been no recorded incidents of falling ice and therefore little regulatory experience exists. At the wind turbine towers at Kincardine on the Bruce Peninsula of Ontario, operators have recorded an average of three icing events each winter. Several proactive actions are available. First, setback criteria can be used to ensure people are kept at a reasonable distance from a wind turbine tower and the rotating blades. Second, temperature sensors as well as sensors to monitor the balance of blades can provide early information about ice buildup. Once ice accumulation is detected, the wind turbine can be shut down and not restarted until operators determine that conditions are safe.

Aesthetics. A wind turbine, or wind turbine field, will be viewed by some as an unwelcome visual intrusion on the landscape, especially if a wind turbine is viewed as not conforming to an area with historical, cultural, or natural values. On the other hand, some people enjoy the appearance of a wind turbine, liking the modern, futuristic appearance along with the symbolism and educational role of an environmentally benign technology located in an easily visible location. The

Furthermore, most birds fly at altitudes much higher than the tops of wind turbines, so they normally fly over rather than through them. The birds at greatest risk are small nocturnal migrants that fly together in large numbers, but even these types usually fly well above the turbine.

Evidence reveals that other urban structures (tall buildings, communication towers) cause much higher rates of mortalities for birds. For example, Dillon Consulting reported that tall communication towers in Canada each often kill more than 1,000 birds annually. A few tall buildings in Toronto are estimated to cause the death of more than 10,000 birds each year. Even normal family houses can be dangerous, with estimates suggesting

SITING WIND TURBINES

The pro-environment view is that small windmill projects are essential in urban areas to satisfy the need to put an energy alternative before the energy consuming public. Under this view, locating windmills at gateways to urban parks, and especially at gateways to natural heritage areas is a major benefit and goal.

— *Dillon Consulting (2000: 44)*

challenge, as Dillon Consulting (2000: 44) observed, is that, 'Given the conformity of view that windmills are a good thing but that they should be placed "somewhere else" and not "here", and the recognition that everyone's "somewhere else" is someone else's "here", a balanced answer is needed.'

Research regarding wind turbines or windmills in Europe and North America has indicated that, prior to their construction, nearby residents usually have some concerns. However, after the wind turbines are operating, the views were either neutral or positive. If the homes receive electricity from the turbines, the attitudes are likely to be more positive.

Summary. Evidence indicates that wind turbines have minimal adverse environmental affects. However, compared to conventional fossil fuel energy sources, they are still relatively expen-

BOX 13.4 WIND POWER FOR FARMS

Wind power is not confined to urban areas and can be generated and used by individuals. To illustrate, two Mennonite farmers in a township just north of Waterloo, Ontario, each have invested more than $35,000 for wind turbines, just under 25 metres high, which they expect will generate enough electricity to supply themselves, with enough electricity left over to sell to the local utility company, Waterloo North Hydro. They anticipate that the wind turbines could pay for themselves in 10–12 years.

Jacob Martin, one of the farmers, has used two smaller windmills for some time to provide power for his home and barn, and has never been connected to the power grid. However, to be able to sell excess power to the grid, he will have to become connected.

One major change has to occur for these initiatives to be a success. Specifically, in Ontario, electricity costs more to provide than what customers are charged for it. The province subsidizes electricity, and therefore farmers such as Jacob Martin and Edwin Martin only can receive the subsidized rate. Since the official subsidized market price is about half the cost of producing wind energy, the potential for the Martins to earn income by selling to the utility will only be realized if the provincial government changes its policy so that consumers are charged the actual price needed to cover the cost of providing electricity.

As Jacob Martin observed, 'If I could get my power and get a cheque for $5,000 a year, that would be good.'

SOURCE: Based on *Kitchener-Waterloo Record* (2004).

Wind turbine on Jacob Martin's farm, Woolwich Township, Ontario (*Bruce Mitchell*).

BOX 13.5   WHAT YOU CAN DO

1. Individuals can modify behaviour to reduce energy and materials use, waste production, and ecosystem degradation. Examples include:
   - Using a bucket, sponge, and trigger nozzle on a hose saves about 300 litres of water each time a car is washed, and also energy used to pump water through a municipal system.
   - Commuting to work by car pool, public transportation, bicycle, or on foot rather than by one person in an automobile will reduce energy consumption and help to reduce greenhouse emissions and other air pollutants.

2. Individuals can use more efficient technology or use products with lower environmental impact throughout their life cycle. Examples include:
   - Use smaller, more efficient automobiles and major appliances with the lowest energy ratings. This, too, will reduce greenhouse gas emissions and other air pollutants.
   - Install water-saving devices in the home, such as low-flow shower heads and toilet dams. Energy will be saved by having to heat less water and also by having to move less

clean and grey water through the water supply system.
   - Replace incandescent light bulbs with fluorescent bulbs, which use about 75 per cent less energy and last 10 times longer.

3. As part of a larger society, individuals can ask for appropriate information and insist that products, services, and planning explicitly address environmental implications. Examples include:
   - Individuals in their communities can promote better planning of urban transit and bicycle routes, and reduced dependency on passenger vehicles.
   - Individuals can lobby their local government, and others, to show leadership in educating the community about types of behaviour and products that are environmentally benign.
   - Individuals can support manufacturers committing to include environmental considerations into their production processes via life-cycle management and environmental management systems.

SOURCE: Based on Environment Canada (2002: 68).

sive, but that could change if the appropriate technology becomes less costly and/or fossil fuel supplies become more expensive. Their increased use in the future will require governments to be proactive, as is occurring in Germany, to create requirements or incentives for energy suppliers to include renewable sources in the mix of sources providing energy to consumers. Furthermore, in establishing the cost of alternative sources, more attention needs to be given to the total costs of each source, including the costs incurred through emissions into the atmosphere. If such comprehensive costing were used, the gap between conventional sources and renewable energy sources would not be as large as it is at the moment (Box 13.4).

IMPLICATIONS

If our ecological footprint is to become lighter related to minerals and energy, it is clear that there will need to be changes by individuals, institutions, and societies. Consumers are heard to complain that manufacturers do not build 'green cars', while

the manufacturers state that customer demands do not indicate that green cars are wanted in sufficient quantity to justify producing them. As a result, if change is to occur, there will need to be adjustments at all levels, with individuals taking initiative to reduce consumption of energy and mineral products, governments providing greater incentives to both individuals and manufacturers to embrace green products, and manufacturers showing leadership to market green products effectively.

In the meantime, what can you do? Box 13.5 identifies some actions or changes in behaviour that, if taken, will help to extend the life of non-renewable or stock resources, as well as encourage greater use of renewable or flow resources. There is no recipe or formula that will lead readily and easily to a society that is less materialistic and energy-intensive. However, small steps such as the ones identified in Box 13.5 can cumulate to cause significant change. Perhaps most importantly, thinking about and taking such actions are first steps in shifting basic beliefs and values.

SUMMARY

1. Non-renewable or stock resources take millions of years to form. Consequently, from a human viewpoint they are for practical purposes fixed in supply and therefore not renewable.

2. If done systematically and correctly, enhanced environmental management practices and extended social responsibility relative to non-renewable resources almost always generate some kind of economic return on investment for business, although usually over the long term.

3. The main environmental issues for the mining and energy sectors include at least : acid mine drainage; sulphur dioxide emissions; metal toxicity; disruption of remote ecosystems due to exploration, test drilling, and operation of oil fields or gas wells; disturbance to aquatic ecosystems from escape of waste heat produced from nuclear energy production; and threat to human and ecosystem health from radioactive waste associated with nuclear energy production over a period of thousands of years.

4. 'Best practice' related to environmental management for mining and fossil fuel firms in Canada should include a combination of basic scientific research to ensure understanding of natural and social systems that can be affected by operations, design of appropriate mitigation measures, environmental impact assessments and reporting, environmental audits, corporate policies that explicitly include environmental aspects, environmental management systems, and life-cycle assessments.

5. Canada is one of the largest exporters of minerals and mineral products in the world, and there are more large mineral exploration companies based in Canada than in any other country. Canada accounts for about one-third of all such companies.

6. The top five minerals in terms of production value in 2002 for Canada were gold, nickel, potash, copper, and cement.

7. In 1998, BHP's Ekati mine, located 200 kilometres south of the Arctic Circle, became the first diamond operation in Canada. A second diamond mine, Diavik, began producing in 2003.

8. The landscape remediation program in Sudbury had a variety of objectives: create a self-sustaining ecosystem with minimal maintenance; use plant species tolerant of acidic soils and low nutrient concentrations; use seed application rates that allow for natural colonization and thus increase species diversity; give preference to the use of native species; restore nutrient cycles and pools by the use of species (legumes) that fix nitrogen; use species that attract and provide cover for wildlife; and undertake initiatives that speed up natural successional changes .

9. The regreening in Sudbury highlights how environmental quality can facilitate economic development.

10. In October 1998, Canada became a major producer of diamonds, when the Ekati Diamond Mine started up. Today, it produces $1.7 million of diamonds each day and Canada is the third largest diamond producer in the world.

11. Mining tailings will be held in a lake at Ekati. As the tailings settle, consolidate, and evolve to permafrost, rocks and soil will be spread over the surface. Revegetation will be started, with the goal of having the entire holding area become a wetland once the mining is completed.

12. Serious challenges were encountered by Ekati in incorporating traditional ecological knowledge into environmental research: (1) two Aboriginal groups were in the midst of land claims negotiations and as a result were reluctant to release traditional knowledge into the public domain because this knowledge was important for their negotiation strategy; (2) concern was expressed by Aboriginal people about using traditional knowledge outside of the context of the cultures and broader system of knowledge that give it meaning and value; (3) there was not one set of traditional knowledge, as the Inuit, Métis, and Dene each have their own traditional knowledge, which did not always coincide; (4) traditional knowledge was viewed by Aboriginal people as their intellectual property, and therefore its use and management had to remain within their control; and (5) there was no documented baseline of traditional knowledge.

13. The Bathurst caribou herd is the largest one in the NWT and includes about 350,000 animals. Since the caribou herd does not follow the same migration route each year and the areas affected by the mining represented less than 0.01 per cent of the range of the herd, it was believed that the mining activity by Ekati would have a very small impact.

14. Beginning in 1992, the mining company responsible for Ekati initiated scientific research to understand the impacts of the proposed mining activity and to develop mitigation measures. In July 1994, the

Minister of Indian Affairs and Northern Development referred the mining project for an environmental assessment, and in July 1995 the company submitted its environmental impact statement. From late January to late February 1996, an Environmental Assessment Panel held public meetings, and in June of that year submitted its report to the federal government. In February 1997, the federal government gave its formal approval, and construction started in May 1997. In January 1999, the first diamonds from Ekati were sold in Antwerp, Belgium.

15. Impact and benefit agreements have been pioneered in Canada, and are intended to ensure that Aboriginal communities benefit from mining projects and, where they contain compensation provisions, that those communities are compensated for the negative impacts of mines on their communities, their land, and their traditional way of life.

16. Canada is ranked as the sixth largest user of primary energy in the world. Fossil fuels are the main type of energy consumed by Canadians, and one of the largest consumers of energy is the transportation sector.

17. The combustion of fossil fuels emits greenhouse gases, such as carbon dioxide and nitrous oxide, which accumulate in the atmosphere and contribute to climate change.

18. Alternative energy sources are considered to be solar, geothermal, hydro, tides, and wind.

19. Wind power is the fastest-growing sector in the world's energy market. Wind turbines have minimal adverse environmental affects. However, compared to conventional fossil fuel energy sources, they are expensive.

KEY TERMS

acid mine drainage	kimberlite pipes	renewable or flow	Whitehorse Mining
frozen core dam	metal toxicity	resources	Initiative
impact and benefit	non-renewable or stock	sulphur dioxide emissions	wind power
agreements	resources		

REVIEW QUESTIONS

1. What are the implications of non-renewable or stock resources for strategies related to 'sustainable development' or for 'sustainable resource management'?

2. How important are non-renewable resources for the Canadian economy?

3. What is the significance of the Whitehorse Mining Initiative?

4. What factors contributed to the successful 'regreening' program of the industrial landscape in Sudbury? What general lessons can be learned from this experience related to landscape remediation?

5. What have been elements of 'best practice' related to the opening of diamond mines in the Canadian North?

6. What was learned from the environmental assessment for the Ekati mine regarding incorporation of local knowledge into scientific understanding of impacts?

7. Why is Canada so dependent on fossil fuels? What would have to change for there to be less dependence?

8. What are the advantages and disadvantages of alternative energy sources?

9. What changes should be made by individuals, organizations, businesses, and governments to reduce energy and mineral use?

RELATED WEBSITES

CANADIAN WIND ENERGY ASSOCIATION (CANWEA):
http://www.canwea.ca/home.html?phpLang'en

ENVIRONMENT CANADA, WIND ATLAS:
http://www.cmc.ec.gc.ca/rpn/modcom/eole/
CanadianAtlas.html

NATURAL RESOURCES CANADA WIND ENERGY
INFORMATION SITE:
http://www.canren.gc.ca/wind/index.asp

EUROPEAN WIND ENERGY ASSOCIATION
(EURWEA):
http://www.ewea.org/

DANISH WINDPOWER ORGANIZATION:
http://www.windpower.org

WINDPOWER MONTHLY:
http://www.wpm.co.nz/

REFERENCES AND SUGGESTED READING

Beckett, P.J. 2000. *A Reflection of Two Landscapes—the Copper and Nickel Mining Region of Sudbury before and after 25 years of Land Reclamation Activity*. Sudbury, Ont.: Laurentian University Library.

Canadian Aboriginal Minerals Association. 2001. *Tapping Aboriginal Resources: Opportunities for Aboriginal Community Development, Mining Metals and Diamonds, Oil and Gas*. Whitehorse, Yukon: Canadian Aboriginal Minerals Association.

Canadian Environmental Assessment Agency, Environmental Assessment Panel. 1996. *NWT Diamonds Project: Report of the Environmental Assessment Panel*. Cat. En105-53 1996.

Chapman, J.D. 1989. *Geography and Energy: Commercial Energy Systems and National Policies*. Harlow, Essex: Longman Scientific and Technical.

Couch, W.J. 2002. 'Strategic resolution of policy, environmental and socio-economic impacts in Canadian Arctic diamond mining: BHP's NWT diamond project', *Impact Assessment and Project Appraisal* 20: 265–78.

Dillon Consulting Limited. 2000. *Wind Turbine Environmental Assessment: Draft Screening Document*. Toronto: Dillon Consulting Limited, Feb.

Environment Canada, National Indicators and Reporting Office. 2003. *Environmental Signals: Canada's National Environmental Indicator Series 2003*. Ottawa; Canada. Cat. En40–775/2002E. <http://www.ec.gc.ca/soer-ree/English/Indicator_series/default.cfm>.

Gunn, J.M., ed. 1995. *Restoration and Recovery of an Industrial Region: Progress in Restoring the Smelter-Damaged Landscape near Sudbury, Ontario*. New York: Springer-Verlag.

Hanks, C., and S. Williams. 2002. 'Perceptions of reality: cumulative effects and the Lac de Gras diamond field', in A.J. Kennedy, ed., *Cumulative Environmental Effects Management*. Edmonton: Alberta Society of Professional Biologists, 411–24.

Hilson, G. 2000. 'Sustainable development policies in Canada's mining sector: an overview of government and industry efforts', *Environmental Science and Policy* 3: 201–11.

———— and V. Nayee. 2002. 'Environmental management system implementation in the mining industry: a key to achieving cleaner production', *International Journal of Minerals and Processes* 64: 19–41.

Hoos, R.A.W., and W.S. Williams. 1999. 'Environmental management of BHP's Ekati Diamond Mine in the western Arctic', in J.E. Udd and A. Keen, eds, *Proceedings of the International Symposium on Mining in the Arctic*, vol. 5. Rotterdam: Balkema, 63–70.

Indigenous Environmental Network. 2002. 'Indigenous Mining Campaign Project', at: <http://www.ienearth.org/mining_campaign.html#project>.

Katary, N. 1982. *Origins and Evolution of 2001: A Developmental Odyssey*. Sudbury, Ont.: Department of Planning and Development, Regional Municipality of Sudbury.

Kennett, S. 1999. *A Guide to Impact and Benefit Agreements*. Calgary: Canadian Institute of Resources Law.

Kitchener-Waterloo Record. 2004. 'Harvesting the wind: local farmers say they can provide power—if regulations let them', 9 Aug., A1–A2.

Land Reclamation Program. 1999. *Land Reclamation Program Annual Report, 1999*. Sudbury, Ont.: Economic Development and Planning Services.

————. 2000. *Land Reclamation Program Annual Report, 2000*. Sudbury, Ont.: Economic Development and Planning Services.

————. 2001. *Land Reclamation Program Annual Report, 2001*. Sudbury, Ont.: Economic Development and Planning Services.

————. 2002. *Land Reclamation Program Annual Report, 2002*. Sudbury, Ont.: Economic Development and Planning Services.

Lautenbach, W.E. 1985. *Land Reclamation Program 1978–1984*. Sudbury, Ont.: Regional Municipality of Sudbury, Vegetation Enhancement Technical Advisory Committee, 15 Apr.

————. 1987. 'The greening of Sudbury', *Journal of Soil and Water Conservation* 42: 228–31.

————, J. Miller, P.J. Beckett, J.J. Negusanti, and K. Winterhalder. 1995. 'Municipal land restoration program: the regreening process', in Gunn (1995: 109–22).

Lees, D. 2000. 'Green rebirth: how three decades of grass-roots determination cleaned up Sudbury's industrial gloom', *Canadian Geographic* 120: 60–70.

Lemieux, A. 2002. 'Canada's global mining presence', *Canadian Minerals Yearbook, 2001*. Ottawa: Minister of Public Works and Government Services.

McAllister, M.L., and C.J. Alexander. 1997. *A Stake in the Future: Redefining the Canadian Mineral Industry*. Vancouver: University of British Columbia Press.

Mining Watch Canada. 2002. 'Aims and Objectives'. At: <http://www.miningwatch.ca/mwC_profile_short.html#anchor28160192>.

Natural Resources Canada. 2000. *Energy in Canada, 2000*. Ottawa: Canada Communication Group.

————. 2003. 'Canadian Mining Facts', *Minerals and Mining Statistics On-Line*. At: <http://mmsd1.mms.nrcan.gc.ca/mmsd/facts/canFact_e.asp?regionId'12>.

————, Minerals and Metals Sector. 1998. *From Mineral Resources to Manufactured Products*. Ottawa: Minister of Public Works and Government Services.

————. 2001a. *Focus 2006: A Strategic Vision for 2001–2006*. Ottawa: Minister of Public Works and Government Services.

————. 2001b. 'Canada's Ranking in World Mining', *Canadian Minerals Yearbook 2000*. Ottawa: Minister of Public Works and Government Services.

O'Reilly, K., and E. Eacott. 1998. *Aboriginal Peoples and Impact and Benefit Agreements: Report of a National Workshop*. Ottawa: Canadian Arctic Resources Committee, Northern Minerals Program Working Paper No. 7.

Rees, J. 1985. *Natural Resources: Allocation, Economics and Policy*. London: Methuen.

Regional Municipality of Sudbury. 1990. 'Five Year Land Reclamation Plan 1990–1994 (Draft)'. Sudbury, Ont.: Regional Municipality of Sudbury, Planning and Development Department, Vegetation Enhancement Technical Advisory Committee, 22 June.

Richardson, N.H. 1991. 'Reshaping a mining town: economic and community development in Sudbury, Ontario', in J. Fox-Przeworski, J. Goddard, and M. de Jong, eds, *Urban Regeneration in a Changing Economy: An International Perspective*. Oxford: Clarendon Press, 164–84.

————, B.I. Savan, and L. Bodnar. 1989. *Economic Benefits of a Clean Environment: Sudbury Case Study. Prepared for the Department of Environment, Canada*. Toronto: N.H. Richardson Consulting.

Ross, N. 2001. *Healing the Landscape: Celebrating Sudbury's Reclamation Success*. Sudbury, Ont.: Vegetation Enhancement Technical Advisory Committee.

Rylatt, M.G. 1999. 'Ekati diamond mine—background and development', *Mining Engineer* 51: 37–43.

Saarinen, O. 1992. 'Creating a sustainable community: the Sudbury case study', in M. Bray and A. Thomson, eds, *At the End of the Shift: Mines and Single Industry Towns in Northern Ontario*. Toronto: Dundurn Press, 165–86.

Sanchez, L.E. 1998. 'Industry response to the challenge of sustainability: the case of the Canadian nonferrous mining sector', *Environmental Management* 22: 521–31.

Strong, M. 1995. 'Planning for the future', in Gunn (1995: 109–22).

Voynick, S. 1999. 'Diamonds on ice', *Compressed Air* 104: 60–8.

Werniuk. J. 1998a. 'Great Canadian diamonds', *Canadian Mining Journal* (Oct.): 8–22.

————. 1998b. 'Where the smart mining is going', *Canadian Mining Journal* (Dec.): 14–18.

Whitehorse Mining Initiative. 1994. *Leadership Council Accord: Final Report*. Nov. (Available from Natural Resources Canada, Minerals and Metals Sector.)

Whiteman, G., and K. Maman. 2001. 'Community consultation in mining: a tool for community empowerment or for public relations', *Cultural Survival Quarterly* 25: 30–5.

Winterhalder, K. 2002. 'The effects of the mining and smelting industry on Sudbury's landscape', in D.H. Rousell and K.J. Jansons, eds, *The Physical Environment of the City of Greater Sudbury*. Ontario Geological Survey Special Volume No. 6. Toronto: Ontario Ministry of Northern Development and Mines, Mines and Minerals Information Centre, 145–73.

Witteman, J., L.M. Davis, and C. Hanks. 1999. 'Regulatory approval process for BHP's Ekati Diamond Mine, Northwest Territories, Canada', in J.E. Udd and A. Keen, eds, *Proceedings of the International Symposium on Mining in the Arctic*, vol. 5. Rotterdam: Balkema, 7–11.

Environmental Change and Challenge Revisited

'The only person who likes change is a wet baby', observes educator Roy Blitzer. 'Two basic rules of life are: 1) change is inevitable; and 2) everybody resists change. Much of the world has its defences up to keep out new ideas.'
— von Oech (1990: 180)

Change is ubiquitous. This book has provided an overview of environmental change and challenge in Canada. Change occurs as a result of both natural and human-induced pressures and it is often difficult to determine the balance between them. However, over time it seems as if human-induced changes are becoming the dominant driving factor for many aspects of environmental change. In many cases, these changes, such as extinction and climate change, are irreversible. Current activities are serving to impoverish the planet for future generations.

The preceding chapters have emphasized the need to understand the ecological aspects of environmental change, along with the various management approaches that may be useful. Reading the book should enable you to understand the background to many of the environmental problems that we face, and also to appreciate the different management approaches to their resolution. In each chapter we have attempted not only to make you aware of the nature of the challenges being faced, but also about some of the solutions that are being tried.

There is one chapter in this final part and its main focus is on solutions. In it, we provide an assessment of the progress at the global and national levels in coming to terms with environmental change. Perhaps more importantly, though, we finish with some suggestions for the kinds of actions that you can take to help contribute to the positive changes that need to be made to ensure that future generations will inherit a planet every bit as beautiful and productive as have the generations before.

We hope that you will not only read the chapter carefully, but also take action to try to improve your balance sheet with or your footprint on the environment! Your efforts, combined with those of thousands of others acting individually, can make substantial changes in the environment of tomorrow.

Reference

von Oech, R. 1990. *A Whack on the Side of the Head*. New York: Warner Books.

Chapter 14

Making It Happen

Learning Objectives

- To identify some of the main global responses to environmental degradation.
- To understand some of the main Canadian responses to environmental degradation.
- To place Canada within the global context for environmental response.
- To assess how important environment is to the administration of your university.
- To make better decisions to minimize your impacts on environment.
- To use your influence more effectively to benefit the environment.
- To clarify what 'the good life' means for you.

When I call to mind my earliest impressions, I wonder whether the process ordinarily referred to as growing up is not actually a process of growing down; whether experience, so much touted among adults as the thing children lack, is not actually a progressive dilution of the essentials by the trivialities of life.

— Aldo Leopold, *A Sand County Almanac*

INTRODUCTION

Aldo Leopold, one of the greatest conservation thinkers and writers, points out that as we get older and our lives get busier we often get distracted from the important things in life, like protecting the environment. Everyone says that environmental protection is important, but most devote minimal effort to actually doing anything about it. We are all members of NATO: No Action, Talk Only. And the same is true of our country. On paper Canada has an impressive raft of legislation, policies, strategies, and action plans regarding the environment. Sadly, the translation of these into 'on-the-ground-improvements' is often chronically under-resourced. Many examples have been cited in this book, ranging from lack of resources to implement Canada's Oceans Strategy as

mandated under the Oceans Act (Chapter 8) through to lack of a detailed plan on how to meet our obligations under the Kyoto Protocol (Chapter 7).

In this chapter we will provide a brief overview of global and Canadian responses to environmental change. But governments are only part of the answer. This final chapter rests on the firm conviction that individuals can make a *significant difference* in how the environmental challenges presented in this book will develop over the next decade, if we are aware of the problems and are willing to do something about them. This chapter provides some ideas about how *you* can become involved in creating change.

GLOBAL PERSPECTIVES

The last century witnessed many changes. It may be characterized as an age of diminishing imperial powers, ongoing wars, atomic bombs, the harnessing of the entire globe into an interconnected economic system, rising consumer demands, and an exploding human population. The next century will witness the continuation of some of these trends, but many scientists seem convinced that global climatic change, water shortages, biological impov-

Resist the temptation to give in to the consumer binge (*Philip Dearden*).

erishment, declining food yields per capita, desertification, pollution, and overpopulation will constitute the backdrop for the events of the next century.

Many of these trends are driven by consumption of material goods, which has emerged over the last couple of decades of the last century as the dominant international ideology (Cross, 2002). From its heartland in Europe, North America, and Japan, the globalization of consumption will be one of the main developments of this century, if not the next couple of decades. Although population growth is still a concern, convincing signs point to falling rates of increase and the stabilization of populations, probably within the next 50 years. In contrast, consumption knows no bounds. Indeed, our whole global economic system is focused on increasing consumption levels. At the individual level our psyches are dominated by images of the consumer goods we hanker for. The shopping mall has become the new place of worship.

Roughly one-quarter of humanity is now within this consumer class, and this number is divided more or less equally between those in developed countries and the rapidly increasing numbers of consumers in developing countries, such as China and India. In the developed nations, demands continue to grow insatiably. In the US, for example, the average size of new houses is 38 per cent larger in 2002 than in 1975, yet the number of people per household is falling. These large houses also have big appetites for furniture, electronics, and all manner of consumer goods. They are also expensive to heat and maintain.

The lesser-developed countries are joining the consumer binge. In 1980, for example, there were virtually no private cars in China. By the year 2000, five million people owned cars. Private automobile ownership is projected to rise to 24 million by 2005, with another one billion potential car buyers waiting to add their impacts to the global environment (WRI, 2003). If the average Chinese consumer developed oil demands similar to Americans, China would consume 90 million barrels of oil per day, 11 million more than the entire global daily production in 2001. The global passenger car fleet is now over 531 million vehicles and increases by 11 million vehicles every year (Sawin, 2004). In Canada there were 220,000 more vehicles on the roads in 2001 than in 2000 (Statistics Canada, 2003). The number of cars produced every year is more than five times what it was in 1950. In one of the most dramatic examples of our lack of progress in the areas where it matters most, the mileage of the Model T Ford, first built nearly a century ago, was better than that of the average new Ford today (Sawin, 2004).

The impacts of growing consumer demands are far-reaching. Consumers have driven the more than threefold increase in water consumption and fivefold rise in fossil fuel consumption over the last 50 years, which has served to deny others—those in desperate need—the use of resources in sufficient quantity and/or quality.

International action to address environmental problems has been sporadic over the last decade. The **World Summit on Sustainable Development (WSSD)** held in Johannesburg in 2002 made various commitments, such as the goal of halving the proportion of people without access to adequate sanitation by the year 2015. This will entail providing service to an additional 125 million people a year. A non-binding agreement was also signed by 192 nations to restore fisheries to their maximum sustainable yield by 2015. Unfortunately, compliance is often marginal at best, even for binding international agreements, let alone non-binding agreements. More general declarations also emerged, such as the need to 'promote public procurement policies that encourage development and diffusion of environmentally sound goods and services'. These declarations have good intentions, but little effort is expended to promote implementation (Box 14.1).

This is also the conclusion reached regarding the follow-up to the predecessor of the WSSD, the first **Earth Summit** held in Rio de Janeiro in 1992. The UN Commission on Sustainable Develop-

ment was struck as the official institution to monitor and pursue implementation. Despite over a decade of talk, little has emerged in the way of concrete action. Yet, the amount of money

---

BOX 14.1 HIGHLIGHTS FROM THE JOHANNESBURG IMPLEMENTATION PLAN

The Johannesburg Plan of Implementation is one of two negotiated documents from the 2002 World Summit on Sustainable Development. It encourages countries to fulfill their commitments from the 1992 Rio Earth Summit by participating in a 10-year framework of programs on sustainable consumption and production. Broad expectations and goals for this framework include the following:

- Have industrial countries take the lead in promoting sustainable production and consumption.
- Through common but differentiated responsibilities, make sure that all countries benefit from the process of shifting to sustainable consumption and production.
- Make sustainable production and consumption cross-cutting issues and include them in sustainable development policies.
- Focus on youth, especially in industrial countries. Use consumer information tools and advertising campaigns to communicate issues of sustainable production and consumption to youth.
- Promote implementation of the polluter pays principle, which internalizes environmental

costs and incorporates the financial burden of pollution into the price of the product.

- Incorporate life-cycle analysis into policy in order to track a product from production to consumption and disposal. Use this approach to improve product efficiency.
- Support public procurement policies that encourage the development of environmentally sensitive goods and services.
- Develop cleaner, more efficient, and more affordable energy sources to diversify the energy supply. Phase out subsidies that inhibit sustainable development.
- Encourage voluntary industry initiatives that promote corporate environmental and social responsibility, especially among financial institutions. Examples include codes of conduct, certification, ISO standards, and global reporting initiative guidelines.
- Collect cost-effective examples of cleaner production and promote cleaner production methods, especially in developing countries and among small and medium-sized enterprises.

SOURCE: Worldwatch Institute (2004a).

| Table 14.1 | Annual Expenditure on Luxury Items compared with Funding Needed to Meet Selected Basic Needs |

Product	Annual Expenditure	Social or Economic Goal	Additional Annual Investment Needed to Achieve Goal
Makeup	$18 billion	Reproductive health care for all women	$12 billion
Pet food in Europe and United States	$17 billion	Elimination of hunger and malnutrition	$19 billion
Perfumes	$15 billion	Universal literacy	$5 billion
Ocean cruises	$14 billion	Clean drinking water for all	$10 billion
Ice cream in Europe	$11 billion	Immunizing every child	$1.3 billion

SOURCE: Worldwatch Institute (2004a: 10).

INTERNATIONAL PROGRESS?

Despite these steps forward, the sobering reality is that the limited gains made since 1992 in shifting towards more sustainable patterns of production and consumption have been overwhelmed by the continued global growth of the consumer society. Delegates spent many hours at the World Summit in Johannesburg debating what they might do to turn this situation around. The power of vested interests and institutional inertia translated into reluctance on the part of many governments to commit to a clear program of action toward this end.

— *French (2004: 154)*

required is not large, especially when compared with the amounts spent on consumer items (Table 14.1).

Some landmark international agreements have been signed in the last decade, such as the Kyoto Protocol, the Convention on Biological Diversity, and the Stockholm Convention on Persistent Organic Pesticides. International bodies such as the International Whaling Commission (Chapter 8) have also made important philosophical changes that emphasize conservation rather than consumption. The European Union has developed a climate emissions trading law that will ascribe a market value to carbon dioxide. Certification is also gaining ground in some areas. The Forest Stewardship Council (FSC) (Chapter 9) has now certified more than 39 million hectares of commercial

BOX 14.2 TRADE AND THE ENVIRONMENT IN CONFLICT?

Meetings of the World Trade Organization (WTO) have generated increasing attention since the organization was first formed in 1995. The last one, in Cancun, Mexico, came to a premature end when no agreements could be reached between an increasingly militant block of lesser-developed countries and key developed countries on several issues relating to protectionism. Environment is a central concern. One of the goals of the WTO is to break down unfair trade barriers that prevent countries from gaining access to the markets of others. This is particularly important to many developing countries that want to market and sell their goods to the rich consumers of the developed world. Many developed nations, over the years, have erected trade barriers to protect their own industries from being undercut by such overseas competition.

However, many countries have also established laws and policies that relate to minimum standards, including environmental standards, that must be met by locally produced and imported goods sold within the country. Lesser-developed countries argue that such laws and policies discriminate against their products, since it is too costly for them to meet these standards. Therefore, setting such standards is protectionist and violates international trade agreements.

This difference of perspective has set the stage for some high-profile international disputes relating to the conflict between environmental protection and trade barriers. The US, for example, placed an embargo on tuna from Mexico following declaration of the domestic Marine Mammal Protection Act. Mexican tuna fishing led to the deaths of thousands of dolphins through the practice of setting tuna nets around the highly visible dolphins that school with the tuna. Mexico claimed, successfully, that this discriminated against their fishers. Several Asian countries recently won a similar case involving the shrimp fishery. US law requires shrimp fishers to have turtle-excluding devices on their nets to prevent the drowning of turtles. All species of marine turtle are listed as endangered.

Canada has also been active, appealing to the WTO to turn over a French ban on Canadian exports of the well-known public health hazard and carcinogen chrysotile asbestos. In a ruling lauded by experts as the first time the WTO had recognized public safety over trade, the WTO upheld the ban.

International trade and the rulings of the WTO obviously have major implications for future environmental goals. How these will be played out is far from certain. On the one hand, there is enormous potential for international trade law to actively encourage a transition to a more sustainable society. Some of the most damaging environmental activities, for example, are the perverse subsidies that many governments use to support unsustainable fishing, agricultural, and forestry activities. On the other, past rulings do not create a sense of optimism that this proactive environmental role will be realized in the near future.

forest in 58 different countries. A Marine Steward-ship Council, modelled on the FSC, was recently initiated, and to date it has certified seven fisheries offering 170 certified seafood products in 14 countries. However, significant barriers to international and national efforts to enact environmentally sound practices are increasingly being challenged as 'protectionist' (Box 14.2).

TREATIES, AGREEMENTS, AND COMMITMENT

The number of international environmental accords has exploded as countries awaken to the seriousness of transboundary and global ecological threats. The un Environment Program (UNEP) estimates that there are now more than 500 international treaties and other agreements related to the environment, more than 300 of them negotiated in the last 30 years.

But reaching agreements is only the first step. The larger challenge is seeing that the ideals expressed in them become reality. What is needed is not more agreements, but a commitment to breathe life into the hundreds of existing accords by implementing and enforcing them.

— *Mastny and French (2002: 13)*

Individual countries have showed marked differences in their willingness to take effective action to reduce environmental degradation. In the US, environmental groups have documented over 100 rule changes in US legislation that have reduced environmental protection since the onset of the George W. Bush administration in 2001. The US has also refused to be party to the Kyoto Protocol, despite the fact that it is by far the largest contributor to greenhouse gas emissions. In contrast, the UK has declared that it will take steps to reduce its emissions of greenhouse gases by 60 per cent by the year 2050, greatly exceeding the Kyoto requirements.

Obviously, given the very different conditions pertaining in different parts of the world, different solutions are in order. For some countries, increased consumption is required to allow people to meet their basic needs. For others, drastic reductions in consumption are necessary. Such reduction, however, does not necessarily imply a reduction in quality of life. The Human Development Index (HDI) of the United Nations, for example, takes into consideration education, longevity, and living standards rather than just GNP (gross national product) as measures of development. For very poor people even a small increase in energy consumption can make a major difference to their living standard. It is difficult to spend long hours studying at night, for example, if there is no electricity available. This retards educational levels, which in turn hold back economic development. The benefits of additional energy availability and income increase up to a certain point. After that point, there is no relationship between consumption and the HDI. North American consumption rates are more than triple this threshold level (Suarez, 1995).

Other research points to similar conclusions. Robert Prescott-Allen measured 87 indicators ranging from life expectancy and school enrolment through to deforestation in 180 countries to develop his Wellbeing Index. He found no relationship between energy consumption and the well-being of a country. In fact, the United Arab Emirates, with the world's second highest per capita energy consumption, was ranked 173rd for well-being. Austria was ranked fifth in well-being and twenty-sixth in terms of energy consumption. Sweden and the Netherlands are ranked very similarly in terms of human well-being, yet the Netherlands has a much lower environmental rating. This illustrates that high ratings of human

Global challenges require different solutions in different parts of the world. While most of us need to cut our consumption, this peasant family in Myanmar has nothing to cut. They need to increase their levels of consumption just to meet what many of us would classify as basic needs (*Philip Dearden*).

well-being can be achieved at different environmental costs.

Imagine if this were not the case, that, rather, a direct correlation existed between energy consumption and societal well-being. This would create the imperative to raise the level of energy consumption in the world at least to North American levels, which would entail a fivefold increase in energy use and a likely fivefold increase in all the attendant problems, including acid rain generation (Chapter 4) and global climatic change (Chapter 7). The scale of disruption to planetary biogeochemical cycles (Chapter 4) would be catastrophic.

Thankfully, this is not the case. The studies cited above confirm what many people already know intuitively, although this truism is something we struggle to address in terms of national policies or individual choices. *The best things in life aren't things.* Consuming more will not, after a certain threshold, improve the overall well-being of society or individuals. The challenge at the international level is to enact policies and programs that will see consumption levels raised in the needy countries but reduced in the over-consuming countries, such as Canada. Calculations suggest that a reduction in material consumption in wealthier nations will have to be in the order of 90 per cent for some semblance of sustainability and equity to emerge (Gardner and Sampat, 1998). In short, there needs to be a rediscovery in the devel-

oped countries of the *quality* of life to replace the current emphasis on the *quantity* of goods that can be acquired. Associated changes in the whole operation of our economic system will be required, 'a new perception of value, a shift from the acquisition of goods as a measure of affluence to an economy where the continuous receipt of quality, utility and performance promotes well-being' (Hawken et al., 1999: 10).

NATIONAL PERSPECTIVES

According to the polls, Canadians care about their environment, with around 90 per cent feeling that environmental problems will have either a great deal or a fair amount of an effect on the health of future generations. This percentage has remained virtually unchanged over the last decade (Environics, 2002: 3). Natural resources are also fundamental to our economy, with forestry, agriculture, and fishing accounting for 13 per cent of GDP. Furthermore, many Canadians feel a close personal identity with the ecosystems and wildlife of this vast

'The best things in life aren't things.' This Bali rice farmer has none of the mechanical aids of our modern farmers, none of the household appliances that you and I have, but he is a happy man (*Philip Dearden*).

Extreme poverty forces people in some parts of the world to consume the resources today that they and later generations will need to survive in the future. Here a poor fisher from Sri Lanka's east coast wheels dynamited coral reef fragments to the brick factory for processing. He knows he is destroying the reef that should feed his children in the future, but the reef has been so overfished that it no longer provides for the needs of today. Intact, the reef could have been a protective barrier against the Asian tsunamis of 26 December 2004 (*Philip Dearden*).

country. Leaving a healthy environment for future generations is the number-one issue by which Canadians define what being a Canadian means to them (Environment Canada, 2003a). However, over the past 50 years population levels have doubled and the economy has expanded almost sevenfold, placing ever greater pressures on the environment (Figure 14.1). Environmental change is ubiquitous, ranging from the earlier breakup of ice on the Arctic Ocean to the increasing numbers of species listed as endangered and the amount of waste generated in our households. Many aspects of these changes have been detailed in the previous chapters and some summary trends are shown in Box 14.3. Here we want to provide a brief overview of some

of the responses to environmental trends.

Following the 1992 Earth Summit in Rio de Janeiro, Canada was one of 178 nations that endorsed **Agenda 21**, a series of actions to move towards greater sustainability. One of these commitments was to develop national strategies for sustainable development. In 1997, the General Assembly of the UN reiterated this need. A target date of 2002 was set for the completion of national strategies. In response, Canada's Parliament amended legislation requiring certain government departments to prepare sustainable development strategies to move Canada towards this goal, and also created the position of Commissioner of the Environment and Sustainable Development to

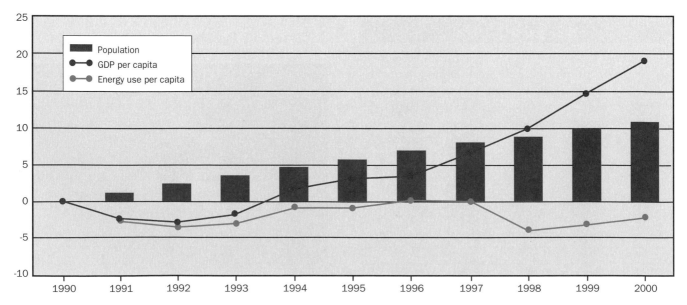

Figure 14.1 Change in population, GDP per capita, and energy use per capita (% change since 1990). SOURCE: Environment Canada (2003b: vi). Reproduced with the permission of the Minister of Public Works and Government Services, 2005.

BOX 14.3 SOME ENVIRONMENTAL TRENDS IN CANADA

- Land area that is strictly protected has doubled since 1992.
- Over 40 per cent of Canada's ecoregions have no strictly protected area.
- Per capita non-hazardous wastes have increased by 10 per cent since 1998.
- Total energy consumption has increased 10 per cent since 1990.
- Automobile use has increased 9 per cent since 1990.
- Fuel efficiency of new vehicles has not improved since 1982.
- Greenhouse gas emissions are up 20 per cent since 1990.
- Stratospheric ozone levels have declined 2–6 per cent since 1980.
- Mercury emissions to air have declined 77 per cent since 1990.
- Total sulphur dioxide emissions are down 19 per cent since 1991.
- Phosphorus loadings as a result of municipal wastes are estimated to have declined by 44 per cent since 1983.

SOURCE: Environment Canada (2003b).

---

monitor and report on the progress of these departments. The Commissioner's office has undertaken reviews of numerous government programs and provides a valuable source of information. The findings of the Commissioner have been instrumental in some cases in encouraging government departments to improve their performance.

The key responsibility for implementing sustainable development in Canada rests with Environment Canada (Box 14.4). Some 28 federal government departments and agencies were required to submit sustainable development strategies, first in 1998 and subsequently in 2001. The Commissioner has reviewed whether the submissions address the right issues and whether significant progress has been made. The 2002 annual report of the Commissioner (Auditor General of Canada, 2002) specifically focused on whether federal departments had met the expectation to change the way they deliver their mandates in light of sustainable development requirements. The audit concluded that significant changes had not

taken place as a result of the Agenda 21 requirement that countries move towards greater sustainability. In other words, it is largely business as usual.

There have, however, been additions to legislation and the development of new programs that should lead to improvements in environmental trends in the future. Recent new federal legislation includes the Oceans Act (Chapter 8), a new National Parks Act, an Act to create National Marine Conservation Areas, the Species at Risk Act (Chapter 11), and the Canadian Environmental Protection Act. The Throne Speech in 2003 announced the government's commitment to create 10 new national parks and five new National Marine Conservation Areas by 2010. This landmark announcement will lead to the essential completion of Canada's national park system after virtually no progress over the last two decades. In addition, the federal government, over opposition from several vocal provinces, especially Alberta, ratified the Kyoto Protocol (Chapter 7), and a commitment was made in the Throne Speech in early 2004 to allocate funds to support action. Many provinces have also enacted legislation that should help improve environmental conditions. Ontario, for example, passed the Safe Drinking Water and Nutrient Control Acts in the aftermath of the Walkerton tragedy (Chapter 12).

These examples indicate a society in a state of transition. Not all environmental indicators are getting worse (see Box 14.3; Environment Canada, 2003b), nor is legislation ignoring the environment. However, there are still some very serious trends of environmental deterioration and major gaps between the requirements and implementation of legislation and policy.

SUSTAINABLE DEVELOPMENT IN CANADA

Sustainable development strategies of federal departments and agencies are not yet fulfilling their potential to influence a change towards sustainable development. The strategies are used as a communication tool, a foundation for future change, and a focal point for managing sustainable development. But currently they are not the strategic documents they were meant to be.

— *Auditor General of Canada (2002: 1)*

dramatic (and often costly) must be the remedial actions when they do occur. In some cases, remedial action is delayed so long that the desired environmental conditions cannot be achieved again. Extinction is the ultimate example, but there are many others. Scientists have found that although politicians have met agreed-upon targets for reduced sulphate depositions, large areas are still receiving excessive sulphates, pH levels are not improving in many lakes, and biological indicators of recovery are even further behind. Reductions were made, but they were too little and too late to halt the acidification of many lakes in eastern Canada. The real challenge is to identify and mitigate these problems before they develop so far that they become either irreversible or so costly to reverse that they cannot be pursued.

The Conference Board of Canada recently reviewed Canada's performance in comparison with other OECD (Organization for Economic Co-operation and Development) countries (23 of the leading nations of the developed world). While praising Canada's economic status, the Conference Board expressed grave concern regarding environmental progress. In five categories—economy, society, health, education, and innovation—Canada was ranked in the top 12 countries. On environment, Canada ranked sixteenth. We ranked particularly poorly in air quality, waste generation and disposal, and water quality. We were second to last in nitrogen oxide emissions and last in carbon dioxide emissions (Conference Board of Canada, 2003).

PERSONAL PERSPECTIVES

Pick Up That Degree

'The future is increasingly a race between education and catastrophe.'
— *H.G. Wells*

The difficulty with many of the environmental challenges discussed above is the enormity of their scale. They are so widespread that most Canadians do not realize they are happening, especially because we are probably more sheltered from their effects than populations of smaller, more densely populated countries. Most of these problems also have long lag times (the period between when the processes are set in motion and when the effects are felt), especially when the effects may have

This mix tells us of a society that has yet to really grasp the significance of the environmental challenges currently faced by Canadians. The 20 per cent increase in Canada's greenhouse gas emissions over the last decade, for example, was greater than both population growth (11 per cent) and total domestic energy consumption (Environment Canada, 2003a). Given Canada's commitment under the Kyoto Protocol to cut emissions by 6 per cent from the 1990 base level (Figure 14.2), this growth is both disturbing and symptomatic of the Canadian approach. It has taken 10 years to formulate a plan on how to deal with climate change and there is still widespread disagreement in Canada on the plan. But over that time period, emissions have continued to increase, making the challenge even more daunting.

Unfortunately the longer we delay action to address environmental degradation, the more

Canada, in many instances, is not winning the battle to maintain and restore wildlife and habitat. A recent assessment of the status of selected species by the Committee on the Status of Endangered Wildlife in Canada (COSEWIC) points to the long-term challenges in protecting species at risk. To monitor progress COSEWIC reassesses the status of select species to determine trends. A recent reassessment showed improvements in status of 20 per cent of the 170 species reassessed compared with the base year of 1985, and no change or a decline in status in the majority of cases. In parallel with the species at risk assessments, recent study of migratory bird populations also shows disturbing trends; approximately 35 per cent of land species and almost one-half of the 56 shorebird species are in decline.
— *Environment Canada (2003a: 17)*

different impacts in different parts of the globe. Global warming, for example, may lead to cooling and increased precipitation in some areas, while the effects may be just the opposite in other areas. Changes of this complexity and magnitude require long-term study before they can be understood. The scale of change also suggests that, within the human lifespan, many of these trends may be irreversible.

Is our educational system preparing you and the students that will follow you to understand and deal with these changes? Urgent actions are required if we are to help diffuse these trends, but actions do not occur in a vacuum. They require some understanding of the road we are on, where it is going, and how we can get on other, more desirable roads. Educational systems should help to bring about this understanding. Currently, many schools, colleges, and universities graduate students who have little or no idea about how the ecosphere functions and how human activities impair those functions. They shop, travel, eat, drink, work, and play in blissful ignorance of the impacts they may be having on life support systems.

Every morning as one of the authors cycles to work, he is reminded of this. His route takes him through the grounds of a local high school. Outside, at the back, numerous boys (yes, always boys!) tinker with their cars. This is not recess time; they are taking a class on cars. A class on cars should, first and foremost, deal with the tremendous impacts that cars have on the environment. How much matter and energy does it take to build and maintain one? What are the impacts on the carbon cycle and global warming? What are the effects of high-temperature combustion on the ozone layer? Astonishingly, he found that none of the students had even considered such questions, let alone knew the answers or the significance of the answers. Our education system is backwards. Before being given time to tinker with cars in school, students should be required to take and pass a course on the environmental implications of cars, but they have no time for such things. Cars are part of their 'real' world.

Colleges and universities are frequently accused of not being part of the 'real' world. By 'real' world, people usually mean the economic realities of today's society. However, this is not the real 'real' world; it is a game that humans introduced to facilitate barter and exchange. Important? Yes! Is this the real world? Only partially! The real world is composed of the air

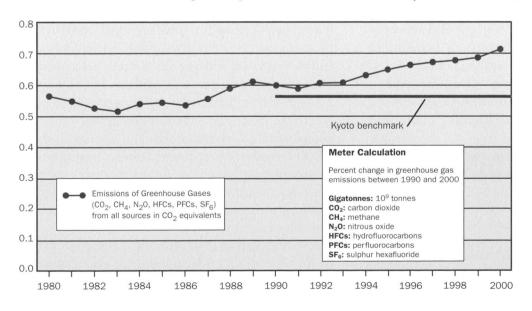

Figure 14.2 Canadian greenhouse gas emissions, 1980–2000 (gigatonnes). SOURCE: Environment Canada (2003b: 20). Reproduced with the permission of the Minister of Public Works and Government Services, 2005.

we breathe, the water we drink, the organisms that keep the life support systems going, and the ground we stand on. Without these things, there can be no other invented 'real' world. We concentrate on balancing these play budgets when, in reality, the more significant budgets of energy throughflow and material balance determine the future of society. The deficit? Yes, Canadians have a huge deficit, but it relates to the 60 million bison that no longer roam the Prairies, the skies cleared of all but a fraction of the birds that used to flock in such large numbers that day turned to night, and the seas that were once home to the largest animals ever to evolve on the planet but where now not even the smallest fish are safe from sonar detection, vacuum trawling, and human consumption. Yes, we have a deficit, most of which remains unrecognized in the routine financial and economic accounting procedures that help shape national policies.

A primary function of our educational system should be to give students a general level of understanding about the nature of the environment and resources. We have requirements for general levels of language and mathematical competence, but require nothing from our students in terms of this most fundamental challenge of the future. Indeed, there are now pressures, especially on universities, to put greater and greater emphasis on meeting the short-term economic demands of society. Business schools and faculties of commerce flourish, yet few additional resources are allocated for programs dealing with the environment.

Even in such programs, colleges and universities have seldom done a good job of instilling in students an appreciation for and love of the planet. Increasingly, science programs have become a process of learning more and more about less and less. They have produced technically competent

ON EDUCATION

... without significant precautions, education can equip people merely to be more effective vandals of the earth. If one listens closely, it may even be possible to hear the Creation groan every year in May when another batch of smart, degree-holding, but ecologically-illiterate, Homo sapiens who are eager to succeed are launched into the biosphere.

— *Orr (1994)*

scientists, but these programs have often missed the mark considerably in terms of maintaining students' wonder about the natural world and combining the rigour of scientific inquiry with deep moral questioning. Science programs often mistake the laboratory for the 'real' world and cut students off from a more comprehensive understanding of, and passion for, their environment.

You can create change on your campus, as the guest statement by Graham Watt-Gremm indicates. Are there sufficient courses on the environment? Do these courses cover a wide spectrum from the technical to the philosophical, and, more important, are students encouraged or even required to select from courses all along this spectrum? You should also remember that campuses are large consumers and processors of matter and energy. How efficient are they? Has anyone undertaken an environmental audit of your campus? How are wastes disposed of? How much recycling occurs? Are chemicals used for landscaping? Does the faculty pension fund invest in businesses with unsound environmental practices? There are many questions that can be investigated through course work, environmental clubs, or individually. If you are interested in pursuing these ideas, you may wish to draw on such experiences elsewhere (e.g., see Smith, 1993; Thompson and van Bakel, 1995) and start some activities on your own campus. For an example of how one interested student managed to get the ball rolling at a Canadian university, see Bardati (1995). Similarly, at the University of Waterloo, engaged students, faculty, and administrators have pursued a multi-pronged 'WATGREEN' program since 1990 that involved surveys and audits—and real change in campus behaviours— with an emphasis on such projects as: solid waste management to reduce waste by composting and recycling; landscape practices, including replacing annual plants with perennials and phasing out the use of biocides; water usage; energy consumption; environmental awareness and habits; and changing what is consumed, which encompassed plans for building retrofitting, as well as reuse of materials and use of biodegradable cleaning products.

So, pick up that degree and encourage others to increase their understanding of environmental challenges. Do not be intimidated by people who think that interest in the environment and higher learning is not the 'real' world. Challenge your teachers to inspire you. Be interested. Apply what

Guest Statement

How Students Brought Sustainability into Campus Planning at the University of Victoria

Graham Watt-Gremm

For most of UVic's four decades, the university planned and developed according to the initial vision and recommendations of the 1960s campus plan: sprawling buildings amid courtyards and gardens, a circular ring road surrounding a segmented academic core. After a period of rapid growth in the 1990s, the university again turned its attention to planning and released a series of planning updates and draft campus plans between 1999 and 2002.

Graham Watt-Gremm

To many in the university community, the new plan signalled business as usual, with no improvements in sustainability, and continued sprawling development, parking lots, and courtyards. Furthermore, the plan was inconsistent, vague, and non-committal, providing no structure for future planning decisions. Student groups such as the UVic Sustainability Project (UVSP) and research centres like the POLIS Project on Ecological Governance responded strongly by rallying the university community and providing a number of recommendations for improvements to the plan itself and to the planning process that produced it.

For the most part, the university recognized these concerns and recommendations, producing a much-improved Campus Plan in 2003. The new plan includes consistent recognition of sustainability principles, a commitment to maintaining and restoring the integrity of campus natural areas and wildlife, protection of aquatic systems through stormwater management, green building practices, and transportation demand management. The university also committed to reviewing and improving processes for planning, communication, and consultation. While the plan is not perfect, and the university faces a number of challenges in implementing its vision, it marks an opportunity for the university to demonstrate leadership in sustainability and environmental management.

I believe that without the dedicated and passionate involvement of students such as those working with the UVSP, the university could not have improved the campus plan to this extent. From the beginning, students at the UVSP established themselves as a resource for faculty, staff, and community by research-

ing aspects of campus ecology and sustainability, including baseline inventories of campus ecosystems, audits of energy, water, and recycling systems, and reports on planning and sustainability.

UVSP staff and volunteers brought their knowledge to administration and management by gaining representation on key decision-making bodies such as the Campus Development Committee, Senate, and student governance committees. Even more important to these contributions were the development of relationships and a new understanding of shared interests among all members of the university community. To this end, the UVSP has engaged in community mapping, facilitated workshops on planning and sustainability issues, led high-profile projects in restoration and transportation, and taken community members on interactive nature walks and campus exploration.

It is not always easy for a small student organization to contribute to planning processes. Relationships take time to build, but staff turnover on student organizations is high, and as new energy and ideas come in older (unfinished) initiatives are left behind. There are also trade-offs between activism and continuing dialogue—the UVSP could not organize protests while keeping a position of influence with decision-makers. This highlights the need to allow a diversity of approaches to environmental activism. Common to all approaches is the need to develop deep knowledge of the issues, to participate as much as possible in consultation and decision-making, and to develop relationships among all stakeholders.

Graham Watt-Gremm is a Master of Science student in Environmental Studies at the University of Victoria and former Ecosystem Planning Leader with the UVic Sustainability Project.

BOX 14.5 DREAMING OF A 'GREEN' CHRISTMAS?

Christmas heralds the biggest consumer bash of the year, although merchants are also trying to persuade us to be equally excessive at other times. Take control of your consumer lifestyle at Christmas and the battle is half won. Consider the following gifts:

1. Arrange an event, an outing, or a personal service rather than giving a material item.
2. Increase 'green' education by giving a book or subscription to a magazine. You could also buy someone a membership in a 'green' organization, such as Pollution Probe, the Canadian Parks and Wilderness Society, or Greenpeace.
3. Give a houseplant, a backyard composting kit, or an unbreakable coffee mug to replace the use of disposable ones. The World Wildlife Fund also enables you to protect an acre of rain forest by making a donation to fund a project that was started by unemployed students.
4. Give, in your gift recipient's name, a goat, a llama, a water buffalo, or any number of other animals through Heifer International, a 60-year-old NGO with a mission of alleviating hunger, poverty, and environmental degradation by providing food-producing animals and sustainable agriculture education to families in need.
5. Give something that conserves energy, such as a bus pass or an energy-saving shower head.
6. Give second-hand items.
7. Make your own gifts, such as a sweater, dried flowers, or jam.
8. Give items that display the EcoLogo of three doves.
9. Choose gifts that require little wrapping. Reuse old wrapping paper or use reusable fabric gift bags instead.
10. Put 'Planet Earth' at the top of your list. If we all gave the planet an offering for Christmas that would make it feel better, the Earth would be a little more loved and a little less stressed.

you learn to your life. Don't be misled, however, into thinking that the formal education system is the only source of learning. Keep on reading. Many of the most inspiring works on the environment do not make their way onto college or university reading lists, and most of the rewarding environmental experiences are certainly not part of the curriculum. Challenge society to change and seek a new kind of relationship between humanity and our home, planet Earth.

Light Living

'Light living' expresses the need to tread as lightly as possible, to minimize our ecological footprints (Chapter 1). Many books and guides, such as *The Canadian Green Consumer Guide*, have been written on the topic. What follows is a brief selection of ideas that you may wish to try. Given the statistics presented elsewhere in the text regarding per capita energy consumption, waste production, and water consumption levels in Canada, you can be assured that Canadians contribute greatly to overconsumption. **Light living** is often characterized by the four R's: refuse, reduce, reuse, and recycle.

Refuse

We live in a society geared towards making consumption easy. Our newspapers are full of advertisements regarding the best buys. Turn on the radio and you hear from the sponsor, or you are bombarded with commercials on the TV. Estimates suggest that the average American will see 35,000 television ads every year. Most of us are surrounded by shopping opportunities on a daily basis. We can drive to one of several mega-malls in most Canadian cities, park with ease and little expense, and consume from a wide variety of stores, pay by credit card, ATM, cheque, or even cash. We frequently shop not to fulfill basic needs but to indulge frivolous and petty whims. Clothes are discarded when no longer fashionable rather than when they lose their durability. Gadgets are discarded in favour of newer and shinier models. Christmas as a religious holiday has been replaced as the time when we pay homage to our greatest god, **consumerism** (Box 14.5).

Resist and *refuse* to buy anything that you do not really, really *need*. If the purchase is necessary, shop carefully. Buy items that are less harmful to the environment during all stages of the product life cycle, from manufacture to consumption and disposal. There are products certified by the government for their low impacts, as discussed later. Buy one quality item rather than a succession of several shoddy ones to fulfill your needs. Buy organic produce wherever available or, better still and if possible, refuse to buy any and grow your own. It is difficult, because consuming is easy. All the messages that we receive from society extol the virtues of buying things.

Reduce

Can you reduce your consumption of certain items? Energy is a good place to start and a major contributor to greenhouse gas (GHG) emissions. More than a quarter of Canada's GHG emissions occur as a result of the everyday activities of Canadians. The government is asking every Canadian to reduce their emissions by one tonne per year, or about 20 per cent of current consumption. A lot of energy is taken up in space heating in Canada, but must you have that thermostat set so high? Canadians tend to keep their houses much warmer inside than Northern Europeans, for example. These cultures (New Zealanders, too) are accustomed to setting a low thermostat and wearing more clothing in the house in winter. They would not expect to be comfortable wearing just a T-shirt. Turn down the thermostat, wear more clothes, and turn the thermostat down further when you go out or go to sleep.

Reduce lighting costs by replacing burnt-out bulbs with long-life bulbs. They may be more expensive to buy, but they last 10 years or longer and use 75 per cent less electricity. When replacing electrical appliances, a major factor in your choice should be energy efficiency. Think before you turn on a light or an appliance about whether you really need it. Every time you flick the switch or plug in an appliance you are sending out the message of demand. Electricity-producing utilities and governments will react to your message by building new production facilities, with all their attendant environmental costs. If you do not want those costs, try not to send along as many messages signalling your demand.

Transport is also a big energy consumer, accounting for one-quarter of all energy used in Canada. Road vehicles are responsible for 83 per

In Europe few students have cars. Cities are designed for people, and it's easy to get to campus by bicycle or mass transit. Even when people buy cars they prefer small ones (*Philip Dearden*).

cent of that share. Canadians' average per capita gasoline consumption is 1,100 litres per year, compared with 350–500 litres in European countries. It is also estimated that each kilometre of road or highway takes up about 6.5 hectares of land. In Ontario, which has 155,000 km of highways, roads, and streets, this would add up to one million hectares for motorized vehicles. Whenever feasible, walk or ride a bicycle. If you have to use motorized transport, use public transport, such as buses and trains. If you must have a car, get a small economical one with a standard transmission, use it sparingly, and try to carpool (Box 14.6). Most people are aware that larger cars (and SUVs) consume more gasoline. However, since it takes 18 litres of water to produce one litre of gasoline, they are also contributing to water deficiencies (Sawin, 2004).

The Union of Concerned Scientists (Brower and Leon, 1999) suggests that our food choices are second only to transport in terms of their environmental impacts. Agriculture covers over 25 per cent of the world's surface and profoundly affects the health of natural ecosystems. Choosing a meat-rich diet with ingredients imported from afar generates as much as nine times the carbon emissions of a vegetarian meal made from local produce (Carlsson-Kanyama, 1998) and requires two to four times the land to produce (FAO, 2003). It is not only what we choose to eat but also how much we eat. In Canada, as in other Western countries and in the consumer classes of the developing countries, growing obesity is one of the main health challenges. Eat only as much as your body needs to function in an effective and healthy manner. Become part of the food democracy movement. This means using your purchasing power to help establish connections between consumers and producers in the local area and making sure that you eat the kinds of food you wish to eat, not the kinds promoted by government subsidies and industrial agriculture.

You can also think about reducing the waste associated with the things you buy. Many products are overpackaged. They look good on the store shelf, but will only add to the amount of waste sent to the landfill. When you can, buy groceries in bulk to help cut down on packaging. Reduce your waste by starting a compost heap for kitchen wastes. Reduce and, if possible, eliminate your use of toxic materials. Products that may seem quite innocuous (paint, solvents, and cleaning agents, for example) become hazardous wastes when disposed of. Try to find alternatives to these products. Don't buy more product than you need, and dispose of them in full accordance with the instructions from your local municipality.

BOX 14.6 THE ENVIRONMENT-FRIENDLY DRIVER

How you drive is as important as what you drive in terms of minimizing negative environmental impacts. Here are some tips:

1. Slow down: Cutting speed from 112 km per hour to 80 km per hour reduces fuel consumption by 30 per cent.
2. Keep your tires inflated. By cutting tire drag, radials give you 6–8 per cent fuel saving.
3. Keep your car tuned. This will ensure maximum efficiency and minimum pollution.
4. Minimize idling times. Even –20° C requires only a couple of minutes of warm-up.

Reuse

Buy products, such as rechargeable batteries, that can be reused. Try to find another use for something no longer useful in its original state. Use plastic food containers to store things in your fridge or workshop. Return with the same plastic bags that you bought your groceries in for your next load of groceries, or better still, use cotton bags to transport your goods. When you're finished with something, it may still be useful to someone else. Organize a garage sale or donate the items to charity rather than throwing them out.

Recycle

Recycling facilities have sprung up all across the country over the last decade. Recyclable materials include newspaper, cardboard, mixed paper, glass, and aluminum. In many areas you can now also recycle plastics, car batteries, tires, and oil. These materials can be reprocessed into new goods. It

Figure 14.3 How not to reduce greenhouse gas emissions. SOURCE: Reprinted by permission of Adrian Raeside.

takes 30–55 per cent less energy, for example, to make new paper from old paper than it does to start fresh from a new tree. Estimates suggest that if we recycled all the paper we used in Canada, we would save 80 million trees. The amount of newsprint is over 50 kg per person per year—enough to account for one whole mature tree. Similar efficiencies can be obtained by recycling other materials where less energy is required for remanufacture. Oil, for example, fuels many industrial processes. Estimates suggest that if 1 per cent of the Canadian population recycled instead of trashing an aluminum can a day, the oil saved would make 21 million litres of gasoline. However, the bottom line remains that while it is better to recycle than not, reducing consumption levels in the first place is still the preferred option.

THE LAW OF EVERYBODY

There are also many other ways in which you can reduce the pressure on the environment. These are just a few suggestions to get you started. The key point to remember is that the accumulated actions of many concerned individuals acting together will make a difference. In this book we have introduced you to many different scientific laws and principles related to understanding and managing the environment. However, if everyone knew and enacted this last one, it would go a long way to ameliorating our environmental challenges. It is, simply, the **law of everybody**.

'Everybody's Got To Do Something'

If we all did a myriad of small things that we could do easily, without having much negative or even noticeable influence on our lifestyle (e.g., drive 10 per cent less a year, buy a few less toys, shower with a friend, etc.), we would find that many of our environmental challenges would be greatly reduced in scale. It is not important that all people do everything, but everyone has got to do something!

There are now a few websites where you can find out more about your own environmental impact and concrete ways in which your impact can be reduced. For example, try the *personal challenges* identified on the website of the famous Canadian environmentalist, David Suzuki: <www.davidsuzuki.org>, and explore some of the Web sites at the end of this and other chapters in the book.

Influence

One of the best ways to influence business is through the purchasing power of consumers. If consumers band together and refuse to buy certain products because of their impacts on the environment, then the manufacturer will either have to respond to these concerns or go out of business. There are many successful examples of these kinds of actions. During the 1970s and 1980s, for example, conservationists were able to exert increasing pressure on hamburger chains to change their source of beef supply to ensure that tropical rain forests were not being cut down to be replaced by grass to graze cattle for hamburgers. Following boycotts, all major chains were persuaded to ensure that their sources did not contribute to the destruction of the rain forests. Similar campaigns have been used on tuna canners to ensure that tuna is not caught by methods that kill dolphins, which often swim with schools of tuna.

Consumer boycotts can be a very effective way of influencing business practices and reducing environmental impacts. When choosing fish to eat, for example, check out the websites at the end of Chapter 8. Download the 'Seafood Watch' choice guide, compiled by the Monterey Aquarium in California: <www.montereybayaquarium.org>. This wallet-sized and regularly updated guide lists

MANAGING PLANET EARTH

A second myth is that with enough knowledge and technology, we can, in the words of Scientific American *(1989), 'manage planet earth.' Higher education has largely been shaped by the drive to extend human domination to its fullest. In this mission, human intelligence may have taken the wrong road. Nonetheless, managing the planet has a nice ring to it. It appeals to our fascination with digital readouts, computers, buttons and dials. But the complexity of earth and its life systems can never be safely managed. The ecology of the top inch of topsoil is still largely unknown as is its relationship to the larger systems of the biosphere. What might be managed, however, is us: human desires, economies, politics, and communities. But our attention is caught by those things that avoid the hard choices implied by politics, morality, ethics, and common sense. It makes far better sense to reshape ourselves to fit a finite planet than to attempt to reshape the planet to fit our infinite wants.*

— Orr (1994: 9)

three categories of seafood: best choices, caution, and avoid. If consumers followed these guidelines it would alleviate much of the pressure on fishery resources.

Not only should we be sending messages to producers about our environmental standards through purchasing power, we should also let non-conforming producers know why we are not buying their products. If you believe in eating wild rather than farmed salmon, always ask the origin of the salmon before you buy it. If you believe in fair-trade coffee or certified wood products, always let retailers know that is what you are shopping for. Home Depot, for example, is the largest retailer of wood products in the world. Following a consumer boycott organized by a non-governmental organization, Home Depot finally made a commitment to phase in sale of only certified wood products in its stores. This decision, rapidly echoed by two other major retailers, probably did more to protect old-growth forests in BC than all the protests of the last 20 years. However, the change was slow in coming and there is still opportunity for improvement in Home Depot's purchasing practices. Home Depot argues that one of the main reasons behind its tardy behaviour is simply that customers do not ask for certified wood products and often do not buy them preferentially (Mastny, 2004).

As a responsible consumer it is important that you let your preferences be known. This also raises the question of corporate responsibility (Box 14.7). Is it the responsibility of the retailer to take part in, or even lead, the transference to a more sustainable way of doing things by informing its customers about the environmental implications of different products? Or is it sufficient merely to react to customer demands once they are manifest in the marketplace?

Life-cycle assessments (LCAs) are now gaining greater support from both government and industry. LCAs identify inputs, outputs, and potential environmental impacts of a product or service throughout its lifetime. The 2002 World Summit on Sustainable Development called for greater investment in LCAs and the United Nations Environment Program is developing a program to assist in the development and dissemination of practical tools to help in life-cycle assessment. One industrial example is from Volvo, which provides LCAs for the various components involved in vehicle manufacture. Various NGOs, such as Green

Seal, provide standards for different products that manufacturers must meet to gain their endorsement. Others have excellent programs to assist consumers, industry, and government in making wise procurement decisions, such as the American-based 'Center for a New American Dream' <www.newdream.org>. In Canada, the Canadian Environmental Choice product-labelling program is based on this approach. This program aims to influence business practices by shifting consumer and institutional purchasing to products and services that conform to environmental guidelines. A board establishes the guidelines that each category of product must achieve to obtain recognition, signified by the three doves logo. These guidelines consider factors such as recyclability, design of packaging material, use of recyclate in manufacturing, effluent toxicity, and energy consumption. In addition to this list, however, the government requires the board to take into account the economic impacts of the guidelines to ensure that they do not impinge on the economic competitiveness of Canadian products. As David Cohen, a member of the board, concludes:

> The insistence on an economic analysis requirement for a non-regulatory program, which attempts to do little more than provide reasonably accurate consumer information, reveals the ambiguity of official commitments to environmental recovery and the depth of government concern that environmental actions might bring economic damage. If governments believe environmental information for consumers is a serious economic threat, we have good reason to be concerned about the design and likely results of governments' more direct regulatory initiatives for the environment. (Cohen, 1994: 27)

Governments make many decisions that can aid or retard sustainability. Public transport can be subsidized instead of private cars, land-use planning and building codes can be designed to minimize energy and material use, and governments can help facilitate shared public consumption, such as provision of libraries and swimming pools, rather than encouraging private acquisition of these services. Many municipalities have inherited what has been called the 'infrastructure of consumption' from previous generations (OECD, 2002). There is now a need to redesign the way we

BOX 14.7 CORPORATE CONTRIBUTIONS

Large corporations and retailers are highly visible targets for environmental action and have attracted a lot of attention. However, many corporations do much better than we do personally, or than our governments do, in taking a systematic look at some of their environmental impacts and moving to address them. For example, the ubiquitous global fast-food corporation, McDonald's, has the Earth Effort program, which embraces the reduce, reuse, and recycle philosophy described in the text. Every year McDonald's is committed to buying at least $100 million worth of recycled products for building, operating, and equipping their facilities. Carry-out bags are made from recycled corrugated boxes and newsprint; take-out drink trays are made from recycled newspapers. New restaurants have been constructed with concrete blocks made from recycled photographic film and roofs made from computer casings. They have also cut down on the amount of waste produced; for example, sandwich packaging has been reduced by over 90 per cent by switching from foam packaging to paper wraps for sandwiches. In 2001, McDonald's adopted compostable food packaging from reclaimed potato starch and other materials.

McDonald's also has a strict policy that they will not buy the beef of cattle raised on land that was converted from rain forests for that purpose. In addition, it has programs to reduce energy consumption and to take part in a wide range of local initiatives ranging from tree planting to local litter drives. For its efforts, McDonald's in the US has won White House awards and the National Recycling Coalition's Award for Outstanding Corporate Leadership.

McDonald's is not, of course, alone in its efforts. Many well-known corporations have similar programs. Eddie Bauer, for example, has joined forces with an NGO, American Forests, in a tree-planting effort on damaged forest ecosystems in eight reforestation sites in the US and Canada. Consumers can opt to add a dollar to the price of their purchase at the store, and in turn the store will plant a tree for each dollar and donate another tree. Nike, the shoe, clothing, and sports equipment giant, has not only boosted the proportion of organic cotton in its own products but in 2001 also helped to launch 'Organic Exchange', a network of 55 businesses all committed to expand significantly the use of organic cotton within the next 10 years. Other companies, such as Texas Instruments, Levi Strauss, and the Ford Motor Company, have banded their purchasing power together to buy recycled paper.

Also consider Valhalla Pure Outfitters of Vernon, BC, makers of high-quality outdoor clothing. They now make all their polyester fleece jackets from recycled materials that look, feel, wear, and last the same as new fabric, even though they are made from recycled pop bottles. Each pop bottle diverted from a landfill site for recycling is sent to a factory in South Carolina, where they are separated by colour and then reprocessed before being sent to finishing plants in the US and Quebec. The final cost is similar to making new material, but has the added environmental advantages. Each jacket is equivalent to about 25 pop bottles.

These are not the only ways in which the corporate and business world is contributing positively towards addressing environmental problems. Some of them also have substantial foundations to which environmental groups can apply for funding for specific problems. As government coffers are drained, more pressure is put on these sources. The Laidlaw Foundation, for example, used to give $50,000 a year for environmental projects, and requests rarely exceeded this amount. The Foundation now disburses $350,000 but has requests exceeding $1 million. The Richard Ivey Foundation in London, Ontario, provides up to 80 per cent of its $2 million budget for environmental projects each year, but still manages to meet only 5 per cent of the amount requested. With such pressure on resources, it is essential to ensure that the money is used effectively to address the most pressing problems.

If there are environmental problems in your area that you think need to be addressed, don't be afraid to approach local businesses for support. They may well agree with you and be happy to make a contribution to a carefully crafted solution.

live in accord with our increased understanding of planetary limits.

What can you do to encourage governments to act? Election times offer a major opportunity. Make a point of asking your local candidates at all levels of government for their views on certain environmental or sustainable development issues. Help publicize these views. It is much more difficult for politicians to change their positions if they are well

A wind farm in southern Alberta (*Philip Dearden*).

known to the public. Make sure that the politicians follow through on their pre-election promises on environmental matters. Make sure you express your views on the environment to politicians between elections. Write letters, call them, go and see them, and organize demonstrations.

Governments are often more than willing to adopt environmentally friendly, voter-popular measures—as long as they do not involve any significant economic costs. Issues are often very complex, too complex for the understanding of many individuals, who nonetheless care about acting in an environmentally responsible fashion. For this reason, concerned people may band together in non-governmental organizations. Such organizations represent the collective concern and

resources of many people, and consequently are in a much better position to attack a problem than an individual. One of the best ways to spend your conservation dollar is to support such a group. They have had significant impacts on government policies in Canada. A list of some organizations is provided in the Appendix.

IMPLICATIONS

The overall implications of this book should be brutally clear by now. Humanity is facing some major challenges over this next decade regarding our relationship with the Earth and its life support systems. So great are our capabilities to affect these systems that human-controlled influences now dominate many natural processes, resulting in the many critical environmental problems described throughout this book. Government programs at the international and national levels continue to emerge, but actions that make a difference on the ground are few and far between. Still, we are not powerless to change the direction of society. As individuals and as concerned individuals banding together, we can effect many of the changes that need to be made and we can strive for greater well-being, not wealth, in the future. So, despite the severity of the challenges, it is important to remain optimistic.

Above all, stay cheerful, stay active, look after yourself, look after others, love this planet, and don't give up! We are on the most beautiful planet we know about. Canada has some of the most breathtaking wonders in the universe. We have a responsibility. We can think of no better advice than that offered by Edward Abbey (1977):

The symbol of three doves in the shape of a maple leaf identifies products certified under Canada's Environmental Choice Program. The three doves represent Canada's consumers, industry, and government working together to improve the environment (*Printed with permission of the Environmental Choice Program*).

One final paragraph of advice: *Do not burn yourselves out. Be as I am—a reluctant enthusiast ... a part-time crusader, a half-hearted fanatic. Save the other half of yourselves and your lives for pleasure and adventure. It is not enough to fight for the land; it is even more important to enjoy it. While you can. While it's still here. So get out there and ... ramble out yonder and explore the forests, encounter the grizz, climb the mountains, bag the parks, run the rivers, breathe deep of that yet sweet and lucid air, sit* quietly for a while and contemplate the precious stillness, that lovely, mysterious and awesome space. Enjoy yourselves, keep your brain in your head and your head firmly attached to the body, the body active and alive, and I promise you this much: I promise you this one sweet victory over our enemies, over those desk-bound people with their ears in a safe deposit box and their eyes hypnotized by desk calculators. I promise you this: you will outlive the bastards.*

SUMMARY

1. The last century witnessed the start of a consumer-dominated society in several parts of the world that is rapidly spreading and threatens to engulf most of the world in the early part of this century.

2. Population growth rates are showing declines, but consumerism continues to expand. Environmental conditions continue to deteriorate and are likely to do so even more as consumerism spreads.

3. A global response to environmental change emerged following the Earth Summit held in Rio de Janeiro in 1992. Evaluations suggest that progress on meeting the targets set out in *Agenda 21*, the official follow-up, has been very slow. An updated agenda was established following the World Summit on Sustainable Development held in Johannesburg in 2002. There is potential for significant conflict between the movements to improve global environmental conditions and those seeking reductions in international trade barriers.

4. Important global agreements have been forged, including the Kyoto Protocol and the UN Convention on Biodiversity, that set frameworks for global actions. The contribution of national governments to these initiatives varies greatly.

5. Research examining the relationship between consumerism and well-being shows that a positive relationship does exist when incomes are small. However, after a certain point is reached there is no relationship. Most of the developed world has already attained triple this income level.

6. Polls indicate that the environment is important to Canadians, yet our response to environmental degradation at both international and national scales continues to be very slow, despite the clear signals we have regarding the state of the environment.

7. Recent federal legislation includes the Oceans Act, a new National Parks Act, an Act to create National Marine Conservation Areas, the Species at Risk Act, and the Canadian Environmental Protection Act. In addition, the federal government ratified the Kyoto Protocol and many provinces have also enacted

legislation that should help improve environmental conditions.

8. The Conference Board of Canada recently reviewed Canada's performance in comparison with other OECD countries. In five categories—economy, society, health, education, and innovation—Canada was ranked in the top 12 (of 23) countries. On environment, Canada ranked sixteenth.

9. Ordinary citizens can have a positive impact in many ways on the environmental challenges facing society. A first step is building awareness of these problems. Universities and colleges should be intimately involved with this process by ensuring that all graduates have a measure of environmental literacy before they graduate. Universities and colleges are also large consumers of matter and energy and should lead by example in reducing their impacts on the environment.

10. Individuals can help by living in accord with the four R's: *refuse* to be goaded into overconsumption, *reduce* consumption of matter and energy where possible, *reuse* materials where possible, and *recycle* those that you cannot reuse.

11. Individuals can also wield influence by banding together and taking collective action, for example, through consumer boycotts of products and companies that engage in environmentally destructive practices. One of the most successful ways of doing this is by joining a non-governmental organization (NGO) composed of and supported by like-minded individuals. As an individual you will probably not be able to afford to support an environmental lobbyist in Ottawa. However, if thousands of people contribute, then this becomes a reality.

12. In spite of the serious nature of many of the environmental problems described in the book, it is important to stay optimistic about their solution and take time to get out and enjoy the beauty and challenge of one of the most splendid parts of the planet, Canada, our home.

KEY TERMS

Agenda 21	law of everybody	World Summit on Sustainable
consumerism	life-cycle assessments (LCAs)	Development (WSSD)
Earth Summit	light living	

REVIEW QUESTIONS

1. What do you think are the main factors driving global environmental degradation?
2. Which factors do you think are the most serious and why?
3. Discuss what you think should be the top priorities for global action.
4. Outline some of the main responses to environmental degradation at both the international and national levels.
5. Discuss Canada's current and potential role at the international level.
6. What do you think are the main barriers to more effective government response in Canada?
7. Discuss your institution's present and potential role in raising environmental awareness.

8. What are three concrete steps towards 'light living' that you are willing to make over the next month?
9. What are some initiatives that you could take as an individual or as part of your community to reduce negative environmental consequences from our activities?
10. Find out the names and mandates of the environmental NGOs in your area. Is there anything you can do to help them?
11. Who is your local MP? What is his or her view on environmental issues? Exercise your democratic right—phone and find out.

RELATED WEBSITES

CITY GREEN:
  http://www.citygreen.ca
DAVID SUZUKI FOUNDATION:
  http://www.davidsuzuki.org/consumer/
ENVIRONMENT CANADA, WHAT YOU CAN DO:
  http://www.ec.gc.ca/eco/main_e.htm
CENTER FOR A NEW AMERICAN DREAM:
  http://www.newdream.org
COMMISSIONER OF THE ENVIRONMENT AND
SUSTAINABLE DEVELOPMENT:
  http://www.oag-bvg.gc.ca
ENVIRONMENTALLY SOUND TRANSPORTATION:
  http://www.best.bc.ca
HEIFER INTERNATIONAL:
  http://www.heifer.org/
  http://www.catalog.heifer.org/index.cfm
NATURAL RESOURCES CANADA, OFFICE OF
ENERGY EFFICIENCY:
  http://oee.nrcan.gc.ca/english

NATURE SERVE:
  http://www.natureserve.ca
NATURE WATCH:
  http://www.naturewatch.ca/english
SIERRA YOUTH COALITION:
  http://www.syc-cjs.grg
UNITED NATIONS ENVIRONMENT PROGRAM:
  http://www.unep.org.
UNITED NATIONS HUMAN DEVELOPMENT INDEX:
  http://hdp.undp.org/reports/global/2003/
UNIVERSITY OF BRITISH COLUMBIA,
SUSTAINABILITY PROJECT:
  http://www.sustain.ubc.ca
UNIVERSITY OF VICTORIA, SUSTAINABILITY PROJECT:
  http://uvsp.uvic.ca/main.php
UNIVERSITY OF WATERLOO, SUSTAINABILITY
PROJECT:
  http://watservl.uwaterloo.ca/~uwsp
WORLD SUMMIT ON SUSTAINABLE DEVELOPMENT:
  http://www.joannesburgsummit.org

REFERENCES AND SUGGESTED READING

Abbey, E. 1977. *The Journey Home: Some Words in Defense of the American West.* New York: E.P. Dutton.

Auditor General of Canada. 2002. *Report of the Commissioner of the Environment and Sustainable Development to the House of Commons*: Chapter 5, Sustainable Development Strategies. Ottawa.

Barber, J. 2003. *Production, Consumption and the World Summit for Sustainable Development.* Rockville, Md: Integrative Strategies Forum.

Bardati, D.R. 1995. 'An environmental action plan for Bishop's University', *Journal of Eastern Townships Studies* 6: 19–38.

Brower, M., and W. Leon. 1999. *The Consumers' Guide to Effective Environmental Choices: Practical Advice from the Union of Concerned Scientists.* New York: Three Rivers Press.

Carlsson-Kanyama, A. 1998. 'Climate change and dietary choices: how can emissions of greenhouse gases from food consumption be reduced?', *Food Policy* (Fall/Winter): 288–9.

Cohen, D.S. 1994. 'Subtle effects: requiring economic assessments in the Environmental Choice Program', *Alternatives* 20: 22–7.

Conference Board of Canada. 2003. *Annual Report 2003.* Ottawa.

Cross, G. 2002. *An All-Consuming Century: Why Commercialism Won in Modern America.* New York: Columbia University Press.

Environics. 2002. *Environmental Monitor.*

Environment Canada. 2003a. *Performance Report, 2003.* Ottawa: Environment Canada.

———. 2003b. *Environmental Signals: Canada's National Environmental Indicator Series 2003.* Ottawa, Canada. Cat. En40-775/2002 E. <http://www.ec.gc.ca/soer-ree/English/Indicator_series/default.cfm>.

Food and Agriculture Organization. 2003. *FAOSTAT Statistical Data Base.* At: <http://www.apps.fao.org>.

French, H. 2004 'Linking globalization, consumption and governance', in Worldwatch Institute (2004a: 144–61).

Gardner, G., and P. Sampat. 1998. *Mind over Matter: Recasting the Role of Materials in Our Lives.* Worldwatch Paper 144. Washington: Worldwatch Institute.

——— and A. Assadourian. 2004. 'Rethinking the good life', in Worldwatch Institute (2004a: 164–80).

Hawkins, P., A. Lovins, and L.H. Lovins. 1999. *Natural Capitalism.* Boston: Little, Brown and Company.

Keating, M., and Canadian Global Change Program. 1997. *Canada and the State of the Planet: The Social, Economic and Environmental Trends That Are Shaping Our Lives*. Toronto: Oxford University Press.

Leopold, A. 1949. *A Sand County Almanac*. Oxford: Oxford University Press

Mastny, L. 2004. 'Purchasing for people and the planet', in Worldwatch Institute (2004a: 122–42).

———— and H. French. 2002. 'Crimes of (a) global nature', *Worldwatch* (Sept.-Oct.): 12–23.

Organization for Economic Co-operation and Development (OECD). 2002. *Towards Sustainable Consumption: An Economic Conceptual Framework*. Paris: OECD Environment Directorate.

Orr, D.W. 1994. *Earth in Mind: On Education, Environment and the Human Prospect*. Covelo, Calif.: Island Press.

Prescott-Allen, R. 2001. *The Well-Being of Nations: A Country by Country Index of Quality of Life and the Environment*. Washington and Ottawa: Island Press and IRDC.

Rowe, S. 1993. 'In search of the holy grass: how to bond with the wilderness in nature and ourselves', *Environment Views* (Winter): 7–11.

Sawin, J.L. 2004. 'Making better energy choices', in Worldwatch Institute (2004a: 24–43).

Smith, A. 1993. *Campus Ecology: A Guide to Assessing Environmental Quality and Creating Strategies for Change*. Los Angeles: Living Planet Press.

Statistics Canada. 2003. *Human Activity and the Environment: Annual Statistics 2003*. Ottawa: Statistics Canada.

Suarez, C. 1995. 'Energy needs for sustainable human development', in J. Goldenberg and T.B. Johansson, eds, *Energy as an Instrument for Socio-Economic Development*. New York: UNDP, 18–27.

Thompson, D., and S. van Bakel. 1995. *A Practical Introduction to Environmental Management on Canadian Campuses*. Ottawa: National Roundtable on the Environment and the Economy.

United Nations. 2003. *Plan of Implementation of the World Summit on Sustainable Development*. New York: UN.

Worldwatch Institute. 2004a. *State of the World 2004*. New York: W.W. Norton. www.worldwatch.org.

————. 2004b. *Vital Signs: The Trends That Are Shaping Our Future*. New York: W.W. Norton.

Conservation Organizations

INTERNATIONAL ORGANIZATIONS

Alliance for the Wild Rockies
P.O. Box 8731
Missoula, MT 59807
Phone: (406) 721–5420
Fax: (406) 721–9917
E-mail: awr@wildrockiesalliance.com
Website: http://www.wildrockiesalliance.com

Antarctic and Southern Ocean Coalition
1630 Connecticut Ave. NW, 3rd Floor
Washington, DC 20009
Phone: (202) 234-2480
Fax: (202) 387-4823
Website: http://www.asoc.org

Center for Marine Conservation
1725 DeSales Street NW
Washington, DC 20036
Phone: (202) 429–5609
Fax: (202) 872–0619
E-mail: dccmc@ix.netcom.com

Center for Plant Conservation
P.O. Box 299
St Louis, MO 63166–0299
Phone: (314) 577–9450
Fax: (314) 577–9465

Clean Water Action Project
4455 Connecticut Avenue NW, Suite A300
Washington, DC 20008
Phone: (202) 895–0420
Fax: (202) 895–0438

Conservation International
1919 M Street NW, Suite 600
Washington, DC 20036
Phone: (202) 912–1000

Cousteau Society
710 Settlers Land Road
Hampton, VA 23669
Phone: (757) 722 9300
Fax: (757) 722–8185
E-mail: cousteau@cousteausociety.org

Earth Council
Phone: (416) 498–3150
Fax: (416) 498–7296
E-mail: ecsondra@web.ca

Earth Island Institute
300 Broadway, Suite 28
San Francisco, CA 94133–3312
Phone: (415) 788–3666
Fax: (415) 788–7324
E-mail: earthisland@earthisland.org

EarthKind
2100 L Street NW
Washington, DC 20037
Phone: (202) 778–6149
Fax: (202) 778–6134
E-mail: earthknd@ix.netcom.com

Earthwatch
3 Clock Tower Place, Suite 100
Box 75
Maynard, MA 01754
Phone: (978) 461–0018
Toll-free: (800) 776–0188
Fax: (978) 461–2332

Ecological Agricultural Projects
Macdonald Campus
McGill University
Ste-Anne-de-Bellevue, QC
H9X 3V9
Phone: (514) 398–7771
Fax: (514) 398–7621
E-mail: ecological.agriculture@mcgill.ca

Environmental Action Foundation
333 John Carlyle Street, Suite 200
Alexandria, VA 22314
Phone: (703) 548–3118
Fax: (703) 548–3119
E-mail: info@agc.org

Friends of Animals
777 Post Road, Suite 205
Darien, CT 06820
Phone: (203) 656–1522
Fax: (203) 656–0267

Friends of the Earth International
International Secretariat
P.O. Box 19199
1000 GD Amsterdam
The Netherlands
Phone: 31–20–622–1369
Fax: 31–20–639–2181

Greenpeace International
Keizersgracht 176
1016 DW
Amsterdam
The Netherlands
Phone: 31–20–523–6222
Fax: 31–20–523–6000
E-mail: supporter.services@int.greenpeace.org
Website: http://www.greenpeace.org

International Water Resources Association
University of New Mexico
1915 Roma NE
Albuquerque, NM 87131–1436
Phone: (505) 277–9400
Fax: (505) 277–9405

International Wildlife Coalition
70 East Falmouth Highway
East Falmouth, MA 02536
Phone: (508) 548–8328
Fax: (508) 548–8542

IUCN (World Conservation Union)
Rue Mauverney 28
CH 1196 Gland
Switzerland
Phone: 41–22–999–0155
Fax: 41–22–999–0015
Website: http://www.iucn.org

**Jane Goodall Institute for Wildlife
　Research, Education, and Conservation**
8700 Georgia Ave., Suite 500
Silver Spring, MD 20910
Phone: (240) 645-4000
Fax: (301) 565-3188
Website: http://www.janegoodall.org.

Jersey Wildlife Preservation Trust
Les Augres Manor
Trinity, Jersey
Channel Islands
JE3 5BP
Phone: 15–34–864–666
Fax: 15–34–865–161
E-mail: ldexec@itl.net

National Audubon Society
700 Broadway
New York, NY 10003
Phone: (212) 979–3000
E-mail: webmaster@audubon.org
Website: http://www.audubon.org

Ocean Alliance and Whale Conservation Institute
191 Weston Road
Lincoln, MA 01773
Phone: (781) 259–0423
Toll-free: (800) 969–4253
Fax: (781) 259–0288
E-mail: question@oceanalliance.org

Ocean Voice International
P.O. Box 37026
3332 McCarthy Road
Ottawa, ON
K1V 0W0
Phone: (613) 264–8986
Fax: (613) 521–4205
E-mail: ah194@freenet.carleton.ca

Rainforest Action Network
450 Sansome Street, Suite 700
San Francisco, CA 94111
Phone: (415) 398–4404
Fax: (415) 398–2732
E-mail: rainforest@ran.org

Sierra Club
85 Second Street, 2nd Floor
San Francisco, CA 94105
Phone: (415) 977–5500
Fax: (415) 977–5799
E-mail: information@sierraclub.org
Website: http://www.sierraclub.org

Soil and Water Conservation Society
7515 North Ankeny Road
Ankeny, IA 50021–9764
Phone: (515) 289–2331
Toll-free: (800) 843–7645
Fax: (515) 289–1227
E-mail: swcs@swcs.org

United Nations Environment Program
Information Officer
Regional Office for North America
2 United Nations Plaza, Room DC 2–803
New York, NY 10017
Phone: (212) 963–8210
Fax: (212) 963–7341
E-mail: uneprona@un.org
Website: http://www.unep.ch/unep.html
Convention on Biodiversity:
http://www.unep.ch/biodiv.html

World Commission on Forests and Sustainable Development
CP 51
CH–1219 Châtelaine
Geneva, Switzerland
Phone: 22–979–9165
Fax: 22–979–9060
E-mail: dameena@1prolink.ch
Website: http://www.iisd.org/wcfsd

World Resources Institute
1709 New York Avenue NW, Suite 700
Washington, DC 20006
Phone: (202) 638–6300
Fax: (202) 638–0036
Website: http://www.wri.org

World Society for the Protection of Animals
29 Perkins Street
P.O. Box 190
Boston, MA 02130
Phone: (617) 522–7000
Fax: (617) 522–7077
E-mail: wspa@world.std.com

Worldwatch Institute
1776 Massachusetts Avenue NW
Washington, DC 20036
Phone: (202) 452–1999
Fax: (202) 296–7365
E-mail: worldwatch@worldwatch.org
Website: http://www.worldwatch.org

CANADIAN NATIONAL ORGANIZATIONS

Animal Alliance of Canada
221 Broadview Avenue, Suite 201
Toronto, ON
M4M 2G3
Phone: (416) 462–9541
Fax: (416) 462–9647
E-mail: aac@inforamp.net

Assembly of First Nations
1 Nicholas Street, Suite 1002
Ottawa, ON
K1N 7B7
Phone: (613) 241–6789
Fax: (613) 241–5808

**Association for the Protection of
  Fur-Bearing Animals**
3727 Renfrew Street
Vancouver, BC
V5M 3L7
Phone: (604) 435–1850
Fax: (604) 435–1840

Canadian Arctic Resources Committee
1276 Wellington Street, 2nd Floor
Ottawa, ON
K1Y 3A7
Phone: (613) 759–4284
Toll-free: (866) 949–9006
Fax: (613) 759–4581

Canadian Council on Ecological Areas
Secretariat/Manager
c/o Leigh Warren
4507 Torbolton Ridge Road
R.R. 2
Woodlawn, ON
K0A 3M0
Phone: (613) 832–1995
E-mail: leigh.warren@ec.gc.ca

Canadian Earth Energy Association
130 Slater Street #1050
Ottawa, ON
K1P 6E2
Phone: (613) 230–2332
Fax: (613) 237–1480

Canadian Environmental Law Association
130 Spadina Avenue, Suite 301
Toronto, ON
M5V 2L4
Phone: (416) 960–2284
Fax: (416) 960–9392

Canadian Global Change Program
University of Victoria
P.O. Box 1700 STN CSC
Victoria, BC
V8W 2Y2
Phone: (250) 472-4337
Fax: (250) 472-4830
E-mail: cgcp@uvic.ca
Website: http://www.globalentres.org/cgcp

Canadian Nature Federation
1 Nicholas Street, Suite 520
Ottawa, ON
K1N 7B7
Phone: (613) 562–3447
Fax: (613) 562–3371

**Canadian Parks and Wilderness Society
  (CPAWS)**
National Office
880 Wellington Street, Suite 506
Ottawa, ON
K1R 6K7
Phone: (613) 569–7226
Toll-free: (800) 333–WILD (9453)
Fax: (613) 569–7098

Canadian Water Resources Association
P.O. Box 1329
Cambridge, ON
N1R 7G6
Phone: (519) 622–4764
Fax: (519) 621–4844
E-Mail: cwranat@grandriver.ca
Website: http://www.cwra.org

**City Farmer—Canada's Office of Urban
  Agriculture**
Box 74561, Kitsilano RPO
Vancouver, BC
V6K 4P4
Phone: (604) 685–5832
Fax: (604) 685–0431
E-mail: cityfarm@unixg.ubc.ca
Website: http://www.cityfarmer.org

Earth Day Canada
111 Peter St., Suite 503
Toronto, ON
M5V 2H1
Phone: (416) 599–1991, ext. 107
Fax: (416) 599–3100

Energy Probe
225 Brunswick Avenue
Toronto, ON
M5S 2M6
Phone: (416) 964–9223
Fax: (416) 964–8239
E-mail: EnergyProbe@nextcity.com

Environment Bureau
Agriculture and Agri-Food Canada
Co-operatives Secretariat
1525 Carling Avenue, Room 367
Ottawa, ON
K1A 0C5
Phone: (613) 759–7309

Environment Canada
Inquiry Centre
70 Crémazie Street
Gatineau, QC
K1A 0H3
Fax: (819) 994–1412
TTY: (819) 994-0736 (Teletype for the hearing
    impaired)
E-mail: enviroinfo@ec.gc.ca
A Guide to a Green Government:
    http://www.ec.gc.ca/sd-dd_consult/factsheet
        Greengov1_e.htm
Site of the Environment InfoBase:
    http://www.ec.gc.ca/soer-ree/
Sustainable Development Strategy:
    http://www.ec.gc.ca/sd-dd_consult/
        SDS2004/index_e.cfm

Fisheries and Oceans Canada
Communications Branch
200 Kent Street
13th Floor, Station 13228
Ottawa, ON
K1A 0E6
Phone: (613) 993–0999
Fax: (613) 990–1866
TDD: (613) 941-6517
E-mail: info@www.ncr.dfo.ca
Website: http://www.ncr.dfo.ca/home_e.htm

Friends of the Earth Canada
206–260 St Patrick Street
Ottawa, ON
K1N 5K5
Phone: (613) 241–0085

Greenpeace Canada
250 Dundas St. W, Suite 605
Toronto, ON
M5T 2Z5
Phone: (416) 597–8408
Toll-free: (800) 320–7183
Fax: (416) 597–8422

International Fund for Animal Welfare
1 Nicholas Street
Ottawa, ON
K1N 7B7
Phone: (613) 241–8996
Toll-free: (888) 500–4329
Fax: (613) 241–0641

**International Institute for Sustainable
    Development**
161 Portage Avenue East, 6th Floor
Winnipeg, MB
R3B 0Y4
Phone: (204) 958–7700
Fax: (204) 958–7710
E-mail: info@iisd.ca
Website: http://www.iisd.org

Marine Protected Areas Network
Commission for Environmental Co-operation
393, rue St-Jacques Ouest
Bureau 200
Montreal, QC
H2Y 1N9
Phone: (514) 350–4300
Fax: (514) 350–4314

National Parks Directorate Parks Canada
25 Eddy Street, 4th Floor
Gatineau, QC
K1A 0M5
Phone: (888) 773–8888
E-mail: National-ParksWebmaster@pch.gc.ca
Website: http://parkscanada.pch.gc.ca

Nature Canada
1 Nicholas Street, Suite 606
Ottawa, ON
K1N 7B7
Toll-free: (800) 267–4088

Nature Conservancy of Canada
110 Eglinton Avenue West, Suite 400
Toronto, ON
M4R 1A3
Phone: (416) 932–3202
Toll-free: (800) 465–0029
Fax: (416) 932–3208
E-mail: nature@natureconservancy.ca

Probe International
225 Brunswick Avenue
Toronto, ON
M5S 2M6
Phone: (416) 964–9223
Fax: (416) 964–8239

Sea Shepherd Conservation Society
P.O. Box 48446
Vancouver, BC
V7X 1A2
Phone: (604) 688–7325

Sierra Club of Canada
1 Nicholas Street, Suite 412
Ottawa, ON
K1N 7B7
Phone: (613) 241–4611
Fax: (613) 241–2292

Wildlife Habitat Canada
1750 Courtwood Crescent, Suite 310
Ottawa, ON
K2C 2B5
Phone: (613) 722–2090
Toll-free: (800) 669–7919
Fax: (613) 722–3318
E-mail: reception@whc.org

Wildlife Preservation Trust Canada
120 King Street
Guelph, ON
Phone: (519) 836–9314
Fax: (519) 824–6776
E-mail: wptc@inforamp.net

World Wildlife Fund Canada
245 Eglinton Ave. East, Suite 410
Toronto, ON
M4P 3J1
Phone: (416) 489–8800
Toll-free: (800) 267–2632
Fax: (416) 489–8055

Zoocheck Canada Inc.
2646 St Clair Avenue East
Toronto, ON
M4B 3M1
Phone: (416) 285–1744
Fax: (416) 285–4670
E-mail: zoocheck@zoocheck.com

BRITISH COLUMBIA

BC Spaces for Nature
Box 673
Gibsons, BC
V0N 1V0
Phone: (604) 886–8605
Fax: (604) 886–3768

BC Wild
Box 2241, Main Post Office
Vancouver, BC
V6B 3W2
Phone: (604) 669–4802
Fax: (604) 669–6833

CPAWS, BC Chapter
610–555 West Georgia Street
Vancouver, BC
V6B 1Z6
Phone: (604) 685–7445
Fax: (604) 685–6449
E-mail: info@cpawsbc.org

Friends of Clayoquot Sound
P.O. Box 489
Tofino, BC
V0R 2Z0
Phone: (604) 725–4218
Fax: (604) 725–2527

Ministry of Sustainable Resource Management
P.O. Box 9054, Stn Prov Govt
Victoria, BC
V8W 9E2
Phone: (250) 356–9076
Fax: (250) 356–8273

Ministry of Water, Land, and Air Protection
P.O. Box 9047, Stn Prov Govt
Room 112, Parliament Buildings
Victoria, BC
V8W 9E2
Phone: (250) 387–1187
Fax: (250) 387–1356

Nature Trust of British Columbia
260–1000 Roosevelt Crescent
North Vancouver, BC
V7P 1M3
Phone: (604) 924–9771
Toll-free: (866) 288–7878
Fax: (604) 924–9772
E-mail: info@naturetrust.bc.ca

Ocean Resource Conservation Alliance
P.O. Box 1189
Sechelt, BC
V0N 3A0
Phone: (604) 885–7518
Fax: (604) 885–2518

Sierra Club of Canada, BC Chapter
302–733 Johnson Street
Victoria, BC
V8W 3C7
Phone: (250) 386–5255
Fax: (250) 386–4453

Western Canada Wilderness Committee
Head Office
227 Abbott Street
Vancouver, BC
V6B 2K7
Phone: (604) 683–8220
Toll-free: (800) 661–9453
Fax: (604) 683–8229
E-mail: info@wildernesscommittee.org
Website: http://www.wildernesscommittee.org

Wildlife Rescue Association of British Columbia
5216 Glencairn Drive
Burnaby, BC
V5B 3C1
Phone: (604) 526–7275
Fax: (604) 524–2890

ALBERTA

Alberta Environmental Protection
Communications
9915–108 Street, 9th Floor
Edmonton, AB
T5K 2G8
Phone: (780) 427–6267 (communications)

General inquiries:
Phone: (780) 427–2700 (toll-free by first dialing 310–0000)
Fax: (780) 422–4086

Alberta Sport, Recreation, Parks, and Wildlife Foundation
Communications Branch
7th Floor, Standard Life Centre
10405 Jasper Avenue
Edmonton, AB
T5J 4R7
Phone: (780) 427–6530
E-mail: Comdev.Communications@gov.ab.ca

Alberta Wilderness Association
Box 6398, Station D
Calgary, AB
T2P 2E1
Phone: (403) 283–2025
Fax: (403) 270–2743

Bow Valley Naturalists
Box 1693
Banff, AB
T0L 0C0
Phone/fax: (403) 762–4160

CPAWS, Calgary/Banff Chapter
1120–1202 Centre St. SE
Calgary, AB
T2G 5A5
Phone: (403) 232–6686
Fax: (403) 232–6988
E-mail: info@cpawscalgary.org

CPAWS, Edmonton Chapter
P.O. Box 52031
8540–109 Street, Suite 202
Edmonton, AB
T6G 2T5
Phone: (780) 432–0967
Fax: (780) 439–4913
E-mail: info@cpaws-edmonton.org

Federation of Alberta Naturalists
11759 Groat Road
Edmonton, AB
T5M 3K6
Phone: (780) 427–8124
Fax: (780) 422–2663 (c/o FAN)
E-mail: info@fanweb.ca

SASKATCHEWAN

CPAWS, Saskatchewan Chapter
203–115 2nd Avenue North
Saskatoon, SK
S7K 2B1
Phone: (306) 955–6197
Fax: (306) 665–2128
E-mail: info@cpaws-sask.org

Environment and Resource Management
Education and Communications
3211 Albert Street
Regina, SK
S4S 5W6
Toll-free: (800) 667–2757

Nature Saskatchewan
1860 Lorne Street, Suite 206
Regina, SK
S4P 3W6
Phone: (306) 780–9273
Toll-free: (800) 667–4668
Fax: (306) 780–9263

Saskatchewan Environmental Society
P.O. Box 1372
Saskatoon, SK
S7K 3N9
Phone: (306) 665–1915
Fax: (306) 665–2128

MANITOBA

CPAWS, Manitoba Chapter
Box 344
Winnipeg, MB
R3C 2H5
Phone: (204) 949–0782
Fax: (204) 949–0783

Department of Conservation
Phone: (204) 945-3744
Fax: (204) 945-4261
E-mail: mgi@gov.mb.ca

Department of Environment
123 Main Street, Suite 160
Winnipeg, MB
R3C 1A5
Phone: (204) 945–7100

Manitoba Naturalists Society
63 Albert Street, Suite 401
Winnipeg, MB
R3B 1G4
Phone/fax: (204) 943–9029

ONTARIO

Citizens Network on Waste Management
17 Major Street
Kitchener, ON
N2H 4R1
Phone: (519) 744–7503
Fax: (519) 744–1546
E-mail: jjackson@web.net

CPAWS, Wildlands League Chapter
401 Richmond Street West, Suite 380
Toronto, ON
M5V 3A8
Phone: (416) 971–9453
E-mail: cpaws@web.net

Earthroots Coalition
401 Richmond Street West, Suite 410
Toronto, ON
M5V 3A8
Phone: (416) 599–0152
Fax: (416) 340–2429

Elora Centre for Environmental Excellence
82 Metcalfe Street
Elora, ON
N0B 1S0
Phone: (519) 846–0841
Fax: (519) 846–2642
E-mail: info@ecee.on.ca

Energy Action Council of Toronto
401 Richmond Street West, Suite 401
Toronto, ON
M5V 3A8
Phone: (416) 488–3966
Fax: (416) 977–2157

Environment North
704 Holly Crescent
Thunder Bay, ON
P7E 2T2
Phone: (807) 475–5267
Fax: (807) 577–6433

Federation of Ontario Naturalists
355 Lesmill Road
Don Mills, ON
M3B 2W8
Phone: (416) 444–8419
Fax: (416) 444–9866

Ministry of Environment and Energy
Communications
135 St Clair Avenue West, 2nd Floor
Toronto, ON
M4V 1P5
Phone: (416) 323–4321
Toll-free: (800) 565–4923

Water Environment Association of Ontario
P.O. Box 176
Milton, ON
L9T 4N9
Phone: (416) 410–6933
Fax: (416) 410–1626

QUEBEC

La Fondation pour la sauvegarde des espèces menacées
Édifice Marie-Guyart, rez-de-chaussée
675 boulevard René-Lévesque est
Quebec City, QC
G1R 5V7
Phone: (418) 521–3830
Toll-free: (800) 561–1616
Fax: (418) 646–5974
E-mail: info@menv.gouv.qc.ca

La Fondation québécoise en environment
1255 carré Phillips, bureau 706
Montreal, QC
H3B 3G1
Phone: (514) 849–3323
Toll-free: (800) 361–2503
Fax: (514) 849–0028
E-mail: info@fqe.qc.ca

L'Union québécoise pour la conservation de la nature (UQCN)
1085 avenue de Salaberry, bureau 300
Quebec City, QC
G1R 2V7
Phone: (418) 648–2104
Fax: (418) 648–0991
E-mail: courrier@uqcn.qc.ca

Ministry of Environment and Wildlife
Institutional Affairs
675 boulevard René-Lévesque est, 8th Floor
Quebec City, QC
G1R 5V7
Phone: (418) 643–1853
Toll-free: (800) 561–1616

Ministry of Natural Resources
Direction des communications
5700 4th Avenue Ouest, B 302
Charlesbourg, QC
G1H 6R1
Phone: (418) 627–8609
Fax: (418) 643–0720
E-mail: communications@mrnfp.gouv.qc.ca

NEW BRUNSWICK

Conservation Council of New Brunswick
180 St John Street
Fredericton, NB
E3B 4A9
Phone: (506) 458–8747
Fax: (506) 458–1047

Department of the Environment
Communications and Environmental Education
P.O. Box 6000
Fredericton, NB
E3B 5H1
Phone: (506) 453–3700

Nature Trust of New Brunswick
P.O. Box 603, Station A
Fredericton, NB
E3B 5A6
Phone: (506) 457–2398
Fax: (506) 450–2137
E-mail: ntnb@nbnet.nb.ca

New Brunswick Federation of Naturalists
924 Prospect Street, Suite 110
Fredericton, NB
E3B 2T9
Phone: (506) 459–4209
E-mail: nbfn@nb.aibn.com

**New Brunswick Protected Natural Areas
  Coalition**
180 St John Street
Fredericton, NB
E3B 4A9
Phone: (506) 451–9902
Fax: (506) 458–1047

PRINCE EDWARD ISLAND

Environment, Energy, and Forestry
Jones Building, 4th and 5th Floors
11 Kent Street
P.O. Box 2000
Charlottetown, PEI
C1A 7N8
Phone: (902) 368–5000
Fax: (902) 368–5830

Island Nature Trust
P.O. Box 265
Charlottetown, PEI
C1A 7K4
Phone: (902) 566–9150
Fax: (902) 628–6331
E-mail: intrust@isn.net

NOVA SCOTIA

CPAWS, Nova Scotia Chapter
4th Floor, 1526 Dresden Row
Halifax, NS
Phone: (902) 446–4155
E-mail: coordinator@cpawsns.org

Department of Environment and Labour
5151 Terminal Road
P.O. Box 697
Halifax, NS
B3J 2T8
Phone: (902) 424–5300
Fax: (902) 424–0503

Ecology Action Centre
1568 Argyle St., Suite 31
Halifax, NS
B3J 2B3
Phone: (902) 429–2202

Federation of Nova Scotia Naturalists
6360 Young Street
Halifax, NS
B3L 2A1

NEWFOUNDLAND AND LABRADOR

**Department of Environment and
  Conservation**
4th Floor, West Block
Confederation Building
P.O. Box 8700
St John's, NL
A1B 4J6
Phone: (709) 729–2664
Fax: (709) 729–6639

**Newfoundland and Labrador
  Environmental Association**
90 O'Leary Ave.
Suite 101, Parsons Building
St John's, NL
A1B 2C7
Phone: (709) 772–3333
Fax: (709) 772–3213
E-mail: info@neia.org

**Newfoundland and Labrador
  Environmental Association**
140 Water Street, Suite 603
St John's, NL
A1C 6H6
Phone: (709) 722–1740
Fax: (709) 726–1813

**Protected Areas Association of
  Newfoundland and Labrador**
Box 1027, Station C
St John's, NL
A1C 5M5
Phone/fax: (709) 726–2603

**Wilderness and Ecological Reserves
  Advisory Council**
c/o Parks and Natural Areas Division
Department Tourism and Culture
P.O. Box 8700
St John's, NL
A1B 4J6
Phone: (709) 729–2421
Fax: (709) 729–1100

YUKON

CPAWS, Yukon Chapter

P.O. Box 31095
211 Main St.
Whitehorse, YT
Y1A 5P7
Phone: (867) 393–8080
Fax: (867) 393–8081

Department of Environment

Government of Yukon
Box 2703
Whitehorse, YT
Y1A 2C6
Phone: (867) 667–5652
Toll-free (in Yukon): (800) 661–0408, local 5652
Fax: (867) 393–6213
E-mail: environmentyukon@gov.yk.ca

Renewable Resources

Communications
P.O. Box 2703
Whitehorse, YT
Y1A 2C6
Phone: (403) 667–5237

Yukon Conservation Society

302 Hawkins Street
Whitehorse, YT
Y1A 1X6
Phone: (867) 668–5678
Fax: (867) 668–6637
E-mail: ycs@ycs.yk.ca

NORTHWEST TERRITORIES

Department of Resources, Wildlife, and Economic Development

Government Information Office
P.O. Box 1320
Yellowknife, NWT
X1A 2L9
Phone: (403) 669–2302

Ecology North

5013 51th Street
Yellowknife, NWT
X1A 1S5
Phone: (867) 873–6019
Fax: (867) 873–9195
E-mail: econorth@ssimicro.com

NUNAVUT

Department of Sustainable Development

Government of Nunavut Communications
Iqaluit, NU
X0A 0H0
Phone: (867) 975–5925
Fax: (867) 975–5980
Website: http://www.gov.nu.ca/sd.htm

Nunavut Parks

Government of Nunavut Communications
Iqaluit, NU
X0A 0H0
Phone: (867) 975–5900
Fax: (867) 975–5990
E-mail: Parks@gov.nu.ca
Website: http://www.nunavutparks.com/
visitorscentre/links.cfm

Glossary

abiotic components: Non-living parts of the ecosystem, including chemical and physical factors, such as light, temperature, wind, water, and soil characteristics.

acid deposition: Rain or snow that has a lower pH than precipitation from unpolluted skies; also includes dry forms of deposition, such as nitrate and sulphate particles.

acid mine drainage: Acidic drainage from waste rock and mine tailings caused by the oxidization of iron sulphides to create sulphuric acid, which in turn dissolves residual metals.

acid shock: The buildup of acids in water bodies and standing water over the winter, resulting in higher acidity than experienced through the rest of the year.

active management: Purposeful interference by resource and environmental managers in ecosystems, which recognizes that the human forces of change are now so ubiquitous that even protected areas are affected; includes habitat restoration, creation of wildlife corridors, reintroduction of extirpated species, prescribed burning, and management of hyper-abundant species.

adaptive environmental management: An approach that develops policies and practices to deal with the uncertain, the unexpected, and the unknown; approaches management as an experiment from which we learn by trial and error.

aerobic: Requiring oxygen.

Agenda 21: A 40-chapter report from the Earth Summit in 1992 that outlines goals and priorities for economic development and environmental protection in the twenty-first century.

albedo effect: The extent to which the surface of the Earth reflects rather than absorbs incoming radiation from the sun. Snow has a high albedo, but as temperatures rise the area covered in snow will be replaced by areas free of snow, uncovering rocks and vegetation with lower albedo values that absorb radiation and thus add to warming.

alien (also known as exotic, introduced, invader, or non-native) species: Any organism, such as zebra mussels, purple loosestrife, and Eurasian water milfoil in Canada, that enters an ecosystem beyond its normal range through deliberate or inadvertent introduction by humans.

alternative dispute resolution: A non-judicial approach to resolving disputes that uses negotiation, mediation, or arbitration, with a focus on reparation for harm done and on improving future conduct.

anadromous: Aquatic life, such as salmon, that spend part of their lives in salt water and part in fresh water.

annual allowable cut (AAC): The amount of timber that is permitted to be cut annually from a specified area.

anthropocentric view: Human-centred, whereby values are defined relative to human interests, wants, and needs.

aquaculture: Seafood farming, the fastest-growing food production sector in the world.

aquifer: A formation of permeable rocks or loose materials that contains usable sources of groundwater and may extend from a few square kilometres to several thousand square kilometres.

arbitration: A procedure for dispute resolution in which a third party is selected to listen to the views and interests of the parties in dispute and develop a solution to be accepted by the participants.

Atlantic Maritime: Ecozone stretching from the mouth of the St Lawrence River across New Brunswick, Nova Scotia, and Prince Edward Island that is heavily influenced by the Atlantic Ocean, which creates a cool, moist maritime climate but with quite variable conditions between the upland masses, such as the Cape

Breton and New Brunswick highlands, and the coastal lowlands that support most of the population.

atmosphere: Layer of air surrounding the Earth.

autotrophs: Organisms, such as plants, that produce their own food, generally via photosynthesis.

benthic: Of or living on or at the bottom of a water body.

bioaccumulation: The storage of chemicals in an organism in higher concentrations than are normally found in the environment.

biocentric (ecocentric) view: A view that values aspects of the environment simply because they exist and accepts that they have the right to exist.

biocides: Chemicals that kill many different kinds of living things; also called pesticides.

bioconcentration: The combined effect of bioaccumulation and biomagnification.

biodiversity: The variety of life forms that inhabit the Earth. Biodiversity includes the genetic diversity among members of a population or species as well as the diversity of species and ecosystems.

biogeochemical cycles: Series of biological, chemical, and geological processes by which materials cycle through ecosystems.

biomagnification: Buildup of chemical elements or substances in organisms in successively higher trophic levels.

biological oxygen demand (BOD): The amount of dissolved oxygen required for the bacterial decomposition of organic waste in water.

biomass: The sum of all living material, or of all living material of particular species, in a given environment.

biomass pyramid: In terrestrial ecosystems, greater biomass generally exists at the level of primary consumers, with the least total biomass at the highest trophic levels; in marine ecosystems, the reverse is true and the pyramid is inverted—greater biomass is at the highest trophic level while the primary consumers, phytoplankton, at any given time comprise much less biomass but reproduce rapidly.

biome: A major ecological community of organisms, both plant and animal, that is usually characterized by the dominant vegetation type, for example, a tundra biome and a tropical rain forest biome.

biosphere: Total of all areas on Earth where organisms are found; includes deep ocean and part of the atmosphere.

biotic components: Those parts of ecosystems that are living; organisms.

biotic potential: The ability of species to reproduce regardless of the level that an environment can support; see *carrying capacity*.

Boreal Cordillera: Ecozone, to the south and west of the Tundra Cordillera, in northern BC and southern Yukon made up of mountains in the west and east, separated by intermontane plains, with a wet climate that gives rise to tree growth.

boreal forest: One of the largest forest belts in the world, extending all across North America and Eurasia, encompassing roughly a third of the Earth's forest land and 14 per cent of the world forest biomass, and separating the treeless tundra regions to the north from the temperate deciduous forests or grasslands to the south.

Boreal Plains: Ecozone extending from the southern part of the Yukon in a wide sweeping band down into southeastern Manitoba consisting of a generally flat to undulating surface similar to the Prairie zone to the south.

Boreal Shield: The largest ecozone in Canada, stretching along the Canadian Shield from Saskatchewan to Newfoundland.

bottom trawling: One of the most destructive means of fishing, where heavy nets are dragged along the sea floor scooping up everything in their path.

bycatch: Non-target organisms caught or captured in the course of catching a target species, as in the fisheries, where estimates suggest that 25 per cent of the world's catch is dumped because it is not the right species or size.

calorie: A unit of heat energy, the amount of heat required to raise 1 g of water by 1° C.

carbon balance: A balance between the amount of CO_2 in the atmosphere and bicarbonate in the water.

carbon sequestration: Reforestation and afforestation to ameliorate carbon dioxide loadings in the atmosphere because trees and shrubs use the excess CO_2.

carnivore: An organism that consumes only animals.

carrying capacity: Maximum population size that a given ecosystem can support for an indefinite period or on a sustainable basis.

CFCs (chlorofluorocarbons): Gaseous synthetic substances composed of chlorine, fluorine, and carbon. They have been used as refrigerants, aerosol propellants, cleaning solvents, and in the manufacture of plastic foam. CFCs cause ozone depletion in the stratosphere.

chemoautotroph: A producer organism that converts inorganic chemical compounds into energy.

chlorophylls: Pigments of plant cells that absorb sunlight, thus enabling plants to capture solar energy.

clear-cutting: A forest harvesting technique in which an entire stand of trees is felled and removed.

climate: The long-term weather pattern of a particular region.

climate change: A long-term alteration in the climate of a particular location or region, or for the entire planet.

climate modelling: Various mathematical and computerized approaches for determining past climate trends in an effort to build scenarios predicting future climate, which use any or all of the following factors in measurement: incoming and outgoing radiation; energy dynamics or flows around the globe; surface processes affecting climate such as snow cover and vegetation; chemical composition of the atmosphere; and time step or resolution (time over which the model runs and the spatial scale to which it applies).

climax community: Last stage of succession; a relatively stable, long-lasting, complex, and interelated community of organisms.

coevolution: Process whereby two species evolve adaptations as a result of extensive interactions with each other.

collaboration: The art of working together.

co-management: An arrangement in which a government agency shares or delegates some of its legal authority regarding a resource or environmental management issue with local inhabitants of an area.

commensalism: An interaction between two species that benefits one species and neither harms nor benefits the other.

competitive exclusion principle: The principle that competition between two species with similar requirements will result in the exclusion of one of the species.

compound: The coming together of two different atoms to form a different substance, such as water (H_2O), a compound made up of two hydrogen atoms (H) and one oxygen atom (O).

consumers: Organisms that cannot produce their own food and must get it by eating or decomposing other organisms; in economics, those who use goods and services.

consumerism: Wasteful consumption of resources to satisfy wants rather than needs.

context: Specific characteristics of a time and place.

contour cultivation: The cultivation and seeding of fields parallel to the contour of the slope, which serves to reduce the speed of runoff by catching soil particles in the plow furrows.

Convention on Biological Diversity (CBD): International treaty that emerged from the World Summit on Sustainable Development, held in Rio de Janeiro in 1992, that requires signatories, including Canada, to develop biodiversity strategies, identify and monitor important components of biodiversity, develop endangered species legislation/protected areas systems, and promote environmentally sound and sustainable development in areas adjacent to protected areas.

Convention on International Trade in Endangered Species of Wild Fauna and Flora (CITES): A 1973 treaty currently ratified by more than 120 countries (including Canada) that establishes lists of species for which international trade is to be controlled or monitored (e.g., orchids, cacti, parrots, large cat species, sea turtles, rhinos, primates, etc.).

co-ordination: The effective or harmonious working together of different departments, groups, and individuals.

coral bleaching: Death of corals caused by water temperatures becoming too warm.

coral polyps: Individual biotic members of a coral reef.

critical population size: Population level below which a species cannot successfully reproduce.

crop rotation: Alternating crops in fields to help restore soil fertility and also control pests.

culmination age: The age of economic maturity of a tree crop, which varies widely but usually falls within the 60–120-year range in Canada.

custom-designed solutions: Management approach in which the specific conditions of a place and time are recognized, and the attempt to ameliorate or resolve a problem takes these specifics into account.

DDT (dichlorodiphenyltrichloroethane): An organo-chlorine insecticide used first to control malaria-carrying mosquitoes and lice and later to control a variety of insect pests, but now banned in Canada because of its persistence in the environment and ability to bioaccumulate.

decomposer food chain: A specific nutrient and energy pathway in an ecosystem in which decomposer organisms (bacteria and fungi) consume dead plants and animals as well as animal wastes; essential for the return of nutrients to soil and carbon dioxide to the atmosphere. Also called detritus food chain.

denitrification: The conversion of nitrate to molecular nitrogen by bacteria in the nitrogen cycle.

detritus: Organic waste, such as fallen leaves.

drought: Condition whereby a combination of lack of precipitation, temperature, evaporation, evapotranspiration, and the inability of soil to retain moisture leads to the loss of resilience among flora and fauna in dry conditions.

dykes: Walls or earth embankments along a watercourse to control flooding, or encircling a town or a property to protect it from flooding (ring dyke), or across a stream whereby the flow of water is stopped from going upstream by a sluice gate (cross dyke).

Earth Summit: The International Conference on Environment and Development held in Rio de Janeiro during June 1992 when most of the nations of the world met to discuss issues related to sustainable development.

ecocentric (biocentric) view: The view that a natural order governs relationships between living things and that a harmony and balance reflect this natural order, which humankind tends to disrupt.

E. coli: *Escherichia coli*, a bacterium present in fecal matter that can get into a water supply and pollute it, as happened in Walkerton, Ontario, in May 2000.

ecologically extinct: When a species exists in such low numbers that it can no longer fulfill its ecological role in the ecosystem.

ecological footprint: The land area a community needs to provide its consumptive requirements for food, water, and other products and to dispose of the wastes from this consumption.

ecological succession: The gradual replacement of one assemblage of species by another as conditions change over time.

ecosphere: Refers to the entire global ecosystem, which comprises atmosphere, lithosphere, hydrosphere, and biosphere as inseparable components.

ecosystem: Short for ecological system; a community of organisms occupying a given region within a biome, including the physical and chemical environment of that community and all the interactions among and between organisms and their environment.

ecosystem-based management (EBM): Holistic management that takes into account the entire ecosystem and emphasizes biodiversity and ecosystem integrity, as opposed to focusing primarily or solely on a resource or resources, such as water or timber, within an ecosystem.

ecosystem homeostasis: A balance, implying not a static state but a dynamic equilibrium, where internal processes adjust for changes in external conditions.

ecotone: The transitional zone of intense competition for resources and space between two communities.

El Niño: A marked warming of the waters in the eastern and central portions of the tropical Pacific that triggers weather changes and events in two-thirds of the world.

emission trading: Under the Kyoto Protocol, a system whereby one country that will exceed its allotted limit of greenhouse gas emissions can buy an amount of greenhouse gas emissions from another country that will not reach its own established emissions limit.

endangered: An official designation assigned by the Committee on the Status of Endangered Wildlife in Canada to any indigenous species or subspecies or geographically separate population of fauna or flora that is threatened with imminent extinction or extirpation throughout all or a significant portion of its Canadian range.

endemic species: A plant or animal species confined to or exclusive to a specific area.

endocrine disruption: The interference of normal bodily processes such as sex, metabolism, and growth by chemicals in such products as soaps and detergents that are released into an ecosystem, as happens among aquatic species, often causing feminization.

energy: The capacity to do work; found in many forms, including heat, light, sound, electricity, coal, oil, and gasoline.

entropy: A measure of disorder. The second law of thermodynamics applied to matter says that all systems proceed to maximum disorder (maximum entropy).

environment: The combination of the atmosphere, hydrosphere, cryosphere, lithosphere, and biosphere in which humans, other living species, and non-animate phenomena exist.

environmental impact assessment: Part of impact assessment that identifies and predicts the impacts from development proposals on both the biophysical environment and on human health and well-being.

environmental justice: The right to a safe, healthy, productive, and sustainable environment for all living things, in which 'environment' is viewed in its totality.

estuary: Coastal regions, such as inlets or mouths of rivers, where salt water and fresh water mix.

euphotic zone: Zone of the ocean to which light from the sun reaches.

eutrophic: Pertaining to a body of water rich in nutrients.

eutrophication (also known as nutrient enrichment): The overfertilization of a body of water by nutrients that produce more organic matter than the water body's self-purification processes can overcome.

evapotranspiration: Evaporation of water from soil and transpiration of water from plants.

evolution: A long-term process of change in organisms caused by random genetic changes that favour the survival and reproduction of those organisms possessing the genetic change; organisms become better adapted to their environment through evolution.

exclusive economic zones (EEZs): Areas off the coasts of a nation that are claimed by that nation for its sole responsibility and exploitation, as permitted by the UN Convention on the Law of the Sea.

ex situ preservation: The conservation of species outside their natural habitat, including breeding in captivity, so that they can be reintroduced to their natural habitat, as has been done, for example, with the black-footed ferret and swift fox.

extinction: The elimination of all the individuals of a species.

extirpated: An official designation assigned by the Committee on the Status of Endangered Wildlife in Canada to any indigenous species or subspecies or geographically separate population of fauna or flora no longer known to exist in the wild in Canada but occurring elsewhere.

extrinsic values: Those values that humans derive from other species, including consumptive and non-consumptive values.

falldown effect: The lower volume of harvestable timber at the culmination age for second growth on sites where old-growth forest was previously harvested.

fishing down the food chain: Harvesting at progressively lower trophic levels as higher trophic levels become depleted.

flood plain: Low-lying land along a river, stream, or creek or around a lake that under normal conditions is flooded from time to time.

floodway or diversion: An excavated channel to divert flood waters away from a population centre.

food chain: A specific nutrient and energy pathway in ecosystems proceeding from producer to consumer; along the pathway, organisms in higher trophic levels gain energy and nutrients by consuming organisms at lower trophic levels.

food web: Complex intermeshing of individual food chains in an ecosystem.

forest tenure: The conditions that govern forest ownership and use.

fossil fuels: Organic fuels (coal, natural gas, oil, tar sands, and oil shale) derived from once-living plants or animals.

fragmentation: The division of an ecosystem or species habitat into small parcels as a result of human activity, such as agriculture, highways, pipelines, population settlements, etc.

frozen core dam: Type of dam used in mining in the North, such as at the Ekati diamond mine in the NWT, with a central core of frozen soil saturated with ice and bonded to the natural permafrost, and surrounded with granular fill to ensure both stability and thermal protection.

gaseous cycles: Cycles of elements that have most of their matter in the atmosphere.

general circulation models (GCMs): The most prominent and most complex type of climate modelling, which takes into account the three-dimensional nature of the Earth's atmosphere, oceans, or both.

generalist species: Species, like the black bear and coyote, with a very broad niche, where few things organic are not considered a potential food item.

genetically modified organisms (GMOs): Organisms created by humans through genetic manipulation combining genes from different and often totally unrelated species to create a different organism that is economically more productive and/or has greater resistance to pathogens.

global warming: Changes in average temperatures of the Earth's surface, although these changes are not uniform, i.e., some regions experience significantly higher temperatures, others only slight changes upward, and still other regions might experience somewhat cooler temperatures.

grasshopper effect: Atmospheric transport and deposition of persistent and volatile chemical pollutants, whereby the pollutants evaporate into the air in warmer climates and travel in the atmosphere towards cooler areas, condensing out again when the temperature drops. The cycle then repeats itself in a series of 'hops' until the pollutants reach climates where they can no longer evaporate.

greenhouse effect: A warming of the Earth's atmosphere caused by the presence of certain gases (e.g., water vapour, carbon dioxide, methane) that absorb radiation emitted by the Earth, thereby retarding the loss of energy to space.

greenhouse gas (GHG): A gas that contributes to the greenhouse effect, such as carbon dioxide.

green manure: Growing plants that are plowed into the soil as fertilizer.

Green Revolution: Development in plant genetics (hybridization) in the late 1950s and early 1960s resulting in high-yield varieties producing three to five times more grain than previous plants but requiring intensive irrigation and fertilizer use.

gross national product (GNP): The total value of all goods and services produced for final consumption in an economy, used by economists as an index or indicator to compare national economies or periods of time within a single national economy.

gross primary productivity (GPP): The total amount of energy produced by autotrophs over a given period of time.

groundwater: Water below the Earth's surface in the saturated zone.

habitat: The environment in which a population or individual lives.

heat: The total energy of all moving atoms.

herbivores: Animals that eat plants, that is, primary consumers.

heritage rivers: Rivers designated for special protection by the Canadian Heritage Rivers Board because of their historical, cultural, ecological, and recreational significance.

heterotroph: An organism that feeds on other organisms.

high-quality energy: Energy that is easy to use, such as a hot fire or coal or gasoline, but that disperses quickly.

hybridization: The crossbreeding of two varieties or species of plants or animals.

hydrological cycle: The circulation of water through bodies of water, the atmosphere, and land.

hydrosphere: One of three main layers of the ecosphere, which contains all the water on Earth.

impact and benefit agreements (IBAs): Voluntary agreements between extractive industries and communities that go beyond formal impact assessment requirements and are intended to facilitate extraction of resources in a way that contributes to the economic and social well-being of local people and communities.

impact assessment: Thorough consideration of the effects of a project that takes into account its potential and probable impacts on the environment and on society or a community, and that assesses the technology proposed for the project as well as the technology available for dealing with any negative impacts.

indicators: Specific facets of a particular system, such as the population of a key species within an ecosystem, that tell us something of the current state of the system but do not help us understand why the system is in that state.

indigenous knowledge: Understanding of climate, animals and animal behaviour, soil, waters, and/or plants within an ecosystem based on experiential knowledge from living or working in a particular area for a long period of time; also referred to as traditional ecological knowledge (TEK) or local knowledge.

inertia: The tendency of a natural system to resist change.

in situ preservation: Conservation strategies that focus on a species within its natural habitat.

integrated pest management (IPM): The avoidance or reduction of yield losses caused by diseases, weeds, insects, etc. while minimizing the negative impacts of chemical pest control.

integrated plant nutrient systems (IPNSs): Maximization of the efficiency of nutrient use by recycling all plant nutrient sources within the farm and by using nitrogen fixation by legumes.

intensive livestock operations (ILOs): Factory farms, feedlots, etc. where large quantities of external energy inputs are required to raise for market larger numbers of animals than the area in which they are raised can support, which can result in problems of disease and in dealing with animal waste.

International Joint Commission: A bilateral institution, consisting of three Canadian commissioners and three American commissioners, established by the 1909 Boundary Waters Treaty to manage interjurisdictional resource issues between Canada and the United States.

interspecific competition: Competition between members of different species for limited resources, such as food, water, or space.

intraspecific competition: Competition between members of the same species for limited resources, such as food, water, or space.

intrinsic value: A belief that nature has value in and of itself, apart from its value to humanity; a central focus for the preservation of species.

James Bay and Northern Quebec Agreement: Treaty signed in 1975 by the James Bay Cree and Quebec Inuit with the Quebec government, permitting the continuance of the James Bay Project and granting to the Natives, among other things, $232.5 million in compensation and outright ownership of 5,543 km^2, as well as exclusive hunting, fishing, and trapping rights to an additional 62,160 km^2, often considered to be the first modern Aboriginal land claims settlement in Canada.

James Bay Project: A hydroelectric megaproject in northern Quebec, begun in the 1970s, that has involved extensive dams on the La Grande and other rivers and has flooded thousands of square kilometres of the James Bay Cree homeland.

keystone species: Critical species in an ecosystem whose loss profoundly affects several or many others.

kimberlite pipes: Rare, carrot-shaped igneous rock formations sometimes containing diamonds and found in parts of northern Canada.

kinetic energy: The energy of objects in motion.

Kyoto Protocol: An international agreement reached in Kyoto, Japan, in 1997 that targets 38 developed nations as well as the European Community to ensure that 'their aggregate anthropocentric carbon dioxide equivalent emissions of the greenhouse gases [e.g., carbon dioxide (CO_2); methane (CH_4); nitrous oxide (N_2O); hydrofluorocarbons (HFCs); perfluorocarbons (PFCs); sulphur hexafluoride (SF_6)]...do not exceed their assigned amounts'. The Protocol came into effect in 2004 when 55 countries accounting for 55 per cent of 1990 global carbon dioxide emissions had ratified it.

law of conservation of energy: Law stating that energy cannot be created or destroyed; it is merely changed from one form into another. Also known as the first law of thermodynamics.

law of conservation of matter: Law that tells us that matter cannot be created or destroyed, but merely transformed from one form into another.

law of everybody: The understanding that if everyone did many small things of a conserving and environmentally aware nature, major environmental problems, threats, dangers would be ameliorated or alleviated.

life-cycle assessments (LCAs): Identification of inputs, outputs, and potential environmental impacts of a product or service throughout its lifetime, from manufacture to use and ultimate disposal.

light living: The need to tread as lightly as possible, to minimize our ecological footprints, often characterized by the four R's: refuse, reduce, reuse, and recycle.

limiting factor: A chemical or physical factor that determines whether an organism can survive in a given ecosystem. In most ecosystems, rainfall is the limiting factor.

lithosphere: The Earth's crust.

Livestock Revolution: The shift in production units from family farms to factory farms and feedlots that depend on outside supplies of feed, energy, and other inputs to produce vastly more livestock, a shift that has fuelled the growth in meat consumption worldwide, which has doubled since 1977.

longline: Type of commercial fishing using lines with many baited hooks.

low-quality energy: Energy that is diffuse, dispersed, at low temperatures, and difficult to gather; most of the energy available to us.

LULU: Locally unwanted land use, often the source of a NIMBY reaction.

macronutrient: A chemical substance needed by living organisms in large quantities (for example, carbon, oxygen, hydrogen, and nitrogen).

marine protected areas (MPAs): Underwater reserves set aside and protected from normal human exploitation because of the fragility, rarity, or valued biodiversity of their ecosystems.

matter: What things are made of—92 natural and 17 synthesized chemical elements such as carbon, oxygen, hydrogen, and calcium.

mediation: A negotiation process guided by a facilitator (mediator).

megaprojects: Large-scale engineering or resource development projects that cost at least $1 billion and take several years to complete.

mesosphere: Layer of the atmosphere extending from the stratosphere, at about 50 km above Earth, to about 80 km above Earth.

metal toxicity: The poisonous or harmful nature of metals and minerals, such as asbestos and lead, both to humans and to ecosystems.

micronutrient: An element needed by organisms, but only in small quantities, such as copper, iron, and zinc.

mineralization: The process in which biomass is converted back to ammonia (NH_3) and ammonium salts (NH_4) by bacterial action and returned to the soil when plants die.

minimum viable population (MVP): The smallest population size of a species that can be predicted to have a very high chance of persisting for the foreseeable future.

Mixed Wood Plains: The most urbanized ecozone in Canada, spreading from the lower Great Lakes north and east through the St Lawrence Valley, with gently rolling topography and a continental climate characterized by warm, humid summers and cool winters.

monoculture cropping: Cultivation of one plant species (such as corn) over a large area, which leaves the crop highly susceptible to disease and insects, especially when all of the individual plants are genetically identical.

Montane Cordillera: Ecozone in the Interior of BC with considerable contrast between the summits of the snow-bound peaks through to high montane valleys, rolling plateaus, and deeply entrenched desert-like conditions and a climate generally characterized by long, cold winters and short, warm summers.

Montreal Protocol: Signed in 1987 by thirty-two nations, this agreement established a schedule for reducing use of chlorofluorocarbons and halons to reduce the rate of depletion of the ozone layer.

multi-barrier approach: A method of ensuring the quality of a water supply by using a series of measures (e.g., system security, source protection through pollution regulations within a watershed, water treatment and filtration, testing), each independently acting as a barrier to water-borne contaminants through the system.

mutualism: Relationship between two organisms, having to do with food supplies, protection, or transport, that is beneficial to both.

negative feedback: Control mechanism present in the ecosystem and in all organisms—information in the form of chemical, physical, and biological agents influences processes, causing them to shut down or reduce their activity.

negotiation: One of the two main types of alternative dispute resolution, when two or more parties involved in a dispute join in a voluntary, joint exploration of issues with the goal of reaching a mutually acceptable agreement.

net primary productivity (NPP): Gross primary productivity (the total amount of energy that plants produce) minus the energy plants use during cellular respiration.

new forestry: A silvicultural approach that mimics natural processes more closely through emphasizing long-term site productivity by maintaining ecological diversity.

niche: An organism's place in the ecosystem: where it lives, what it consumes, and how it interacts with all biotic and abiotic factors.

NIMBY: 'Not in my backyard', a phrase used to describe local people's reactions when a noxious or undesired facility—for example, a landfill site, a sand and gravel pit, or an expressway—is proposed to be built adjacent to or near their property.

nitrogen fixation: Conversion of gaseous (atmospheric) nitrogen (N_2) into ammonia (NH_3) by bacteria, such as those that grow on the root nodules of legumes.

non-point sources: Sources of pollution in which pollutants are discharged over a widespread area or from a number of small inputs rather than from distinct, identifiable sources.

non-renewable or stock resources: Those resources, such as oil, coal, and minerals, that take millions of years to form, and thus, for practical purposes, are fixed in supply and therefore not renewable.

non-timber forest products (NTFPs): Forest resources of economic value but not related to the lumber and pulp and paper industries, such as wild rice, mushrooms and berries, maple syrup, edible nuts, furs and hides, medicines, ornamental cuttings, which currently contribute an estimated $442 million annually to the Canadian economy.

no-till/conservation agriculture (NT/CA): Zero, minimum, or low tillage to protect and stimulate the biological functioning of the soil while maintaining and improving crop yields, which includes direct sowing or drilling of seeds instead of plowing, maintenance of permanent cover of plant material on the soil, and crop rotation.

nutrients: Elements or compounds that an organism must take in from its environment because it cannot produce them or cannot produce them as fast as needed.

old-growth forests: Forests that generally have a significant number of huge, long-lived trees; many large, standing dead trees; numerous logs lying about the forest floor; and multiple layers of canopy created by the crowns of trees of various ages and species.

oligotrophic: Nutrient poor.

omnivores: Organisms that eat both plants and animals.

optimal foraging theory: The relationship between the benefit of making a kill and feeding against the cost of the energy expended to make the kill.

oxygen sag curve: The drop in oxygen levels in a body of water when organic wastes are added and the number of bacteria rises to help break down the waste.

ozone: An atmospheric gas (O_3) that, when present in the stratosphere, helps protect the Earth from ultraviolet rays. However, when it is present near the Earth's surface, it is a primary component of urban smog and has detrimental effects on both vegetation and human respiratory systems.

ozone layer: Thin layer of ozone molecules in the stratosphere that absorbs ultraviolet light and converts it to infrared radiation, effectively screening out 99 per cent of the ultraviolet light.

Pacific Maritime: Ecozone characterized by the influence of the Pacific Ocean, with the highest rainfall figures in Canada (up to 3,000 mm) and the warmest average temperatures.

Palliser's Triangle: Semi-arid area of southern Alberta and Saskatchewan and extreme southwest Manitoba first identified by Captain John Palliser during an expedition to the Canadian West in 1857–60 sponsored by the Royal Geographical Society and the British Colonial Office.

parasitism: Relationship in which one species lives in or on another, which acts as its host.

partnerships: A sharing of responsibility and power between two or more groups, especially a government agency and a second party, regarding a resource or environmental issue; co-management is an example of a partnership.

photosynthesis: A two-part process in plants and algae involving: (1) the capture of sunlight and its conversion into cellular energy, and (2) the production of organic molecules, such as glucose and amino acids from carbon dioxide, water, and energy from the sun.

phototrophs: Organisms that produce complex chemicals through photosynthesis.

phytoplankton: Single-celled algae and other free-floating photosynthetic organisms.

placebo policies: Government and management policies designed to play down the salience of environmental concerns and to sidestep issues by addressing the symptoms of a problem rather than its causes.

point sources: Easily discernible 'end-of-pipe' sources of pollution, such as a factory or a town sewage system.

pollutant: A substance that adversely alters the physical, chemical, or biological quality of the Earth's living systems or that accumulates in the cells or tissues of living organisms in amounts that threaten their health or survival.

polynyas: Ice-free areas of permanent open water in the Arctic surrounded by ice, created by tides, currents, ocean-bottom upwellings, and winds, that are biologically productive and vary greatly in size from 60–90 metres in diameter to as large as the North Water polynya between Ellesmere Island and Greenland that may cover as much as 130,000 km^2.

potential energy: Stored energy that is available for later use.

population: A group of organisms of the same species living within a specified region.

positive feedback loop: A situation in which a change in a system in one direction provides the conditions to cause the system to change further in the same direction.

precautionary principle: A guideline stating that when there is a possibility of serious or irreversible environmental damage resulting from a course of action, such as a development project, lack of scientific certainty is not an acceptable reason for postponing a measure to prevent environmental degradation, or for assuming that damage in the future can be rectified by some kind of technological fix.

predator: An organism that actively hunts its prey.

prey: An organism (e.g., deer) that is attacked and killed by a predator.

prey switching: A familiar foraging behaviour whereby a predator shifts from its target species, after it is depleted or not available in an area, to the next most preferred or profitable species, until that too is depleted, and then continuing to move down the food chain, as wolves will do in moving from caribou to Arctic hare to small

rodents, or as humans have done in fishing down the food chain in commercial fisheries.

primary consumers: The first consuming organisms in a given food chain, such as a grazer in grazer food chains or a decomposer organism or insect in decomposer food chains; primary consumers belong to the second trophic level.

primary succession: The development of a biotic community in an area previously devoid of organisms.

producers: Autotrophs capable of synthesizing organic material, thus forming the basis of the food web.

protected areas: Areas such as national and provincial parks, wildlife sanctuaries, and game preserves established to protect species and ecosystems.

rainshadow effect: The decrease in precipitation levels as the air warms up in its descent from the mountains and it can hold more moisture, i.e., there is considerably less precipitation on the leeward side of a mountain or a mountain range than on the windward side.

range of tolerance: Range of abiotic factors within which an organism can survive, from the minimum amount of a limiting factor that the organism requires to the maximum amount that it can withstand.

reclamation: The process of bringing back an area to a useful, good condition—similar to rehabilitation.

Red List: An annual listing of species at risk prepared by the IUCN and produced by thousands of scientific experts, which is the best source of knowledge on the status of global biodiversity.

renewable or flow resources: Those resources that are renewed naturally within a relatively short period of time, such as water, air, animals, and plants, as well as solar radiation, wind power, and tidal energy.

resilience: Ability of an ecosystem to return to normal after a disturbance.

resource partitioning: A situation in which the resources are used at different times or in different ways by species with overlap of fundamental niches, such as owls and hawks, which seek the same prey but at different times during the day.

resources: Such things as forests, wildlife, oceans, rivers and lakes, minerals, and petroleum.

Rio Declaration: The 27 principles related to sustainable development, developed at the Earth Summit in Rio de Janeiro during June 1992.

risk assessment: Determining the probability or likelihood of an environmentally or socially negative event of some specified magnitude.

salinization: Deposition of salts in irrigated soils, making soil unfit for most crops. It is caused by a rising water table due to inadequate drainage of irrigated soils.

secondary consumers: Second consuming organisms in a food chain; belong to the third trophic level.

secondary succession: The sequential development of biotic communities after the complete or partial destruction of an existing community by natural or anthropogenic forces.

second growth: A second forest that develops after harvest of the original forest.

second law of thermodynamics: When energy is converted from one form to another, it is degraded; that is, it is converted from a concentrated to a less concentrated form. The amount of useful energy decreases during such conversions.

sedimentary cycles: Those cycles of elements, such as the phosphorus and sulphur cycles, that hold most of their matter in the lithosphere.

serial depletion: When one stock after another becomes progressively depleted as a result of prey switching, even if the total catch remains the same.

shifting baseline: When scientists have no other option than to take the current or recent degraded state as the baseline for stock biomass rather than the historical ecological abundance.

Silent Spring: Rachel Carson book published in 1962 that detailed the disastrous effects of biocides on the environment.

silviculture: The practice of directing the establishment, composition, growth, and quality of forest stands through a variety of activities, including harvesting, reforestation, and site preparation.

soil compaction: The compression of soil as a result of frequent heavy machinery use on wet soils or overstocking of cattle on the land.

soil erosion: A natural process whereby soil is removed from its place of formation by gravitational, water, and wind processes.

soil horizons: Layers found in most soils.

specialist: Organism that has a narrow niche, usually feeding on one or a few food materials and adapted to a particular habitat.

speciation: Formation of new species.

species: A group of individuals that share certain identical physical characteristics and are capable of producing fertile offspring.

Species at Risk Act (SARA): Canadian legislation passed in 2002 that mandates the Committee on the Status of Endangered Wildlife in Canada to maintain lists of species at risk and to recommend to the minister responsible that particular species be given special protection in their environment.

stakeholders: Persons or groups with a legal responsibility relative to a problem or issue, or likely to be affected by decisions or actions regarding the problem or issue, or able to provide an obstacle to a solution of the problem or issue.

stewardship: Activities undertaken towards caring for the Earth.

stratosphere: The layer of the atmosphere (about 10–50 km above the Earth's surface) in which temperatures rise with increasing altitude.

strip cropping: A technique similar to contour cultivation in which different crops are planted in strips parallel to the slope.

subsidiarity: A policy and management approach stipulating that decisions should be taken at the level closest to where consequences are most noticeable or have the most direct impact.

subsistence farming: The production of food and other necessities to satisfy the needs of the farm household.

sulphur dioxide emissions: Release into the atmosphere of huge quantities of sulphur, mainly in the form of dioxides, as a result of smelting sulphide ores and from the burning of fossil fuels, which causes air pollution and climate change.

summer fallow: A practice common on the Prairies in which land is plowed and kept bare to minimize moisture losses through evapotranspiration, but which leads to increased salinization.

sustainable development: Economic development that meets current needs without compromising the ability of future generations to meet their needs.

sustainable livelihoods: A human-centred approach to broad environmental management directed towards ways for local people to meet basic needs (food, housing), as well as other needs related to security and dignity, through meaningful work, at the same time minimizing environmental degradation, rehabilitating damaged environments, and addressing concerns about social justice.

sustained yield: The amount of harvestable material that can be removed from an ecosystem over a long period of time with no apparent deleterious effects on the system.

synergism: An interaction between two substances that produces a greater effect than the effect of either one alone; an interaction between two relatively harmless components in the environment.

taiga: The portion of the boreal forest lying between the southern boundary of the tundra and the closed-crown coniferous forest to the south, characterized by coniferous forests, soil that thaws during the summer months, abundant precipitation, and high species diversity.

Taiga Plains: Gently rolling northern ecozone with a high proportion of surface water storage, wetlands, and organic soils, a cold and relatively dry climate, and, on better-drained localities and uplands, mixed coniferous-deciduous forests.

technocentric perspective: The assumption that humankind is able to understand, control, and manipulate nature to suit its purposes, and that nature and other living and non-living things exist to meet human needs and wants.

tertiary consumers: In a food chain, organisms at the top that consume other organisms.

thermocline: Sharp transition in temperature between the warmer surface waters of the ocean and the cooler waters underneath, generally occurring between 120 and 240 metres in depth.

thermohaline circulation: The movement of carbon-saturated water around the globe, mainly as a result of differing water densities.

thermosphere: Uppermost layer of the atmosphere, beyond the mesosphere.

threatened species: A species designated by COSEWIC as likely to become endangered in Canada if factors threatening its vulnerability are not reversed.

threshold: A point or limit beyond which something is unsatisfactory relative to a consideration, such as health, welfare, or ecological integrity.

total allowable catch (TAC): The amount, in tonnage, of a particular aquatic species that the federal Department of Fisheries and Oceans, for example, determines can be landed within a particular fishery in a given year.

transpiration: The loss of water vapour through the pores of a plant.

trophic: Relating to processes of energy and nutrient transfer from one or more organisms to others in an ecosystem.

trophic level: Functional classification of organisms in a community according to feeding relationships: the first trophic level includes green plants, the second level includes herbivores, and so on.

troposphere: Innermost layer of the atmosphere that contains 99 per cent of the water vapour, up to 90 per cent of the Earth's air, and is responsible for our weather, extending about 6–17 km up from the Earth, depending on latitude and season.

uncertainty: A situation in which the probability or odds of a future event are not known and therefore indicates the presence of doubt.

vision: A view for the future for a region, community, or group that is realistic, credible, attractive, and attainable.

vulnerable species: An official designation assigned by the Committee on the Status of Endangered Wildlife in Canada to any indigenous species or subspecies or geographically separate population of fauna or flora that is particularly at risk because of low or declining numbers, because it occurs at the fringe of its range or in restricted areas, or for some other reason, but which is not a threatened species.

water table: The top of the zone of saturation.

weather: The sum total of atmospheric conditions (temperature, pressure, winds, moisture, and precipitation) in a particular place for a short period of time.

wetlands: Areas that are hybrid aquatic and terrestrial systems, such as swamps and marshes, where the ground is saturated with water much or all of the time.

Whitehorse Mining Initiative: Multi-stakeholder accord agreed to by governments, the mining industry, labour, Aboriginal groups, and environmental organizations designed to support an overall vision of a 'socially, economically, and environmentally sustainable and prosperous mining industry, underpinned by political and community consensus'.

wind power: The fastest-growing sector in the world's energy market, which uses wind turbines that have minimal adverse environmental affects to generate electricity.

windthrow: Uprooting and blowing down of trees by wind.

World Summit on Sustainable Development (WSSD): Conference held in Johannesburg in 2002, 10 years after the Earth Summit in Rio, that made various commitments, such as the goal of halving the proportion of people without access to adequate sanitation by the year 2015.

zone of physiological stress: Upper and lower limits of the range of tolerance in which organisms have difficulty surviving.

zooplankton: Non-photosynthetic, single-celled aquatic organisms.

zooxanthellae: Unicellular algae.

Index

Abbey, Edward, 31, 514–15
Abbotsford Aquifer, 121
abiotic components, 56–60, 530
Aboriginal peoples: bioaccumulation and, 260, 350, 353; mining and, 463, 464, 474–5, 475, 482; parks and, 415, 418; *see also* First Nations; Inuit; land claims
accommodation: climate change and, 202
acid, 130; deposition, 112, 130–40, 530; mine drainage, 461, 530; precipitation, 278, 341; shock, 132, 530
acidity: measuring, 131; soil, 58, 109, 341–2
Acres International, 12–13
adaptation measures, 202
Advisory Committee on Watershed-based Source Protection Planning (Ont.), 440–1
advocate: role of, 434
aerobic organisms, 46, 530
aesthetics: forestry and, 311–12; wind power and, 486–7
afforestation, 212
Africa, 329
Agenda 21, 242, 502, 530
Agricultural Land Preservation Act (Que.), 327
Agricultural Land Reserve Act (BC), 327
agriculture, 325–370; acid deposition and, 133; Canadian challenges in, 339–59; Canadian trends in, 338–9; as ecological process, 330; habitat change and, 385; impact of, 337–8; industrialized, 331–8; monoculture cropping, 337, 537; no-till/conservation (NT/CA), 362–3, 538; organic, 363–6; pollution by, 430–7; social implications of, 326; subsidies for, 328, 357–8; subsistence, 333, 533; wind power for, 487
albatross, 237, 238
albedo effect, 86, 520
Alberta: endangered species in, 386, 395; Kyoto and, 160, 214–15
Algonquin Provincial Park, 386
alien species, 79–85, 520
alkalinity, soil, 58
alternative dispute resolution (ADR), 176–7, 178, 520
ammonia, 356, 430–1
anadromous species, 111, 259, 520
Anderson, David, 213
Andrey, J., and L. Mortsch, 169, 187, 208
annual allowable cut (AAC), 290–1, 520

antelope, 65, 385
anthropocentric view, 5, 376, 520
'Anthropozoic' age, 376
aquaculture, 265–70, 520
aquifers, 121, 520
arbitration, 177, 520
Arctic: biocides in, 349, 350; climate change and, 190, 216; nitrogen in, 115; ocean ecosystems in, 228; species in, 49, 53, 89
Arctic Archipelago ecozone, 246
Arctic Basin ecozone, 244, 246
Arnstein, S., 166–7
Asia, 329
Atlantic Canada: fisheries and, 248–56
Atlantic ecozone, 247–8
Atlantic Groundfish Strategy (TAGS), 251
Atlantic Maritime ecozone, 283, 530–1
Atlantic Seal Hunt Management Plan, 247
atmosphere, 40, 524; climate change and, 189; water in, 118–20
atmospheric fixation, 115
Auditor General of Canada, 85
Audubon, John James, 382
auk, great, 379
Austria, 500
automobiles, 481, 483; in China, 497; driving, 510; and energy use, 513
autotrophs, 45, 531

Babcock, Tim, 16–17
Bacillus thuringiensis (Bt), 297–8, 332
backcasting, 153
bacteria, 112, 115, 359
ballast water, 83–4
Banff-Bow Valley Task Force, 408
Banff National Park, 386, 397, 406–9
Bangladesh, 18
Barlow, M., and E. May, 10, 14
base (chemical), 131
BC Hydro, 172
Beanlands, G.E., and P.N. Duinker, 171
bears: grizzly, 90, 305, 308, 395–96, 417, 478; polar, 90, 240

Beaufort Sea, 258
beaver, 64–5, 66
Beazley, K., 376
Beazley, K., and R. Boardman, 416
beetle, mountain pine, 303
Bell, Warren, 353
benthic plants, 126, 531
Berger, Thomas, 434
Berkes, F., 428, 429
'best practice', 145–7
BHP World Minerals, 473, 480
bioaccumulation, 260, 347–9, 531
biocentric view, 5, 531
biocides: 294–9, 333, 335–6, 343–56, 531; 'biocide
    treadmill', 343; domestic use of, 351, 352; mobility
    of, 345, 347, 350; non-selective, 345; persistence of,
    347–9; regulation of, 353–6; see also herbicides;
    pesticides
bioconcentration, 240, 350, 352–3, 531
biodiversity, 79, 85, 94–100, 531; 'crisis', 371; forests
    and, 276–83, 299–305; response to loss of, 392–98;
    valuing, 372–6; see also diversity
Biodiversity Management Program, 217–20
biogeochemical cycles, 45, 105–16, 531
biological controls, 298
biological nitrogen fixation (BNF), 115
biological oxygen demand (BOD), 126, 531
biomagnification, 260, 349, 351–2, 531
biomass, 45, 52, 531; energy, 481; potential, 531;
    pyramids, 230, 531
biomes, 56, 531
biosphere, 531
Biosphere Reserve Program, 416
biotic: components, 56, 60–5, 531; potential, 89, 531;
    pyramids, 51–2; relationships, 62, 64–5
birds: biocides and, 346, 349, 351–2; endangered, 346;
    extinctions of, 382, 383, 387, 390–1; migratory,
    387–9; wind power and, 485–6; see also specific
    species
bison, 385; wood, 411–13
Blusson, Stewart, 464, 472–3
Boreal Cordillera ecozone, 278, 531
boreal forest, 276, 282, 531
Boreal Plains ecozone, 281–2, 531
Boreal Shield ecozone, 276–8, 282–3, 531
Boundary Waters Treaty, 166
Bourassa, Robert, 426, 427
Brandon, Man., 168
Bregha, F., et al., 172
British Columbia, 85; aquaculture in, 265–9; fires in,
    300; fisheries in, 257–8; forestry in, 280, 287, 295–6,
    303, 313–14; Kyoto and, 215; offshore oil and gas
    and, 258, 260; protected areas in, 403–04
British Columbia Ministry of Environment, Lands and
    Parks, 168
Brklacich, M., C.R. Bryant, and B. Smit, 360
Browne, C.L. and S.J. Hecnar, 371–2
Browning, M.H.R., and T.C. Hutchinson, 135

'brownlash', 208
Bruner, A.G., et al., 400
budworm, spruce, 298, 299
buffering capacity, 133
Bunnell, F., and L.L. Kremsater, 305
Burgess Shales, 94
Burn, D.H., and N.K. Goel, 443
Bush, George W., 151, 210–11
'business model', 151
butterfly effect, 86
bycatch, 235, 238, 531

calorie, 42, 531
Campbell, Gordon, 260
Canada: alien species in, 81–5; biodiversity in, 97–8,
    100; Carolinian zone in, 95; debt and deficit
    reduction in, 150; endangered species in, 371–2;
    energy use in, 20, 23, 460, 485, 481, 483;
    environmental response in, 501–4; as forest nation,
    276; international rank of, 504; jurisdictions in,
    23–4; Kyoto and, 213–15, 217–20; loss of
    biodiversity and, 394–7; oceans and, 242–8, 261–5;
    population in, 19, 22; precipitation in, 119
Canada Land Inventory (CLI), 327
Canada–US Air Quality Agreement, 138
Canada-Wide Acid Rain Strategy for Post-2000, 137
Canadian Climate Change Model, 200–1
Canadian Council of Forest Ministers (CCFM), 314
Canadian Council of Ministers of Environment, 190
Canadian Endangered Species Conservation Council
    (CESCC), 394–95, 396
Canadian Environmental Assessment Research
    Council, 171
Canadian Environmental Choice, 512
Canadian Environmental Protection Act, 261
Canadian Environmental Protection Agency, 261
Canadian Food Inspection Agency (CFIA), 354–5
Canadian Heritage Rivers Program (CHRP), 448–54
Canadian Pacific Railway, 465
Canadian Parks and Wilderness Society (CPAWS), 276
Canadian Regional Climate Model (CRCM), 195
Canadian System of Soil Classification, 59
Canadian Water and Wastewater Association, 430
Canadian Wildlife Service, 263
carbofuran, 346–7
carbon, 105, 107, 116, 117; balance, 230–2, 524;
    sequestration, 219–20, 531
carbon dioxide, 116, 188, 193, 195–6, 217, 219, 230–2
caribou, 90, 473, 477–8
carnivores, 46, 531
carp, Asian big head, 85
Carpenter, R.A., 169, 207, 208–9
carrying capacity, 21, 29–30, 87–8, 532
Carson, Rachel, 343
'Catastrophozoic' age, 376
Centre for Marine Conservation, 238
Chalecki, E.L., 207

change: ecosystem, 69–103; natural, 4–5; responses to, 495–518; societal, 4–5; threshold of, 66, 541; *see also* climate change; environmental change

chaos theory, 87

Chapman, J.D., 483, 485

Charlottetown, PEI, 201

chemicals: as pollutants, 430–1

chemoautotrophs, 45, 532

Chevalier, G., et al., 428–9

China, 18, 210, 497

Chisasibi, Que., 427–8

chlorinated hydrocarbons, 344

chloroflurocarbons (CFCs), 532

chlorophylls, 45, 532

Chrétien, Jean, 213–14, 265, 403

Clarke, A., and C.M. Harris, 246

Clayoquot Sound, 175

clear-cutting, 291–3, 295, 302, 532

climate, 187–8, 532; modelling, 193, 532; and vegetation, 73

climate change, 183, 186–225, 532; adaptation measures, 202; and atmosphere, 188; communication and, 205–10; definition of, 187; forests and, 278; freshwater systems and, 205; human activities and, 196–8; less-developed countries and, 209–10; lifestyles and, 210; measuring, 194; media and, 209; modelling, 192–4; ocean systems and, 205, 230–2, 241; science and, 188–90; and sea levels, 190, 197–202, 241; and snow cover, 190; terrestrial systems and, 204–5; and tourism, 33, 202–3; and uncertainty, 194, 207, 208–10; *see also* global warming

Climate Change and Emissions Management Act (Alta), 214–15

climatic climax, 73

climax community, 70, 73, 532

coastal areas, 33, 196, 241

cod, 38, 39, 226, 248, 249–57, 395; estimated stocks of, 254–5; seals and, 247

coevolution, 90, 112, 532

Cohen, David, 511

collaboration, 1, 164–5, 466–8, 532

Collins, Harry, 449

colonization, 70–4

colonizers, primary, 70

co-management, 168, 176, 418, 532

commensalism, 64, 532

commercialization, 150, 151

Commissioner of the Environment and Sustainable Development, 502–3

Committee on the Status of Endangered Wildlife in Canada (COSEWIC), 226, 307, 373, 379, 380, 391, 395, 396, 505

communication, 168–9, 178, 184, 536; climate change and, 205–10

communities, 54–5, 71, 73

competition, species, 61–2, 536

competitive exclusion principle, 61, 532

composite indices, 28–9

compounds, chemical, 105, 532

Conestoga-Rovers and Associates, 14

Conference Board of Canada, 504

conservation: genetic, 338; organizations for, 413–14

consultation, 462–3; public, 151, 165, 175–6

consumerism, 515, 532; responsible, 319

consumers: energy, 45–6, 532; human, 2, 515–18, 532; primary, 46, 539; secondary, 46, 540; tertiary, 47, 539

consumption: global, 21; globalization of, 497; reduction, 500, 501, 509

contamination: endangered species and, 372; soil, 356, 359; water, 356, 359, 479; *see also* pollution

context (planning and management), 149–52, 534

contour cultivation, 363, 534

control, top-down/bottom-up, 51

Convention on Biological Diversity (CBD), 393–4, 499, 532

Convention on International Trade in Endangered Species of Wild Fauna and Flora (CITES), 382, 392–3, 532

Convention on Long-Range Trans-Boundary Air Pollution, 138

conversion rate, 290–1

co-ordination, 165–6, 532

coral: bleaching, 229, 532; polyps, 229, 525; reefs, 229–30

cormorants, 383

corporations, 513

Couch, W.J., 472, 479

cougars, 387

courts: dispute resolution and, 175–7; First Nations' fishing rights and, 256–8; pesticides and, 352

Coward, H., R. Ommer, and T. Pitcher, 422

cowbird, 387, 389

crane, whooping, 412–13

Cree, 426–9

Cressman, D.R., 434–5

Crombie, David, 434

crop(s): genetically modified (GM), 332, 364; rotation, 363, 532

cryptosporidium, 439

culmination age (forestry), 290, 533

cumulative effects, 172

dams, 171–2, 424, 475; frozen core, 477, 479, 534–5

Darwin, Charles, 38, 99

data: baseline, 475–9; quantitative/qualitative, 39

DDT (dichlorodiphenyltrichloroethane), 260, 347–9, 533

decision-making, shared, 166–70

'decision-regulators', 55

decomposers, 108

deer, 302, 383

deforestation, 196, 212, 377–9

Dene, 475

denitrification, 112, 116, 533

Denny, Cecil, 411

Department of Fisheries and Oceans (DFO), 244

detritivores, 47
detritus, 47, 528; nutrients, 108
developing countries: Kyoto and, 213
diamonds, 464–5, 472–80
Diavik diamond mine, 464
Diduck, A., and B. Mitchell, 168
diffuse pollution, *see* non-point sources
Dillon Consulting Limited, 485–8
dinosaurs, 93
dioxins, 260
discharge areas, 121
diseases, 11–12, 349; aquaculture and, 266–7; biocides
    and, 345; hunger and, 329; manure and, 356; *see also*
    health
dispute resolution, 147, 174–7
diversity, 78; farm animal, 338; forest ecosystem,
    276–83; genetic, 95; *see also* biodiversity
Doer, Gary, 215
dominant limiting factor, 60
Dominion Coal Company (DOMCO), 10
Dominion Iron and Steel Company (DISCO), 10
downloading, 24, 150
Drivers-Pressures-State-Impact-Response (DPSIR)
    indicator, 26–7
driving forces (DPSIR), 26
droughts, 446–8, 533
ducks, 388–9
Ducks Unlimited, 413
Duinker, P.N., et al., 313
Dumont, C., et al., 428–9
dykes, 444, 533

eagles, 389
earthquakes, 32
Earth Summit (Rio de Janeiro), 21, 150, 171, 332, 394,
    498, 533
ecocentric: values, 155–6, 157; view, 5, 376, 531, 533
E. coli, 334, 356, 438, 533
Ecological Integrity Panel, 408
ecological: 'footprint', 24–5, 488, 533; knowledge,
    traditional (TEK), 38, 173, 191, 474–5, 475, 536;
    ratios, 96
ecology, landscape, 63–4
economy: mining and, 464, 472; world, 5
ecosphere, 37–143, 533; definition of, 40
ecosystem-based management (EBM), 303, 315, 533
ecosystems: aquatic, 78, 82–4, 334; change in, 69–103;
    definition of, 56, 533; disruption of, 461, 465;
    diversity in, 95; ecosystems perspective, 156, 158–60,
    183; energy flows and, 41–69; forest, 276–87;
    homeostasis in, 79–87, 535; marine, 49–50; matter
    cycling and, 104–43; mature/immature, 78; oceanic,
    227–33; oligotrophic, 125, 531; productivity in,
    53–4; structure of, 56–65
ecotones, 73, 533
edaphic climax, 73, 77
Eddie Bauer, 513
education: environment and, 504–11

efficiency, 150
Ehrlich, P., and A., Ehrlich, 208
Ekati diamond mine, 464, 472–80
electricity, 20; cost v. price, 484, 487, 488; *see also*
    hydroelectricity
El Niño, 188, 200, 535
emission: credits, 212–13; trading, 213, 533
Emond, D.P., 176
employment: fisheries, 249, 255; mining, 466–8
empowerment, 154–5, 165–6
Endeavour Hydrothermal Vents Marine Protected
    Area, 267
endemism, 97–8, 534
endocrine disruption, 240, 261, 534
energetics: aquaculture and, 268
energy, 42, 459–61, 463, 480–1, 483–8, 534; biomass,
    481; Canada's use of, 460, 481, 483; consumption
    of, 18, 20, 21, 23; ecosystems and, 41–68; flows of,
    45–56, 333; geothermal, 480–1, 483; high-quality,
    42, 535; hydroelectric, 172, 425–30, 461; kinetic, 42,
    538; low-quality, 42, 539; non-renewable sources of,
    480–1; nuclear, 461, 464, 483; potential, 42, 541;
    pyramid, 53; radiant, 45; resources, 480–1, 483–8;
    solar, 480–1; well-being and, 501
energy balance models (EBMs), 193
entropy, 43–4, 534
environment, 5, 534; and education, 504–11; global,
    15, 17–22; science-based management of, 7–9
Environmental Assessment Panel, 172
environmental change: Canadian responses to, 501–4;
    global responses to, 496–501; individual responses
    to, 504–11
Environmental Impact Assessment Review Panel,
    475–7, 478–9
environmental justice, 146, 154–5, 534
Environmental Protection Agency (US), 129
Environmental Sustainability Index (ESI), 29
Environment Canada, 15, 17, 29, 129, 263, 312, 365,
    430, 436–7, 481, 503–4; indicator metres, 27
epidemiology, environmental, 12
Erie, Lake, 128–9, 432, 434
erosion: control, 362–3; forestry and, 311; soil, 339–41,
    434–5, 540; 'sustainable', 366
escapement, 266–7
estuaries, 54, 110, 524
ethics, 155–6
euphotic zone, 49, 110 , 228, 524
European Union, 211, 252, 484, 499
eutrophication, 78, 109, 125–9, 359, 432–3, 524; control
    of, 128–9; cultural, 124, 126; effects of, 126–8
evapotranspiration, 118, 524
evolution, 90–3, 529; phyletic, 91
exclusive economic zones (EEZs), 242, 524
exotic species, 530; *see also* alien species
Experimental Lakes Research Area (ELA), 174
exports: electricity, 427; grain, 357–8; mineral, 464
ex situ preservation, 384, 534

extinction, 91–3, 410, 472, 534; causes, 376–89; ecological, 373, 528; mass, 91–3; rate of, 92; in tropics, 376–7, 379–80; vulnerability to, 389–92
extirpation, 65, 92, 373, 524

'fact', 38
'factory farms', 337
Falconbridge Nickel Mines, 468
falcons, 382, 390–1
falldown effect, 291, 524
fallow, summer, 342, 540
Federation of Canadian Municipalities, 25
feedback, 86–7; loops, 233; negative, 86, 539; positive, 74, 78, 85, 539
ferret, black-footed, 383
fertilizers, 109, 110, 115, 196, 333, 335–6, 341, 356, 359
Finland, 286–7
Fipke, Chuck, 464, 472
fires, forest, 72, 74, 98, 293–4, 299–300
First Nations, 23, 155; Clayoquot Sound and, 175; climate change and, 192; co-management and, 168; fisheries and, 256–8; forestry and, 288, 290, 313–14, 315; see also Aboriginal peoples; land claims
fish: acid deposition and, 132; extinctions, 382; feminization of, 240; fish kills, 356, 359
fisheries: acid deposition and, 135; and alien species, 85; Canadian, 248–56; capelin, 41–2; environmental change and, 255; 'fishing down the food chain', 235, 256, 524; incentives for, 255; inshore, 252, 253; lobster, 11, 12, 257; longline fishing, 238, 242, 537; management, 226–7, 233–40; offshore, 253–4; overfishing, 252–4; serial depletion of stocks, 235, 540; sport, 60; total allowable catches (TACs), 249, 250, 254, 541; and water diversion, 428–9; see also aquaculture; cod; salmon
Fisheries Research Centre, 234
Fisheries Resource Conservation Council (FRCC), 250
Flanagan, K., and S. Rasheed, 410
flood(s), 183, 201, 441–6; control of, 424; major Canadian, 441
flood plains, 441–6, 534
floodways, 444–5, 534
food: affordable, 329; consumption, 340; future need for, 328; population and, 19; potential new sources, 374–5; production, 340, 359–65
food chains, 42, 46–52, 330, 534; decomposer, 47, 526; grazing, 47, 49
food webs, 49, 534
'footprint', ecological, 24–5, 488, 533
'forcings', radiative, 195–7
forecasting, 153
forest(s), 275–324; and biodiversity, 273–9, 301–7; boreal, 276, 282–3, 531; culmination age, 291, 533; disturbance, 300, 303, 304; global perspective on, 289; global warming and, 75–6; health, 312; management, intensive, 299–301, 313; old-growth, 291, 296, 302, 304, 306–7, 538; ownership of, 286–7, 290, 314–15; private, 286–7, 315, 317;

products of, 284–90, 317; reclamation, 290, 539; and recreation, 284; second growth, 540; urban, 315, 316; see also forestry industry
Forest Practices Act (BC), 293
forestry industry, 276, 277; acid deposition and, 135; and aesthetics, 312–13; certification of, 320; clear-cutting, 292–4, 295, 304, 532; competitiveness of, 287, 316; conflict in, 288, 290; fire and, 73, 74, 98, 294, 300–1; harvesting methods, 292–6, 307–9; impacts of, 301–13; 'new', 312–15; non-timber forest products (NTFPs), 284–5, 311–13, 538; and site fertility, 307–11; sustainability of, 291–3, 314–15; value of, 285
Forest Stewardship Council (FSC), 320, 499
fossil fuels, 116, 117, 240–1, 465, 487, 534; climate change and, 195, 214; see also oil and gas
fossils, 90, 92, 93
Foster, Bristol, 49
Foster, H.D., and W.R.D. Sewell, 423
four Rs, 508–11
fox, swift, 384, 385
Francis, Charles, 97
Fraser River, 430, 452, 454
French, H., 499
Freedman, B., et al., 307, 309
fungi, alien, 83
furans, 260
future: alternative possible, 169–70; desirable, 146, 152–3; feasible, 153; likely, 153; probable, 153

Gabriel, A.O., and R.D. Kreutzwiser, 446
Gaia hypothesis, 80
Gardner, G., and A. Assadourian, 502, 515
Garipey, M., 173
Garrison Diversion project, 166
gaseous cycles, 107, 112–16, 535
Gedalof, Ze'ev, 75–6
gene pools, 375
general circulation models (GCMs), 193–4, 535
genetic diversity, 94
genetically engineered (GE) wheat, 364
genetically modified (GM crops, 332, 364
genetically modified organisms (GMOs), 332, 333, 535; labelling, 332
Georgian Lakeland Tourism Region, 203–4
glaciers, 122–3, 190
Global POPs Protocol, 350–1
global warming, 74, 75–6, 187, 231–3, 535; see also climate change
Goodarzi, E., and M. Mukhopadhyay, 12
Gorrie, P., 432
Gould, Stephen Jay, 93, 94
governments: biocides and, 353; Canadian, 23–4; context and, 150–2; and endangered species, 394–8; forests and, 286–7, 290; influencing, 512–14; Kyoto and, 213–14; leadership of, 436; negotiations among, 434–5; oceans and, 243–4; see also jurisdiction; specific jurisdictions

Grand River, 449–52, 453–4

grass, 74, 362

grasshopper effect, 350, 535

Gray, B., 165

Great Bear Rainforest, 417

Great Lakes: alien species in, 83–5; pollution, 430, 432–7; water levels, 446–7

Great Lakes Water Quality Agreement, 129, 174, 433, 435

Great Whale River, 427

greenhouse effect, 189, 535

greenhouse gases (GHGs), 74, 87, 186, 188–9, 195–7, 210–12, 504, 509, 535; 'greenhouse gas intensity', 210

Green Plan, 435

Green Revolution, 328, 330–36, 357–8, 535

Green Seal, 512

gross national product (GNP), 20, 29, 535

gross primary productivity (GPP), 53–4, 535

groundwater, 118, 120–5, 424, 438, 439, 446–7, 535

habitat, 61, 535; alien species and, 389, 391; chemical changes in, 387, 389; physical changes to, 385–92

Halifax, 261–3

Hall, Joseph, 442

Hanna, Kevin, 286–7

Harper, Stephen, 213

Harris, Leslie, 254

harvesting methods (forest), 305, 307

Harvey, Harold, 129

health, human: acid deposition and, 135–6; aquaculture and, 268; pollution and, 11–12; see also diseases

Health Canada, 353

heat: as energy, 42

Helsinki Protocol, 138

Helzer, Chris, 63

Hengeveld, H., et al., 188

herbicides, 294, 296, 302, 332; see also biocides; pesticides

herbivores, 46, 535

heterotrophs, 45, 535

Hilson, G., 460–1

Himmelman, A.T., 164–5

Hitch, Michael, 480, 482

Ho, Elise, 191

Hobson, K.A., et al., 385

hog-processing, 168

Holling, C.S., 170

Home Depot, 516

homeostasis, ecosystem, 79–87

Homer-Dixon, T., 150

'hot air' credit, 213

Hounsel, Steve, 217

Hudson, Que., 352

human activities, 15, 17–22; biogeochemical cycles and, 109, 112, 115, 117, 123, 125–40; and climate change, 196–7; extinctions and, 91; hydrological cycle and, 424–30; succession and, 78–9

Human Development Index (HDI), 29, 499

hunger, 329, 357–8

hurricanes: Hazel, 446; Juan, 3–4

Hutchings, J.A., and R.A. Myers, 255

Huxley, Thomas, 233

hybridization, 333, 535

hydroelectricity, 174, 425–30, 461; boreal forest and, 276, 277; water diversion for, 425–30; see also electricity

hydrogen, 105

hydrological cycle, 107–8, 116–25, 424, 535; forestry and, 310; human intervention in, 424–30; see also water

hydrosphere, 39, 535

hydrothermal venting systems, 267

hypotheses, 38

ice cover, 199–200

ideologies, 151–2

impact assessment, 14, 147, 168, 171–4, 178, 429–30, 535; environmental (EIA), 14, 171, 177, 426–7, 527; social, 171, 426–7

impact and benefit agreements (IBAs), 480, 535

impact indicators (DPSIR), 26

impacts, cumulative, 429–30

Independent Environmental Monitoring Agency, 480

India, 18, 210

indicators, 2, 25–9, 78, 365, 536; indicator species, 72, 259

indices, 28–9

indigenous knowledge, 38, 173, 536; see also ecological knowledge, traditional

inertia, 80, 536

'ingenuity gap', 150

insecticides, 297–300; see also biocides; pesticides

insects, rain-forest, 96

in situ preservation, 384, 536

integrated: database, 27–8; integrated management, 262; integrated pest management (IPM), 361, 363, 536; integrated plant nutrient systems (IPNSs), 363, 536

intensification, agricultural, 327–8

intensive forest management, 297–8, 311

intensive livestock operations (ILOs), 356–9, 536

Intergovernmental Panel on Climate Change, 190, 195–6, 212

intermediary disturbance hypothesis, 78

International Drinking Water Supply and Sanitation Decade, 437

International Food Policy Research Institute (IFPRI), 328, 343

International Joint Commission (IJC), 166, 432–4, 435–7, 445, 536

International Maritime Organization (IMO), 84–5

International Model Forest Program, 317–18

International Nickel Company (Inco), 466, 468

International Reference Group on Great Lakes Pollution from Land Use Activities (PLUARG), 432–4

International Union for the Conservation of Nature and Natural Resources (IUCN), 32–3, 238, 239, 378

International Whaling Commission, 245, 499

Inuit, 173, 245, 246, 426, 475; bioconcentration and, 350–1, 352
irrigation, 341–2, 424
IUCN (World Conservation Union), 32–3, 238, 239, 378

Jaccard, M., et al., 186
Jacques Whitford, 13
James Bay and Northern Quebec Agreement, 430–1, 536
James Bay Hydroelectric Project, 191, 425–30, 483, 536
James Bay II, 427–9
Japan, 18, 19
Jelinski, Dennis, 63–4
Jessen, Sabine, 264
Johannesburg Plan of Implementation, 498
Joint Action Group (JAG), 14
jurisdiction: Canadian oceans and, 243–4; context and, 150–2; see also governments; specific jurisdictions

Kay, Charles, 92
Keith, L.B., et al., 53
Kennedy, Robert Jr, 175
kimberlites, 473, 479, 536
Klein, Ralph, 214
Kluane National Park Reserve, 386
knapweed, 81, 83
knowledge: indigenous 38, 173, 536; traditional ecological (TEK), 38, 173, 191, 474–5, 475, 536
Krajnc, A., 151
Kreutzwiser, R., 150
K-strategists, 89–90
Kyoto Protocol, 151, 160, 209–19, 499, 500, 503, 536; assigned amounts, 211–12; clean development mechanism, 213–14; and developing countries, 214; emission credits and trading under, 213–14; greenhouse gases identified under, 213; legal basis for, 211; targets under, 210

labelling: GMOs, 332; pesticide, 353–5; product, 512
La Grande River, 426–9
Lahey, A., 10
Laidlaw Foundation, 513
lakes, 125, 424; acid, 139; water levels in, 446–7; water storage and, 123; see also eutrophication
land: agricultural, 325, 327–8; degradation of, 339–42; use, locally unwanted (LULU), 537
land claims, 175, 287–88, 415, 418, 426–7, 474–5, 479–80
Land Quality Index, 28
Landry, M., et al., 410
landscape ecology, 63–4
La Niña, 200
Lautenbach, W.E., 469
Lautenschlager, R.A., and T.P. Sullivan, 296
laws: conservation of energy, 536; conservation of matter, 105, 536; 'law of everybody', 511–14, 536
Lee, K.N., 170

Lemeiux, A., 464
Leopold, Aldo, 373, 496
Le Prestre, P.G., and P. Stoett, 392–3
lichens, 70, 72
life: definition of, 43
lifestyles: climate change and, 208
'light living', 512–15, 537
limiting factor principle, 58, 60, 433, 537
List of Wildlife Species at Risk (Red List), 235, 238, 378, 396, 397, 539
lithosphere, 39, 537
livestock, 336–7, 338, 355–6, 359
Livestock Revolution, 336–7, 537
Living Planet Index, 2, 15, 28
Long Lake, NWT, 475, 477, 479
long-run sustained yield (LRSY), 290–1
long-term view, 160
loons, 135
loosestrife, purple, 81
Lord, G.R., 24
Louise, Lake, 404, 406
Lovelock, James, 80
LULU (locally unwanted land use), 537
Lumsden Lake, 129

McAllister, Ian, 417
Macdonald, John A., 406
McDonald's, 517
McGinnity, P., et al., 266–7
Mackenzie River, 424
McIntosh, R.P., 173
McLeman, R., and B. Smit, 205
McLuhan, Marshall, 37
McNeely, Jeff, 33
macronutrients, 105, 537
Makah people, 245
Malthus, Thomas, 19
management: active, 409–10, 530; adaptive environmental, 147, 159, 170–1, 173, 178, 530; environmental, 32; fisheries, 226–7, 233–40; forestry, 74, 290–319; guidelines for science-based, 7–9; and long-term view, 160; non-renewable resources, 460–1; oceans, 226–74; protected areas, 403–18; trial-and-error, 171
management and planning: philosophy of, 145–63; process, method, and product of, 164–7
Manitoba: Kyoto and, 215
Manitoba Floodway Expansion Authority, 445
Manitoba Round Table on Environment and Economy, 165
manure, 356, 359, 439; green, 363–4, 535
marine: marine agreements, international, 241–2; ecozones, Canadian, 244–8; protected areas (MPAs), 242, 263–5, 267
Marine Stewardship Council, 501–2
marmot, 387
Marshall, Donald Jr, 257
Martin, Jacob, 487

Martin, Paul, 92
Mastny, L., and H. French, 500
matter, 104–5, 537
May, E., and M. Barlow, 11
meat, consumption of, 330, 336
media, 207
mediation, 176, 530
megaprojects, 434–5, 537
Mercredi, Ovide, 155, 176
mercury, 428–9, 430–1
mesosphere, 40, 537
metals: consumption of, 20; heavy, 12; recycled, 460; toxic, 133, 136, 465, 537; *see also* minerals; mining
methane, 189, 196
Métis, 475
micronutrients, 105, 537
Miller, A., and P. Rusnock, 298
Miller, Gord, 152
Mills, T.J., T.M. Quigley, and F.J. Everest, 7–8
mineralization, 115–16, 537
minerals, 459–93; *see also* metals; mining
Minerals and Metals Policy, 463
minimum viable population (MVP), 410, 537
mining, 9–15, 459–93; boreal forest and, 276, 277; coal, 9–11, 131; mine tailings, 475, 477; *see also* diamonds; energy; metals; mining
Mining Association of Canada, 462
'miracle seeds', 333
Mitchell, B., 23
Mixed Wood Plains ecozone, 283, 537
Model Forest Program, 314, 317–18
monitoring, 175, 428–9, 436
monoculture cropping, 337, 537
Montane Cordillera ecozone, 280–1, 537
Montreal Protocol, 537
Mosquin, T., 97
mountain lion, 373
Mulkins, Lindsay, 63
multi-barrier approach, 439–40, 537
murrelets, 389, 391
muskox, 90, 384
mutualism, 48, 62, 64, 72, 537
Myers, Ransom, 235
'myth of superabundance', 423

Nahanni National Park, 275
Nakashima, D.J., 173
Nanus, B., 152
National Forest Strategies, 313, 315–19; Coalition (NFSC), 314
National Marine Conservation Areas (NMCAs), 263–5, 503
National Marine Fisheries Service (US), 269
National Parks Act, 403–4, 406, 408, 415
National Parks Association, 404
National Pollutant Release Inventory, 430
National Research Council (US), 195, 197

National Roundtable on the Environment and Economy (NRTEE), 29
National Soil Conservation Program, 434–5
National Wildlife Areas, 263
Natural Resources Canada, 464–5
natural selection, 91
Nature Conservancy of Canada, 413
negotiation, 176, 538
neo-conservatism, 151
NEPTUNE (North-East Pacific Time-series Undersea Networked Experiment), 232
net community productivity (NCP), 54–5
Netherlands, 500
net primary productivity (NPP), 54–5, 376–7, 537–8
New Brunswick: budworm in, 298; fisheries in, 257
Newfoundland: offshore oil and gas and, 260
niche, 61, 538
Nigeria, 18
Nike, 513
NIMBY, 538
nitrates, 112, 116, 121
nitrification, 116
nitrogen, 105, 131, 342–3, 356, 359, 430–1; residual, 365; fixation of, 112, 115–16, 538; oxides, 138, 196
non-renewable resources, 459–93, 538
non-timber forest products (NTFPs), 284–5, 311–13, 538
Norse, Elliott, 238
North America: extinctions in, 92, 379–92
North American Waterfowl Management Plan, 388–9
North Atlantic Oscillation, 200
North Battleford, Sask., 439
Northwest Atlantic ecozone, 246–8
Northwest Atlantic Fisheries Organization, 250, 252–3
Northwest Territories, 464–5, 474–80
no-till/conservation agriculture (NT/CA), 362–3, 538
'not satisfactorily restocked' (NSR), 296
Nottaway-Broadback-Rupert river system, 427
Nova Scotia: forestry in, 292, 298; offshore oil and gas and, 260; tar pools in, 4, 9–15, 16–17
Nunavut: diamonds in, 468–9; Kyoto and, 215
nutrient(s), 105, 538; enrichment, 534; loss of, 342–3; movement of, 105–40; ocean ecosystems and, 227–8; as pollutant, 240, 334; removal of, 307–11; storage of, 105–7

oceans: Canada's, 242–8; ecosystems, 49, 227–33; management of, 226–74; sulphur in, 110–11
Oceans Act, 261–2
Oceans Strategy, 244, 262–3
O'Connor, Dennis, 439–41
oil and gas development, 240–1, 248, 258; *see also* fossil fuels
oil spills, 239, 240–1
Okalik, Paul, 215
Okanagan Valley, BC, 79–80
omnivores, 46, 538

one-dimensional radiative-convective (RC) climate models, 194

Ontario: air pollution in, 135; budget-cutting in, 150; climate change and, 202–4; drought in, 446–7; Environmental Commissioner of, 152; Kyoto and, 215; pest management in, 361; wind power in, 483–8

Ontario Conservation Authority, 24, 159–60

Ontario Land Stewardship Program, 434

Ontario Power Generation, 84, 217, 219–20

optimal foraging theory, 6, 538

organic matter, loss of, 342–3

O'Riordan, T., 155–6

Orr, D.W., 506, 511

oscillation (air pressure), 200

overharvesting, 459

owls: burrowing, 346; spotted, 305–6

oxygen, 45–6, 49–50, 105; depletion of, 126; oxygen sag curve, 126, 538

ozone, 538; layer, 196, 538

Pacific Marine ecozone, 244

Pacific Maritime ecozone, 279–80, 538

Pacific Rim National Park Reserve, 415

packaging, 510

PAHs (polycyclic aromatic hydrocarbons), 10–13

Palliser's Triangle, 447, 538

pandas, 61

parasitism, 62, 538

parks: development within, 403–8; external threats to, 409–10, 416; and fragmentation, 410–14, 534; management challenges for, 403–18; marine, 263–5; national, 204–5, 396, 397, 503; stakeholder interests and, 415, 418; system of, 371–2, 398, 399, 402–18

Parks Canada, 263–5, 398, 403

Parson, E.A., 148, 166

participation, public, 165–9, 173–4

partnerships, 168, 177, 436, 462–3, 538

pathogens, 334

Pauly, Daniel, 256

PCBs (polychlorinated biphenyls), 11, 12–13, 260, 433

Pest Control Products Act, 354, 355

pesticides: 334, 335; in Arctic, 350; cosmetic use of, 352; report card on, 354–5; types of, 344; see also biocides; herbicides

Pest Management Regulatory Agency (PMRA), 353

Peterson, C.H., et al., 240

pharmaceuticals, 375

philosophy, planning and management, 145–62

phosphorus, 105, 108–11, 129, 343, 356, 359, 432–5

photosynthesis, 45, 538

phototrophs, 45–6, 538

pH scale, 109, 131

phyla, 49

phytoplankton, 47–49, 126, 228, 230, 539

pigeon, passenger, 379, 381–2

'placebo policies', 151, 539

planning and management: context, 149–52, 534; land-use, 177; philosophy, 145–64; process, method,

and product of, 164–82; regional and land-use, 177; vision, 152–3, 539

plants: benthic, 126, 531; carnivorous, 48; see also specific species

PLUARG (International Reference Group on Great Lakes Pollution from Land Use Activities), 432–4

poaching, 399

Point Pelee National Park, 371–2

pollination, 375

pollution, 105, 532; agricultural, 430, 431–6; air, 356, 466; aquaculture and, 268; custom-designed solutions, 149–50, 533; extinctions and, 387, 389; forests and, 277–78; health and, 135–6; industrial, 9–15, 430–1; oceanic, 239–40, 260–1; sources (point and non-point), 128, 240, 430–7, 538; prevention v. control, 154; water, 118, 334, 430–7; see also PCBs; POPs

polynyas, 246, 539

POPs (persistent organic pollutants), 260, 261, 353, 354

population, 539; critical size, 532; human, 2, 18–20, 54, 328, 330; species, 52, 56, 87–94

Portage La Prairie, Man., 444

Portugal, 252

poverty, 154; global, 20

power: delegated, 176; dispute resolution and, 176; sharing, 166

precautionary principle, 171, 263, 539

precipitation, 118–20; acidity of, 131

predators, 49, 51, 62, 91, 256, 539; control of, 268, 382–3, 385, 386–7

Prescott-Allen, Robert, 500

pressure indicators (DPSIR), 26

prey, 49, 51, 62, 91, 539; switching, 235, 539

Prince Edward Island, 4; biocides in, 345; rising sea level and, 197–202

privatization, 150, 151

producers, 539; energy, 45–6

productivity: ecosystem, 53–5; gross primary (GPP), 53–4, 535; net community (NCP), 54; primary, 228, 230; net primary (NPP), 53, 376–7, 537–8

protected areas, 397–418, 539; Canadian, 402–18; classification of, 398, 402; development within, 403–8; external threats to, 409–10, 416; forest, 276, 311–12; global perspective on, 398–400; management challenges for, 403–18; provincial, 402–3; roles of, 400; and stakeholder interests, 416, 415

Protected Marine Areas, 265

provinces: endangered species and, 394–5; Kyoto and, 214–16, 218–19; see also governments; jurisdictions; specific provinces

puffins, 41–2

pulp and paper industry, 261

Putin, Vladimir, 211

Quebec: climate change and, 203; hydroelectricity in, 425–30; Kyoto and, 215; wind power in, 483

Quinn, F., et al., 425

radiation, adaptive, 91
'radiative forcings', 195–7
Rafferty-Alameda Dam, 172
Raincoast Conservation Society, 417
Rainham, D., 11, 12
rainshadow effect, 119, 539
Ramsar site, 411
raptors, 390
recharge areas, 121
reclamation, forest, 290, 539
recycling, 469, 510–11
Red List, 238, 240, 378, 396, 397, 539
Red River, 430; flood, 442–6; Floodway, 444
reforestation, 212, 294, 296, 470
regional planning, 177–8
'regreening recipe', 471
Reimchen, Tom, 259
Report on International Obligations (RIO), 215, 217–19
reproduction, 89–90
research: types of, 6–7
reservoirs, 428–9
residence time: carbon cycle, 107, 116; hydrological cycle, 118
resilience (ecosystem), 80, 539
resource(s), 5, 539; consumed, 460; definition of, 157; flow, 459–60, 539; non-renewable, 459–93; recycled, 460; renewable, 459–60 539; management of, 7–9, 177–8; partitioning, 61–2, 539; stock, 459–93, 538
respiration: autotrophic, 54; cellular, 46, 52–5; heterotrophic, 55
response indicators (DPSIR), 26
reuse ('light living'), 510
revegetation, 468–9
Richard Ivey Foundation, 513
Richardson, N.H., 472
Rideau River, 452
Riding Mountain National Park, 410, 416
Rio Declaration, 211, 540
Rio Summit, see Earth Summit
risk assessment, 171–4, 178, 209–11, 540
rivers, 424–5; heritage, 448–54, 535
roadkill, 407
rock cycle, 111
rodents, 66
Rohde, K., 96
Ross, N., 465, 471
Rowe, Stan, 105, 515
Royal Commission on the Future of the Toronto Waterfront, 156, 158
r-strategists, 89–90
runoff, 356–7; agricultural, 430, 431–7; urban, 430, 431, 439
Russia, 211

St Clair River-Lake St Clair-Detroit River system, 436–7
St Lawrence, Gulf of, 200–1
St Lawrence River, 430, 436–7

salinization, soil, 341–2, 540
salmon, 226, 257–58; Atlantic, 259, 266–7
Sanchez, L.E., 461
sand dunes, 73
saprotrophs, 47
Saskatchewan: Kyoto and, 215
Saskatchewan Round Table on Environment and Economy, 156, 168
Saskatchewan Watershed Authority, 441
Schindler, D.W., 275
Schroeder, Peter, 357–8
Schueler, F.W., and D.E. McAllister, 283
science: credible, 432–3; decision-making and, 29–32; links with policy, 436; management and, 184; natural, 37–40; social, 39
scientific method, 37–9
Scott, D.N., 203–4
sea level: climate change and, 190, 197–203, 241
sea lions, 228–39
seals, 236–7, 244, 246, 383; cod and, 256; hunting, 247
sea otters, 85–6, 244
sea urchins, 85
sedimentary cycles, 107, 108–12, 540
sedimentation, 334
selection (silviculture practice), 295–6
Selin, S., and D. Chavez, 165
serial depletion (fish stocks), 235, 540
sewage: as pollution, 261; treatment, 202
Shaw, R.W., 196, 199, 201, 202
shelterwood, 295–6
shifting baseline (fisheries), 235, 540
Shiklomanov, I.A., 424
short-term view, 160
Shrubsole, Dan, 156, 157, 445–6
Sierra Club of Canada, 215, 217–19
Silent Spring, 540
silo effect, 146
silviculture, 291–302, 540
Slocombe, S.D., 159
Smeesters, Edith, 355
Smith, Dan J., 75–6
Smith, L.G., 171
snow cover, 190
soil(s), 56–60; acid deposition and, 132–3, 135; agricultural, 365; alkalinity, 58; in Canada, 59; compaction of, 341, 540; erosion of, 341–2, 434–5, 540; horizons, 57, 540; moisture of, 121; permeability of, 57–8; profile, 57, 58; salinization of, 341–2, 540
Soil and Water Environmental Enhancement Program (SWEEP), 438
Soule, M., 376
South America, 99
Southeast Asia, 229
Southern Oscillation, 200
'Spaceship Earth', 37
Spain, 252
Sparrow, Herbert, 434
speciation, 91–3, 540

species: alien, 79–85, 384, 530; anadromous, 111, 259, 530; biodiversity and, 94–100; diversity, 95; endangered, 27, 371–98, 533; endemic, 97–8; exotic, 530; forests and, 278–84; indicator, 72, 259; interactions among, 60–5; introduced, 530; invader, 79–85, 530; keystone, 64–5, 85–6, 536; non-native, 530; number of, 47–52, 94–100; reintroduction, 401; removal of, 85–6; at risk, 303; specialist, 60–1, 540; threatened, 379; vulnerable, 541; *see also* alien species
species-area curves, 96
Species at Risk Act (SARA), 394–8, 540
sponge reefs, 238
Sri Lanka, 32
stakeholders, 165–69, 540
Standing Committee on Environment and Sustainable Development, 352, 353
state indicators (DPSIR), 26
stewardship, 413–15, 540
Stewart, J.M., 441
Stockholm Convention on Persistent Organic Pollutants, 261, 499
storms, 4, 201–2
storm surges, 201–2
stormwater systems, 431
'strategic assessment', 172
stratosphere, 40, 540
stress, physiological, zone of, 60, 542
strip cropping, 362, 540
Stuart-Smith, J., et al., 83
Study of Environmental Arctic Change (SEARCH), 216
sublimation, 120
subsidiarity, 24, 150, 540
subsidies, agricultural, 328, 357–8
succession: cyclic, 73; ecological, 70–9, 533; primary, 70–3, 539; secondary, 73, 79, 540
Sudbury, Ont., 465–72
Suffling, R., and D. Scott, 409
sulphates, 10, 111–12, 136–8
sulphur, 105, 111–12, 131; sulphur dioxide, 459, 540
Sumatra, 32
'superabundance, myth of', 423
'Superstack', 468
sustainability, 44; on campus, 506, 507; food production and, 359–66; in forestry, 293–4, 313, 314
sustainable: community, 472; consumption, 498; development, 7, 146, 153–4, 157, 262–4, 472, 502–3, 540; livelihoods, 146, 154, 540
Sustainable Development Index, 29
Sustainable Development Technology Canada, 214
Suzuki, David, 511
Sweden, 288–9, 500
Swinnerton, G.S., 407
Sydney Steel Corporation (SYSCO), 12
Sydney Tar Ponds, 1–2, 4, 9–15, 16–17
synergism, 87, 109, 541
systems, 30–1; control, 31, 55; natural, 4, 6–7, 31; socio-economic, 4, 6–7
systems perspective, 156, 158–60, 183

Table Mountain, Nfld, 77
taiga, 281, 541
Taiga Plains ecozone, 281, 541
tariffs, softwood, 318
technocentric perspective, 156, 157, 541
technology assessment, 171
Tembec Inc., 319
temperature: global average, 187, 189, 195; ocean, 255–6
tenure, forest, 286–7, 290, 313–14, 534
Thailand, 33
Thelon Game Sanctuary, 384
'theory', 38
thermocline, 230, 541
thermodynamics: laws of, 42–4; second law of, 541
thermohaline circulation, 231–3, 541
thermosphere, 40, 541
Thompson, I.D., et al., 299
'threatened status', 306
timber forest products, 285–90
Timoney, K.P., 318
Tinker, J., 153
tolerance, range of, 60, 539
Toronto, 352, 446; wind power in, 483–8
Toronto Hydro, 484
Toronto Renewable Energy Cooperative (TREC), 484
total accumulated ice coverage (TAIC), 200
total allowable catches (TACs), 249, 250, 254, 541
tourism: climate change and, 33, 202–3; nature-based, 375–60, 403–8; winter, 203–4
toxic materials, 240, 260, 433, 435–7, 461, 537
trade: endangered species, 382, 392–3; environment and, 499; ivory, 392
traditional ecological knowledge (TEK), 38, 173, 191, 474, 475, 536
transpiration, 118, 541
transport: energy and, 513–14; public, 509–10
trawling, bottom, 238, 242, 531
treaties, environmental, 392–4, 499–500
trees: acid deposition and, 132–3; carbon sequestration and, 219–20; seed, 295
Trist, E., 170
trophic levels, 46–7, 541
tropics: exinctions in, 376–7, 379–80
troposphere, 40, 541
Trudeau, Justin and Pierre, 275
tsunamis, 32–3
tundra: mining in, 473–4
Turpel, Mary Ellen, 155,
Turtle Island, 155
turtles, 238, 371–2
two-dimensional statistical-dynamic (SD) climate models, 193

uncertainty, 541; climate change and, 195, 207, 208–10; communicating, 169
unemployment: fishery workers and, 249, 250; insurance program, 255

Union of Concerned Students, 509–10
United Arab Emirates, 500
United Kingdom, 500
United Nations Commission on Sustainable
    Development, 498
United Nations Convention on Biological
    Diversity, 85
United Nations Convention on the Law of
    the Sea, 241
United Nations Development Program, 28, 29, 331
United Nations Environmental Program, 28, 500
United Nations Food and Agricultural Organization
    (FAO), 28, 226, 325, 329, 343, 363
United States: acid deposition by, 131, 132, 136–9;
    agreements with, 23, 27, 129, 138, 166, 432–7;
    energy use in, 20; international agreements and,
    151; Kyoto and, 210–11; population in, 19; Red
    River flood in, 442; reduced protection in, 500;
    softwood tariffs and, 317–18; wind power in, 483
University of Victoria Sustainability Project, 507
University of Waterloo, 506
upwellings, nutrient, 228
urbanization, 327

Valhalla Pure Outfitters, 513
values: ecocentric, 155–6; ecological, 373–4; economic,
    374–50; ethical, 376; evolutionary, 374; extrinsic,
    372–6, 534; intrinsic, 372–6, 536
Vancouver, 3
Vancouver Island, 175
Veale, Barbara, 451, 453–4
Vegetation Enhancement Technical Advisory
    Committee (VETAC), 471–2
VENUS (Victoria Experimental Network Under the
    Sea), 232
Victoria, BC, 261
vision (planning and management), 152–3, 541
Vitousek, P., et al., 15
volcanoes, 188
Volvo, 516
von Oech, R., 495

Walkerton, Ont., 442–5, 507; Inquiry, 439–41
Wallace, Alfred Russel, 38
walruses, 236
Warnock, R.G., and M.A. Skeel, 397
wastewater, 430–1
water, 423–58; acidified, 136; agriculture and, 333–4,
    359; availability of, 60, 424; ballast, 84–5; diamond
    mining and, 478–9; diversion, 424–30; fresh, 120–1;
    groundwater, 118, 121–5, 424, 438, 439, 446–7,
    535; as hazard, 441–7; human requirements for, 438;
    properties of, 123; quality, 434–41; safety of,
    439–41, 537; security of, 437–41 storage of, 118,
    120–2; water table, 446, 541; treatment, 439; use by
    Canadians, 423, 437–8; see also hydrological cycle

waterfowl, 388
watershed protection, 440–1
Watt-Cloutier, Sheila, 350
Watt-Gremm, Graham, 506, 507
weather, 187–8, 541
weathering, 111
Weber, M.G., and S.W. Taylor, 742
well-being, 504–5, 510; Wellbeing Index, 500–1
wells, 121, 439
Wernuik, J., 472
wetlands, 424, 541; drainage of, 334; loss of, 385, 387,
    388
whales, 228, 236–7, 239, 244–6; beluga, 349; bottlenose,
    396; great, 89; hunting, 245; killer, 238–9; pollution
    and, 260; right, 393; watching, 245, 382
wheat, genetically modified, 365–6
White Dog Reserve, 14
Whitehorse Mining Initiative, 462–3, 541–2
Whiteman, G. and K. Mamen, 463, 464
Wielgus, R.B., and P.R. Vernier, 305
wildlife: protection of, 505; wind power and, 485–6;
    see also specific species
Wilkinson, Merv, 313
wind power, 483–83, 542
WindShare, 484
windthrow, 312, 542
Winnipeg, 442–6; Floodway, 445
wolves, 386–7
Wong, P.Y., and I. Brodo, 72
Wood Buffalo National Park, 411–13
Woodley, S., 409
World Bank, 28
World Commission on Environment and
    Development, 7, 146, 154, 403
World Conservation Union (IUCN), 32–3, 238, 239,
    378
World Food Summit, 329
World Heritage sites, 403, 406, 411
World Summit on Sustainable Development (WSSD),
    21, 242, 265, 394, 497–9, 412, 442
World Trade Organization, 211, 499
World Watch Institute, 17
World Wildlife Fund, 15, 398, 403

Yellowknife, NWT, 14, 480
Yellowstone National Park, 98, 410
yew, western, 375
yield, sustained, 313, 541

zebra mussel, 83–4, 88
Zimmermann, E., 157
zooplankton, 49–50, 542
zooxanthellae, 229, 542